D0224538

Get the Most Out of Your Liberal Arts Mathematics MyMathLab Courses

Pearson

MyMathLab has helped millions of students succeed in their math courses. As classrooms and students have evolved in mathematics, MyMathLab has evolved as well, with enhancements that make it easy for you to support your students.

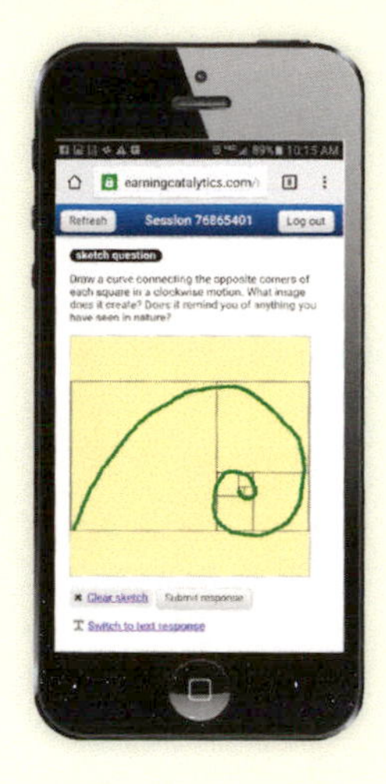

Get Students Engaged

Learning Catalytics™—a student response tool that uses students' smartphones, tablets, or laptops to engage them in more interactive tasks and thinking—is available through any MyMathLab course to foster student engagement and peer-to-peer learning. You can generate class discussion, guide your lecture, and promote peer-to-peer learning with real-time analytics.

Personalize Learning for Each Student

MyMathLab can personalize homework assignments for students based on their performance on a test or quiz. This way, students can focus on just the topics they have not yet mastered.

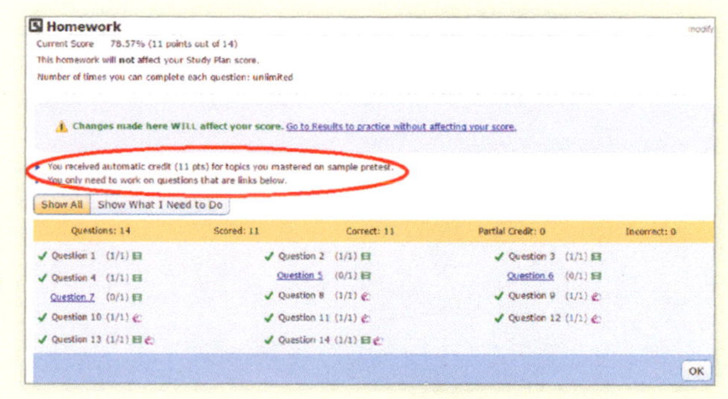

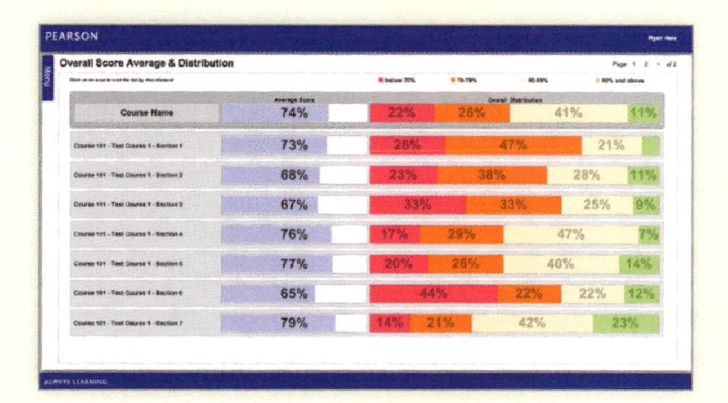

Use Data to Tailor Your Course

A comprehensive gradebook with enhanced reporting functionality helps you efficiently manage your course. The new Reporting Dashboard presents student performance data at the class, section, and program levels in an accessible, visual manner. Item Analysis allows you to track class-wide understanding at the exercise level, so that you can refine your lectures or assignments to address just-in-time student needs.

www.mymathlab.com

Pearson

Get the Most Out of Your Liberal Arts Mathematics MyMathLab Courses

Excursions in Modern Mathematics shows that math is more than a set of formulas— it's a powerful and beautiful tool that can be applicable and interesting to anyone. With chapters categorized by social choice, management science, growth, shape and form, and statistics, you can dig into topics that are beyond what your students have seen before, and give them a new appreciation for this subject.

Spark Your Students' Interest with:

...Math That's Applicable

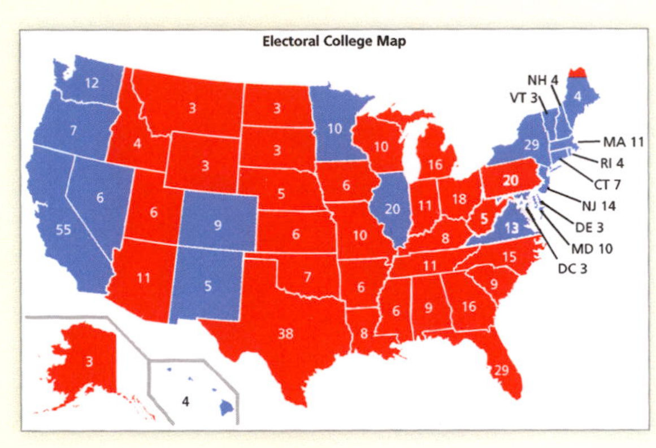

The math of politics, government, and social science is more important than ever for students to think about and is increasingly relevant to their civic lives. How do we elect our leaders? How do we measure the power of individuals and groups when it comes to social choice? These are key questions for students to explore.

...Math That's Beautiful

Exploring the connections between math and shapes—both natural and man-made—gives students a chance to see that there is so much more to math than algebra. It truly can be beautiful.

...Math That's Modern

Much of the math in this text was discovered in the last century. Show your students how math is an evolving tool that is extremely useful to today's society. For example, in a world where social networks are a fact of life, the math of how people and places connect to each other is more relevant than ever.

www.mymathlab.com

Excursions in Modern Mathematics

9th edition

Excursions in Modern Mathematics

Peter Tannenbaum

California State University—Fresno

Director, Portfolio Management: *Anne Kelly*
Courseware Portfolio Manager: *Marnie Greenhut*
Courseware Portfolio Assistant: *Stacey Miller*
Content Producer: *Patty Bergin*
Managing Producer: *Karen Wernholm*
Media Producer: *Nick Sweeny*
TestGen Content Manager: *Mary Durnwald*
MathXL Content Developer: *Robert Carroll*
Product Marketing Manager: *Alicia Frankel*
Product Marketing Assistant: *Hanna Lafferty*
Senior Author Support/Technology Specialist: *Joe Vetere*
Rights and Permissions Project Manager: *Gina Cheselka*
Manufacturing Buyer: *Carol Melville, LSC Communications*
Associate Director of Design: *Blair Brown*
Text Design, Production Coordination, Composition, and Illustrations: *Cenveo Publisher Services*
Cover Design: *Barbara T. Atkinson*
Cover Image: *Elisabeta Stan/Shutterstock*

Acknowledgments of third-party content appear on page 569, which constitutes an extension of this copyright page.

Library of Congress Cataloging-in-Publication Data

Names: Tannenbaum,
Peter, 1946–
Title: Excursions in modern mathematics / Peter Tannenbaum,
California State
University, Fresno.
Description: 9th edition. | Boston
: Pearson, [2018] | Includes index.
Identifiers: LCCN 2016042060 | ISBN
9780134468372 (hardcover) | ISBN
0134468376 (hardcover)
Subjects: LCSH:
Mathematics. | Mathematics--Textbooks. |
Probabilities--Textbooks. |
Mathematical statistics--Textbooks.
Classification: LCC QA36 .T35 2018 | DDC
510--dc23
LC record available at https://lccn.loc.gov/2016042060

KV 12 20 2018 0809

ISBN 13: 978-0-13-446837-2
ISBN 10: 0-13-446837-6

To the members of the board of Last Tango

Contents

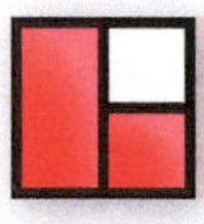

PART 2 MANAGEMENT SCIENCE

PART 5 STATISTICS

Preface

FROM THE AUTHOR

"To most outsiders, modern mathematics is unknown territory. Its borders are protected by dense thickets of technical terms; its landscapes are a mass of indecipherable equations and incomprehensible concepts. Few realize that the world of modern mathematics is rich with vivid images and provocative ideas."

– *Ivars Peterson,*
The Mathematical Tourist

This text started more than 20 years ago as a set of lecture notes for a new, experimental "math appreciation" course (these types of courses are described, sometimes a bit derisively, as "math for poets"). Over time, the lecture notes grew into a text and the "poets" turned out to be social scientists, political scientists, economists, psychologists, environmentalists, and many other "ists." Over time, and with the input of many users, the contents have been expanded and improved, but the underlying philosophy of the text has remained the same since those handwritten lecture notes were handed out to my first group of students.

Excursions in Modern Mathematics is a travelogue into that vast and alien frontier that many people perceive mathematics to be. My goal is to show the open-minded reader that mathematics is a lively, interesting, useful, and surprisingly rich human activity.

The "excursions" in *Excursions* represent a collection of topics chosen to meet the following simple criteria.

- **Applicability.** There is no need to worry here about that great existential question of college mathematics: What is this stuff good for? The connection between the mathematics presented in these excursions and down-to-earth, concrete real-life problems is transparent and immediate.
- **Accessibility.** As a general rule, the excursions in this text do not presume a background beyond standard high school mathematics—by and large, intermediate algebra and a little Euclidean geometry are appropriate and sufficient prerequisites. (In the few instances in which more advanced concepts are unavoidable, an effort has been made to provide enough background to make the material self-contained.) A word of caution—this does not mean that the excursions in this book are easy! In mathematics, as in many other walks of life, simple and basic are not synonymous with easy and superficial.
- **Modernity.** Unlike much of traditional mathematics, which is often hundreds of years old, most of the mathematics in this text has been discovered within the last 100 years, and in some cases within the last couple of decades. Modern mathematical discoveries do not have to be the exclusive province of professional mathematicians.
- **Aesthetics.** The notion that there is such a thing as beauty in mathematics is surprising to most casual observers. There is an important aesthetic component in mathematics, and just as in art and music (which mathematics very much resembles), it often surfaces in the simplest ideas. A fundamental objective of this text is to develop an appreciation of the aesthetic elements of mathematics.

Outline

The excursions are organized into five independent parts, each touching on a different area where mathematics and the real world interface.

- PART 1 **Social Choice.** This part deals with mathematical applications to politics, social science, and government. How are *elections* decided? (Chapter 1);

How can the *power* of individuals, groups, or voting blocs be measured? (Chapter 2); How can assets commonly owned be *divided* in a *fair* and equitable manner? (Chapter 3); How are seats *apportioned* in a legislative body? (Chapter 4).

- **PART 2 Management Science.** This part deals with questions of efficiency—how to manage some valuable resource (time, money, energy) so that utility is maximized. How do we sweep over a network with the least amount of backtracking? (Chapter 5); How do we find the shortest or least expensive route that *visits* a specified set of locations? (Chapter 6); How do we create efficient networks that *connect* people or things? (Chapter 7); How do we schedule a project so that it is completed as early as possible? (Chapter 8).
- **PART 3 Growth.** In this part we discuss, in very broad terms, the mathematics of growth and decay, profit and loss. In Chapter 9 we cover mathematical models of *population growth*, mostly biological and human populations but also populations of inanimate "things" such as garbage and pollution. Since money plays such an important role in our lives, it deserves a chapter of its own. In Chapter 10 we discuss a few of the key concepts of *financial mathematics*: interest, investments, retirement savings, and consumer debt.
- **PART 4 Shape and Form.** In this part we cover a few connections between mathematics and the shape and form of objects—natural or human-made. What is *symmetry*? What *types* of symmetries exist in nature and art? (Chapter 11); What kind of geometry lies hidden behind the *kinkiness* of the many irregular shapes we find in nature? (Chapter 12); What is the connection between the *Fibonacci numbers* and the *golden ratio* (two abstract mathematical constructs) and the *spiral* forms that we regularly find in nature? (Chapter 13).
- **PART 5 Statistics.** In one way or another, statistics affects all our lives. Government policy, insurance rates, our health, our diet, and our political lives are all governed by statistical information. This part deals with how the statistical information that affects our lives is collected, organized, and interpreted. What are the purposes and strategies of *data collection*? (Chapter 14); How is data *organized, presented,* and *summarized*? (Chapter 15); How do we use mathematics to measure *uncertainty* and *risk*? (Chapter 16); How do we use mathematics to model, analyze, and make predictions about *real-life, bell-shaped* data sets? (Chapter 17).

Acknowledgments

A large number of colleagues have contributed both formally and informally to the evolution of this text. (My apologies to anyone whose name has inadvertently been left out.)

The following reviewers contributed to this edition (thank you for your great comments and suggestions):

Tamara Carter, *Texas A&M University*
Bruce Corrigan-Salter, *Wayne State University*
Barbara Hess, *California University of Pennsylvania*
Jennifer L. Jameson, *Coconino Community College*
Stephanie Lafortune, *College of Charleston*
Christine Latulippe, *Norwich University*
Jill Rafael, *Sierra College*
Robin Rufatto, *Ball State University*
Dawn Slavens, *Midwestern State University*
Cindy Vanderlaan, *Indiana University-Purdue University*

In addition to the above, special thanks to those who contributed to specific aspects of this project: Dale Buske, who produced the Student and Instructor's Solutions Manuals; Katie Tannenbaum, my favorite indexer; and Nick Sweeny and the team at LearningMate for producing the new and much improved version of the Applets.

The following is a list of reviewers of older editions: Lowell Abrams, Diane Allen, Teri Anderson, Carmen Artino, Erol Barbut, Donald Beaton, Gregory Budzban, Guanghwa Andy Chang, Lynn Clark, Terry L. Cleveland, Leslie Cobar, Crista Lynn Coles, Kimberly A. Conti, Irene C. Corriette, Ronald Czochor, Robert V. DeLiberato, Nancy Eaton, Lily Eidswick, Lauren Fern, Kathryn E. Fink, Stephen I. Gendler, Marc Goldstein, Josephine Guglielmi, Abdi Hajikandi, William S. Hamilton, Cynthia Harris, Harold Jacobs, Peter D. Johnson, Karla Karstens, Lynne H. Kendall, Stephen Kenton, Tom Kiley, Katalin Kolossa, Randa Lee Kress, Jean Krichbaum, Thomas Lada, Diana Lee, Kim L. Luna, Mike Martin, Margaret A. Michener, Mika Munakata, Thomas O'Bryan, Daniel E. Otero, Philip J. Owens, Matthew Pickard, Kenneth Pothoven, Lana Rhoads, David E. Rush, Shelley Russell, Kathleen C. Salter, Theresa M. Sandifer, Paul Schembari, Salvatore Sciandra Jr., Deirdre Smith, Marguerite V. Smith, William W. Smith, Hilary Spriggs, David Stacy, Zoran Sunik, Paul K. Swets, W. D. Wallis, John Watson, Sarah N. Ziesler, and Cathleen M. Zucco–Teveloff.

Last, but not least, the *real movers and shakers* in the editorial staff that made this edition possible and deserve special recognition: mover and shaker in-chief (and Senior Acquisitions Editor) Marnie Greenhut, the voice of reason and calm whenever the project hit rough waters; Content Producer Patty Bergin and Project Manager at Cenveo Publisher Services Marilyn Dwyer, both of whose patience, good humor, and attention to detail made the logistics of producing this edition the smoothest ever; Product Marketing Manager Alicia Frankel; Field Marketing Manager Andrew Noble; Designer Barbara Atkinson; and Media Producer Nick Sweeny.

A Final Word to the Reader

> "It's not what you look at that matters, it's what you see."
> – *Henry David Thoreau*

My goal in writing this text is to shine a small light on all that mathematics can be when looked at in the right way—useful, interesting, subtle, beautiful, and accessible. I hope that you will see something of that in this text.

Peter Tannenbaum

New in This Edition

- New and updated examples from pop culture, sports, politics, and science.
- New material on **Retirement Savings** in Chapter 10.
- New **Applet Bytes** exercises in the exercise sets require the use of the new applets in MyMathLab and encourage students to delve deeper into the concepts using the applets.
- New and updated exercises have been informed by MyMathLab data analytics including level of difficulty and appropriateness.
- New **Annotated Instructor's Edition** provides annotations indicating where Applets, Animated Whiteboard Concept Videos, and Learning Catalytics are relevant, in addition to Discussion Ideas and Teaching Tips. Answers to exercises are still in the back of the book.
- New in MyMathLab for *Excursions in Modern Mathematics*, Ninth Edition
    - New and improved **applets** designed by the author help students visualize the more difficult concepts and develop deeper understanding:
        - Voting Methods
        - Banzhaf and Shapley-Shubik Power
        - Method of Sealed Bids
        - Method of Markers
        - Apportionment Methods
        - Euler Paths and Circuits: Fleury's Algorithm
        - Hamilton Paths and Circuits
        - Traveling Salesman
        - Kruskal's Algorithm
        - Priority List Scheduling
        - Finance Calculator
        - Rigid Motions
        - Geometric Fractals
        - Numerical Summaries of Data

      All applets are assignable in MyMathLab and exercises have been written to guide students.
    - Engaging **Animated Whiteboard Concept Videos** bring concepts to life in an exciting and interesting fashion using narration and animated drawing. Videos cover topics such as Fair Division, Eulerizing Graphs, Self-Similarity, The Golden Ratio, and Normal Curves. Students will see math in a fresh, new way!
    - **Learning Catalytics**, a "bring your own device" student engagement, assessment, and classroom intelligence system, is available in MyMathLab with annotations at point-of-use for instructors in the Annotated Instructor's Edition. LC annotations provide a corresponding code for each question as it becomes relevant to integrate into the classroom. Within Learning Catalytics, simply search for the question using the code in the text's annotation.
    - **StatCrunch,** a powerful, Web-based statistical software that allows users to perform complex analyses, share data sets, and generate compelling reports, has been integrated into the MyMathLab for the first time. The vibrant online community offers tens of thousands of data sets shared by users.

Resources for Success

Pearson

MyMathLab® Online Course for *Excursions in Modern Mathematics,* 9th Edition by Peter Tannenbaum (access code required)

MyMathLab is available to accompany Pearson's market-leading text offerings. To give students a consistent tone, voice, and teaching method, each text's flavor and approach are tightly integrated into the accompanying MyMathLab course, making learning the material as seamless as possible.

New Applets

New and improved applets help students explore concepts more deeply. Assignable in MyMathLab, they encourage students to visualize and interact with concepts such as apportionment, Euler circuits, and geometric fractals. Applet Bytes (exercises and explorations based on the applets) are available for some chapters.

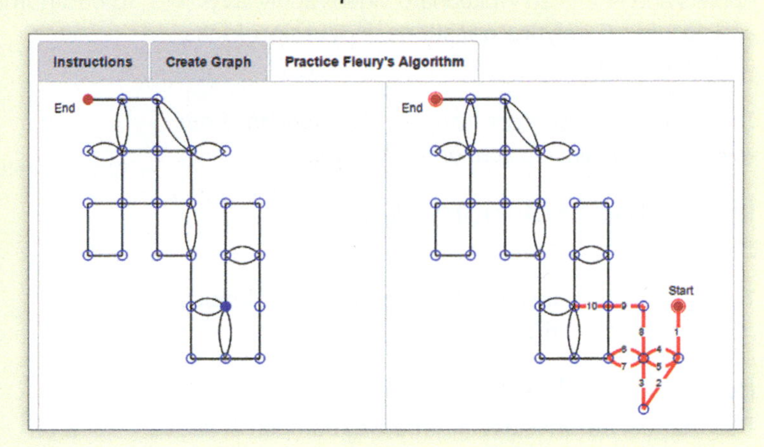

Animated Concept Videos

New Animated Whiteboard Concept videos use narration and animated drawing to bring concepts to life in a more engaging manner for students. Videos cover topics such as Fair Division, Eulerizing Graphs, Self-Similarity, The Golden Ratio, and Normal Curves.

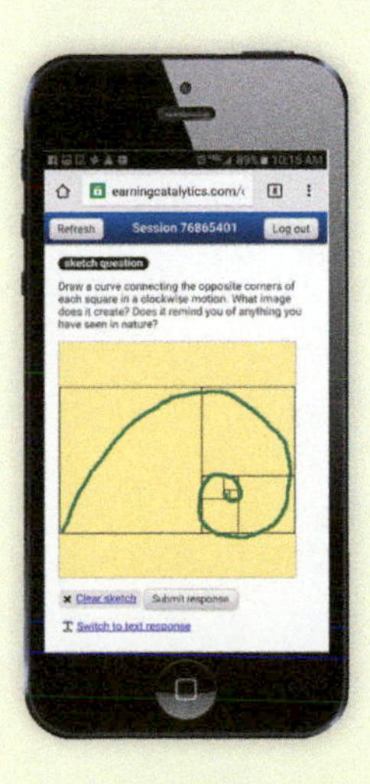

Learning Catalytics

Integrated into MyMathLab, Learning Catalytics uses students' mobile devices for an engagement, assessment, and classroom intelligence system that gives instructors real-time feedback on student learning. LC annotations in the Annotated Instructor's Edition provide a corresponding tag to search for when a Learning Catalytics question is relevant to the topic at hand.

StatCrunch

Newly integrated StatCrunch allows students to harness technology to perform complex analysis on data.

www.mymathlab.com

Resources for Success

Instructor Resources

NEW! Annotated Instructor's Edition

ISBN 10: 0-13-446908-9 **ISBN 13:** 978-0-13-446908-9

The AIE provides annotations for instructors, including suggestions about where media resources like Applets and Animated Whiteboard Videos apply, as well as Learning Catalytics questions, discussion ideas, and teaching tips.

The following resources can be downloaded from www.pearsonhighered.com or in MyMathLab.

Instructor's Solutions Manual

Dale R. Buske, St. Cloud State University

This manual contains detailed, worked-out solutions to all exercises in the text.

Instructor's Testing Manual

This manual includes two alternative multiple-choice tests per chapter.

Image Resources Library

This resource in MyMathLab contains all art from the text for instructors to use in their own presentations and handouts.

PowerPoints

These editable slides present key concepts and definitions from the text. You can add art from the Image Resource Library in MyMathLab® or slides that you develop on your own.

TestGen

TestGen® (www.pearsoned.com/testgen) enables instructors to build, edit, print, and administer tests using a computerized bank of questions developed to cover all the objectives of the text.

Student Resources

Student's Solutions Manual

Dale R. Buske, St. Cloud State University

ISBN 10: 0-13-446913-5 **ISBN 13:** 978-0-13-446913-3

This manual provides detailed worked-out solutions to odd-numbered walking and jogging exercises.

www.mymathlab.com

PART 1

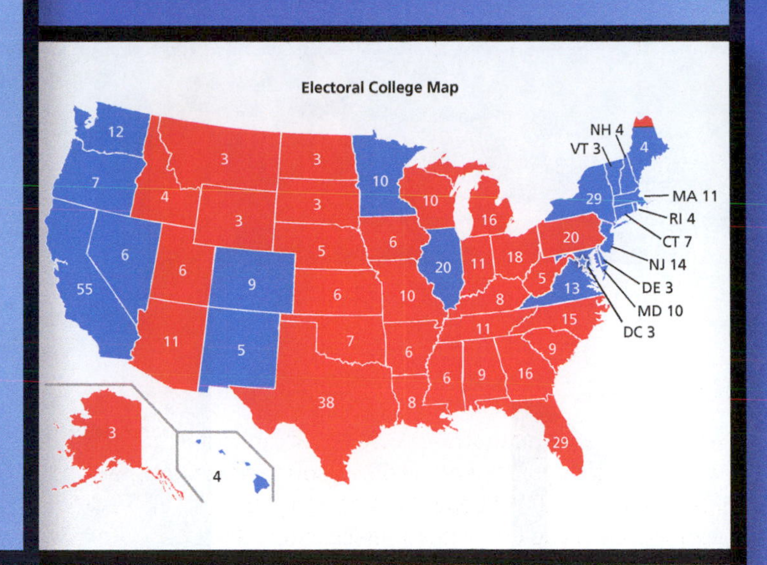

Social Choice

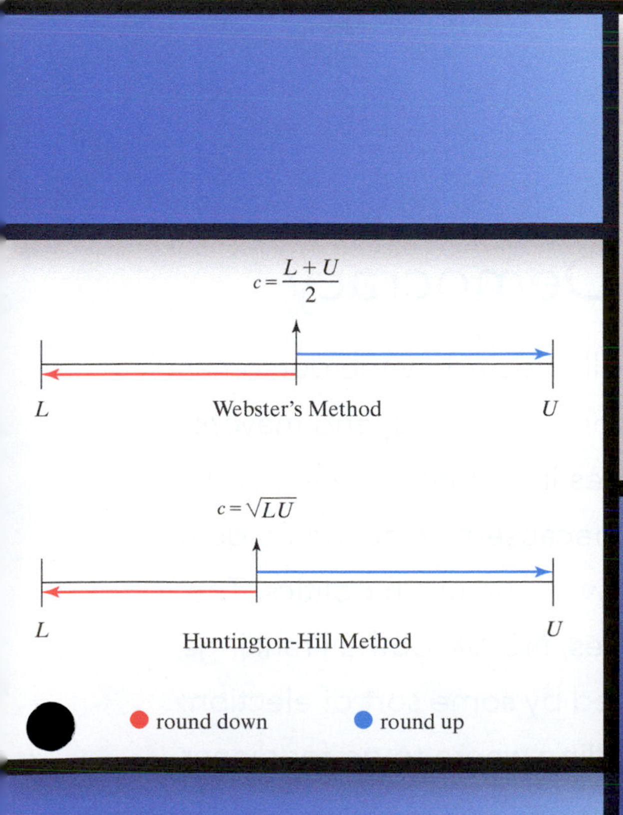

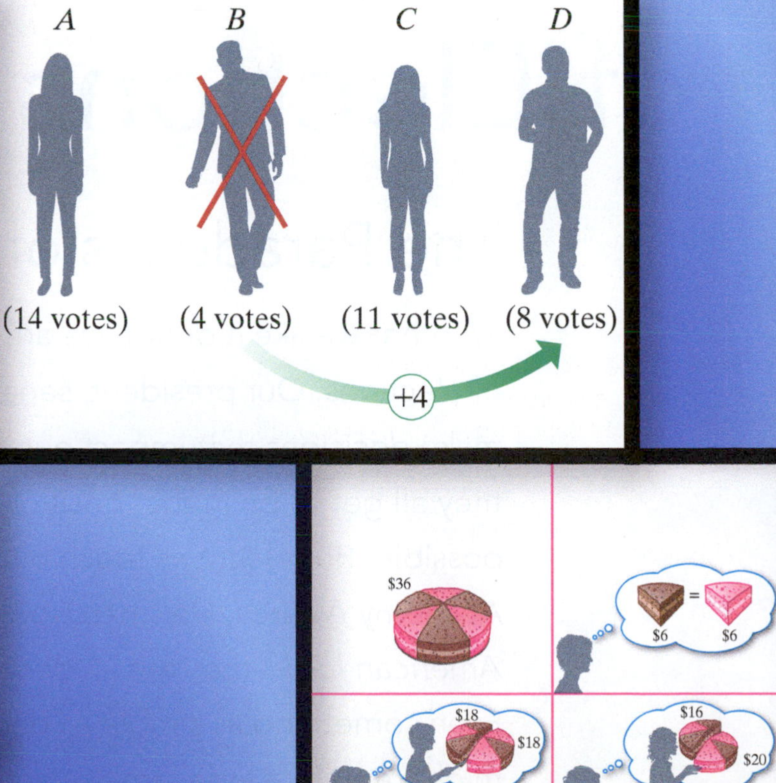

2015 Heisman Trophy finalists: Derrick Henry of the University of Alabama, Christian McCaffrey of Stanford University, and Deshaun Watson of Clemson University. (For the full story, see page 13.)

The Mathematics of Elections

The Paradoxes of Democracy

Whether we like it or not, we are all affected by the outcomes of elections. Our president, senators, governors, and mayors make decisions that impact our lives in significant ways, and they all get to be in that position because an election made it possible. But elections touch our lives not just in politics. The Academy Awards, Heisman trophies, NCAA football rankings, *American Idol*—they are all decided by some sort of election. Even something as simple as deciding where to go for dinner might require a little family election.

We have elections because we don't all think alike. Since we cannot all have things our way, we vote. But *voting* is only the first half of the story, the one we are most familiar with. As playwright Tom Stoppard suggests, it's the second half of the story—the *counting*—that is at the heart of the democratic process. How do we sift through the many choices of individual voters to find the collective choice of the group? More important, how well does the process work? Is the process always fair? Answering these questions and explaining a few of the many intricacies and subtleties of *voting theory* are the purpose of this chapter.

But wait just a second! Voting theory? Why do we need a fancy theory to figure out how to count the votes? It all sounds pretty simple: We have an election; we count the ballots. Based on that count, we decide the outcome of the election in a consistent and fair manner. Surely, there must be a reasonable way to accomplish this. Surprisingly, there isn't!

In the late 1940's the American economist Kenneth Arrow discovered a remarkable fact: For elections involving three or more candidates, there is no consistently fair democratic method for choosing a winner. In fact, Arrow demonstrated that *a method for determining election results that is always fair is a mathematical impossibility*. This fact, the most famous in voting theory, is known as *Arrow's Impossibility Theorem.*

> "It's not the voting that's democracy; it's the counting."
>
> *– Tom Stoppard*

This chapter is organized as follows. We will start with a general discussion of *elections* and *ballots* in Section 1.1. This discussion provides the backdrop for the remaining sections, which are the heart of the chapter. In Sections 1.2 through 1.5 we will explore four of the most commonly used *voting methods*—how they work and how they are used in real-life applications. In Section 1.6 we will introduce some basic principles of fairness for voting methods and apply these *fairness criteria* to the voting methods discussed in Sections 1.2 through 1.5. The section concludes with a discussion of the meaning and significance of Arrow's Impossibility Theorem.

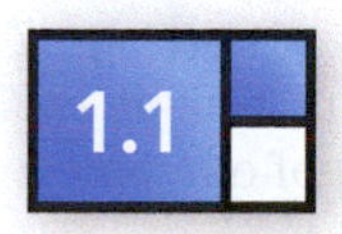

The Basic Elements of an Election

Big or small, important or trivial, *all* elections share a common set of elements.

- **The candidates.** The purpose of an election is to choose from a set of *candidates* or *alternatives* (at least two—otherwise it is not a real election). Typically, the word *candidate* is used for people and the word *alternative* is used for other things (movies, football teams, pizza toppings, etc.), but it is acceptable to use the two terms interchangeably. In the case of a generic choice (when we don't know if we are referring to a person or a thing), we will use the term *candidate*. While in theory there is no upper limit on the number of candidates, for most elections (in particular the ones we will discuss in this chapter) the number of candidates is small.
- **The voters.** These are the people who get a say in the outcome of the election. In most democratic elections the presumption is that all voters have an equal say, and we will assume this to be the case in this chapter. (This is not always true, as we will see in great detail in Chapter 2.) The number of voters in an election can range from very small (as few as 3 or 4) to very large (hundreds of millions). In this section we will see examples of both.
- **The ballots.** A ballot is the device by means of which a voter gets to express his or her opinion of the candidates. The most common type is a paper ballot, but a

voice vote, a text message, or a phone call can also serve as a "ballot" (see Example 1.5 *American Idol*). There are many different forms of ballots that can be used in an election, and Fig. 1-1 shows a few common examples. The simplest form is the **single-choice ballot**, shown in Fig. 1-1(a). Here very little is being asked of the voter ("pick the candidate you like best, and keep the rest of your opinions to yourself!"). At the other end of the spectrum is the **preference ballot**, where the voter is asked to rank *all* the candidates in order of preference. Figure 1-1(b) shows a typical preference ballot in an election with five candidates. In this ballot, the voter has entered the candidates' names in order of preference. An alternative version of the same preference ballot is shown in Fig. 1-1(c). Here the names of the candidates are already printed on the ballot and the voter simply has to mark first, second, third, etc. In elections where there are a large number of candidates, a **truncated preference ballot** is often used. In a truncated preference ballot the voter is asked to rank some, but not all, of the candidates. Figure 1-1(d) shows a truncated preference ballot for an election with dozens of candidates.

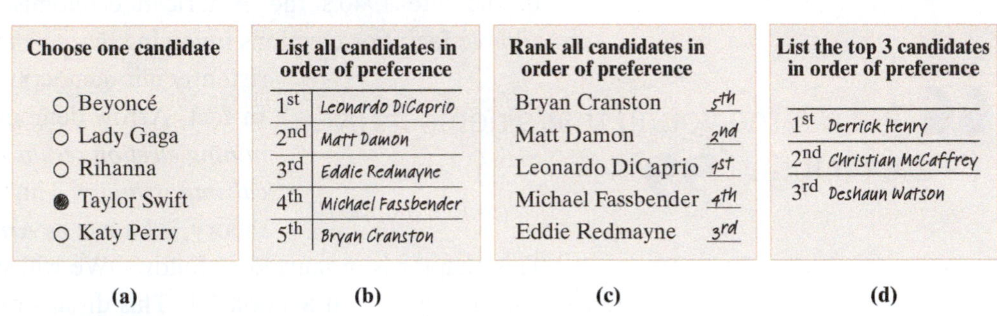

FIGURE 1-1 (a) Single-choice ballot, (b) preference ballot, (c) a different version of the same preference ballot, and (d) truncated preference ballot.

- **The outcome.** The purpose of an election is to use the information provided by the ballots to produce some type of outcome. But what types of outcomes are possible? The most common is **winner-only**. As the name indicates, in a winner-only election all we want is to find a winner. We don't distinguish among the nonwinners. There are, however, situations where we want a broader outcome than just a winner—say we want to determine a first-place, second-place, and third-place candidate from a set of many candidates (but we don't care about fourth place, fifth place, etc.). We call this type of outcome a **partial ranking**. Finally, there are some situations where we want to rank *all* the candidates in order: first, second, third, . . . , last. We call this type of outcome a **full ranking**, or just a **ranking** for short.
- **The voting method.** The final piece of the puzzle is the method that we use to tabulate the ballots and produce the outcome. This is the most interesting (and complicated) part of the story, but we will not dwell on the topic here, as we will discuss voting methods throughout the rest of the chapter.

It is now time to illustrate and clarify the above concepts with some examples. We start with a simple example of a fictitious election. This is an important example, and we will revisit it many times throughout the chapter. You may want to think of Example 1.1 as a mathematical parable, its importance being not in the story itself but in what lies hidden behind it. (As you will soon see, there is a lot more to Example 1.1 than first meets the eye.)

EXAMPLE 1.1 THE MATH CLUB ELECTION (WINNER-ONLY)

The Math Appreciation Society (MAS) is a student club dedicated to an unsung but worthy cause: that of fostering the enjoyment and appreciation of mathematics among college students. The MAS chapter at Tasmania State University is holding its annual election for club president, and there are four *candidates* running: Alisha, Boris, Carmen, and Dave (A, B, C, and D for short).

Every member of the club is eligible to vote, and the vote takes the form of a *preference ballot*. Each voter is asked to rank each of the four candidates in order of preference. There are 37 *voters* who submit their ballots, and the 37 *preference ballots* submitted are shown in Fig. 1-2.

Ballot	Ballot	Ballot	Ballot	Ballot	Ballot	Ballot	Ballot	Ballot	Ballot	Ballot	Ballot	Ballot	Ballot	Ballot	Ballot	Ballot	Ballot
1st *A*	1st *B*	1st *A*	1st *C*	1st *B*	1st *C*	1st *A*	1st *B*	1st *C*	1st *A*	1st *C*	1st *D*	1st *A*	1st *A*	1st *C*	1st *A*	1st *C*	1st *D*
2nd *B*	2nd *D*	2nd *B*	2nd *B*	2nd *D*	2nd *B*	2nd *B*	2nd *D*	2nd *B*	2nd *B*	2nd *B*	2nd *C*	2nd *B*	2nd *B*	2nd *B*	2nd *B*	2nd *B*	2nd *C*
3rd *C*	3rd *C*	3rd *C*	3rd *D*	3rd *C*	3rd *D*	3rd *C*	3rd *C*	3rd *D*	3rd *C*	3rd *D*	3rd *B*	3rd *C*	3rd *C*	3rd *D*	3rd *C*	3rd *D*	3rd *B*
4th *D*	4th *A*	4th *D*	4th *A*	4th *A*	4th *A*	4th *D*	4th *A*	4th *A*	4th *D*	4th *A*	4th *A*	4th *D*	4th *D*	4th *A*	4th *D*	4th *A*	4th *A*

Ballot	Ballot	Ballot	Ballot	Ballot	Ballot	Ballot	Ballot	Ballot	Ballot	Ballot	Ballot	Ballot	Ballot	Ballot	Ballot	Ballot	Ballot	Ballot
1st *C*	1st *A*	1st *D*	1st *D*	1st *C*	1st *C*	1st *D*	1st *A*	1st *D*	1st *C*	1st *A*	1st *D*	1st *B*	1st *A*	1st *C*	1st *A*	1st *A*	1st *D*	1st *A*
2nd *B*	2nd *B*	2nd *C*	2nd *C*	2nd *B*	2nd *B*	2nd *C*	2nd *B*	2nd *C*	2nd *B*	2nd *B*	2nd *C*	2nd *D*	2nd *B*	2nd *D*	2nd *B*	2nd *B*	2nd *C*	2nd *B*
3rd *D*	3rd *C*	3rd *B*	3rd *B*	3rd *D*	3rd *D*	3rd *B*	3rd *C*	3rd *B*	3rd *D*	3rd *C*	3rd *B*	3rd *C*	3rd *C*	3rd *B*	3rd *C*	3rd *C*	3rd *B*	3rd *C*
4th *A*	4th *D*	4th *A*	4th *A*	4th *A*	4th *A*	4th *A*	4th *D*	4th *A*	4th *A*	4th *D*	4th *A*	4th *A*	4th *D*	4th *A*	4th *D*	4th *D*	4th *A*	4th *D*

FIGURE 1-2 The 37 preference ballots for the Math Club election.

Last but not least, what about the *outcome* of the election? Since the purpose of the election is to choose a club president, it is pointless to discuss or consider which candidate comes in second place, third place, etc. This is a *winner-only* election.

EXAMPLE 1.2 THE MATH CLUB ELECTION (FULL RANKING)

Suppose now that we have pretty much the same situation as in Example 1.1 (same candidates, same voters, same preference ballots), but in this election we have to choose not only a president but also a vice-president, a treasurer, and a secretary. According to the club bylaws, the president is the candidate who comes in first, the vice-president is the candidate who comes in second, the treasurer is the candidate who comes in third, and the secretary is the candidate who comes in fourth. Given that there are four candidates, each candidate will get to be an officer, but there is a big difference between being elected president and being elected treasurer (the president gets status and perks; the treasurer gets to collect the dues and balance the budget). In this version how you place matters, and the outcome should be a full *ranking* of the candidates.

EXAMPLE 1.3 THE ACADEMY AWARDS

The Academy Awards (also known as the Oscars) are given out each year by the Academy of Motion Picture Arts and Sciences for Best Picture, Best Actress, Best Actor, Best Director, and many other, lesser categories (Sound Mixing, Makeup, etc.). The election process is not the same for all awards, so for the sake of simplicity we will just discuss the selection of Best Picture.

The *voters* in this election are all the eligible members of the Academy (a tad over 6000 voting members for the 2016 Academy Awards). After a complicated preliminary round (a process that we won't discuss here) somewhere between eight and ten films are selected as the nominees—these are our *candidates*. (For most other awards there are only five nominees.) Each voter is asked to submit a preference ballot ranking the ten candidates. There is only a winner (the other candidates are not ranked), with the winner determined by a voting method called plurality-with-elimination that we will discuss in detail in Section 1.4. (The winner of the 2016 Best Picture Award was *Spotlight*.)

The part with which people are most familiar comes after the ballots are submitted and tabulated—the annual Academy Awards ceremony, held each year in late February. How many movie fans realize that behind one of the most extravagant and glamorous events in pop culture lies an election?

EXAMPLE 1.4 THE HEISMAN TROPHY

The Heisman Memorial Trophy Award is given annually to the "most outstanding player in collegiate football." The Heisman, as it is usually known, is not only a very prestigious award but also a very controversial award. With so many players playing so many different positions, how do you determine who is the most "outstanding"?

In theory, any football player in any division of college football is a potential *candidate* for the award. In practice, the real candidates are players from Division I programs and are almost always in the glamour positions—quarterback or running back. (Since its inception in 1935, only once has the award gone to a defensive player—Charles Woodson of Michigan.)

The *voters* are members of the media plus all past Heisman award winners still living, plus one vote from the public (as determined by a survey conducted by ESPN). There are approximately 930 *voters* (the exact number of voters varies each year). Each voter submits a *truncated preference ballot* consisting of a first, second, and third choice (see Fig. 1-1[d]). A first-place vote is worth 3 points, a second-place vote 2 points, and a third-place vote 1 point. The candidate with the most total points from all the ballots is awarded the Heisman trophy in a televised ceremony held each December at the Downtown Athletic Club in New York.

While only one player gets the award, the finalists are ranked by the number of total points received, in effect making the *outcome* of the Heisman trophy a *partial ranking* of the top candidates. (For the 2015 season, the winner was Derrick Henry of the University of Alabama, second place went to Christian McCaffrey of Stanford, and third place went to Deshaun Watson of Clemson.)

EXAMPLE 1.5 *AMERICAN IDOL*

American Idol is a popular reality TV singing competition for individuals. Each year, the winner of *American Idol* gets a big recording contract, and many past winners have gone on to become famous recording artists (Kelly Clarkson, Carrie Underwood, Taylor Hicks). While there is a lot at stake and a big reward for winning, *American Idol* is not a winner-only competition, and there is indeed a ranking of all the finalists. In fact, some nonwinners (Clay Aiken, Jennifer Hudson) have gone on to become great recording artists in their own right.

The 12 (sometimes 13) candidates who reach the final rounds of the competition compete in a weekly televised show. During and immediately after each

weekly show the voters cast their votes. The candidate with the fewest number of votes gets eliminated from the competition, and the following week the process starts all over again with one fewer candidate (occasionally two candidates are eliminated in the same week—see Table 1-11). And who are the *voters* responsible for deciding the fate of these candidates? Anyone and everyone—you, me, Aunt Betsie—we are all potential voters. All one has to do to vote for a particular candidate is to text or call a toll-free number specific to that candidate ("to vote for Clark, call 1-866-IDOLS09," etc.). *American Idol* voting is an example of democracy run amok—you can vote for a candidate even if you never heard her sing, and you can vote as many times as you want.

VOTE
CLARK
TONIGHT
5 WAYS TO VOTE
1. SuperVote Online - http://www.americanidol.com/vote ends 8 am PT next day
2. SuperVote on FoxNow App - download in iTunes or Google Play
3. Google Search - google "American Idol Vote" or "Idol Vote"
4. Text Voting - text the "9" to 21523 ends 11pm PT
5. Toll-Free Voting - call 1-866-IDOLS09

By the final week of the competition there are only two finalists left, and after one last frenzied round of voting, the winner is determined. (For the final 2016 season of *American Idol* the two finalists were La'Porsha Renae and Trent Harmon, the eventual winner. Table 1-11 shows a detailed summary of the results.)

Examples 1.1 through 1.5 represent just a small sample of how elections can be structured, both in terms of the ballots (think of these as the *inputs* to the election) and the types of outcomes we look for (the *outputs* of the election). We will revisit some of these examples and many others as we wind our way through the chapter.

Preference Ballots and Preference Schedules

Let's focus now on elections where the balloting is done by means of preference ballots, as in Examples 1.1 and 1.2. The great advantage of preference ballots (compared with, for example, single-choice ballots) is that they provide a great deal of useful information about an individual voter's preferences—in both direct and indirect ways.

To illustrate what we mean, consider the preference ballot shown in Fig. 1-3. This ballot directly tells us that the voter likes candidate C best, B second best, D third best, and A last. But, in fact, this ballot tells us a lot more—it tells us unequivocally which candidate the voter would choose if it came down to a choice between just two candidates. For example, if it came down to a choice between, say, A and B, which one would this voter choose? Of course she would choose B—she has B above A in her ranking. Thus, a preference ballot allows us to make relative comparisons between any two candidates—*the candidate higher on the ballot is always preferred over the candidate in the lower position.* Please take note of this simple but important idea, as we will use it repeatedly later in the chapter.

Ballot
1st C
2nd B
3rd D
4th A

FIGURE 1-3

The second important idea we will use later is the assumption that the relative preferences in a preference ballot do not change if one of the candidates withdraws or is eliminated. Once again, we can illustrate this using Fig. 1-3. What would happen if for some unforeseen reason candidate B drops out of the race right before the ballots are tabulated? Do we have to have a new election? Absolutely not—the old ballot simply becomes the ballot shown on the right side of Fig. 1-4. The candidates above B stay put and each of the candidates below B moves up a spot.

Ballot		Ballot
1st C		1st C
~~2nd B~~	→	2nd D
3rd D		3rd A
4th A		

FIGURE 1-4

In an election with many voters, some voters will vote exactly the same way—for the same candidates in the same order of preference. If we take a careful look at the 37 ballots submitted for the Math Club election shown in Fig. 1-2, we see that 14 ballots look exactly the same (*A* first, *B* second, *C* third, *D* fourth), another 10 ballots look the same, and so on. So, if you were going to tabulate the 37 ballots, it might make sense to put all the *A*-*B*-*C*-*D* ballots in one pile, all the *C*-*B*-*D*-*A* ballots in another pile, and so on. If you were to do this you would get the five piles shown in Fig. 1-5 (the order in which you list the piles from left to right is irrelevant). Better yet, you can make the whole idea a little more formal by putting all the ballot information in a table such as Table 1-1, called the **preference schedule** for the election.

Ballot	Ballot	Ballot	Ballot	Ballot
1st *A*	1st *C*	1st *D*	1st *B*	1st *C*
2nd *B*	2nd *B*	2nd *C*	2nd *D*	2nd *D*
3rd *C*	3rd *D*	3rd *B*	3rd *C*	3rd *B*
4th *D*	4th *A*	4th *A*	4th *A*	4th *A*
14	10	8	4	1

FIGURE 1-5 The 37 Math Club election ballots organized into piles.

Number of voters	14	10	8	4	1
1st	*A*	*C*	*D*	*B*	*C*
2nd	*B*	*B*	*C*	*D*	*D*
3rd	*C*	*D*	*B*	*C*	*B*
4th	*D*	*A*	*A*	*A*	*A*

TABLE 1-1 ■ Preference schedule for the Math Club election

We will be working with preference schedules throughout the chapter, so it is important to emphasize that a preference schedule is nothing more than a convenient bookkeeping tool—it summarizes all the elements that constitute the input to an election: the candidates, the voters, and the balloting. Just to make sure this is clear, we conclude this section with a quick example of how to read a preference schedule.

EXAMPLE 1.6 THE CITY OF KINGSBURG MAYORAL ELECTION

Number of voters	93	44	10	30	42	81
1st	*A*	*B*	*C*	*C*	*D*	*E*
2nd	*B*	*D*	*A*	*E*	*C*	*D*
3rd	*C*	*E*	*E*	*B*	*E*	*C*
4th	*D*	*C*	*B*	*A*	*A*	*B*
5th	*E*	*A*	*D*	*D*	*B*	*A*

TABLE 1-2 ■ Preference schedule for the Kingsburg mayoral election

Table 1-2 shows the preference schedule summarizing the results of the most recent election for mayor of the city of Kingsburg (there actually is a city by that name, but the election is fictitious). Just by looking at the preference schedule we can answer all of the relevant input questions:

- *Candidates*: there were five candidates (*A*, *B*, *C*, *D*, and *E*, which are just abbreviations for their real names).
- *Voters*: there were 300 voters that submitted ballots (add the numbers at the head of each column: $93 + 44 + 10 + 30 + 42 + 81 = 300$).
- *Balloting*: the 300 preference ballots were organized into six piles as shown in Table 1-2.

The question that still remains unanswered: Who is the winner of the election? In the next four sections we will discuss different ways in which such output questions can be answered.

Ties

In any election, be it a *winner-only* election or a *ranking* of the candidates, ties can occur. What happens then?

In some elections (for example, Academy Awards, sports awards, and reality TV competitions) ties are allowed to stand and need not be broken. Here are a few interesting examples:

- **Academy Awards:** In 1932, Frederic March and Wallace Beery tied for Best Actor; in 1968, Katharine Hepburn and Barbra Streisand tied for Best Actress. A few more ties for lesser awards have occurred over the years. When ties occur, both winners receive the Oscar.
- **Grammys:** In 1992, Lisa Fischer and Patti LaBelle tied for Best Female R&B Vocal Performance and shared the award. More recently (2007), Aretha Franklin and Mary J. Blige shared the award for Best Gospel Performance with the Clark Sisters.
- **NFL Most Valuable Player:** In 2004, Peyton Manning and Steve McNair shared the MVP award. (So did Brett Favre and Barry Sanders in 1997, as well as Norm van Brocklin and Joe Schmidt in 1960.)
- **Cy Young Award:** Mike Cuellar of the Baltimore Orioles and Denny McLain of the Detroit Tigers tied for the American League award in 1969. There have been no other ties since.
- ***American Idol*:** In 2016 Gianna Isabella and Olivia Rox tied for 9th place in the finals. They were declared as sharing the 9th-10th position. Surprisingly, this tie was followed by a second tie for the 7th-8th position between Lee Jean and Avalon Young. (see Table 1-11).

In other situations, especially in elections for political office (president, senator, mayor, city council, etc.), ties cannot be allowed (can you imagine having co-presidents or co-mayors?), and then a tie-breaking rule must be specified. The Constitution, for example, stipulates how a tie in the Electoral College is broken, and most elections have a set rule for breaking ties. The most common method for breaking a tie for political office is through a runoff election, but runoff elections are expensive and take time, so many other tie-breaking procedures are used. Here are a few interesting examples:

- In the 2009 election for a seat in the Cave Creek, Arizona, city council, Thomas McGuire and Adam Trenk tied with 660 votes each. The winner was decided by drawing from a deck of cards. Mr. McGuire drew first—a six of hearts. Mr. Trenk (the young man with the silver belt buckle) drew next and drew a king of hearts. This is how Mr. Trenk became a city councilman.

- In the 2010 election for mayor of Jefferson City, Tennessee, Rocky Melton and Mark Potts tied with 623 votes each. The decision then went to a vote of the city council. Mr. Potts became the mayor.
- In the 2011 election for trustees of the Island Lake, Illinois, village board, Allen Murvine and Charles Cernak tied for one of the three seats with 576 votes each. The winner was decided by a coin toss. Mr. Cernak called tails and won the seat.

Ties and tie-breaking procedures add another layer of complexity to an already rich subject. To simplify our presentation, in this chapter we will stay away from ties as much as possible.

1.2 The Plurality Method

The **plurality method** is arguably the simplest and most commonly used method for determining the outcome of an election. With the plurality method, all that matters is how many first-place votes each candidate gets: In a *winner-only* election the candidate with the most first-place votes is the winner; in a *ranked* election the candidate with the most first-place votes is first, the candidate with the second most is second, and so on.

For an election decided under the plurality method, *preference ballots* are not needed, since the voter's second, third, etc. choices are not used. But, since we already have the preference schedule for the Math Club election (Examples 1.1 and 1.2) let's use it to determine the outcome under the plurality method.

EXAMPLE 1.7 THE MATH CLUB ELECTION UNDER THE PLURALITY METHOD

Number of voters	14	10	8	4	1
1st	*A*	*C*	*D*	*B*	*C*
2nd	*B*	*B*	*C*	*D*	*D*
3rd	*C*	*D*	*B*	*C*	*B*
4th	*D*	*A*	*A*	*A*	*A*

TABLE 1-3 ■ Preference schedule for the Math Club election

We discussed the Math Club election in Section 1.1. Table 1-3 shows once again the preference schedule for the election. Counting only first-place votes, we can see that *A* gets 14, *B* gets 4, *C* gets 11, and *D* gets 8. So there you have it: In the case of a *winner-only* election (see Example 1.1) the winner is *A* (Headline: "Alisha wins presidency of the Math Club"); in the case of a *ranked election* (see Example 1.2) the results are: *A* first (14 votes); *C* second (11 votes); *D* third (8 votes); *B* fourth (4 votes). (Headline "New board of MAS elected! President: Alisha; VP: Carmen; Treasurer: Dave; Secretary: Boris.")

The vast majority of elections for political office in the United States are decided using the plurality method. The main appeal of the plurality method is its simplicity, but as we will see in our next example, the plurality method has many drawbacks.

EXAMPLE 1.8 THE 2010 MAINE GOVERNOR'S ELECTION

Like many states, Maine chooses its governor using the plurality method. In the 2010 election there were five candidates: Eliot Cutler (Independent), Paul LePage (Republican), Libby Mitchell (Democrat), Shawn Moody (Independent), and Kevin Scott (Independent). Table 1-4 shows the results of the election. Before reading on, take a close look at the numbers in Table 1-4 and draw your own conclusions.

Candidate	Votes	Percent
Eliot Cutler (Independent)	208,270	36.5%
Paul LePage (Republican)	218,065	38.2%
Libby Mitchell (Democrat)	109,937	19.3%
Shawn Moody (Independent)	28,756	5.0%
Kevin Scott (Independent)	5,664	1.0%

Source: The New York Times, www.elections.nytimes.com/2010/results/governor.

TABLE 1-4 ■ Results of 2010 Maine gubernatorial election

A big problem with the plurality method is that when there are more than two candidates we can end up with a winner that does not have a *majority* (i.e., more than 50%) of the votes. The 2010 Maine gubernatorial election is a case in point. As Table 1-4 shows, Paul LePage became governor with the support of only 38.2% of the voters (which means, of course, that 61.8% of the voters in Maine wanted someone else). A few days after the election, an editorial piece in the *Portland Press Herald* expressed the public concern about the outcome.

> *The election of Paul LePage with 38 percent of the vote means Maine's next governor won't take office with the support of the majority of voters—a situation that has occurred in six of the last seven gubernatorial elections. . . . Some people . . . say it's time to reform the system so Maine's next governor can better represent the consensus of voters.* (From *Is Winning An Election Enough?* by Tom Bell in Portland Press Herald. Copyright © 2010 by MaineToday Media, Inc. Used by permission of MaineToday Media, Inc.)

The second problem with the Maine governor's election is the closeness of the election: Out of roughly 571,000 votes cast, less than 10,000 votes separated the winner and the runner-up. This is not the plurality method's fault, but it does raise the possibility that the results of the election could have been *manipulated* by a small number of voters. Imagine for a minute being inside the mind of a voter we call Mr. Insincere: "Of all these candidates, I like Kevin Scott the best. But if I vote for Scott I'm just wasting my vote—he doesn't have a chance. All the polls say that it really is a tight race between LePage and Cutler. I don't much care for either one, but LePage is the better of two evils. I'd better vote for LePage." The same thinking, of course, can be applied in the other direction—voters afraid to "waste" their vote on Scott (or Moody, or Mitchell) and insincerely voting for Cutler over Le Page. The problem is that we don't know how many *insincere votes* went one way or the other, and the possibility that there were enough insincere votes to tip the results of the election cannot be ruled out.

While all voting methods can be manipulated by insincere voters, the plurality method is the one that can be most easily manipulated, and insincere voting is quite common in real-world elections. For Americans, the most significant cases of insincere voting occur in close presidential or gubernatorial races between the two major party candidates and a third candidate ("the spoiler") who has little or no chance of winning. Insincere voting not only hurts small parties and fringe candidates, it has unintended and often negative consequences for the political system itself. The history of American political elections is littered with examples of independent candidates and small parties that never get a fair voice or a fair level of funding (it takes 5% of the vote to qualify for federal funds for the next election) because of the "let's not waste our vote" philosophy of insincere voters. The ultimate consequence of the plurality method is an entrenched two-party system that often gives the voters little real choice.

The last, but not least, of the problems with the plurality method is that a candidate may be preferred by the voters over all other candidates and yet not win the election. We will illustrate how this can happen with the example of the fabulous Tasmania State University marching band.

EXAMPLE 1.9 THE FABULOUS TSU BAND GOES BOWLING

Tasmania State University has a superb marching band. They are so good that this coming bowl season they have invitations to perform at five different bowl games: the Rose Bowl (R), the Hula Bowl (H), the Fiesta Bowl (F), the Orange Bowl (O), and the Sugar Bowl (S). An election is held among the 100 band members to decide in which of the five bowl games they will perform. Each band member submits a preference ballot ranking the five choices. The results of the election are shown in Table 1-5 on the next page.

Number of voters	49	48	3
1st	*R*	*H*	*F*
2nd	*H*	*S*	*H*
3rd	*F*	*O*	*S*
4th	*O*	*F*	*O*
5th	*S*	*R*	*R*

TABLE 1-5 ■ Preference schedule for the band election

Under the plurality method the winner of the election is the Rose Bowl (*R*), with 49 first-place votes. It's hard not to notice that this is a rather bad outcome, as there are 51 voters that have the Rose Bowl as their last choice. By contrast, the Hula Bowl (*H*) has 48 first-place votes and 52 second-place votes. Simple common sense tells us that the Hula Bowl is a far better choice to represent the wishes of the entire band. In fact, we can make the following persuasive argument in favor of the Hula Bowl: If we compare the Hula Bowl with any other bowl on a *head-to-head* basis, the Hula Bowl is always the preferred choice. Take, for example, a comparison between the Hula Bowl and the Rose Bowl. There are 51 votes for the Hula Bowl (48 from the second column plus the 3 votes in the last column) versus 49 votes for the Rose Bowl. Likewise, a comparison between the Hula Bowl and the Fiesta Bowl would result in 97 votes for the Hula Bowl (first and second columns) and 3 votes for the Fiesta Bowl. And when the Hula Bowl is compared with either the Orange Bowl or the Sugar Bowl, it gets all 100 votes. Thus, no matter with which bowl we compare the Hula Bowl, there is always a majority of the band that prefers the Hula Bowl.

Marie Jean Antoine Nicolas Caritat, Marquis de Condorcet (1743–1794)

A candidate preferred by a majority of the voters over every other candidate when the candidates are compared in head-to-head comparisons is called a **Condorcet candidate** (named after the Marquis de Condorcet, an eighteenth-century French mathematician and philosopher). Not every election has a Condorcet candidate, but if there is one, it is a good sign that this candidate represents the voice of the voters better than any other candidate. In Example 1.9 the Hula Bowl is the Condorcet candidate—it is not unreasonable to expect that it should be the winner of the election. We will return to this topic in Section 1.6.

1.3 The Borda Count Method

Jean-Charles de Borda (1733–1799)

The second most commonly used method for determining the winner of an election is the **Borda count method**, named after the Frenchman Jean-Charles de Borda. In this method each place on a ballot is assigned points as follows: 1 point for *last* place, 2 points for *second from last* place, and so on. At the top of the ballot, a *first-place* vote is worth N points, where N represents the number of candidates. The points are tallied for each candidate separately, and the candidate with the highest total is the winner. If we are ranking the candidates, the candidate with the second-most points comes in second, the candidate with the third-most points comes in third, and so on. We will start our discussion of the Borda count method by revisiting the Math Club election.

EXAMPLE 1.10 THE MATH CLUB ELECTION (BORDA COUNT)

Table 1-6 shows the preference schedule for the Math Club election with the Borda points for the candidates shown in parentheses to the right of their names. For example, the 14 voters in the first column ranked A first (giving A $14 \times 4 = 56$ points), B second ($14 \times 3 = 42$ points), and so on.

Number of voters	14	10	8	4	1
1st (4 points)	A (56)	C (40)	D (32)	B (16)	C (4)
2nd (3 points)	B (42)	B (30)	C (24)	D (12)	D (3)
3rd (2 points)	C (28)	D (20)	B (16)	C (8)	B (2)
4th (1 point)	D (14)	A (10)	A (8)	A (4)	A (1)

TABLE 1-6 ■ Borda points for the Math Club election

When we tally the points,

A gets $56 + 10 + 8 + 4 + 1 = 79$ points,

B gets $42 + 30 + 16 + 16 + 2 = 106$ points,

C gets $28 + 40 + 24 + 8 + 4 = 104$ points,

D gets $14 + 20 + 32 + 12 + 3 = 81$ points.

The Borda winner of this election is Boris! (Wasn't Alisha the winner of this election under the plurality method?)

If we have to rank the candidates, B is first, C second, D third, and A fourth. To see what a difference the voting method makes, compare this ranking with the ranking obtained under the plurality method (Example 1.7).

EXAMPLE 1.11 THE 2015 HEISMAN AWARD

For general details on the Heisman Award, see Example 1.4. The Heisman is determined using a Borda count, but with *truncated preference ballots*: each voter chooses a first, second, and third choice out of a large list of candidates, with a first-place vote worth 3 points, a second-place vote worth 2 points, and a third-place vote worth 1 point.

Table 1-7 shows a summary of the balloting for the three 2015 finalists. The table shows the number of first-, second-, and third-place votes for each of the three finalists; the last column shows the total point tally for each. Notice that Table 1-7 is not a preference schedule. Because the Heisman uses truncated preference ballots and many candidates get votes, it is easier and more convenient to summarize the balloting this way.

Player	1st (3 pts.)	2nd (2 pts.)	3rd (1 pt.)	Total points
Derrick Henry	378	277	144	1832
Christian McCaffrey	290	246	177	1539
Deshaun Watson	148	240	241	1165

Source: Heisman Award, *www.heisman.com*

TABLE 1-7 ■ 2015 Heisman Award: top three finalists

The last column of Table 1-7 shows the total number of points received by each finalist: Derrick Henry of the University of Alabama was the winner with 1832 points, Christian McCaffrey of Stanford University came in second with 1539 points, and Deshaun Watson of Clemson University came in third with 1165 points.

Many variations of the standard Borda count method are possible, the most common being a change in the values assigned to the various positions on the ballot. We will call these **modified Borda count** methods. Example 1.12 illustrates one situation where a modified Borda count is used.

EXAMPLE 1.12 THE 2015 NATIONAL LEAGUE CY YOUNG AWARD

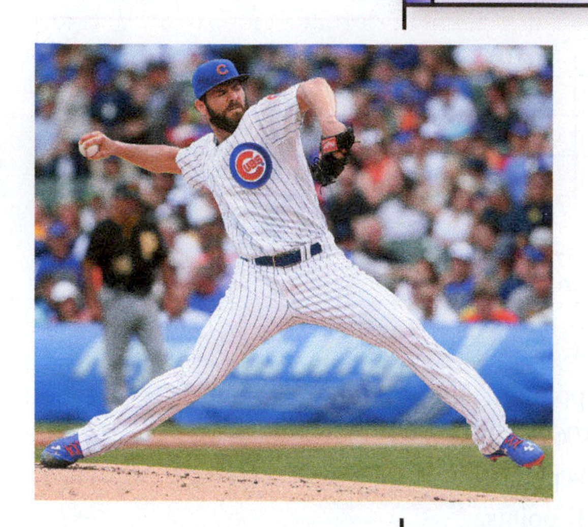

The Cy Young Award is an annual award given by Major League baseball for "the best pitcher" in each league (one award for the American League and one for the National League). For the National League award there are 30 *voters* (they are baseball writers—two from each of the 15 cities having a National League team), and each voter submits a truncated preference ballot with a first, second, third, fourth, and fifth choice. The modification in the Cy Young calculations (in effect for the first time with the 2010 award) is that first place is worth 7 points (rather than 5). The other places in the ballot count just as in a regular Borda count: 4 points for second, 3 points for third, 2 points for fourth, and 1 point for fifth. The idea here is to give extra weight to first-place votes—the gap between a first and a second place is bigger than the gap between a second and a third place, a third and fourth place, and so on.

Table 1-8 shows the top five finalists for the 2015 National League Cy Young award.

Pitcher	1st (7 pts.)	2nd (4 pts.)	3rd (3 pts.)	4th (2 pts.)	5th (1 pt.)	Total points
Jake Arrieta (Cubs)	17	11	2	0	0	169
Zack Greinke (Dodgers)	10	17	3	0	0	147
Clayton Kershaw (Dodgers)	3	2	23	1	1	101
Gerrit Cole (Pirates)	0	0	2	13	8	40
Max Scherzer (Nationals)	0	0	0	13	6	32

Source: CBS Sports, *www.CBSSports.com*

TABLE 1-8 ■ 2015 National League Cy Young Award: top five finalists

In real life, the Borda count method (or some variation of it) is widely used in a variety of settings, from individual sport awards to music industry awards to the hiring of school principals, university presidents, and corporate executives. It is generally considered to be a much better method for determining the outcome of an election than the plurality method. In contrast to the plurality method, it takes into account the voter's preferences not just for first place but also for second, third, etc., and then chooses as the winner the candidate with the best average ranking—the best compromise candidate, if you will.

1.4 The Plurality-with-Elimination Method

In the United States most municipal and local elections have a majority requirement—a candidate needs a majority of the votes to get elected. With only two candidates this is rarely a problem (unless they tie, one of the two candidates must have a majority of the votes). When there are three or more candidates running, it can easily happen that no candidate has a majority. Typically, the candidate or candidates with the fewest first-place votes are eliminated, and a runoff election is held. But runoff elections are expensive, and in these times of tight budgets more efficient ways to accomplish the "runoff" are highly desirable.

A very efficient way to implement the runoff process without needing runoff elections is to use preference ballots, since a preference ballot tells us not only which candidate the voter wants to win but also which candidate the voter would choose in a runoff (with one important caveat—we assume the voters are consistent in their preferences and would stick with their original ranking of the candidates). The idea is to use the information in the preference schedule to eliminate the candidates with the *fewest* first-place votes one at a time until one of them gets a majority. This method has become increasingly popular and is now known under several other names, including, *instant-runoff voting* (IRV), *ranked-choice voting* (RCV), and the *Hare method*. For the sake of clarity, we will call it the *plurality-with-elimination* method—it is the most descriptive of all the names.

Here is a formal description of the **plurality-with-elimination method**:

- **Round 1.** Count the first-place votes for each candidate, just as you would in the plurality method. If a candidate has a majority of the first-place votes, then that candidate is automatically declared the winner. If no candidate has a majority of the first-place votes, eliminate the candidate (or candidates if there is a tie) with the *fewest* first-place votes and transfer (pass down) those first-place votes to the next eligible candidate(s) on those ballots. Cross out the name(s) of the eliminated candidate(s) from the preference schedule.
- **Round 2.** Recount the votes. If a candidate now has a majority of the first-place votes, declare that candidate the winner. Otherwise, eliminate the candidate(s) with the fewest votes and transfer (pass down) those first-place votes to the next eligible candidate(s) on those ballots. Cross out the name(s) of the eliminated candidate(s) from the preference schedule.
- **Rounds 3, 4, . . .** Repeat the process, each time eliminating the candidate with the fewest first-place votes and transferring those first-place votes to the next eligible candidates on those ballots. Continue until there is a candidate with a majority of the first-place votes. That candidate is the winner of the election.

  In a ranked election the candidates should be ranked in reverse order of elimination: the candidate eliminated in the last round gets second place, the candidate eliminated in the second-to-last round gets third place, and so on.

EXAMPLE 1.13 THE MATH CLUB ELECTION (PLURALITY-WITH-ELIMINATION)

Let's see how the plurality-with-elimination method works when applied to the Math Club election. For the reader's convenience Table 1-9 shows the preference schedule again.

Number of voters	14	10	8	4	1
1st	*A*	*C*	*D*	*B*	*C*
2nd	*B*	*B*	*C*	*D*	*D*
3rd	*C*	*D*	*B*	*C*	*B*
4th	*D*	*A*	*A*	*A*	*A*

TABLE 1-9 ■ Preference schedule for the Math Club election

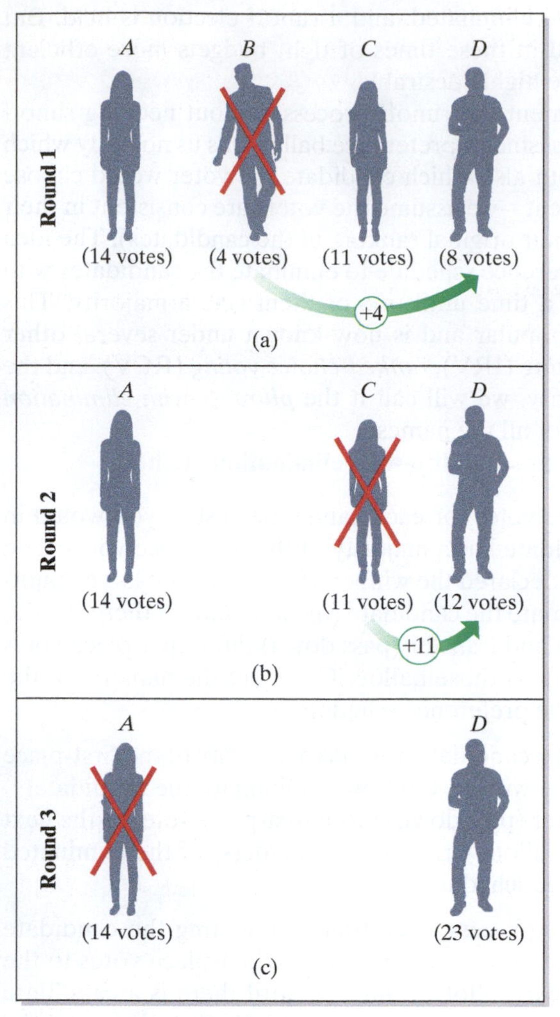

FIGURE 1-6 Boris is eliminated first, then Carmen, and then Alisha. The last one standing is Dave.

Round 1.

Candidate	A	~~B~~	C	D
First-place Votes	14	④	11	8

+4

B has the fewest first-place votes and is eliminated first [Fig. 1-6(a)]. After B is eliminated, the four votes that originally went to B are transferred to D (per column 4 of Table 1-9).

Round 2. We now recount the first-place votes. The new tally is

Candidate	A	~~C~~	D
First-place Votes	14	⑪	12

+11

In this round C has the fewest first-place votes and is eliminated [Fig. 1-6(b)]. The 11 votes that went to C in round 2 are all transferred to D (per columns 2 and 5 of Table 1-9).

Round 3. Once again we recount the first-place votes and end up with

Candidate	A	D
First-place votes	14	23

Now D has a majority of the first-place votes and is declared the winner [Fig. 1-6(c)].

In the case of a ranked election we have D first, A second (eliminated in round 3), C third (eliminated in round 2), and B last (eliminated in round 1).

Our next example illustrates a few subtleties that can come up when applying the plurality-with-elimination method.

EXAMPLE 1.14 THE CITY OF KINGSBURG MAYORAL ELECTION

Table 1-10 shows the preference schedule for the Kingsburg mayoral election first introduced in Example 1.6. To save money Kingsburg has done away with runoff elections and now uses plurality-with-elimination for all local elections. (Notice that since there are 300 voters voting in this election, a candidate needs 151 or more votes to win.)

Number of voters	93	44	10	30	42	81
1st	A	B	C	C	D	E
2nd	B	D	A	E	C	D
3rd	C	E	E	B	E	C
4th	D	C	B	A	A	B
5th	E	A	D	D	B	A

TABLE 1-10 ■ Preference schedule for the Kingsburg mayoral election

Round 1. Here C has the fewest number of first-place votes and is eliminated first. Of the 40 votes originally cast for C, 10 are transferred to A (per column 3 of Table 1-10) and 30 are transferred to E (per column 4 of Table 1-10).

Candidate	A	B	~~C~~	D	E
Votes	93	44	40	42	81

+10 (C → A) +30 (C → E)

Round 2. After a recount of the first-place votes, D has the fewest and is eliminated. The 42 votes originally cast for D are now transferred to E, the next eligible candidate since C has already been eliminated (see column 5 of Table 1-10).

Candidate	A	B	~~D~~	E
Number of first-place votes	103	44	42	111

+42 (D → E)

Round 3.

Candidate	A	B	E
Number of first-place votes	103	44	153

After a recount of the first-place votes, E has a majority and is declared the winner. (If this were a ranked election, we would continue on to Round 4, only to determine second place between A and B, but this is an election for mayor, and second place, third place, etc. are meaningless.)

Several variations of the plurality-with-elimination method are used in real-life elections. One of the most popular goes by the name *instant-runoff voting* (also called *ranked-choice voting* in some places). Instant-runoff voting uses a truncated preference ballot (typically asking for just first, second, and third choice). Once the ballots are cast the process works very much like plurality-with-elimination: the candidate(s) with the fewest first-place votes are eliminated and those votes are transferred to the second-place candidates in those ballots; in the next round the candidate(s) with the fewest votes are eliminated and those votes are transferred to the next eligible candidate, and so on. There is one important difference: unlike regular plurality-with-elimination there is a point at which some votes can no longer be transferred (say your vote was for candidates X, Y, and Z—if and when all three of them are eliminated there is no one to transfer your vote to). Such votes are called *exhausted votes* and although perfectly legal, they don't count in the final analysis.

Ranked-choice voting is used in several U.S. cities in elections for mayor and city council, including San Francisco, Minneapolis, St. Paul, and Oakland, California, as well as in elections for political office in Australia, Canada, Ireland, and New Zealand. We will illustrate how ranked-choice voting works with the 2014 election for mayor of Oakland.

EXAMPLE 1.15 THE 2014 OAKLAND MAYORAL ELECTION

In 2014 a total of 16 candidates were running for mayor of Oakland, an inordinately large number for an election for political office, and it took 15 rounds of elimination before a winner emerged. But Oakland has been using ranked-choice voting since 2010, so the entire elimination process took place inside a computer, and the final results were known without delay.

Of the 16 candidates, only four had a realistic chance of winning, so to keep things simple we will show how the vote count evolved once the other candidates were eliminated (rounds 1 through 12) and just the main four candidates were left.

- **Round 13.**

Dan Siegel	Jean Quan	Rebecca Kaplan	Elizabeth Schaaf
17,402 votes	18,049 votes	18,662 votes	39,941 votes

Dan Siegel has the fewest number of first-place votes and is eliminated. His 17,402 are transferred as follows: 2476 to Quan, 4679 to Kaplan, and 3877 to Schaaf. The remaining 6370 ballots are "exhausted" because all the candidates in those ballots are gone. At this point, the exhausted ballots are discarded [Fig. 1-7(a)].

- **Round 14.**

Jean Quan	Rebecca Kaplan	Elizabeth Schaaf
20,525 votes	23,341 votes	43,818 votes

Now Jean Quan has the fewest number of first-place votes and is eliminated. Her 20,525 are transferred as follows: 5080 go to Kaplan, 4988 go to Schaaf, and 10,457 are exhausted [Fig. 1-7(b)].

- **Round 15.**

Rebecca Kaplan	Elizabeth Schaaf
28,421 votes	48,806 votes

We are now down to Kaplan and Schaaf, and Schaaf has the majority of the votes.

So this is how Elizabeth "Libby" Schaaf got to be the current mayor of Oakland (at least through 2018). It took 15 rounds but only one set of ballots—the heavy lifting was done by ranked-choice voting.

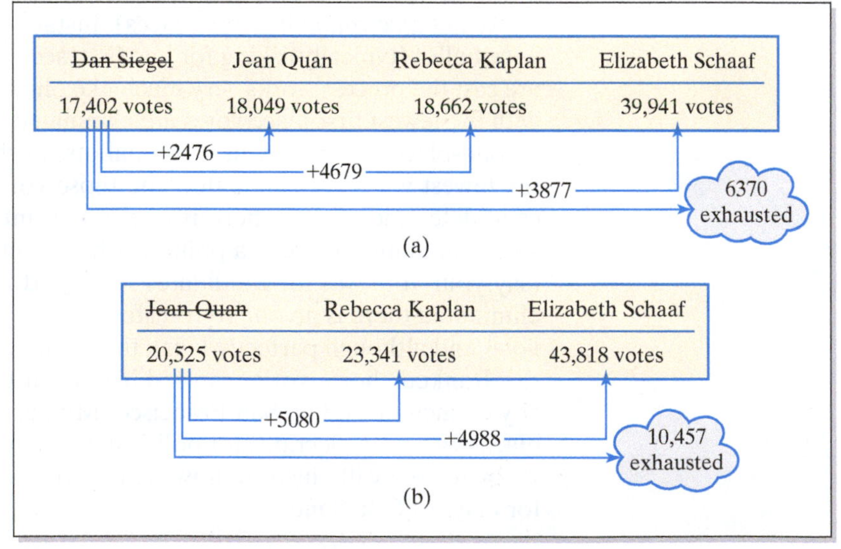

Source: Alameda County Registrar of Voters, ACGov.org

FIGURE 1-7 Results of 2014 Oakland mayoral election

As mentioned earlier in this section, the practical advantage of plurality-with-elimination is that it does away with expensive and time-consuming runoff elections. There is one situation, however, where expense is not an issue and delaying the process is part of the game: televised competitions such as *Dancing with the Stars*, *The Voice*, and

American Idol. The longer the competition goes, the higher the ratings—having many runoffs accomplishes this goal. All of these "elections" work under the same variation of the plurality-with-elimination method: have a round of competition, vote, eliminate the candidate (or candidates) with the fewest votes. The following week have another round of competition and repeat the process. This builds up to the last round of competition, when there are two finalists left. Millions of us get caught up in the hoopla. We will illustrate one such election using the 2016 *American Idol* competition (by the way, 2016 was officially the final year of American idol, although there are plenty of rumors that it will come back sometime in the future).

EXAMPLE 1.16 THE 2016 *AMERICAN IDOL* COMPETITION

We discussed *American Idol* as an election in Example 1.5. Table 1-11 shows the evolution of the 2016 competition. As noted in Example 1.5, the winner is the big deal, but how the candidates place in the competition is also of some relevance, so we consider *American Idol* a ranked election. Working our way up from the bottom of Table 1-11, we see how the process of elimination played out: Gianna Isabella and Olivia Rox were eliminated in the first round and tied for 9th-10th place; Lee Jean and Avalon Young were eliminated in the second round and tied for 7th-8th place; Tristan MacIntosh was eliminated in the third round and placed in 6th place . . . and so it went for a total of seven rounds. In the final round it came down to a showdown between La'Porsha Renae and Trent Harmon. La'Porsha was eliminated and Trent became the 2016 *American Idol.*

Winner	→	Trent Harmon		
Runner-up	→	La'Porsha Renae	→	eliminated in the final round
3rd place	→	Dalton Rapattoni	→	eliminated in the 6th round
4th place	→	MacKenzie Bourg	→	eliminated in the 5th round
5th place	→	Sonika Vaid	→	eliminated in the 4th round
6th place	→	Tristan McIntosh	→	eliminated in the 3rd round
7th place 8th place	→	Lee Jean Avalon Young (tie)	→	eliminated in the 2nd round
9th place 10th place	→	Gianna Isabella Olivia Rox (tie)	→	eliminated in the 1st round

TABLE 1-11 ■ 2016 American Idol results

1.5 The Method of Pairwise Comparisons

One of the most useful features of a preference schedule is that it allows us to find the winner of any **pairwise comparison** between candidates. Specifically, given any two candidates—call them X and Y—we can count how many voters rank X above Y and how many rank Y above X. The one with the most votes wins the pairwise comparison. This is the basis for a method called the **method of pairwise comparisons** (sometimes also called *Copeland's method*). For each possible pairwise comparison between candidates, give 1 point to the winner, give 0 points to the loser, and when the pairwise comparison ends up in a tie give each candidate $\frac{1}{2}$ point. The candidate with the most points is the winner. (If we are ranking the candidates, the candidate with the second-most points is second, and so on.)

The method of pairwise comparisons is very much like a round-robin tournament: every player plays every other player once; the winner of each "match" gets a point and the loser gets no points (if there is a tie each gets $\frac{1}{2}$ point); and the player with the most points wins the tournament. As usual, we will start with the Math Club election as our first example.

EXAMPLE 1.17 THE MATH CLUB ELECTION (PAIRWISE COMPARISONS)

Number of voters	14	10	8	4	1
1st	*A*	*C*	*D*	*B*	*C*
2nd	*B*	*B*	*C*	*D*	*D*
3rd	*C*	*D*	*B*	*C*	*B*
4th	*D*	*A*	*A*	*A*	*A*

TABLE 1-12 ■ Preference schedule for the Math Club election

Table 1-12 shows, once again, the preference schedule for the Math Club election. With four candidates, there are six possible pairwise comparisons to consider (see the first column of Table 1-13). For the sake of brevity, we will go over a couple of these pairwise comparisons in detail and leave the details of the other four to the reader.

- *A* v *B*: The first column of Table 1-12 represents 14 votes for *A* (*A* is ranked higher than *B*); the remaining 23 votes are for *B* (*B* is ranked higher than *A* in the last four columns of the table). The winner of this comparison is *B*.
- *C* v *D*: The first, second, and last columns of Table 1-12 represent votes for *C* (*C* is ranked higher than *D*); the third and fourth columns represent votes for *D* (*D* is ranked higher than *C*). Thus, *C* has 25 votes to *D*'s 12 votes. The winner of this comparison is *C*.

Pairwise comparison	Votes	Winner
(1) *A* v *B*	*A*: 14 votes *B*: 23 votes	*B*
(2) *A* v *C*	*A*: 14 votes *C*: 23 votes	*C*
(3) *A* v *D*	*A*: 14 votes *D*: 23 votes	*D*
(4) *B* v *C*	*B*: 18 votes *C*: 19 votes	*C*
(5) *B* v *D*	*B*: 28 votes *D*: 9 votes	*B*
(6) *C* v *D*	*C*: 25 votes *D*: 12 votes	*C*

Total points: $C = 3, B = 2, D = 1, A = 0$

TABLE 1-13 ■ Pairwise comparisons for the Math Club election

We continue this way, checking the results of all six possible comparisons (try it now on your own, before you read on!). Once you are done, you should get something along the lines of Table 1-13. A tally of the point totals (shown at the bottom of the table) gives us the outcome of the election: In a winner-only election the winner is *C* (with 3 points); in a ranked election *C* is first (3 points), *B* second (2 points), *D* third (1 point), and *A* fourth (no points).

Method	Winner only	Ranking 1st	2nd	3rd	4th
Plurality	*A*	*A*	*C*	*D*	*B*
Borda count	*B*	*B*	*C*	*D*	*A*
Plurality with elimination	*D*	*D*	*A*	*C*	*B*
Pairwise comparisons	*C*	*C*	*B*	*D*	*A*

TABLE 1-14 ■ The outcome of the Math Club election under four different voting methods

If you have been paying close attention, you may have noticed that the results of the Math Club election have been different under each of the voting methods we have discussed—both in terms of the winner and in terms of the ranking of the candidates. This can be seen quite clearly in the summary results shown in Table 1-14. It is amazing how much the outcome of an election can depend on the voting method used!

One more important comment about Example 1.17: Notice that *C* was the *undefeated* champion, as *C* won each of the pairwise comparisons against the other candidates. (We already saw that there is a name for a candidate that beats all the other candidates in pairwise comparisons—we call such a candidate a *Condorcet* candidate.) The method of pairwise comparisons always chooses the Condorcet candidate (when there is one) as the winner of the election, but this is not true with all methods. Under the plurality method, for example, you can have a Condorcet candidate that does not win the election (see Example 1.9).

Although the method of pairwise comparisons is a pretty good method, in real-life elections it is not used as much as the other three methods we discussed. In the next example we will illustrate one interesting and meaningful (if you are a football fan) application of the method—the selection of draft choices in the National Football League. Because NFL teams are extremely secretive about how they make their draft decisions, we will illustrate the general idea with a made-up example.

EXAMPLE 1.18 THE NFL DRAFT

The Los Angeles LAXers are the newest expansion team in the NFL and are awarded the first pick in the upcoming draft. The team's draft committee (made up of coaches, scouts, and team executives) has narrowed down the list to five candidates: Allen, Byers, Castillo, Dixon, and Evans. After many meetings, the draft committee is ready to vote for the team's first pick in the draft. The election is to be decided using the method of pairwise comparisons.

Table 1-15 shows the preference schedule obtained after each of the 22 members of the draft committee submits a preference ballot ranking the five candidates. There is a total of 10 separate pairwise comparisons to be looked at, and the results are shown in Table 1-16 (we leave it to the reader to check the details).

Number of voters	2	6	4	1	1	4	4
1st	A	B	B	C	C	D	E
2nd	D	A	A	B	D	A	C
3rd	C	C	D	A	A	E	D
4th	B	D	E	D	B	C	B
5th	E	E	C	E	E	B	A

TABLE 1-15 ■ LAXer's draft choice election

Pairwise comparison	Votes	Winner (points)
A v B	A: 7 votes B: 15 votes	B (1)
A v C	A: 16 votes C: 6 votes	A (1)
A v D	A: 13 votes D: 9 votes	A (1)
A v E	A: 18 votes E: 4 votes	A (1)
B v C	B: 10 votes C: 12 votes	C (1)
B v D	B: 11 votes D: 11 votes	$B\left(\frac{1}{2}\right); D\left(\frac{1}{2}\right)$
B v E	B: 14 votes E: 8 votes	B (1)
C v D	C: 12 votes D: 10 votes	C (1)
C v E	C: 10 votes E: 12 votes	E (1)
D v E	D: 18 votes E: 4 votes	D (1)
Total points: $A = 3, B = 2\frac{1}{2}, C = 2, D = 1\frac{1}{2}, E = 1$		

TABLE 1-16 ■ Pairwise comparisons for Example 1.18

We can see from Table 1-16 that the winner of the election is Allen with 3 points. Notice that things are a little trickier here: Allen is the winner of the election even though the draft committee prefers Byers to Allen in a pairwise comparison between the two. We will return to this example in Section 1.6.

You probably noticed in Examples 1-17 and 1-18 that, compared with the other methods, pairwise comparisons takes a lot more work. Each comparison requires a separate calculation, and there seems to be a lot of comparisons that need to be checked. How many? We saw that with 4 candidates there are 6 separate comparisons and with 5 candidates there are 10. As the number of candidates grows, the number of comparisons grows even more. Table 1-17 illustrates the relation between the number of candidates and the number of pairwise comparisons.

Number of candidates	4	5	6	7	8	9	10	...	N
Number of pairwise comparisons	6	10	15	21	28	36	45	...	$\frac{N(N-1)}{2}$

TABLE 1-17 ■ The number of pairwise comparisons

A nice formula for counting the number of pairwise comparisons in an election is given below.

In an election with N candidates, the number of pairwise comparisons is

$$N(N-1)/2$$

1.6 Fairness Criteria and Arrow's Impossibility Theorem

Kenneth J. Arrow (1921–)
Nobel Prize (Economics), 1972

So far, this is what we learned: There are many different types of elections and there are different ways to decide their outcome. We examined four different voting methods in some detail, but there are many others that we don't have time to discuss here (see Exercises 68–70 for a small sample). So now comes a different but fundamental question (that may have already crossed your mind): *Of all those voting methods out there, which one is the best?* As simple as it sounds, this question has vexed social scientists and mathematicians for centuries, going back to Condorcet and Borda in the mid 1700s. For multi-candidate elections (three or more candidates) there is no good answer. In fact, we now know that there are limitations to *all* voting methods. This is a very important and famous discovery known as **Arrow's Impossibility Theorem**. In this section we will discuss the basic ideas behind this theorem.

In the late 1940s the American economist Kenneth Arrow turned the question of finding an ideal voting method on its head and asked himself the following: *What would it take for a voting method to at least be a fair voting method*? To answer this question Arrow set forth a minimum set of requirements that we will call Arrow's **fairness criteria**. (In all fairness, Arrow's original formulation was quite a bit more complicated than the one we present here. The list below is a simplified version.)

- **The majority criterion.** If there is a majority candidate (i.e., a candidate with a majority of the first place votes), then that candidate should be the winner of the election.
- **The Condorcet criterion.** If there is a Condorcet candidate (i.e., a candidate who beats each of the other candidates in a pairwise comparison), then that candidate should be the winner of the election.
- **The monotonicity criterion.** If candidate X is the winner of an election, then X would still be the winner had a voter ranked X higher in his preference ballot. (In other words, a voter should not be able to hurt the winner by moving her up in his ballot.)

- **The independence-of-irrelevant-alternatives (IIA) criterion.** If candidate X is the winner of an election, then X would still be the winner had one or more of the *irrelevant alternatives* (i.e., losing candidates) not been in the race. (In other words, the winner should not be hurt by the elimination from the election of irrelevant alternatives.)

The above fairness criteria represent some (not necessarily all) of the basic principles we expect a democratic election to satisfy and can be used as a benchmark by which we can measure any voting method. If a method violates any one of these criteria, then there is the potential for unfair results under that method.

The next set of examples illustrates how *violations* of the different fairness criteria might occur.

EXAMPLE 1.19 THE BORDA COUNT VIOLATES THE MAJORITY CRITERION

Number of voters	6	2	3
1st	A	B	C
2nd	B	C	D
3rd	C	D	B
4th	D	A	A

TABLE 1-18 ■ Preference schedule for Example 1.19

Table 1-18 shows the preference schedule for a small election. The majority candidate in this election is A with 6 out of 11 first-place votes. However, when we use the Borda count we get A: 29 points, B: 32 points, C: 30 points, D: 19 points, so B is the winner!

So here we have a rather messy situation: A has a majority of the first-place votes, and yet A is *not the winner* under the Borda count method. This is what we mean by "*a violation of the Majority Criterion*"!

Essentially what happened in Example 1.19 is that in a very small scale, A was a "polarizing" candidate—a majority of the voters had A as their first choice, but at the same time there were many voters who had A as their last choice. Candidate B, on the other hand, was more of a compromise candidate—few first-place votes but enough second- and third-place votes to make a difference and beat A. Polarizing candidates (voters either love them or hate them) are quite common in elections for public office (just look at the 2016 presidential election for a case in point), and it is not hard to imagine how real world violations of the majority criterion could easily occur.

EXAMPLE 1.20 THE PLURALITY METHOD VIOLATES THE CONDORCET CRITERION

Number of voters	49	48	3
1st	R	H	F
2nd	H	S	H
3rd	F	O	S
4th	O	F	O
5th	S	R	R

TABLE 1-19 ■ Preference schedule for Example 1.20

Let's revisit Example 1.9 (The Fabulous TSU Band Goes Bowling). Table 1-19 shows the preference schedule once again. In this election the Hula Bowl (H) is the *Condorcet candidate* (see Example 1.9 for the details), and yet the *winner* under the plurality method is the Rose Bowl (R).

Example 1.20 illustrates how, by disregarding the voters' preferences other than first choice, the plurality method can end up choosing a clearly inferior candidate (the Rose Bowl) over a clearly superior Condorcet candidate (the Hula Bowl).

EXAMPLE 1.21 PLURALITY-WITH-ELIMINATION VIOLATES THE MONOTONICITY CRITERION

This example comes in two parts—a before and an after. The "before" part shows how the voters intend to vote just before they cast their ballots. Table 1-20(a) shows the preference schedule for the "before" election. If all voters vote as shown in Table 1-20(a), *C* will be the winner under the plurality-with-elimination method (*B* is eliminated in the first round and the 8 votes for *B* get transferred to *C* in the second round).

Now imagine that just *before* the ballots are cast the two voters represented by the last column of Table 1-20(a) decide that they really like *C* better than *A* and switch *C* from second place to first place on their ballots. Since this is a change favorable to *C*, and *C* was going to win before, we would expect *C* to remain the winner. Surprisingly, this is not the case—just check it out: the preference schedule after the switch is given by Table 1-20(b). Now *A* is eliminated in the first round, the 7 votes for *A* are transferred to *B*, and *B* becomes the winner of the election!

Number of voters	7	8	10	2
1st	*A*	*B*	*C*	*A*
2nd	*B*	*C*	*A*	*C*
3rd	*C*	*A*	*B*	*B*

(a)

Number of voters	7	8	10	2
1st	*A*	*B*	*C*	*C*
2nd	*B*	*C*	*A*	*A*
3rd	*C*	*A*	*B*	*B*

(b)

TABLE 1-20 ■ Preference schedules for Example 1.21 (a) before the change and (b) after the change

Looking at Example 1.21 in retrospect we can say that *C* lost the election because of *too many* first-place votes! (Had *C* been able to convince the two voters in the last column not to switch their ballots in her favor, she would have won!) The monotonicity criterion essentially says that this kind of perverse reversal of electoral fortunes represents a violation of a key principle of fairness: *A candidate should never be penalized for getting more votes.* (Just imagine how bizarre it would be to see your typical politician campaigning *not* to get too many first-place votes!)

EXAMPLE 1.22 PAIRWISE COMPARISONS VIOLATES THE IIA

This example is a continuation of Example 1.18 (The NFL Draft). Table 1-21 is a repeat of Table 1-15. We saw in Example 1.18 that the winner of the election under the method of pairwise comparisons is *A* (you may want to go back and refresh your memory). The LAXers are prepared to make *A* their number-one draft choice and offer him a big contract. *A* is happy. End of story? Not quite.

Number of voters	2	6	4	1	1	4	4
1st	*A*	*B*	*B*	*C*	*C*	*D*	*E*
2nd	*D*	*A*	*A*	*B*	*D*	*A*	*C*
3rd	*C*	*C*	*D*	*A*	*A*	*E*	*D*
4th	*B*	*D*	*E*	*D*	*B*	*C*	*B*
5th	*E*	*E*	*C*	*E*	*E*	*B*	*A*

TABLE 1-21 ■ LAXer's draft choice election: Original preference schedule

Just before the announcement is made, it is discovered that one of the irrelevant alternatives (C) should not have been included in the list of candidates. (Nobody had bothered to tell the draft committee that C had failed the team physical!) So, C is removed from the preference schedule and everything is recalculated. The new preference schedule is now shown in Table 1-22 (it is Table 1-21 after C is removed). Table 1-23 shows the result of the six pairwise comparisons between A, B, D, and E. We can see that now the winner of the election is B! Other than B, nobody is happy!

Number of voters	2	6	4	1	1	4	4
1st choice	A	B	B	B	D	D	E
2nd choice	D	A	A	A	A	A	D
3rd choice	B	D	D	D	B	E	B
4th choice	E	E	E	E	E	B	A

TABLE 1-22 ■ Preference schedule after C is removed

Pairwise comparison	Votes	Winner
A v B	A: 7 votes B: 15 votes	B
A v D	A: 13 votes D: 9 votes	A
A v E	A: 18 votes E: 4 votes	A
B v D	B: 11 votes D: 11 votes	tie
B v E	B: 14 votes E: 8 votes	B
D v E	D: 18 votes E: 4 votes	D

Total points: $B = 2\frac{1}{2}$, $A = 2$, $D = 1\frac{1}{2}$, $E = 0$

TABLE 1-23 ■ Pairwise comparisons for Table 1.22

Example 1.22 illustrates a typical violation of the IIA: The elimination of an irrelevant alternative (C) penalized A and made him lose an election he would have otherwise won and in turn allowed B to win an election he would have otherwise lost. Clearly this is not fair, and the independence of irrelevant alternatives criterion aims to prevent these types of situations.

The point of the four preceding examples is to illustrate the fact that each of the voting methods we studied in this chapter violates one of Arrow's fairness criteria. In fact, some of the voting methods violate more than one criterion. The full story of which fairness criteria are violated by each voting method is summarized in Table 1-24.

If you are looking at Table 1-24 and asking yourself "So what's the point? Why did we spend so much time learning about voting methods that are so flawed?" you have a legitimate gripe. The problem is that we don't have better options: *every voting method—whether already known or yet to be invented—is flawed.* This remarkable fact was discovered in 1949 by Kenneth Arrow and is known as *Arrow's Impossibility Theorem.* To be more precise, Arrow demonstrated that for elections involving three or more candidates *it is mathematically impossible for a voting method to satisfy all four of his fairness criteria.*

Criterion	Plurality	Borda count	Plurality-with-elimination	Pairwise comparisons
Majority	✔	*Yes*	✔	✔
Condorcet	*Yes*	*Yes*	*Yes*	✔
Monotonicity	✔	✔	*Yes*	✔
IIA	*Yes*	*Yes*	*Yes*	*Yes*

TABLE 1-24 ■ Summary of violations of Arrow's fairness criteria: *Yes* indicates that the method *could violate* the criterion; ✔ indicates that the method *satisfies* the criterion.

In one sense, Arrow's Impossibility Theorem is a bit of a downer. It tells us that no matter how hard we try, democracy will never have a perfectly fair voting method and that the potential for some form of unfairness is built into every election. This does not mean that every election is unfair or that every voting method is equally bad, nor does it mean that we should stop trying to improve the quality of our voting experience.

Conclusion

Elections are the mechanism that allows us to make social decisions in a *democracy*. (In contrast to a *dictatorship*, where social decisions are made by one individual and elections are either meaningless or nonexistent.) The purpose of this chapter is to help you see elections in a new light.

In this chapter we discussed many important concepts, including *preference ballots*, *preference schedules*, *winner-only* versus *ranked elections*, *voting methods*, and *fairness criteria*. We saw plenty of examples of elections—some made-up, some real. We learned some specific skills such as interpreting a preference schedule and calculating the outcome of an election under four different voting methods and variations thereof.

Beyond the specific concepts and skills, there were several important general themes that ran through the chapter:

- *Elections are ubiquitous.* The general public tends to think of elections mostly in terms of political decisions (president, governor, mayor, city council, etc.), but elections are behind almost every meaningful social decision made outside the political arena—Academy Awards, *American Idol*, Heisman Trophy, MVP awards, Homecoming Queen, where to go to dinner, etc.
- *There are many different voting methods.* The outcome of an election can be determined in many different ways. In this chapter we discussed in some detail only four voting methods: *plurality, Borda count, plurality-with-elimination*, and *pairwise comparisons*. By no means do these four exhaust the list—there are many other voting methods, some quite elaborate and exotic.

- *Different voting methods can produce different outcomes.* We saw an extreme illustration of this with the Math Club election: each of the four voting methods produced a different winner. Since there were four candidates, we can say that each of them won the election (just pick the "right" voting method). Of course, this doesn't happen all the time and there are many situations where different voting methods produce the same outcome.
- *Fairness in voting is elusive.* For a voting method to be considered fair there are certain basic criteria that it should consistently satisfy. These are called *fairness criteria*. We introduced four in this chapter (*majority, Condorcet, monotonicity*, and *independence of irrelevant alternatives*), but there are others. Each fairness criterion represents a basic principle we expect a democratic election to satisfy. When a voting method violates any one of these criteria then there is the potential for unfair results under that method. All of the voting methods we discussed in this chapter violate at least one (sometimes several) of the criteria, and there is a good reason why: for elections with three or more candidates it is mathematically impossible for any voting method to satisfy all four fairness criteria. This is a simplified version of *Arrow's Impossibility Theorem.*

One concluding thought about this chapter: One should not interpret Arrow's Impossibility Theorem to mean that democracy is bad and that elections are pointless. The lesson to be learned from Arrow's Impossibility Theorem is that no voting system is perfect, because there are some built-in limitations to the process of making decisions in a democracy. This is a good thing to know, and somewhat surprisingly, it is a knowledge made possible through the power of mathematical ideas.

"The search of the great minds of recorded history for the perfect democracy, it turns out, is the search for a chimera, a logical self-contradiction."

– *Paul Samuelson*

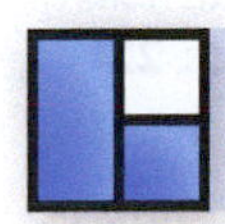

KEY CONCEPTS

1.1 The Basic Elements of an Election

- **single-choice ballot:** a ballot in which a voter only has to choose one candidate, **4**
- **preference ballot:** a ballot in which the voter has to rank all candidates in order of preference, **4**
- **truncated preference ballot:** a ballot in which a voter only has to rank the top *k* choices rather than all the choices, **4**
- **ranking (full ranking):** in an election, an outcome that lists *all* the candidates in order of preference (first, second, . . . , last), **4**
- **partial ranking:** in an election, an outcome where just the top *k* candidates are ranked, **4**
- **preference schedule:** a table that summarizes the preference ballots of all the voters, **8**

1.2 The Plurality Method

- **plurality method:** a voting method that ranks candidates based on the number of first-place votes they receive, **10**
- **insincere voting:** voting for candidates in a manner other than the voter's real preference with the purpose of manipulating the outcome of the election, **11**
- **Condorcet candidate:** a candidate that beats all the other candidates in pairwise comparisons, **12**

1.3 The Borda Count Method

- **Borda count method:** a voting method that assigns points to positions on the ballot and ranks candidates according to the number of points, **12**

1.4 The Plurality-with-Elimination Method

- **plurality-with-elimination method:** a voting method that chooses the candidate with a majority of the votes; when there isn't one it eliminates the candidate(s) with the least votes and transfers those votes to the next highest candidate on those ballots, continuing this way until there is a majority candidate, **15**
- **ranked-choice voting (instant-runoff voting):** a variation of the plurality-with-elimination method based on truncated preference ballots, **17**

1.5 The Method of Pairwise Comparisons

- **method of pairwise comparisons:** a voting method based on head-to-head comparisons between candidates that assigns one point to the winner of each comparison, none to the loser, and $\frac{1}{2}$ point to each of the two candidates in case of a tie, **19**

1.6 Fairness Criteria and Arrow's Impossibility Theorem

- **fairness criteria:** basic rules that define formal requirements for fairness—a fair voting method is expected to always satisfy these basic rules, **22**
- **majority criterion:** a fairness criterion that says that if a candidate receives a majority of the first-place votes, then that candidate should be the winner of the election, **22**
- **Condorcet criterion:** a fairness criterion that says that if there is a Condorcet candidate, then that candidate should be the winner of the election, **22**
- **monotonicity criterion:** a fairness criterion that says that a candidate who would otherwise win an election should not lose the election merely because some voters changed their ballots in a manner that favors that candidate, **22**
- **independence-of-irrelevant-alternatives criterion:** a criterion that says that a candidate who would otherwise win an election should not lose the election merely because one of the losing candidates withdraws from the race, **23**
- **Arrow's Impossibility Theorem:** a theorem that proves that it is mathematically impossible for a voting method to satisfy all of the fairness criteria, **22, 26**

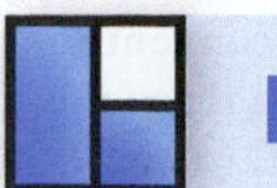

EXERCISES

WALKING

1.1 Ballots and Preference Schedules

1. Figure 1-8 shows the preference ballots for an election with 21 voters and 5 candidates. Write out the preference schedule for this election.

Ballot	Ballot	Ballot	Ballot	Ballot	Ballot	Ballot
1st *C*	1st *A*	1st *B*	1st *A*	1st *C*	1st *D*	1st *A*
2nd *E*	2nd *D*	2nd *E*	2nd *B*	2nd *E*	2nd *C*	2nd *B*
3rd *D*	3rd *B*	3rd *A*	3rd *C*	3rd *D*	3rd *B*	3rd *C*
4th *A*	4th *C*	4th *C*	4th *D*	4th *A*	4th *E*	4th *D*
5th *B*	5th *E*	5th *D*	5th *E*	5th *B*	5th *A*	5th *E*

Ballot	Ballot	Ballot	Ballot	Ballot	Ballot	Ballot
1st *B*	1st *A*	1st *D*	1st *D*	1st *A*	1st *C*	1st *A*
2nd *E*	2nd *B*	2nd *C*	2nd *C*	2nd *B*	2nd *E*	2nd *D*
3rd *A*	3rd *C*	3rd *B*	3rd *B*	3rd *C*	3rd *D*	3rd *B*
4th *C*	4th *D*	4th *A*	4th *E*	4th *D*	4th *A*	4th *C*
5th *D*	5th *E*	5th *E*	5th *A*	5th *E*	5th *B*	5th *E*

Ballot	Ballot	Ballot	Ballot	Ballot	Ballot	Ballot
1st *B*	1st *C*	1st *A*	1st *C*	1st *A*	1st *D*	1st *D*
2nd *E*	2nd *E*	2nd *B*	2nd *E*	2nd *D*	2nd *C*	2nd *C*
3rd *A*	3rd *D*	3rd *C*	3rd *D*	3rd *B*	3rd *B*	3rd *B*
4th *C*	4th *A*	4th *D*	4th *A*	4th *C*	4th *A*	4th *E*
5th *D*	5th *B*	5th *E*	5th *B*	5th *E*	5th *E*	5th *A*

FIGURE 1-8

2. Figure 1-9 shows the preference ballots for an election with 17 voters and 4 candidates. Write out the preference schedule for this election.

Ballot	Ballot	Ballot	Ballot	Ballot
1st *C*	1st *B*	1st *A*	1st *C*	1st *B*
2nd *A*	2nd *C*	2nd *D*	2nd *A*	2nd *C*
3rd *D*	3rd *D*	3rd *B*	3rd *D*	3rd *D*
4th *B*	4th *A*	4th *C*	4th *B*	4th *A*

Ballot	Ballot	Ballot	Ballot	Ballot	Ballot
1st *A*	1st *A*	1st *B*	1st *B*	1st *C*	1st *C*
2nd *D*	2nd *C*	2nd *C*	2nd *C*	2nd *A*	2nd *A*
3rd *B*	3rd *D*	3rd *D*	3rd *D*	3rd *D*	3rd *D*
4th *C*	4th *B*	4th *A*	4th *A*	4th *B*	4th *B*

Ballot	Ballot	Ballot	Ballot	Ballot	Ballot
1st *A*	1st *A*	1st *C*	1st *B*	1st *A*	1st *C*
2nd *C*	2nd *D*	2nd *A*	2nd *C*	2nd *D*	2nd *A*
3rd *D*	3rd *B*	3rd *D*	3rd *D*	3rd *B*	3rd *D*
4th *B*	4th *C*	4th *B*	4th *A*	4th *C*	4th *B*

FIGURE 1-9

3. An election is held to choose the Chair of the Mathematics Department at Tasmania State University. The candidates are Professors Argand, Brandt, Chavez, Dietz, and Epstein (*A*, *B*, *C*, *D*, and *E* for short). Table 1-25 shows the preference schedule for the election.

Number of voters	5	5	3	3	3	2
1st	*A*	*C*	*A*	*B*	*D*	*D*
2nd	*B*	*E*	*D*	*E*	*C*	*C*
3rd	*C*	*D*	*B*	*A*	*B*	*B*
4th	*D*	*A*	*C*	*C*	*E*	*A*
5th	*E*	*B*	*E*	*D*	*A*	*E*

TABLE 1-25

(a) How many people voted in this election?

(b) How many first-place votes are needed for a majority?

(c) Which candidate had the fewest last-place votes?

4. The student body at Eureka High School is having an election for Homecoming Queen. The candidates are Alicia, Brandy, Cleo, and Dionne (*A*, *B*, *C*, and *D* for short). Table 1-26 shows the preference schedule for the election.

Number of voters	202	160	153	145	125	110	108	102	55
1st	*B*	*C*	*A*	*D*	*D*	*C*	*B*	*A*	*A*
2nd	*D*	*B*	*C*	*B*	*A*	*A*	*C*	*B*	*D*
3rd	*A*	*A*	*B*	*A*	*C*	*D*	*A*	*D*	*C*
4th	*C*	*D*	*D*	*C*	*B*	*B*	*D*	*C*	*B*

TABLE 1-26

(a) How many students voted in this election?

(b) How many first-place votes are needed for a majority?

(c) Which candidate had the fewest last-place votes?

Exercises 5 through 8 refer to the following format for a preference ballot: The names of the candidates are printed on the ballot in some random order, and the voter is simply asked to rank the candidates [for example, see Fig. 1-1(c)]. For ease of reference we call this the "printed-names" format. (This format makes it easier on the voters and is useful when the names are long or when a misspelled name invalidates the ballot. The main disadvantage is that it tends to favor the candidates who are listed first.)

5. An election is held using the "printed-names" format for the preference ballots. Table 1-27 shows the results of the election. Rewrite Table 1-27 in the conventional preference schedule format used in the text. (Use *A*, *B*, *C*, *D*, and *E* as shorthand for the names of the candidates.)

Number of voters	37	36	24	13	5
Alvarez	3rd	1st	2nd	4th	3rd
Brownstein	1st	2nd	1st	2nd	5th
Clarkson	4th	4th	5th	3rd	1st
Dax	5th	3rd	3rd	5th	4th
Easton	2nd	5th	4th	1st	2nd

TABLE 1-27

6. An election is held using the "printed-names" format for the preference ballots. Table 1-28 shows the results of the election. Rewrite Table 1-28 in the conventional preference schedule format used in the text. (Use A, B, C, D, and E as shorthand for the names of the candidates.)

Number of voters	14	10	8	7	4
Andersson	2nd	3rd	1st	5th	3rd
Broderick	1st	1st	2nd	3rd	2nd
Clapton	4th	5th	5th	2nd	4th
Dutkiewicz	5th	2nd	4th	1st	5th
Eklundh	3rd	4th	3rd	4th	1st

TABLE 1-28

7. Table 1-29 shows a conventional preference schedule for an election. Rewrite Table 1-29 using a format like that in Table 1-27 (as if the ballots were "printed-names" ballots).

Number of voters	14	10	8	7	4
1st	Bob	Bob	Ana	Dee	Eli
2nd	Ana	Dee	Bob	Cat	Bob
3rd	Eli	Ana	Eli	Bob	Ana
4th	Dee	Eli	Dee	Eli	Cat
5th	Cat	Cat	Cat	Ana	Dee

TABLE 1-29

8. Table 1-30 shows a conventional preference schedule for an election. Rewrite Table 1-30 using a format like that in Table 1-28 (as if the ballots were "printed-names" ballots).

Number of voters	37	36	24	13	5
1st	Ada	Bo	Dina	Ceci	Bo
2nd	Ceci	Ada	Bo	Ada	Dina
3rd	Bo	Dina	Ceci	Eva	Eva
4th	Eva	Ceci	Eva	Bo	Ada
5th	Dina	Eva	Ada	Dina	Ceci

TABLE 1-30

9. The Demublican Party is holding its annual convention. The 1500 voting delegates are choosing among three possible party platforms: L (a liberal platform), C (a conservative platform), and M (a moderate platform). Seventeen percent of the delegates prefer L to M and M to C. Thirty-two percent of the delegates like C the most and L the least. The rest of the delegates like M the most and C the least. Write out the preference schedule for this election.

10. The Epicurean Society is holding its annual election for president. The three candidates are A, B, and C. Twenty percent of the voters like A the most and B the least. Forty percent of the voters like B the most and A the least. Of the remaining voters 225 prefer C to B and B to A, and 675 prefer C to A and A to B. Write out the preference schedule for this election.

1.2 Plurality Method

11. Table 1-31 shows the preference schedule for an election with four candidates (A, B, C, and D). Use the plurality method to

(a) find the winner of the election.

(b) find the complete ranking of the candidates.

Number of voters	27	15	11	9	8	1
1st	C	A	B	D	B	B
2nd	D	B	D	A	A	A
3rd	B	D	A	B	C	D
4th	A	C	C	C	D	C

TABLE 1-31

12. Table 1-32 shows the preference schedule for an election with four candidates (A, B, C, and D). Use the plurality method to

(a) find the winner of the election.

(b) find the complete ranking of the candidates.

Number of voters	29	21	18	10	1
1st	D	A	B	C	C
2nd	C	C	A	B	B
3rd	A	B	C	A	D
4th	B	D	D	D	A

TABLE 1-32

13. Table 1-33 shows the preference schedule for an election with four candidates (A, B, C, and D). Use the plurality method to

(a) find the winner of the election.

(b) find the complete ranking of the candidates.

Number of voters	6	5	4	2	2	2	2
1st	C	A	B	B	C	C	C
2nd	D	D	D	A	B	B	D
3rd	A	C	C	C	A	D	B
4th	B	B	A	D	D	A	A

TABLE 1-33

14. Table 1-34 shows the preference schedule for an election with four candidates (A, B, C, and D). Use the plurality method to

(a) find the winner of the election.

(b) find the complete ranking of the candidates.

Number of voters	6	6	5	4	3	3
1st	A	B	B	D	A	B
2nd	C	C	C	A	B	A
3rd	D	A	D	C	C	C
4th	B	D	A	B	D	D

TABLE 1-34

15. Table 1-35 shows the preference schedule for an election with five candidates (A, B, C, D, and E). The number of voters in this election was very large, so the columns of the preference schedule show percentages rather than actual numbers of voters. Use the plurality method to

(a) find the winner of the election.

(b) find the complete ranking of the candidates.

Percent of voters	24	23	19	14	11	9
1st	C	D	D	B	A	D
2nd	A	A	A	C	C	C
3rd	B	C	E	A	B	A
4th	E	B	C	D	E	E
5th	D	E	B	E	D	B

TABLE 1-35

16. Table 1-36 shows the preference schedule for an election with five candidates (A, B, C, D, and E). The number of voters in this election was very large, so the columns of the preference schedule show percentages rather than numbers of voters. Use the plurality method to

(a) find the winner of the election.

(b) find the complete ranking of the candidates.

Percent of voters	25	21	15	12	10	9	8
1st	C	E	B	A	C	C	C
2nd	E	D	D	D	D	B	E
3rd	D	B	E	B	E	A	D
4th	A	A	C	E	A	E	B
5th	B	C	A	C	B	D	A

TABLE 1-36

17. Table 1-25 (see Exercise 3) shows the preference schedule for an election with five candidates (A, B, C, D, and E). In this election ties are not allowed to stand, and the following tie-breaking rule is used: *Whenever there is a tie between candidates, the tie is broken in favor of the candidate with the fewer last-place votes.* Use the plurality method to

(a) find the winner of the election.

(b) find the complete ranking of the candidates.

18. Table 1-26 (see Exercise 4) shows the preference schedule for an election with four candidates (A, B, C, and D). In this election ties are not allowed to stand, and the following tie-breaking rule is used: *Whenever there is a tie between candidates, the tie is broken in favor of the candidate with the fewer last-place votes.* Use the plurality method to

(a) find the winner of the election.

(b) find the complete ranking of the candidates.

19. Table 1-25 (see Exercise 3) shows the preference schedule for an election with five candidates (A, B, C, D, and E). In this election ties are not allowed to stand, and the following tie-breaking rule is used: *Whenever there is a tie between two candidates, the tie is broken in favor of the winner of a head-to-head comparison between the candidates.* Use the plurality method to

(a) find the winner of the election.

(b) find the complete ranking of the candidates.

20. Table 1-26 (see Exercise 4) shows the preference schedule for an election with four candidates (A, B, C, and D). In this election ties are not allowed to stand, and the following tie-breaking rule is used: *Whenever there is a tie between two candidates, the tie is broken in favor of the winner of a head-to-head comparison between the candidates.* Use the plurality method to

(a) find the winner of the election.

(b) find the complete ranking of the candidates.

1.3 Borda Count

21. Table 1-31 (see Exercise 11) shows the preference schedule for an election with four candidates (A, B, C, and D). Use the Borda count method to

(a) find the winner of the election.

(b) find the complete ranking of the candidates.

22. Table 1-32 (see Exercise 12) shows the preference schedule for an election with four candidates (A, B, C, and D). Use the Borda count method to
   (a) find the winner of the election.
   (b) find the complete ranking of the candidates.

23. Table 1-33 (see Exercise 13) shows the preference schedule for an election with four candidates (A, B, C, and D). Use the Borda count method to
   (a) find the winner of the election.
   (b) find the complete ranking of the candidates.

24. Table 1-34 (see Exercise 14) shows the preference schedule for an election with four candidates (A, B, C, and D). Use the Borda count method to
   (a) find the winner of the election.
   (b) find the complete ranking of the candidates.

25. Table 1-35 (see Exercise 15) shows the preference schedule for an election with five candidates (A, B, C, D, and E). The total number of people that voted in this election was very large, so the columns of the preference schedule show percentages rather than actual numbers of voters. Use the Borda count method to find the complete ranking of the candidates. (*Hint*: The ranking is determined by the percentages and does not depend on the number of voters, so you can pick any number to use for the number of voters. Pick a nice round one.)

26. Table 1-36 (see Exercise 16) shows the preference schedule for an election with five candidates (A, B, C, D, and E). The total number of people that voted in this election was very large, so the columns of the preference schedule show percentages rather than actual numbers of voters. Use the Borda count method to find the complete ranking of the candidates. (*Hint*: The ranking is determined by the percentages and does not depend on the number of voters, so you can pick any number to use for the number of voters. Pick a nice round one.)

27. **The 2014 Heisman Award.** Table 1-37 shows the results of the balloting for the 2014 Heisman Award. Find the ranking of the top three finalists and the number of points each one received (see Example 1.11).

Player	School	1st	2nd	3rd
Amari Cooper	Alabama	49	280	316
Melvin Gordon	Wisconsin	37	432	275
Marcus Mariota	Oregon	788	74	22

Source: Heisman Award, *www.heisman.com*

TABLE 1-37

28. **The 2014 AL Cy Young Award.** Table 1-38 shows the top 5 finalists for the 2014 American League Cy Young Award. Find the ranking of the top 5 finalists and the number of points each one received (the point values are the same as those used for the National League Cy Young—see Example 1.12).

Pitcher	1st	2nd	3rd	4th	5th
Felix Hernandez	13	17	0	0	0
Corey Kluber	17	11	2	0	0
Jon Lester	0	0	3	15	7
Chris Sale	0	2	19	5	3
Max Scherzer	0	0	4	6	8

Source: Baseball Writers Association of America, *bbwaa.com/14-al-cy/*

TABLE 1-38

29. An election was held using the conventional Borda count method. There were four candidates (A, B, C, and D) and 110 voters. When the points were tallied (using 4 points for first, 3 points for second, 2 points for third, and 1 point for fourth), A had 320 points, B had 290 points, and C had 180 points. Find how many points D had and give the ranking of the candidates. (*Hint*: Each of the 110 ballots hands out a fixed number of points. Figure out how many, and take it from there.)

30. Imagine that in the voting for the American League Cy Young Award (7 points for first place, 4 points for second, 3 points for third, 2 points for fourth, and 1 point for fifth) there were five candidates (A, B, C, D, and E) and 50 voters. When the points were tallied A had 152 points, B had 133 points, C had 191 points, and D had 175 points. Find how many points E had and give the ranking of the candidates. (*Hint*: Each of the 50 ballots hands out a fixed number of points. Figure out how many, and take it from there.)

1.4 Plurality-with-Elimination

31. Table 1-31 (see Exercise 11) shows the preference schedule for an election with four candidates (A, B, C, and D). Use the plurality-with-elimination method to
   (a) find the winner of the election.
   (b) find the complete ranking of the candidates.

32. Table 1-32 (see Exercise 12) shows the preference schedule for an election with four candidates (A, B, C, and D). Use the plurality-with-elimination method to
   (a) find the winner of the election.
   (b) find the complete ranking of the candidates.

33. Table 1-33 (see Exercise 13) shows the preference schedule for an election with four candidates (A, B, C, and D). Use the plurality-with-elimination method to
   (a) find the winner of the election.
   (b) find the complete ranking of the candidates.

34. Table 1-34 (see Exercise 14) shows the preference schedule for an election with four candidates (A, B, C, and D). Use the plurality-with-elimination method to
   (a) find the winner of the election.
   (b) find the complete ranking of the candidates.

35. Table 1-39 shows the preference schedule for an election with five candidates (A, B, C, D, and E). Find the complete ranking of the candidates using the plurality-with-elimination method.

Number of voters	8	7	5	4	3	2
1st	*B*	*C*	*A*	*D*	*A*	*D*
2nd	*E*	*E*	*B*	*C*	*D*	*B*
3rd	*A*	*D*	*C*	*B*	*E*	*C*
4th	*C*	*A*	*D*	*E*	*C*	*A*
5th	*D*	*B*	*E*	*A*	*B*	*E*

TABLE 1-39

36. Table 1-40 shows the preference schedule for an election with five candidates (A, B, C, D, and E). Find the complete ranking of the candidates using the plurality-with-elimination method.

Number of voters	7	6	5	5	5	5	4	2	1
1st	*D*	*C*	*A*	*C*	*D*	*E*	*B*	*A*	*A*
2nd	*B*	*A*	*B*	*A*	*C*	*A*	*E*	*B*	*C*
3rd	*A*	*E*	*E*	*B*	*A*	*D*	*C*	*D*	*E*
4th	*C*	*B*	*C*	*D*	*E*	*B*	*D*	*E*	*B*
5th	*E*	*D*	*D*	*E*	*B*	*C*	*A*	*C*	*D*

TABLE 1-40

37. Table 1-35 (see Exercise 15) shows the preference schedule for an election with five candidates (A, B, C, D, and E). The number of voters in this election was very large, so the columns of the preference schedule show percentages rather than actual numbers of voters. Use the plurality-with-elimination method to

(a) find the winner of the election.

(b) find the complete ranking of the candidates.

38. Table 1-36 (see Exercise 16) shows the preference schedule for an election with five candidates (A, B, C, D, and E). The number of voters in this election was very large, so the columns of the preference schedule show percentages rather than actual numbers of voters. Use the plurality-with-elimination method to

(a) find the winner of the election.

(b) find the complete ranking of the candidates.

Top-Two Instant-Runoff Voting. *Exercises 39 and 40 refer to a simple variation of the plurality-with-elimination method called top-two IRV. This method works for winner-only elections. In top-two IRV, instead of eliminating candidates one at a time, we eliminate all the candidates except the top two in the first round and transfer their votes to the two remaining candidates.*

39. Find the winner of the election given in Table 1-39 using the *top-two IRV* method.

40. Find the winner of the election given in Table 1-40 using the *top-two IRV* method.

1.5 Pairwise Comparisons

41. Table 1-31 (see Exercise 11) shows the preference schedule for an election with four candidates (A, B, C, and D). Use the method of pairwise comparisons to

(a) find the winner of the election.

(b) find the complete ranking of the candidates.

42. Table 1-32 (see Exercise 12) shows the preference schedule for an election with four candidates (A, B, C, and D). Use the method of pairwise comparisons to

(a) find the winner of the election.

(b) find the complete ranking of the candidates.

43. Table 1-33 (see Exercise 13) shows the preference schedule for an election with four candidates (A, B, C, and D). Use the method of pairwise comparisons to

(a) find the winner of the election.

(b) find the complete ranking of the candidates.

44. Table 1-34 (see Exercise 14) shows the preference schedule for an election with four candidates (A, B, C, and D). Use the method of pairwise comparisons to

(a) find the winner of the election.

(b) find the complete ranking of the candidates.

45. Table 1-35 (see Exercise 15) shows the preference schedule for an election with five candidates (A, B, C, D, and E). The number of voters in this election was very large, so the columns of the preference schedule give the percent of voters instead of the number of voters. Find the winner of the election using the method of pairwise comparisons.

46. Table 1-36 (see Exercise 16) shows the preference schedule for an election with five candidates (A, B, C, D, and E). The number of voters in this election was very large, so the columns of the preference schedule give the percent of voters instead of the number of voters. Find the winner of the election using the method of pairwise comparisons.

47. Table 1-39 (see Exercise 35) shows the preference schedule for an election with 5 candidates. Find the complete ranking of the candidates using the method of pairwise comparisons. (Assume that ties are broken using the results of the pairwise comparisons between the tying candidates.)

48. Table 1-40 (see Exercise 36) shows the preference schedule for an election with 5 candidates. Find the complete ranking of the candidates using the method of pairwise comparisons.

49. An election with five candidates (A, B, C, D, and E) is decided using the method of pairwise comparisons. If B loses two pairwise comparisons, C loses one, D loses one and ties one, and E loses two and ties one,

(a) find how many pairwise comparisons A loses. (*Hint*: First compute the total number of pairwise comparisons for five candidates.)

(b) find the winner of the election.

50. An election with six candidates (A, B, C, D, E, and F) is decided using the method of pairwise comparisons. If A loses four pairwise comparisons, B and C both lose three, D loses one and ties one, and E loses two and ties one,

(a) find how many pairwise comparisons F loses. (*Hint*: First compute the total number of pairwise comparisons for six candidates.)

(b) find the winner of the election.

1.6 Fairness Criteria

51. Use Table 1-41 to illustrate why the Borda count method violates the Condorcet criterion.

Number of voters	6	2	3
1st	A	B	C
2nd	B	C	D
3rd	C	D	B
4th	D	A	A

TABLE 1-41

52. Use Table 1-32 to illustrate why the plurality-with-elimination method violates the Condorcet criterion.

53. Use Table 1-42 to illustrate why the plurality method violates the IIA criterion. (*Hint*: Find the winner, then eliminate F and see what happens.)

Number of voters	49	48	3
1st	R	H	F
2nd	H	S	H
3rd	F	O	S
4th	O	F	O
5th	S	R	R

TABLE 1-42

54. Use the Math Club election (Example 1.10) to illustrate why the Borda count method violates the IIA criterion. (*Hint*: Find the winner, then eliminate D and see what happens.)

55. Use Table 1-43 to illustrate why the plurality-with-elimination method violates the IIA criterion. (*Hint*: Find the winner, then eliminate C and see what happens.)

Number of voters	5	5	3	3	3	2
1st	A	C	A	D	B	D
2nd	B	E	D	C	E	C
3rd	C	D	B	B	A	B
4th	D	B	C	E	C	A
5th	E	A	E	A	D	E

TABLE 1-43

56. Explain why the method of pairwise comparisons satisfies the majority criterion.

57. Explain why the method of pairwise comparisons satisfies the Condorcet criterion.

58. Explain why the plurality method satisfies the monotonicity criterion.

59. Explain why the Borda count method satisfies the monotonicity criterion.

60. Explain why the method of pairwise comparisons satisfies the monotonicity criterion.

JOGGING

61. Two-candidate elections. Explain why when there are only two candidates, the four voting methods we discussed in this chapter give the same winner and the winner is determined by straight majority. (Assume that there are no ties.)

62. Alternative version of the Borda count. The following simple variation of the conventional Borda count method is sometimes used: last place is worth 0 points, second to last is worth 1 point, . . . , first place is worth $N - 1$ points (where N is the number of candidates). Explain why this variation is equivalent to the conventional Borda count described in this chapter (i.e., it produces exactly the same winner and the same ranking of the candidates).

63. Reverse Borda count. Another commonly used variation of the conventional Borda count method is the following: A first place is worth 1 point, second place is worth 2 points, . . . , last place is worth N points (where N is the number of candidates). The candidate with the *fewest* points is the winner, second *fewest* points is second, and so on. Explain why this variation is equivalent to the original Borda count described in this chapter (i.e., it produces exactly the same winner and the same ranking of the candidates).

64. The average ranking. The average ranking of a candidate is obtained by taking the place of the candidate on each of the ballots, adding these numbers, and dividing by the number of ballots. Explain why the candidate with the best (lowest) average ranking is the Borda winner.

65. The 2006 Associated Press college football poll. The AP college football poll is a ranking of the top 25 college football teams in the country. The voters in the AP poll are a group of sportswriters and broadcasters chosen from across the country. The top 25 teams are ranked using a conventional Borda count: a first-place vote is worth 25 points, a second-place vote is worth 24 points, a third-place vote is worth 23 points, and so on. A last-place vote is worth 1 point. Table 1-44 shows the ranking and total points for each of the top three teams at the end of the 2006 regular season. (The remaining 22 teams are not shown here because they are irrelevant to this exercise.)

Team	Points
1. Ohio State	1625
2. Florida	1529
3. Michigan	1526

TABLE 1-44

(a) Given that Ohio State was the unanimous first-place choice of all the voters, find the number of voters that participated in the poll.

(b) Given that all the voters had Florida in either second or third place, find the number of second-place and the number of third-place votes for Florida.

(c) Given that all the voters had Michigan in either second or third place, find the number of second-place and the number of third-place votes for Michigan.

66. The Pareto criterion. The following fairness criterion was proposed by Italian economist Vilfredo Pareto (1848–1923): *If every voter prefers candidate X to candidate Y, then X should be ranked above Y.*

(a) Explain why the Borda count method satisfies the Pareto criterion.

(b) Explain why the pairwise-comparisons method satisfies the Pareto criterion.

67. The 2003–2004 NBA Rookie of the Year vote. Each year, a panel of broadcasters and sportswriters selects an NBA rookie of the year using a modified Borda count. Table 1-45 shows the results of the balloting for the top three finalists of the 2003–2004 season.

Player	1st place	2nd place	3rd place	Total points
LeBron James	78	39	1	508
Carmelo Anthony	40	76	2	430
Dwayne Wade	0	3	108	117

Source: From *LeBron James wins NBA Rookie of Year*. Copyright © 2004 by InsideHoops.com. Used by permission of InsideHoops.com.

TABLE 1-45

Determine how many points are given for each first-, second-, and third-place vote in this election.

68. Top-two IRV is a variation of the plurality-with-elimination method in which all the candidates except the top two are eliminated in the first round and their votes transferred to the top two. (see Exercises 39 and 40).

(a) Use the Math Club election to show that top-two IRV can produce a different outcome than plurality-with-elimination.

(b) Give an example that illustrates why top-two IRV violates the monotonicity criterion.

(c) Give an example that illustrates why top-two IRV violates the Condorcet criterion.

69. The Coombs method. This method is just like the plurality-with-elimination method except that in each round we eliminate the candidate with the *largest number of last-place votes* (instead of the one with the fewest first-place votes).

(a) Find the winner of the Math Club election using the Coombs method.

(b) Give an example that illustrates why the Coombs method violates the Condorcet criterion.

(c) Give an example that illustrates why the Coombs method violates the monotonicity criterion.

70. Bucklin voting. (This method was used in the early part of the 20th century to determine winners of many elections for political office in the United States.) The method proceeds in rounds. **Round 1**: Count first-place votes only. If a candidate has a majority of the first-place votes, that candidate wins. Otherwise, go to the next round. **Round 2**: Count *first- and second-place* votes only. If there are any candidates with a majority of votes, the candidate with the most votes wins. Otherwise, go to the next round. **Round 3**: Count *first-, second-, and third-place* votes only. If there are any candidates with a majority of votes, the candidate with the most votes wins. Otherwise, go to the next round. Repeat for as many rounds as necessary.

(a) Find the winner of the Math Club election using the Bucklin method.

(b) Give an example that illustrates why the Bucklin method violates the Condorcet criterion.

(c) Explain why the Bucklin method satisfies the monotonicity criterion.

RUNNING

71. The 2016 NBA MVP vote. The National Basketball Association Most Valuable Player is chosen using a modified Borda count. Each of the 131 voters (130 sportswriters from the U.S and Canada plus one aggregate vote from the fans) submits ballots ranking the top five players from 1st through 5th place. Table 1-46 shows the results of the 2016 vote. (For the first time in NBA history a single player—Stephen Curry of the Golden State Warriors—was the unanimous choice for first place.) Using the results shown in Table 1-46, determine the point value of each place on the ballot, and (this is the most important part) explain how you came up with the numbers (no looking it up on the web please!). [As usual, assume the point values are all positive integers and that 1st place is worth more than 2nd, 2nd is worth more than 3rd, and so on down the line.]

Player (team)	1st place	2nd place	3rd place	4th place	5th place	Total points
Stephen Curry (Golden State)	131	0	0	0	0	1310
Kawhi Leonard (San Antonio)	0	54	34	26	8	634
LeBron James (Cleveland)	0	40	48	34	9	631
Russell Westbrook (Oklahoma City)	0	29	37	28	14	486
Kevin Durant (Oklahoma City)	0	2	7	22	32	147
Chris Paul (Los Angeles)	0	4	3	9	37	107
Draymond Green (Golden State)	0	2	0	6	18	50
Damian Lillard (Portland)	0	0	1	4	9	26
James Harden (Houston)	0	0	1	1	1	9
Kyle Lowry (Toronto)	0	0	0	1	3	6

Source: National Basketball Association, *nba.com*

TABLE 1-46

72. The Condorcet loser criterion. *If there is a candidate who loses in a one-to-one comparison to each of the other candidates, then that candidate should not be the winner of the election.* (This fairness criterion is a sort of mirror image of the regular Condorcet criterion.)

(a) Give an example that illustrates why the plurality method violates the Condorcet loser criterion.

(b) Give an example that illustrates why the plurality-with-elimination method violates the Condorcet loser criterion.

(c) Explain why the Borda count method satisfies the Condorcet loser criterion.

73. Consider the following fairness criterion: *If a majority of the voters have candidate X ranked last, then candidate X should not be a winner of the election.*

(a) Give an example to illustrate why the plurality method violates this criterion.

(b) Give an example to illustrate why the plurality-with-elimination method violates this criterion.

(c) Explain why the method of pairwise comparisons satisfies this criterion.

(d) Explain why the Borda count method satisfies this criterion.

74. Suppose that the following was proposed as a fairness criterion: *If a majority of the voters rank X above Y, then the results of the election should have X ranked above Y.* Give an example to illustrate why all four voting methods discussed in the chapter can violate this criterion. (*Hint*: Consider an example with no Condorcet candidate.)

75. Consider a modified Borda count where a first-place vote is worth F points ($F > N$ where N denotes the number of candidates) and all other places in the ballot are the same as in the ordinary Borda count: $N - 1$ points for second place, $N - 2$ points for third place, . . . , 1 point for last place. By choosing F large enough, we can make this variation of the Borda count method satisfy the majority criterion. Find the smallest value of F (expressed in terms of N) for which this happens.

APPLET BYTES MyMathLab®

These Applet Bytes are short projects or mini-explorations built around the applet **Voting Methods** (available in MyMathLab in the Multimedia Library or Tools for Success). This applet allows the user to create a preference schedule and then find the results of the election under each of the four voting methods discussed in the chapter. Eliminating the drudgery of having to do the computations by hand allows one to handle much larger elections (the applet can handle elections with up to 26 candidates) as well as explore more interesting and meaningful questions.*

76. Consider the election given by the preference schedule shown in Table 1-47.

(a) Using the *Voting Methods* applet verify that in this election different methods give different results (sometimes the winner is *A*, sometimes the winner is *B* and sometimes there is a tie for first-place between *A* and *B*).

(b) Imagine that you are an "election fixer." Your job is to manipulate (i.e. "tweak") some of the ballots so that the election is **consistent**: the results of the election *are the same* (same winner and same ranking) under each of the four voting methods. Find a modified preference schedule for the election that accomplishes this. (With the Voting Methods applet you can make changes to the election in the Create Preference Schedule tab, check the results under the other tabs and keep going back and forth until you get the results you want.) The goal here is to accomplish a *consistent* election by manipulating as few ballots as possible. (*Hint*: It is possible to do so by changing just *two* ballots.)

*MyMathLab code required.

	49	40	40	51	43	30
1st	*A*	*A*	*B*	*B*	*C*	*C*
2nd	*B*	*C*	*A*	*C*	*A*	*B*
3rd	*C*	*B*	*C*	*A*	*B*	*A*

TABLE 1-47

77. Consider once again the election given by the preference schedule shown in Table 1-47. Manipulate the results of this election (changing as few ballots as possible) so that each of the three candidates is a winner under one of the methods.

78. Consider the election given by the preference schedule shown in Table 1-48.

Votes	93	44	10	30	42	81
1st	*A*	*B*	*C*	*C*	*D*	*E*
2nd	*B*	*D*	*A*	*E*	*C*	*D*
3rd	*C*	*E*	*E*	*D*	*E*	*C*
4th	*D*	*C*	*B*	*A*	*A*	*B*
5th	*E*	*A*	*D*	*B*	*B*	*A*

TABLE 1-48

(a) Using the Voting Methods applet find the results of this election under each of the four voting methods. Determine which of the four voting methods violates the Condorcet criterion in this election.

(b) Manipulate the election (changing as few ballots as possible) so that in the modified election there are no violations of the Condorcet criterion.

79. Consider the following fairness criterion: *If a majority of the voters rank candidate X above candidate Y, then the results of the election should rank X above Y*. Use the Voting Methods applet to find an election in which all four of the voting methods violate this criterion.

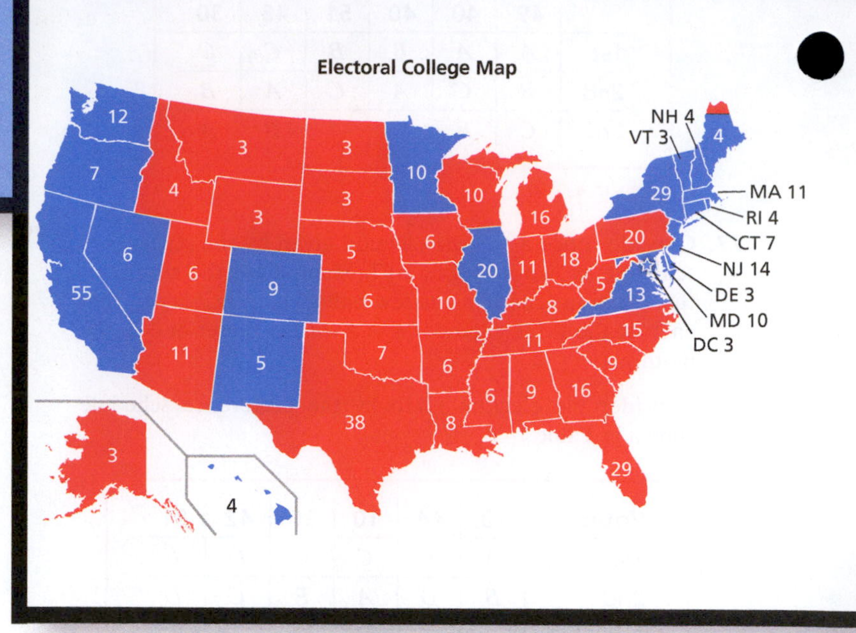

The Electoral College and the 2016 Presidential Election. (For more details see Example 2.19.)

2 The Mathematics of Power

Weighted Voting

In a democracy we take many things for granted, not the least of which is the idea that we are all equal. When it comes to voting, the democratic ideal of equality translates into the principle of *one person–one vote*. (Our entire discussion of elections in Chapter 1 was based on this premise.) But is the principle of *one person–one vote* always fair? Should *one person–one vote* apply when the *voters are* institutions or governments, rather than individuals?

In a diverse society, voters—be they individuals or institutions—are sometimes not equal, and sometimes it is actually desirable to recognize their differences by giving them different amounts of say over the outcome of the voting. This is the exact opposite of the principle of *one voter–one vote*, a principle best described as *one voter–x votes* and formally called *weighted voting* (the word "weight" here refers to the number of votes controlled by a voter, and not to pounds or kilograms).

Weighted voting is not uncommon—we see examples of weighted voting in shareholder votes in corporations (where votes are based on the number of shares owned), in business partnerships (where partners vote according to their percent of ownership), in county boards (where large districts have more votes than small districts), in the United Nations Security Council (where some countries have veto power and others don't), and in the Electoral College.

The Electoral College, that uniquely American institution used to elect the President of the United States, offers a classic illustration of weighted voting. The Electoral College consists of 51 "voters" (each of the 50 states plus the District of Columbia), each with a weight determined by the size of its Congressional delegation (number of Representatives and Senators). At one end of the spectrum is heavyweight California with 55 electoral votes; at the other end of the spectrum are small population states like Wyoming, Montana, North Dakota, and Vermont with just 3 electoral votes.

> "Each State shall appoint a Number of Electors equal to the whole Number of Senators and Representatives to which the state may be entitled in the Congress."
>
> *– Article II, Section 1, U.S. Constitution*

Since the point of weighted voting is to give different voters different amounts of influence in the outcome of the voting (i.e., different amounts of power), the key question we will discuss in this chapter is *how does one go about measuring* a voter's power in a weighted voting situation? We will answer this question not once but twice, and in so doing cover some new and interesting mathematics.

Section 2.1 will introduce and illustrate the concept of a *weighted voting system*, a formal way to describe most weighted voting situations. We will learn about *weights*, *quotas*, *dictators*, and *veto power*. Section 2.2 deals with what is known as the *Banzhaf* measure of power. Here we will learn about *coalitions*, *critical players*, and *dummies*. Section 2.3 deals with what is known as the *Shapley-Shubik* measure of power. Here we will learn about *sequential coalitions*, *pivotal players*, and *factorials*. Section 2.4 is essentially a mathematical detour where we will learn about counting *subsets* and *permutations*.

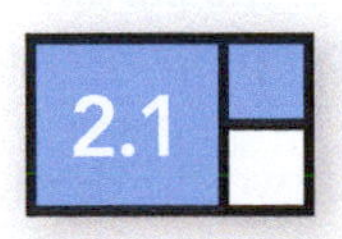

An Introduction to Weighted Voting

We will use the term **weighted voting system** to describe any formal voting arrangement in which voters are not necessarily equal in terms of the number of votes they control. Unlike Chapter 1, where the discussion focused primarily on elections involving three or more choices, in this chapter we will only consider voting on *yes–no* votes, known as **motions**. [Note that any vote between two choices (say *A* or *B*) can be recast as a *yes–no* vote, as in *Yes* is a vote for *A*, *No* is a vote for *B*].

Every weighted voting system is characterized by three elements:

- **The players.** We will refer to the voters in a weighted voting system as the **players**. The players can be persons, but they could also be institutions (corporations, municipalities, districts, states, or countries). We will use N to denote the number of players in a weighted voting system, and typically (unless there is a good reason not to) we will call the players $P_1, P_2, \ldots, P_N$. (Think of P_1 as short for "player 1," P_2 as short for "player 2," etc.)

- **The weights.** The hallmark of a weighted voting system is that each player controls a certain number of votes, and this number is called the **weight** of the player. We will assume that the weights are all positive integers, and we will use $w_1, w_2, \ldots, w_N$ to denote the weights of $P_1, P_2, \ldots, P_N$, respectively. We will use V to denote the total number of votes in the system $(V = w_1 + w_2 + \cdots + w_N)$. (A weighted voting system is only interesting when the weights of the players vary, but in principle we cannot rule out the possibility that all players have the same weight—somewhat boring but technically legal.)
- **The quota.** In addition to the players' weights, every weighted voting system has a *quota*. The **quota** is the *minimum number of votes required to pass a motion*, and is denoted by the letter q. While the most common standard for the quota is a *simple majority* of the votes, the quota may very well be something else. In the U.S. Senate, for example, it takes a simple majority to pass an ordinary law, but it takes a minimum of 60 votes to stop a filibuster, and it takes a minimum of two-thirds of the votes to override a presidential veto. In other weighted voting systems the rules may stipulate a quota of three-fourths of the votes, or four-fifths, or even *unanimity* (100% of the votes).

Notation and Examples

The standard notation used to describe a weighted voting system is to use square brackets and inside the square brackets to write the quota q first (followed by a colon) and then the respective weights of the individual players separated by commas. It is convenient and customary to list the weights in numerical order, starting with the highest, and we will adhere to this convention throughout the chapter.

Thus, a generic weighted voting system with N players can be written as:

$$[q: w_1, w_2, \ldots, w_N] \text{ (with } w_1 \geq w_2 \geq \cdots \geq w_N)$$

We will now look at a few examples to illustrate some basic concepts in weighted voting.

EXAMPLE 2.1 VENTURE CAPITALISM

Four partners (P_1, P_2, P_3, and P_4) decide to start a new business venture. In order to raise the $200,000 venture capital needed for startup money, they issue 20 shares worth $10,000 each. Suppose that P_1 buys 8 shares, P_2 buys 7 shares, P_3 buys 3 shares, and P_4 buys 2 shares, with the usual agreement that one share equals one vote in the partnership. Suppose also that the bylaws say that two-thirds of the partnership votes are required to pass a motion.

Since the total number of votes in the partnership is $V = 20$, and two-thirds of 20 is $13\frac{1}{3}$, the actual quota turns out to be 14 votes $(q = 14)$. Using the weighted voting system notation we just introduced, the partnership can be described mathematically as $[14: 8, 7, 3, 2]$.

EXAMPLE 2.2 ANARCHY

Imagine the same partnership discussed in Example 2.1, with the only difference being that the quota is changed to 10 votes. We might be tempted to think of this partnership as the weighted voting system $[10: 8, 7, 3, 2]$, but there is a problem here: the quota is too small, making it possible for both Yes's and No's to have enough votes to carry a particular motion. (Imagine, for example, that an important decision needs to be made and P_1 and P_4 vote yes and P_2 and P_3 vote no. Now we have a stalemate, since both the Yes's and the No's have enough votes to meet the quota.)

In general, when the quota requirement is less than simple majority we have the potential for both Yes's and No's to win—a mathematical version of anarchy.

EXAMPLE 2.3 GRIDLOCK

Once again, let's look at the partnership introduced in Example 2.1, but suppose now that the quota is set to $q = 21$, that is, more votes than the total number of votes in the system. This would not make much sense. Under these conditions no motion would ever pass and nothing could ever get done—a mathematical version of gridlock.

Given that we expect a weighted voting systems to operate without anarchy or gridlock, from here on we will require that the quota be *more than half the total number of votes* but *never more than the total*; in other words,

$$V/2 < q \leq V \text{ (where } V = w_1 + w_2 + \cdots + w_N)$$

EXAMPLE 2.4 FEW VOTES, MUCH POWER

Let's consider the partnership introduced in Example 2.1 one final time. This time the quota is set to be $q = 19$. Here we can describe the partnership as the weighted voting system $[19: 8, 7, 3, 2]$. What's interesting about this weighted voting system is that the *only way a motion can pass is by the unanimous support of all the players*. (Note that P_1, P_2, and P_3 together have 18 votes—they still need P_4's votes to pass a motion.) In this weighted voting system all four players have the same power. Even though P_4 has only 2 votes, his ability to influence the outcome of the vote is exactly the same as that of P_1, with 8 votes. In a practical sense this weighted voting system is no different from a weighted voting system in which each partner has just one vote and it takes the unanimous agreement of the four partners to pass a motion (i.e., $[4: 1, 1, 1, 1]$).

The surprising conclusion of Example 2.4 is that the weighted voting system $[19: 8, 7, 3, 2]$ represents a "one person–one vote" situation in disguise. This seems like a contradiction only if we think of "*one person–one vote*" as implying that all players have an *equal number of votes rather than an equal say in the outcome of the election*. As Example 2.4 makes abundantly clear, these two things are not the same—just looking at the number of votes a player owns can be very deceptive!

EXAMPLE 2.5 MANY VOTES, LITTLE POWER

Four college friends (P_1, P_2, P_3, and P_4) decide to go into business together. Three of the four (P_1, P_2, and P_3) invest \$10,000 each, and each gets 10 shares in the partnership. The fourth partner (P_4) is a little short on cash, so he invests only \$9,000 and gets 9 shares. As usual, one share equals one vote. The quota is set at 75%, which here means $q = 30$ out of a total of $V = 39$ votes. Mathematically (i.e., stripped of all the irrelevant details of the story), this partnership is just the weighted voting system $[30: 10, 10, 10, 9]$.

Everything seems fine with the partnership until one day P_4 wakes up to the realization that with the quota set at $q = 30$ he is completely out of the decision-making loop: For a motion to pass P_1, P_2, and P_3 all must vote Yes, and at that point it makes no difference how P_4 votes. Thus, P_4's votes will never make a difference in the final outcome of a business decision—in effect, they are worthless!

EXAMPLE 2.6 DICTATORS

Consider the weighted voting system $[11: 12, 5, 4]$. Here one of the players (P_1) owns enough votes to carry a motion singlehandedly. In this situation P_1 is in complete control—if P_1 is for the motion, then the motion will pass; if P_1 is against it, then the motion will fail. Clearly, in terms of the power to influence decisions, P_1 has *all* of it. Not surprisingly, we will say that P_1 is a *dictator*.

- **Dictator.** A player is a **dictator** if and only if *the player's weight is bigger than or equal to the quota*. Mathematically speaking, a weighted voting system with a dictator is not very interesting, since all the power is concentrated in the hands of one player. For the rest of this chapter we will restrict our attention to weighted voting systems where there are no dictators.

EXAMPLE 2.7 VETO POWER

Consider the weighted voting system $[12: 9, 5, 4, 2]$. Here P_1 plays the role of a "spoiler"—without the support of P_1 a motion can't pass. This happens because if we add all of the other players' votes $(5 + 4 + 2 = 11)$, we are still short of the 12 votes needed to pass a motion. Thus, even if all the players except P_1 voted for the motion, the motion still would not pass. In a situation like this we say that P_1 has *veto power.*

In general we will say that a player who is not a dictator has *veto power* if a *motion cannot pass unless the player votes in favor of the motion*. In other words, a player with veto power cannot force a motion to *pass* (the player is not a dictator) but can force a motion to *fail*. If we let w denote the weight of a player with veto power, then the two conditions can be expressed mathematically by the inequalities $w < q$ (the player is not a dictator) and $V - w < q$ (the remaining votes in the system are not enough to pass a motion).

- **Veto power.** A player with weight w has **veto power** if and only if $w < q$ and $V - w < q$ (where $V = w_1 + w_2 + \cdots + w_N$).

2.2 Banzhaf Power

John F. Banzhaf III (1940–)

From the preceding set of examples we can already draw an important lesson: In weighted voting the players' weights can be deceiving. Sometimes a player can have a few votes and yet have as much power as a player with many more votes (see Example 2.4); sometimes two players have almost an equal number of votes, and yet one player has a lot of power and the other one has none (see Example 2.5).

To pursue these ideas further we will need a formal definition of what "power" means and how it can be measured. In this section we will introduce a mathematical method for measuring the power of the players in a weighted voting system. This method was first proposed in 1965 by, of all people, a legal scholar named John Banzhaf III. (Banzhaf is a Professor of Public Interest Law at George Washington University. He was just 25 years old when he first came up with the ideas we will discuss in this section.)

We start our discussion of Banzhaf's method with an application to the U.S. Senate. The circumstances are fictitious, but given how the Senate works these days, not very far-fetched.

EXAMPLE 2.8(a) THE U.S. SENATE

The U.S. Senate has 100 members, and a simple majority of 51 votes is required to pass a bill. Suppose the Senate is composed of 49 Republicans, 48 Democrats, and 3 Independents, and that Senators vote strictly along party lines—Republicans all vote the same way, so do Democrats, and even the Independents are sticking together.

Under this scenario the Senate behaves as the weighted voting system $[51: 49, 48, 3]$. The three players are the Republican party (49 votes), the Democratic party (48 votes), and the Independents (3 votes).

There are only four ways that a bill can pass:

- All players vote Yes. The bill passes unanimously, 100 to 0.
- Republicans and Democrats vote Yes; Independents vote No. The bill passes 97 to 3.
- Republicans and Independents vote Yes; Democrats vote No. The bill passes 52 to 48.
- Democrats and Independents vote Yes; Republicans vote No. The bill passes 51 to 49.

That's it! There is no other way that a bill can pass this Senate. The key observation now is that the Independents have as much influence on the outcome of the vote as the Republicans or the Democrats (as long as they stick together): To pass, a bill needs the support of *any* two out of the three parties.

Before we continue with Example 2.8, we will introduce some important new concepts.

- **Coalition.** We will use the term **coalition** to describe *any* set of players who decide to join forces and vote the same way. In principle, we can have a coalition with as few as *one* player and as many as *all* players. The coalition consisting of all the players is called the **grand coalition**. Since coalitions are just sets of players, the most convenient way to describe coalitions mathematically is to use *set* notation. For example, the coalition consisting of players P_1, P_2, and P_3 can be written as the set $\{P_1, P_2, P_3\}$ (but also as $\{P_3, P_1, P_2\}$, $\{P_2, P_1, P_3\}$, etc.—the order in which the members of a coalition are listed is irrelevant).
- **Winning coalition.** Some coalitions have enough votes to win and some don't. Quite naturally, we call the former **winning coalitions** and the latter **losing coalitions**. Since our focus is on weighted voting systems with no dictators, winning coalitions must have at least two players (only a dictator can be in a single-player winning coalition). At the other end of the spectrum, the grand coalition is always a winning coalition, since it controls all the votes. In some weighted voting systems (see Example 2.4) the grand coalition is the only winning coalition.
- **Critical player.** In a winning coalition a player is said to be a *critical player* for the coalition if the coalition needs that player's votes to win. In other words, if we subtract a critical player's weight from the total weight of the coalition, the number of remaining votes drops below the quota. Here is a more formal way to say the same thing: P is a **critical player** in a winning coalition if and only if $W - w < q$ (where W denotes the weight of the coalition and w denotes the weight of P).
- **Critical count.** For each player we can count the number of times that player is a critical player (i.e., in how many different winning coalitions the player is critical). This number is called the **critical count** for the player.

EXAMPLE 2.8(b) THE U.S. SENATE (*CONTINUED*)

We now recast the analysis of the U.S. Senate [Example 2.8(a)] in the language of coalitions, critical players, and critical counts. Table 2-1 shows the four possible winning coalitions, their weights, and the critical players in each winning coalition underlined. (R represents the Republican Party with 49 votes, D represents the Democratic Party with 48 votes, and I represents the Independents with 3 votes.) Note that in the coalition $\{R, D, I\}$ there are no critical players (any two of the three parties can pass a bill by themselves), but in the two-player coalitions both players are critical. The critical count for R is 2 (notice that R is underlined twice in Table 2-1). In a similar manner we can check the critical counts for D and I—they are also 2. In short, all players have a critical count of 2.

Coalition	Weight
$\{R, D, I\}$	100
$\{\underline{R}, \underline{D}\}$	97
$\{\underline{R}, \underline{I}\}$	52
$\{\underline{D}, \underline{I}\}$	51

TABLE 2-1 ■ Winning coalitions for [51: 49, 48, 3] with critical players underlined

We will now introduce the key concept of this section—the *Banzhaf power index*. John Banzhaf introduced this concept in 1965 in a legal dispute involving the Nassau County (New York) Board of Supervisors (more details in Example 2.13). Essentially, Banzhaf argued that it's the *critical count*, and not the *weight*, that truly measures a player's power—if the critical count of X is the same as the critical count of Y then X and Y have the same power; if the critical count of X is double that of Y then X has twice as much power as Y, and so on. That's the basic idea. We formalize it with the following two definitions.

- **The Banzhaf power index (BPI).** Let $P_1, P_2, \ldots, P_N$ be the players in a weighted voting system, and $B_1, B_2, \ldots, B_N$ denote their respective *critical counts*. Let $T = B_1 + B_2 + \cdots + B_N$ be the *total critical count*. The **Banzhaf power index (BPI)** of a player is the ratio of the player's critical count over the total critical count T. Using the greek letter β ("beta") to indicate Banzhaf power indexes, we have that the BPI of P_1 is $\beta_1 = \frac{B_1}{T}$, the BPI of P_2 is $\beta_2 = \frac{B_2}{T}$, and so on.

  Each Banzhaf power index is a fraction between 0 and 1 and can be expressed either as a fraction, a decimal, or a percent between 0 and 100. You can think of it as a measure of the size of a player's share of the power.
- **The Banzhaf power distribution.** The complete division of the "power pie" among the players based on their Banzhaf power indexes is called the **Banzhaf power distribution** of the weighted voting system. We describe a Banzhaf power distribution by simply listing the power indexes $\beta_1, \beta_2, \ldots, \beta_N$ in order.

  Some comments about Banzhaf power indexes:

- Banzhaf power indexes start out as fractions all having a common denominator T. It is not necessarily a good idea to reduce the fractions to their simplest form—for comparison purposes it is preferable to list all the fractions using a common denominator.
- As fractions, the numbers in the power distribution must always add up to 1. When T is a large number, decimals or percents provide a more convenient numerical description than fractions, but there is a small price to pay: when converted to decimals or percents, the numbers may not add up to 1 because of round-off errors.

The following is a step-by-step recipe for computing the Banzhaf power distribution of a weighted voting system with N players.

COMPUTING THE BANZHAF POWER DISTRIBUTION

- **Step 1.** Make a list of all possible *winning* coalitions.
- **Step 2.** Within each winning coalition determine which are the *critical* players. (For record-keeping purposes, it is a good idea to underline each critical player.)
- **Step 3.** Find the *critical counts* $B_1, B_2, \ldots, B_N$.
- **Step 4.** Find $T = B_1 + B_2 + \cdots + B_N$.
- **Step 5.** Compute the Banzhaf power indexes: $\beta_1 = \frac{B_1}{T}, \beta_2 = \frac{B_2}{T}, \ldots, \beta_N = \frac{B_N}{T}$.

In the next set of examples we will illustrate how to carry out the above sequence of steps.

EXAMPLE 2.9 BANZHAF POWER IN [4: 3, 2, 1]

Let's find the Banzhaf power distribution of the weighted voting system [4: 3, 2, 1] using the five-step procedure described above. The steps are also summarized in Table 2-2.

Winning coalitions	Weight	Critical players
$\{\underline{P_1}, \underline{P_2}\}$	5	$\underline{P_1}, \underline{P_2}$
$\{\underline{P_1}, \underline{P_3}\}$	4	$\underline{P_1}, \underline{P_3}$
$\{\underline{P_1}, P_2, P_3\}$	6	$\underline{P_1}$
Critical counts: $B_1 = 3, B_2 = 1, B_3 = 1$		
Banzhaf power: $\beta_1 = \frac{3}{5}, \beta_2 = \frac{1}{5}, \beta_3 = \frac{1}{5}$		

TABLE 2-2 ■ Winning coalitions in [4: 3, 2, 1] with critical players underlined

Step 1. There are three winning coalitions in this weighted voting system. They are $\{P_1, P_2\}$ with 5 votes, $\{P_1, P_3\}$ with 4 votes, and the grand coalition $\{P_1, P_2, P_3\}$ with 6 votes.

Step 2. In $\{P_1, P_2\}$ both P_1 and P_2 are critical; in $\{P_1, P_3\}$ both P_1 and P_3 are critical; in $\{P_1, P_2, P_3\}$ only P_1 is critical.

Step 3. $B_1 = 3$ (P_1 is critical in three coalitions); $B_2 = 1$ and $B_3 = 1$ (P_2 and P_3 are critical once each).

Step 4. $T = 3 + 1 + 1 = 5$.

Step 5. $\beta_1 = \frac{B_1}{T} = \frac{3}{5}; \beta_2 = \frac{B_2}{T} = \frac{1}{5}; \beta_3 = \frac{B_3}{T} = \frac{1}{5}$. (If we want to express the β's in terms of percentages, then $\beta_1 = 60\%$, $\beta_2 = 20\%$, and $\beta_3 = 20\%$.)

Of all the steps we must carry out in the process of computing Banzhaf power, by far the most demanding is Step 1. When we have only three players, as in Example 2.9, we can list the winning coalitions on the fly—there simply aren't that many—but as the number of players increases, the number of possible winning coalitions grows rapidly, and it becomes necessary to adopt some form of strategy to come up with a list of *all* the winning coalitions. This is important because if we miss a single one, we are in all likelihood going to get the wrong Banzhaf power distribution. One conservative strategy is to make a list of *all* possible coalitions and then cross out the losing ones. The next example illustrates how one would use this approach.

EXAMPLE 2.10 THE NBA DRAFT

When NBA teams prepare for the annual draft of college players, the decision on which college basketball players to draft may involve many people, including the management, the coaches, and the scouting staff. Typically, not all these people have an equal voice in the process—the head coach's opinion is worth more than that of an assistant coach, and the general manager's opinion is worth more than that of a scout. In some cases this arrangement is formalized in the form of a weighted voting system. Let's use a fictitious team—the Flyers—for the purposes of illustration.

In the Flyers' draft system the head coach (P_1) has 4 votes, the general manager (P_2) has 3 votes, the director of scouting operations (P_3) has 2 votes, and the assistant coach (P_4) has 1 vote. Of the 10 votes cast, a simple majority of 6 votes is required for a Yes vote on a player to be drafted. In essence, the Flyers operate as the weighted voting system $[6: 4, 3, 2, 1]$.

We will now find the Banzhaf power distribution of this weighted voting system using Steps 1 through 5.

Coalition	Weight	Coalition	Weight
$\{\underline{P_1}, \underline{P_2}\}$	7 ✓	$\{\underline{P_1}, P_2, P_3\}$	9 ✓
$\{\underline{P_1}, \underline{P_3}\}$	6 ✓	$\{\underline{P_1}, \underline{P_2}, P_4\}$	8 ✓
$\{P_1, P_4\}$	5	$\{\underline{P_1}, \underline{P_3}, P_4\}$	7 ✓
$\{P_2, P_3\}$	5	$\{\underline{P_2}, \underline{P_3}, \underline{P_4}\}$	6 ✓
$\{P_2, P_4\}$	4	$\{P_1, P_2, P_3, P_4\}$	10 ✓
$\{P_3, P_4\}$	3		
Critical counts: $B_1 = 5, B_2 = 3, B_3 = 3, B_4 = 1$			
Banzhaf power: $\beta_1 = \frac{5}{12}, \beta_2 = \frac{3}{12}, \beta_3 = \frac{3}{12}, \beta_4 = \frac{1}{12}$			

TABLE 2-3 ■ Banzhaf power in [6: 4, 3, 2, 1] (✓ indicates a winning coalition)

Step 1. Table 2-3 shows a list of *all* possible coalitions and their weights with the winning coalitions followed by a check mark. (Note that the list of coalitions is organized systematically—two-player coalitions first, three-player coalitions next, etc. Also note that within each coalition the players are listed in numerical order from left to right. Both of these are good bookkeeping strategies, and you are encouraged to use them when you do your own work.)

Step 2. Now we disregard the losing coalitions and go to work on the winning coalitions only. For each winning coalition we determine which players are critical. The critical players are underlined in Table 2-3. (Don't take someone else's word for it—please check that these are indeed the critical players!)

Step 3. We now tally how many times each player is underlined in Table 2-3. These are the critical counts: $B_1 = 5, B_2 = 3, B_3 = 3$, and $B_4 = 1$.

Step 4. $T = 5 + 3 + 3 + 1 = 12$.

Step 5. $\beta_1 = \frac{5}{12} = 41\frac{2}{3}\%; \beta_2 = \frac{3}{12} = 25\%; \beta_3 = \frac{3}{12} = 25\%$, and $\beta_4 = \frac{1}{12} = 8\frac{1}{3}\%$.

An interesting and unexpected result of these calculations is that the team's general manager (P_2) and the director of scouting operations (P_3) have the same Banzhaf power index—not exactly the arrangement originally intended when one was given three votes and the other one two votes.

Shortcuts for Computing Banzhaf Power Distributions

As the number of players grows, listing all the coalitions takes a lot of effort. Sometimes we can save ourselves some of the work by figuring out directly which are the winning coalitions.

EXAMPLE 2.11 [5: 3, 2, 1, 1, 1]

In a weighted voting system with 5 players and no dictators there is a total of 26 possible coalitions (we will discuss this in more detail in Section 2.4). We can save a fair amount of effort by listing only the winning coalitions. A little organization will help: We will go through the winning coalitions systematically according to the number of players in the coalition. There is only one two-player winning coalition, namely $\{P_1, P_2\}$. The only three-player winning coalitions are those that include P_1. All four-player coalitions are winning coalitions, and so is the grand coalition.

Steps 1 and 2. Table 2-4 shows *all* the winning coalitions, with the critical players underlined. (You should double check to make sure that these are the right critical players in each coalition—it's good practice!)

Step 3. The critical counts are $B_1 = 11, B_2 = 5, B_3 = 3, B_4 = 3$, and $B_5 = 3$.

Step 4. $T = 25$.

Step 5. $\beta_1 = \frac{11}{25} = 44\%; \beta_2 = \frac{5}{25} = 20\%; \beta_3 = \beta_4 = \beta_5 = \frac{3}{25} = 12\%$.

Winning coalition	Weight	Winning coalition	Weight
$\{\underline{P_1}, \underline{P_2}\}$	5	$\{\underline{P_1}, P_2, P_3, P_4\}$	7
$\{\underline{P_1}, \underline{P_2}, P_3\}$	6	$\{\underline{P_1}, P_2, P_3, P_5\}$	7
$\{\underline{P_1}, \underline{P_2}, P_4\}$	6	$\{\underline{P_1}, P_2, P_4, P_5\}$	7
$\{\underline{P_1}, \underline{P_2}, P_5\}$	6	$\{\underline{P_1}, P_3, P_4, P_5\}$	6
$\{\underline{P_1}, \underline{P_3}, \underline{P_4}\}$	5	$\{\underline{P_2}, \underline{P_3}, \underline{P_4}, \underline{P_5}\}$	5
$\{\underline{P_1}, \underline{P_3}, \underline{P_5}\}$	5	$\{P_1, P_2, P_3, P_4, P_5\}$	8
$\{\underline{P_1}, \underline{P_4}, \underline{P_5}\}$	5		
Critical counts: $B_1 = 11, B_2 = 5, B_3 = 3, B_4 = 3, B_5 = 3$			
Banzhaf power: $\beta_1 = \frac{11}{25}, \beta_2 = \frac{5}{25}, \beta_3 = \frac{3}{25}, \beta_4 = \frac{3}{25}, \beta_5 = \frac{3}{25}$			

TABLE 2-4 ■ Banzhaf power in [5: 3, 2, 1, 1, 1]

EXAMPLE 2.12 THE POWER OF TIEBREAKERS

The Tasmania State University Promotion and Tenure committee consists of five members: the dean (D) and four other faculty members of equal standing (F_1, F_2, F_3, and F_4). (For convenience we are using a slightly different notation for the players.) In this committee faculty members vote first, and motions are carried by simple majority. The dean votes *only* to break a 2–2 tie. Is this a weighted voting system? If so, what is the Banzhaf power distribution?

The answer to the first question is Yes, and the answer to the second question is that we can apply the same steps we used before even though we don't have specific weights for the players.

Winning coalitions (faculty only)	Winning coalitions (dean breaks tie)
$\{\underline{F_1}, \underline{F_2}, \underline{F_3}\}$	$\{\underline{F_1}, \underline{F_2}, \underline{D}\}$
$\{\underline{F_1}, \underline{F_2}, \underline{F_4}\}$	$\{\underline{F_1}, \underline{F_3}, \underline{D}\}$
$\{\underline{F_1}, \underline{F_3}, \underline{F_4}\}$	$\{\underline{F_1}, \underline{F_4}, \underline{D}\}$
$\{\underline{F_2}, \underline{F_3}, \underline{F_4}\}$	$\{\underline{F_2}, \underline{F_3}, \underline{D}\}$
$\{F_1, F_2, F_3, F_4\}$	$\{\underline{F_2}, \underline{F_4}, \underline{D}\}$
	$\{\underline{F_3}, \underline{F_4}, \underline{D}\}$
Critical counts: 6 for each F; 6 for D	
Banzhaf power: Same for all players $\left(\frac{1}{5} = 20\%\right)$	

TABLE 2-5 ■ Banzhaf power in TSU promotion and tenure committee

Step 1. The possible winning coalitions fall into two groups: (1) A majority (three or more) faculty members vote Yes. In this case the dean does not vote. These winning coalitions are listed in the first column of Table 2-5. (2) Only two faculty members vote yes, and the dean breaks the tie with a Yes vote. These winning coalitions are listed in the second column of Table 2-5.

Step 2. In the winning coalitions consisting of just three faculty members, all faculty members are critical. In the coalition consisting of all four faculty members, no faculty member is critical. In the coalitions consisting of two faculty members plus the dean, all players, including the dean, are critical. Table 2-5 shows the critical players underlined in each winning coalition.

Step 3. The dean is critical six times (in each of the six coalitions in the second column). Each of the faculty members is critical three times in the first column and three times in the second column. Thus, all players have a critical count of 6.

Steps 4 and 5. Since all five players are critical an equal number of times, they share power equally (i.e., the Banzhaf power index of each player is $\frac{1}{5} = 20\%$). ■

"The Vice President of the United States shall be the President of the Senate, but shall have no Vote, unless they be equally divided."

– Article I, Section 3, U.S. Constitution

The surprising conclusion of Example 2.12 is that although the rules appear to set up a special role for the dean (the role of tiebreaker), in practice the dean is no different than any of the faculty members. A larger-scale version of Example 2.12 occurs in the U.S. Senate, where the vice president of the United States can only vote to break a tie. An analysis similar to the one used in Example 2.12 shows that as a member of the Senate the vice president has the same Banzhaf power index as an ordinary senator.

Our next example is of interest for both historical and mathematical reasons.

EXAMPLE 2.13 THE NASSAU COUNTY (N.Y.) BOARD OF SUPERVISORS (1960s)

	Name (symbol)	Weight
"Big 3"	Hempstead 1 (H1)	31
	Hempstead 2 (H2)	31
	Oyster Bay (OB)	28
"Lesser 3"	North Hempstead (NH)	21
	Long Beach (LB)	2
	Glen Cove (GC)	2
		$V = 115$

TABLE 2-6 ■ Nassau County districts and weights (1960s)

Throughout the 1900's county boards in the state of New York operated as weighted voting systems. The reasoning behind weighted voting was that counties are often divided into districts of uneven size and it seemed unfair to give an equal vote to both large and small districts. To eliminate this type of unfairness a system of *proportional representation* was used: Each district would have a number of votes roughly proportional to its population. Every 10 years, after the Census, the allocation of votes could change if the population changed, but the principle of proportional representation remained.

In this example we will focus on one specific case: the Nassau County Board of Supervisors in the 1960s. At the time, Nassau County was divided into six districts. Table 2-6 shows the names of the six districts and the number of votes each district had on the Board of Supervisors (based on population data from the 1960 Census). The total number of votes was $V = 115$, and the quota was $q = 58$ (simple majority). Thus, from a mathematical point of view the Nassau County Board of Supervisors could be simply described as $[58: 31, 31, 28, 21, 2, 2]$. We use H1, H2, OB, NH, LB, and GC as shorthand for the six districts.

To simplify the computations that follow we find it convenient to divide the six districts into the "Big 3" (H1, H2, and OB) and the "Lesser 3"(NH, LB, and GC). The winning coalitions will be organized into four groups:

1. Coalitions of the form $\{H1, H2, OB, *\}$. These are coalitions that include each of the "Big 3" and some subset of the "Lesser 3" (i.e., all, just two, just one, or none of the "Lesser 3"). There are eight such coalitions. In these coalitions, none of the players are critical.
2. Coalitions of the form $\{H1, H2, *\}$. These are coalitions that include H1 and H2 and some subset of the "Lesser 3." OB is not included in these coalitions. There are eight such coalitions. In these coalitions H1 and H2 are always critical; the remaining players are never critical.
3. Coalitions of the form $\{H1, OB, *\}$. These are coalitions that include H1 and OB and some subset of the "Lesser 3." H2 is not included in these coalitions. There are eight such coalitions. In these coalitions H1 and OB are always critical; the remaining players are never critical.
4. Coalitions of the form $\{H2, OB, *\}$. These are coalitions that include H2 and OB and some subset of the "Lesser 3." H1 is not included in these coalitions. There are eight such coalitions. In these coalitions H2 and OB are always critical; the remaining players are never critical.

The list above covers all possible winning coalitions. It follows from all of the preceding observations (summarized in Table 2-7) that the critical count for each of the "Big 3" is 16, and the critical count for each of the "Lesser 3" is 0. The Banzhaf power index of each of the "Big 3," therefore, is $\frac{1}{3} = 33\frac{1}{3}\%$, and the Banzhaf power index of each of the "Lesser 3" is 0!

Coalition type	Number of coalitions	Weight
(1) $\{H_1, H_2, OB, *\}$	8	90 or more
(2) $\{\underline{H_1}, \underline{H_2}, *\}$	8	62 or more
(3) $\{\underline{H_1}, \underline{OB}, *\}$	8	59 or more
(4) $\{\underline{H_2}, \underline{OB}, *\}$	8	59 or more
Critical counts: 16 for H1, H2 and OB; 0 for NH, LB and GC.		
Banzhaf power: $\frac{16}{48} = \frac{1}{3} = 33\frac{1}{3}\%$ for H1, H2 and OB; 0 for NH, LB and GC.		

TABLE 2-7 ■ Banzhaf power in the Nassau County Board of Supervisors (1960s)

Example 2.13 reinforces the fact that in weighted voting you can have a lot of votes and yet have zero power. A player with zero power is called a **dummy**. In the Nassau County Board of Supervisors Glen Cove, Long Beach, and North Hempstead were all dummies. The supervisors for these three districts attended meetings, participated in discussions, and voted in earnest not realizing that their votes were never relevant. The case of North Hempstead is especially serious—a district with a large population but no representation.

In a series of articles and law filings, John Banzhaf made the case that what was happening in Nassau County to North Hempstead, Long Beach, and Glen Cove (the three dummies) violated the "equal protection" clause guaranteed by the Fourteenth Amendment of the Constitution. Because of Banzhaf's mathematical analysis and legal persistence, in 1993 a federal court made it unconstitutional for county boards to use proportional representation as the basis for weighted voting.

EXAMPLE 2.14 THE UNITED NATIONS SECURITY COUNCIL

As presently constituted, the United Nations Security Council is made up of 15 member nations. Five of the member nations are permanent members: China (CH), France (FR), the Russian Federation (RU), the United Kingdom (UK), and the United States (US). The remaining 10 member nations are nonpermanent members appointed for two-year terms on a rotating basis. We will use a generic symbol NP to describe any nonpermanent member. To pass a resolution in the Security Council requires a minimum of nine Yes votes—a Yes vote from each of the five permanent members plus at least four additional Yes votes from nonpermanent members. If any one of the permanent members votes No, the motion fails regardless of how many Yes votes it gets. In other words, each of the permanent members has *veto power*. It's clear that the permanent members have more power than the nonpermanent members—but how much more?

Even though we don't have the weights of the players, with the help of a little mathematics we will be able to compute the Banzhaf power distribution. This is a very large weighted voting system, with 15 players and hundreds of winning coalitions, so we don't want to list the winning coalitions.

- **Step 1.** For our purposes, the winning coalitions can be divided into two types:
  1. Coalitions with the 5 permanent members plus 4 nonpermanent members (first row of Table 2-8). These are coalitions that just make the cut, and every member of such a coalition is critical. There are 210 coalitions in this group.
  2. Coalitions with the 5 permanent members plus 5 or more nonpermanent members (second row of Table 2-8). In these coalitions none of the nonpermanent members is critical (there is slack now for the requirement of a minimum of 4 nonpermanent members). The permanent members continue to be critical. There are 638 coalitions in this group.

We now have enough information to find the Banzhaf power distribution in the Security Council, but we will have to do so in a slightly roundabout way (Step 4 before Steps 2 and 3).

- **Step 4.** We will first find the total critical count T. Each of the 210 coalitions of type (1) has 9 critical players. This contributes $210 \times 9 = 1890$ to T. Each of the 638 coalitions of type (2) has 5 critical players. This contributes another $638 \times 5 = 3190$ to T. Adding the two gives $T = 1890 + 3190 = 5080$.
- **Steps 2 and 3.** Each permanent member is critical in each of the 848 winning coalitions. This gives each permanent member a critical count of 848. Combining the critical counts of the five permanent members gives us $848 \times 5 = 4240$. The remaining 840 $(5080 - 4240)$ are divided equally among the 10 nonpermanent members, so each nonpermanent member has a critical count of 84.
- **Step 5.** The Banzhaf power index of each permanent member is $\frac{848}{5080} = 16.7\%$; the Banzhaf power index of each nonpermanent member is $\frac{84}{5080} = 1.65\%$.

Coalition type	Number of coalitions
(1) {CH, FR, RU, UK, US, NP_1, NP_2, NP_3, NP_4}	210
(2) {CH, FR, RU, UK, US + 5 or more NPs}	638

- Total critical count: $T = 210 \times 9 + 638 \times 5 = 5080$
- Critical count for CH, FR, RU, UK, and US: 848 each
- Critical count for all 10 NPs together: $5080 - 848 \times 5 = 840$
- Critical count for each NP $= \frac{840}{10} = 84$

- BPI of each of the 5 permanent members: $\frac{848}{5080} \cong 16.7\%$
- BPI of each of the 10 nonpermanent members: $\frac{84}{5080} \cong 1.65\%$

TABLE 2-8 ■ Banzhaf power in the United Nations Security Council

Example 2.14 has an interesting aftermath. The power of the veto gives the permanent members roughly 10 times as much power as that of the nonpermanent members, and this was not the way that the Security Council was originally intended to work. It took Banzhaf's interpretation of power together with some nice mathematics to figure this out. To make the power distribution more balanced, there are plans in the works to change the current 15-nation arrangement by adding additional permanent members and changing the voting rules.

2.3 Shapley-Shubik Power

Lloyd Shapley (1923 – 2016)
Nobel Prize in Economics (2012)

Martin Shubik (1926 –)

In this section we will discuss a different approach to measuring power, first proposed by the American mathematician Lloyd Shapley and the economist Martin Shubik in 1954. The key difference between the Shapley-Shubik measure of power and the Banzhaf measure of power centers on the concept of a *sequential coalition*. Under the Shapley-Shubik method the assumption is that coalitions are formed sequentially: some player starts the coalition, a second player then joins it, then a third joins, and so on.

We will illustrate the main idea with a simple example.

EXAMPLE 2.15 [4: 3, 2, 1] REVISITED

We discussed this weighted voting system in Section 2.2. We will now look at things from a slightly different point of view. Suppose that players join coalitions one at a time and that we want to consider the order in which the players join the coalition. With three players there are six possibilities. (Just for now we will list them in some random order. Later we will try to organize ourselves a little better.)

1. P_2 goes first, P_1 goes second, P_3 goes last. We describe this using the notation $\langle P_2, P_1, P_3 \rangle$. (The $\langle\ \rangle$ is a convenient way to indicate that the players are listed in order from left to right.) This coalition starts with 2 votes, picks up 3 more when P_1 joins (now it has 5 votes), and picks up 1 more vote when P_3 joins. The key observation is that the player who had the pivotal role in making the coalition a winning coalition was P_1. We will call P_1 the *pivotal* player of that coalition.
2. P_3 goes first, P_2 goes second, P_1 goes last. We write this as $\langle P_3, P_2, P_1 \rangle$. This coalition starts with 1 vote, picks up 2 more when P_2 joins (now it has 3 votes), and picks up 3 more when P_1 joins. Once again, it's not until P_1 joins the coalition that there are enough votes to meet the quota. Score another one for P_1!
3. $\langle P_1, P_2, P_3 \rangle$. This coalition starts with 3 votes and gets 2 more votes when P_2 joins. No need to go any further. The pivotal player in this coalition is P_2.
4. $\langle P_1, P_3, P_2 \rangle$. This coalition starts with 3 votes and gets 1 more vote when P_3 joins. That's 4 votes, enough to win. The pivotal player in this coalition is P_3.
5. $\langle P_2, P_3, P_1 \rangle$. This coalition starts with 2 votes and gets 1 more vote when P_3 joins. That's 3 votes, not enough to win. It takes P_1 joining at the end to turn the coalition into a winning coalition. The pivotal player in this coalition is P_1.

6. $\langle P_3, P_1, P_2 \rangle$. This coalition starts with 1 vote and gets 3 more votes when P_1 joins. That's 4 votes, enough to win. The pivotal player in this coalition is P_1.

The six scenarios above cover all possibilities. Going down the list, we can see that in these six scenarios P_1 was the pivotal player four times, whereas P_2 and P_3 were pivotal once each. One could reasonably argue that P_1 has four times as much power as either P_2 or P_3.

We will now formally define some of the ideas introduced in Example 2.15 and end with a definition of Shapley-Shubik power. We will also introduce an important mathematical concept—the factorial of a number. We will discuss factorials in greater detail in Section 2.4, but for now we just need to know what the term *factorial* means.

- **Sequential coalition.** A **sequential coalition** is an *ordered list* of the players. We write sequential coalitions in the form $\langle P_1, P_2, \ldots, P_N \rangle$. The order of the players is from left to right. A good way to think of a sequential coalition is as a line of people with the head of the line on the left.
- **Factorial of N.** For any positive integer N, the product of all the integers between 1 and N is called the **factorial** of N and written $N!$.

$$N! = 1 \times 2 \times 3 \times \cdots \times N.$$

  The reason factorials are relevant to our discussion here is the following: *A weighted voting system with N players has N! sequential coalitions.*
- **Pivotal player.** When looking at a sequential coalition start counting votes from left to right. At some point (possibly not until the very end) there are enough votes to win. The first player that makes this possible is the **pivotal player** in that coalition. Every sequential coalition has *one and only one* pivotal player.
- **Pivotal count.** For each player we count how many times the player is a pivotal player as we run over all possible sequential coalitions. This gives the **pivotal count** for that player. We use the notation $SS_1, SS_2, \ldots, SS_N$ to denote the pivotal counts of $P_1, P_2, \ldots, P_N$ respectively.
- **The Shapley-Shubik power index (SSPI).** The **Shapley-Shubik power index** of a player is the ratio of the player's pivotal count over the total pivotal count for all the players. (Note that with N players there are $N!$ sequential coalitions and since there is exactly one pivotal player in each, the total pivotal count for all players is $N!$.) Thus, we can rephrase the definition as: The Shapley-Shubik power index of a player is *the ratio of the player's pivotal count over N!*. We will use the Greek letter σ (sigma) to represent the Shapley-Shubik power, so that $\sigma_1, \sigma_2, \ldots, \sigma_N$ denote the Shapley-Shubik power indexes of $P_1, P_2, \ldots, P_N$ respectively.
- **The Shapley-Shubik power distribution.** The list $\sigma_1, \sigma_2, \ldots, \sigma_N$ of all the Shapley-Shubik power indexes gives the **Shapley-Shubik power distribution** of the weighted voting system. It essentially describes the division of the "power pie" among the players based on their Shapley-Shubik power indexes.

The following is a summary of the steps needed to compute the Shapley-Shubik power distribution of a weighted voting system with N players.

COMPUTING A SHAPLEY-SHUBIK POWER DISTRIBUTION

- **Step 1.** Make a list of the $N!$ sequential coalitions with the N players.
- **Step 2.** In each sequential coalition determine *the* pivotal player. (For bookkeeping purposes underline the pivotal players.)
- **Step 3.** Find the *pivotal counts* $SS_1, SS_2, \ldots, SS_N$.
- **Step 4.** Compute the SSPI's: $\sigma_1 = \frac{SS_1}{N!}, \sigma_2 = \frac{SS_2}{N!}, \ldots, \sigma_N = \frac{SS_N}{N!}$.

EXAMPLE 2.16 THE NBA DRAFT REVISITED

In Example 2.10 (The NBA Draft) we computed the Banzhaf power distribution of the weighted voting system $[6: 4, 3, 2, 1]$. Now we will find the Shapley-Shubik power distribution.

Steps 1 and 2. Table 2-9 shows the 24 sequential coalitions with P_1, P_2, P_3, and P_4. In each sequential coalition the pivotal player is underlined. (Note that the 24 sequential coalitions in Table 2-9 are not randomly listed—each column corresponds to the sequential coalitions with a given first player. You may want to use the same or a similar pattern when you do the exercises on Shapley-Shubik power distributions.)

$\langle P_1, \underline{P_2}, P_3, P_4 \rangle$	$\langle P_2, \underline{P_1}, P_3, P_4 \rangle$	$\langle P_3, \underline{P_1}, P_2, P_4 \rangle$	$\langle P_4, P_1, \underline{P_2}, P_3 \rangle$
$\langle P_1, \underline{P_2}, P_4, P_3 \rangle$	$\langle P_2, \underline{P_1}, P_4, P_3 \rangle$	$\langle P_3, \underline{P_1}, P_4, P_2 \rangle$	$\langle P_4, P_1, \underline{P_3}, P_2 \rangle$
$\langle P_1, \underline{P_3}, P_2, P_4 \rangle$	$\langle P_2, P_3, \underline{P_1}, P_4 \rangle$	$\langle P_3, P_2, \underline{P_1}, P_4 \rangle$	$\langle P_4, P_2, \underline{P_1}, P_3 \rangle$
$\langle P_1, \underline{P_3}, P_4, P_2 \rangle$	$\langle P_2, P_3, \underline{P_4}, P_1 \rangle$	$\langle P_3, P_2, \underline{P_4}, P_1 \rangle$	$\langle P_4, P_2, \underline{P_3}, P_1 \rangle$
$\langle P_1, P_4, \underline{P_2}, P_3 \rangle$	$\langle P_2, P_4, \underline{P_1}, P_3 \rangle$	$\langle P_3, P_4, \underline{P_1}, P_2 \rangle$	$\langle P_4, P_3, \underline{P_1}, P_2 \rangle$
$\langle P_1, P_4, \underline{P_3}, P_2 \rangle$	$\langle P_2, P_4, \underline{P_3}, P_1 \rangle$	$\langle P_3, P_4, \underline{P_2}, P_1 \rangle$	$\langle P_4, P_3, \underline{P_2}, P_1 \rangle$

TABLE 2-9 ■ Sequential coalitions for [6: 4, 3, 2, 1] (pivotal players underlined)

Step 3. The pivotal counts are $SS_1 = 10$, $SS_2 = 6$, $SS_3 = 6$, and $SS_4 = 2$.

Step 4. The Shapley-Shubik power distribution is given by $\sigma_1 = \frac{10}{24} = 41\frac{2}{3}\%$, $\sigma_2 = \frac{6}{24} = 25\%$, $\sigma_3 = \frac{6}{24} = 25\%$, and $\sigma_4 = \frac{2}{24} = 8\frac{1}{3}\%$.

If you compare this result with the Banzhaf power distribution obtained in Example 2.10, you will notice that here the two power distributions are the same. If nothing else, this shows that it is not impossible for the Banzhaf and Shapley-Shubik power distributions to agree. In general, however, for randomly chosen real-life situations it is very unlikely that the Banzhaf and Shapley-Shubik methods will give the same answer. ■

EXAMPLE 2.17 A CITY COUNCIL WITH A "STRONG MAYOR"

In some cities the city council operates under what is known as the "strong-mayor" system. Under this system the city council can pass a motion under simple majority, but the mayor has the power to veto the decision. The mayor's veto can then be overruled by a "supermajority" of the council members. As an illustration of the strong-mayor system we will consider the city council of Cleansburg. In Cleansburg the city council has four members plus a strong mayor who has a vote as well as the power to veto motions supported by a simple majority of the council members. On the other hand, the mayor cannot veto motions supported by all four council members. Thus, a motion can pass if the mayor plus two or more council members support it or, alternatively, if the mayor is against it but the four council members support it.

Common sense tells us that under these rules, the four council members have the same amount of power but the mayor has more. We will compute the Shapley-Shubik power distribution of this weighted voting system to figure out exactly how much more. For convenience, we will let P_1 denote the mayor and P_2, P_3, P_4, P_5 denote the four council members.

With five players, this weighted voting system has $5! = 120$ sequential coalitions to consider. Obviously, we would prefer not to have to write them all down. Depending on where the mayor sits on the sequential coalition, we have 5 different scenarios, each consisting of 24 sequential coalitions (Table 2-10).

Type of sequential coalition	Number
1. $\langle P_1, *, \underline{*}, *, * \rangle$	24
2. $\langle *, P_1, \underline{*}, *, * \rangle$	24
3. $\langle *, *, \underline{P_1}, *, * \rangle$	24
4. $\langle *, *, *, \underline{P_1}, * \rangle$	24
5. $\langle *, *, *, \underline{*}, P_1 \rangle$	24
Pivotal counts: $SS_1 = 48$; $SS_2 = SS_3 = SS_4 = SS_5 = 18$	
Shapley-Shubik power: $\sigma_1 = \frac{48}{120} = 40\%$; $\sigma_2 = \sigma_3 = \sigma_4 = \sigma_5 = \frac{18}{120} = 15\%$	

TABLE 2-10 ■ Shapley-Shubik power in Cleansburg City Council (P_1 = mayor, * = city council member)

Notice that the mayor is the pivotal player when he is in the third or fourth position (rows 3 and 4 in Table 2-10). This gives the mayor a pivotal count of 48. The remaining 72 sequential coalitions have city council members as pivotal players (in equal numbers). This gives each city council member a pivotal count of $\frac{72}{4} = 18$.

It follows that the Shapley-Shubik power index of the mayor is $\sigma_1 = \frac{48}{120} = 40\%$ and each of the four council members has a Shapley-Shubik power index of 15%. In conclusion, the Shapley-Shubik power distribution of the Cleansburg City Council is $\sigma_1 = 40\%$, $\sigma_2 = \sigma_3 = \sigma_4 = \sigma_5 = 15\%$.

For the purposes of comparison, the reader is encouraged to calculate the Banzhaf power distribution of the Cleansburg City Council (see Exercise 76). ■

EXAMPLE 2.18 THE UNITED NATIONS SECURITY COUNCIL REVISITED

In Example 2.14, we computed the Banzhaf power distribution in the United Nations Security Council. In this example we will outline how one might be able to find the Shapley-Shubik power distribution. We will skip a few of the mathematical details as they go beyond the scope of this book. (The complete calculations can be found in the applet ***Banzhaf and Shapley-Shubik Power*** available in *MyMathLab* in the *Multimedia Library* or in *Tools for Success*. Just click on the "Create WVS" tab and then choose "U.N. Security Council" from the Custom menu at the bottom of the window.)

Recall that the U.N. Security Council has 5 permanent members and 10 nonpermanent members. A minimum of 9 votes is needed to carry a resolution, but the permanent members have veto power—a resolution carries only if all 5 permanent members and at least 4 nonpermanent members vote for it. The following is an outline of the computations:

Step 1. There are 15! (about *1.3 trillion*) sequential coalitions with 15 players.

Steps 2 and 3. A nonpermanent member can be pivotal only if that member is the 9th player in the coalition, preceded by all five of the permanent members and three nonpermanent members (see Fig. 2-1). For *each* nonpermanent member, there are approximately *2.44 billion* sequential coalitions that meet this description.

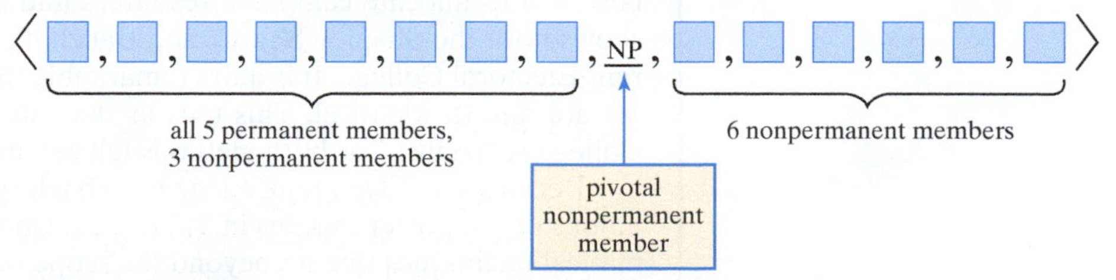

FIGURE 2-1 Sequential coalitions in the U.N. Security Council with pivotal nonpermanent member.

Step 4. From the numbers in Steps 1 through 3 above we can conclude that the Shapley-Shubik power index of each nonpermanent member is approximately $\frac{2.44\text{ billion}}{1.3\text{ trillion}} \approx 0.001865 = 0.1865\%$. Together, the 10 nonpermanent members (each with a Shapley-Shubik power index of 0.1865%) have 1.865% of the power pie, leaving the remaining 98.135% to be divided equally among the 5 permanent members. Thus, the Shapley-Shubik power index of each permanent member is approximately 19.627%.

This analysis shows the enormous difference between the Shapley-Shubik power of the permanent and nonpermanent members of the Security Council—permanent members have roughly 100 times the Shapley-Shubik power of nonpermanent members!

EXAMPLE 2.19 THE ELECTORAL COLLEGE (2012, 2016, 2020)

The president of the United States is allegedly the most powerful person in the world. To get elected president, a candidate must get a majority (270 or more) of the 538 electoral votes in the Electoral College. Electoral votes are assigned to the 50 states and the District of Columbia based on their representation in Congress (i.e., number of electoral votes = number of representatives + 2; the District of Columbia gets 3 votes by law). The first column of Table 2-11 shows the number of electoral votes for each state based on the 2010 Census. These numbers apply to the 2012, 2016, and 2020 presidential elections (they will probably change after the 2020 Census).

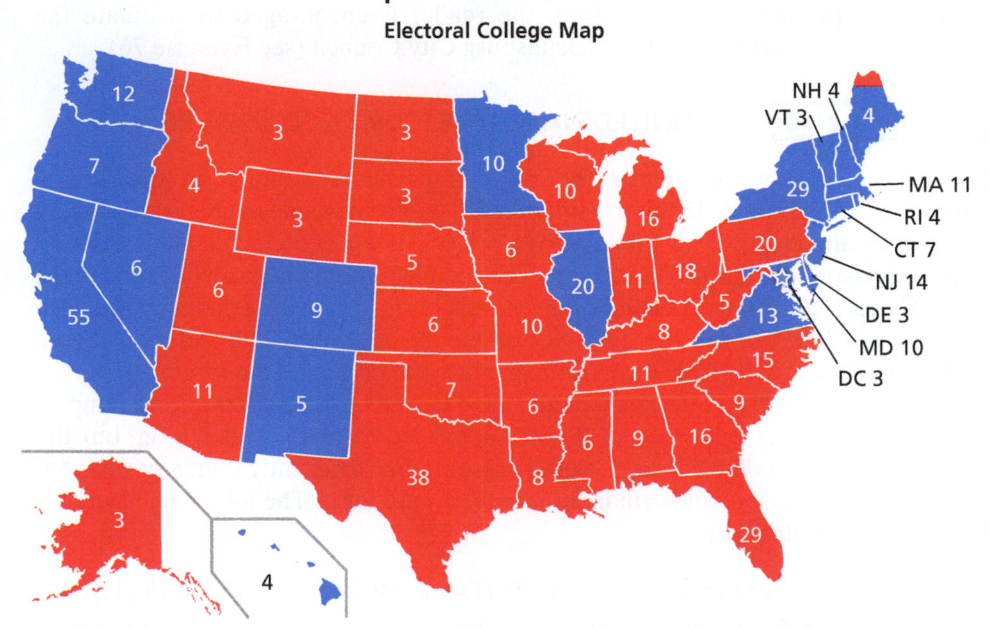

2016 Presidential Election Results (blue = Hillary Clinton, red = Donald Trump).

Each state casts all its electoral votes to the winner of the popular vote in that state (there are two small exceptions to this "winner-take-all" rule—Maine and Nebraska allow their electoral votes to be split), and voting typically boils down to two options—the Republican candidate or the Democratic candidate (the last time a third-party candidate received any electoral votes was 1968). If we assume that there are only two choices (Republican or Democrat) and make a slight concession to expediency and assume that Maine and Nebraska do not split their votes, then the Electoral College becomes a very large and complicated weighted voting system (538 total votes, *quota* $q = 270$, and 51 *players* with *weights* as shown in the first column of Table 2-11).

The power distribution in the Electoral College is one of the many important variables to consider in a presidential campaign (for example in determining the best way to allocate campaign resources and effort among the states). Table 2-11 shows both the Shapley-Shubik and Banzhaf power index of each state in the current Electoral College. It is quite remarkable that the numbers all the way down the list are almost identical. This tells us that the power distribution in the Electoral College is "robust," as both methods tell essentially the same story.

Because the Electoral College is such a large weighted voting system, the calculations of the power indexes in Table 2-11 can only be done using advanced mathematical techniques that go beyond the scope of this book and require some sophisticated software. (Some of the details can be found in the applet ***Banzhaf and Shapley-Shubik Power*** available in *MyMathLab* in the *Multimedia Library* or in *Tools for Success.* Click on the "Create WVS" tab and choose "Electoral College" in the Custom menu at the bottom of the window.)

State	Weight	Shapley-Shubik	Banzhaf	State	Weight	Shapley-Shubik	Banzhaf
Alabama	9	1.64%	1.64%	Montana	3	0.54%	0.55%
Alaska	3	0.54%	0.55%	Nebraska*	5	0.9%	0.91%
Arizona	11	2.0%	2.0%	Nevada	6	1.1%	1.1%
Arkansas	6	1.1%	1.1%	New Hampshire	4	0.72%	0.73%
California	55	11.0%	11.4%	New Jersey	14	2.6%	2.6%
Colorado	9	1.64%	1.64%	New Mexico	5	0.9%	0.91%
Connecticut	7	1.3%	1.3%	New York	29	5.5%	5.4%
Delaware	3	0.54%	0.55%	North Carolina	15	2.8%	2.7%
District of Columbia	3	0.54%	0.55%	North Dakota	3	0.54%	0.55%
Florida	29	5.5%	5.4%	Ohio	18	3.3%	3.3%
Georgia	16	2.9%	2.9%	Oklahoma	7	1.3%	1.3%
Hawaii	4	0.72%	0.73%	Oregon	7	1.3%	1.3%
Idaho	4	0.72%	0.73%	Pennsylvania	20	3.7%	3.7%
Illinois	20	3.7%	3.7%	Rhode Island	4	0.72%	0.73%
Indiana	11	2.0%	2.0%	South Carolina	9	1.64%	1.64%
Iowa	6	1.1%	1.1%	South Dakota	3	0.54%	0.55%
Kansas	6	1.1%	1.1%	Tennessee	11	2.0%	2.0%
Kentucky	8	1.5%	1.5%	Texas	38	7.3%	7.2%
Louisiana	8	1.5%	1.5%	Utah	6	1.1%	1.1%
Maine*	4	0.72%	0.73%	Vermont	3	0.54%	0.55%
Maryland	10	1.8%	1.8%	Virginia	13	2.4%	2.4%
Massachusetts	11	2.0%	2.0%	Washington	12	2.2%	2.2%
Michigan	16	2.9%	2.9%	West Virginia	5	0.9%	0.91%
Minnesota	10	1.8%	1.8%	Wisconsin	10	1.8%	1.8%
Mississippi	6	1.1%	1.1%	Wyoming	3	0.54%	0.55%
Missouri	10	1.8%	1.8%				

*Not bound by "winner-take-all" rule.

TABLE 2-11 ■ Shapley-Shubik and Banzhaf Power in the Electoral College (2012, 2016, and 2020 Presidential Elections)

2.4 Subsets and Permutations

The goal of this section is to provide a quick primer of basic facts about two very important mathematical concepts: *subsets* and *permutations*. These basic facts turn out to be quite useful when working with coalitions and sequential coalitions. (Some of the ideas introduced in this section will be covered again in future chapters.)

Subsets and Coalitions

Coalitions are essentially sets of players that join forces to vote the same way on a motion. (This is the reason we used set notation in Section 2.2 to work with coalitions.) By definition, a **subset** of a set is *any* combination of elements from the set. This includes the set with nothing in it (called the *empty set* and denoted by { }) as well as the set with all the elements (the original set itself).

EXAMPLE 2.20 SUBSETS OF $\{P_1, P_2, P_3\}$ AND $\{P_1, P_2, P_3, P_4\}$

Table 2-12 shows the eight subsets of the set $\{P_1, P_2, P_3\}$ and the 16 subsets of the set $\{P_1, P_2, P_3, P_4\}$. The subsets of $\{P_1, P_2, P_3, P_4\}$ are organized into two groups—the "no P_4" group and the "add P_4" group. The subsets in the "no P_4" group are exactly the 8 subsets of $\{P_1, P_2, P_3\}$; the subsets of the "add P_4" group are obtained by adding P_4 to each of the 8 subsets of $\{P_1, P_2, P_3\}$. Thus, adding one more element to the set $\{P_1, P_2, P_3\}$ doubled the number of subsets. Note that the number of subsets of a set is completely independent of the names of its elements, so any set with 3 elements, regardless of what they are called, has 8 subsets, and any set with four elements has 16 subsets.

Subsets of $\{P_1, P_2, P_3\}$	Subsets of $\{P_1, P_2, P_3, P_4\}$	
	"no P_4"	"add P_4"
$\{\ \}$	$\{\ \}$	$\{P_4\}$
$\{P_1\}$	$\{P_1\}$	$\{P_1, P_4\}$
$\{P_2\}$	$\{P_2\}$	$\{P_2, P_4\}$
$\{P_1, P_2\}$	$\{P_1, P_2\}$	$\{P_1, P_2, P_4\}$
$\{P_3\}$	$\{P_3\}$	$\{P_3, P_4\}$
$\{P_1, P_3\}$	$\{P_1, P_3\}$	$\{P_1, P_3, P_4\}$
$\{P_2, P_3\}$	$\{P_2, P_3\}$	$\{P_2, P_3, P_4\}$
$\{P_1, P_2, P_3\}$	$\{P_1, P_2, P_3\}$	$\{P_1, P_2, P_3, P_4\}$

TABLE 2-12

The key observation in Example 2.20 is that by adding one more element to a set we double the number of subsets, and this principle applies to sets of any size. Since a set with 4 elements has 16 subsets, a set with 5 elements must have 32 subsets and a set with 6 elements must have 64 subsets. This leads to the following key fact:

NUMBER OF SUBSETS

A set with N elements has 2^N subsets.

Now that we can count subsets, we can also count coalitions. The only difference between coalitions and subsets is that we don't consider the empty set a coalition (the purpose of a coalition is to cast votes, so you need at least one player to have a coalition).

NUMBER OF COALITIONS

A weighted voting system with N players has $2^N - 1$ coalitions.

Finally, when computing critical counts and Banzhaf power, we are interested in listing just the winning coalitions. If we assume that there are no dictators, then winning coalitions have to have at least two players and we can, therefore, rule out all the single-player coalitions from our list. There are N different single-player coalitions, which leads to our next fact:

NUMBER OF COALITIONS OF TWO OR MORE PLAYERS

A weighted voting system with N players has $2^N - N - 1$ coalitions of two or more players.

Permutations and Sequential Coalitions

- **Permutation.** A **permutation** of a set of objects is an ordered list of the objects. When you change the order of the objects you get a different *permutation* of those objects.

When the objects are people, a permutation is analogous to having the people form a line: first person, second person, etc. Rearrange the folks in the line and you get a different permutation.

EXAMPLE 2.21 **PERMUTATIONS OF P_1, P_2, P_3 AND P_1, P_2, P_3, P_4**

The six permutations of P_1, P_2, P_3 are shown at the top of Fig. 2-2. Now imagine that P_4 shows up and we have to add her to the line. We can position P_4 at the back of the line, or in third place, or second place, or at the front of the line (Fig. 2-2). For each of the six permutations of P_1, P_2, P_3 we get four permutations of P_1, P_2, P_3, P_4 depending on where we choose to put P_4. Thus, we have a total of 24 permutations of the 4 objects.

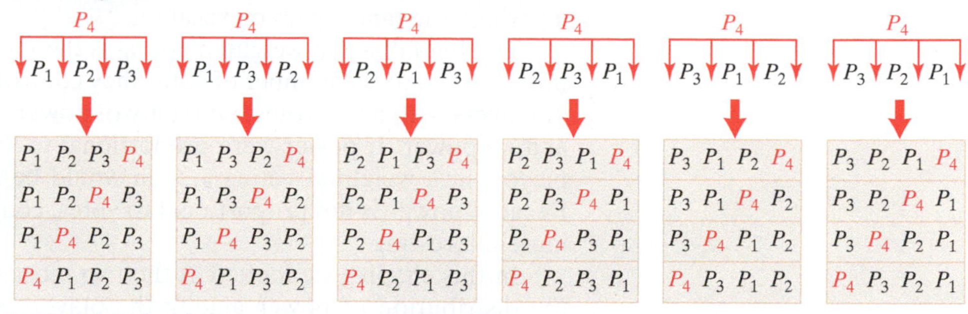

FIGURE 2-2 The 24 permutations of P_1, P_2, P_3, and P_4.

Note that $6 = 3 \times 2 \times 1 = 3!$ and $24 = 4 \times 3 \times 2 \times 1 = 4!$. Generalizing the idea in Example 2.21 gives the following important fact:

NUMBER OF PERMUTATIONS

There are $N!$ different *permutations* of N objects ($N! = 1 \times 2 \times \ldots \times N$).

The connection between sequential coalitions and permutations is pretty straightforward: A sequential coalition of N players is just a permutation of the players. This leads to the following:

NUMBER OF SEQUENTIAL COALITIONS

A weighted voting system with N players has $N!$ sequential coalitions.

EXAMPLE 2.22 **PERMUTATIONS WITH A PLAYER IN A SPECIFIED POSITION**

In Example 2.21 we generated the 24 permutations of P_1, P_2, P_3, P_4 by putting P_4 in each of 4 possible slots in a permutation of P_1, P_2, P_3. This means that there are 6 coalitions with P_4 in last place, 6 more with P_4 in third place, 6 with P_4 in second place, and 6 with P_4 at the head of the line (read Fig. 2-2 horizontally). What is true for P_4 is true for each of the other players: There are 6 permutations with P_3 in last place, 6 more with P_3 in third place, 6 with P_3 in second place, and 6 with P_3 at the head of the line. Same for P_2 and P_1.

The observations made in Example 2.22 can be generalized to any player in any position in any weighted voting system.

NUMBER OF SEQUENTIAL COALITIONS WITH P IN THE kTH POSITION

Let P be a player in a weighted voting system with N players and k an arbitrary position between first and last. There are $(N-1)!$ sequential coalitions with P in the kth position.

Conclusion

In this chapter we discussed weighted voting—voting situations in which the voters (called players) control different numbers of votes. The idea behind weighted voting is that sometimes we want to give some players more power than others over the outcome of the voting. Weighted voting is used in partnerships where some partners own a larger share than others; in international bodies where some countries are more influential than others; in county boards where some districts are bigger than others; in the Electoral College where the big states have more votes than the small states; and even at home, where Mom has more votes than anyone else when deciding where the family goes on vacation.

The big fallacy in weighted voting is the idea that the power of the players is proportional to the number of votes they control. In this chapter we saw examples of players with a few votes but plenty of power as well as players with many votes and no power. If power were proportional to votes, computing the power of the players in a weighted voting system would be easy but also very uninteresting. Because power is not proportional to votes, computing power is harder but much more interesting.

In this chapter we looked at the two most commonly used ways to compute the distribution of power among the players in a weighted voting system: the *Banzhaf power distribution* and *the Shapley-Shubik power distribution*. These two approaches provide two different ways to measure power, and, while they occasionally agree (see Examples 2.10 and 2.16), they can also differ significantly (see Examples 2.14 and 2.18). Of the two, which one is better?

Unfortunately, there is no simple answer. Both are useful, and in some sense the choice is subjective. Perhaps the best way to evaluate them is to think of them as being based on a slightly different set of assumptions. The idea behind Banzhaf power is that players are free to come and go, negotiating their allegiance for power. Underlying Shapley-Shubik power is the assumption that when a player joins a coalition the player is making a commitment to stay. In the latter case a player's power is generated by its ability to be in the right place at the right time.

In practice the choice of which method to use for measuring power is based on which of the assumptions better fits the specifics of the situation. Contrary to what we've often come to expect, mathematics does not give us the answer, just the tools that might help us make an informed decision.

KEY CONCEPTS

2.1 An Introduction to Weighted Voting

- **weighted voting system:** a formal voting arrangement where each player controls a given number of votes, **39**
- **motion:** a vote between two options (Yes and No), **39**
- **player:** a voter in a weighted voting system, **39**
- **weight:** of a player, the number of votes controlled by the player, **40**
- **quota:** the minimum number of votes required to pass a motion, **40**
- **dictator:** a player whose weight is bigger or equal to the quota, **42**

- **veto power:** the power to keep the remaining players from passing a motion, **42**

2.2 Banzhaf Power

- **coalition:** a set of players that join forces and agree to vote together, **43**
- **grand coalition:** the coalition consisting of all the players, **43**
- **winning coalition:** a coalition with enough votes to carry a motion, **43**
- **losing coalition:** a coalition that doesn't have enough votes to carry a motion, **43**
- **critical player:** in a winning coalition, a player without whom the coalition would be a losing coalition, **43**
- **critical count:** for each player, the number of winning coalitions in which the player is a critical player, **43**
- **Banzhaf power index:** for each player, the ratio B/T, where B is the player's critical count and T is the total critical count for all the players, **44**
- **Banzhaf power distribution:** a list consisting of the Banzhaf power indexes of all the players, **44**

2.3 Shapley-Shubik Power

- **sequential coalition:** an ordered listing of all the players, **51**
- **factorial:** for a positive integer N, the product of all the integers from 1 to N, $(N! = 1 \times 2 \times 3 \times \cdots \times N)$, **51**
- **pivotal player:** in a sequential coalition, as the votes are tallied from left to right, the first player whose votes make the total equal to or higher than the quota, **51**
- **pivotal count:** for each player, the number of sequential coalitions in which the player is a pivotal player, **51**
- **Shapley-Shubik power index:** for each player, the ratio $SS/N!$, where SS is the player's pivotal count and N is the number of players, **51**
- **Shapley-Shubik power distribution:** a list consisting of the Shapley-Shubik power indexes of all the players, **51**

2.4 Subsets and Permutations

- **subset:** of a set S, any combination of elements from S (including the empty subset { } as well as the entire set S), **55**
- **permutation:** an ordered arrangement of a set of objects, **56**
- **number of subsets:** 2^N (where N is the number of elements in the set), **56**
- **number of coalitions:** $2^N - 1$ (where N is the number of players in the weighted voting system), **56**
- **number of coalitions with 2 or more players:** $2^N - N - 1$, **56**
- **number of sequential coalitions (permutations):** $N!$ (where N is the number of players in the weighted voting system), **57**
- **number of sequential coalitions with player P in a fixed position:** $(N-1)!$, **57**

EXERCISES

WALKING

2.1 Weighted Voting

1. Five partners (P_1, P_2, P_3, P_4, and P_5) jointly own the Gaussian Electric Company. P_1 owns 15 shares of the company, P_2 owns 12 shares, P_3 and P_4 each owns 10 shares and P_5 owns 3 shares, with the usual agreement that one share equals one vote. Describe the partnership as a weighted voting system using the standard notation $[q: w_1, w_2, \ldots, w_N]$ if
   - **(a)** decisions in the partnership are made by simple majority.
   - **(b)** decisions in the partnership require two-thirds of the votes.

2. Five partners (P_1, P_2, P_3, P_4, and P_5) jointly own the Euler Moving Company. P_1 owns 30%, P_2 owns 25%, P_3 owns 20%, P_4 owns 16% and P_5 owns 9% of the company. Suppose that each 1% of ownership equals one vote. Describe the partnership as a weighted voting system using the standard notation $[q: w_1, w_2, \ldots, w_N]$ if
   - **(a)** decisions in the partnership are made by simple majority.
   - **(b)** decisions in the partnership require two-thirds of the votes.

3. Consider the weighted voting system $[q: 6, 4, 3, 3, 2, 2]$.
   - **(a)** What is the smallest value that the quota q can take?
   - **(b)** What is the largest value that the quota q can take?
   - **(c)** What is the value of the quota if *at least* three-fourths of the votes are required to pass a motion?
   - **(d)** What is the value of the quota if *more* than three-fourths of the votes are required to pass a motion?

4. Consider the weighted voting system $[q: 10, 6, 5, 4, 2]$.
   - **(a)** What is the smallest value that the quota q can take?
   - **(b)** What is the largest value that the quota q can take?
   - **(c)** What is the value of the quota if *at least* two-thirds of the votes are required to pass a motion?
   - **(d)** What is the value of the quota if *more* than two-thirds of the votes are required to pass a motion?

5. In each of the following weighted voting systems, determine which players, if any, have veto power.
   - **(a)** $[7: 4, 3, 3, 2]$
   - **(b)** $[9: 4, 3, 3, 2]$
   - **(c)** $[10: 4, 3, 3, 2]$
   - **(d)** $[11: 4, 3, 3, 2]$

6. In each of the following weighted voting systems, determine which players, if any, have veto power.
   - **(a)** $[9: 8, 4, 2, 1]$
   - **(b)** $[12: 8, 4, 2, 1]$
   - **(c)** $[14: 8, 4, 2, 1]$
   - **(d)** $[15: 8, 4, 2, 1]$

7. Consider the weighted voting system $[q: 7, 5, 3]$. Find the *smallest* value of q for which
   - **(a)** all three players have veto power.
   - **(b)** P_2 has veto power but P_3 does not.

8. Consider the weighted voting system $[q: 10, 8, 6, 4, 2]$. Find the *smallest* value of q for which
   - **(a)** all five players have veto power.
   - **(b)** P_3 has veto power but P_4 does not.

9. A committee has four members (P_1, P_2, P_3, and P_4). In this committee P_1 has twice as many votes as P_2; P_2 has twice as many votes as P_3; P_3 and P_4 have the same number of votes. The quota is $q = 49$. For each of the given definitions of the *quota*, describe the committee using the notation $[q: w_1, w_2, w_3, w_4]$. (*Hint*: Write the weighted voting system as $[49: 4x, 2x, x, x]$, and then solve for x.)
   - **(a)** The quota is defined as a *simple majority* of the votes.
   - **(b)** The quota is defined as *more than two-thirds* of the votes.
   - **(c)** The quota is defined as *more than three-fourths* of the votes.

10. A committee has six members (P_1, P_2, P_3, P_4, P_5, and P_6). In this committee P_1 has twice as many votes as P_2; P_2 and P_3 each has twice as many votes as P_4; P_4 has twice as many votes as P_5; P_5 and P_6 have the same number of votes. The quota is $q = 121$. For each of the given definitions of the *quota*, describe the committee using the notation $[q: w_1, w_2, w_3, w_4, w_5, w_6]$. (*Hint*: Write the weighted voting system as $[121: 8x, 4x, 4x, 2x, x, x]$, and then solve for x.)
    - **(a)** The quota is defined as a *simple majority* of the votes.
    - **(b)** The quota is defined as *more than two-thirds* of the votes.
    - **(c)** The quota is defined as *more than three-fourths* of the votes.

2.2 Banzhaf Power

11. Consider the weighted voting system $[q: 7, 5, 3]$.
    - **(a)** What is the weight of the coalition formed by P_1 and P_3?
    - **(b)** For what values of the quota q is the coalition formed by P_1 and P_3 a winning coalition?
    - **(c)** For what values of the quota q is the coalition formed by P_1 and P_3 a losing coalition?

12. Consider the weighted voting system $[q: 10, 8, 6, 4, 2]$.
    - **(a)** What is the weight of the coalition formed by P_2, P_3, and P_4?
    - **(b)** For what values of the quota q is the coalition formed by P_2, P_3, and P_4 a winning coalition?
    - **(c)** For what values of the quota q is the coalition formed by P_2, P_3, and P_4 a losing coalition?

13. Find the Banzhaf power distribution of a weighted voting system with four players in which the winning coalitions (with critical players underlined) are

$$\{\underline{P_1}, \underline{P_2}, \underline{P_3}\}, \{\underline{P_1}, \underline{P_2}, \underline{P_4}\}, \{\underline{P_1}, \underline{P_2}, P_3, P_4\}.$$

14. Find the Banzhaf power distribution of a weighted voting system with five players in which the winning coalitions (with critical players underlined) are

$\{\underline{P_1}, \underline{P_2}, \underline{P_3}\}$, $\{\underline{P_1}, \underline{P_2}, \underline{P_4}\}$, $\{\underline{P_1}, \underline{P_2}, P_3, P_4\}$, $\{\underline{P_1}, \underline{P_2}, \underline{P_3}, P_5\}$, $\{\underline{P_1}, \underline{P_2}, \underline{P_4}, P_5\}$, $\{\underline{P_1}, \underline{P_3}, \underline{P_4}, \underline{P_5}\}$, $\{\underline{P_1}, P_2, P_3, P_4, P_5\}$.

15. Consider the weighted voting system $[10: 6, 5, 4, 2]$.

(a) Which players are critical in the winning coalition $\{P_1, P_2, P_4\}$?

(b) Write down all winning coalitions.

(c) Find the Banzhaf power distribution of this weighted voting system.

16. Consider the weighted voting system $[5: 3, 2, 1, 1]$.

(a) Which players are critical in the winning coalition $\{P_1, P_3, P_4\}$?

(b) Write down all winning coalitions.

(c) Find the Banzhaf power distribution of this weighted voting system.

17. (a) Find the Banzhaf power distribution of the weighted voting system $[6: 5, 2, 1]$.

(b) Find the Banzhaf power distribution of the weighted voting system $[3: 2, 1, 1]$. Compare your answers in (a) and (b).

18. (a) Find the Banzhaf power distribution of the weighted voting system $[7: 5, 2, 1]$.

(b) Find the Banzhaf power distribution of the weighted voting system $[5: 3, 2, 1]$. Compare your answers in (a) and (b).

19. Consider the weighted voting system $[q: 5, 4, 3, 2, 1]$. Find the Banzhaf power distribution of this weighted voting system when

(a) $q = 10$

(b) $q = 11$

(c) $q = 12$

(d) $q = 15$

20. Consider the weighted voting system $[q: 8, 4, 2, 1]$. Find the Banzhaf power distribution of this weighted voting system when

(a) $q = 8$

(b) $q = 9$

(c) $q = 10$

(d) $q = 14$

21. In a weighted voting system with three players the winning coalitions are: $\{P_1, P_2\}$, $\{P_1, P_3\}$, and $\{P_1, P_2, P_3\}$.

(a) Find the critical players in each winning coalition.

(b) Find the Banzhaf power distribution of the weighted voting system.

22. In a weighted voting system with four players the winning coalitions are: $\{P_1, P_2\}$, $\{P_1, P_2, P_3\}$, $\{P_1, P_2, P_4\}$, and $\{P_1, P_2, P_3, P_4\}$.

(a) Find the critical players in each winning coalition.

(b) Find the Banzhaf power distribution of the weighted voting system.

23. The Nassau County (N.Y.) Board of Supervisors (1960's version). In the 1960's, the voting in the Nassau County Board of Supervisors was represented by the weighted voting system $[58: 31, 31, 28, 21, 2, 2]$. Assume that the players are denoted by P_1 through P_6.

(a) List all the *two-* and *three-player* winning coalitions and find the critical players in each coalition.

(b) List all the winning coalitions that have P_4 as a member and find the critical players in each coalition.

(c) Use the results in (b) to find the Banzhaf power index of P_4.

(d) Use the results in (a) and (c) to find the Banzhaf power distribution of the weighted voting system.

24. The Nassau County Board of Supervisors (1990s version). In the 1990s, after a series of legal challenges, the Nassau County Board of Supervisors was redesigned to operate as the weighted voting system $[65: 30, 28, 22, 15, 7, 6]$.

(a) List all the *three-player* winning coalitions and find the critical players in each coalition.

(b) List all the *four-player* winning coalitions and find the critical players in each coalition. (*Hint*: There are 11 four-player winning coalitions.)

(c) List all the *five-player* winning coalitions and find the critical players in each coalition.

(d) Use the results in (a), (b), and (c) to find the Banzhaf power distribution of the weighted voting system.

25. A law firm is run by four partners $(A, B, C, \text{and } D)$. Each partner has one vote and decisions are made by majority rule, but in the case of a 2-2 tie, the coalition with D (the junior partner) loses. (For example, $\{A, B\}$ wins, but $\{A, D\}$ loses.)

(a) List all the winning coalitions in this voting system and find the critical players in each.

(b) Find the Banzhaf power distribution in this law firm.

26. A law firm is run by four partners $(A, B, C, \text{and } D)$. Each partner has one vote and decisions are made by majority rule, but in the case of a 2-2 tie, the coalition with A (the senior partner) wins.

(a) List all the winning coalitions in this voting system and the critical players in each.

(b) Find the Banzhaf power index of this law firm.

2.3 Shapley-Shubik Power

27. Table 2-13 shows the 24 sequential coalitions (with pivotal players underlined) in a weighted voting system with four players. Find the Shapley-Shubik power distribution of this weighted voting system.

$\langle P_1, P_2, \underline{P_3}, P_4\rangle$	$\langle P_1, P_2, \underline{P_4}, P_3\rangle$	$\langle P_1, P_3, \underline{P_2}, P_4\rangle$	$\langle P_1, P_3, P_4, \underline{P_2}\rangle$
$\langle P_1, P_4, \underline{P_2}, P_3\rangle$	$\langle P_1, P_4, P_3, \underline{P_2}\rangle$	$\langle P_2, P_1, \underline{P_3}, P_4\rangle$	$\langle P_2, P_1, \underline{P_4}, P_3\rangle$
$\langle P_2, P_3, \underline{P_1}, P_4\rangle$	$\langle P_2, P_3, P_4, \underline{P_1}\rangle$	$\langle P_2, P_4, \underline{P_1}, P_3\rangle$	$\langle P_2, P_4, P_3, \underline{P_1}\rangle$
$\langle P_3, P_1, \underline{P_2}, P_4\rangle$	$\langle P_3, P_1, P_4, \underline{P_2}\rangle$	$\langle P_3, P_2, \underline{P_1}, P_4\rangle$	$\langle P_3, P_2, P_4, \underline{P_1}\rangle$
$\langle P_3, P_4, P_1, \underline{P_2}\rangle$	$\langle P_3, P_4, P_2, \underline{P_1}\rangle$	$\langle P_4, P_1, \underline{P_2}, P_3\rangle$	$\langle P_4, P_1, P_3, \underline{P_2}\rangle$
$\langle P_4, P_2, \underline{P_1}, P_3\rangle$	$\langle P_4, P_2, P_3, \underline{P_1}\rangle$	$\langle P_4, P_3, P_1, \underline{P_2}\rangle$	$\langle P_4, P_3, P_2, \underline{P_1}\rangle$

TABLE 2-13

28. Table 2-14 shows the 24 sequential coalitions (with pivotal players underlined) in a weighted voting system with four players. Find the Shapley-Shubik power distribution of this weighted voting system.

$\langle P_1, \underline{P_2}, P_3, P_4 \rangle$	$\langle P_1, \underline{P_2}, P_4, P_3 \rangle$	$\langle P_1, \underline{P_3}, P_2, P_4 \rangle$	$\langle P_1, \underline{P_3}, P_4, P_2 \rangle$
$\langle P_1, P_4, \underline{P_2}, P_3 \rangle$	$\langle P_1, P_4, \underline{P_3}, P_2 \rangle$	$\langle P_2, \underline{P_1}, P_3, P_4 \rangle$	$\langle P_2, \underline{P_1}, P_4, P_3 \rangle$
$\langle P_2, P_3, \underline{P_1}, P_4 \rangle$	$\langle P_2, P_3, \underline{P_4}, P_1 \rangle$	$\langle P_2, P_4, \underline{P_1}, P_3 \rangle$	$\langle P_2, P_4, \underline{P_3}, P_1 \rangle$
$\langle P_3, \underline{P_1}, P_2, P_4 \rangle$	$\langle P_3, \underline{P_1}, P_4, P_2 \rangle$	$\langle P_3, P_2, \underline{P_1}, P_4 \rangle$	$\langle P_3, P_2, \underline{P_4}, P_1 \rangle$
$\langle P_3, P_4, \underline{P_1}, P_2 \rangle$	$\langle P_3, P_4, \underline{P_2}, P_1 \rangle$	$\langle P_4, P_1, \underline{P_2}, P_3 \rangle$	$\langle P_4, P_1, \underline{P_3}, P_2 \rangle$
$\langle P_4, P_2, \underline{P_1}, P_3 \rangle$	$\langle P_4, P_2, \underline{P_3}, P_1 \rangle$	$\langle P_4, P_3, \underline{P_1}, P_2 \rangle$	$\langle P_4, P_3, \underline{P_2}, P_1 \rangle$

TABLE 2-14

29. Consider the weighted voting system $[16: 9, 8, 7]$.

(a) Write down all the sequential coalitions, and in each sequential coalition identify the pivotal player.

(b) Find the Shapley-Shubik power distribution of this weighted voting system.

30. Consider the weighted voting system $[8: 7, 6, 2]$.

(a) Write down all the sequential coalitions, and in each sequential coalition identify the pivotal player.

(b) Find the Shapley-Shubik power distribution of this weighted voting system.

31. Find the Shapley-Shubik power distribution of each of the following weighted voting systems.

(a) $[15: 16, 8, 4, 1]$

(b) $[18: 16, 8, 4, 1]$

(c) $[24: 16, 8, 4, 1]$

(d) $[28: 16, 8, 4, 1]$

32. Find the Shapley-Shubik power distribution of each of the following weighted voting systems.

(a) $[8: 8, 4, 2, 1]$

(b) $[9: 8, 4, 2, 1]$

(c) $[12: 8, 4, 2, 1]$

(d) $[14: 8, 4, 2, 1]$

33. Find the Shapley-Shubik power distribution of each of the following weighted voting systems.

(a) $[51: 40, 30, 20, 10]$

(b) $[59: 40, 30, 20, 10]$ (*Hint*: Compare this situation with the one in (a).)

(c) $[60: 40, 30, 20, 10]$

34. Find the Shapley-Shubik power distribution of each of the following weighted voting systems.

(a) $[41: 40, 10, 10, 10]$

(b) $[49: 40, 10, 10, 10]$ (*Hint*: Compare this situation with the one in (a).)

(c) $[50: 40, 10, 10, 10]$

35. In a weighted voting system with three players the winning coalitions are: $\{P_1, P_2\}$, $\{P_1, P_3\}$, and $\{P_1, P_2, P_3\}$.

(a) List the sequential coalitions and identify the pivotal player in each one.

(b) Find the Shapley-Shubik power distribution of the weighted voting system.

36. In a weighted voting system with three players the winning coalitions are: $\{P_1, P_2\}$ and $\{P_1, P_2, P_3\}$.

(a) List the sequential coalitions and identify the pivotal player in each sequential coalition.

(b) Find the Shapley-Shubik power distribution of the weighted voting system.

37. Table 2-15 shows the 24 sequential coalitions in a weighted voting system with four players. In some cases the pivotal player is underlined, and in some cases it isn't. Find the Shapley-Shubik power distribution of this weighted voting system. (*Hint*: First find the pivotal player in the remaining sequential coalitions.)

$\langle P_1, \underline{P_2}, P_3, P_4 \rangle$	$\langle P_2, P_1, P_3, P_4 \rangle$	$\langle P_3, P_1, P_2, P_4 \rangle$	$\langle P_4, P_1, P_2, P_3 \rangle$
$\langle P_1, P_2, P_4, P_3 \rangle$	$\langle P_2, P_1, P_4, P_3 \rangle$	$\langle P_3, P_1, P_4, P_2 \rangle$	$\langle P_4, P_1, P_3, P_2 \rangle$
$\langle P_1, \underline{P_3}, P_2, P_4 \rangle$	$\langle P_2, P_3, \underline{P_1}, P_4 \rangle$	$\langle P_3, P_2, P_1, P_4 \rangle$	$\langle P_4, P_2, P_1, P_3 \rangle$
$\langle P_1, P_3, P_4, P_2 \rangle$	$\langle P_2, P_3, \underline{P_4}, P_1 \rangle$	$\langle P_3, P_2, P_4, P_1 \rangle$	$\langle P_4, P_2, P_3, P_1 \rangle$
$\langle P_1, P_4, \underline{P_2}, P_3 \rangle$	$\langle P_2, P_4, \underline{P_1}, P_3 \rangle$	$\langle P_3, P_4, \underline{P_1}, P_2 \rangle$	$\langle P_4, P_3, P_1, P_2 \rangle$
$\langle P_1, P_4, \underline{P_3}, P_2 \rangle$	$\langle P_2, P_4, \underline{P_3}, P_1 \rangle$	$\langle P_3, P_4, \underline{P_2}, P_1 \rangle$	$\langle P_4, P_3, P_2, P_1 \rangle$

TABLE 2-15

38. Table 2-16 shows the 24 sequential coalitions in a weighted voting system with four players. In some cases the pivotal player is underlined, and in some cases it isn't. Find the Shapley-Shubik power distribution of this weighted voting system. (*Hint*: First find the pivotal player in the remaining sequential coalitions.)

$\langle P_1, \underline{P_2}, P_3, P_4 \rangle$	$\langle P_2, P_1, P_3, P_4 \rangle$	$\langle P_3, P_1, P_2, P_4 \rangle$	$\langle P_4, P_1, P_2, P_3 \rangle$
$\langle P_1, P_2, P_4, P_3 \rangle$	$\langle P_2, P_1, P_4, P_3 \rangle$	$\langle P_3, P_1, P_4, P_2 \rangle$	$\langle P_4, P_1, P_3, P_2 \rangle$
$\langle P_1, P_3, \underline{P_2}, P_4 \rangle$	$\langle P_2, P_3, \underline{P_1}, P_4 \rangle$	$\langle P_3, P_2, P_1, P_4 \rangle$	$\langle P_4, P_2, P_1, P_3 \rangle$
$\langle P_1, P_3, \underline{P_4}, P_2 \rangle$	$\langle P_2, P_3, P_4, \underline{P_1}, \rangle$	$\langle P_3, P_2, P_4, P_1 \rangle$	$\langle P_4, P_2, P_3, P_1 \rangle$
$\langle P_1, P_4, \underline{P_2}, P_3 \rangle$	$\langle P_2, P_4, \underline{P_1}, P_3 \rangle$	$\langle P_3, P_4, \underline{P_1}, P_2 \rangle$	$\langle P_4, P_3, P_1, P_2 \rangle$
$\langle P_1, P_4, \underline{P_3}, P_2 \rangle$	$\langle P_2, P_4, P_3, P_1 \rangle$	$\langle P_3, P_4, P_2, P_1 \rangle$	$\langle P_4, P_3, P_2, P_1 \rangle$

TABLE 2-16

2.4 Subsets and Permutations

39. Let A be a set with 10 elements.

(a) Find the number of subsets of A.

(b) Find the number of subsets of A having one or more elements.

(c) Find the number of subsets of A having exactly one element.

(d) Find the number of subsets of A having two or more elements. [*Hint*: Use the answers to parts (b) and (c).]

40. Let A be a set with 12 elements.

(a) Find the number of subsets of A.

(b) Find the number of subsets of A having one or more elements.

(c) Find the number of subsets of A having exactly one element.

(d) Find the number of subsets of A having two or more elements. [*Hint*: Use the answers to parts (b) and (c).]

41. For a weighted voting system with 10 players,

(a) find the total number of coalitions.

(b) find the number of coalitions with two or more players.

42. Consider a weighted voting system with 12 players.

(a) Find the total number of coalitions in this weighted voting system.

(b) Find the number of coalitions with two or more players.

43. Consider a weighted voting system with six players (P_1 through P_6).

(a) Find the total number of coalitions in this weighted voting system.

(b) How many coalitions in this weighted voting system do not include P_1? (*Hint*: Think of all the possible coalitions of the remaining players.)

(c) How many coalitions in this weighted voting system do not include P_3? [*Hint*: Is this really different from (b)?]

(d) How many coalitions in this weighted voting system do not include both P_1 and P_3?

(e) How many coalitions in this weighted voting system include both P_1 and P_3? [*Hint*: Use your answers for (a) and (d).]

44. Consider a weighted voting system with five players (P_1 through P_5).

(a) Find the total number of coalitions in this weighted voting system.

(b) How many coalitions in this weighted voting system do not include P_1? (*Hint*: Think of all the possible coalitions of the remaining players.)

(c) How many coalitions in this weighted voting system do not include P_5? [*Hint*: Is this really different from (b)?]

(d) How many coalitions in this weighted voting system do not include P_1 or P_5?

(e) How many coalitions in this weighted voting system include both P_1 and P_5? [*Hint*: Use your answers for (a) and (d).]

For Exercises 45 through 48 you should use a calculator with a factorial key (typically, it's a key labeled either x! or n!). All scientific calculators and most business calculators have such a key. There are also many free scientific calculators available online. Just Google "online scientific calculator" and pick one you like.

45. Use a calculator to compute each of the following.

(a) 13!

(b) 18!

(c) 25!

(d) Suppose that you have a supercomputer that can list *one trillion* (10^{12}) sequential coalitions per second. Estimate (in years) how long it would take the computer to list all the sequential coalitions of 25 players.

46. Use a calculator to compute each of the following.

(a) 12!

(b) 15!

(c) 20!

(d) Suppose that you have a supercomputer that can list one billion (10^9) sequential coalitions per second. Estimate (in years) how long it would take the computer to list all the sequential coalitions of 20 players.

47. Use a calculator to compute each of the following.

(a) $\frac{13!}{3!}$ (b) $\frac{13!}{3!10!}$ (c) $\frac{13!}{4!9!}$ (d) $\frac{13!}{5!8!}$

48. Use a calculator to compute each of the following.

(a) $\frac{12!}{2!}$ (b) $\frac{12!}{2!10!}$ (c) $\frac{12!}{3!9!}$ (d) $\frac{12!}{4!8!}$

The purpose of Exercises 49 and 50 is for you to learn how to numerically manipulate factorials. If you use a calculator to answer these questions, you are defeating the purpose of the exercise. Please try Exercises 49 and 50 without using a calculator.

49. (a) Given that 10! = 3,628,800, find 9!

(b) Find $\frac{11!}{10!}$

(c) Find $\frac{11!}{9!}$

(d) Find $\frac{9!}{6!}$

(e) Find $\frac{101!}{99!}$

50. (a) Given that 20! = 2,432,902,008,176,640,000, find 19!

(b) Find $\frac{20!}{19!}$

(c) Find $\frac{201!}{199!}$

(d) Find $\frac{11!}{8!}$

51. Consider a weighted voting system with seven players (P_1 through P_7).

(a) Find the number of sequential coalitions in this weighted voting system.

(b) How many sequential coalitions in this weighted voting system have P_7 as the first player?

(c) How many sequential coalitions in this weighted voting system have P_7 as the last player?

(d) How many sequential coalitions in this weighted voting system do not have P_1 as the *first* player?

52. Consider a weighted voting system with six players (P_1 through P_6).

(a) Find the number of sequential coalitions in this weighted voting system.

(b) How many sequential coalitions in this weighted voting system have P_4 as the last player?

(c) How many sequential coalitions in this weighted voting system have P_4 as the *third* player?

(d) How many sequential coalitions in this weighted voting system do not have P_1 as the first player?

53. A law firm has seven partners: a senior partner (P_1) with 6 votes and six junior partners (P_2 through P_7) with 1 vote each. The quota is a *simple majority* of the votes. (This law firm operates as the weighted voting system [7: 6, 1, 1, 1, 1, 1, 1].)

(a) In how many sequential coalitions is the senior partner P_1 the pivotal player? (*Hint*: First note that P_1 is the pivotal player in all sequential coalitions except those in which he is the first player.)

(b) Using your answer in (a), find the Shapley-Shubik power index of the senior partner P_1.

(c) Using your answer in (b), find the Shapley-Shubik power distribution in this law firm.

54. A law firm has six partners: a senior partner (P_1) with 5 votes and five junior partners (P_2 through P_6) with 1 vote each. The quota is a *simple majority* of the votes. (This law firm operates as the weighted voting system [6: 5, 1,1, 1, 1, 1].)

(a) In how many sequential coalitions is the senior partner P_1 the pivotal player? (*Hint*: First note that P_1 is the pivotal player in all sequential coalitions except those in which he is the first player.)

(b) Using your answer in (a), find the Shapley-Shubik power index of the senior partner P_1.

(c) Using your answer in (b), find the Shapley-Shubik power distribution in this law firm.

JOGGING

55. Dummies. We defined a *dummy* as a player that is never critical. Explain why each of the following is true:

(a) If P is a dummy, then any winning coalition that contains P would also be a winning coalition without P.

(b) P is a dummy if and only if the Banzhaf power index of P is 0.

(c) P is a dummy if and only if the Shapley-Shubik power index of P is 0.

56. (a) Consider the weighted voting system [22: 10,10, 10, 10, 1]. Are there any dummies? Explain your answer.

(b) Without doing any work [but using your answer for (a)], find the Banzhaf and Shapley-Shubik power distributions of this weighted voting system.

(c) Consider the weighted voting system [q: 10, 10, 10, 10, 1]. Find all the possible values of q for which P_5 is not a dummy.

(d) Consider the weighted voting system [34: 10, 10,10, 10, w]. Find all positive integers w which make P_5 a dummy.

57. Consider the weighted voting system [q: 8, 4, 1].

(a) What are the possible values of q?

(b) Which values of q result in a dictator? (Who? Why?)

(c) Which values of q result in exactly one player with veto power? (Who? Why?)

(d) Which values of q result in more than one player with veto power? (Who? Why?)

(e) Which values of q result in one or more dummies? (Who? Why?)

58. Consider the weighted voting system [9: w, 5, 2, 1].

(a) What are the possible values of w?

(b) Which values of w result in a dictator? (Who? Why?)

(c) Which values of w result in a player with veto power? (Who? Why?)

(d) Which values of w result in one or more dummies? (Who? Why?)

59. Equivalent voting systems. Two weighted voting systems are *equivalent* if they have the same number of players and exactly the same winning coalitions.

(a) Show that the weighted voting systems [8: 5, 3, 2] and [2: 1, 1, 0] are equivalent.

(b) Show that the weighted voting systems [7: 4, 3, 2, 1] and [5: 3, 2, 1, 1] are equivalent.

(c) Show that the weighted voting system discussed in Example 2.12 is equivalent to [3: 1, 1, 1, 1, 1].

(d) Explain why equivalent weighted voting systems must have the same Banzhaf power distribution.

(e) Explain why equivalent weighted voting systems must have the same Shapley-Shubik power distribution.

60. Veto power. A player P with weight w is said to have *veto power* if and only if $w < q$, and $V - w < q$ (where V denotes the total number of votes in the weighted voting system). Explain why each of the following is true:

(a) A player has veto power if and only if the player is a member of every winning coalition.

(b) A player has veto power if and only if the player is a critical player in the *grand* coalition.

61. Consider the generic weighted voting system [q: w_1, w_2, . . . , w_N]. (Assume $w_1 \geq w_2 \geq \cdots \geq w_N$.)

(a) Find all the possible values of q for which no player has veto power.

(b) Find all the possible values of q for which every player has veto power.

(c) Find all the possible values of q for which P_i has veto power but P_{i+1} does not. (*Hint*: See Exercise 60.)

62. The Smith family has two parents (P_1 and P_2) and three children (c_1, c_2, and c_3). Family vacations are decided by a majority of the votes, but at least one parent must vote Yes (i.e., the three children don't have enough weight to carry the motion).

(a) If we use [q: p, p, c, c, c] to describe this weighted voting system, find q, p, and c.

(b) Find the Banzhaf Power distribution of this weighted voting system.

63. The weighted voting system [8: 6, 4, 2, 1] represents a partnership among four partners (P_1, P_2, P_3, and you!). You are the partner with just one vote, and in this situation you have no power (you dummy!). Not wanting to remain a dummy, you offer to buy one vote. Each of the other four partners is willing to sell you one of their votes, and they are all asking the same price. From which partner should you buy in order to get as much power for your buck as possible? Use the Banzhaf power index for your calculations. Explain your answer.

64. The weighted voting system $[27: 10, 8, 6, 4, 2]$ represents a partnership among five people (P_1, P_2, P_3, P_4, and P_5). You are P_5, the one with two votes. You want to increase your power in the partnership and are prepared to buy one share (one share equals one vote) from any of the other partners. Partners P_1, P_2, and P_3 are each willing to sell cheap ($1000 for one share), but P_4 is not being quite as cooperative—she wants $5000 for one of her shares. Given that you still want to buy one share, from whom should you buy it? Use the Banzhaf power index for your calculations. Explain your answer.

65. The weighted voting system $[18: 10, 8, 6, 4, 2]$ represents a partnership among five people (P_1, P_2, P_3, P_4, and P_5). You are P_5, the one with two votes. You want to increase your power in the partnership and are prepared to buy shares (one share equals one vote) from any of the other partners.

(a) Suppose that each partner is willing to sell one share and that they are all asking the same price. Assuming that you decide to buy only one share, from which partner should you buy? Use the Banzhaf power index for your calculations.

(b) Suppose that each partner is willing to sell two shares and that they are all asking the same price. Assuming that you decide to buy two shares from a single partner, from which partner should you buy? Use the Banzhaf power index for your calculations.

(c) If you have the money and the cost per share is fixed, should you buy one share or two shares (from a single person)? Explain.

66. Mergers. Sometimes in a weighted voting system two or more players decide to merge—that is to say, to combine their votes and always vote the same way. (Note that a merger is different from a coalition—coalitions are temporary, whereas mergers are permanent.) For example, if in the weighted voting system $[7: 5, 3, 1]$ P_2 and P_3 were to merge, the weighted voting system would then become $[7: 5, 4]$. In this exercise we explore the effects of mergers on a player's power.

(a) Consider the weighted voting system $[4: 3, 2, 1]$. In Example 2.9 we saw that P_2 and P_3 each have a Banzhaf power index of $1/5$. Suppose that P_2 and P_3 merge and become a single player P^*. What is the Banzhaf power index of P^*?

(b) Consider the weighted voting system $[5: 3, 2, 1]$. Find first the Banzhaf power indexes of players P_2 and P_3 and then the Banzhaf power index of P^* (the merger of P_2 and P_3). Compare.

(c) Rework the problem in (b) for the weighted voting system $[6: 3, 2, 1]$.

(d) What are your conclusions from (a), (b), and (c)?

67. (a) Verify that the weighted voting systems $[12: 7, 4, 3, 2]$ and $[24: 14, 8, 6, 4]$ result in exactly the same Banzhaf power distribution. (If you need to make calculations, do them for both systems side by side and look for patterns.)

(b) Based on your work in (a), explain why the two proportional weighted voting systems $[q: w_1, w_2, \ldots, w_N]$ and $[cq: cw_1, cw_2, \ldots, cw_N]$ always have the same Banzhaf power distribution.

68. (a) Verify that the weighted voting systems $[12: 7, 4, 3, 2]$ and $[24: 14, 8, 6, 4]$ result in exactly the same Shapley-Shubik power distribution. (If you need to make calculations, do them for both systems side by side and look for patterns.)

(b) Based on your work in (a), explain why the two proportional weighted voting systems $[q: w_1, w_2, \ldots, w_N]$ and $[cq: cw_1, cw_2, \ldots, cw_N]$ always have the same Shapley-Shubik power distribution.

69. In a weighted voting system with four players the winning coalitions are: $\{P_1, P_2, P_3\}$, $\{P_1, P_2, P_4\}$, $\{P_1, P_3, P_4\}$, and $\{P_1, P_2, P_3, P_4\}$.

(a) Find the Banzhaf power distribution of the weighted voting system.

(b) Find the Shapley-Shubik power distribution of the weighted voting system.

70. The Fresno City Council. In Fresno, California, the city council consists of seven members (the mayor and six other council members). A motion can be passed by the mayor and at least three other council members, or by at least five of the six ordinary council members.

(a) Describe the Fresno City Council as a weighted voting system.

(b) Find the Shapley-Shubik power distribution for the Fresno City Council. (*Hint*: See Example 2.17 for some useful ideas.)

RUNNING

71. A partnership has four partners (P_1, P_2, P_3, and P_4). In this partnership P_1 has twice as many votes as P_2; P_2 has twice as many votes as P_3; P_3 has twice as many votes as P_4. The quota is a *simple majority* of the votes. Show that P_1 is always a *dictator*. (*Hint*: Write the weighted voting system in the form $[q: 8x, 4x, 2x, x]$, and express q in terms of x. Consider separately the case when x is even and the case when x is odd.)

72. In a weighted voting system with three players, the six sequential coalitions (each with the pivotal player underlined) are: $\langle P_1, \underline{P_2}, P_3\rangle$, $\langle P_1, \underline{P_3}, P_2\rangle$, $\langle P_2, \underline{P_1}, P_3\rangle$, $\langle P_2, P_3, \underline{P_1}\rangle$, $\langle P_3, \underline{P_1}, P_2\rangle$, and $\langle P_3, P_2, \underline{P_1}\rangle$. Find the Banzhaf power distribution of the weighted voting system.

73. A professional basketball team has four coaches, a head coach (H) and three assistant coaches (A_1, A_2, A_3). Player personnel decisions require at least three Yes votes, one of which must be H's.

(a) If we use $[q: h, a, a, a]$ to describe this weighted voting system, find q, h, and a.

(b) Find the Shapley-Shubik power distribution of the weighted voting system.

74. A law firm has $N+1$ partners: the senior partner with N votes, and N junior partners with one vote each. The quota is a simple majority of the votes. Find the Shapley-Shubik power distribution in this weighted voting system. (*Hint:* Try Exercise 53 or 54 first.)

75. Decisive voting systems. A weighted voting system is called *decisive* if for every losing coalition, the coalition consisting of the remaining players (called the *complement*) must be a winning coalition.

(a) Show that the weighted voting system $[5: 4, 3, 2]$ is decisive.

(b) Show that the weighted voting system $[3: 2, 1, 1, 1]$ is decisive.

(c) Explain why any weighted voting system with a dictator is decisive.

(d) Find the number of winning coalitions in a decisive voting system with N players.

76. The Cleansburg City Council. Find the Banzhaf power distribution in the Cleansburg City Council discussed in Example 2.17.

77. Relative voting power. The *relative voting weight* w_i of a player P_i is the fraction of votes controlled by that player. A player's Banzhaf power index β_i can differ considerably from his relative voting weight w_i. One indicator of the relation between Banzhaf power and relative voting weight is the ratio between the two (called the *relative Banzhaf voting power*): $\pi_i = \frac{\beta_i}{w_i}$.

(a) Compute the relative Banzhaf voting power of California in the Electoral College (see Table 2-11).

(b) Compute the relative Banzhaf voting power of each player in Example 2.13.

78. Suppose that in a weighted voting system there is a player A who hates another player P so much that he will always vote the opposite way of P, regardless of the issue. We will call A the *antagonist* of P.

(a) Suppose that in the weighted voting system $[8: 5, 4, 3, 2]$, P is the player with two votes and his antagonist A is the player with five votes. The other two players we'll call P_2 and P_3. What are the possible coalitions under these circumstances? What is the Banzhaf power distribution under these circumstances?

(b) Suppose that in a generic weighted voting system with N players there is a player P who has an antagonist A. How many coalitions are there under these circumstances?

(c) Give examples of weighted voting systems where a player A can

(i) increase his Banzhaf power index by becoming an antagonist of another player.

(ii) decrease his Banzhaf power index by becoming an antagonist of another player.

(d) Suppose that the antagonist A has more votes than his enemy P. What is a strategy that P can use to gain power at the expense of A?

79. (a) Give an example of a weighted voting system with four players and such that the Shapley-Shubik power index of P_1 is $\frac{3}{4}$.

(b) Show that in any weighted voting system with four players a player cannot have a Shapley-Shubik power index of more than $\frac{3}{4}$ unless he or she is a dictator.

(c) Show that in any weighted voting system with N players a player cannot have a Shapley-Shubik power index of more than $\frac{(N-1)}{N}$ unless he or she is a dictator.

(d) Give an example of a weighted voting system with N players and such that P_1 has a Shapley-Shubik power index of $\frac{(N-1)}{N}$.

80. (a) Explain why in any weighted voting system with N players a player with veto power must have a Banzhaf power index bigger than or equal to $\frac{1}{N}$.

(b) Explain why in any weighted voting system with N players a player with veto power must have a Shapley-Shubik power index bigger than or equal to $\frac{1}{N}$.

APPLET BYTES MyMathLab®

These Applet Bytes are short projects or mini-explorations built around the applet **Banzhaf and Shapley-Shubik Power** (available in MyMathLab in the Multimedia Library or Tools for Success.) This applet allows the user to create a weighted voting system by entering as inputs the number of players, the weights of the players and the quota. The applet then computes the Banzhaf and Shapley-Shubik power of the players in the weighted voting system. The applet can handle very large weighted voting systems (up to 50 players) and thus makes it possible to do computations that would be impossible by hand, freeing the user to explore more interesting and meaningful questions.*

81. The First Electoral College (1792). The election of 1792 was the first presidential election decided by electors assigned to each state based on the Census of 1790. There were 15 states at that time—the original 13 colonies plus Vermont (admitted in 1791) and Kentucky (admitted in 1792). The number of electors from each state is shown in the second column of Table 2-17. Using the applet *Banzhaf and Shapley-Shubik Power* and the number of electors as a proxy for votes in the Electoral College†, compute the Banzhaf and Shapley power indexes of each of the 15 states. (Assume the quota is given by strict majority.)

*MyMathLab code required.

†In reality, at that time the Electoral College was a little more complicated than that: each elector had *two* votes to cast for two different candidates, and the electors from each state did not have to vote for the same two candidates. The candidate with the most electoral votes became President, the candidate with the second-most Vice President. For the purposes of this exercise we will assume the more modern version of the Electoral College, where each elector has just one vote and all the electors from a given state vote for the same candidate.

State	Number of Electors	Banzhaf Power	Shapley-Shubik Power
Virginia	21		
Massachusetts	16		
Pennsylvania	15		
New York	12		
North Carolina	12		
Connecticut	9		
Maryland	8		
South Carolina	8		
New Jersey	7		
New Hampshire	6		
Georgia	4		
Kentucky	4		
Rhode Island	4		
Vermont	3		
Delaware	3		
Total	**132**		

TABLE 2-17

82. The Nassau County (N.Y.) Board of Supervisors in the 1960's. In Example 2.13 we discussed the Nassau County Board of Supervisors in the 1960's. (See Example 2.13 for the historical details). Table 2-18 shows:

Column 1. The players (districts),

Column 2. The weights of the players (number of votes on the board), which are roughly proportional to the district populations,

Column 3. The weights expressed as a percentage of the total number of votes. (For example, N. Hempstead with 21 out of 115 votes has $21/115 \approx 0.1826 = 18.26\%$ of the votes.) We can take this number to roughly represent the percent of the population in each district.

Column 4. The Banzhaf power index of each player expressed as a percent,

Column 5. The "gap" between weight and power computed as the absolute value of the difference between the percentages in columns 3 and 4.

Player	Weight	Weight as a %	Banzhaf Power	Gap
Hempstead 1	31	26.96%	33.33%	6.37%
Hempstead 2	31	26.96%	33.33%	6.37%
Oyster Bay	28	24.35%	33.33%	8.98%
N. Hempstead	21	18.26%	0%	18.26%
Long Beach	2	1.74%	0%	1.74%
Glen Cove	2	1.74%	0%	1.74%
Total	**115**			
Quota	**58**			

TABLE 2-18

In a good weighted voting system we want the gaps between weights and power to be small—the smaller the gaps the better the voting system accomplishes what it is supposed to do, namely, *allocate power to each district in proportion to its population*. By this measure, the 1960's Nassau County Board was a terrible weighted voting system with huge gaps (just look at N. Hempstead with 18.26% of the population and 0% of the power). In a sense this is the argument used by the courts to declare the composition of the Nassau County board illegal and to remand the board to reconstitute itself with different numbers.

Suppose the standard set by the court for a legally acceptable board is that the gaps should all be under 4% (not exactly true, but just imagine this for the purposes of the exercise). Using the applet *Banzhaf and Shapley-Shubik Power* try different values of the quota (but keep the original weights of the districts the same) until you find one that meets the above requirement.

83. The Nassau County (N.Y.) Board of Supervisors in the 1990's. (You should first try your hand at Exercise 82 before you work on this one.) By the 1990's, following a series of court cases and population changes the Nassau County Board of Supervisors was reconstituted as the weighted voting system [65: 30, 28, 22, 15, 7, 6] (Table 2-19).

Player	Weight	Weight as a %	Banzhaf Power	Gap
Hempstead 1	30			
Hempstead 2	28			
Oyster Bay	22			
N. Hempstead	15			
Long Beach	7			
Glen Cove	6			
Total	**108**			
Quota	**65**			

TABLE 2-19

(a) Complete the remaining three columns of Table 2-19. Use the applet to do the Banzhaf Power computations in the fourth column of the table.

(b) What is the size of the largest gap for $q = 65$? What is the average value of the gaps for $q = 65$?

(c) Imagine that you are a lawyer arguing in front of the court that the weighted voting system [65: 30, 28, 22, 15, 7, 6] is not good enough, and that you can provide several values of the quota that produce a much better weighted voting system (with smaller values for both the largest gap and the average gap). Show which values of the quota do this and compute the *best* largest gap and average gaps possible. [*Hint*: There are several such values and they are all less than $q = 65$.]

3

Brothers (and business partners) Tom, Fred, and Henry Elghanayan. Breaking-up was a piece of cake. (For the details, see Example 3.13.)

The Mathematics of Sharing

Fair-Division Games

As children, the Elghanayan brothers probably did not give much thought to the question of how to best divide their Halloween stash of candy. Fifty years later, when Tom, Fred, and Henry Elghanayan decided to break up their forty-year partnership and split up the assets, they did indeed give a great deal of thought to the question of division. After all, they were about to divide $3 billion worth of New York City real estate holdings.

The fact that the Elghanayan brothers were able to split up so much valuable property (8000 apartments, 9 office buildings, and 9 development projects) without lawsuits or bad blood is a testament to the power of a surprising idea: When different parties—with different preferences and values—have to divide commonly owned assets, it is actually possible to carry out the division in a way that is more than just fair. Indeed, it is so fair that *all* parties come out of the deal feeling that they got more than they deserved.

The problem of dividing a real estate empire worth $3 billion and the problem of dividing a stash of Halloween candy are not all that different. Other than the stakes involved, they share the same basic elements: a set of assets, a group of individuals who share ownership of those assets, and a willingness to divide the assets in a fair way. These types of problems are part of an interesting and important class of problems known as *fair-division problems*.

Fairness is an innate human value, based on empathy, the instinct for cooperation, and, to some extent, self-interest. A theory for sharing things fairly based on cooperation, reason, and logic is one of the great achievements of social science, and, once again, we can trace the roots of this achievement to a branch of mathematics known as game theory. In this chapter we will give a basic introduction to the subject.

> "I'm happy with the assets I ended up with, and my guess is that they are happy also. Assets have different values to different people. There is no such thing as a fair market value; it's a fair market value in the mind of each person."
>
> – *Henry Elghanayan*

The chapter is organized in a manner very similar to that of Chapters 1 and 2. We will start (Section 3.1) with some general background: What are the elements of a *fair-division game* and what are the requirements for a *fair-division method*? In the remaining five sections we introduce five different fair-division methods. In Section 3.2 we discuss the *divider-chooser* method, a fair-division method that applies when dividing a *continuous* (i.e., infinitely divisible) asset between two players. In Sections 3.3 and 3.4 we discuss two different extensions of the divider-chooser method that can be used to divide a continuous asset among three or more players: the *lone-divider* method and the *lone-chooser* method. In Sections 3.5 and 3.6 we discuss two different fair-division methods that can be used when dividing a set of discrete (i.e., indivisible) assets: the method of *sealed bids* and the method of *markers*.

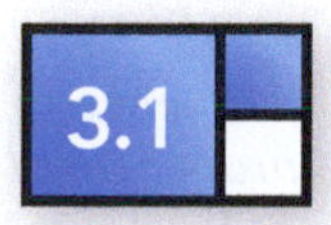

Fair-Division Games

In this section we will introduce some of the basic concepts and terminology of fair division. Much as we did in Chapter 2 when we studied weighted voting systems, we will think of fair-division problems in terms of games—with players, goals, rules, and strategies.

Basic Elements of a Fair-Division Game

The underlying elements of every *fair-division game* are as follows:

- **The assets.** This is the formal name we will give to the "goodies" being divided. Typically, the assets are tangible physical objects such as real estate, jewelry, art, candy, cake, and so on. In some situations the assets being divided may be intangible things such as rights—water rights, drilling rights, broadcast licenses, and so on. (While the term "assets" has a connotation of positive value, sometimes the assets being divided have negative values—chores, obligations, liabilities, etc. In this chapter we will focus primarily on the division of positive assets, but in Example 3.12 we discuss a situation in which negative-valued assets are being divided.) Regardless of the nature of the assets, we will use the symbol S throughout this chapter to denote the set of *all* assets being divided.

- **The players.** In every fair-division game there is a set of parties with the right (or in some cases the duty) to divide among themselves a set of jointly owned assets. They are the *players* in the game. Most of the time the players in a fair-division game are individuals, but it is worth noting that some of the most significant applications of fair division occur when the players are *institutions* (ethnic groups, political parties, states, and even nations).
- **The value systems.** The fundamental assumption we will make is that each player has the ability to give a value to any part of the assets. Specifically, this means that each player can look at the set S or any subset of S and assign to it a value—either in absolute terms ("to me, that plot of land is worth $875,000") or in relative terms ("to me, that plot of land is worth 30% of the total value of the assets").
- **A fair-division method.** These are the rules that govern the way the game is played. Much like the voting methods discussed in Chapter 1, fair-division methods are very specific and leave no room for ambiguity. We will discuss in much greater detail various fair-division methods in the remaining sections of the chapter.

Like most games, fair-division games are predicated on certain assumptions about the players. For the purposes of our discussion, we will make the following four assumptions:

- **Rationality.** Each of the players is a thinking, rational entity seeking to maximize his or her share of the assets. We will further assume that in the pursuit of this goal, a player's moves are based on reason alone (we are taking emotion, psychology, undue risk-taking, and all other nonrational elements out of the picture).
- **Cooperation.** The players are willing participants and accept the rules of the game as binding. The rules are such that after a *finite* number of moves by the players, the game terminates with a division of the assets. (There are no outsiders such as judges or referees involved in these games—just the players and the rules.)
- **Privacy.** Players have *no* useful information on the other players' value systems and thus of what kinds of moves they are going to make in the game. (This assumption does not always hold in real life, especially if the players are siblings or friends.)
- **Symmetry.** Players have *equal* rights in sharing the assets. A consequence of this assumption is that, at a minimum, each player is entitled to a *proportional* share of the assets—when there are two players, each is entitled to at least one-half of the assets, when there are three players each is entitled to at least one-third of the assets, and in general, when there are N players each is entitled to at least $1/N$ th of the assets.

Fair Shares and Fair Divisions

Given a set of players $P_1, P_2, \ldots, P_N$ and a set of assets S, the ultimate purpose of a fair-division game is to produce a *fair division* of S. But what does this mean? To formalize the precise meaning of this seemingly subjective idea, we introduce two very important definitions.

- **Fair share (to a player P).** Suppose that s denotes a share of the set of assets S and that P is one of the players in a fair-division game with N players. The share

s is called a **proportional fair share** (or just simply a **fair share**) *to P*, if, in *P*'s opinion, the value of s is at least $\frac{1}{N}$th of the value of S.

- **Fair division (of the set of assets *S*).** Suppose we are able to divide the set of assets S into N shares (call them $s_1, s_2, \ldots, s_N$) and assign each of these shares to one of the players. If each player considers the share he or she received to be a fair share (i.e., worth at least $\frac{1}{N}$th of the value of S), then we have achieved a **fair division** of the set of assets S.

EXAMPLE 3.1 FAIR DIVISIONS

Three brothers—say Henry, Tom, and Fred—are splitting up their partnership and dividing a bunch of assets of unspecified value. The set of assets S is divided (at this point we don't care how) into three shares: s_1, s_2, and s_3. In Henry's opinion, the values of the three shares (expressed as a percentage of the value of S) are, respectively, 32%, 31% and 37%. This information is displayed in the first row of Table 3-1. Likewise, the second and third rows of Table 3-1 show the values that Tom and Fred assign, respectively, to each of the three shares.

	s_1	s_2	s_3
Henry	32%	31%	37%
Tom	34%	31%	35%
Fred	$33\frac{1}{3}$%	$33\frac{1}{3}$%	$33\frac{1}{3}$%

TABLE 3-1 ■ Players' valuation of shares for Examples 3.1 and 3.2

- **Fair shares.** With three players, the threshold for a fair share is $33\frac{1}{3}$%. To Henry, s_3 is the only fair share; to Tom, s_1 and s_3 are both fair shares; to Fred, who values all three shares equally, they are all fair shares.
- **Fair divisions.** To assign each player a fair share we start with Henry because with him we have only one option—we have to assign him s_3. With Tom we seem to have, at least in principle, two options—s_1 and s_3 are both fair shares—but s_3 is gone, so our only choice is to assign him s_1. This leaves the final share s_2 for Fred. Since all three players received a fair share, we have achieved a fair division. (Notice that this does not mean that all players are equally happy—Tom would have been happier with s_3 rather than s_1, but such is life. All that a fair division promises the players is that they will get a *fair share*, not the *best share*.)

Fair-Division Methods

A **fair-division method** is a set of rules that, when properly used by the players, *guarantees* that at the end of the game each player will have received a fair share of the assets. The key requirement is the guarantee—no matter what the circumstances, the method should produce a fair division of the assets. The next example illustrates a seemingly reasonable method that fails this test.

EXAMPLE 3.2 DRAWING LOTS

Once again we go back to Henry, Tom, and Fred and their intention to split their partnership fairly. The method they adopt for the split seems pretty reasonable: One of the brothers divides the assets into three shares (s_1, s_2, and s_3), and then they draw lots to determine the order in which they get to choose. The values of the shares, once again, are shown in Table 3-1. Say the order is Henry first, Tom second, and Fred last. In this case Henry will undoubtedly choose s_3, Tom will then choose s_1, and Fred will end up with s_2. This works out fine, as this gives a fair division of S. So far, so good.

Suppose, however, that the when they draw lots the order is Tom first, Henry second, and Fred last. In this case Tom will start by choosing s_3. Now Henry has to choose between s_1 and s_2, neither of which in his eyes is a fair share. In this case the method fails to produce a fair division. Since drawing lots sometimes works and other times doesn't, we can't call it a fair-division method.

There are many different fair-division methods known, but in this chapter we will discuss only a few of the classic ones. Depending on the nature of the set *S*, a fair-division game can be classified as one of three types: *continuous, discrete*, or *mixed*, and the fair-division methods used depend on which of these types we are facing.

- **Continuous fair division.** Here the set *S* is divisible in infinitely many ways, and shares can be increased or decreased by arbitrarily small amounts. Typical examples of continuous fair-division games involve the division of land, a cake, a pizza, and so forth.
- **Discrete fair division.** Here the set *S* is made up of objects that are indivisible, like paintings, houses, cars, boats, jewelry, and so on. (What about pieces of candy? One might argue that with a sharp enough knife a piece of candy could be chopped up into smaller and smaller pieces, but nobody really does that—it's messy. As a semantic convenience let's agree that a piece of candy is indivisible, and, therefore, dividing candy is a discrete fair-division game.)
- **Mixed fair division.** This is the case where some of the assets are continuous and some are discrete. Dividing an estate consisting of jewelry, a house, and a parcel of land is a mixed fair-division game.

Fair-division methods are classified according to the nature of the problem involved. Thus, there are *discrete fair-division* methods (used when the set *S* is made up of indivisible, discrete objects), and there are *continuous fair-division* methods (used when the set *S* is an infinitely divisible, continuous set). Mixed fair-division games can usually be solved by dividing the continuous and discrete parts separately, so we will not discuss them in this chapter. We will start our excursion into fair-division methods with a classic method for continuous fair division.

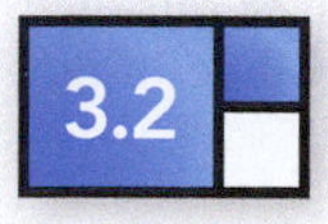

3.2 The Divider-Chooser Method

When two players are dividing a continuous asset, the standard method used is the **divider-chooser method**. This method is also known as "you cut; I choose," and most of us have used it at some time or another when splitting with someone else a piece of cake, a sandwich, or a pint of ice cream. As the name suggests, one player, called the *divider*, divides *S* into two shares, and the second player, called the *chooser*, picks the share he or she wants, leaving the other share to the divider.

Under the *rationality* and *privacy* assumptions introduced in Section 3.1, this method guarantees that both divider and chooser will get a share worth 50% or more of the total value of *S*. Why? Not knowing the chooser's likes and dislikes (privacy assumption), the divider can only guarantee himself a 50% share by dividing *S* into two halves of equal value (rationality assumption); the chooser is guaranteed a 50% or better share by choosing the piece she likes best.

EXAMPLE 3.3 DIVIDING A CHEESECAKE

After meeting online through a computer dating website, Angie and Brad decide to go to the county fair on their first date. They buy jointly a raffle ticket, and, as luck would have it, they win a fancy chocolate-strawberry cheesecake. The retail value of this cheesecake is $36.

Brad has no preference between chocolate and strawberry—he values them the same. It follows that in his eyes, each of the six sections of the cake, either chocolate or strawberry, has a value of $6 [Fig. 3-1(b)].

Angie, on the other hand, is partial to strawberry over chocolate, and, being hyper-rational, she can quantify this preference: she claims she values a strawberry

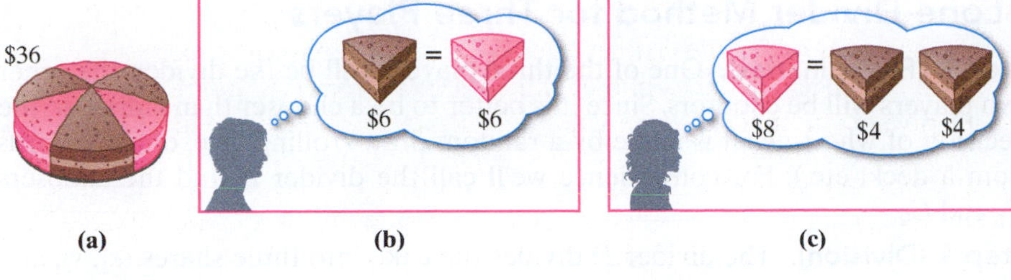

FIGURE 3-1

piece *twice as much* as she values a chocolate piece of the same size. (In a somewhat informal shorthand one could say that to Angie, $2C$'s $= 1S$, meaning that it takes two parts chocolate to equal in value one part strawberry). It follows that to Angie the value of each chocolate section is $4 while the value of each strawberry section is $8 [Fig. 3-1(c)]. (Here are the details of the calculation: If the value of a chocolate section is x, then the three chocolate sections are worth $3x$, the three strawberry sections are worth $6x$, and the whole cake is worth $9x$. Setting $9x = 36$ gives $x = 4$.)

FIGURE 3-2 (a) Brad cuts. (b) Angie picks.

Let's now see how Angie and Brad might divide this cake using the divider-chooser method. Brad volunteers to go first and be the divider. He cuts the cake in half as shown in Fig. 3-2(a). It is now Angie's turn to choose, and for obvious reasons she chooses the half with two parts strawberry and one part chocolate, which in her eyes has a value of $20 [Fig. 3-2(b)].

The final outcome is that Brad gets a piece that in his eyes is worth exactly half of the cake ($18), while Angie ends up with a piece that in her eyes is worth a bit more ($20). This is, nonetheless, a fair division of the cake—both players get pieces worth 50% or more. Mission accomplished!

Example 3.3 illustrates why, given a choice, it is always *better to be the chooser than the divider*—the divider is guaranteed a share worth exactly 50% of the total value of S, but the chooser can do better and end up with a share worth more than 50%. Since a fair-division method should treat all players equally, both players should have an equal chance of being the chooser. This is best done by means of a coin toss, with the winner of the coin toss getting the privilege of making the choice.

The basic idea behind the divider-chooser method is that when an asset is divided between two people, the same person should not be both divider and chooser. This is a very old idea, and we can find examples of it as far back as the Old Testament: When Lot and Abraham argued over grazing rights for their sheep, Abraham proposed, "Let us divide the land into left and right. If you go left, I will go right; and if you go right, I will go left" [that is, "you choose first, I'll take the other half"] (Genesis 13:1–9).

But how do we implement the divider-chooser idea when the division is among three, or four, or N players? In the next two sections we will discuss two different ways to extend the divider-chooser method to the case of three or more players.

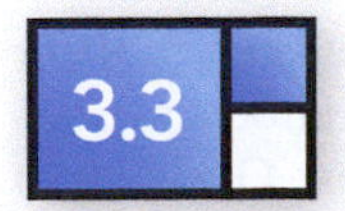

3.3 The Lone-Divider Method

The first important breakthrough in the mathematics of fair division came in 1943, when the Polish mathematician Hugo Steinhaus came up with a clever way to extend the divider-chooser method to the case of *three* players, one of whom plays the role of the *divider* while the other two play the role of *choosers*. Steinhaus's approach was subsequently generalized to any number of players N (one divider and $N - 1$ choosers) by Princeton mathematician Harold Kuhn. In either case we will refer to this method as the **lone-divider method**.

We start this section with a description of Steinhaus's *lone-divider method* for the case of $N = 3$ players. We will describe the process in terms of dividing a cake, a commonly used and convenient metaphor for a continuous asset.

The Lone-Divider Method for Three Players

- **Step 0 (Preliminaries).** One of the three players will be the divider; the other two players will be choosers. Since it is better to be a chooser than a divider, the decision of who is what is made by a random draw (rolling dice, drawing cards from a deck, etc.). For convenience we'll call the divider D and the choosers C_1 and C_2.
- **Step 1 (Division).** The divider D divides the cake into three shares (s_1, s_2, and s_3). D will get one of these shares, but at this point does not know which one. Not knowing which share will be his (privacy assumption) forces D to divide the cake into three shares of equal value (rationality assumption). It follows that each of the three shares is a fair share to D, and that D is indifferent as to which of the three he gets.
- **Step 2 (Bidding).** C_1 declares (usually by writing on a slip of paper) which of the three pieces are fair shares to her. Independently, C_2 does the same. These are the *bids*. A chooser's bid *must list every single piece that he or she considers to be a fair share* (i.e., worth one-third or more of the cake)—it may be tempting to bid only for the very best piece, but this is a strategy that can easily backfire. To preserve the privacy requirement, it is important that the bids be made independently, without the choosers being privy to each other's bids.
- **Step 3 (Distribution).** Who gets which piece? The answer, of course, depends on which pieces are listed in the bids. For convenience, we will separate the pieces into two types: C-pieces (these are pieces *chosen* by one or both of the choosers) and U-pieces (these are *unwanted* pieces that did not appear in either of the bids). Expressed in terms of value, a U-piece is a piece that *both* choosers value at less than $33\frac{1}{3}\%$ of the cake, and a C-piece is a piece that *at least* one of the choosers (maybe both) values at $33\frac{1}{3}\%$ or more. Depending on the number of C-pieces, there are two separate cases to consider.

  Case 1. When there are two or more C-pieces, there is always a way to assign to each chooser a fair share chosen from her bid list. (The details will be covered in Examples 3.4 and 3.5.) Once each chooser gets her fair share, the divider gets the last remaining piece and a fair division has been accomplished. (Sometimes we might end up in a situation where C_1 likes C_2's piece better than her own and vice versa. In that case it is perfectly reasonable to let them swap pieces—this would make each of them happier than they already were, and who could be against that?)

  Case 2. When there is only one C-piece, we have a bit of a problem because it means that both choosers are bidding for the very same piece [Fig. 3-3(a)]. The solution here requires a little more creativity. First, we take care of the divider D—to whom each of the three pieces is a fair share—by giving him one of the two pieces that neither chooser wants [Fig. 3-3(b).] (If the two choosers can agree on the least desirable piece, then so much the better; if they can't agree, then the choice of which piece to give the divider can be made randomly.)

  After D gets his piece, the two pieces left (the C-piece and the remaining U-piece) are recombined into one piece that we call the B-piece [Fig. 3-3(c)]. Now that we have one piece (the B-piece) and two players (the choosers), we can revert to the *divider-chooser method* to finish the fair division: One player cuts the B-piece into two halves; the other player chooses the half she likes best [Fig. 3-3(d)].

  This process results in a fair division of the cake because it guarantees fair shares for all players. We know that D ends up with a fair share by the very fact that D did the original division, but what about C_1 and C_2? The key observation is that in the eyes of both C_1 and C_2 the B-piece is worth *more than two-thirds of the value of the original cake* (think of the B-piece as 100% of the original cake minus a U-piece worth less than $33\frac{1}{3}\%$), so when we divide *it* fairly into *two* shares, each party is guaranteed *more than one-third* of the original cake. We will come back to this point in Example 3.6.

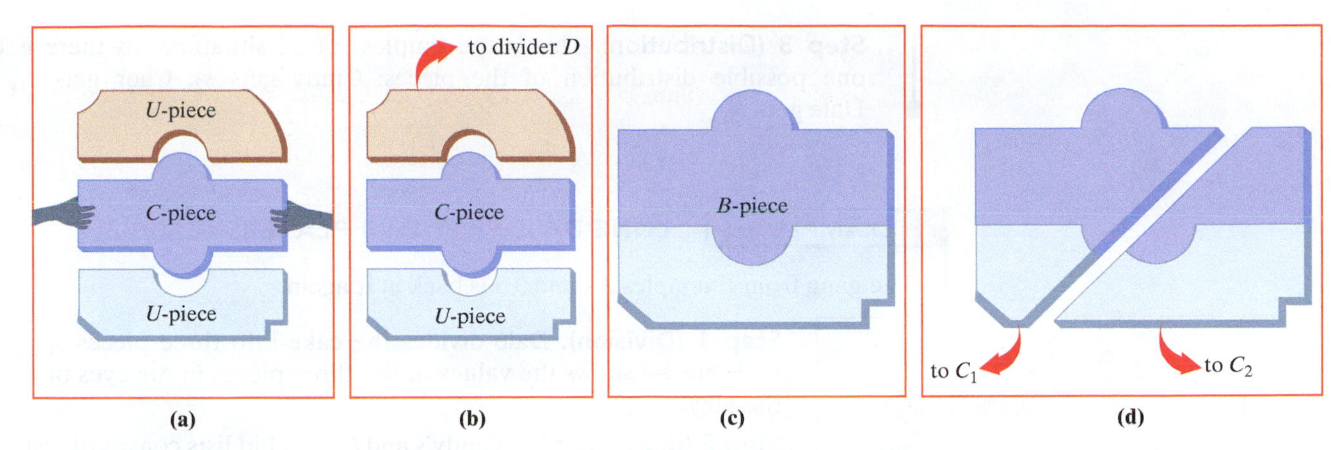

FIGURE 3-3 Case 2 in the lone-divider method (three players). (a) Both choosers covet the same piece. (b) The divider is assigned one of the *U*-pieces. (c) The two remaining pieces are recombined into the *B*-piece. (d) The *B*-piece is divided by the remaining two players using the divider-chooser method.

We will now illustrate the details of Steinhaus's lone-divider method for three players with several examples. In all these examples we will assume that the divider has already divided the cake into three shares s_1, s_2, and s_3. In each example the values that each of the three players assigns to the shares, expressed as percentages of the total value of the cake, are shown in a table. The reader should remember, however, that this information is never available in full to the players—an individual player only knows the percentages on his or her row.

EXAMPLE 3.4 LONE DIVIDER WITH 3 PLAYERS (CASE 1, VERSION 1)

Dale, Cindy, and Cher are dividing a cake using Steinhaus's lone-divider method. They draw cards from a well-shuffled deck of cards, and Dale draws the low card (bad luck!) and has to be the divider.

	s_1	s_2	s_3
Dale	$33\frac{1}{3}\%$	$33\frac{1}{3}\%$	$33\frac{1}{3}\%$
Cindy	35%	10%	55%
Cher	40%	25%	35%

TABLE 3-2 ■ Players' valuation of shares for Example 3.4

Step 1 (Division). Dale divides the cake into three pieces s_1, s_2, and s_3. Table 3-2 shows the values of the three pieces in the eyes of each of the players.

Step 2 (Bidding). From Table 3-2 we can assume that Cindy's bid list is $\{s_1, s_3\}$ and Cher's bid list is also $\{s_1, s_3\}$.

Step 3 (Distribution). The *C*-pieces are s_1 and s_3. There are two possible distributions. One distribution would be: Cindy gets s_1, Cher gets s_3, and Dale gets s_2. An even better distribution (the *optimal* distribution) would be: Cindy gets s_3, Cher gets s_1, and Dale gets s_2. In the case of the first distribution, both Cindy and Cher would benefit by swapping pieces, and there is no rational reason why they would not do so. Thus, using the rationality assumption, we can conclude that in either case the final result will be the same: Cindy gets s_3, Cher gets s_1, and Dale gets s_2. ■

EXAMPLE 3.5 LONE DIVIDER WITH 3 PLAYERS (CASE 1, VERSION 2)

We'll use the same setup as in Example 3.4—Dale is the divider, Cindy and Cher are the choosers.

	s_1	s_2	s_3
Dale	$33\frac{1}{3}\%$	$33\frac{1}{3}\%$	$33\frac{1}{3}\%$
Cindy	30%	40%	30%
Cher	60%	15%	25%

TABLE 3-3 ■ Players' valuation of shares for Example 3.5

Step 1 (Division). Dale divides the cake into three pieces s_1, s_2, and s_3. Table 3-3 shows the values of the three pieces in the eyes of each of the players.

Step 2 (Bidding). Here Cindy's bid list is $\{s_2\}$ only, and Cher's bid list is $\{s_1\}$ only.

Step 3 (Distribution). This is the simplest of all situations, as there is only one possible distribution of the pieces: Cindy gets s_2, Cher gets s_1, and Dale gets s_3.

EXAMPLE 3.6 LONE DIVIDER WITH 3 PLAYERS (CASE 2)

The gang from Examples 3.4 and 3.5 is back at it again.

	s_1	s_2	s_3
Dale	$33\frac{1}{3}\%$	$33\frac{1}{3}\%$	$33\frac{1}{3}\%$
Cindy	20%	30%	50%
Cher	10%	20%	70%

TABLE 3-4 ■ Players' valuation of shares for Example 3.6

Step 1 (Division). Dale divides the cake into three pieces s_1, s_2, and s_3. Table 3-4 shows the values of the three pieces in the eyes of each of the players.

Step 2 (Bidding). Here Cindy's and Cher's bid lists consist of just $\{s_3\}$.

Step 3 (Distribution). The only C-piece is s_3. Cindy and Cher talk it over, and without giving away any other information agree that of the two U-pieces, s_1 is the least desirable, so they all agree that Dale should get s_1. (Dale doesn't care which of the three pieces he gets, so he has no rational objection.) The remaining pieces (s_2 and s_3) are then recombined to form the B-piece, to be divided between Cindy and Cher using the divider-chooser method (one of them divides the B-piece into two shares, the other one chooses the share she likes better). Regardless of how this plays out, both of them will get a very healthy share of the cake: Cindy will end up with a piece worth at least 40% of the original cake (the B-piece is worth 80% of the original cake to Cindy), and Cher will end up with a piece worth at least 45% of the original cake (the B-piece is worth 90% of the original cake to Cher).

The Lone-Divider Method for More Than Three Players

Harold Kuhn (1925–2014)

In 1967 Harold Kuhn, a mathematician at Princeton University, extended Steinhaus's lone-divider method to more than three players. The first two steps of Kuhn's method are a straightforward generalization of Steinhaus's lone-divider method for three players, but the distribution step requires some fairly sophisticated mathematical ideas and is rather difficult to describe in full generality. We will only give an outline here and will illustrate the details with a couple of examples.

- **Step 0 (Preliminaries).** One of the players is chosen to be the divider D, and the remaining $N-1$ players are all choosers. As always, it's better to be a chooser than a divider, so the selection of the divider D should be made by a random draw.
- **Step 1 (Division).** The divider D divides the set S into N shares $s_1, s_2, s_3, \ldots, s_N$. D is guaranteed of getting one of these shares, but doesn't know which one.
- **Step 2 (Bidding).** Each of the $N-1$ choosers independently submits a bid list consisting of *every share that he or she considers to be a fair share* (in this case, this means any share worth $\frac{1}{N}$*th* or more of the total).
- **Step 3 (Distribution).** The bid lists are opened. Much as we did in the case of three players, we will have to consider two separate cases, depending on how these bid lists turn out.

  Case 1. If there is a way to assign a different share to each of the $N-1$ choosers, then that should be done. (Needless to say, the share assigned to a chooser should be from his or her bid list.) The divider, to whom all shares are presumed to be of equal value, gets the last unassigned share. At the end, players may choose to swap pieces if they want.

  Case 2. There is a *standoff*—in other words, there are two choosers both bidding for just one share, or three choosers bidding for just two shares, or, in more general terms, K choosers bidding for less than K shares. This is a much more complicated case, and what follows is a rough sketch of what to do. To resolve a standoff, we first separate the players involved in the standoff (let's call them the "standoff players") from the rest ("non-standoff players"). We also separate the shares that are part of the standoff ("standoff shares) from the rest ("non-standoff shares"). Each of the non-standoff players, including the divider, can be assigned a fair share from among the non-standoff shares and sent packing. All the shares left are recombined into a new piece to be divided among the standoff players, and the process starts all over again.

The following two examples will illustrate some of the ideas behind the lone-divider method in the case of four players. The first example is one without a standoff; the second example involves a standoff.

EXAMPLE 3.7 **LONE DIVIDER WITH 4 PLAYERS (CASE 1)**

We have one divider, Demi, and three choosers, Chan, Chloe, and Chris.

Step 1 (Division). Demi divides the cake into four shares s_1, s_2, s_3, and s_4. Table 3-5 shows how each of the players values each of the four shares. Remember that the information on each row of Table 3-5 is private and known only to that player.

Step 2 (Bidding). Chan's bid list is $\{s_1, s_3\}$; Chloe's bid list is $\{s_3\}$ only; Chris's bid list is $\{s_1, s_4\}$. (Keep in mind that with 4 players the threshold for a fair share is 25%.)

	s_1	s_2	s_3	s_4
Demi	25%	25%	25%	25%
Chan	30%	20%	35%	15%
Chloe	20%	20%	40%	20%
Chris	25%	20%	20%	35%

TABLE 3-5 ■ Players' valuation of shares for Example 3.7

Step 3 (Distribution). The bid lists are opened. It is clear that for starters Chloe must get s_3—there is no other option. This forces the rest of the distribution: s_1 must then go to Chan, and s_4 goes to Chris. Finally, we give the last remaining piece, s_2, to Demi.

This distribution results in a fair division of the cake, although it is not entirely "envy-free"—Chan wishes he had Chloe's piece (35% is better than 30%) but Chloe is not about to trade pieces with him, so he is stuck with s_1. (From a strictly rational point of view, Chan has no reason to gripe—he did not get the best piece, but got a piece worth 30% of the total, better than the 25% he is entitled to.) ■

EXAMPLE 3.8 **LONE DIVIDER WITH 4 PLAYERS (CASE 2)**

Once again, we will let Demi be the divider and Chan, Chloe, and Chris be the three choosers (same players, different game).

Step 1 (Division). Demi divides the cake into four shares s_1, s_2, s_3, and s_4. Table 3-6 shows how each of the players values each of the four shares.

Step 2 (Bidding). Chan's bid list is $\{s_4\}$; Chloe's bid list is $\{s_2, s_3\}$; Chris's bid list is $\{s_4\}$.

	s_1	s_2	s_3	s_4
Demi	25%	25%	25%	25%
Chan	20%	20%	20%	40%
Chloe	15%	35%	30%	20%
Chris	22%	23%	20%	35%

TABLE 3-6 ■ Players' valuation of shares for Example 3.8

Step 3 (Distribution). The bid lists are opened, and the players can see that there is a standoff brewing on the horizon—Chan and Chris are both bidding for s_4. First we separate Chan and Chris (the standoff players) from Chloe and Demi (the non-standoff players). We also separate s_4 (the standoff piece) from the rest. Next, we assign fair shares to Chloe and Demi chosen from the non-standoff set of pieces s_1, s_2, and s_3. Chloe could be given either s_2 or s_3. (She would rather have s_2, of course, but it's not for her to decide.) A coin toss is used to determine which one. Let's say Chloe ends up with s_3.

Demi could now be given either s_1 or s_2. Another coin toss, and Demi ends up with s_1. The final move is . . . you guessed it!—recombine s_2 and s_4 into a single piece to be divided between Chan and Chris using the divider-chooser method. Since $(s_2 + s_4)$ is worth 60% to Chan and 58% to Chris (you can check it out in Table 3-6), regardless of how this final division plays out they are both guaranteed a final share worth more than 25% of the cake.

Mission accomplished! We have produced a fair division of the cake.

3.4 The Lone-Chooser Method

A. M. Fink (1932—)

A completely different approach for extending the divider-chooser method was proposed in 1964 by A. M. Fink, a mathematician at Iowa State University. In this method one player plays the role of chooser, all the other players start out playing the role of dividers. For this reason, the method is known as the **lone-chooser method**. Once again, we will start with a description of the method for the case of three players.

The Lone-Chooser Method for Three Players

- **Step 0 (Preliminaries).** We have one chooser and two dividers. Let's call the chooser C and the dividers D_1 and D_2. As before, it is better to be the chooser, so C is the winner of some random draw.
- **Step 1 (First Division).** D_1 and D_2 divide S between themselves into *two* fair pieces using the divider-chooser method. Let's say that D_1 ends up with s_1 and D_2 ends up with s_2 [Fig. 3-4(a)].
- **Step 2 (Subdivision).** Each divider divides his or her piece into three subshares. Thus, D_1 divides s_1 into three subshares, which we will call s_{1a}, s_{1b}, and s_{1c}. Likewise, D_2 divides s_2 into three subshares, which we will call s_{2a}, s_{2b}, and s_{2c} [Fig. 3-4(b)].
- **Step 3 (Selection).** The chooser C now selects one of D_1's three subshares and one of D_2's three subshares [Fig. 3-4(c)]. These two subshares make up C's final share. D_1 then keeps the remaining two subshares from s_1, and D_2 keeps the remaining two subshares from s_2.

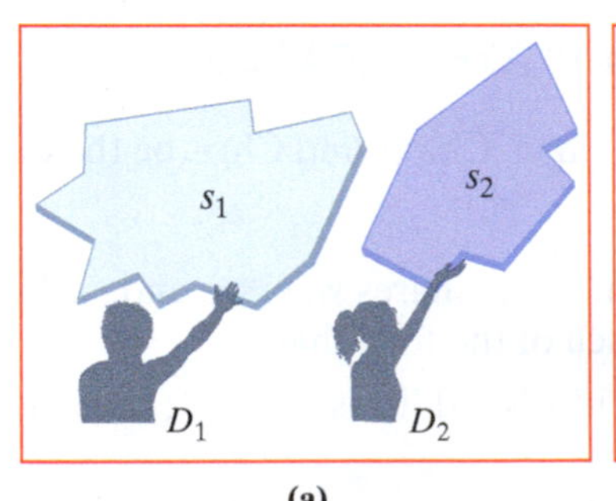

(a)

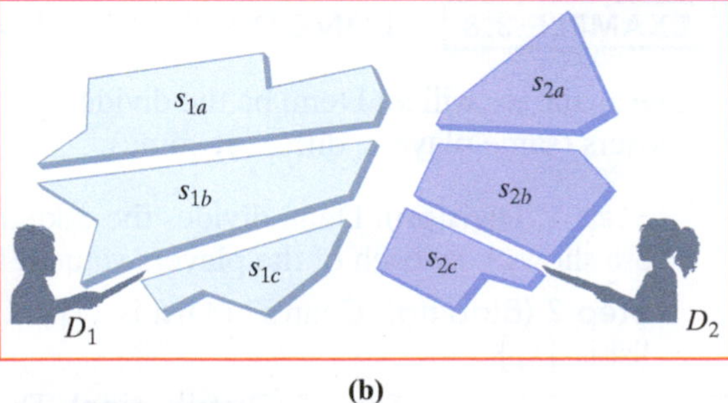

(b)

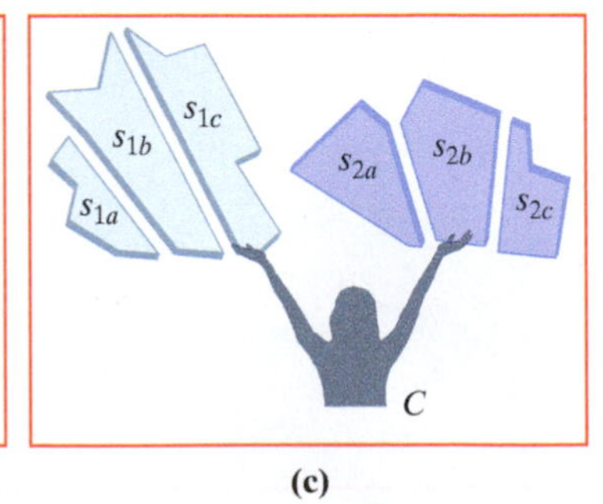

(c)

FIGURE 3-4 (a) First division. (b) Subdivision. (c) Selection.

Why is this a fair division of S? D_1 ends up with two-thirds of s_1. To D_1, s_1 is worth at least one-half of the total value of S, so two-thirds of s_1 is at least one-third—a fair share. The same argument applies to D_2. What about the chooser's share? We don't know what s_1 and s_2 are each worth to C, but it really doesn't matter—a one-third or better subshare of s_1 plus a one-third or better subshare of s_2 equals a one-third or better share of $(s_1 + s_2)$ and, thus, a fair share of S.

The following example illustrates in detail how the lone-chooser method works with three players.

EXAMPLE 3.9 LONE CHOOSER WITH 3 PLAYERS

David, Dinah, and Cher are dividing a fancy mango-raspberry cake [Fig. 3-5(a)] among themselves using the lone-chooser method. The value of this cake is $54 (it is indeed a fancy cake), so for a fair division, each of them should end up with a share worth $18 or more.

Their individual value systems (not known to one another, but available to us as outside observers) are as follows:

- David likes mango and raspberry the same, so to him each of the six wedges—either mango or raspberry—has a value of $9 [Fig. 3-5(b)].
- Dinah likes mango *twice* as much as she likes raspberry, so to her the mango wedges are worth $12 each and the raspberry wedges are worth $6 each ($1M = 2R$; $3M + 3R = 54$ implies $R = 6$ and $M = 12$) [Fig. 3-5(c)].
- Cher likes raspberry *three* times as much as she likes mango, so to her the mango wedges are worth $4.50 each and the raspberry wedges are worth $13.50 each ($1R = 3M$; $3M + 3R = 54$ implies $M = 4.50$ and $R = 13.50$) [Fig. 3-5(d)].

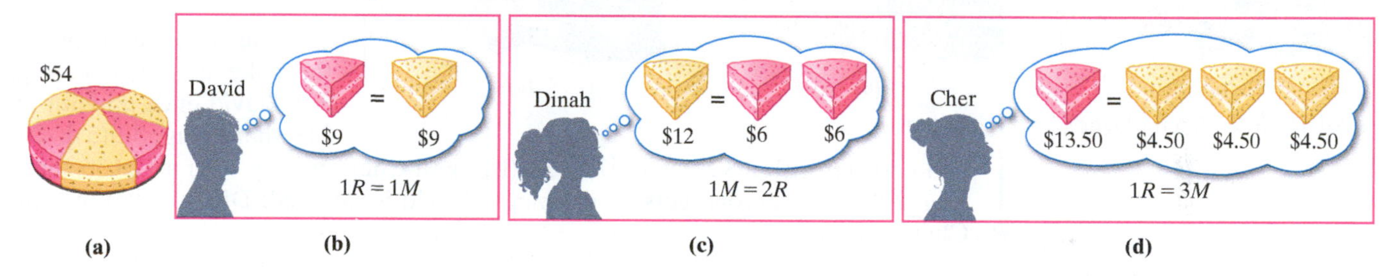

FIGURE 3-5

When the three draw cards from a deck, Cher gets the high card and chooses to be the chooser. Thus, she gets to sit out Steps 1 and 2.

Step 1 (First Division). David and Dinah start by dividing the cake between themselves using the divider-chooser method. After a coin flip, David has to do the first division and, being indifferent between mango and raspberry, cuts the cake into two halves of equal value [Fig. 3-6(a)]. Dinah, who prefers mango over raspberry, chooses the half with the most mango (to her worth $30) [Fig. 3-6(b)].

FIGURE 3-6 (a) David cuts. (b) Dinah chooses.

Step 2 (Subdivision). David divides his half into three subshares that in his opinion are of equal value. In his case, the three subshares are all the same size [Fig. 3-7(a)]. Dinah also divides her half into three subshares that in her opinion are of equal value ($10 per share). She does this by adding to the raspberry wedge a small sliver (10°) from each of the mango wedges [Fig. 3-7(b)]. This works because to her a 10° sliver is worth $2 (it is, after all, one-sixth of a $12 wedge).

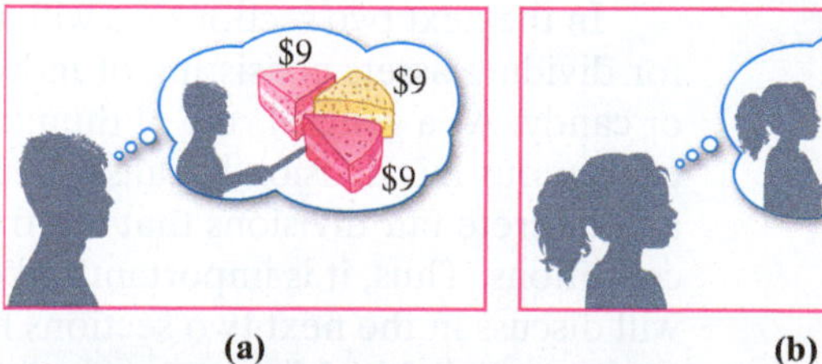

FIGURE 3-7 (a) David subdivides his share. (b) Dinah subdivides her share.

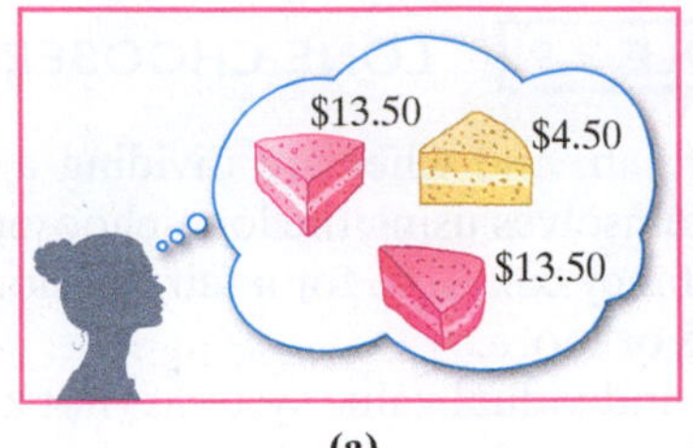

(a)

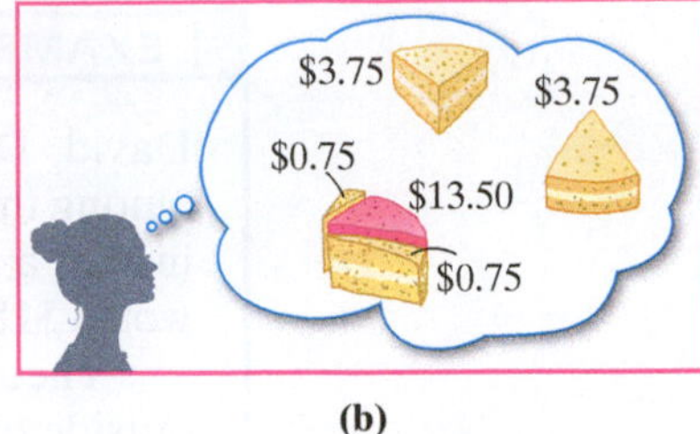

(b)

FIGURE 3-8 The values of the subshares in Cher's eyes.

Step 3 (Selection). It's now Cher's turn to choose one subshare from David's three and one subshare from Dinah's three. Figure 3-8 shows the values of the subshares in Cher's eyes. It's clear what her choices will be: She will take one of the two raspberry wedges from David and the raspberry wedge with a little mango on the sides from Dinah.

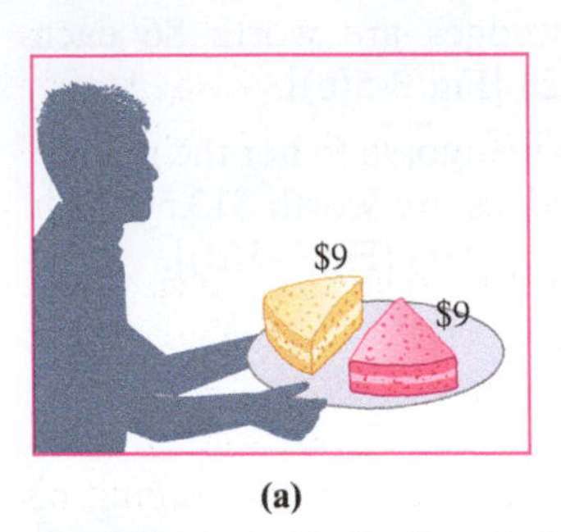

(a)

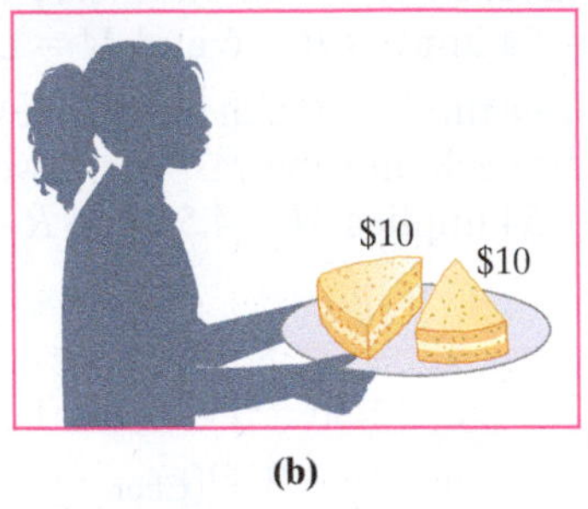

(b)

(c)

FIGURE 3-9 Each player's final fair share.

The final division of the cake is shown in Fig. 3-9: David ends up with an \$18 share [Fig. 3-9(a)], Dinah ends up with a \$20 share [Fig. 3-9(b)], and Cher ends up with a \$28.50 share [Fig. 3-9(c)]. Everyone gets a fair share—David is satisfied, Dinah is happy, and Cher is ecstatic. ■

The Lone-Chooser Method for *N* Players

For the general case of N players, the lone-chooser method involves one chooser C and $N-1$ dividers $D_1, D_2, \ldots, D_{N-1}$. As always, it is preferable to be a chooser rather than a divider, so the chooser is determined by a random draw. The method is based on an inductive strategy—if you can do it for three players, then you can do it for four players; if you can do it for four, then you can do it for five; and so on. Thus, when we get to N players, we can assume that we can use the lone-chooser method to divide the set S among $N-1$ players.

- **Step 1 (First Division).** $D_1, D_2, \ldots, D_{N-1}$ divide fairly the set S among themselves, acting as if C didn't exist. This is a fair division among $N-1$ players, so each one gets a piece he or she considers worth at least $\frac{1}{(N-1)}$th of S.
- **Step 2 (Subdivision).** Each divider subdivides his or her piece into N subshares.
- **Step 3 (Selection).** The chooser C finally gets to play. C selects one subshare from each divider—one subshare from D_1, one from D_2, and so on. At the end, C ends up with $N-1$ subshares, which make up C's final share, and each divider gets to keep the remaining $N-1$ subshares in his or her subdivision.

When properly played, the lone-chooser method guarantees that everyone, dividers and chooser alike, ends up with a fair share.

In the next two sections we will discuss *discrete* fair-division methods—methods for dividing assets consisting of indivisible objects such as houses, cars, art, jewels, or candy. As a general rule of thumb, discrete fair division is harder to achieve than continuous fair division because there is a lot less flexibility in the division process, and discrete fair divisions that are truly fair are only possible under a limited set of conditions. Thus, it is important to keep in mind that while both of the methods we will discuss in the next two sections have limitations, they still are the best methods we have available. Moreover, when they work, both methods work remarkably well and produce surprisingly good fair divisions.

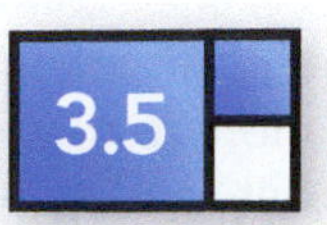

3.5 The Method of Sealed Bids

The **method of sealed bids** was originally proposed by the Polish mathematicians Hugo Steinhaus (1887–1972) and Bronislaw Knaster (1893–1980) around 1948. The best way to illustrate how this method works is by means of an example.

EXAMPLE 3.10 SETTLING GRANDMA'S ESTATE

In her last will and testament, Grandma plays a little joke on her four grandchildren (Art, Betty, Carla, and Dave) by leaving just three items—a cabin in the mountains, a vintage 1955 Rolls Royce, and a Picasso painting—with the stipulation that the items must remain with the grandchildren (not sold to outsiders) and must be divided fairly among them. How can we possibly resolve this conundrum? The method of sealed bids will give an ingenious and elegant solution.

- **Step 1 (Bidding).** Each of the players makes a bid (in dollars) for each of the items in the estate, giving his or her honest assessment of the actual value of each item. To satisfy the privacy assumption, it is important that the bids are done independently, and no player should be privy to another player's bids before making his or her own. The easiest way to accomplish this is for each player to submit his or her bid in a sealed envelope. When all the bids are in, they are unsealed. The first three rows in Table 3-7 show the player's bids for each of the items.

  Once the bids are unsealed we can calculate the fair-share value for each of the players. We first add a player's bids for all of the items in the estate (Total Value row in Table 3-7) and then divide this total by the number of players (Fair-share value row in Table 3-7).

  $$\text{Fair-share value} = \text{Total value}/N$$

	Art	Betty	Carla	Dave
Cabin	420,000	450,000	411,000	398,000
Vintage Rolls	80,000	70,000	87,000	92,000
Painting	680,000	640,000	634,000	590,000
Total value	1,180,000	1,160,000	1,132,000	1,080,000
Fair-share value	295,000	290,000	283,000	270,000

TABLE 3-7 ■ The original bids and fair-share values for each player. Winning bids are circled in red.

- **Step 2 (Allocation).** Each item will go to the highest bidder for that item. (If there is a tie, the tie can be broken with a coin flip.) In this example the cabin goes to Betty, the vintage Rolls Royce goes to Dave, and the Picasso painting goes to Art. Notice that Carla gets nothing. Not to worry—it all works out at the end! (In this method it is possible for one player to get none of the items and another player to get many or all of the items. Much like in a silent auction, it's a matter of who bids the highest.)
- **Step 3 (Initial Settlement).** It's now time to settle things up. Depending on what items (if any) a player gets in Step 2, he or she will owe money to the estate or be owed money by the estate.

  The fair-share values are the baseline for the initial settlements—if the total value of the items that the player gets in Step 2 is more than his or her fair-share value, then the player *pays* the estate the difference. If the total value of the items that the player gets is *less* than his or her fair-share value, then the player *gets* the difference in cash.

  Table 3-8 illustrates how the first settlements are calculated. Take Art, for example. Art receives the Picasso painting, which he valued at $680,000. But his fair-share value is only $295,000, so Art must make up the difference of $385,000 by paying it to the estate. Likewise, Betty gets the cabin for $450,000 but has to pay $160,000 to the estate. Carla is not getting any items, so she receives the full value of her share ($283,000) in cash *from* the estate. Dave gets the vintage Rolls, which he values at $92,000. Since he is entitled to a fair share valued at $270,000, the difference is given to him in cash from the estate.

At this point each of the four heirs has received a fair share, and we might consider our job done, but this is not the end of the story—there is more to come (good news mostly!). If we add Art and Betty's payments to the estate and subtract the payments made by the estate to Carla and Dave, we discover that there is a *surplus* of $84,000! ($385,000 and $160,000 came in from Art and Betty respectively; $283,000 and $178,000 went out to Carla and Dave.)

	Art	Betty	Carla	Dave	
Item(s) received	Picasso	Cabin	none	Rolls	
Value received	$680,000	$450,000	0	$92,000	
Fair-share value	$295,000	$290,000	$283,000	$270,000	Surplus
To (from) estate	$385,000	$160,000	($283,000)	($178,000)	$84,000

TABLE 3-8 ■ Initial settlement of the estate leaves an $84,000 surplus

- **Step 4 (Division of the Surplus).** The surplus is common money that belongs to the estate, and thus is to be divided equally among the players. In our example each player's share of the $84,000 surplus is $21,000.
- **Step 5 (Final Settlement).** Everything done up to this point could be done on paper, but now, finally, real money needs to change hands! Art gets the Picasso painting and pays the estate $364,000 ($385,000 – $21,000). Betty gets the cabin and has to pay the estate $139,000 ($160,000 – $21,000). Carla gets $304,000 in cash ($283,000 + $21,000). Dave gets the vintage Rolls Royce plus $199,000 ($178,000 + $21,000).

When two players are dividing a commonly owned discrete asset, one of them will end up with the asset and the other one must get an equivalent share of something, usually money. The method of sealed bids handles this situation nicely and provides a solution that makes both players happy (assuming, of course, that they are rational players). Consider the following example.

EXAMPLE 3.11 SPLITTING UP THE HOUSE

Al and Betty are getting a divorce. The only joint property of any value is their house. Rather than hiring attorneys and going to court to figure out how to split up the house, they agree to give the method of sealed bids a try.

Al's bid on the house is $340,000; Betty's bid is $364,000. The fair-share value of the house is $170,000 to Al and $182,000 to Betty. Since Betty is the higher bidder, she gets to keep the house and must pay Al cash for his share. The computation of how much cash Betty pays Al can be done in two steps. In the first settlement, Betty owes the estate $182,000. Of this money, $170,000 pays for Al's fair share, leaving a surplus of $12,000 to be split equally between them. The bottom line is that Betty ends up paying $176,000 to Al for his share of the house. Table 3-9 summarizes the split.

	Al	Betty
Bid for house	$340,000	$364,000
Fair-share value	$170,000	$182,000
To (from) estate	($170,000)	$182,000
Share of surplus	$6,000	$6,000
Final settlement	gets $176,000	gets house, pays $176,000

TABLE 3-9 ■ Method of sealed bids for Example 3.11

The method of sealed bids can provide an excellent solution not only to settlements of property in a divorce but also to the equally difficult and often contentious issue of splitting up a partnership. The catch is that in these kinds of splits we can rarely count on the rationality assumption to hold. A divorce or partnership split devoid of emotion, spite, and hard feelings is a rare thing indeed!

Auctions, Reverse Auctions, and Negative Bidding

Most people are familiar with the principle behind a standard **auction** (we can thank eBay for that): A seller has an item of *positive* value (say a vintage car) that he is willing to give up in exchange for money, and there are bidders who are willing to part with their money to get the item. The *highest* bidder gets the item. In some situations the bidders are allowed to make only one secret bid (usually in writing) and the highest bidder gets the item. These types of auctions are called **sealed-bid auctions**. A good way to think of the method of sealed bids is as a variation of a sealed-bid auction.

The key idea behind the method of sealed bids is that for each item there is a sealed-bid auction in which the players are simultaneously sellers and bidders. The highest bidder for the item ends up being the buyer (of the other players' shares), the other players are all sellers (of their shares). The twist is that a player doesn't know whether he will end up a buyer or a seller until the bids are opened. Not knowing forces the player to make honest bids (if he bids too high he runs the risk of buying an overpriced item; if he bids too low he runs the risk of selling an item too cheaply).

The above interpretation of the method of sealed bids helps us understand how the method can also be used to divide negative-valued items such as chores and other unpleasant responsibilities (for lack of a better word let's just call such items "baddies"). The only difference is that now we will model the process after a *reverse auction*. A **reverse auction** is the flip side of a regular auction: There is a buyer who wants to buy a service (for example repairing a leaky roof) and bidders who are willing to provide the service in exchange for cash. Here the *lowest* bidder gets the job (i.e., gets to sell his services to the buyer). The buyer pays because he doesn't want to climb on the roof and risk breaking his neck, and the low bidder does the job because he wants the cash.

When we combine the basic rules for the method of sealed bids with the idea of a reverse auction, we have a method for dividing any discrete set of baddies. We will illustrate how this works in our next example.

EXAMPLE 3.12 APARTMENT CLEANUP 101

Anne, Belinda, and Clara are college roommates. The school year just ended, and they are getting ready to move out of their apartment. They talk to their landlord about getting their $1200 cleaning deposit back, and he tells them that there are five major chores that they need to do to get their cleaning deposit (or at least most of it) back. Table 3-10 lists the five chores as well as the bids that each roommate made for each chore. Remember that now the bids represent bids for services provided. (When Anne bids $90 for washing the windows she is saying something like "I am willing to wash the windows if I can get $90 in credit toward the final settlement on the cleaning deposit. If someone else wants to do it for less that's fine—let them do it.")

	Anne	Belinda	Clara
1. Clean bathrooms	$120	$80	$75
2. Clean kitchen	$85	$80	$70
3. Patch walls and paint	$180	$170	$210
4. Shampoo carpets	$80	$110	$100
5. Wash windows	$90	$70	$100
Total value	$555	$510	$555
Fair-share value	$185	$170	$185

TABLE 3-10 ■ Bids and fair-share values for each player in Example 3.12

The fair division of the chores is summarized in Table 3-11. Since this is a reverse auction, each chore is assigned to the lowest bidder. Anne ends up with chore 4, Belinda ends up with chores 3 and 5, and Clara ends up with chores 1 and 2. In the first settlement Anne has to "pay" $105 (no money is changing hands yet—this is just a debit on the final settlement of the cleaning deposit); Belinda gets a credit of $70; and Clara has a debit of $40. The details of the calculations are shown in Table 3-11. There is a surplus of $75, so each roommate gets a surplus credit of $25.

	Anne	Belinda	Clara	
Chores assigned (low bidder)	4	3 and 5	1 and 2	
Value of service provided	$80	$240	$145	
Fair-share value of duties	$185	$170	$185	Surplus
Debit (credit) toward final settlement	$105	($70)	$40	$75

TABLE 3-11 ■ First settlement and surplus for Example 3.12

The final settlement: Because they did such a good job, Anne, Belinda, and Clara get their full $1200 cleaning deposit back. The money is divided as follows:

- Anne, who shampooed the carpets, gets $320.
  [$400 (share of cleaning deposit) + $25 (surplus credit) − $105 (settlement debit)]
- Belinda, who painted walls and cleaned windows, gets $495.
  [$400 (share of cleaning deposit) + $25 (surplus credit) + $70 (settlement credit)]
- Clara, who cleaned the bathrooms and the kitchen, gets $385.
  [$400 (share of cleaning deposit) + $25 (surplus credit) − $40 (settlement debit)]

Whether we are dividing assets ("goodies") or chores ("baddies"), the method of sealed bids offers an elegant and effective solution to the problem of fair division. There are, however, two limitations to the method of sealed bids:

1. Players must have enough money to properly play the game. A player that is cash strapped is at a disadvantage—he may not be able to make the right bids if he cannot back them up. (So, if you are the poor grandson dividing grandma's estate with a bunch of fat-cat cousins, you may want to argue for something other than the method of sealed bids.)
2. Players must accept the fact that all assets are marketable commodities (i.e., the principle that "everything has a price"). When there is an item that more than one player considers "priceless" (some kind of family heirloom, for example), then there are serious issues that the method of sealed bids cannot resolve.

Fine-Grained, Discrete Fair Division

Under the right set of circumstances, a discrete fair-division problem can be solved using continuous fair-division methods. This happens when the set of assets consists of many items with plenty of relatively low-valued items to "smooth" out the division. We call these situations "fine-grained" fair-division games. To clarify the idea, imagine that you have a large pile of rocks and you want to split the rocks into three piles of equal weight. If all you have is big boulders then it is not very likely you will be able to do it, but if you also have a big supply of small rocks (the fine grain, so to speak) then your chances of making three piles of equal weight are quite good.

Using continuous fair-division methods to solve discrete fair-division problems is very helpful—continuous fair-division is easier and has fewer restrictions than discrete fair division. Our next example illustrates how we can use a continuous fair-division method to solve a fine-grained, discrete fair-division problem. It also brings us back full circle to the story of the Elghanayan brothers.

EXAMPLE 3.13 DIVIDING $3 BILLION WORTH OF REAL ESTATE

In 2009, Henry, Tom and Fred Elghanayan agreed to break up Rockrose Development Corporation, the $3-billion New York City real estate empire they jointly owned (about 8000 apartments, 9 office towers, and 9 major development projects). The breakup was reasonably amicable (there were a few bruised egos but no hard feelings), and each of the three brothers ended up with a share that he considered fair. How did they do it?

First, notice that because of the large number of apartments (the small rocks), this is an example of a fine-grained, discrete, fair-division problem. Thus, it was possible to divide the assets into three shares using a variation of the lone-divider method. Fred ended up as the divider (how he ended up with that dubious honor will be explained soon), and his task was to divide the assets into three "piles" of equal value. After Fred was finished with the division (it took him about 60 days to make up the three piles) a coin flip was used to determine which of the other two brothers (the choosers) would get to choose first. Henry won the coin flip, and this determined the order of choice: Henry first, Tom second, Fred last.

As in any lone-divider strategy, being the divider is the least desirable role, and to determine who would end up in that role the brothers used a variation of the method of sealed bids: a reverse auction in which the lowest bidder for the assets (Fred) got to be the divider. This insured that both choosers valued the assets at a higher rate than the divider and guaranteed that Fred, knowing that he would choose last, would divide the assets into three approximately equal shares.

The most important postscript to this story is that all three brothers were happy with the outcome of the split, a very uncommon ending to such a big stakes division. ■

3.6 The Method of Markers

The **method of markers** is a discrete fair-division method proposed in 1975 by William F. Lucas, a mathematician at the Claremont Graduate School. The method has the great virtue that it does not require the players to put up any of their own money. On the other hand, unlike the method of sealed bids, this method can only be used effectively in the case of a fine-grained, discrete, fair-division game.

In this method we start with the items lined up in a random but fixed sequence called an *array*. Each of the players then gets to make an independent *bid* on the items in the array. A player's bid consists of dividing the array into segments of consecutive items (as many segments as there are players) so that each of the segments represents a fair share of the entire set of items.

For convenience, we might think of the array as a string. Each player then "cuts" the string into N segments, each of which he or she considers an acceptable share. (Notice that to cut a string into N sections, we need $N-1$ cuts.) In practice, one way to make the "cuts" is to lay markers in the places where the cuts are made. Thus, each player can make his or her bids by placing $N-1$ markers so that they divide the array into N segments. To ensure privacy, no player should see the markers of another player before laying down his or her own.

The final step is to give to each player one of the segments in his or her bid. The easiest way to explain how this can be done is with an example.

EXAMPLE 3.14 DIVIDING THE POST-HALLOWEEN STASH

Alice, Bianca, Carla, and Dana want to divide their jointly owned post-Halloween stash of candy. Their teacher, Mrs. Jones, offers to divide the candy for them, but the students reply that they just learned about a cool fair-division game they want to try, and they can do it themselves, thank you.

As a preliminary step, the 20 pieces are lined up in an array. The order in which the pieces are lined up should be random (the easiest way to do this is to dump the pieces into a paper bag, shake the bag, and take the pieces out, as shown in Fig. 3-10). (For convenience we are using colored squares with labels 1 through 20 to represent the various pieces of candy. Imagine if you will that #1 is a lollypop, #2 is a peppermint patty, and so on.)

FIGURE 3-10 The items lined up in an array.

- **Step 1 (Bidding).** Each student writes down independently on a piece of paper exactly where she wants to place her three markers. (Three markers divide the array into four sections.) The bids are opened, and the results are shown in Fig. 3-11. The A-labels indicate the position of Alice's markers (A_1 denotes her first marker, A_2 her second marker, and A_3 her third and last marker). Alice's bid means that she is willing to accept one of the following as a fair share of the candy: pieces 1 through 5 (first segment), pieces 6 through 11 (second segment), pieces 12 through 16 (third segment), or pieces 17 through 20 (last segment). Bianca's bid is shown by the B-markers and indicates how she would break up the array into four segments that are fair shares; same for Carla's bid (shown by the C-markers) and Dana's bid (shown by the D-markers).

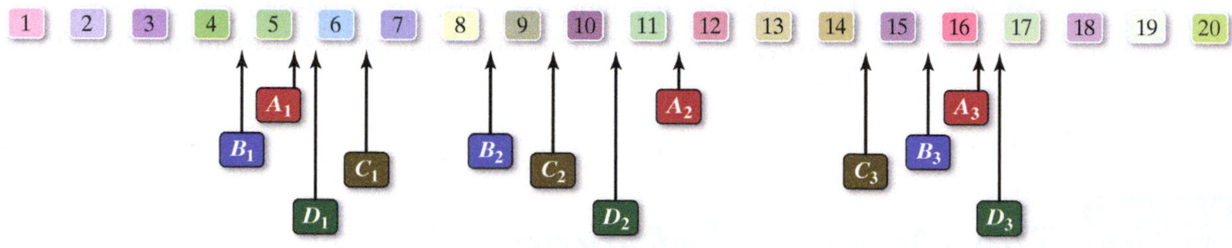

FIGURE 3-11 The bids.

- **Step 2 (Allocations).** This is the tricky part, where we are going to give to each player one of the segments in her bid. Here is how to do it: Scan the array from left to right until the first *first marker* comes up. Here the first *first marker* is Bianca's B_1. This means that Bianca will be the first player to get her fair share, consisting of the first segment in her bid (pieces 1 through 4, Fig. 3-12).

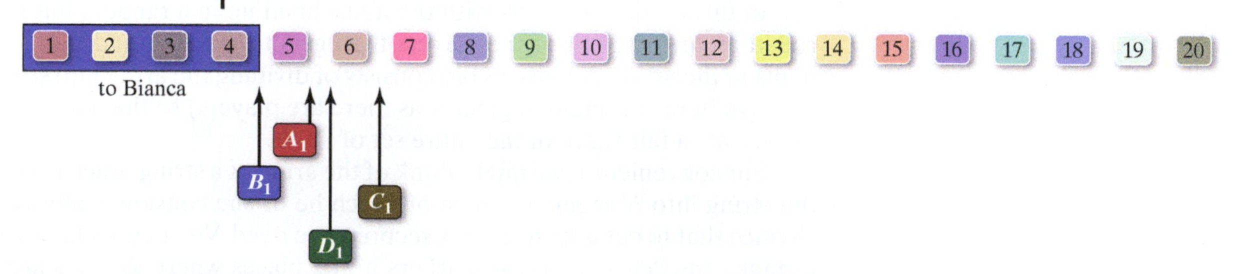

FIGURE 3-12 B_1 is the first 1-marker. Bianca goes first, gets her first segment.

Bianca is done now, and her markers can be removed since they are no longer needed. Continue scanning from left to right looking for the first *second marker*. Here the first second marker is Carla's C_2, so Carla will be the second player taken care of. Carla gets the second segment in her bid (pieces 7 through 9, Fig. 3-13). Carla's remaining markers can now be removed.

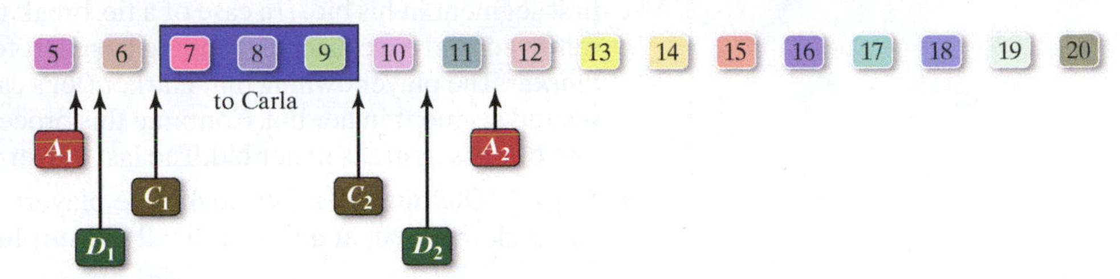

FIGURE 3-13 C_2 is the first 2-marker (among *A*'s, *C*'s, and *D*'s). Carla goes second, gets her second segment.

Continue scanning from left to right looking for the first *third marker*. Here there is a tie between Alice's A_3 and Dana's D_3. As usual, a coin toss is used to break the tie and Alice will be the third player to go—she will get the third segment in her bid (pieces 12 through 16, Fig. 3-14). Dana is the last player and gets the last segment in her bid (pieces 17 through 20, Fig. 3-15). At this point each player has gotten a fair share of the 20 pieces of candy. The amazing part is that there is *leftover candy*!

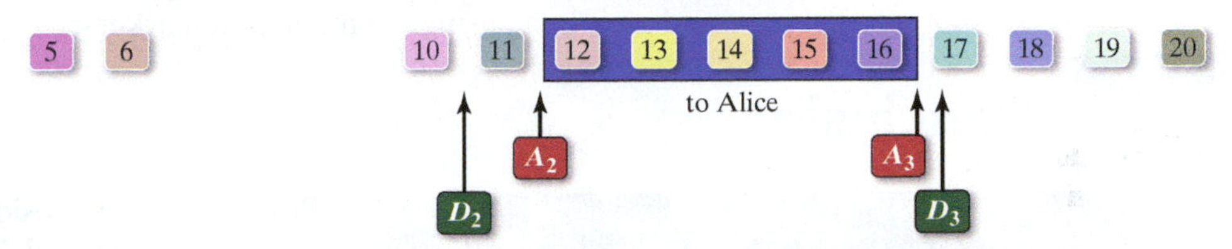

FIGURE 3-14 A_3 and D_3 are tied as the first 3-marker. After a coin toss, Alice gets her third segment.

FIGURE 3-15 Dana is last, gets her last segment.

- **Step 3 (Dividing the Surplus).** The easiest way to divide the surplus is to randomly draw lots and let the players take turns choosing one piece at a time until there are no more pieces left. Here the leftover pieces are 5, 6, 10, and 11 (Fig. 3-16). The players now draw lots; Carla gets to choose first and takes piece 11. Dana chooses next and takes piece 5. Bianca and Alice receive pieces 6 and 10, respectively.

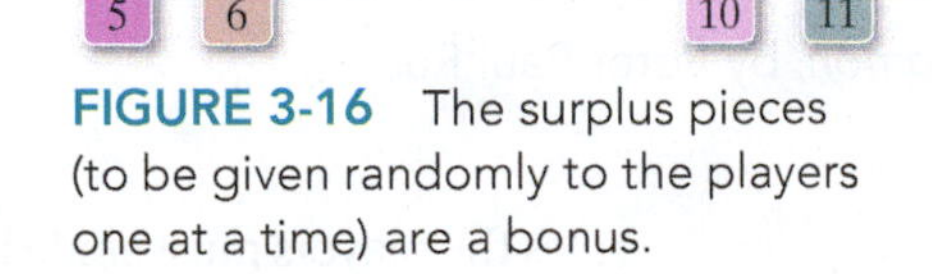

FIGURE 3-16 The surplus pieces (to be given randomly to the players one at a time) are a bonus.

The ideas behind Example 3.14 can be generalized to any number of players. We now give the general description of the method of markers with *N* players and *M* indivisible items.

- **Preliminaries.** The items are arranged randomly into an array. For convenience, label the items 1 through *M*, going from left to right.
- **Step 1 (Bidding).** Each player independently divides the array into *N* segments (segments 1, 2, . . . , *N*) by placing $N - 1$ markers along the array. These segments are assumed to represent the fair shares of the array in the opinion of that player.

- **Step 2 (Allocations).** Scan the array from left to right until the first *first marker* is located. The player owning that marker (let's call him P_1) goes first and gets the first segment in his bid. (In case of a tie, break the tie randomly.) P_1's markers are removed, and we continue scanning from left to right, looking for the first *second marker*. The player owning that marker (let's call her P_2) goes second and gets the second segment in her bid. Continue this process, assigning to each player in turn one of the segments in her bid. The last player gets the last segment in her bid.
- **Step 3 (Dividing the Surplus).** The players take turns in some random order and pick one item at a time until all the surplus items are given out.

(The applet ***Method of Markers***, available in *MyMathLab* in the *Multimedia Library* or in *Tools for Success* allows the user to walk through a step-by-step implementation of the method of markers for problems involving up to 6 players and up to 40 items.)

Despite its simple elegance, the method of markers can be used only under some fairly restrictive conditions. In particular, the method assumes that every player is able to divide the array of items into segments in such a way that each of the segments has approximately equal value. This is usually possible in a fine-grained, fair-division game, but almost impossible to accomplish when there is a combination of expensive and inexpensive items (good luck trying to divide fairly 19 candy bars and an iPhone using the method of markers!).

Conclusion

The Judgment of Solomon, by Peter Paul Rubens, 1577–1640

Problems of fair division are as old as humankind. One of the best-known and best-loved biblical stories is built around one such problem. Two women, both claiming to be the mother of the same baby, make their case to King Solomon. As a solution King Solomon proposes to cut the baby in two and give each woman a share. (Basically, King Solomon was proposing a continuous solution to a discrete fair-division problem!) This solution is totally unacceptable to the true mother, who would rather see the baby go to the other woman than have it "divided" into two shares. The final settlement, of course, is that the baby is returned to its rightful mother.

The problem of dividing an object or a set of objects among the members of a group is a practical problem that comes up regularly in our daily lives. When we are dividing a pizza, a cake, or a bunch of candy, we don't always pay a great deal of attention to the issue of fairness, but when we are dividing real estate, jewelry, or some other valuable asset, dividing things fairly becomes a critical issue.

In this chapter we approached fair division as a formal game. Like many other formal games, fair division involves rules the players must abide by (each *fair-division method* has its own set of rules) as well as fundamental assumptions about the players (*rationality*, *privacy*, *cooperation*, and *symmetry*). We discussed three *continuous* fair-division methods—the *divider-chooser* method (only works for divisions between two players) and the *lone-divider* and *lone-chooser* methods (they work for any number of players)—and two *discrete* fair-division methods—the *method of sealed bids* (produces great results but requires players to have plenty of cash to sit at the table) and the *method of markers* (requires no cash but only works with a fine-grained set of assets).

There are several other fair-division methods that we did not cover in this chapter, and what we offered here was a fairly basic overview. Nonetheless, we were able to get a glimpse of an interesting (and surprising) application of mathematics and game theory to one of the most important questions of social science—how to get humans to share their common assets in a rational and fair way.

KEY CONCEPTS

3.1 Fair-Division Games

- **assets:** the items or things being divided; usually of positive value, but can also be things with negative value such as liabilities, obligations, or chores, **69**
- **players:** the parties that jointly own the assets and wish to divide them among themselves, **70**
- **value system:** the opinion that each player has regarding the value of the assets or any part thereof, **70**
- **fair share (proportional fair share):** to a player, a share that in the player's opinion is worth at least $(\frac{1}{N})$ of the total value of the assets (where N denotes the number of players), **71**
- **fair division:** a division of the assets that gives each player a fair share, **71**
- **fair-division method:** a procedure that guarantees as its outcome a fair division of the assets, **71**
- **continuous fair division:** a division involving assets that can be divided in infinite ways and by making arbitrarily small changes, **72**
- **discrete fair division:** a division involving assets consisting of indivisible objects or objects that can only be divided up to a point, **72**

3.2 The Divider-Chooser Method

- **divider-chooser method:** two players; one cuts the assets into two shares, and the other one chooses one of the shares, **72**

3.3 The Lone-Divider Method

- **lone-divider method:** N players $(N \geq 2)$; the lone divider cuts the assets into N shares; the others (choosers) declare which shares they consider to be fair, **73**

3.4 The Lone-Chooser Method

- **lone-chooser method:** N players $(N \geq 2)$; all but one player (the dividers) divide the assets fairly among themselves, and each then subdivides his or her share into N sub-shares; the remaining player (chooser) picks one sub-share from each divider, **78**

3.5 The Method of Sealed Bids

- **method of sealed bids:** each player bids a dollar value for each item with the item going to the highest bidder; the other players get cash from the winning bid for their equity on the item, **81**
- **sealed-bid auction:** an auction of an item of positive value; only one secret bid allowed per bidder; highest bidder gets the item, **83**
- **reverse auction:** an auction in which the item being auctioned is a job or a chore; the lowest bidder gets the amount of his or her bid as payment for doing that job or chore, **83**

3.6 The Method of Markers

- **method of markers:** each player bids on how to split an array of the items into N sections; the player with the "smallest" bid for the first section gets it; among the remaining players, the player with the "smallest" bid for the second section gets it, and so on until every player gets one of his or her sections, **85**

EXERCISES

WALKING

3.1 Fair-Division Games

1. Henry, Tom, and Fred are breaking up their partnership and dividing among themselves the partnership's real estate assets equally owned by the three of them. The assets are divided into three shares (s_1, s_2, and s_3). Table 3-12 shows the values of the shares to each player expressed as a percent of the total value of the assets.
   - **(a)** Which of the shares are fair shares to Henry?
   - **(b)** Which of the shares are fair shares to Tom?
   - **(c)** Which of the shares are fair shares to Fred?
   - **(d)** Find all possible fair divisions of the assets using s_1, s_2, and s_3 as shares.
   - **(e)** Of the fair divisions found in (d), which one is the best?

	s_1	s_2	s_3
Henry	25%	40%	35%
Tom	28%	35%	37%
Fred	$33\frac{1}{3}\%$	$33\frac{1}{3}\%$	$33\frac{1}{3}\%$

TABLE 3-12

2. Alice, Bob, and Carlos are dividing among themselves the family farm equally owned by the three of them. The property is divided into three shares (s_1, s_2, and s_3). Table 3-13 shows the values of the shares to each player expressed as a percent of the total value of the property.
   - **(a)** Which of the shares are fair shares to Alice?
   - **(b)** Which of the shares are fair shares to Bob?
   - **(c)** Which of the shares are fair shares to Carlos?
   - **(d)** Find all possible fair divisions of the assets using s_1, s_2, and s_3 as shares.
   - **(e)** Of the fair divisions found in (d), which one is the best?

	s_1	s_2	s_3
Alice	38%	28%	34%
Bob	$33\frac{1}{3}\%$	$33\frac{1}{3}\%$	$33\frac{1}{3}\%$
Carlos	34%	40%	26%

TABLE 3-13

3. Angie, Bev, Ceci, and Dina are dividing among themselves a set of common assets equally owned by the four of them. The assets are divided into four shares (s_1, s_2, s_3, and s_4). Table 3-14 shows the values of the shares to each player expressed as a percent of the total value of the assets.
   - **(a)** Which of the shares are fair shares to Angie?
   - **(b)** Which of the shares are fair shares to Bev?
   - **(c)** Which of the shares are fair shares to Ceci?
   - **(d)** Which of the shares are fair shares to Dina?
   - **(e)** Find all possible fair divisions of the assets using s_1, s_2, s_3, and s_4 as shares.

	s_1	s_2	s_3	s_4
Angie	22%	26%	28%	24%
Bev	25%	26%	22%	27%
Ceci	20%	30%	27%	23%
Dina	25%	25%	25%	25%

TABLE 3-14

4. Mark, Tim, Maia, and Kelly are dividing among themselves a set of common assets equally owned by the four of them. The assets are divided into four shares (s_1, s_2, s_3, and s_4). Table 3-15 shows the values of the shares to each player expressed as a percent of the total value of the assets.
   - **(a)** Which of the shares are fair shares to Mark?
   - **(b)** Which of the shares are fair shares to Tim?
   - **(c)** Which of the shares are fair shares to Maia?
   - **(d)** Which of the shares are fair shares to Kelly?
   - **(e)** Find all possible fair divisions of the assets using s_1, s_2, s_3, and s_4 as shares.

	s_1	s_2	s_3	s_4
Mark	20%	32%	28%	20%
Tim	25%	25%	25%	25%
Maia	15%	15%	30%	40%
Kelly	24%	26%	24%	26%

TABLE 3-15

5. Allen, Brady, Cody, and Diane are sharing a cake. The cake had previously been divided into four slices (s_1, s_2, s_3, and s_4). Table 3-16 shows the values of the slices in the eyes of each player.
   - **(a)** Which of the slices are fair shares to Allen?
   - **(b)** Which of the slices are fair shares to Brady?
   - **(c)** Which of the slices are fair shares to Cody?
   - **(d)** Which of the slices are fair shares to Diane?
   - **(e)** Find all possible fair divisions of the cake using s_1, s_2, s_3, and s_4 as shares.

	s_1	s_2	s_3	s_4
Allen	$4.00	$5.00	$6.00	$5.00
Brady	$3.00	$3.50	$4.00	$5.50
Cody	$6.00	$4.50	$3.50	$4.00
Diane	$7.00	$4.00	$4.00	$5.00

TABLE 3-16

6. Carlos, Sonya, Tanner, and Wen are sharing a cake. The cake had previously been divided into four slices (s_1, s_2, s_3, and s_4). Table 3-17 shows the values of the slices in the eyes of each player.
   (a) Which of the slices are fair shares to Carlos?
   (b) Which of the slices are fair shares to Sonya?
   (c) Which of the slices are fair shares to Tanner?
   (d) Which of the slices are fair shares to Wen?
   (e) Find all possible fair divisions of the cake using s_1, s_2, s_3, and s_4 as shares.

	s_1	s_2	s_3	s_4
Carlos	$3.00	$5.00	$5.00	$3.00
Sonya	$4.50	$3.50	$4.50	$5.50
Tanner	$4.25	$4.50	$3.50	$3.75
Wen	$5.50	$4.00	$4.50	$6.00

TABLE 3-17

7. Four partners (Adams, Benson, Cagle, and Duncan) jointly own a piece of land with a market value of $400,000. Suppose that the land is subdivided into four parcels s_1, s_2, s_3, and s_4. The partners are planning to split up, with each partner getting one of the four parcels.
   (a) To Adams, s_1 is worth $40,000 more than s_2, s_2 and s_3 are equal in value, and s_4 is worth $20,000 more than s_1. Determine which of the four parcels are fair shares to Adams.
   (b) To Benson, s_1 is worth $40,000 more than s_2, s_4 is $8,000 more than s_3, and together s_4 and s_3 have a combined value equal to 40% of the value of the land. Determine which of the four parcels are fair shares to Benson.
   (c) To Cagle, s_1 is worth $40,000 more than s_2 and $20,000 more than s_4, and s_3 is worth twice as much as s_4. Determine which of the four parcels are fair shares to Cagle.
   (d) To Duncan, s_1 is worth $4,000 more than s_2; s_2 and s_3 have equal value; and s_1, s_2, and s_3 have a combined value equal to 70% of the value of the land. Determine which of the four parcels are fair shares to Duncan.
   (e) Find a fair division of the land using the parcels s_1, s_2, s_3, and s_4 as fair shares.
8. Four players (Abe, Betty, Cory, and Dana) are sharing a cake. Suppose that the cake is divided into four slices s_1, s_2, s_3, and s_4.
   (a) To Abe, s_1 is worth $3.60, s_4 is worth $3.50, s_2 and s_3 have equal value, and the entire cake is worth $15.00. Determine which of the four slices are fair shares to Abe.
   (b) To Betty, s_2 is worth twice as much as s_1, s_3 is worth three times as much as s_1, and s_4 is worth four times as much as s_1. Determine which of the four slices are fair shares to Betty.
   (c) To Cory, s_1, s_2, and s_4 have equal value, and s_3 is worth as much as s_1, s_2, and s_4 combined. Determine which of the four slices are fair shares to Cory.
   (d) To Dana, s_1 is worth $1.00 more than s_2, s_3 is worth $1.00 more than s_1, s_4 is worth $3.00, and the entire cake is worth $18.00. Determine which of the four slices are fair shares to Dana.
   (e) Find a fair division of the cake using s_1, s_2, s_3, and s_4 as fair shares.

Exercises 9 through 12 refer to the following situation: Angelina and Brad jointly buy the gourmet chocolate-strawberry cake shown in Fig. 3-17. The price of this fancy cake is $72.

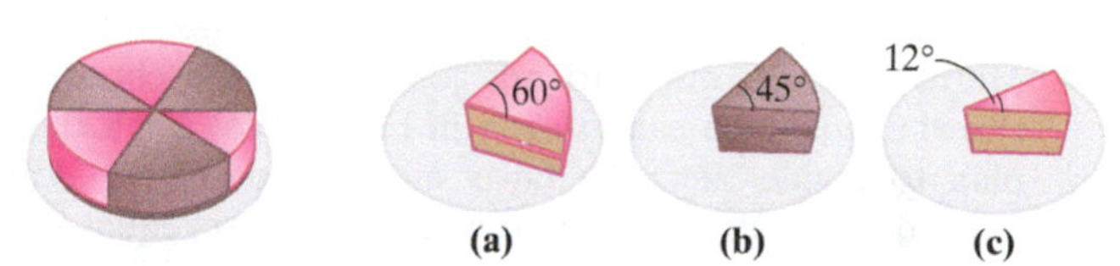

FIGURE 3-17

9. Suppose that Angelina values strawberry cake *twice* as much as she values chocolate cake. Find the dollar value to Angelina of each of the following pieces:
   (a) the strawberry wedge shown in Fig. 3-17(a)
   (b) the chocolate slice shown in Fig. 3-17(b)
   (c) the strawberry sliver shown in Fig. 3-17(c)
10. Suppose that Brad values chocolate cake *three* times as much as he values strawberry cake. Find the dollar value to Brad of each of the following pieces:
   (a) the strawberry wedge shown in Fig. 3-17(a)
   (b) the chocolate slice shown in Fig. 3-17(b)
   (c) the strawberry sliver shown in Fig. 3-17(c)
11. Suppose that Brad values chocolate cake *four* times as much as he values strawberry cake. Find the dollar value to Brad of each of the following pieces:
   (a) the strawberry wedge shown in Fig. 3-17(a)
   (b) the chocolate slice shown in Fig. 3-17(b)
   (c) the strawberry sliver shown in Fig. 3-17(c)
12. Suppose that Angelina values strawberry cake *five* times as much as she values chocolate cake. Find the dollar value to Angelina of each of the following pieces:
   (a) the strawberry wedge shown in Fig. 3-17(a)
   (b) the chocolate slice shown in Fig. 3-17(b)
   (c) the strawberry sliver shown in Fig. 3-17(c)

13. Karla and five other friends jointly buy the chocolate-strawberry-vanilla cake shown in Fig. 3-18(a) for $30 and plan to divide the cake fairly among themselves. After much discussion, the cake is divided into the six equal-sized slices $s_1, s_2, \ldots, s_6$ shown in Fig. 3-18(b). Suppose that Karla values strawberry cake *twice* as much as vanilla cake and chocolate cake *three* times as much as vanilla cake.

(a) Find the dollar value to Karla of each of the slices s_1 through s_6.

(b) Which of the slices s_1 through s_6 are fair shares to Karla?

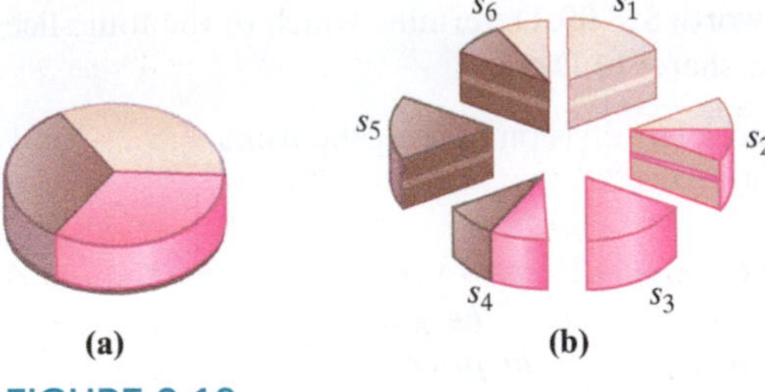

FIGURE 3-18

14. Marla and five other friends jointly buy the chocolate-strawberry-vanilla cake shown in Fig. 3-18(a) for $30 and plan to divide the cake fairly among themselves. After much discussion, the cake is divided into the six equal-sized slices $s_1, s_2, \ldots, s_6$ shown in Fig. 3-18(b). Suppose that Marla values vanilla cake *twice* as much as chocolate cake and chocolate cake *three* times as much as strawberry cake.

(a) Find the dollar value to Marla of each of the slices s_1 through s_6.

(b) Which of the slices s_1 through s_6 are fair shares to Marla?

3.2 The Divider-Chooser Method

Exercises 15 and 16 refer to the following situation: Jackie and Karla jointly bought the half meatball–half vegetarian foot-long sub shown in Fig. 3-19 for $8.00. They plan to divide the sandwich fairly using the divider-chooser method. Jackie likes meatball subs three times as much as vegetarian subs; Karla is a strict vegetarian and does not eat meat at all. Assume that Jackie just met Karla and has no idea that she is a vegetarian. Assume also that when the sandwich is cut, the cut is made perpendicular to the length of the sandwich. (You can describe different shares of the sandwich using the ruler and interval notation. For example, [0, 6] describes the vegetarian half, [6, 8] describes one-third of the meatball half, etc.).

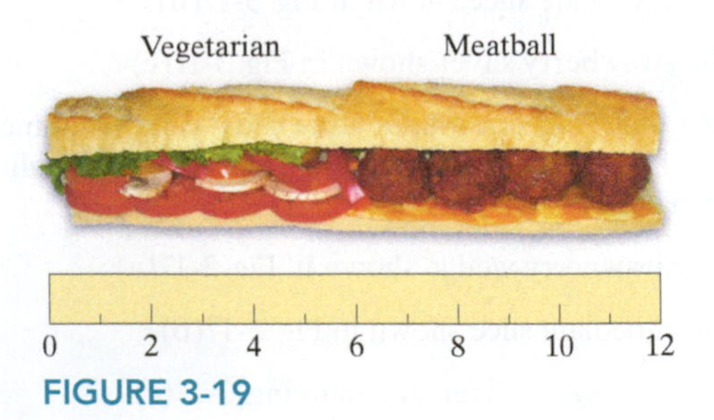

FIGURE 3-19

15. Suppose that they flip a coin and Jackie ends up being the divider.

(a) Describe how Jackie should cut the sandwich into two shares s_1 and s_2.

(b) After Jackie cuts, Karla gets to choose. Specify which of the two shares Karla should choose and give the value of the share to Karla.

16. Suppose they flip a coin and Karla ends up being the divider.

(a) Describe how Karla should cut the sandwich into two shares s_1 and s_2.

(b) After Karla cuts, Jackie gets to choose. Specify which of the two shares Jackie should choose and give the value of the share to Jackie.

Exercises 17 and 18 refer to the following situation: Martha and Nick jointly bought the giant 28-in. sub sandwich shown in Fig. 3-20 for $9. They plan to divide the sandwich fairly using the divider-chooser method. Martha likes ham subs twice as much as she likes turkey subs, and she likes turkey and roast beef subs the same. Nick likes roast beef subs twice as much as he likes ham subs, and he likes ham and turkey subs the same. Assume that Nick and Martha just met and know nothing of each other's likes and dislikes. Assume also that when the sandwich is cut, the cut is made perpendicular to the length of the sandwich. (You can describe different shares of the sandwich using the ruler and interval notation. For example, [0, 8] describes the ham part, [8, 12] describes one-third of the turkey part, etc.).

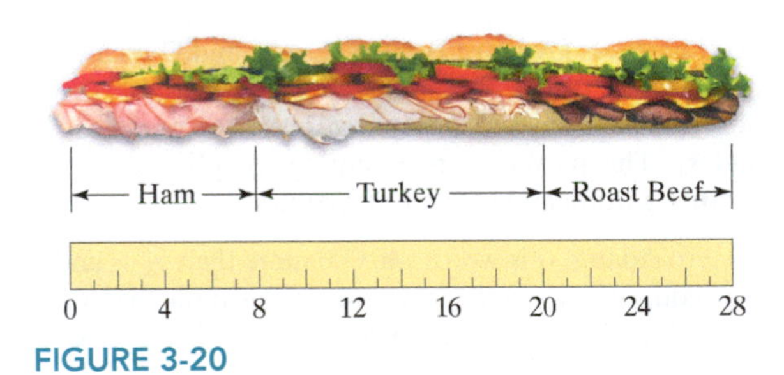

FIGURE 3-20

17. Suppose that they flip a coin and Martha ends up being the divider.

(a) Describe how Martha would cut the sandwich into two shares s_1 and s_2.

(b) After Martha cuts, Nick gets to choose. Specify which of the two shares Nick should choose, and give the value of the share to Nick.

18. Suppose that they flip a coin and Nick ends up being the divider.

(a) Describe how Nick would cut the sandwich into two shares s_1 and s_2.

(b) After Nick cuts, Martha gets to choose. Specify which of the two shares Martha should choose and give the value of the share to Martha.

Exercises 19 and 20 refer to the following situation: David and Paula are planning to divide the chocolate-vanilla-strawberry cake shown in Fig. 3-21(a) using the divider-chooser method. David likes chocolate cake and vanilla cake the same, and he likes both of them twice as much as strawberry cake. Paula likes vanilla cake and strawberry cake the same, but she is allergic to chocolate cake.

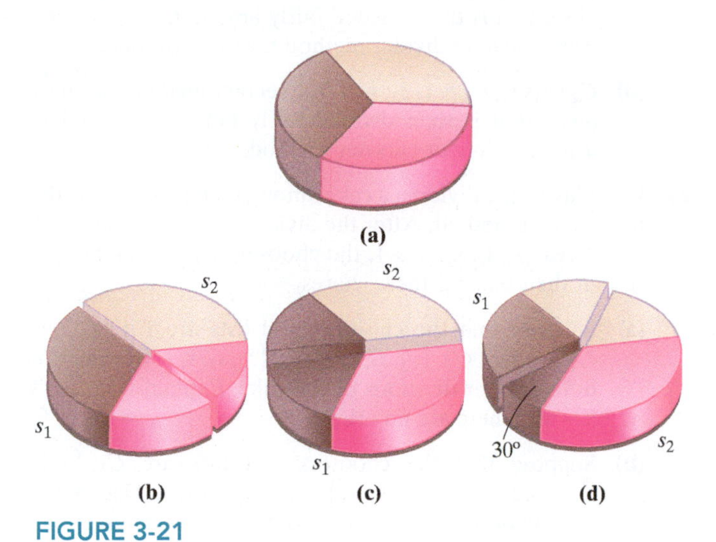

FIGURE 3-21

19. Suppose that David is the divider and Paula is the chooser.

(a) Is the cut shown in Fig. 3-21(b) a possible 50-50 cut that David might have made as the divider? If so, describe the share Paula should choose and give the value (as a percent) of that share to Paula. If the cut is not a 50-50 cut, give the values of the two shares to David.

(b) Is the cut shown in Fig. 3-21(c) a possible 50-50 cut that David might have made as the divider? If so, describe the share Paula should choose and give the value (as a percent) of that share to Paula. If the cut is not a 50-50 cut, give the values of the two shares to David.

(c) Is the cut shown in Fig. 3-21(d) a possible 50-50 cut that David might have made as the divider? If so, describe the share Paula should choose and give the value (as a percent) of that share to Paula. If the cut is not a 50-50 cut, give the values of the two shares to David.

20. Suppose that Paula is the divider and David is the chooser.

(a) Is the cut shown in Fig. 3-21(b) a possible 50-50 cut that Paula might have made as the divider? If so, describe the share David should choose and give the value (as a percent) of that share to David. If the cut is not a 50-50 cut, give the values of the two shares to Paula.

(b) Is the cut shown in Fig. 3-21(c) a possible 50-50 cut that Paula might have made as the divider? If so, describe the share David should choose and give the value (as a percent) of that share to David. If the cut is not a 50-50 cut, give the values of the two shares to Paula.

(c) Is the cut shown in Fig. 3-21(d) a possible 50-50 cut that Paula might have made as the divider? If so, describe the share David should choose and give the value (as a percent) of that share to David. If the cut is not a 50-50 cut, give the values of the two shares to Paula.

3.3 The Lone-Divider Method

21. Three partners are dividing a plot of land among themselves using the lone-divider method. After the divider D divides the land into three shares s_1, s_2, and s_3, the choosers C_1 and C_2 submit their bids for these shares.

(a) Suppose that the choosers' bid lists are C_1: $\{s_2, s_3\}$; C_2: $\{s_1, s_3\}$. Describe *two* different fair divisions of the land.

(b) Suppose that the choosers' bid lists are C_1: $\{s_2, s_3\}$; C_2: $\{s_1, s_3\}$. Describe *three* different fair divisions of the land.

22. Three partners are dividing a plot of land among themselves using the lone-divider method. After the divider D divides the land into three shares s_1, s_2, and s_3, the choosers C_1 and C_2 submit their bids for these shares.

(a) Suppose that the choosers' bid lists are C_1: $\{s_2\}$; C_2: $\{s_1, s_3\}$. Describe *two* different fair divisions of the land.

(b) Suppose that the choosers' bid lists are C_1: $\{s_1, s_2\}$; C_2: $\{s_2, s_3\}$. Describe *three* different fair divisions of the land.

23. Four partners are dividing a plot of land among themselves using the lone-divider method. After the divider D divides the land into four shares s_1, s_2, s_3, and s_4, the choosers C_1, C_2, and C_3 submit their bids for these shares.

(a) Suppose that the choosers' bid lists are C_1: $\{s_2\}$; C_2: $\{s_1, s_3\}$; C_3: $\{s_2, s_3\}$. Find a fair division of the land. Explain why this is the only possible fair division.

(b) Suppose that the choosers' bid lists are C_1: $\{s_2, s_3\}$; C_2: $\{s_1, s_3\}$; C_3: $\{s_1, s_2\}$. Describe *two* different fair divisions of the land.

(c) Suppose that the choosers' bid lists are C_1: $\{s_2\}$; C_2: $\{s_1, s_3\}$; C_3: $\{s_1, s_4\}$. Describe *three* different fair divisions of the land.

24. Four partners are dividing a plot of land among themselves using the lone-divider method. After the divider D divides the land into four shares s_1, s_2, s_3, and s_4, the choosers C_1, C_2, and C_3 submit their bids for these shares.

(a) Suppose that the choosers' bid lists are C_1: $\{s_2\}$; C_2: $\{s_1, s_3\}$; C_3: $\{s_2, s_3\}$. Find a fair division of the land. Explain why this is the only possible fair division.

(b) Suppose that the choosers' bid lists are C_1: $\{s_2\}$; C_2: $\{s_1, s_3\}$; C_3: $\{s_1, s_4\}$. Describe *three* different fair divisions of the land.

(c) Suppose that the choosers' bid lists are C_1: $\{s_2\}$; C_2: $\{s_1, s_2, s_3\}$; C_3: $\{s_2, s_3, s_4\}$. Describe *three* different fair divisions of the land.

25. Mark, Tim, Maia, and Kelly are dividing a cake among themselves using the lone-divider method. The divider divides the cake into four slices (s_1, s_2, s_3, and s_4). Table 3-18 shows the values of the slices to each player expressed as a percent of the total value of the cake.

(a) Who was the divider?

(b) Find a fair division of the cake.

	s_1	s_2	s_3	s_4
Mark	20%	32%	28%	20%
Tim	25%	25%	25%	25%
Maia	15%	15%	30%	40%
Kelly	24%	24%	24%	28%

TABLE 3-18

26. Allen, Brady, Cody, and Diane are sharing a cake valued at \$20 using the lone-divider method. The divider divides the cake into four slices (s_1, s_2, s_3, and s_4). Table 3-19 shows the values of the slices in the eyes of each player.

(a) Who was the divider?

(b) Find a fair division of the cake.

	s_1	s_2	s_3	s_4
Allen	\$4.00	\$5.00	\$4.00	\$7.00
Brady	\$6.00	\$6.50	\$4.00	\$3.50
Cody	\$5.00	\$5.00	\$5.00	\$5.00
Diane	\$7.00	\$4.50	\$4.00	\$4.50

TABLE 3-19

27. Four partners are dividing a plot of land among themselves using the lone-divider method. After the divider D divides the land into four shares s_1, s_2, s_3, and s_4, the choosers C_1, C_2, and C_3 submit the following bids: C_1: $\{s_2\}$; C_2: $\{s_1, s_2\}$; C_3: $\{s_1, s_2\}$. For each of the following possible divisions, determine if it is a fair division or not. If not, explain why not.

(a) D gets s_3; s_1, s_2, and s_4 are recombined into a single piece that is then divided fairly among C_1, C_2, and C_3 using the lone-divider method for three players.

(b) D gets s_1; s_2, s_3, and s_4 are recombined into a single piece that is then divided fairly among C_1, C_2, and C_3 using the lone-divider method for three players.

(c) D gets s_4; s_1, s_2, and s_3 are recombined into a single piece that is then divided fairly among C_1, C_2, and C_3 using the lone-divider method for three players.

(d) D gets s_3; C_1 gets s_2; and s_1, s_4 are recombined into a single piece that is then divided fairly between C_2 and C_3 using the divider-chooser method.

28. Four partners are dividing a plot of land among themselves using the lone-divider method. After the divider D divides the land into four shares s_1, s_2, s_3, and s_4, the choosers C_1, C_2, and C_3 submit the following bids: C_1: $\{s_3, s_4\}$; C_2: $\{s_4\}$; C_3: $\{s_3\}$. For each of the following possible divisions, determine if it is a fair division or not. If not, explain why not.

(a) D gets s_1; s_2, s_3, and s_4 are recombined into a single piece that is then divided fairly among C_1, C_2, and C_3 using the lone-divider method for three players.

(b) D gets s_3; s_1, s_2, and s_4 are recombined into a single piece that is then divided fairly among C_1, C_2, and C_3 using the lone-divider method for three players.

(c) D gets s_2; s_1, s_3, and s_4 are recombined into a single piece that is then divided fairly among C_1, C_2, and C_3 using the lone-divider method for three players.

(d) C_2 gets s_4; C_3 gets s_3; s_1, s_2 are recombined into a single piece that is then divided fairly between C_1 and D using the divider-chooser method.

29. Five players are dividing a cake among themselves using the lone-divider method. After the divider D cuts the cake into five slices (s_1, s_2, s_3, s_4, s_5), the choosers C_1, C_2, C_3, and C_4 submit their bids for these shares.

(a) Suppose that the choosers' bid lists are C_1: $\{s_2, s_4\}$; C_2: $\{s_2, s_4\}$; C_3: $\{s_2, s_3, s_5\}$; C_4: $\{s_2, s_3, s_4\}$. Describe *two* different fair divisions of the cake. Explain why that's it—why there are no others.

(b) Suppose that the choosers' bid lists are C_1: $\{s_2\}$; C_2: $\{s_2, s_4\}$; C_3: $\{s_2, s_3, s_5\}$; C_4: $\{s_2, s_3, s_4\}$. Find a fair division of the cake. Explain why that's it—there are no others.

30. Five players are dividing a cake among themselves using the lone-divider method. After the divider D cuts the cake into five slices (s_1, s_2, s_3, s_4, s_5), the choosers C_1, C_2, C_3, and C_4 submit their bids for these shares.

(a) Suppose that the choosers' bid lists are C_1: $\{s_2, s_3\}$; C_2: $\{s_2, s_4\}$; C_3: $\{s_1, s_2\}$; C_4: $\{s_1, s_3, s_4\}$. Describe *three* different fair divisions of the land. Explain why that's it—why there are no others.

(b) Suppose that the choosers' bid lists are C_1: $\{s_1, s_4\}$; C_2: $\{s_2, s_4\}$; C_3: $\{s_2, s_4, s_5\}$; C_4: $\{s_2\}$. Find a fair division of the land. Explain why that's it—why there are no others.

31. Four partners (Egan, Fine, Gong, and Hart) jointly own a piece of land with a market value of \$480,000. The partnership is breaking up, and the partners decide to divide the land among themselves using the lone-divider method. Using a map, the divider divides the property into four parcels s_1, s_2, s_3, and s_4. Table 3-20 shows the value of some of the parcels in the eyes of each partner.

	s_1	s_2	s_3	s_4
Egan	\$80,000	\$85,000		\$195,000
Fine		\$100,000	\$135,000	\$120,000
Gong			\$120,000	
Hart	\$95,000	\$100,000		\$110,000

TABLE 3-20

(a) Who was the divider? Explain.

(b) Determine each chooser's bid.

(c) Find a fair division of the property.

32. Four players (Abe, Betty, Cory, and Dana) are dividing a pizza worth $18.00 among themselves using the lone-divider method. The divider divides the pizza into four shares s_1, s_2, s_3, and s_4. Table 3-21 shows the value of some of the slices in the eyes of each player.

	s_1	s_2	s_3	s_4
Abe	$5.00	$5.00	$3.50	
Betty		$4.50		
Cory	$4.80	$4.20	$4.00	
Dana	$4.00	$3.75	$4.25	

TABLE 3-21

(a) Who was the divider? Explain.

(b) Determine each chooser's bid.

(c) Find a fair division of the pizza.

Exercises 33 and 34 refer to the following situation: Jackie, Karla, and Lori are planning to divide the half vegetarian–half meatball foot-long sub sandwich shown in Fig. 3-22 among themselves using the lone-divider method. Jackie likes the meatball and vegetarian parts equally well; Karla is a strict vegetarian and does not eat meat at all; Lori likes the meatball part twice as much as the vegetarian part. (Assume that when the sandwich is cut, the cuts are always made perpendicular to the length of the sandwich. You can describe different shares of the sandwich using the ruler and interval notation—for example, [0, 6] describes the vegetarian half, [6, 8] describes one-third of the meatball half, etc.)

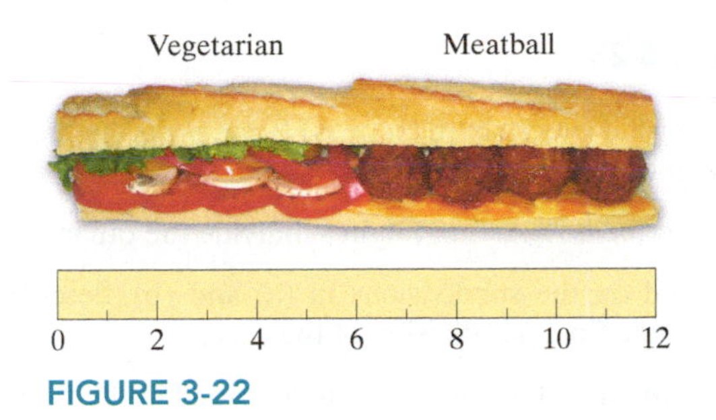

FIGURE 3-22

33. Suppose that Jackie is the divider.

(a) Describe how Jackie should cut the sandwich into three shares. Label the three shares s_1 for the leftmost piece, s_2 for the middle piece, and s_3 for the rightmost piece. Use the ruler and interval notation to describe the three shares. (Assume that Jackie knows nothing about Karla and Lori's likes and dislikes.)

(b) Which of the three shares are fair shares to Karla?

(c) Which of the three shares are fair shares to Lori?

(d) Find three different fair divisions of the sandwich.

34. Suppose that Lori ends up being the divider.

(a) Describe how Lori should cut the sandwich into three shares. Label the three shares s_1 for the leftmost piece, s_2 for the middle piece, and s_3 for the rightmost piece. Use the ruler and interval notation to describe the three shares. (Assume that Lori knows nothing about Karla and Jackie's likes and dislikes.)

(b) Which of the three shares are fair shares to Jackie?

(c) Which of the three shares are fair shares to Karla?

(d) Suppose that Lori gets s_3. Describe how to proceed to find a fair division of the sandwich.

3.4 The Lone-Chooser Method

Exercises 35 through 38 refer to the following situation: Angela, Boris, and Carlos are dividing the vanilla-strawberry cake shown in Fig. 3-23(a) using the lone-chooser method. Figure 3-23(b) shows how each player values each half of the cake. In your answers assume that all cuts are normal "cake cuts" from the center to the edge of the cake. You can describe each piece of cake by giving the angles of the vanilla and strawberry parts, as in "15° strawberry–40° vanilla" or "60° vanilla only."

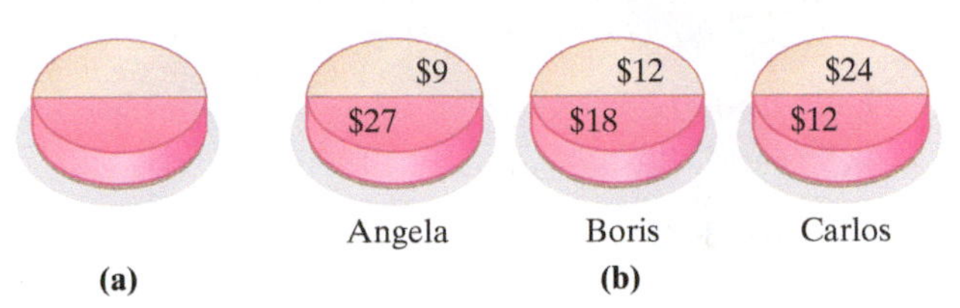

FIGURE 3-23

35. Suppose that Angela and Boris are the dividers and Carlos is the chooser. In the first division, Boris cuts the cake vertically through the center as shown in Fig. 3-24, with Angela choosing s_1 (the left half) and Boris s_2 (the right half). In the second division, Angela subdivides s_1 into three pieces and Boris subdivides s_2 into three pieces.

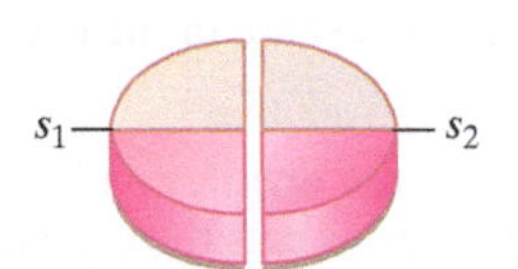

FIGURE 3-24

(a) Describe how Angela would subdivide s_1 into three pieces.

(b) Describe how Boris would subdivide s_2 into three pieces.

(c) Based on the subdivisions in (a) and (b), describe a possible final fair division of the cake.

(d) For the final fair division you described in (c), find the value (in dollars and cents) of each share in the eyes of the player receiving it.

36. Suppose that Carlos and Angela are the dividers and Boris is the chooser. In the first division, Carlos cuts the cake vertically through the center as shown in Fig. 3-24, with Angela choosing s_1 (the left half) and Carlos s_2 (the right half). In the second division, Angela subdivides s_1 into three pieces and Carlos subdivides s_2 into three pieces.

(a) Describe how Carlos would subdivide s_2 into three pieces.

(b) Describe how Angela would subdivide s_1 into three pieces.

(c) Based on the subdivisions in (a) and (b), describe a possible final fair division of the cake.

(d) For the final fair division you described in (c), find the value (in dollars and cents) of each share in the eyes of the player receiving it.

37. Suppose that Angela and Boris are the dividers and Carlos is the chooser. In the first division, Angela cuts the cake into two shares: s_1 (a 120° strawberry-only piece) and s_2 (a 60° strawberry–180° vanilla piece) as shown in Fig. 3-25. Boris picks the share he likes best, and Angela gets the other share. In the second division, Angela subdivides her share of the cake into three pieces and Boris subdivides his share of the cake into three pieces.

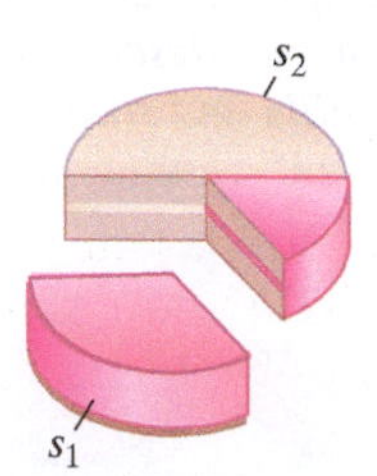

FIGURE 3-25

(a) Describe which share (s_1 or s_2) Boris picks and how he might subdivide it.

(b) Describe how Angela would subdivide her share of the cake.

(c) Based on the subdivisions in (a) and (b), describe a possible final fair division of the cake.

(d) For the final fair division you described in (c), find the value (in dollars and cents) of each share in the eyes of the player receiving it.

38. Suppose that Carlos and Angela are the dividers and Boris is the chooser. In the first division, Carlos cuts the cake into two shares: s_1 (a 135° vanilla-only piece) and s_2 (a 45° vanilla–180° strawberry piece) as shown in Fig. 3-26. Angela picks the share she likes better and Carlos gets the other share. In the second division, Angela subdivides her share of the cake into three pieces and Carlos subdivides his share of the cake into three pieces.

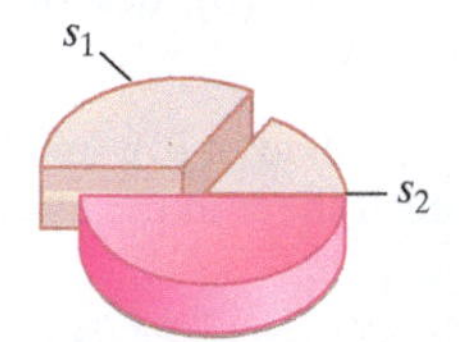

FIGURE 3-26

(a) Describe which share (s_1 or s_2) Angela picks and how she might subdivide it.

(b) Describe how Carlos might subdivide his share of the cake.

(c) Based on the subdivisions in (a) and (b), describe a possible final fair division of the cake.

(d) For the final fair division you described in (c), find the value (in dollars and cents) of each share in the eyes of the player receiving it.

Exercises 39 and 40 refer to the following: Arthur, Brian, and Carl are dividing the cake shown in Fig. 3-27 using the lone-chooser method. Arthur loves chocolate cake and orange cake equally but hates strawberry cake and vanilla cake. Brian loves chocolate cake and strawberry cake equally but hates orange cake and vanilla cake. Carl loves chocolate cake and vanilla cake equally but hates orange cake and strawberry cake. In your answers, assume all cuts are normal "cake cuts" from the center to the edge of the cake. You can describe each piece of cake by giving the angles of its parts, as in "15° strawberry–40° chocolate" or "60° orange only."

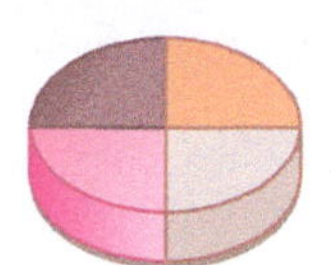

FIGURE 3-27

39. Suppose that Arthur and Brian are the dividers and Carl is the chooser. In the first division, Arthur cuts the cake vertically through the center as shown in Fig. 3-28 and Brian picks the share he likes better. In the second division, Brian subdivides the share he chose into three pieces and Arthur subdivides the other share into three pieces.

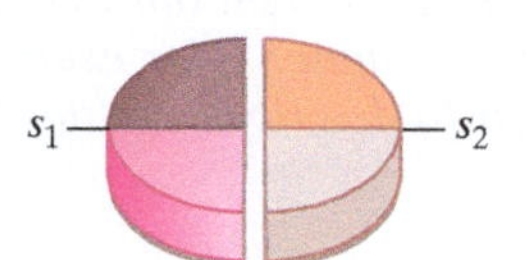

FIGURE 3-28

(a) Describe which share (s_1 or s_2) Brian picks and how he might subdivide it.

(b) Describe how Arthur might subdivide the other share.

(c) Based on the subdivisions in (a) and (b), describe a possible final fair division of the cake.

(d) For the final fair division you described in (c), find the value of each share (as a percentage of the total value of the cake) in the eyes of the player receiving it.

40. Suppose that Carl and Arthur are the dividers and Brian is the chooser. In the first division, Carl makes the cut shown in Fig. 3-29 and Arthur picks the share he likes better. In the second division, Arthur subdivides the share he chose into three pieces and Carl subdivides the other share into three pieces.

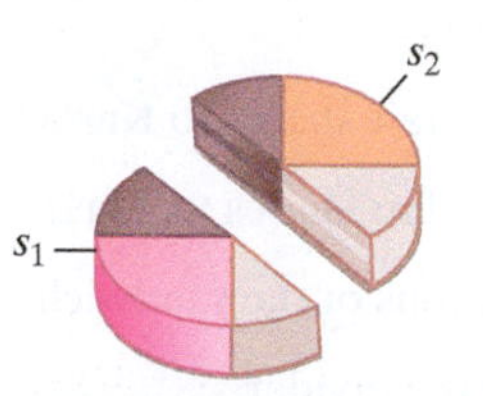

FIGURE 3-29

(a) Describe which share (s_1 or s_2) Arthur picks and how he might subdivide it.

(b) Describe how Carl might subdivide the other share.

(c) Based on the subdivisions in (a) and (b), describe a possible final fair division of the cake.

(d) For the final fair division you described in (c), find the value of each share (as a percentage of the total value of the cake) in the eyes of the player receiving it.

41. Jackie, Karla, and Lori are dividing the foot-long half meatball–half vegetarian sub shown in Fig. 3-30 using the lone-chooser method. Jackie likes the vegetarian and meatball parts equally well, Karla is a strict vegetarian and does not eat meat at all, and Lori likes the meatball part twice as much as she likes the vegetarian part. Suppose that Karla and Jackie are the dividers and Lori is the chooser. In the first division, Karla divides the sub into two shares (a left share s_1 and a right share s_2) and Jackie picks the share he likes better. In the second division, Jackie subdivides the share he picks into three pieces (a "left" piece J_1, a "middle" piece J_2, and a "right" piece J_3) and Karla subdivides the other share into three pieces (a "left" piece K_1, a "middle" piece K_2, and a "right" piece K_3). Assume that all cuts are perpendicular to the length of the sub. (You can describe the pieces of sub using the ruler and interval notation, as in [3, 7] for the piece that starts at inch 3 and ends at inch 7.)

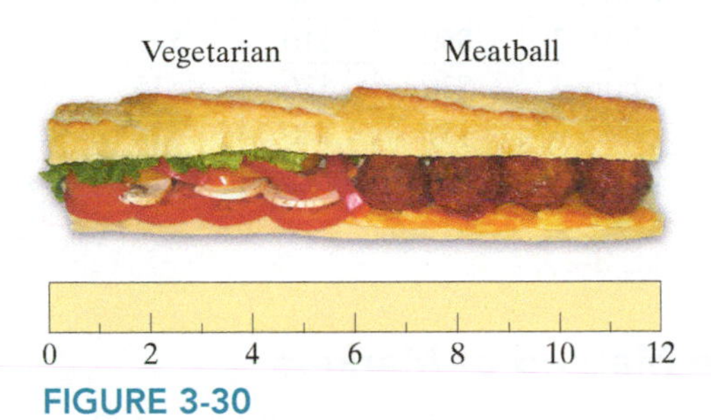

FIGURE 3-30

(a) Describe Karla's first division into s_1 and s_2.

(b) Describe which share (s_1 or s_2) Jackie picks and how he would then subdivide it into the three pieces J_1, J_2, and J_3.

(c) Describe how Karla would subdivide her share into three pieces K_1, K_2, and K_3.

(d) Based on the subdivisions in (a), (b), and (c), describe the final fair division of the sub and give the value of each player's share (as a percentage of the total value of the sub) in the eyes of the player receiving it.

42. Jackie, Karla, and Lori are dividing the foot-long half meatball–half vegetarian sub shown in Fig. 3-30 using the lone-chooser method. Jackie likes the vegetarian and meatball parts equally well, Karla is a strict vegetarian and does not eat meat at all, and Lori likes the meatball part twice as much as she likes the vegetarian part. Suppose that Karla and Lori are the dividers and Jackie is the chooser. In the first division, Lori divides the sub into two shares (a left share s_1 and a right share s_2) and Karla picks the share she likes better. In the second division, Karla subdivides the share she picks into three pieces (a "left" piece K_1, a "middle" piece K_2, and a "right" piece K_3) and Lori subdivides the other share into three pieces (a "left" piece L_1, a "middle" piece L_2, and a "right" piece L_3). Assume that all cuts are perpendicular to the length of the sub. (You can describe the pieces of sub using the ruler and interval notation, as in [3, 7] for the piece that starts at inch 3 and ends at inch 7.)

(a) Describe Lori's first division into s_1 and s_2.

(b) Describe which share (s_1 or s_2) Karla picks and how she would then subdivide it into the three pieces K_1, K_2, and K_3.

(c) Describe how Lori would subdivide her share into three pieces L_1, L_2, and L_3.

(d) Based on the subdivisions in (a), (b), and (c), describe the final fair division of the sub and give the value of each player's share (as a percentage of the total value of the sub) in the eyes of the player receiving it.

3.5 The Method of Sealed Bids

43. Ana, Belle, and Chloe are dividing four pieces of furniture using the method of sealed bids. Table 3-22 shows the players' bids on each of the items.

	Ana	Belle	Chloe
Dresser	$150	$300	$275
Desk	$180	$150	$165
Vanity	$170	$200	$260
Tapestry	$400	$250	$500

TABLE 3-22

(a) Find the value of each player's fair share.

(b) Describe the first settlement (who gets which item and how much do they pay or get in cash).

(c) Find the surplus after the first settlement is over.

(d) Describe the final settlement (who gets which item and how much do they pay or get in cash).

44. Andre, Bea, and Chad are dividing an estate consisting of a house, a small farm, and a painting using the method of sealed bids. Table 3-23 shows the players' bids on each of the items.

	Andre	Bea	Chad
House	$150,000	$146,000	$175,000
Farm	$430,000	$425,000	$428,000
Painting	$50,000	$59,000	$57,000

TABLE 3-23

(a) Describe the first settlement of this fair division and compute the surplus.

(b) Describe the final settlement of this fair-division problem.

45. Five heirs (A, B, C, D, and E) are dividing an estate consisting of six items using the method of sealed bids. The heirs' bids on each of the items are given in Table 3-24.

	A	B	C	D	E
Item 1	$352	$295	$395	$368	$324
Item 2	$98	$102	$98	$95	$105
Item 3	$460	$449	$510	$501	$476
Item 4	$852	$825	$832	$817	$843
Item 5	$513	$501	$505	$505	$491
Item 6	$725	$738	$750	$744	$761

TABLE 3-24

(a) Find the value of each player's fair share.

(b) Describe the first settlement (who gets which item and how much do they pay or get in cash).

(c) Find the surplus after the first settlement is over.

(d) Describe the final settlement (who gets which item and how much do they pay or get in cash).

46. Oscar, Bert, and Ernie are using the method of sealed bids to divide among themselves four items they commonly own. Table 3-25 shows the bids that each player makes for each item.

	Oscar	Bert	Ernie
Item 1	$8600	$5500	$3700
Item 2	$3500	$4200	$5000
Item 3	$2300	$4400	$3400
Item 4	$4800	$2700	$2300

TABLE 3-25

(a) Find the value of each player's fair share.

(b) Describe the first settlement (who gets which item and how much do they pay or get in cash).

(c) Find the surplus after the first settlement is over.

(d) Describe the final settlement (who gets which item and how much do they pay or get in cash).

47. Anne, Bette, and Chia jointly own a flower shop. They can't get along anymore and decide to break up the partnership using the method of sealed bids, with the understanding that one of them will get the flower shop and the other two will get cash. Anne bids $210,000, Bette bids $240,000, and Chia bids $225,000. How much money do Anne and Chia each get from Bette for their third share of the flower shop?

48. Al, Ben, and Cal jointly own a fruit stand. They can't get along anymore and decide to break up the partnership using the method of sealed bids, with the understanding that one of them will get the fruit stand and the other two will get cash. Al bids $156,000, Ben bids $150,000, and Cal bids $171,000. How much money do Al and Ben each get from Cal for their one-third share of the fruit stand?

49. Ali, Briana, and Caren are roommates planning to move out of their apartment. They identify four major chores that need to be done before moving out and decide to use the method of sealed bids to reverse auction the chores. Table 3-26 shows the bids that each roommate made for each chore. Describe the final outcome of the division (which chores are done by each roommate and how much each roommate pays or gets paid.)

	Ali	Briana	Caren
Chore 1	$65	$70	$55
Chore 2	$100	$85	$95
Chore 3	$60	$50	$45
Chore 4	$75	$80	$90

TABLE 3-26

50. Anne, Bess, and Cindy are roommates planning to move out of their apartment. They identify five major chores that need to be done before moving out and decide to use the method of sealed bids to reverse auction the chores. Table 3-27 shows the bids that each roommate made for each chore. Describe the final outcome of the division (which chores are done by each roommate and how much each roommate pays or gets paid.)

	Anne	Bess	Cindy
Chore 1	$20	$30	$40
Chore 2	$50	$10	$22
Chore 3	$30	$20	$15
Chore 4	$30	$20	$10
Chore 5	$20	$40	$15

TABLE 3-27

3.6 The Method of Markers

51. Three players (A, B, and C) are dividing the array of 13 items shown in Fig. 3-31 using the method of markers. The players' bids are as indicated in the figure.

(a) Which items go to A?

(b) Which items go to B?

(c) Which items go to C?

(d) Which items are left over?

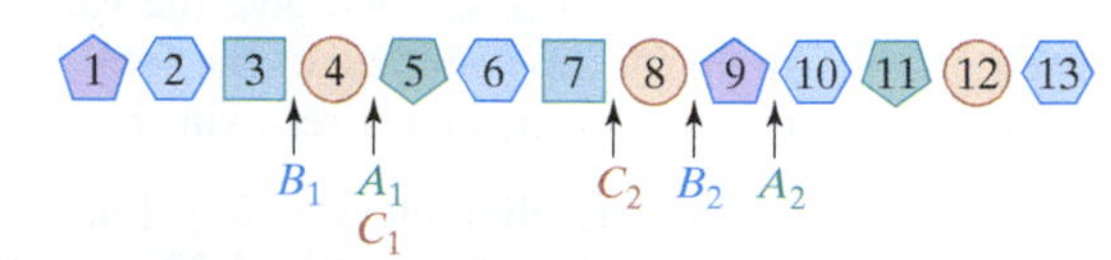

FIGURE 3-31

52. Three players (A, B, and C) are dividing the array of 13 items shown in Fig. 3-32 using the method of markers. The players' bids are as indicated in the figure.

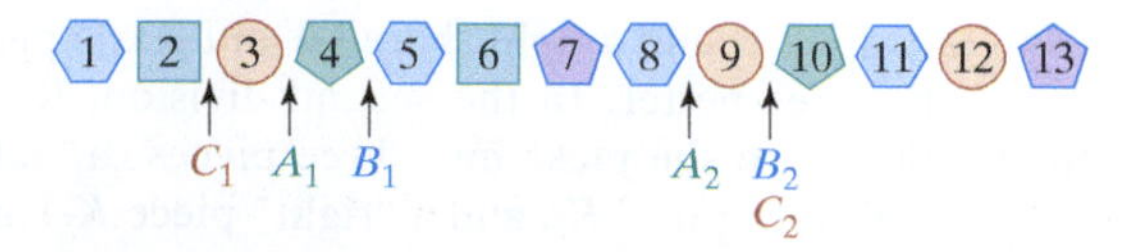

FIGURE 3-32

(a) Which items go to A?

(b) Which items go to B?

(c) Which items go to C?

(d) Which items are left over?

53. Three players (A, B, and C) are dividing the array of 12 items shown in Fig. 3-33 using the method of markers. The players' bids are as indicated in the figure.

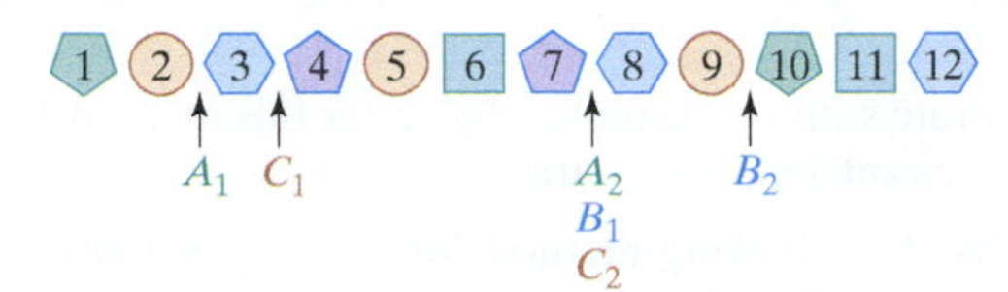

FIGURE 3-33

(a) Which items go to A?

(b) Which items go to B?

(c) Which items go to C?

(d) Which items are left over?

54. Three players (A, B, and C) are dividing the array of 12 items shown in Fig. 3-34 using the method of markers. The players' bids are indicated in the figure.

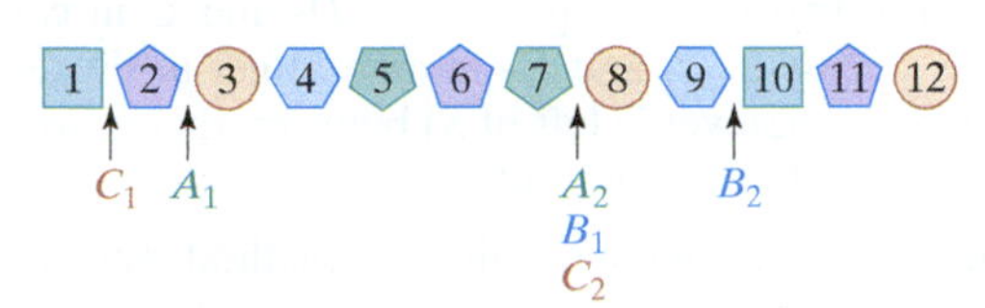

FIGURE 3-34

(a) Which items go to A?

(b) Which items go to B?

(c) Which items go to C?

(d) Which items are left over?

55. Five players (A, B, C, D, and E) are dividing the array of 20 items shown in Fig. 3-35 using the method of markers. The players' bids are as indicated in the figure.

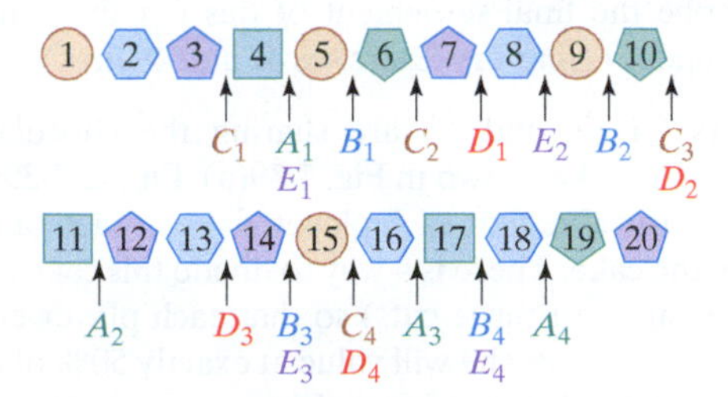

FIGURE 3-35

(a) Describe the allocation of items to each player.

(b) Which items are left over?

56. Four players (A, B, C, and D) are dividing the array of 15 items shown in Fig. 3-36 using the method of markers. The players' bids are as indicated in the figure.

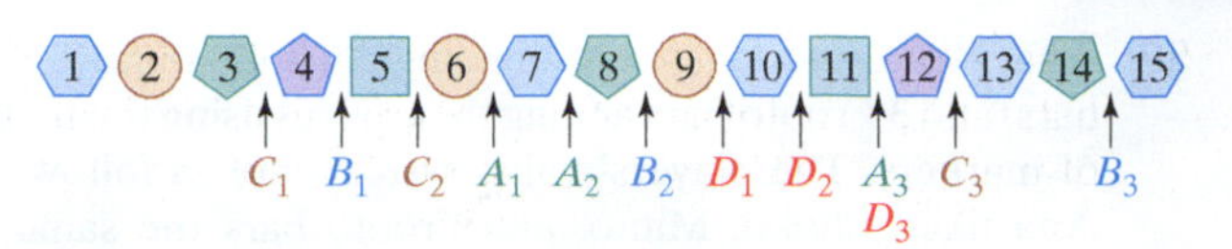

FIGURE 3-36

(a) Describe the allocation of items to each player.

(b) Which items are left over?

57. Quintin, Ramon, Stephone, and Tim are dividing a collection of 18 classic superhero comic books using the method of markers. The comic books are randomly lined up in the array shown below. (The W's are Wonder Woman comic books, the S's are Spider-Man comic books, the G's are Green Lantern comic books, and the B's are Batman comic books.)

W S S G S W W B G G G S G S G S B B

The value of the comic books in the eyes of each player is shown in Table 3-28.

	Quintin	Ramon	Stephone	Tim
Each W is worth	\$12	\$9	\$8	\$5
Each S is worth	\$7	\$5	\$7	\$4
Each G is worth	\$4	\$5	\$6	\$4
Each B is worth	\$6	\$11	\$14	\$7

TABLE 3-28

(a) Describe the placement of each player's markers. (Use Q_1, Q_2, and Q_3 for Quintin's markers, R_1, R_2, and R_3 for Ramon's markers, and so on.)

(b) Describe the allocation of comic books to each player and describe what comic books are left over.

58. Queenie, Roxy, and Sophie are dividing a set of 15 CDs—3 Beach Boys CDs, 6 Grateful Dead CDs, and 6 opera CDs using the method of markers. Queenie loves the Beach Boys but hates the Grateful Dead and opera. Roxy loves the Grateful Dead and the Beach Boys equally well but hates opera. Sophie loves the Grateful Dead and opera equally well but hates the Beach Boys. The CDs are lined up in an array as follows:

O O O GD GD GD BB BB BB GD GD GD O O O

(O represents the opera CDs, GD the Grateful Dead CDs, and BB the Beach Boys CDs.)

(a) Describe the placement of each player's markers. (Use Q_1 and Q_2 for Queenie's markers, R_1 and R_2 for Roxy's markers, etc.)

(b) Describe the allocation of CDs to each player and describe what CDs are left over.

(c) Suppose that the players agree that each one gets to pick an extra CD from the leftover CDs. Suppose that Queenie picks first, Sophie picks second, and Roxy picks third. Describe which leftover CDs each one would pick.

59. Ana, Belle, and Chloe are dividing 3 Choko bars, 3 Minto bars, and 3 Frooto bars among themselves using the method of markers. The players' value systems are as follows: (1) Ana likes Choko, Minto, and Frooto bars the same; (2) Belle loves Minto bars but hates Choko and Frooto bars; (3) Chloe likes Frooto bars twice as much as she likes Choko or Minto bars. Suppose that the candy is lined up exactly as shown in Fig. 3-37.

FIGURE 3-37

(a) Describe the placement of each player's markers. (Use A_1 and A_2 for Ana's markers, B_1 and B_2 for Belle's markers, and C_1 and C_2 for Chloe's markers.)

(*Hint*: For each player, compute the value of each piece as a fraction of the value of the booty first. This will help you figure out where the players would place their markers.)

(b) Describe the allocation of candy to each player and which pieces of candy are left over.

(c) Suppose that the players decide to divide the leftover pieces by a random lottery in which each player gets to choose one piece. Suppose that Belle gets to choose first, Chloe second, and Ana last. Describe the division of the leftover pieces.

60. Arne, Bruno, Chloe, and Daphne are dividing 3 Choko bars, 3 Minto bars, 3 Frooto bars, and 6 Rollo bars among themselves using the method of markers. Arne hates Choko Bars but likes Minto bars, Frooto bars, and Rollo bars equally well. Bruno hates Minto bars but likes Choko bars, Frooto bars, and Rollo bars equally well. Chloe hates Frooto bars and Rollo bars and likes Choko three times as much as Minto bars. Daphne hates Choko and Minto bars and values a Frooto bars as equal to two-thirds the value of a Rollo bars (i.e., 2 Rollo bars equal 3 Frooto bars). Suppose the candy is lined up exactly as shown in Fig. 3-38.

FIGURE 3-38

(a) Describe the placement of each player's markers. (*Hint*: For each player, compute the value of each piece as a fraction of the value of the booty first. This will help you figure out where the players would place their markers.)

(b) Describe the allocation of candy to each player and which pieces of candy are left over.

(c) After the allocation, each player is allowed to pick one piece of candy. Will there be any arguments?

JOGGING

61. Consider the following method for dividing a continuous asset S among three players (two dividers and one chooser):

Step 1. Divider 1 (D_1) cuts S into two pieces s_1 and s_2 that he considers to be worth, $\frac{1}{3}$ and $\frac{2}{3}$ of the value of S, respectively.

Step 2. Divider 2 (D_2) cuts the second piece s_2 into two halves s_{21} and s_{22} that she considers to be of equal value.

Step 3. The chooser C chooses one of the three pieces (s_1, s_{21}, or s_{22}), D_1 chooses next, and D_2 gets the last piece.

(a) Explain why under this method C is guaranteed a fair share.

(b) Explain why under this method D_1 is guaranteed a fair share.

(c) Illustrate with an example why under this method D_2 is not guaranteed a fair share.

62. Consider the following method for dividing a continuous asset S among three players:

Step 1. Divider 1 (D_1) cuts S into two pieces s_1 and s_2 that he considers to be worth, $\frac{1}{3}$ and $\frac{2}{3}$ of the value of S, respectively.

Step 2. Divider 2 (D_2) gets a shot at s_1. If he thinks that s_1 is worth $\frac{1}{3}$ of S or less, he can pass (case 1); if he thinks that s_1 is worth more than $\frac{1}{3}$, he can trim the piece to a smaller piece s_{11} that he considers to be worth exactly $\frac{1}{3}$ of S (case 2).

Step 3. The chooser C gets a shot at either s_1 (in case 1) or at s_{11} (in case 2). If she thinks the piece is a fair share, she gets to keep it (case 3). Otherwise, the piece goes to the divider that considers it to be worth $\frac{1}{3}$ (D_1 in case 1, D_2 in case 2).

Step 4. The two remaining players (D_2 and C in case 1, D_1 and C in case 2, D_1 and D_2 in case 3) get to divide the "remainder" (whatever is left of S) between themselves using the divider-chooser method.

Explain why the above is a fair-division method that guarantees that if played properly, each player will get a fair share.

63. Two partners (A and B) jointly own a business but wish to dissolve the partnership using the method of sealed bids. One of the partners will keep the business; the other will get cash for his half of the business. Suppose that A bids \$$x$ and B bids \$$y$. Assume that B is the high bidder. Describe the final settlement of this fair division in terms of x and y.

64. Three partners (A, B, and C) jointly own a business but wish to dissolve the partnership using the method of sealed bids. One of the partners will keep the business; the other two will each get cash for their one-third share of the business. Suppose that A bids \$$x$, B bids \$$y$, and C bids \$$z$. Assume that C is the high bidder. Describe the final settlement of this fair division in terms of x, y, and z. (*Hint*: Try Exercises 47 and 48 first.)

65. Three players (A, B, and C) are sharing the chocolate-strawberry-vanilla cake shown in Fig. 3-39(a). Figure 3-39(b) shows the relative value that each player gives to each of the three parts of the cake. There is a way to divide this cake into three pieces (using just three cuts) so that each player ends up with a piece that he or she will value at exactly 50% of the value of the cake. Find such a fair division.

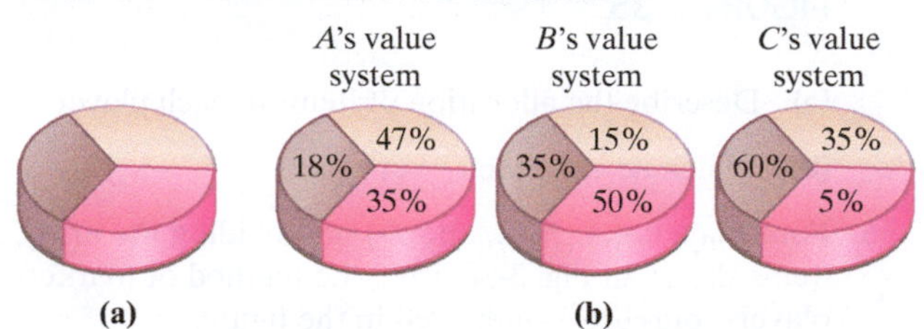

FIGURE 3-39

66. Angelina and Brad are planning to divide the chocolate-strawberry cake shown in Fig. 3-40(a) using the divider-chooser method, with Angelina being the divider. Suppose that Angelina values chocolate cake *three* times as much as she values strawberry cake. Figure 3-40(b) shows a generic cut made by Angelina dividing the cake into two shares s_1 and s_2 of equal value to her. Think of s_1 as an "$x°$ chocolate–$y°$ strawberry" share. For each given measure of the angle x, find the corresponding measure of the angle y.

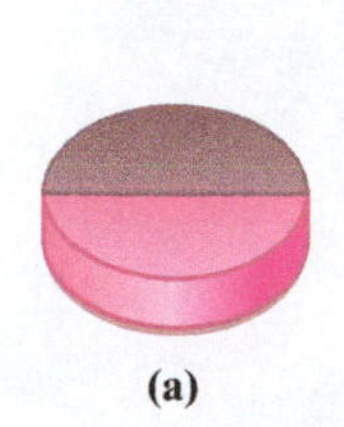
(a)

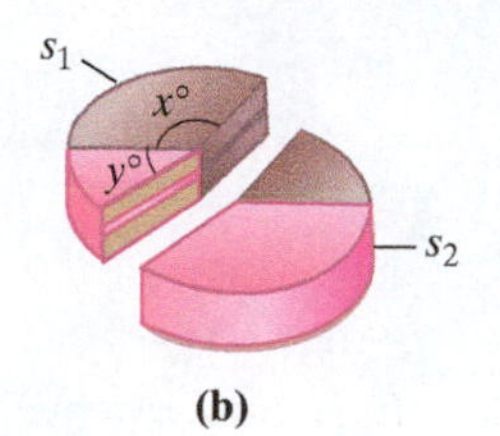

(b)

FIGURE 3-40

(a) $x = 60°$.

(b) $x = 72°$.

(c) $x = 108°$.

(d) $x = 120°$.

67. Standoffs in the lone-divider method. In the lone-divider method, a *standoff* occurs when a set of k choosers are bidding for less than k items. The types of *standoffs* possible depend on the number of players. With $N = 3$ players, there is only *one* type of standoff—the two choosers are bidding for the same item. With $N = 4$ players, there are *three* possible types of standoffs: two choosers are bidding on the same item, or three choosers are bidding on the same item, or three choosers are bidding on just two items. With $N = 5$ players, there are *six* possible types of standoffs, and the number of possible types of standoffs increases rapidly as the number of players increases.

(a) List the six possible types of *standoffs* with $N = 5$ players.

(b) What is the number of possible types of *standoffs* with $N = 6$ players?

(c) What is the number of possible types of *standoffs* with N players? Give your answer in terms of N. (*Hint*: You will need to use the formula given in Chapter 1 for the number of pairwise comparisons among a set of objects.)

68. Efficient and envy-free fair divisions. A fair division is called **efficient** if there is no other fair division that gives *every* player a share that is as good or better (i.e., any other fair division that gives some players a better share must give some other players a worse share). A fair division is called **envy-free** if every player ends up with a share that he or she feels is *as good as or better than* that of any other player.

Suppose that three partners (A, B, and C) jointly own a piece of land that has been subdivided into six parcels ($s_1, s_2, \ldots, s_6$). The partnership is splitting up, and the partners are going to divide fairly the six parcels among themselves. Table 3-29 shows the value of each parcel in the eyes of each partner.

(a) Find a fair division of the six parcels among the three partners that is *efficient*.

(b) Find a fair division of the six parcels among the three partners that is *envy-free*.

	A	B	C
s_1	\$20,000	\$16,000	\$19,000
s_2	\$19,000	\$18,000	\$18,000
s_3	\$18,000	\$19,000	\$15,000
s_4	\$16,000	\$20,000	\$12,000
s_5	\$15,000	\$15,000	\$20,000
s_6	\$12,000	\$12,000	\$16,000

TABLE 3-29

(c) Find a fair division of the six parcels among the three partners that is *efficient but not envy-free.*

(d) Find a fair division of the six parcels among the three partners that is *envy-free but not efficient.*

RUNNING

69. Suppose that N players bid on M items using the method of sealed bids. Let T denote the table with M rows (one for each item) and N columns (one for each player) containing all the players' bids (i.e., the entry in column j, row k represents player j's bid for item k). Let $c_1, c_2, \ldots, c_N$ denote, respectively, the sum of the entries in column 1, column 2, $\ldots$, column N of T, and let $r_1, r_2, \ldots, r_M$ denote, respectively, the sum of the entries in row 1, row 2, $\ldots$, row M of T. Let $w_1, w_2, \ldots, w_M$ denote the winning bids for items 1, 2, $\ldots$, M, respectively (i.e., w_1 is the largest entry in row 1 of T, w_2 is the largest entry in row 2, etc.). Let S denote the surplus money left after the first settlement.

(a) Show that

$$S = (w_1 + w_2 + \cdots + w_M) - \frac{(c_1 + c_2 + \cdots + c_N)}{N}.$$

(b) Using (a), show that

$$S = \left(w_1 - \frac{r_1}{N}\right) + \left(w_2 - \frac{r_2}{N}\right) + \cdots + \left(w_M - \frac{r_M}{N}\right).$$

(c) Using (b), show that $S \geq 0$.

(d) Describe the conditions under which $S = 0$.

70. Asymmetric method of sealed bids. Suppose that an estate consisting of M indivisible items is to be divided among N heirs ($P_1, P_2, \ldots, P_N$) using the method of sealed bids. Suppose that Grandma's will stipulates that P_1 is entitled to x_1% of the estate, P_2 is entitled to x_2% of the estate, $\ldots$, P_N is entitled to x_N% of the estate. The percentages add up to 100%, but they are not all equal (Grandma loved some grandchildren more than others). Describe a variation of the method of sealed bids that ensures that each player receives a "fair share" (i.e., P_1 receives a share that she considers to be worth at least x_1% of the estate, P_2 receives a share that he considers to be worth at least x_2% of the estate, etc.).

71. Lone-chooser is a fair-division method. Suppose that N players divide a cake using the *lone-chooser method.* The chooser is C and the dividers are $D_1, D_2, \ldots, D_{N-1}$. Explain why, when properly played, the method guarantees to each player a fair share. (You will need one argument for the dividers and a different argument for the chooser.)

"Representatives . . . shall be apportioned among the several States . . . according to their respective Numbers." *Article 1, Section 2, U.S. Constitution.* (For details, read "A Brief History of Apportionment in the United States," in the Conclusion.)

The Mathematics of Apportionment

Making the Rounds

April 1, 2010 was not just any ordinary April Fool's Day. With an exquisite sense of irony, the United States Census Bureau designated April 1 as "Census Day" for the 2010 Census—the day when the population of the United States was officially counted. And indeed it was—there seemed to be around 308,745,538 of us going about our business that day.

The Constitution mandates that every 10 years—on years ending with a 0—the government produce a "head count" of the U.S.

population, broken down by state.(The modern version of the U.S.Census does a lot more than count heads, but at the moment we will focus on just the state population counts—we will discuss other aspects of the U.S. Census in greater detail in Chapter 14.) The key purpose of the state population numbers is to meet the constitutional requirement of "proportional representation." The Constitution requires that seats in the U.S. House of Representatives be "apportioned among the... States according to their respective Numbers," so every 10 years two things must happen: (a) the state populations must be determined (that's the Census part), and (b) the seats in the House of Representatives must be apportioned to the states based on their populations (that's the "apportionment" part). It doesn't sound all that complicated, does it? You will be surprised!

> "Representatives . . . shall be apportioned among the several States . . . according to their respective Numbers. The actual Enumeration shall be made . . . every ten Years . . . "
>
> – *Article I, Section 2, U.S. Constitution*

This chapter is an excursion into the mathematics behind the apportionment part, and it contains many interesting twists and turns. There are also a lot of parallels between apportionment and voting, and the organization of the chapter is very similar to that of Chapter 1. We will start the chapter with an introduction to the basic concepts of apportionment: What is an *apportionment problem*? What are the elements common to all apportionment problems? What is an *apportionment method*? In Sections 4.2 through 4.4 we will introduce four classic apportionment methods named after four famous figures in American history: *Hamilton's method*, *Jefferson's method*, *Adams's method*, and *Webster's method*. In Section 4.5 we will discuss the *Huntington-Hill apportionment method*, the method currently used for the apportionment of the U.S. House of Representatives. Finally, in Section 4.6 we will discuss the flaws and paradoxes of the various apportionment methods, concluding with an important result known as the *Balinski-Young Impossibility Theorem.*

4.1 Apportionment Problems and Apportionment Methods

> "ap·por·tion: to *divide* and *share* out according to a *plan*; to make a *proportionate* division or distribution of . . . "
>
> – *Merriam-Webster Dictionary**

Obviously, the verb *apportion* is the key word in this chapter. There are two critical elements in the dictionary definition of the word: (1) We are *dividing* and *sharing* things, and (2) we are doing this on a *proportional* basis and in a *planned*, organized fashion.

We will start this section with a pair of examples that illustrate the nature of the problem we are dealing with. These examples will raise several questions, but the answers will come later.

---

*By permission. From *Merriam-Webster's Collegiate® Dictionary, 11th Edition ©2012* by Merriam-Webster, Incorporated (www.Merriam-Webster.com).

EXAMPLE 4.1 KITCHEN CAPITALISM

Mom has a total of 50 identical pieces of candy (let's say caramels), which she is planning to divide among her five children (this is the "divide and share" part). Like any good mom, she is intent on doing this fairly. Of course, the easiest thing to do would be to give each child 10 caramels—by most standards, that would be fair. Mom, however, is thinking of the long-term picture—she wants to teach her children about the value of work and about the relationship between work and reward. This leads her to the following idea: She announces to the kids that the candy is going to be divided at the end of the week in proportion to the amount of time each of them spends helping with the weekly kitchen chores—if you worked twice as long as your brother you get twice as much candy, and so on (this is the "proportionate" division part). Unwittingly, mom has turned this division problem into an apportionment problem.

At the end of the week, the numbers are in. Table 4-1 shows the amount of work done by each child during the week. (Yes, mom did keep up-to-the-minute records!)

Child	Alan	Betty	Connie	Doug	Ellie	Total
Minutes worked	150	78	173	204	295	900

TABLE 4-1 ■ Work (in minutes) per child

According to the ground rules, Alan, who worked 150 out of a total of 900 minutes, should get $8\frac{1}{3}$ pieces. (Setting up the proportion $\frac{150}{900} = \frac{x}{50}$ and solving for x gives $x = 8\frac{1}{3}$.) Here comes the problem: Since the pieces of candy are indivisible, it is impossible for Alan to get his $8\frac{1}{3}$ pieces—he can get 8 pieces (and get shorted) or he can get 9 pieces (and someone else will get shorted). A similar problem occurs with each of the other children. Betty's exact fair share should be $4\frac{1}{3}$ pieces; Connie's should be $9\frac{11}{18}$ pieces; Doug's, $11\frac{1}{3}$ pieces; and Ellie's, $16\frac{7}{18}$ pieces. (Be sure to double-check these figures!) Because none of these shares can be realized, an absolutely fair apportionment of the candy is going to be impossible. What should mom do?

Example 4.1 shows all the elements of an apportionment problem—there are discrete identical objects to be divided (the pieces of candy), and there is a proportionality criterion for the division (number of minutes worked during the week). We will say that the pieces of candy are *apportioned* to the kids, and we will describe the final solution as an *apportionment* (Alan's *apportionment* is x pieces, Betty's *apportionment* is y pieces, etc.).

Let's now consider a seemingly different apportionment problem.

EXAMPLE 4.2 THE INTERGALACTIC CONGRESS OF UTOPIA

It is the year 2525, and all the planets in the Utopia galaxy have finally signed a peace treaty. Five of the planets (Alanos, Betta, Conii, Dugos, and Ellisium) decide to join forces and form an Intergalactic Federation. The Federation will be ruled by an Intergalactic Congress consisting of 50 "elders," and the 50 seats in the Intergalactic Congress are to be *apportioned* among the planets according to their respective populations.

The population data for each of the planets (in billions) are shown in Table 4-2. Based on these population figures, what is the correct *apportionment* of seats to each planet?

Planet	Alanos	Betta	Conii	Dugos	Ellisium	Total
Population	150	78	173	204	295	900

Source: Intergalactic Census Bureau.

TABLE 4-2 ■ Intergalactic Federation: Population figures (in billions) for 2525

Example 4.2 is another example of an apportionment problem—here the objects being divided are the 50 seats in the Intergalactic Congress, and the proportionality criterion is population. But there is more to it than that. The observant reader may have noticed that the numbers in Examples 4.1 and 4.2 are identical—it is only the framing of the problem that has changed. Mathematically speaking, Examples 4.1 and 4.2 are one and the same apportionment problem—solve either one and you have solved both!

Apportionment: Basic Concepts and Terminology

Between the extremes of apportioning the seats in the Intergalactic Congress of Utopia (important, but too far away!) and apportioning the candy among the children in the house (closer to home, but the galaxy will not come to an end if the apportionments aren't done right!) fall many other, real-life problems that are both important and relevant: assigning classrooms to departments in a university, assigning nurses to shifts in a hospital, etc. But the gold standard for apportionment applications is the allocation of seats to "states" in a legislature, and, thus, it is standard practice to borrow the terminology of legislative apportionment and apply it to apportionment problems in general.

The basic elements of every apportionment problem are as follows:

- **The "states."** This is the term we will use to describe the parties having a stake in the apportionment. Unless they have specific names (Alanos, Betta, etc.), we will use A, B, C, etc. to denote the states. Note that we are giving the term a very broad meaning—the "states" can be districts, counties, planets, and even people (as in Example 4.1). As usual, we will use N to denote the number of states.
- **The "seats."** This term describes the set of M *identical, indivisible objects* that are being divided among the N states. Because the seats are indivisible objects, the number of seats apportioned to each state must be a whole number—no fractions allowed! For convenience, we will assume that there are more seats than there are states, thus ensuring that every state can potentially get a seat. (This assumption does not imply that every state *must* get a seat! Such a requirement is not part of the general apportionment problem, although it is part of the constitutional requirements for the apportionment of the U.S. House of Representatives.)
- **The "populations."** This is a set of N positive numbers (for simplicity we will assume that they are whole numbers) that are used as the basis for the apportionment of the seats to the states. We will use $p_1, p_2, \ldots, p_N$ to denote the states' respective populations and P to denote the total population $(P = p_1 + p_2 + \cdots + p_N)$.
- **An apportionment method.** A systematic procedure that *guarantees* a division of the M seats (no more and no less) to the N states using some formula that is based on the state populations.

We now formally define several important concepts that we will use throughout this chapter.

- **The standard divisor.** The **standard divisor (SD)** is the ratio of total population to seats $\left(SD = \frac{P}{M}\right)$. It gives us a unit of measurement (SD *people* $= 1$ seat) for our apportionment calculations.
- **The standard quotas.** The **standard quota** of a state (sometimes called the *fair quota*) is the exact fractional number of seats that the state would get if fractional seats were allowed. We will use the notation $q_1, q_2, \ldots, q_N$ to denote the standard quotas of the respective states. To find a state's standard quota, we divide the state's population by the standard divisor (for a state with population p, its standard quota is $q = \frac{p}{SD}$). In general, the standard quotas can be expressed as decimals or fractions—it would be almost a miracle if one of them turned out to

be a whole number—and when using the decimal form, it's customary to round them to two or three decimal places.

- **Upper and lower quotas.** Associated with each standard quota are two other important numbers—the **lower quota** (the standard quota rounded down) and the **upper quota** (the standard quota rounded up). In the unlikely event that the standard quota is a whole number, the lower and upper quotas are the same. We will use L's to denote lower quotas and U's to denote upper quotas. For example, the standard quota $q = 32.92$ has lower quota $L = 32$ and upper quota $U = 33$.

We will now introduce the most important apportionment example of this chapter. We will return to this example many times. While the story behind the example is obviously fiction, the issues it raises are very real.

EXAMPLE 4.3 THE CONGRESS OF PARADOR

Parador is a small republic located in Central America and consists of six states: Azucar, Bahia, Cafe, Diamante, Esmeralda, and Felicidad (A, B, C, D, E, and F for short). There are 250 seats in the Congress, which, according to the laws of Parador, are to be apportioned among the six states in proportion to their respective populations. What is the "correct" apportionment?

Table 4-3 shows the population figures for the six states according to the most recent census.

State	A	B	C	D	E	F	Total
Population	1,646,000	6,936,000	154,000	2,091,000	685,000	988,000	12,500,000

TABLE 4-3 ■ Republic of Parador (populations by state)

The first step we will take to tackle this apportionment problem is to compute the standard divisor (SD). In the case of Parador, the standard divisor is $SD = \frac{12{,}500{,}000}{250} = 50{,}000$. (Typically, the standard divisor is not going to turn out to be such a nice whole number, but it is always going to be a rational number.)

The standard divisor $SD = 50{,}000$ tells us that in Parador, each seat in the Congress corresponds to 50,000 people. We can now use this yardstick to find the number of seats that each state *should get by the proportionality criterion*—all we have to do is divide the state's population by 50,000. These are the standard quotas. For example, take state A. If we divide the population of A by 50,000, we get $q_1 = \frac{1{,}646{,}000}{50{,}000} = 32.92$. If seats in the Congress could be apportioned in fractional parts, then the fair apportionment to A would be 32.92 seats, but, of course, this is impossible—seats in the Congress are indivisible objects! Thus, the standard quota of 32.92 is a number that we cannot meet, although in this case we might be able to come close to it: The next best solution might be to apportion 33 seats to state A. Hold that thought!

Table 4-4 shows the standard quotas for each of the states in Parador (rounded to two decimal places). Notice that the sum of the standard quotas equals 250, the number of seats being apportioned.

State	A	B	C	D	E	F	Total
Population	1,646,000	6,936,000	154,000	2,091,000	685,000	988,000	12,500,000
Standard quota	32.92	138.72	3.08	41.82	13.70	19.76	250

TABLE 4-4 ■ Republic of Parador: Standard quotas for each state ($SD = 50{,}000$)

Standard quotas are the benchmark for a *fair apportionment*, but we must somehow convert these standard quotas to whole numbers. This is where the rubber meets the road: How do we round the quotas into whole numbers?

At first glance, this seems like a dumb question. After all, we all learned in school how to round decimals to whole numbers—round down if the fractional part is less than 0.5, round up otherwise. This kind of rounding is called *rounding to the nearest integer*, or simply *conventional rounding*. Unfortunately, conventional rounding will not work in this example, and Table 4-5 shows why not—we would end up giving out 251 seats in the Congress, and there are only 250 seats to give out! (Add an extra seat, you say? Sorry, this is not a banquet. The number of seats is fixed—we don't have the luxury of adding or subtracting seats!)

State	*A*	*B*	*C*	*D*	*E*	*F*	**Total**
Standard quota	32.92	138.72	3.08	41.82	13.70	19.76	250.00
Nearest integer	33	139	3	42	14	20	**251**

TABLE 4-5 ■ Conventional rounding of the standard quotas

The most important moral of Example 4.3 is that conventional rounding of the standard quotas is *not* an apportionment method. It can end up apportioning more than M seats (as in Example 4.3, where 251 seats were needed but only 250 were available) or less than M seats. Sometimes things do fall into place, and it ends up apportioning exactly M seats, but since this is not guaranteed, we can't count on it as an apportionment method. So, what do we try next? For now, we leave this example—a simple apportionment problem looking for a simple solution—with no answers (but, we hope, with some gained insight!). Our search for a good solution to this problem will be the theme of our journey through the rest of this chapter.

4.2 Hamilton's Method

Alexander Hamilton (1757–1804)

Alexander Hamilton first proposed this method in 1792 as a way to apportion the U.S. House of Representatives (for more details see "A Brief History of Apportionment in the United States" in the Conclusion). Hamilton's method is also known as *Vinton's method* and *the method of largest remainders*.

Hamilton's method can be described quite briefly: *Every state gets at least its lower quota. As many states as possible get their upper quota, with the one with highest residue (i.e., fractional part) having first priority, the one with second highest residue second priority, and so on.* A little more formally, it goes like this:

HAMILTON'S METHOD

- **Step 1.** Calculate each state's standard quota.
- **Step 2.** Round the standard quotas down and give to each state its *lower quota*.
- **Step 3.** Give the surplus seats (one at a time) to the states with the largest *residues* (fractional parts) until there are no more surplus seats.

EXAMPLE 4.4 PARADOR'S CONGRESS (HAMILTON'S METHOD)

As promised, we are revisiting the Parador Congress apportionment problem. We are now going to find our first real solution to this problem—the solution given by Hamilton's method (we will call it the *Hamilton apportionment*). Table 4-6 shows all the details. (Reminder to the reader: The standard quotas in the second row of the table were originally computed in Example 4.3.) After each state receives its lower quota, there are 4 surplus seats left. These go to the states with the largest residues: *A* first (0.92), *D* second (0.82), *F* third (0.76), and *B* last (0.72). *E* and *C* get no extra seats.

State	*A*	*B*	*C*	*D*	*E*	*F*	Total
Standard quota	32.92	138.72	3.08	41.82	13.70	19.76	250.00
Lower quota	32	138	3	41	13	19	246
Residue	0.92	0.72	0.08	0.82	0.70	0.76	4.00
Order of Surplus	First	Fourth		Second		Third	
Apportionment	33	139	3	42	13	20	250

TABLE 4-6 ■ Parador's Congress: The Hamilton apportionment

At first glance, Hamilton's method appears to be a good method for apportioning seats, but a careful look at Table 4-6 already shows hints of possible unfairness. Compare the fates of state *B*, with a residue of 0.72, and state *E*, with a residue of 0.70. State *B* gets the last surplus seat; state *E* gets nothing! Sure enough, 0.72 is more than 0.70, so following the rules of Hamilton's method, we give priority to *B*. By the same token, *B* is a huge state, and as a percentage of its population the 0.72 represents an insignificant amount, whereas state *E* is a relatively small state, and its residue of 0.70 (for which it gets nothing) represents more than 5% of its population $\left(\frac{0.70}{13.70} \approx 0.051 = 5.1\%\right)$.

It could be reasonably argued that Hamilton's method has a major flaw in the way it relies entirely on the size of the residues without consideration of what those residues represent as a percent of the state's population. In so doing, Hamilton's method creates a systematic bias in favor of larger states over smaller ones. This is bad—a good apportionment method should be *population neutral*, meaning that it should not be biased in favor of large states over small ones or vice versa. In Section 4.6 we will see that bias in favor of large states is only one of the many serious flaws of Hamilton's method.

Our next example gives us a preview of some of the issues that can arise when using Hamilton's method. The numbers used in the example are clearly made up, but the situation the example illustrates is very real.

EXAMPLE 4.5 THE BAMA PARADOX

The small country of Calavos consists of three states: Bama, Tecos, and Ilnos. With a total population of $P = 200{,}000$ and $M = 200$ seats in the House of Representatives, we have an easy calculation of the standard divisor $(SD = 1000)$. With such a beautiful standard divisor (of course, the numbers are rigged!), we can quickly find the standard quotas (just divide the populations by 1000). Table 4-7 shows the details of the apportionment of the 200 seats under Hamilton's method: Bama gets 10 seats, Tecos get 90 seats, Ilnos gets 100 seats.

State	Bama	Tecos	Ilnos	Total
Population	9400	90,300	100,300	200,000
Standard quota	9.4	90.3	100.3	200
Lower quota	9	90	100	199
Surplus	1	0	0	1
Hamilton apportionment	10	90	100	200

TABLE 4-7 ■ Hamilton's method with $M = 200$, $SD = 1000$

Now we will ask ourselves what happens if the number of seats to be apportioned is increased to 201, but nothing else changes—the populations remain the same. Since there is one more seat to give out, the apportionment has to be recomputed. Table 4-8 shows the new apportionment for a House with 201 seats. First, we need to find the standard divisor — not such a nice round number anymore! Now $SD = \frac{200{,}000}{201} \approx 995.02$. The standard quotas (rounded to two decimal places) are, respectively, $\frac{9400}{995.02} \approx 9.45$, $\frac{90{,}300}{995.02} \approx 90.75$, and $\frac{100{,}300}{995.02} \approx 100.8$ (third row of Table 4-8). We now have two surplus seats, and they go to Ilnos first and Tecos second. Bama's apportionment is now down to 9 seats (last row of Table 4-8).

State	Bama	Tecos	Ilnos	Total
Population	9400	90,300	100,300	200,000
Standard quota	9.45	90.75	100.8	201
Lower quota	9	90	100	199
Surplus	0	1	1	2
Hamilton apportionment	9	91	101	201

TABLE 4-8 ■ Hamilton's method with $M = 201$, $SD \approx 995.02$

Strangely, when there were 200 seats to be apportioned, Hamilton's method gave Bama 10 seats, but when the number of seats was increased to 201 Hamilton's method gave Bama 9 seats. How could this happen? Notice the effect that the increase in the number of seats has on the size of the residues: In a House with 200 seats, Bama is at the head of the priority line for surplus seats, but when the number of seats goes up to 201, Bama gets shuffled to the back of the line. Welcome to the wacky arithmetic of Hamilton's method!

4.3 Jefferson's Method

Thomas Jefferson (1743–1826)

Our next method is attributed to Thomas Jefferson, who proposed it as the right apportionment method for the U.S. House of Representatives in 1792. Jefferson's method, as it is known in the United States, is also known as *d'Hondt's method* and *the method of greatest divisors*. (A more detailed account of the use of Jefferson's method to apportion the U.S. House of Representatives can be found in the section "A Brief History of Apportionment in the United States" in the Conclusion.)

Jefferson's method is based on an approach very different from Hamilton's method, so there is some irony in the fact that we will explain the idea behind Jefferson's method by taking one more look at Hamilton's method.

Recall that under Hamilton's method we start by dividing every state's population by the standard divisor. This gives us the standard quotas. Step 2 is then to round *every* state's quota down. Notice that up to this point Hamilton's method uses a uniform policy for all states—every state is treated in exactly the same way. If you are looking for fairness, this is obviously good. But now comes the bad part: we have some leftover seats which we need to distribute, but not enough for every state. Thus, we must choose some states over others for preferential treatment. No matter how fair we try to be about it, there is no getting around the fact that some states get extra seats and others don't.

We saw in Example 4.5 that it is in the handling of the surplus seats (Step 3) that Hamilton's method runs into trouble. So here is an interesting idea: Let's tweak things so that when the quotas are rounded down, *there are no surplus seats*! (The idea, in other words, is to get rid of Step 3.) But under Hamilton's method there is always going to be at least one surplus seat, so how do we work this bit of magic?

The answer is by changing the divisor, which then changes the quotas. The idea is that by using a *smaller* divisor, we make the quotas bigger. If we hit it just right, when we round the slightly bigger quotas down . . . poof! The surplus seats are gone. When we pull that bit of numerical magic, we have in fact implemented *Jefferson's method.*

The best way to get a good feel for the inner workings of Jefferson's method is through an example, so we will do that next.

EXAMPLE 4.6 PARADOR'S CONGRESS (JEFFERSON'S METHOD)

Once again we are going to apportion the 250 seats in Parador's Congress, this time using Jefferson's method. For comparison purposes, we start with a review of how Hamilton's method tackles this problem. The first two steps in Hamilton's method are to compute the standard quotas (by dividing the populations by the standard divisor $SD = 50{,}000$) and then round them down to the lower quotas. These calculations are shown in Table 4-9. We know that there is a surplus of 4 seats and that Step 3 of Hamilton's method deals with that issue (but not too well!).

State	Population	Standard quota	Lower quota
A	1,646,000	32.92	32
B	6,936,000	138.72	138
C	154,000	3.08	3
D	2,091,000	41.82	41
E	685,000	13.70	13
F	988,000	19.76	19
Total	12,500,000	250	246

TABLE 4-9 ■ Calculations for Example 4.6 with the standard divisor $SD = 50{,}000$

We will now try analogous calculations using a *modified divisor* $d = 49{,}500$ to divide the populations. This gives us different quotas, which we call *modified quotas*. The second column of Table 4-10 shows these modified quotas rounded to two decimal places (using a calculator it's easy to check these numbers). When we round these modified quotas down (last column of Table 4-10) we end up with 250 seats—no leftovers! This is the apportionment according to Jefferson's method.

State	Population	Modified quota ($d = 49{,}500$)	Rounded down to
A	1,646,000	33.25	33
B	6,936,000	140.12	140
C	154,000	3.11	3
D	2,091,000	42.24	42
E	685,000	13.84	13
F	988,000	19.96	19
Total	12,500,000		**250**

TABLE 4-10 ■ Jefferson's method: Calculations for Example 4.6 with divisor $d = 49{,}500$

Two questions come to mind when looking at Example 4.6: (1) Can we really substitute the standard divisor by some other random divisor?, and (2) if so, where did the divisor $d = 49{,}500$ come from? We will address each of these questions in order.

The answer to the first question is straightforward: Yes, we can use divisors other than the standard divisor! The standard divisor is a benchmark for a good apportionment, but we are not necessarily bound to it. If the intent is to round the quotas down and use the lower quotas for the apportionments, then the standard divisor is not a good choice because we will have surplus seats to deal with. By choosing the right divisor we can make the problem of surplus seats disappear. One useful way to think about this change in divisors is as *a change in the units of measurement*. Instead of using 1 seat $= SD$ people as our unit, we use 1 seat $= d$ people.

As far as the second question (how did we come up with the divisor $d = 49{,}500$ in Example 4.6?), the short answer is that we used educated trial and error. Since our goal is to have no surplus seats, we must make the modified quotas *bigger* than the standard quotas. To do this we must divide the populations by a *smaller* number than the

State	Population	Modified quota (d = 49,000)	Rounded down to
A	1,646,000	33.59	33
B	6,936,000	141.55	141
C	154,000	3.14	3
D	2,091,000	42.67	42
E	685,000	13.97	13
F	988,000	20.16	20
Total	12,500,000		252

TABLE 4-11 ■ Jefferson's method: Calculations for Example 4.6 with divisor $d = 49{,}000$

standard divisor. So we start with some number $d < SD$ that is our first guess. Let's say we make our first guess $d = 49{,}000$. Table 4-11 shows the calculation of the modified and lower quotas for this divisor. Using this divisor we see that we overshot our target of 250 by 2 seats. This means that we made our divisor *too small*! So we try another number somewhere between 49,000 and 50,000. A reasonable second guess then is $d = 49{,}500$. This one we know works! (Fortunately, there is a full range of numbers that work as divisors, so it's not like we have to hit a bullseye. In Example 4.6 any divisor d between 49,401 and 49,542 also works.)

The formal description of Jefferson's method is surprisingly short. The devil is in the details.

JEFFERSON'S METHOD

- **Step 1.** Find a "suitable" divisor d.
- **Step 2.** Using d as the divisor, compute each state's modified quota (modified quota = state population/d).
- **Step 3.** Each state is apportioned its modified *lower quota*.

The hard part of implementing Jefferson's method is Step 1. A flowchart outlining the trial-and-error approach for finding a divisor that works under Jefferson's method is shown in Fig. 4-1. (There can be a fair amount of drudgery involved in implementing this flowchart by hand or with a calculator. The applet *Apportionment Methods*, available in *MyMathLab* in the *Multimedia Library* or in *Tools for Success* eliminates the mindless calculations: you just play with the values of the divisor d; the applet takes care of the rest.)

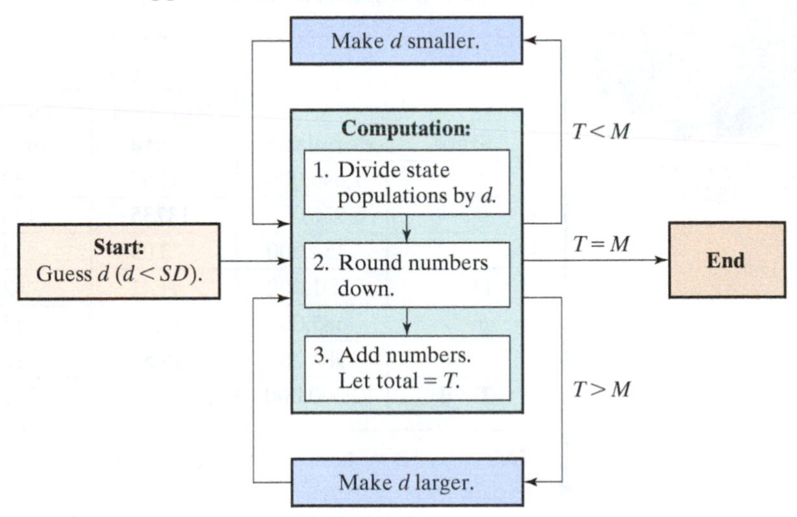

FIGURE 4-1 Flowchart for trial-and-error implementation of Jefferson's method.

Jefferson's method is but one of a group of apportionment methods based on the principle that the standard unit (1 seat = SD people) is not an absolute requirement and that, if necessary, we can change to a different unit: 1 seat = d people, where d is a suitably chosen number. The number d is called a **divisor** (sometimes we use the term **modified divisor** to emphasize the point that it is not the standard divisor), and apportionment methods that use modified divisors are called **divisor methods**. Different divisor methods are based on different philosophies of how the **modified quotas** *should be rounded to whole numbers*, but they all follow one script: When you are done rounding the modified quotas, all M seats have been apportioned (no more, and no less). To be able to do this, you just need to find a suitable divisor d. For Jefferson's method, the rounding rule is to always round the modified quotas down. In the next two sections we will discuss other divisor methods based on different rounding rules.

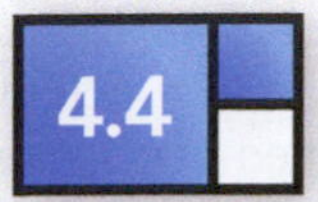

4.4 Adams's and Webster's Methods

Adams's Method

Adams's method (also known as *the method of smallest divisors*) was first proposed by John Quincy Adams in 1832 as a counterpoint to Jefferson's method. In essence, Adams's method is the mirror image of Jefferson's method: Instead of rounding all quotas down to their lower quotas it rounds them *up to their upper quotas*. For this to work we must make the modified quotas *smaller* than the standard quotas, which in turn means that the modified divisor must be *larger* than the standard divisor. As with any divisor method, the trick is to find a divisor that works.

Once again, we will illustrate Adams's method using the apportionment of Parador's Congress.

EXAMPLE 4.7 PARADOR'S CONGRESS (ADAMS'S METHOD)

John Quincy Adams (1767–1848)

Our starting point is to find a suitable divisor d. In Adams's method, the divisor d has to be bigger than the standard divisor $SD = 50{,}000$. Let's make our first guess $d = 50{,}500$. Table 4-12 (third and fourth columns) shows the computations with this divisor. The total number of seats apportioned would be 251. This total is just one seat above our target of 250, so we need to make the quotas just a tad smaller. To do this, we must increase the divisor a little bit.

Let's make our second guess $d = 50{,}700$. The new computations are shown in the last two columns of Table 4-12. This divisor works! The apportionment under Adams's method is given in the last column of Table 4-12. The good news is that there are plenty of other d's that work just as well (any number between 50,628 and 50,999 will work as a divisor).

		First guess: $d = 50{,}500$		Second guess: $d = 50{,}700$	
State	Population	Modified quota	rounded up to	Modified quota	rounded up to
A	1,646,000	32.59	33	32.47	33
B	6,936,000	137.35	138	136.8	137
C	154,000	3.05	4	3.04	4
D	2,091,000	41.41	42	41.24	42
E	685,000	13.56	14	13.51	14
F	988,000	19.56	20	19.49	20
Total	12,500,000		251		250

TABLE 4-12 ■ Adams's method: Calculations for Example 4.7 with divisors 50,500 and 50,700

A very short description of Adams's method (remarkably similar to that of Jefferson's method) follows. It looks harmless enough, but Step 1 can take several trials. (The applet *Apportionment Methods* available in *MyMathLab* provides a seamless way to carry out these trials.)

ADAMS'S METHOD

- **Step 1.** Find a "suitable" divisor d.
- **Step 2.** Using d as the divisor, compute each state's modified quota (modified quota = state population/d).
- **Step 3.** Each state is apportioned its modified *upper quota*.

Webster's Method

Daniel Webster (1782–1852)

What is the obvious compromise between rounding all the quotas down (Jefferson's method) and rounding all the quotas up (Adams's method)? What about *conventional rounding* (round the quotas down when the fractional part is less than 0.5 and up otherwise)?

It makes a lot of sense, but haven't we been there before (Example 4.3) and found that conventional rounding is *not* an apportionment method? Yes, we have tried conventional rounding of the *standard quotas* and found it didn't work, but we haven't considered conventional rounding using *modified quotas.* Now that we know that we can use modified divisors to manipulate the quotas, it is always possible to find a suitable divisor that will make conventional rounding work. This is the idea behind Webster's method (also known as the *Webster-Willcox method* and *the method of major fractions*). Daniel Webster first proposed this method in 1832 (for details on the use of Webster's method to apportion the U.S. House of Representatives see the section "A Brief History of Apportionment in the United States." in the Conclusion).

WEBSTER'S METHOD

- **Step 1.** Find a "suitable" divisor d.
- **Step 2.** Using d as the divisor, compute each state's modified quota (modified quota = state population/d).
- **Step 3.** Find the apportionments by rounding each modified quota to the nearest integer (conventional rounding).

EXAMPLE 4.8 PARADOR'S CONGRESS (WEBSTER'S METHOD)

To find a suitable divisor for Webster's method we start with the standard divisor $SD = 50{,}000$ and compute the *standard quotas* (we already did this more than once, but there they are, once again, in the third column of Table 4-13). When we round the standard quotas to the nearest integer (fourth column of Table 4-13) we are unlucky and overshoot our target of 250 seats by just one seat. This means that we must bring the quotas down (by just a little) by using a slightly bigger divisor d. Let's try $d = 50{,}100$. The modified quotas are now shown in the fifth column of Table 4-13, and when we round these to the nearest integer (last column of Table 4-13) we get what we want—a Webster apportionment of Parador's Congress.

		First guess: $SD = 50{,}000$		Second guess: $d = 50{,}100$	
State	**Population**	**Standard quota**	**Nearest integer**	**Modified quota**	**Nearest integer**
A	1,646,000	32.92	33	32.85	33
B	6,936,000	138.72	139	138.44	138
C	154,000	3.08	3	3.07	3
D	2,091,000	41.82	42	41.74	42
E	685,000	13.70	14	13.67	14
F	988,000	19.76	20	19.72	20
Total	12,500,000		251		250

TABLE 4-13 ■ Webster's method: Calculations for Example 4.8 with the standard divisor $SD = 50{,}000$ and the modified divisor $d = 50{,}100$

Notice that the change that made the difference was very subtle and happened with state *B*, the largest state. The change of divisors dropped *B*'s standard quota from 138.72 to a new quota of 138.44. Dropping B's decimal part to be under 0.5 made all the difference—now B's quota gets rounded down instead of up!

As with the other divisor methods, 50,100 was just one of a range of numbers that work in this example. In fact, any number between 50,080 and 50,385 will work just as well.

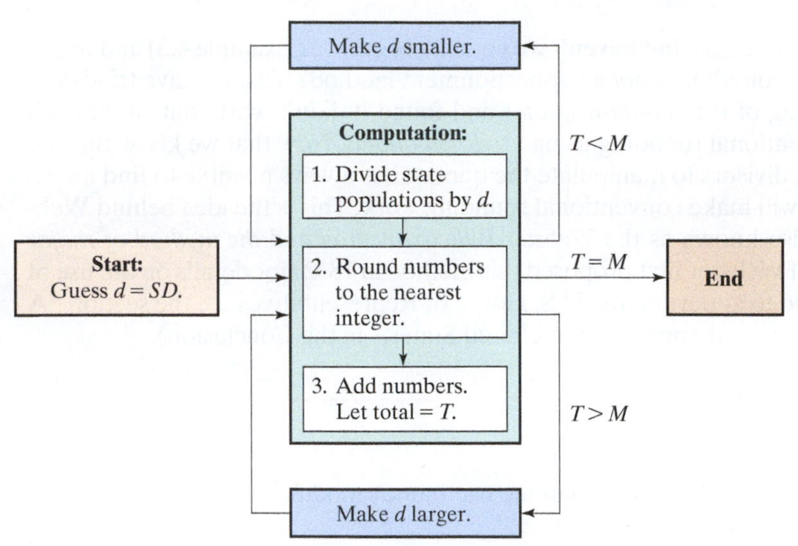

FIGURE 4-2 Flowchart for trial-and-error implementation of Webster's method.

Figure 4-2 shows a flowchart illustrating how to find a divisor d that works for Webster's method. The main difference when we use trial and error to implement Webster's method as opposed to Jefferson's method (compare Figs. 4-1 and 4-2) is the choice of the first guess for d. With Webster's method *the first guess should be the standard divisor.* Often (but not always) rounding the standard quotas to the nearest integer does work out. In such cases we have an easy apportionment using Webster's method. Unfortunately, there are times when the standard divisor SD doesn't quite do the job. In such cases we proceed with the trial-and-error approach until we find a divisor that works. (As with the other methods, a no-hassle implementation of the trial-and-error approach for Webster's method is available in the applet *Apportionment Methods*.)

4.5 The Huntington-Hill Method

Edward V. Huntington (1874–1952)

Until 1940, the method used to apportion the U.S. House of Representatives was chosen by Congress every 10 years, typically for political reasons and with little consideration given to the mathematical subtleties of apportionment or to consistency. Hamilton's method, Jefferson's method, and Webster's method were all used at some time or another. Finally, in 1941 Congress passed, and President Franklin D. Roosevelt signed, "An Act to Provide for Apportioning Representatives in Congress among the Several States by the Equal Proportions Method," generally known as *the 1941 apportionment act.* This act made two important changes in the process of apportioning the seats in the House of Representatives every 10 years: (1) the number of seats is permanently set at $M = 435$, and (2) the method of apportionment is permanently set to be the *Equal Proportions method*, more commonly known as the *Huntington-Hill method.* (The method was first proposed by the American statistician Joseph A. Hill and later revised by the mathematician Edward V. Huntington. Thus the name.)

If for no other reason than the fact that the Huntington-Hill method is the current method used to apportion the U.S. House of Representatives, we will discuss the method in some detail in this section. The Huntington-Hill method is a sophisticated variation of Webster's method, and in many cases it produces exactly the same apportionment as Webster's method. At the same time, the two methods are *not* equivalent, and in some cases they do produce different apportionments. We will give examples of both of these situations in this section.

But, first things first: The key to understanding the Huntington-Hill method is to understand how the quotas are rounded.

The Geometric Mean and The Huntington-Hill Rounding Rule

Under the Huntington-Hill method, quotas are rounded in a manner that is very similar to but not quite the same as the conventional rounding used in Webster's method. For example, a quota of 2.48 would be rounded down to 2 in Webster's method, but under the Huntington-Hill method 2.48 would be rounded up to 3. How come? Under the Huntington-Hill method, the *cutoff point* for rounding a quota down or up is a tad to the left of 0.5, and it varies with the size of the quota. For example, if the quota is between 2 and 3, the cutoff point is $c = \sqrt{6} \approx 2.4495$—quotas smaller than c get rounded down, quotas bigger than c get rounded up (see Fig. 4-3). Similarly, for a quota between 3 and 4 the cutoff point is $c = \sqrt{12} \approx 3.4641$.

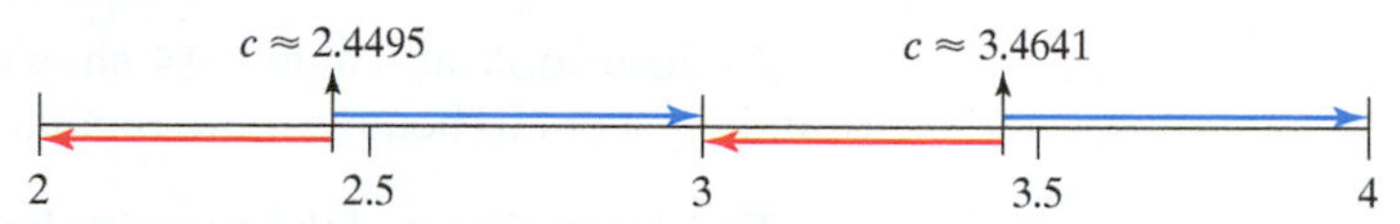

FIGURE 4-3 Quotas in the red zones are rounded down; quotas in the blue zones are rounded up.

To understand these unusual cutoff points for rounding we introduce the two key concepts in this section.

- **Geometric mean** The geometric mean G of two positive numbers a and b is the square root of their product: $G = \sqrt{ab}$. (*Note*: The geometric mean of a and b always falls somewhere between a and b.)
- **Huntington-Hill rounding rule** Given a quota q, the cutoff for rounding it is given by $c = \sqrt{LU}$ (i.e., the geometric mean of the lower quota L and the upper quota U). In other words, if $q < c$, then round q down to L, and if $q > c$ then round q up to U [Fig. 4-4(b)]. (There is no need to worry about the case $q = c$. This can never happen because c is always an irrational number and q is always a rational number.)

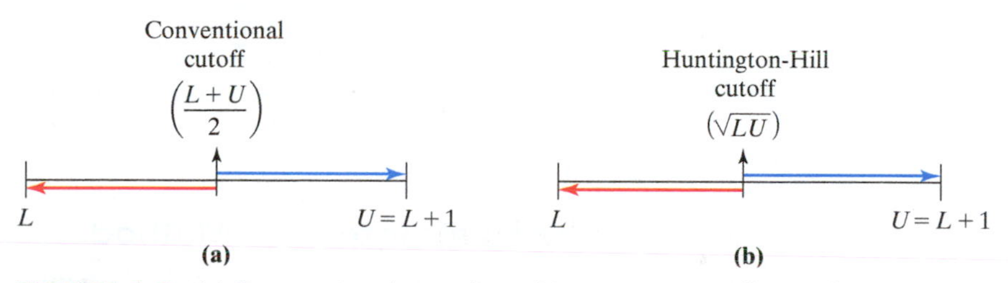

FIGURE 4-4 (a) Conventional rounding, (b) Huntington-Hill rounding rule.

The best way to understand how the Huntington-Hill rounding rule works is through an example.

EXAMPLE 4.9 HUNTINGTON-HILL ROUNDING

In each case we will round the quota q up or down using the Huntington-Hill rounding rule. [An (*) indicates when the rounding deviates from conventional rounding.] For convenience Table 4-14 shows cutoff points (and approximations to three decimal places) for all quotas between 1 and 10.

- $q = 1.51$. This quota gets rounded up to 2, just as it would under Webster's method.
- $q = 1.41$. This quota gets rounded down to 1, just as it would under Webster's method.
- $q = 1.42$. (*) This quota gets rounded up to 2, because it is above the cutoff point $c = \sqrt{2} \approx 1.414$.
- $q = 3.48$. (*) This quota gets rounded up to 4, because it is above the cutoff point $c = \sqrt{12} \approx 3.464$.
- $q = 5.48$. (*) This quota gets rounded up to 6, because it is above the cutoff point $c = \sqrt{30} \approx 5.477$. Notice that here we are just barely above the cutoff point, and we had to go to three decimal places to see the difference between the two numbers.

Quota (q)	Cutoff point c is
$1 < q < 2$	$\sqrt{2} \approx 1.414$
$2 < q < 3$	$\sqrt{6} \approx 2.449$
$3 < q < 4$	$\sqrt{12} \approx 3.464$
$4 < q < 5$	$\sqrt{20} \approx 4.472$
$5 < q < 6$	$\sqrt{30} \approx 5.477$
$6 < q < 7$	$\sqrt{42} \approx 6.481$
$7 < q < 8$	$\sqrt{56} \approx 7.483$
$8 < q < 9$	$\sqrt{72} \approx 8.485$
$9 < q < 10$	$\sqrt{90} \approx 9.487$

TABLE 4-14 ■ Cutoff points for Huntington-Hill rounding of quotas between 1 and 10

- $q = 6.48$. This quota gets rounded down to 6, just as under conventional rounding. It is below the cutoff point $c = \sqrt{42} \approx 6.481$.
- $q = 42.49$. Here we need to first calculate the cutoff point c. It is given by the geometric mean of 42 and 43: $c = \sqrt{(42)(43)} = \sqrt{1806} \approx 42.497$. Since q is below the cutoff point, it gets rounded down to 42.

A close look at Table 4-14 shows that the cutoff points for Huntington-Hill rounding seem to behave quite regularly. The following two facts are pretty useful:

- The decimal part of the cutoff point is always smaller than 0.5, but not by much—it always falls between 0.41 and 0.5.
- The bigger the quota, the closer the decimal part of the cutoff point is to 0.5 (and, thus, the more the rounding rule behaves like conventional rounding).

These two facts reinforce the observation that Huntington-Hill rounding is only a little bit "off center": Quotas with decimal parts above 0.5 always get rounded up, quotas with decimal parts below 0.41 always get rounded down. There is only a very narrow window to the left of 0.5 where Huntington-Hill rounding is going to be different from conventional rounding (Fig. 4-5).

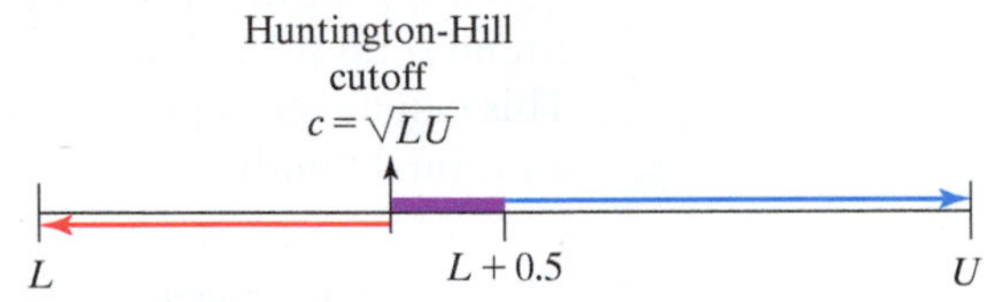

FIGURE 4-5 The Huntington-Hill rounding rule differs from conventional rounding only when the quota is in the purple zone.

The Huntington-Hill Method

Once we understand the Huntington-Hill rounding rule, the description of the Huntington-Hill method is straightforward. In fact, it is very similar to the description of Webster's method.

HUNTINGTON-HILL METHOD

- **Step 1.** Find a "suitable" divisor d. [Here a suitable divisor means a divisor that produces an apportionment of exactly M seats when the quotas (populations divided by d) are rounded using the *Huntington-Hill rounding rule*.]
- **Step 2.** Find the apportionment of each state by rounding its quota using the Huntington-Hill rounding rule.

EXAMPLE 4.10 PARADOR'S CONGRESS (HUNTINGTON-HILL METHOD)

We saw in Example 4.8 (Parador's Congress using Webster's method) that the standard divisor $SD = 50{,}000$ did not work, but that the modified divisor $d = 50{,}100$ did. The modified quotas for this divisor are shown in the third column of Table 4-15 (same quotas as in the fifth column of Table 4-13). We now need to round these quotas up or down using the Huntington-Hill rounding rules. The cutoff points for rounding are shown in the fourth column of Table 4-15, and, as is often the case, the rounding turns out to be exactly the same as under conventional rounding. The apportionment under Huntington-Hill method is shown in the last column of Table 4-15 and is exactly the same apportionment as under Webster's method (see Example 4.8).

State	Population	Modified quota	Cutoff point	Round quota to
A	1,646,000	32.85	32.496	33
B	6,936,000	138.44	138.499	138
C	154,000	3.07	3.464	3
D	2,091,000	41.74	41.497	42
E	685,000	13.67	13.491	14
F	988,000	19.72	19.494	20
Total	12,500,000			250

TABLE 4-15 ▪ Huntington-Hill method: Calculations for Example 4.10 with divisor $d = 50{,}100$

Example 4.10 illustrates a fairly common situation: the same divisors that work for Webster's method work for the Huntington-Hill method. In this case both methods produce exactly the same apportionment.

Our next two examples illustrate situations where the Huntington-Hill method and Webster's method produce different apportionments. The numbers have been carefully rigged to illustrate the mathematical subtlety of how this can happen.

EXAMPLE 4.11 *SD* WORKS FOR HUNTINGTON-HILL BUT NOT FOR WEBSTER

Parador's neighbor is the small republic of Guayuru. Guayuru consists of five states (A, B, C, D, and E for short). The populations of the five states are shown in the first row of Table 4-16. The Guayuru House of Representatives has a total of 40 seats that are to be apportioned to the states using the Huntington-Hill method. Our starting point is to compute the standard divisor and the standard quotas, just as we do with Webster's method. As luck would have it, the total population of Guayuru is 400,000, which makes the standard divisor $SD = 10{,}000$ and the calculation of the standard quotas particularly nice. The standard quotas are shown in the second row of Table 4-16. The third row of Table 4-16 shows the rounding of the standard quotas under the Huntington-Hill rounding rules. (Notice that both state A and state E get rounded up—the decimal parts of their quotas are under 0.5 but above their respective Huntington-Hill cutoff points). We got lucky once again and these numbers work! The total is 40, and we have an honest to goodness Huntington-Hill apportionment! On the other hand, when we use conventional rounding of the standard quotas, the numbers don't work (last row of Table 4-16). This apportionment is different from the one we would get using Webster's method (see Exercise 40).

State	*A*	*B*	*C*	*D*	*E*	Total
Population	34,800	104,800	64,800	140,800	54,800	400,000
Standard quota	3.48	10.48	6.48	14.08	5.48	40
Huntington-Hill rounding	4	10	6	14	6	40 ✓
Conventional rounding	3	10	6	14	5	38 ✗

TABLE 4-16 ▪ $SD = 10{,}000$ works for Huntington-Hill rounding but not for conventional rounding

EXAMPLE 4.12 ***SD* WORKS FOR WEBSTER BUT NOT FOR HUNTINGTON-HILL**

This example is a follow-up to Example 4.11. Following a new census we see a few small changes in the populations of the states, but the rest of the story is unchanged. The first row of Table 4-17 shows the new state populations: the populations of B and C have increased a little from what they were in Example 4.11, the population of D has decreased. The total population is still 400,000, there are still 40 seats to be apportioned, and the standard divisor is still $SD = 10{,}000$. The standard quotas are shown in the second row of Table 4-17. If we round the quotas using conventional rounding (third row of Table 4-17), we get a perfectly good apportionment. This would be the apportionment under Webster's method. Unfortunately, this is not the apportionment under the Huntington-Hill method. When we round the standard quotas using conventional rounding the Huntington-Hill rounding rules (fourth row of Table 4-17), states A and E get rounded up and the total number of seats is 42. This does not work, so we have to look for a different divisor. The divisor $d = 10{,}030$ brings all the quotas down a little (fifth row of Table 4-17). In particular, the quotas of B and E move to the left of their cutoff points and we get the Huntington-Hill apportionment shown in the last row of Table 4-17.

State	A	B	C	D	E	Total
Population	34,800	105,100	65,100	140,200	54,800	400,000
Standard quota	3.48	10.51	6.51	14.02	5.48	40
Conventional rounding	3	11	7	14	5	40 ✓
Huntington-Hill rounding	4	11	7	14	6	42 ✗
Modified quota ($d = 10{,}030$)	3.47	10.479	6.491	13.978	5.464	
Conventional rounding	3	10	6	14	5	38 ✗
Huntington-Hill rounding	4	10	7	14	5	40 ✓

TABLE 4-17 ■ Conventional and Huntington-Hill rounding for Example 4.12 using standard quotas ($SD = 10{,}000$) and modified quotas ($d = 10{,}030$)

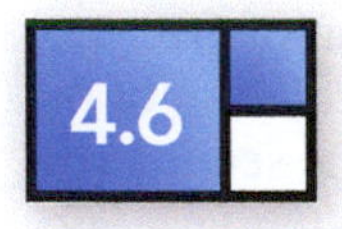

4.6 The Quota Rule and Apportionment Paradoxes

By and large, each of the apportionment methods we discussed in this chapter has both positive and negative features. Some methods are better than others, but none is flawless. In this section we will discuss the curious flaws that apportionment methods can exhibit and how they are reflected in the methods we studied.

The following discussion can be summarized by three key facts:

- Except for Hamilton's method, all of the other methods discussed in this chapter violate the *quota rule.*
- Hamilton's method can produce *apportionment paradoxes.*
- It is *impossible* for an apportionment method to both *satisfy the quota rule* and not produce *apportionment paradoxes.*

So, let's clarify what these facts really mean, starting with the **quota rule**.

- **The Quota rule.** Every state should be apportioned either its lower quota L or its upper quota U.

The basic idea behind the quota rule is that the standard quota is the true measure of a fair apportionment, and a fair apportionment method, therefore, should not differ from the standard quota by more than one seat. For example, if a state has a standard quota of 138.72 then it is reasonable to apportion to that state 139 or possibly 138 seats, but never 140 seats or 137 seats. (Here is another way you might think about it: If someone owes you $138.72 but only has dollar bills to pay you with, you would accept $139 and grudgingly $138, but you would not expect $140 or be happy with $137.)

Any apportionment method that might apportion more than U seats or less than L seats to any state (even if it happens just once) is said to *violate the quota rule*. There are two types of quota rule violations:

- **Upper-quota violations.** If an apportionment method apportions to a state more than U seats we say that it *violates the upper quota*. (For example, if the standard quota of a state is 138.72 and the state gets 140 seats, then we have an upper-quota violation.)
- **Lower-quota violations.** If an apportionment method apportions to a state less than L seats we say that it *violates the lower quota*. (For example, if the standard quota of a state is 138.72 and the state gets 137 seats, then we have a lower-quota violation.)

We will now consider how the various methods we learned in this chapter fare with respect to the quota rule.

- *Hamilton's method satisfies the quota rule.* The reason for this is that Hamilton's method starts with the standard quotas, rounds all of them down, and then gives surplus seats to some states and not to others. The states that get a surplus seat end up with their upper quota U and the states that don't end up with their lower quota L. In either case, the quota rule will be satisfied.
- *Jefferson's method violates the upper quota.* Go back and take another look at Example 4.6 Parador's Congress (Jefferson's method), where state B gets apportioned 140 seats even though its standard quota is 138.72. That one violation is enough—Jefferson's method *violates* the upper quota. Moreover, under Jefferson's method all quota violations are *upper-quota* violations [Exercise 71(a)], and they tend to happen with the larger states.
- *Adams's method violates the lower quota.* Now go take a look at Example 4.7 Parador's Congress (Adams's method). Here state B gets apportioned 137 seats, far short of its standard quota of 138.72. This is an egregious violation of the lower quota. Under Adams's method all quota violations are *lower-quota* violations [Exercise 71(b)], and they also tend to happen with the larger states.
- *Webster's method and the Huntington-Hill method violate the quota rule* (here the violations can be either upper- or lower-quota violations). Quota violations under either method are pretty rare, but they can happen (see Exercise 65).

Apportionment Paradoxes

Next, we will consider three important paradoxes that sometimes pop up in apportionment problems: the *Alabama paradox*, the *population paradox*, and the *new-states paradox*.

- **Alabama paradox.** This paradox occurs when a state gets apportioned fewer seats simply by virtue of the fact that the total number of seats being apportioned has increased.

  We already saw an example of the Alabama paradox in Example 4.5: Under Hamilton's method a fictitious state called Bama gets 10 seats when the total number of seats being apportioned is $M = 200$, but only 9 seats when the total number of seats is increased to $M = 201$.

  Of the apportionment methods we discussed in this chapter, only Hamilton's method can produce the quirky arithmetic behind the Alabama paradox: When we increase the number of seats to be apportioned, each state's standard quota goes up, but not by the same amount. This means that the order of the residues may get scrambled, moving some states ahead of others in the priority order for the surplus seats. This can result in some state or states losing seats they already had.

[**Historical note:** The origin of the name *Alabama paradox* can be traced to the events surrounding the 1882 apportionment of the U.S. House of Representatives. As different apportionment bills for the House were being debated, it came to light that in a House with 299 seats, Alabama's apportionment would be 8 seats, whereas in a House with 300 seats, Alabama's apportionment would be 7 seats. The effect of adding that extra seat would be to give one more seat to Texas and one more seat to Illinois. To do this, a seat would have to be taken away from Alabama!]

- **Population paradox.** This paradox occurs when a state gets apportioned fewer seats simply by virtue of the fact that its population has increased. More specifically, the population paradox represents situations in which state X loses a seat to state Y even though X grew in population at a higher rate than Y.

Our next example illustrates how the population paradox can occur under Hamilton's method.

EXAMPLE 4.13 A TALE OF TWO PLANETS

In the year 2525 the five planets in the Utopia galaxy finally signed a peace treaty and agreed to form an Intergalactic Federation governed by an Intergalactic Congress. This is the story of the two apportionments that broke up the Federation.

Part I. The Apportionment of 2525. The first Intergalactic Congress was apportioned using Hamilton's method, based on the population figures (in billions) shown in the second column of Table 4-18. (This is the apportionment problem introduced in Example 4.2.) There were 50 seats apportioned.

Planet	Population	Standard quota	Lower quota	Surplus	Apportionment
Alanos	150	8.333 . . .	8	0	8
Betta	**78**	4.333 . . .	4	0	**4**
Conii	173	9.6111 . . .	9	1	10
Dugos	204	11.333 . . .	11	0	11
Ellisium	**295**	16.3888 . . .	16	1	**17**
Total	900	50.00	48	2	50

Source: Intergalactic Census Bureau.

TABLE 4-18 ■ Intergalactic Congress: Apportionment of 2525

Let's go over the calculations. Since the total population of the galaxy is 900 billion, the standard divisor is $SD = \frac{900}{50} = 18$ billion. Dividing the planet populations by this standard divisor gives the standard quotas shown in the third column of Table 4-18. After the lower quotas are handed out (column 4), there are two surplus seats. The first surplus seat goes to Conii (with the highest decimal part of .6111...), and the other one to Ellisium (with the second highest decimal part of .3888...). The last column shows the apportionments. (Keep an eye on the apportionments of Betta and Ellisium—they are central to how this story unfolds.)

Part II. The Apportionment of 2535. After 10 years of peace, all was well in the Intergalactic Federation. The Intergalactic Census of 2535 showed only a few changes in the planets' populations—an 8 billion increase in the population of Conii, and a 1 billion increase in the population of Ellisium. The populations of the other planets remained unchanged from 2525. Nonetheless, a new apportionment was required. Table 4-19 shows the details of the 2535 apportionment under Hamilton's method. Notice that the total population increased to 909 billion, so the standard divisor for this apportionment was $SD = \frac{909}{50} = 18.18$. (You might want to check all the calculations in Table 4-19 for yourself.)

Planet	Population	Standard quota	Lower quota	Surplus	Apportionment
Alanos	150	8.25	8	0	8
Betta	**78**	4.29	4	1	**5**
Conii	181	9.96	9	1	10
Dugos	204	11.22	11	0	11
Ellisium	**296**	16.28	16	0	**16**
Total	909	50.00	48	2	50

Source: Intergalactic Census Bureau.

TABLE 4-19 ■ Intergalactic Congress: Apportionment of 2535

The one remarkable thing about the 2535 apportionment is that Ellisium went down from 17 to 16 seats as its *population went up* and Betta *gained that seat* even though its *population remained unchanged*! The epilogue to this story is that Ellisium used the 2535 apportionment as an excuse to invade Betta, and the years of peace and prosperity in Utopia were over. ■

- **New-states paradox.** This paradox occurs when a state's apportionment is affected simply by virtue of the fact that a new state, together with its fair share of seats, has been added to the apportionment calculations.

  [**Historical note:** In 1907 Oklahoma joined the Union. Prior to Oklahoma becoming a state, there were 386 seats in the House of Representatives. At the time the fair apportionment to Oklahoma was five seats, so the size of the House of Representatives was changed from 386 to 391. The point of adding these five seats was to give Oklahoma its fair share of seats and leave the apportionments of the other states unchanged. However, when the new apportionments were calculated, Maine's apportionment went up from three to four seats and New York's went down from 38 to 37 seats.]

Our last example gives a simple illustration of how the new-states paradox can occur under Hamilton's method. For a change of pace, we will discuss something other than legislatures.

EXAMPLE 4.14 GARBAGE TIME

The Metro Garbage Company has a contract to provide garbage collection and recycling services in two districts of Metropolis, Northtown (with 10,450 homes) and the much larger Southtown (89,550 homes). The company runs 100 garbage trucks, which are apportioned under Hamilton's method according to the number of homes in the district. A quick calculation shows that the standard divisor is $SD = 1000$ homes, a nice, round number which makes the rest of the calculations (shown in Table 4-20) easy. As a result of the apportionment, 10 garbage trucks are assigned to service Northtown and 90 garbage trucks to service Southtown.

District	Homes	Quota	Apportionment
Northtown	10,450	10.45	**10**
Southtown	89,550	89.55	**90**
Total	100,000	100.00	100

TABLE 4-20 ■ Metro garbage truck apportionments

Now imagine that the Metro Garbage Company is bidding to expand its territory by adding the district of

Newtown (5250 homes) to its service area. In its bid to the City Council the company promises to buy five additional garbage trucks for the Newtown run so that its service to the other two districts is not affected. But when the new calculations (shown in Table 4-21) are carried out, there is a surprise: One of the garbage trucks assigned to Southtown has to be reassigned to Northtown! (You should check these calculations for yourself. Notice that the standard divisor has gone up a little and is now approximately 1002.38.)

District	Homes	Quota	Apportionment
Northtown	10,450	10.43	**11**
Southtown	89,550	89.34	**89**
Newtown	5,250	5.24	5
Total	105,250	105.00	105

TABLE 4-21 ■ Revised Metro garbage truck apportionments

Michel Balinski [1933–]

H. Peyton Young [1945–]

Table 4-22 summarizes the five apportionment methods of this chapter in terms of apportionment paradoxes and the quota rule. Of the five methods, Webster's method and the Huntington-Hill method are in some sense the best because while in theory they can violate the quota rule, in practice this happens very rarely. From the purely mathematical point of view, the ideal apportionment method would be one that does *not violate the quota rule* and does *not produce any of the apportionment paradoxes.* In 1980 two mathematicians—Michel Balinski of the State University of New York at Stony Brook and H. Peyton Young of Johns Hopkins University—were able to prove that such an apportionment method is a mathematical impossibility: *Any apportionment method that satisfies the quota rule can produce paradoxes, and any apportionment method that does not produce* paradoxes violates the quota rule. This fact, known as **Balinski and Young's Impossibility Theorem**, forces us to accept the fact that all apportionment methods have one flaw or another.

Method	Quota Rule	Paradoxes		
		Alabama	Population	New-States
Hamilton	No violations	Yes	Yes	Yes
Jefferson	Upper-quota violations	No	No	No
Adams	Lower-quota violations	No	No	No
Webster	Lower- and upper-quota violations	No	No	No
Huntington-Hill	Lower- and upper-quota violations	No	No	No

TABLE 4-22 ■ Summary of five apportionment methods

Conclusion

Apportionment is a problem of surprising mathematical depth—to say nothing of its political and historical complexities. For a change of pace, in this conclusion we will take a quick look at the intriguing and convoluted history of the apportionment of the United States Congress (which essentially means the House of Representatives, since in the Senate every state gets two seats).

A Brief History of Apportionment in the United States

As mandated by the United States Constitution, apportionments of the House are to take place every 10 years, following each census of the population. The real problem was that the Constitution left the method of apportionment and the number of seats to be apportioned essentially up to Congress. The only two restrictions stated in the Constitution were that (1) "each State shall have at least one Representative," and (2) "The number of Representatives shall not exceed one for every thirty thousand" (Article I, Section 2).

Following the 1790 Census, and after considerable and heated debate, Congress passed the first "act of apportionment" in 1792. The bill, sponsored by Alexander Hamilton (then Secretary of the Treasury), established a House of Representatives with $M = 120$ seats and apportioned under the method we now call *Hamilton's method*. In April 1792, at the urging of Hamilton's archenemy and then Secretary of State Thomas Jefferson, President George Washington vetoed the bill. (Incidentally, this was the first presidential veto in U.S. history.) Jefferson convinced Washington to support a different apportionment bill, based on a House of Representatives with $M = 105$ seats and apportioned under the method we now call *Jefferson's method.* Jefferson made the case for the superiority of his method over Hamilton's based on unbiased mathematical arguments, but politics was involved as well—under Jefferson's method Virginia would get an extra seat. And it didn't hurt that both Jefferson and Washington were Virginians. Unable to override the president's veto and facing a damaging political stalemate, Congress finally adopted Jefferson's apportionment bill. This is how the first House of Representatives came to be constituted. (For more on this first apportionment, see Exercise 80.)

Jefferson's method remained in use for five decades, up to and including the apportionment of 1832. The great controversy during the 1832 apportionment debate centered on New York's apportionment. Under Jefferson's method, New York would get 40 seats even though its *standard quota* was only 38.59. This apportionment exposed, in a very dramatic way, the critical weakness of Jefferson's method—it produces *upper-quota violations*. Two alternative apportionment bills were considered during the 1832 apportionment debate, one proposed by John Quincy Adams that would apportion the House using the method we now call *Adams's method* and a second one, sponsored by Daniel Webster, that would do the same using the method we now call *Webster's method*. Both of these proposals were defeated and the original apportionment bill passed, but the 1832 apportionment was the last gasp for Jefferson's method.

Webster's method was adopted for the 1842 apportionment, but in 1852 Congress passed a law making Hamilton's method the "official" apportionment method for the House of Representatives. Since it is not unusual for Hamilton's method and Webster's method to produce exactly the same apportionment, an "unofficial" compromise was also adopted in 1852: Choose the number of seats in the House so that the apportionment is the same under either method. This compromise was used again for the apportionment act of 1862.

In 1872, as a result of a power grab among states, an apportionment bill was passed that can only be described as a total mess—it was based on no particular method and produced an apportionment that was inconsistent with both Hamilton's method and Webster's method. The apportionment of 1872 was in violation of both the Constitution (which requires that some method be used) and the 1852 law (which designated Hamilton's method as the method of choice).

In 1876 Rutherford B. Hayes defeated Samuel L. Tilden in one of the most controversial and disputed presidential elections in U.S. history. Hayes won in the Electoral College (despite having lost the popular vote) after Congress awarded him the disputed electoral votes from three southern states—Florida, Louisiana, and South Carolina. One of the many dark sidebars of the 1876 election was that, had the House of Representatives been legally apportioned, Tilden would have been the clear-cut winner in the Electoral College.

In 1882 the *Alabama paradox* first surfaced. In looking at possible apportionments for different House sizes, it was discovered that for a House with 299 seats Alabama would get 8 seats, but if the House size were increased to 300 seats Alabama's apportionment would decrease to 7 seats. So how did Congress deal with this disturbing discovery? It essentially glossed it over, choosing a House with $M = 325$ seats, a number for which Hamilton's method and Webster's method would give the same apportionment. The same strategy was adopted in the apportionment bill of 1892.

In 1901 the Alabama paradox finally caught up with Congress. When the Census Bureau presented to Congress tables showing the possible apportionments under Hamilton's method for all House sizes between 350 and 400 seats, it was pointed out that two states—Maine and Colorado—were impacted by the Alabama paradox: For most all House sizes between 350 and 400 Maine would get 4 seats, except for five sizes where Maine would end up with only 3 seats: 357, 382, 386, 389, and 390. Colorado would get 3 seats for all possible House sizes except 357, for which it would only get 2 seats. The 1901 apportionment bill presented to Congress was for a House with 357 seats—the exact sweet spot if you wanted to rob both Maine and Colorado of a seat! This was not a coincidence, and faster than you can say "we are being robbed," the debate in Congress escalated into a frenzy of name-calling and accusations, with the end result being that the bill was defeated and Hamilton's method was scratched for good. The final apportionment of 1901 used Webster's method and a House with $M = 386$ seats.

Webster's method remained in use for the apportionments of 1901, 1911, and 1931. (No apportionment bill was passed following the 1920 Census, in direct violation of the Constitution.)

In 1941 Congress passed a law that established a fixed size for the House of Representatives (435 seats) and a permanent method of apportionment known as the *method of equal proportions*, or the *Huntington-Hill method*. The 1941 law (*Public Law 291, H.R. 2665, 55 Stat 761: An Act to Provide for Apportioning Representatives in Congress among the Several States by the Equal Proportions Method*) represented a realization by Congress that politics should be taken out of the apportionment debate and that the apportionment of the House of Representatives should be purely a mathematical issue.

Are apportionment controversies then over? Not a chance. With a fixed-size House, one state's gain has to be another state's loss. In the 1990 apportionment Montana was facing the prospect of losing one of its two seats—seats it had held in the House for 80 years. Not liking the message, Montana tried to kill the messenger. In 1991 Montana filed a lawsuit in federal District Court (*Montana v. United States Department of Commerce*) in which it argued that the Huntington-Hill method is unconstitutional. A panel of three federal judges ruled by a 2-to-1 vote in favor of Montana. The case then went on appeal to the Supreme Court, which overturned the decision of the lower federal court and upheld the constitutionality of the Huntington-Hill method. So, until either Congress or a different Supreme Court say otherwise, the Huntington-Hill method remains the chosen method for apportioning the U.S. House of Representatives.

KEY CONCEPTS

4.1 Apportionment Problems and Apportionment Methods

- **apportionment problem:** any problem where M identical, indivisible objects (the "seats") must be divided among N parties (the "states") using some proportionality criterion (the "populations"), **103**
- **states:** a metaphor for the parties in an apportionment problem, **105**
- **seats:** a metaphor for the identical, indivisible objects being apportioned, **105**
- **apportionment method:** given M seats to be apportioned among N states, a method that guarantees a division of the M seats (no more and no less) to the N states based on their populations, **105**
- **standard divisor:** the ratio of total population to number of seats being apportioned (i.e., $SD = P/M$), **105**
- **standard quota (quota):** for a state with population p, the ratio p/SD; it represents what the fair apportionment to the state would be if seats were divisible into fractional parts, **105**
- **lower quota:** the largest integer smaller or equal to the quota, **106**
- **upper quota:** the smallest integer bigger or equal to the quota, **106**

4.2 Hamilton's Method

- **Hamilton's method:** uses the standard divisor and standard quotas; first gives each state its lower quota, then gives out surplus seats according to the "remainders," highest remainder first and so on, **107**

4.3 Jefferson's Method

- **Jefferson's method:** a divisor method that produces an apportionment by using modified quotas that are always rounded down, **111**
- **divisor (modified divisor):** a number d representing the yardstick for measuring seats in terms of populations (1 seat = d population members), **111**
- **divisor method:** any apportionment method that uses modified divisors and modified quotas for the apportionment calculations, **111**
- **modified quota:** a quota obtained using a modified divisor, **111**

4.4 Adams's and Webster's Methods

- **Adams's method:** a divisor method that produces an apportionment by using modified quotas that are always rounded up, **112**
- **Webster's method:** a divisor method that produces an apportionment by using modified quotas that are rounded using conventional rounding, **113**

4.5 The Huntington-Hill Method

- **geometric mean:** for two positive numbers, the square root of their product (i.e., the geometric mean of a and b is $\sqrt{ab}$), **115**
- **Huntington-Hill rounding rule:** given a quota q with lower quota L and upper quota U, q is rounded down to L if $q < \sqrt{LU}$, and rounded up to U if $q > \sqrt{LU}$, **115**

- **Huntington-Hill method:** a divisor method that produces an apportionment by using modified quotas that are rounded according to the Huntington-Hill rounding rule, **116**

4.6 The Quota Rule and Apportionment Paradoxes

- **quota rule:** every state should be apportioned either its lower quota L or its upper quota U, **118**
- **upper-quota violation:** an apportionment of seats to a state that is larger than the state's upper quota, **119**
- **lower-quota violation:** an apportionment of seats to a state that is smaller than the state's lower quota, **119**
- **Alabama paradox:** an apportionment paradox where a state may lose seats to another state merely because of an increase in the number of seats being apportioned, **119**
- **population paradox:** an apportionment paradox where a state may lose seats to another state merely because its population increased at a higher rate, **120**
- **new-states paradox:** an apportionment paradox where a state may lose seats to another state merely because of the introduction of a new state—together with its fair number of seats—into the apportionment calculations, **121**
- **Balinski and Young's Impossibility Theorem:** a perfect apportionment method (no violations of the quota rule and no apportionment paradoxes) is a mathematical impossibility, **122**

EXERCISES

WALKING

4.1 Apportionment Problems and Apportionment Methods

1. The Bandana Republic is a small country consisting of four states: Apure (population 3,310,000), Barinas (population 2,670,000), Carabobo (population 1,330,000), and Dolores (population 690,000). Suppose that there are $M = 160$ seats in the Bandana Congress, to be apportioned among the four states based on their respective populations.

   (a) Find the standard divisor.

   (b) Find each state's standard quota.

2. The Republic of Wadiya is a small country consisting of four provinces: A (population 4,360,000), B (population 2,280,000), C (population 729,000), and D (population 2,631,000). Suppose that there are $M = 200$ seats in the Wadiya Congress, to be apportioned among the four provinces based on their respective populations.

   (a) Find the standard divisor.

   (b) Find each province's standard quota.

3. The Scotia Metropolitan Area Rapid Transit Service (SMARTS) operates six bus routes (A, B, C, D, E, and F) and 125 buses. The number of buses apportioned to each route is based on the number of passengers riding that route. Table 4-23 shows the daily average ridership on each route.

   (a) Find the standard divisor.

   (b) Explain what the standard divisor represents in this problem.

   (c) Find the standard quotas (round your answers to three decimal places).

Route	A	B	C	D	E	F
Ridership	45,300	31,070	20,490	14,160	10,260	8,720

TABLE 4-23

4. University Hospital has five major units: Emergency Care (ECU), Intensive Care (ICU), Maternity (MU), Pediatrics (PU), and Surgery (SU). There are 250 nurses working in these five units and they are apportioned to the units based on the number of beds in each unit, shown in Table 4-24.

   (a) Find the standard divisor.

   (b) Explain what the standard divisor represents in this problem.

   (c) Find the standard quotas (rounded to three decimal places).

Unit	*ECU*	*ICU*	*MU*	*PU*	*SU*
Beds	21	19	35	30	25

TABLE 4-24

5. The Republic of Tropicana is a small country consisting of five states (*A*, *B*, *C*, *D*, and *E*). The total population of Tropicana is 27.4 million. According to the Tropicana constitution, the seats in the legislature are apportioned to the states according to their populations. Table 4-25 shows each state's standard quota:

   (a) Find the number of seats in the Tropicana legislature.

   (b) Find the standard divisor.

   (c) Find the population of each state.

State	*A*	*B*	*C*	*D*	*E*
Standard quota	41.2	31.9	24.8	22.6	16.5

TABLE 4-25

6. Tasmania State University is made up of five different schools: Agriculture, Business, Education, Humanities, and Science (*A*, *B*, *E*, *H*, and *S* for short). The total number of students at TSU is 12,500. The faculty positions at TSU are apportioned to the various schools based on the schools' respective enrollments. Table 4-26 shows each school's standard quota:

   (a) Find the number of faculty positions at TSU.

   (b) Find the standard divisor.

   (c) Find the number of students enrolled in each school.

School	*A*	*B*	*E*	*H*	*S*
Standard quota	32.92	15.24	41.62	21.32	138.90

TABLE 4-26

7. There are 435 seats in the U.S. House of Representatives. According to the 2010 U.S. Census, 8.14% of the U.S. population lived in Texas. Compute Texas's standard quota in 2010 (rounded to two decimal places).

8. There are 435 seats in the U.S. House of Representatives. According to the 2010 U.S. Census, 12.07% of the U.S. population lived in California. Compute California's standard quota in 2010 (rounded to two decimal places).

9. The Interplanetary Federation of Fraternia consists of six planets: Alpha Kappa, Beta Theta, Chi Omega, Delta Gamma, Epsilon Tau, and Phi Sigma (*A*, *B*, *C*, *D*, *E*, and *F* for short). The federation is governed by the Inter-Fraternia Congress, consisting of 200 seats apportioned among the planets according to their populations. Table 4-27 gives the planet populations as percentages of the total population of Fraternia:

   (a) Find the standard divisor (expressed as a percent of the total population).

   (b) Find the standard quota for each planet.

Planet	*A*	*B*	*C*	*D*	*E*	*F*
Population percentage	11.37	8.07	38.62	14.98	10.42	16.54

TABLE 4-27

10. The small island nation of Margarita is made up of four islands: Aleta, Bonita, Corona, and Doritos (*A*, *B*, *C*, and *D* for short). There are 125 seats in the Margarita Congress, which are apportioned among the islands according to their populations. Table 4-28 gives the island populations as percentages of the total population of Margarita:

    (a) Find the standard divisor (expressed as a percent of the total population).

    (b) Find the standard quota for each island.

Island	*A*	*B*	*C*	*D*
Population percentage	6.24	26.16	28.48	39.12

TABLE 4-28

4.2 Hamilton's Method

11. Find the apportionment under Hamilton's method of the Bandana Republic Congress described in Exercise 1.

12. Find the apportionment under Hamilton's method of the Wadiya Congress described in Exercise 2.

13. Find the apportionment under Hamilton's method of the SMARTS buses to the six bus routes described in Exercise 3.

14. Find the apportionment under Hamilton's method of the University Hospital nurses to the five units described in Exercise 4.

15. Find the apportionment under Hamilton's method of the Republic of Tropicana legislature described in Exercise 5.

16. Find the apportionment under Hamilton's method of the faculty at Tasmania State University to the five schools described in Exercise 6.

17. Find the apportionment under Hamilton's method of the Inter-Fraternia Congress described in Exercise 9.

18. Find the apportionment under Hamilton's method of the Margarita Congress described in Exercise 10.

19. Happy Rivers County consists of three towns: Dunes, Smithville, and Johnstown. Each year the social workers employed by the county are apportioned among the three towns based on the number of cases in each town over the previous calendar year. The number of cases in each town in 2016 is shown in Table 4-29.

(a) Suppose there are 24 social workers employed by the county. Use Hamilton's method to apportion the social workers to the towns based on the caseloads shown in Table 4-29.

(b) Suppose there are 25 social workers employed by the county. Use Hamilton's method to apportion the social workers to the towns based on the caseloads shown in Table 4-29.

(c) Compare your answers in (a) and (b). What is strange about the two apportionments?

Town	Dunes	Smithville	Johnstown
Number of cases	41	106	253

TABLE 4-29

20. Plainville Hospital has three wings (A, B, and C). The nurses in the hospital are assigned to the three wings based on the number of beds in each wing, shown in Table 4-30.

(a) Suppose there are 20 nurses working at the hospital. Use Hamilton's method to apportion the nurses to the wings based on Table 4-30.

(b) Suppose an additional nurse is hired at the hospital, bringing the total number of nurses to 21. Use Hamilton's method to apportion the nurses based on Table 4-30.

(c) Compare your answers in (a) and (b). What is strange about the two apportionments?

Wing	A	B	C
Number of beds	154	66	30

TABLE 4-30

4.3 Jefferson's Method

21. The small nation of Fireland is divided into four counties: Arcadia, Belarmine, Crowley, and Dandia. Fireland uses Jefferson's method to apportion the 100 seats in the Chamber of Deputies among the four counties. Table 4-31 shows the populations of the four counties after the most recent census.

(a) Find the standard divisor and the standard quotas for each county.

(b) Determine how many seats would be apportioned if each county was given its lower quota.

(c) Determine how many seats would be apportioned if the divisor $d = 197{,}000$ is used to compute the modified quotas and then all of them are rounded down.

(d) Determine how many seats would be apportioned if the divisor $d = 195{,}000$ is used to compute the modified quotas and then all of them are rounded down.

(e) Determine how many seats would be apportioned if the divisor $d = 195{,}800$ is used to compute the modified quotas and then all of them are rounded down.

(f) Determine how many seats would be apportioned if the divisor $d = 196{,}000$ is used to compute the modified quotas and then all of them are rounded down.

(g) Without doing any additional computations, find three different divisors that would work under Jefferson's method.

County	Arcadia	Belarmine	Crowley	Dandia
Population	4,500,000	4,900,000	3,900,000	6,700,000

TABLE 4-31

22. The Republic of Galatia is divided into four provinces: Anline, Brock, Clanwin, and Drundell. Galatia uses Jefferson's method to apportion the 50 seats in its House of Representatives among the four provinces. Table 4-32 shows the populations of the four provinces (in millions) after the most recent census.

(a) Find the standard divisor and the standard quotas for each province.

(b) Determine how many seats would be apportioned if each province was given its lower quota.

(c) Determine how many seats would be apportioned if the divisor $d = 500{,}000$ is used to compute the modified quotas and then all of them are rounded down.

(d) Determine how many seats would be apportioned if the divisor $d = 530{,}000$ is used to compute the modified quotas and then all of them are rounded down.

(e) Determine how many seats would be apportioned if the divisor $d = 520{,}000$ is used to compute the modified quotas and then all of them are rounded down.

(f) Determine how many seats would be apportioned if the divisor $d = 510{,}000$ is used to compute the modified quotas and then all of them are rounded down.

(g) Without doing any additional computations, find three different divisors that would work under Jefferson's method.

Province	Anline	Brock	Clanwin	Drundell
Population (in millions)	5.9	7.8	6.1	6.9

TABLE 4-32

23. Find the apportionment under Jefferson's method of the Bandana Republic Congress described in Exercise 1. (*Hint*: Look for suitable divisors in the interval 49,250 to 49,500.)

24. Find the apportionment under Jefferson's method of the Wadiya Congress described in Exercise 2. (*Hint*: Look for suitable divisors in the interval 49,400 to 49,600.)

25. Find the apportionment under Jefferson's method of the SMARTS buses to the six bus routes described in Exercise 3. (*Hint*: Look for suitable divisors in the interval 1000 to 1035.)

26. The small republic of Guayuru (see Example 4.11) consists of five states (A, B, C, D, and E for short). The populations of the five states are shown in Table 4-33. Find the apportionment under Jefferson's method of the $M = 40$ seats in the Guayuru House of Representatives. (*Hint*: Look for suitable divisors in the interval 9000 to 9500.)

State	A	B	C	D	E
Population	34,800	104,800	64,800	140,800	54,800

TABLE 4-33

27. Find the apportionment under Jefferson's method of the Republic of Tropicana legislature described in Exercise 5. (*Hint*: Look for suitable divisors in the interval 196,000 to 197,000.)

28. Find the apportionment under Jefferson's method of the faculty at Tasmania State University described in Exercise 6. (*Hint*: Look for suitable divisors in the interval 49.2 to 49.8.)

29. Find the apportionment under Jefferson's method of the Inter-Fraternia Congress described in Exercise 9. (*Hint*: Express the modified divisors in terms of percents of the total population and look for suitable divisors in the interval 0.49% to 0.5%.)

30. Find the apportionment under Jefferson's method of the Margarita Congress described in Exercise 10. (*Hint*: Express the modified divisors in terms of percents of the total population and look for suitable divisors in the interval 0.7% to 0.8%.)

4.4 Adams's and Webster's Methods

31. Find the apportionment under Adams's method of the Republic of Tropicana legislature described in Exercise 5. (*Hint*: Look for suitable divisors in the interval 205,000 to 206,000.)

32. Find the apportionment under Adams's method of the faculty at Tasmania State University described in Exercise 6. (*Hint*: Look for suitable divisors in the interval 50.6 to 51.0.)

33. Find the apportionment under Adams's method of the Inter-Fraternia Congress described in Exercise 9. (*Hint*: Express the modified divisors in terms of percents of the total population and look for suitable divisors in the interval 0.5% to 0.51%.)

34. Find the apportionment under Adams's method of the Margarita Congress described in Exercise 10. (*Hint*: Express the modified divisors in terms of percents of the total population and look for suitable divisors in the interval 0.81% to 0.82%.)

35. Find the apportionment under Webster's method of the Bandana Republic Congress described in Exercise 1.

36. Find the apportionment under Webster's method of the Wadiya Congress described in Exercise 2. (*Hint*: Look for suitable divisors in the interval 50,100 to 50,150.)

37. Find the apportionment under Webster's method of the Republic of Tropicana legislature described in Exercise 5.

38. Find the apportionment under Webster's method of the faculty at Tasmania State University described in Exercise 6.

39. Find the apportionment under Webster's method of the Inter-Fraternia Congress described in Exercise 9. (*Hint*: Express the modified divisors in terms of percents of the total population.)

40. Find the apportionment under Webster's method of the Guayuru House of Representatives described in Exercise 26 (also in Example 4.11).

4.5 The Huntington-Hill Method

41. Round each number using the Huntington-Hill rounding rules.
(a) 1.514 **(b)** 1.4 **(c)** 1.414
(d) 1.415 **(e)** 1.449

42. Round each number using the Huntington-Hill rounding rules.
(a) 8.585 **(b)** 8.485 **(c)** 8.484 **(d)** 8.486

43. In the 2010 apportionment of the U.S. House of Representatives, Rhode Island had a standard quota of 1.488879 and a modified quota of 1.485313. How many seats were apportioned to Rhode Island? Explain your answer.

44. In the 2010 apportionment of the U.S. House of Representatives, Missouri had a standard quota of 8.458641 and a modified quota of 8.483. How many seats were apportioned to Missouri? Explain your answer.

45. A small country consists of five states: A, B, C, D, and E. The standard quotas for each state are given in Table 4.34.
(a) Find the number of seats being apportioned.
(b) Find the apportionment under the Huntington-Hill method.

State	A	B	C	D	E
Standard quota	3.52	10.48	1.41	12.51	12.08

TABLE 4-34

46. A small country consists of five states: A, B, C, D, and E. The standard quotas for each state are given in Table 4.35.
(a) Find the number of seats being apportioned.
(b) Find the apportionment under the Huntington-Hill method.

State	A	B	C	D	E
Standard quota	25.49	14.52	8.48	30.71	20.8

TABLE 4-35

47. A small country consists of five states: A, B, C, D, and E. The standard quotas for each state are given in Table 4.36.

(a) Find the number of seats being apportioned.

(b) Find the apportionment under the Huntington-Hill method.

State	A	B	C	D	E
Standard quota	3.46	10.49	1.42	12.45	12.18

TABLE 4-36

48. A small country consists of five states: A, B, C, D, and E. The standard quotas for each state are given in Table 4.37.

(a) Find the number of seats being apportioned.

(b) Find the apportionment under the Huntington-Hill method.

State	A	B	C	D	E
Standard quota	25.496	14.491	8.486	30.449	21.078

TABLE 4-37

49. A country consists of six states, with the state's populations given in Table 4-38. The number of seats to be apportioned is $M = 200$.

(a) Find the apportionment under Webster's method.

(b) Find the apportionment under the Huntington-Hill method.

(c) Compare the apportionments found in (a) and (b).

State	A	B	C	D	E	F
Population	344,970	408,700	219,200	587,210	154,920	285,000

TABLE 4-38

50. A country consists of six states, with the state's populations given in Table 4-39. The number of seats to be apportioned is $M = 200$.

(a) Find the apportionment under Webster's method.

(b) Find the apportionment under the Huntington-Hill method.

(c) Compare the apportionments found in (a) and (b).

State	A	B	C	D	E	F
Population	344,970	204,950	515,100	84,860	154,960	695,160

TABLE 4-39

4.6 The Quota Rule and Apportionment Paradoxes

51. If the standard quota of state X is 35.41, then which of the following apportionments to state X is (or are) possible under Hamilton's method?

(a) 35.4 or 35.5

(b) any positive integer less than 36

(c) 35 or 36

(d) 35 only

(e) 36 only

52. If the standard quota of state Y is 78.24, then which of the following apportionments to state Y is (or are) possible under Hamilton's method?

(a) 78.2 or 78.3

(b) 78 or 79

(c) 78 only

(d) 79 only

(e) any positive integer less than 79

53. If the standard quota of state X is 35.41, then which of the following apportionments to state X is (or are) possible under Jefferson's method?

(a) 34, 35 or 36

(b) 34, 35, 36 or 37

(c) 35 only

(d) 36 only

(e) 35, 36 or 37

54. If the standard quota of state Y is 78.24, then which of the following apportionments to state Y is (or are) possible under Adams's method?

(a) 77, 78 or 79

(b) 77, 78, 79 or 80

(c) 78, 79, 80 or 81

(d) 79 only

(e) 78 only

55. If the standard quota of state X is 35.41, then which of the following are possible apportionments to state X under Webster's method?

(a) 35 or 36

(b) 34, 35, 36 or 37

(c) 33 or 34

(d) 37 or 38

(e) All of the above are possible

56. If the standard quota of state Y is 78.24, then which of the following are possible apportionments to state Y under Webster's method?

(a) 78 or 79

(b) 76 or 77

(c) 80 or 81

(d) 77, 78, 79 or 80

(e) All of the above are possible

57. At the time of the 2000 Census, California's standard quota was 52.45. Under Jefferson's method, California would get an apportionment of 55 seats. What does this say about Jefferson's method?

58. At the time of the 2000 Census, California's standard quota was 52.45. Under Adams's method, California would get an apportionment of 50 seats. What does this say about Adams's method?

59. This exercise refers to the apportionment of social workers in Happy Rivers County introduced in Exercise 19. The answers to parts (a) and (b) in Exercise 19 are an illustration of which paradox? Explain. [You will need to work out parts (a) and (b) of Exercise 19 if you haven't done so yet.]

60. This exercise refers to the apportionment of nurses to wings in Plainsville Hospital introduced in Exercise 20. The answers to parts (a) and (b) in Exercise 20 are an illustration of which paradox? [You will need to work out parts (a) and (b) of Exercise 20 if you haven't done so yet.]

Exercises 61 and 62 are based on the following story: Mom found an open box of her children's favorite candy bars. She decides to apportion the candy bars among her three youngest children according to the number of minutes each child spent doing homework during the week.

61. (a) Suppose that there were 11 candy bars in the box. Given that Bob did homework for a total of 54 minutes, Peter did homework for a total of 243 minutes, and Ron did homework for a total of 703 minutes, apportion the 11 candy bars among the children using Hamilton's method.

(b) Suppose that before mom hands out the candy bars, the children decide to spend a "little" extra time on homework. Bob puts in an extra 2 minutes (for a total of 56 minutes), Peter an extra 12 minutes (for a total of 255 minutes), and Ron an extra 86 minutes (for a total of 789 minutes). Using these new totals, apportion the 11 candy bars among the children using Hamilton's method.

(c) The results of (a) and (b) illustrate one of the paradoxes of Hamilton's method. Which one? Explain.

62. (a) Suppose that there were 10 candy bars in the box. Given that Bob did homework for a total of 54 minutes, Peter did homework for a total of 243 minutes, and Ron did homework for a total of 703 minutes, apportion the 10 candy bars among the children using Hamilton's method.

(b) Suppose that just before she hands out the candy bars, mom finds one extra candy bar. Using the same total minutes as in (a), apportion now the 11 candy bars among the children using Hamilton's method.

(c) The results of (a) and (b) illustrate one of the paradoxes of Hamilton's method. Which one? Explain.

63. This exercise comes in two parts. Read Part I and answer (a) and (b), then read Part II and answer (c) and (d).

Part I. The Intergalactic Federation consists of three sovereign planets: Aila, with a population of 5.2 million, Balin, with a population of 15.1 million, and Cona, with a population of 10.6 million. The Intergalactic Parliament has 50 seats that are apportioned among the three planets based on their populations.

(a) Find the standard divisor in the Intergalactic Parliament.

(b) Find the apportionment of the 50 seats to the three planets under Hamilton's method.

Part II. Based on the results of a referendum, the federation expands to include a fourth planet, Dent, with a population of 9.5 million. To account for the additional population the number of seats in the Intergalactic Parliament is increased by 15 to a total of 65. [9.5 million individuals represent approximately 15 seats based on the standard divisor found in (a).]

(c) Find the apportionment of the 65 seats to the four planets using Hamilton's method.

(d) Which paradox is illustrated by the results of (b) and (c)? Explain.

64. This exercise comes in two parts. Read Part I and answer (a) and (b), then read Part II and answer (c) and (d).

Part I. A catering company contracts to provide catering services to three schools: Alexdale, with 617 students, Bromville, with 1,292 students, and Canley, with 981 students. The 30 food-service workers employed by the catering company are apportioned among the schools based on student enrollments.

(a) Find the standard divisor, rounded to the nearest integer.

(b) Find the apportionment of the 30 workers to the three schools under Hamilton's method.

Part II. The catering company gets a contract to service one additional school—Dillwood, with 885 students. To account for the additional students, the company hires 9 additional food-service workers. [885 students represent approximately 9 workers based on the standard divisor found in (a).]

(c) Find the apportionment of the 39 workers to the four schools under Hamilton's method.

(d) Which paradox is illustrated by the results of (b) and (c)? Explain.

65. The small island nation of Margarita consists of four islands: Aleta, Bonita, Corona, and Doritos. The number of seats to be apportioned is $M = 100$. The population of each island is given in Table 4-40.

(a) Find the apportionment under Webster's method. (*Hint*: Look for suitable divisors in the interval 900 to 1100.)

(b) Find the apportionment under the Huntington-Hill method. [*Hint*: Try the divisor that worked in (a).]

(c) What do the answers in parts (a) and (b) say about Webster's method and the Huntington-Hill apportionment method?

State	*A*	*B*	*C*	*D*
Population	86,915	4,325	5,400	3,360

TABLE 4-40

JOGGING

66. Consider an apportionment problem with *N* states. The populations of the states are given by $p_1, p_2, \ldots, p_N$, and the standard quotas are $q_1, q_2, \ldots, q_N$, respectively. Describe in words what each of the following quantities represents.

(a) $q_1 + q_2 + \cdots + q_N$

(b) $\dfrac{p_1 + p_2 + \cdots + p_N}{q_1 + q_2 + \cdots + q_N}$

(c) $\left(\dfrac{p_N}{p_1 + p_2 + \cdots + p_N}\right) \times 100$

67. For an arbitrary state *X*, let *q* represent its standard quota and *s* represent the number of seats apportioned to *X* under some unspecified apportionment method. Interpret in words the meaning of each of the following mathematical statements.

(a) $s - q \geq 1$ **(b)** $q - s \geq 1$

(c) $|s - q| \leq 0.5$ **(d)** $0.5 < |s - q| < 1$

68. Consider the problem of apportioning $M = 3$ seats between two states, *A* and *B*, using Jefferson's method. Let p_A and p_B denote the populations of *A* and *B*, respectively. Show that if the apportionment under Jefferson's method gives all three seats to *A* and none to *B*, then more than 75% of the country's population must live in state *A*. (*Hint*: Show that a Jefferson apportionment of 3 and 0 seats implies that $p_A > 3p_B$.)

69. Consider the problem of apportioning $M = 3$ seats between two states, *A* and *B*, using Webster's method. Let p_A and p_B denote the populations of *A* and *B*, respectively. Show that if the apportionment under Webster's method gives all three seats to *A* and none to *B*, then more than $83\frac{1}{3}\%$ of the country's population must live in state *A*. (*Hint*: Show that a Webster apportionment of 3 and 0 seats implies that $p_A > 5p_B$.)

70. Consider the problem of apportioning *M* seats between two states, *A* and *B*. Let q_A and q_B denote the standard quotas of *A* and *B*, respectively, and assume that these quotas have decimal parts that are not equal to 0.5. Explain why in this case

(a) Hamilton's and Webster's methods must give the same apportionment.

(b) the Alabama or population paradoxes cannot occur under Hamilton's method. [*Hint*: Use the result of (a).]

(c) violations of the quota rule cannot occur under Webster's method. [*Hint*: Use the result of (a).]

71. (a) Explain why, when Jefferson's method is used, any violations of the quota rule must be upper-quota violations.

(b) Explain why, when Adams's method is used, any violations of the quota rule must be lower-quota violations.

(c) Explain why, in the case of an apportionment problem with two states, violations of the quota rule cannot occur under either Jefferson's or Adams's method. [*Hint*: Use the results of (a) and (b).]

72. Alternative version of Hamilton's method. Consider the following description of an apportionment method:

- **Step 1.** Find each state's standard quota.
- **Step 2.** Give each state (temporarily) its *upper quota* of seats. (You have now given away more seats than the number of seats available.)
- **Step 3.** Let *K* denote the number of extra seats you have given away in Step 2. Take away the *K* extra seats from the *K* states with the smallest fractional parts in their standard quotas.

Explain why this method produces exactly the same apportionment as Hamilton's method.

Lowndes's Method. *Exercises 73 and 74 refer to a variation of Hamilton's method known as Lowndes's method, first proposed in 1822 by South Carolina Representative William Lowndes. The basic difference between Hamilton's and Lowndes's methods is that in Lowndes's method, after each state is assigned the lower quota, the surplus seats are handed out in order of relative fractional parts. (The relative fractional part of a number is the fractional part divided by the integer part. For example, the* ***relative fractional part*** *of 41.82 is* $\frac{0.82}{41} = 0.02$*, and the relative fractional part of 3.08 is* $\frac{0.08}{3} = 0.027$*. Notice that while 41.82 would have priority over 3.08 under Hamilton's method, 3.08 has priority over 41.82 under Lowndes's method because 0.027 is greater than 0.02.)*

73. (a) Find the apportionment of Parador's Congress (Example 4.3) under Lowndes's method.

(b) Verify that the resulting apportionment is different from each of the apportionments found under the other methods discussed in the chapter. In particular, list which states do better under Lowndes's method than under Hamilton's method.

74. Consider an apportionment problem with two states, *A* and *B*. Suppose that state *A* has standard quota q_1 and state *B* has standard quota q_2, neither of which is a whole number.

(Of course, $q_1 + q_2 = M$ must be a whole number.) Let f_1 represent the fractional part of q_1 and f_2 the fractional part of q_2.

(a) Find values q_1 and q_2 such that Lowndes's method and Hamilton's method result in the same apportionment.

(b) Find values q_1 and q_2 such that Lowndes's method and Hamilton's method result in different apportionments.

(c) Write an inequality involving q_1, q_2, f_1, and f_2 that would guarantee that Lowndes' method and Hamilton's method result in different apportionments.

RUNNING

75. **The Hamilton-Jefferson hybrid method.** The Hamilton-Jefferson hybrid method starts by giving each state its lower quota (as per Hamilton's method) and then apportioning the surplus seats using Jefferson's method.

(a) Use the Hamilton-Jefferson hybrid method to apportion $M = 22$ seats among four states according to the following populations: A (population 18,000), B (population 18,179), C (population 40,950), and D (population 122,871).

(b) Explain why the Hamilton-Jefferson hybrid method can produce apportionments that are different from both Hamilton and Jefferson apportionments.

(c) Explain why the Hamilton-Jefferson hybrid method can violate the quota rule.

76. Explain why Jefferson's method cannot produce

(a) the Alabama paradox

(b) the new-states paradox

77. Explain why Adams's method cannot produce

(a) the Alabama paradox

(b) the new-states paradox

78. Explain why Webster's method cannot produce

(a) the Alabama paradox

(b) the new-states paradox

APPLET BYTES MyMathLab®

These Applet Bytes are short projects or mini-explorations built around the applet **Apportionment Methods**, available in MyMathLab in the Multimedia Library or Tools for Success. This applet allows the user to create an apportionment problem by entering as inputs the number of states, the number of seats that are to be apportioned and the respective states' populations. The applet then gives the user the tools to compute the apportionments under each of the four methods discussed in the chapter and to compare the results. The applet carries out all the necessary calculations, freeing the user to explore more interesting and meaningful questions.*

79. Consider the apportionment of Parador's Congress discussed extensively in this chapter (Examples 4.4, 4.6, 4.7, 4.8, and 4.10). Use the applet to:

(a) Verify that all integers between 49,401 and 49,542 work as suitable modified divisors for Jefferson's method, and that no other integers do. [See Example 4.6.] (*Hint*: You don't have to try each and every one of the integers in the range.)

(b) Verify that all integers between 50,628 and 50,999 work as suitable modified divisors for Adams' method, and that no other integers do. [See Example 4.7.]

(c) Find the range of integers that work as suitable modified divisors for Webster's method. [See Example 4.8.]

(d) Find the range of integers that work as suitable modified divisors for the Huntington-Hill method. [See Example 4.10.]

(e) Find all integers that work as suitable modified divisors for the Huntington-Hill method but **do not** work for Webster's method.

(f) Explain why some integers work as suitable divisors for both the Huntington-Hill method and Webster's method while some work only for the Huntington-Hill method. (*Hint*: Choose an integer that works for both methods and an integer that only works for Huntington-Hill and compare what happens with each when you use the two methods.)

80. **The First Apportionment of the U.S. House of Representatives.** The first mathematical apportionment of the U.S. House of Representatives was done in 1792 based on the population figures obtained in the 1790 Census. The population figures are shown in Table 4.41. The number of seats to be apportioned was $M = 105$. The two methods being considered were Hamilton's method and Jefferson's method. Find the apportionments under both of these methods and compare their differences.

State	Population
Connecticut	236,841
Delaware	55,540
Georgia	70,835
Kentucky	68,705
Maryland	278,514
Massachusetts	475,327
New Hampshire	141,822
New Jersey	179,570
New York	331,589
North Carolina	353,523
Pennsylvania	432,879
Rhode Island	68,446
South Carolina	206,236
Vermont	85,533
Virginia	630,560
Total	3,615,920

TABLE 4-41

* MyMathLab code required.

In Applet Bytes 81 and 82 you are asked to create examples of apportionment problems on your own. The idea here is to experiment with the applet and play around with the numbers until you find what you want.

81. **(a)** Use the applet to find an example where Hamilton's method and Webster's method produce the same apportionment.

(b) Use the applet to find an example where Hamilton's method and Webster's method produce different apportionments. (Yes, the Parador Congress example does this but you need to come up with your own example).

82. Use the applet to find an example of an apportionment problem where there is both an upper quota violation under Jefferson's method and a lower quota violation under Adams' method. (Once again, the Parador Congress example fits the bill here but you need to come up with your own example).

PART 2

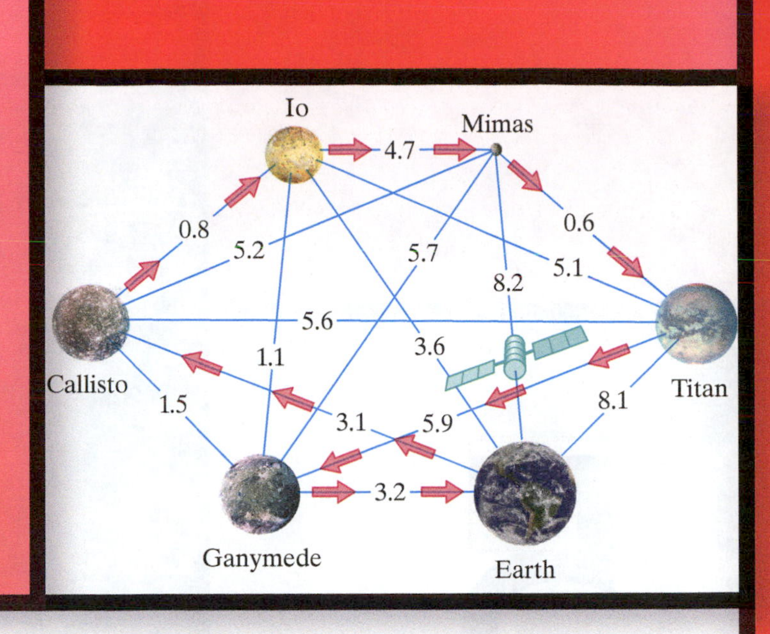

Management Science

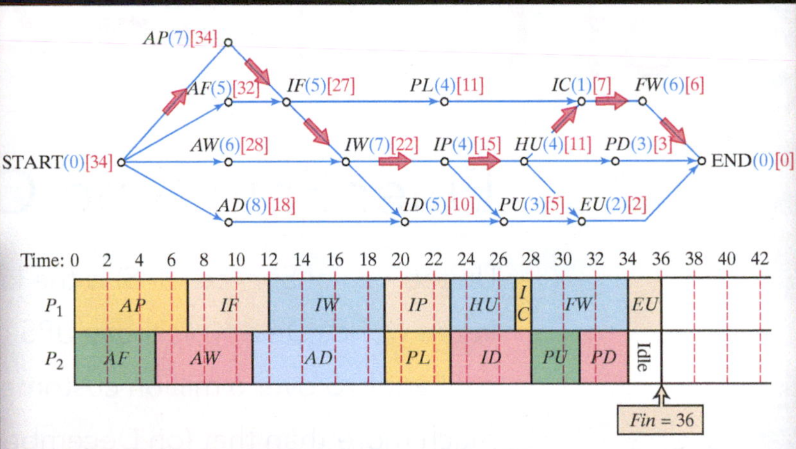

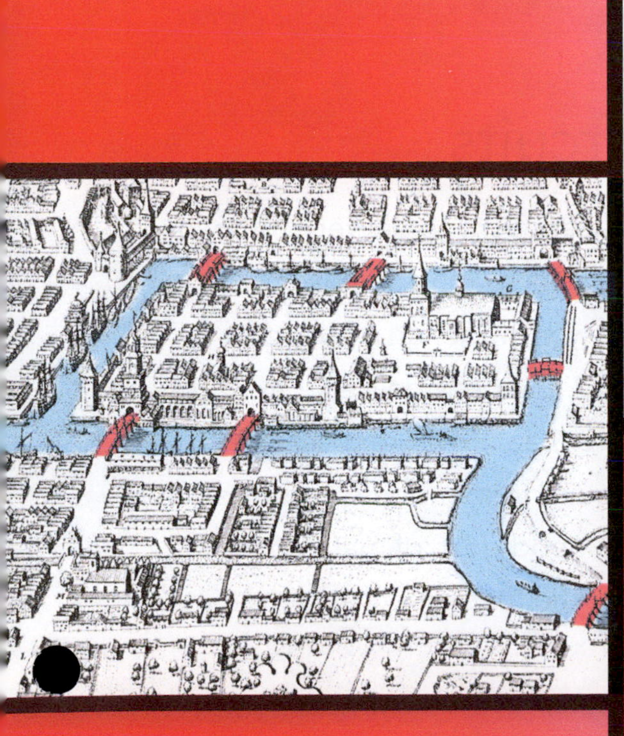

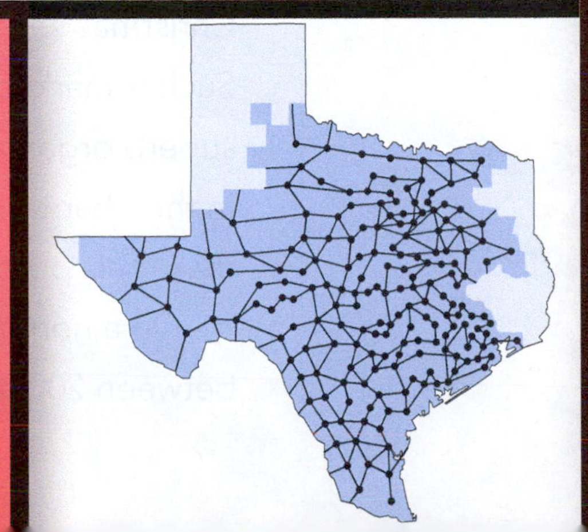

5

It takes more than a strong back to keep those UPS deliveries flowing. For more on the routing of UPS trucks, see Examples 5.3 and 5.24.

The Mathematics of Getting Around

Euler Paths and Circuits

United Parcel Service (UPS) is the largest package delivery company in the world. On a typical day UPS delivers roughly 18 million packages to over 6 million customers worldwide; on a busy day much more than that (on December 22, the peak delivery day of the Christmas season, UPS delivered more than 36 million packages). Such remarkable feats of logistics require tremendous resources, superb organization, and (surprise!) a good dose of mathematics. In this chapter we will discuss some of the mathematical ideas that make this possible.

On a normal day a single UPS driver delivers somewhere between 200 and 500 packages. In rural areas where there may be considerable distances between delivery points, the number is closer to 200; in highly populated urban areas where the delivery points are close to each other the number is closer to 500. But in both cases one of the keys to success in delivering all the packages is the *efficiency* of the delivery route. Efficient routing means, among other things, keeping *deadheading* (the term used to describe retracing parts of the route) at a minimum, and this is where the mathematics comes in: How do you design a route that minimizes the total amount of "wasted" travel?

Problems like those faced by a UPS driver (or a FedEx driver, or a mail carrier for that matter) trying to minimize the total length of his or her route are known as *street-routing problems*, and they have applications to other types of situations such as routing garbage trucks, security patrols, tourist buses, and even late-night pizza deliveries.

Section 5.1 introduces the concept of a *street-routing problem* and shows examples in several different settings and applications. Keep in mind that many of the examples in this section are scaled down in size and scope to keep things simple—in real life the same application will occur on a much larger scale. Regardless of the scale, the mathematical theory behind street-routing problems is one and the same, and we owe much of this theory to the genius of one man—the Swiss mathematician Leonhard Euler. [Euler's role was so significant that in this chapter we will see his last name—pronounced "oiler," by the way—used both as a noun (*Euler circuit*, *Euler path*) and as a verb (*eulerizing*, *semi-eulerizing*).] Section 5.2 introduces *graphs*—the key mathematical tool that allows us to tackle street-routing problems—and some of the concepts and terminology associated with graphs. Section 5.3 gives the mathematical infrastructure needed to solve street-routing problems, consisting of three key facts (*Euler's Circuit Theorem*, *Euler's Path Theorem*, and the *Sum of Degrees Theorem*) and an algorithm known as *Fleury's Algorithm*. In Section 5.4 we will learn how to combine all the preceding ideas to develop the strategies needed to solve street-routing problems in general.

5.1 Street-Routing Problems

We will start this section with a brief discussion of routing problems. What is a **routing problem**? To put it in the most general way, routing problems are concerned with finding ways to route the delivery of *goods* and/or *services* to an assortment of *destinations*. The goods or services in question could be packages, mail, newspapers, pizzas, garbage collection, bus service, and so on. The delivery destinations could be homes, warehouses, distribution centers, terminals, and the like.

There are two basic questions that we are typically interested in when dealing with a routing problem. The first is called the *existence* question. The existence question is simple: Is an actual route possible? For most routing problems, the existence question is easy to answer, and the answer takes the form of a simple yes or no. When the answer to the existence question is yes, then a second question—the *optimization question*—comes into play. Of all the possible routes, which one is the *optimal route*? (*Optimal* here means "the best" when measured against some predetermined variable such as *cost*, *distance*, or *time*.) In most management science problems, the optimization question is where the action is.

In this chapter we will learn how to answer both the existence and optimization questions for a special class of routing problems known as **street-routing problems**.

The common thread in all street-routing problems is what we might call, for lack of a better term, the *sweep-through requirement*—the requirement that the route must sweep at least once through some specified set of connections. In other words, in a street-routing problem the sweep-through requirement means that each street (or bridge, or lane, or highway) within a defined area (be it a town, an area of town, or a subdivision) must be covered by the route. We will refer to these types of routes as *exhaustive routes*. The most common services that typically require exhaustive routing are mail delivery, police patrols, garbage collection, street sweeping, and snow removal. More exotic examples can be census taking, precinct walking, electric meter reading, routing parades, tour buses, and so on.

To clarify some of the ideas we will introduce several examples of street-routing problems (just the problems for now—their solutions will come later in the chapter).

EXAMPLE 5.1 THE SECURITY GUARD PROBLEM

After a rash of burglaries, a private security guard is hired to patrol the streets of the Sunnyside neighborhood shown in Fig. 5-1. The security guard's assignment is to sweep, on foot, the entire neighborhood. Obviously, he doesn't want to walk any more than what is necessary. His starting point is the corner of Elm and J streets across from the school (*S* in Fig. 5-1)—that's where he usually parks his car. (This is relevant because at the end of his patrol he needs to come back to *S* to pick up his car.) Being a practical person, the security guard would like the answers to the following questions:

1. Is it possible to start and end at *S*, cover every block of the neighborhood, and pass through each block *just once*?
2. If some of the blocks will have to be covered more than once, what is an *optimal* route that covers the entire neighborhood? ("Optimal" here means "with the minimal amount of walking.")
3. Can a better route (i.e., less walking) be found by choosing a different starting and ending point? We will answer all of these questions in Section 5.4.

FIGURE 5-1 The Sunnyside neighborhood.

EXAMPLE 5.2 THE MAIL CARRIER PROBLEM

A mail carrier has to deliver mail in the same Sunnyside neighborhood (Fig. 5-1). The difference between the mail carrier's route and the security guard's route is that the mail carrier must make *two* passes through blocks with buildings on both sides of the street and only one pass through blocks with buildings on only one side of the street (and where there are no buildings on either side of the street, the mail carrier does not have to walk at all). In addition, the mail carrier has no choice as to her starting and ending points—she has to start and end her route at the local post office (*P* in Fig. 5-1). Much like the security guard, the mail carrier wants to find the optimal route that would allow her to sweep the neighborhood with the least amount of walking. (Put yourself in her shoes and you would do the same—good weather or bad, she walks this route 300 days a year!)

EXAMPLE 5.3 THE UPS DRIVER PROBLEM

Now we consider the case of a UPS driver who must deliver packages around the Sunnyside neighborhood. The red crosses in Fig. 5-2 indicate the locations (homes or businesses) where packages are to be delivered (it's the week before Christmas so there is an unusually large number of packages to be delivered).

FIGURE 5-2 UPS delivery locations marked with a red cross.

Unlike the mail carrier (required to sweep through every block of the neighborhood where there are homes or businesses), the UPS driver has to sweep only through those blocks where there are red crosses, and only once through such blocks (if the delivery is on the opposite side of the street he just crosses the street on foot). In addition, because of other deliveries, the UPS driver must enter the neighborhood through Fir St. (shown on the right side of Fig. 5-2) and exit the neighborhood through First Ave. (upper left of Fig. 5-2). Once again, we want to determine (and we will in Section 5.4) the most *efficient* route that will allow the UPS driver to deliver all those packages. The requirements for the route are (a) enter and exit the neighborhood where indicated in Fig. 5.2 and (b) sweep through every block where there are delivery locations marked with red crosses. [*Note:* In general, UPS drivers are mostly concerned with the total time it takes to complete their package deliveries rather than the total distance traveled, and often go out of their way to avoid left-turn signals or streets with a lot of traffic. This makes the routing problem more complicated; therefore, to keep things simple we will assume that in this example driving time is proportional to distance traveled (this happens, for example, when there is little or no traffic or when traffic moves evenly throughout the neighborhood). Under this assumption the most efficient route is still the shortest route.]

Leonhard Euler (1707–1783)

Our next example is primarily of historical interest. In the early 1700s, while visiting the medieval town of Königsberg (now in Russia but at the time part of Prussia), Leonhard Euler was introduced to the Königsberg bridges puzzle—a recreational puzzle that might potentially be solved using mathematical ideas, but nobody was clear as to what kind of mathematics was needed. Euler's great contribution was in developing a new mathematical theory (now known as *graph theory*) that could be used to solve the bridges puzzle as well as much more practical and complex problems. Euler was just 28 years old at the time. We will be introduced to Euler's ideas in the next three sections.

EXAMPLE 5.4 THE KÖNIGSBERG BRIDGES PUZZLE

Figure 5-3 shows the layout of the town of Königsberg in the 1700s, with the river Pregel running through town, the two islands on the river, and the north and south banks all connected by the seven bridges shown in red. (*Note*: Present-day Königsberg has a different layout, as two of the bridges no longer exist.) A little

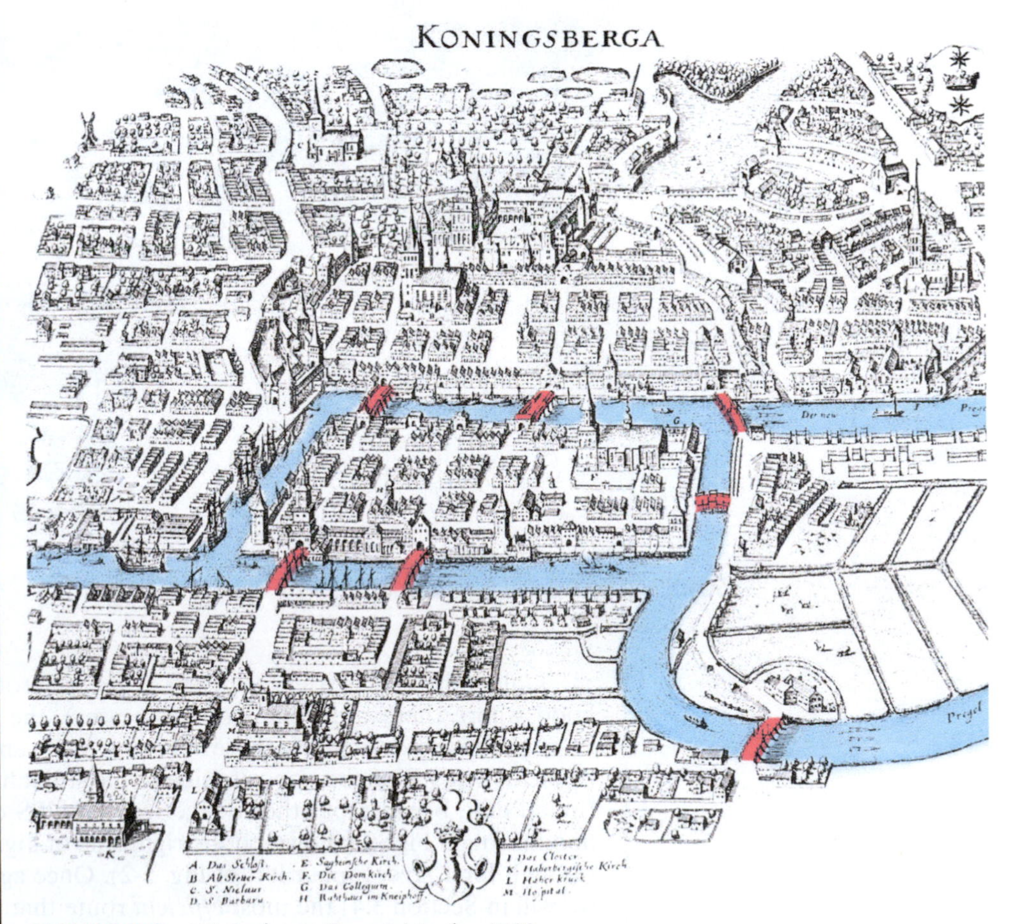

FIGURE 5-3 Königsberg (Prussia) in the 1700s.

game played by the locals at the time was to try to take a walk around town fully crossing every bridge once and only once (i.e., no bridge could be skipped and no bridge could be crossed twice). Nobody was able to do this successfully and a widely held belief in the town was that such a walk was indeed impossible. What was asked of Euler was to rigorously prove this. He did this, and much more. We will soon learn how he did it.

If we think of the bridges as playing a role analogous to that of the streets in the Sunnyside neighborhood, the Königsberg bridges puzzle becomes another street-routing problem. The solution—which Euler was able to demonstrate with a simple mathematical argument—is that *there is no route* that satisfies the requirements of the puzzle (pass through each bridge once and only once).

Our next example is an expanded and modernized version of the Königsberg bridges problem.

EXAMPLE 5.5 THE BRIDGES OF MADISON COUNTY

Madison County is a quaint old place, famous for its beautiful bridges. The Madison River runs through the county, and there are four islands (A, B, C, and D) and 11 bridges joining the islands to both banks of the river (R and L) and one another (Fig. 5-4). A famous photographer is hired to take pictures of each of the 11 bridges for a national magazine. The photographer needs to drive across each bridge once for the photo shoot. The problem is that there is a $25 toll (the locals call it a "maintenance tax") every time an out-of-town visitor drives across a bridge, and the photographer wants to minimize the total cost of the trip. The street-routing problem here is to find a route that passes through each bridge at least once and recrosses as few bridges as possible. Moreover, the photographer can start the route on either bank of the river and, likewise, end it on either bank of the river.

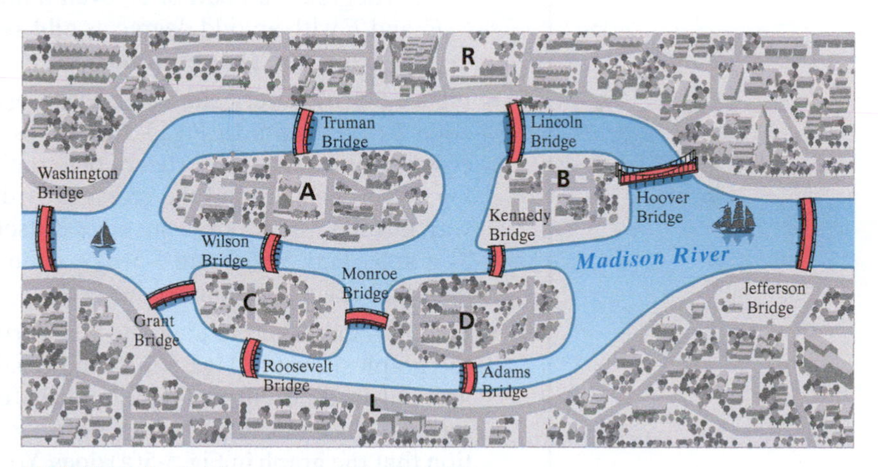

FIGURE 5-4 Bridges on the Madison River.

5.2 An Introduction to Graphs

The key tool we will use to tackle the street-routing problems introduced in Section 5.1 is the notion of a **graph**. (*Note*: The *graphs* we will be discussing here have no relation to the graphs of functions you may have studied in algebra or calculus.) The most common way to describe a *graph* is by means of a picture. The basic elements of such a picture are a set of "dots" called the **vertices** of the graph and a collection of "lines" called the **edges** of the graph. (Unfortunately, this terminology is not universal. In some applications the word "nodes" is used for the vertices and the word "links" is used for the edges. We will stick to vertices and edges as much as possible.) On the surface, that's all there is to it—edges connecting vertices. Below the surface there is a surprisingly rich theory. Let's look at a few examples first.

EXAMPLE 5.6 BASIC GRAPH CONCEPTS

Figure 5-5 shows several examples of graphs. We will discuss each separately.

- Figure 5-5(a) shows a graph with six vertices labeled *A, B, C, D, E,* and *F* (it is customary to use capital letters to label the vertices of a graph). For convenience we refer to the set of vertices of a graph as the **vertex set**. In this graph, the vertex set is $\{A, B, C, D, E, F\}$. The graph has 11 edges (described by listing, in any order, the two vertices that are connected by the edge): AB, AD, BC, etc.
    - When two vertices are connected by an edge we say that they are **adjacent vertices**. Thus, A and B are adjacent vertices, but A and E are not adjacent. The edge connecting B with itself is written as BB and is called a **loop**. Vertices C and D are connected twice (i.e., by two separate edges), so when we list the edges we include CD twice. Similarly, vertices E and F are connected by three edges, so we list EF three times. We refer to edges that appear more than once as **multiple edges**.
    - The complete list of edges of the graph, the **edge list**, is $AB, AD, BB, BC, BE, CD, CD, DE, EF, EF, EF$.
    - The number of edges that meet at each vertex is called the **degree** of the vertex and is denoted by $\deg(X)$. In this graph we have $\deg(A) = 2$, $\deg(B) = 5$ (please note that the loop contributes 2 to the degree of the vertex), $\deg(C) = 3$, $\deg(D) = 4$, $\deg(E) = 5$, and $\deg(F) = 3$. It will be important in the next section to distinguish between vertices depending on whether their degree is an odd or an even number. We will refer to vertices like B, C, E, and F with an odd degree as **odd vertices** and to vertices with an even degree like A and D as **even vertices**.
- Figure 5-5(b) is very similar to Fig. 5-5(a)—the only difference is the way the edge BE is drawn. In Fig. 5-5(a) edges AD and BE cross each other, but the crossing point is not a vertex of the graph—it's just an irrelevant crossing point. Fig. 5-5(b) gets around the crossing by drawing the edge in a more convoluted way, but the way we draw an edge is itself irrelevant. The key point here is that as graphs, Figs. 5-5(a) and 5-5(b) are the same. Both have exactly the same vertices and exactly the same edge list.
- Figure 5-5(c) take the idea one step further—it is in fact, another rendering of the graph shown in Figs. 5-5(a) and (b). The vertices have been moved around and put in different positions, and the edges are funky—no other way to describe it. Despite all the funkiness, this graph conveys exactly the same information that the graph in Fig. 5-5(a) does. You can check it out— same set of vertices and same edge list. The moral here is that while graphs are indeed pictures connecting "dots" with "lines," it is not the specific picture that matters but the story that the picture tells: which dots are connected to each other and which aren't. We can move the vertices around, and we can draw the edges any funky way we want (straight line, curved line, wavy line, etc.)—none of that matters. The only thing that matters is the set of vertices and the list of edges.
- Figure 5-5(d) shows a graph with six vertices. Vertices $A, B, C, D,$ and E form what is known as a **clique**—each vertex is connected to each of the other four. Vertex F, on the other hand, is connected to only one other vertex. This graph has no loops or multiple edges. Graphs without loops or multiple edges are called **simple graphs**. There are many applications of graphs where loops and multiple edges cannot occur, and we have to deal only with simple graphs. (In Examples 5.7 and 5.8 we will see two applications where only simple graphs occur.)

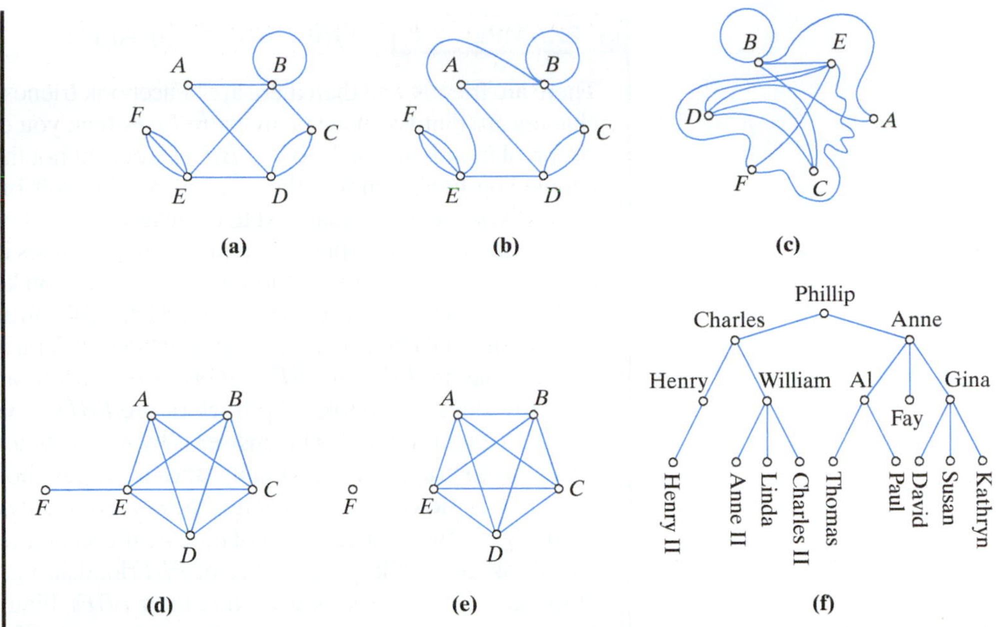

FIGURE 5-5 (a), (b), and (c) are all pictures of the same graph; (d) is a simple, connected graph; (e) is a simple, disconnected graph; and (f) is a graph labeled with names instead of letters.

- Figure 5-5(e) shows a graph very similar to the one in Fig. 5-5(d). The only difference between the two is the absence of the edge *EF*. In this graph there are no edges connecting *F* to any other vertex. For obvious reasons, *F* is called an **isolated** vertex. This graph is made up of two separate and disconnected "pieces"—the clique formed by the vertices *A*, *B*, *C*, *D*, and *E* and the isolated vertex *F*. Because the graph is not made of a single "piece," we say that the graph is *disconnected*, and the separate pieces that make up the graph are called the *components* of the graph. In contrast, a graph consisting of a single "piece" (i.e. component) is *connected*.
- Figure 5-5(f) shows a connected simple graph. The vertices of this graph are names (there is no rule about what the labels of a vertex can be). Can you guess what this graph might possibly represent?

EXAMPLE 5.7 AIRLINE ROUTE MAPS

Figure 5-6 shows the route map for a very small airline called Air Rarotonga. Air Rarotonga serves just 10 islands in the South Pacific, and the route map shows the non-stop flights that are available between the various islands. In essence, the route map is a graph whose vertices are the islands. An edge connects two islands if there is a direct flight between them. No direct flight, no edge. The picture makes a slight attempt to respect geographical facts (the bigger islands are drawn larger but certainly not to scale, and they sit in the ocean more or less as shown), but the point of an airline route map is to show if there is a direct flight from point *X* to point *Y*, and, in that regard, accurate geography is not all that important.

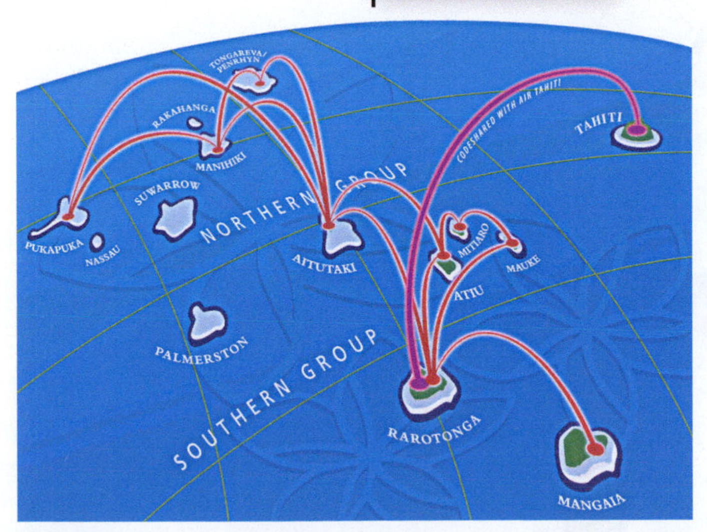

FIGURE 5-6 Air Rarotonga route map.

EXAMPLE 5.8 FRIENDSHIP GRAPHS

There are friends, and then there are "Facebook friends" (*FBF*s). If you are on Facebook, chances are that you have many more *FBF*s than you do real friends. Of course, most of your real friends are likely to be *FBF*s as well, but not the other way around. *FBF*s may be people you hardly know—the sister of cousin Joe's brother-in-law, your college roommate's ex, someone you sat next to on an airplane. All it takes to strike up a *FBF*riendship is an invite and an acceptance—and Facebook makes it a point to facilitate the development of *FBF*riendships with those notorious "Do you know Joe Schmoe?" emails. And if that is not enough, just click on the "Find Friends" tab and you will be provided with a list of new potential *FBF*s under the heading "People You May Know." Typically, the people on these lists are *FBF*s of *FBF*s, *FBF*s of *FBF*s' *FBF*s, and so on.

How does Facebook keep track of the *FBF*s of your *FBF*s, and the *FBF*s of your *FBF*s' *FBF*s, and so on? The answer is by means of a **friendship graph**. The vertices of a friendship graph are individuals, and the edges show friendships among pairs of individuals: If the vertices are friends they are joined by an edge, if they aren't then there is no edge between them. And of course, the same idea can be used for *FBF*riendship. Facebook keeps a large database of *FBF*riendship graphs connecting clusters of tens of thousands of Facebook users and their *FBF*s. Figure 5-7 shows a close-up view of a small cluster of individuals, part of a much larger *FBF*riendship graph.

Even in this small cluster, a careful look reveals a great deal of useful information about the complex web of relations within the group. Take *A* and *B*. We can tell right away that they are not *FBF*'s, as there is no edge directly connecting them. A little closer look also tells us that they have some common *FBF*s (*C*, *D*, and *E*). They will surely show up in each other's "People You May Know" lists, and soon enough a new *FBF*riendship is bound to be born ("Kaching!"—add another edge to the *FBF*riendship graph). Now look at *F* and *G*. They are not *FBF*s and have no common *FBF*s, so maybe not. But wait a second—isn't *G* a *FBF* of a *FBF* of *F*'s? Two degrees of separation is not a problem for Facebook. There is a good chance *F* and *G* will show up in each other's "People You May Know" lists and encouraged to become *FBF*s themselves. After all, *FBF*riendship is cheap. And if things don't work out, a bad *FBF*riendship—unlike a bad friendship—is easy to break off. Just click on the "Unfriend" button and delete the edge from the graph.

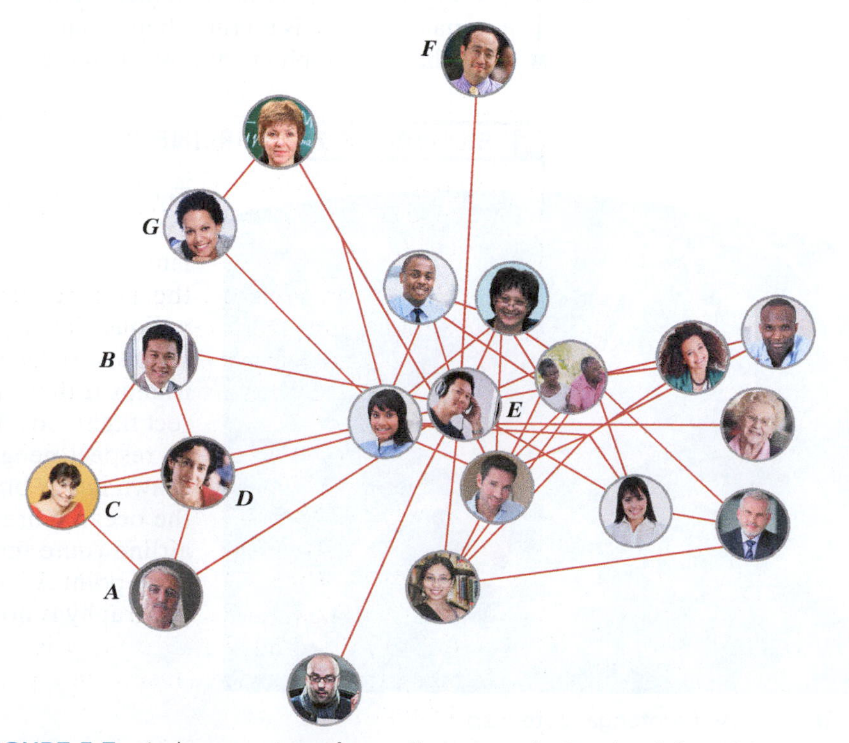

FIGURE 5-7 A close-up view of a small section of a Facebook friendship graph.

The main point of Examples 5.7 and 5.8 is to highlight how powerful (and useful) the concept of a graph can be. Granted, the Facebook friendship graph in Fig. 5-7 and the Air Rarotonga route map in Fig. 5-6 are small (just big enough to make the point), but if you think big you can imagine a United Airlines route map instead, with hundreds of destinations and thousands of flights connecting them. The fundamental idea is still the same—graphs convey visually a tremendous amount of information that would be hard to convey in any other form. Can you imagine describing the complex web of relationships in a friendship graph or in a United Airlines route map any other way?

EXAMPLE 5.9 PATHS AND CIRCUITS

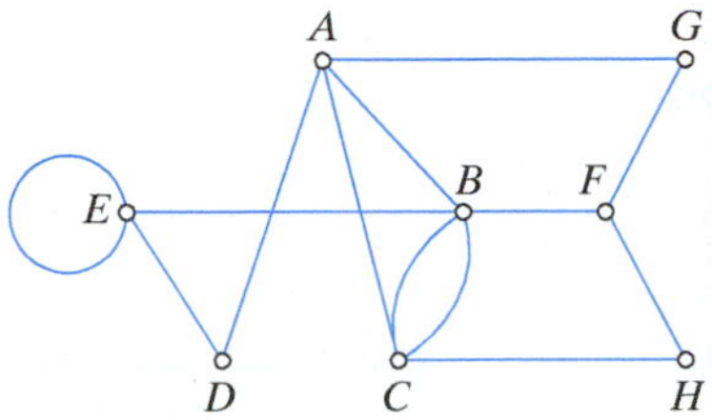

FIGURE 5-8 Graph for Example 5.9.

We say that two edges are **adjacent edges** when they share a common vertex. In Fig. 5-8 for example, AB is adjacent to AC and AD (they share vertex A), as well as to BC, BF, and BE (they share vertex B). A sequence of *distinct* edges each adjacent to the next is called a **path**, and the number of edges in the path is called the **length** of the path. For example in Fig. 5-8, the edges AB, BF, and FG form a path of length 3. A good way to think of a path is as a real-world path—a way to "hike" along the edges of the graph, traveling along the first edge, then the next, and so on. To shorten the notation, we describe the path by just listing the vertices in sequence separated by commas. For example A, B, F, G describes the path formed by the edges AB, BF, and FG.

Here are a few more examples of paths in Fig. 5-8:

- A, B is a path of length 1. Any edge can be thought of as a path of length 1—not very interesting, but it allows us to apply the concept of a path even to single edges.
- A, B, C, A, D, E is a path of length 5 starting at A and ending at E. The path goes through vertex A a second time, but that's OK. It is permissible for a path to revisit some of the vertices. On the other hand, A, C, B, A, C does not meet the definition of a path because the edges of the path have to be distinct (i.e., cannot be revisited) and here AC is traveled twice. So, in a path it's OK to revisit some of the vertices but not OK to revisit any edges.
- A, B, C, A, D, E, E, B is a path of length 7. Notice that this "trip" is possible because of the loop at E.
- A, B, C, A, D, E, B, C is also a legal path of length 7. Here we can use the edge BC twice because there are in fact two distinct edges connecting B and C.

When a trip along the edges of the graph closes back on itself (i.e., starts and ends with the same vertex) we specifically call it a **circuit** rather than a path. Thus, we will restrict the term *path* to open-ended trips and the word *circuit* to closed trips.

Here are a few examples of circuits in Fig. 5-8:

- A, D, E, B, A is a circuit of length 4. Even though it appears like the circuit designates A as the starting (and ending) vertex, a circuit is independent of where we designate the start. In other words, the same circuit can be written as D, E, B, A, D or E, B, A, D, E, etc. They are all the same circuit, but we have to choose one (arbitrary) vertex to start the list.
- B, C, B is a circuit of length 2. This is possible because of the double edge BC. On the other hand, B, A, B is not a circuit because the edge AB is being traveled twice. (Just as in a path, the edges of a circuit have to be distinct.)
- E, E is a circuit of length 1. A loop is the only way to have a circuit of length 1.

In Example 5.9 we saw several examples of paths (and circuits) that are part of the graph in Fig. 5-8, but the important idea we will discuss next in this: Can the path (or circuit) be the entire graph, not just a part of it? In other words, we want to consider the possibility of a path (or a circuit) that *exhausts* all the edges of the graph.

An **Euler path** is a path that covers *all* the edges of the graph. Likewise, an **Euler circuit** is a circuit that covers all the edges of the graph. In other words, we have an Euler path (or circuit) when the entire graph can be written as a path (or circuit).

EXAMPLE 5.10 EULER PATHS AND EULER CIRCUITS

When it comes to having Euler paths or Euler circuits, three scenarios are possible in a connected graph. Figures 5-9, 5-10, and 5-11 illustrate the three possibilities:

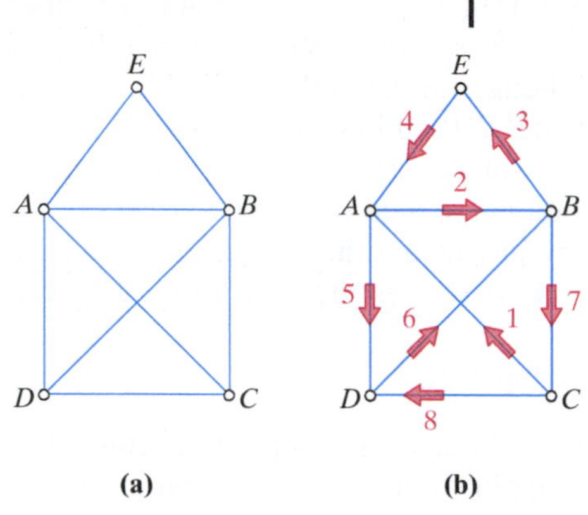

FIGURE 5-9 An Euler path starting at *C* and ending at *D*.

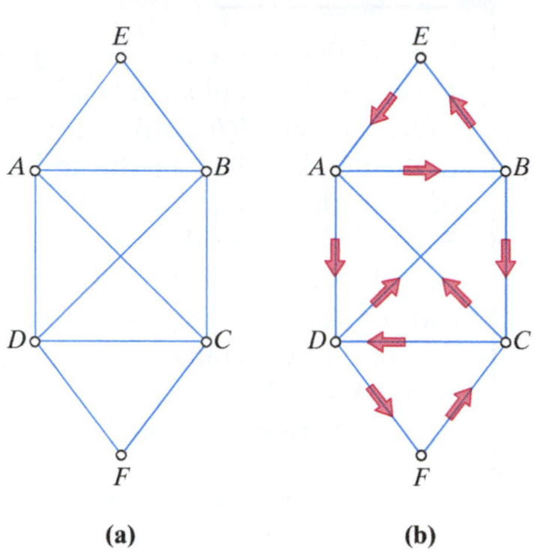

FIGURE 5-10 An Euler circuit.

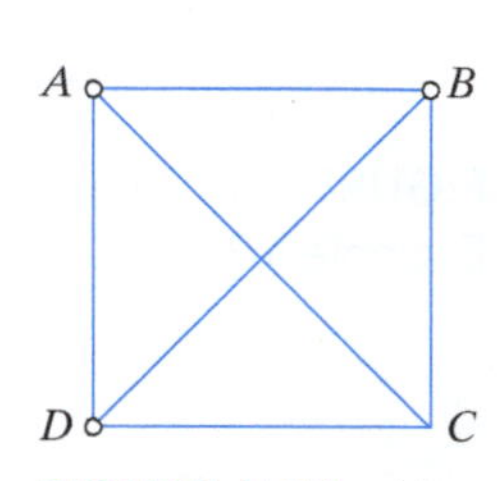

FIGURE 5-11 No Euler path or circuit.

- The graph in Fig. 5-9(a) has an Euler path—in fact, it has several. One of the possible Euler paths is shown in Fig. 5-9(b). The path starts at C and ends at D—just follow the arrows and you will be able to "trace" the edges of the graph without retracing any (just like in elementary school).
- The graph in Fig. 5-10(a) has many possible Euler circuits. One of them is shown in Fig. 5-10(b). Just follow the arrows. Unlike the Euler path in Fig. 5-9(b), the arrows are not numbered. You can start this circuit at any vertex of the graph, follow the arrows, and you will return to the starting vertex having covered all the edges once.
- The graph in Fig. 5-11 has neither an Euler path nor an Euler circuit. That's the way it goes sometimes—some graphs just don't have it!

We introduced the idea of a *connected* or *disconnected* graph in Example 5.6. Formally, we say that a graph is **connected** if you can get from any vertex to any other vertex along some path of the graph. Informally, this says that you can get from any point to any other point by "hiking" along the edges of the graph. Even more informally, it means that the graph is made of one "piece." A graph that is not connected is called **disconnected** and consists of at least two (maybe more) separate "pieces" we call the **components** of the graph.

EXAMPLE 5.11 BRIDGES

Figure 5-12 shows three different graphs. The graph in Fig. 5-12(a) is connected. The graph in Fig. 5-12(b) is disconnected and has two components: *A*, *B*, *C*, *D*, and *E* in one component and *F*, *G*, and *H* in the second component. The graph in Fig. 5-12(c)

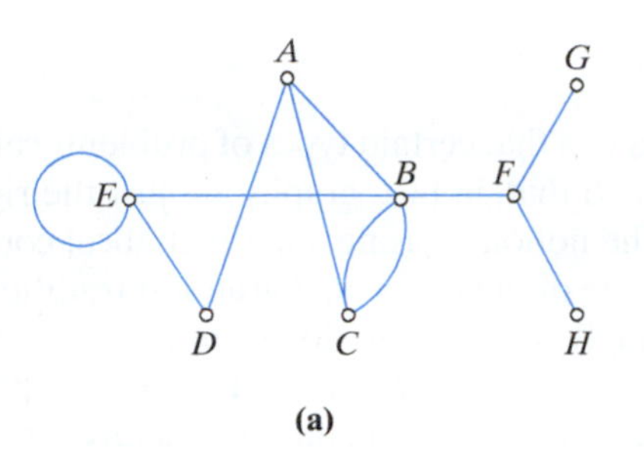

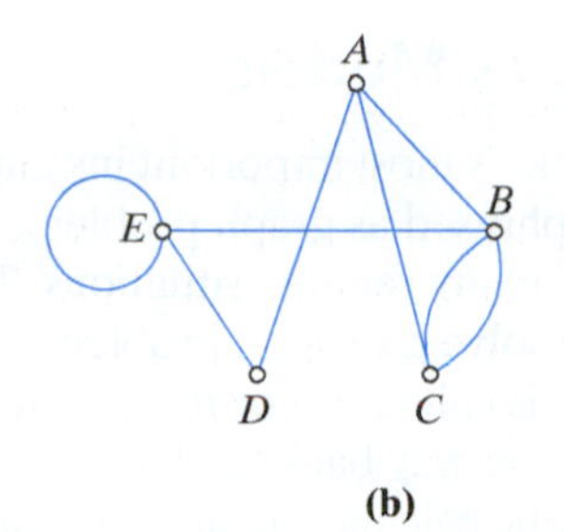

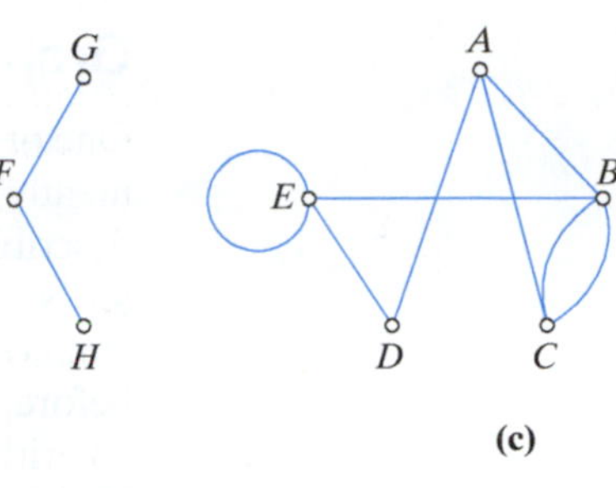

FIGURE 5-12 (a) A connected graph, (b) two components, and (c) three components.

is disconnected and has three components: A, B, C, D, and E in one component, F and H in the second component, and the isolated vertex G by itself as the third component.

Notice that the only difference between the *disconnected* graph in Fig. 5-12(b) and the *connected* graph in Fig. 5-12(a) is the edge BF. Think of BF as a "bridge" that connects the two components shown in Fig. 5-12(b). Not surprisingly, we call such an edge a *bridge*. A **bridge** in a connected graph is an edge that keeps the graph connected—if the bridge were not there, the graph would be disconnected. The graph in Fig. 5-12(a) has three bridges: BF, FG, and FH.

For the reader's convenience, Table 5-1 shows a summary of the basic graph concepts we have seen so far.

Vertices

- **adjacent:** any two vertices connected by an edge
- **vertex set:** the set of vertices in a graph
- **degree:** number of edges meeting at the vertex
- **odd (even):** degree is an odd (even) number
- **isolated:** no edges connecting the vertex (i.e., degree is 0)

Edges

- **adjacent:** two edges that share a vertex
- **loop:** an edge that connects a vertex with itself
- **multiple edges:** more than one edge connecting the same two vertices
- **edge list:** a list of all the edges in a graph
- **bridge:** an edge in a connected graph without which the graph would be disconnected

Paths and circuits

- **path:** a sequence of edges each adjacent to the next, with no edge included more than once, and starting and ending at different vertices
- **circuit:** same as a path, but starting and ending at the same vertex
- **Euler path:** a path that covers all the edges of the graph
- **Euler circuit:** a circuit that covers all the edges of the graph
- **length:** number of edges in a path or a circuit

Graphs

- **simple:** a graph with no loops or multiple edges
- **connected:** there is a path going from any vertex to any other vertex
- **disconnected:** not connected; consisting of two or more components
- **clique:** a set of completely interconnected vertices in the graph (every vertex is connected to every other vertex by an edge)

TABLE 5-1 ■ Glossary of basic graph concepts

Graphs as Models

One of Euler's most important insights was that certain types of problems can be conveniently rephrased as graph problems and that, in fact, graphs are just the right tool for describing many real-life situations. The notion of using a mathematical concept to describe and solve a real-life problem is one of the oldest and grandest traditions in mathematics. It is called *modeling*. Unwittingly, we have all done simple forms of modeling before, all the way back to elementary school. Every time we turn a word problem into an arithmetic calculation, an algebraic equation, or a geometric picture, we are modeling. We can now add to our repertoire one more tool for modeling: graph models.

In the next set of examples we are going to illustrate how we can use graphs to *model* some of the street-routing problems introduced in Section 5.1.

EXAMPLE 5.12 MODELING THE BRIDGES OF KÖNIGSBERG PUZZLE

The Königsberg bridges puzzle introduced in Example 5.4 asked whether it was possible to take a stroll through the old city of Königsberg and cross each of the seven bridges once and only once. Figure 5-13 shows the evolution of a graph model that we can use to answer this question. Figure 5-13(a) shows the original map of the city. Figure 5-13(b) is a "leaner" version of the map, with lots of obviously irrelevant details removed. A little further reflection should convince us that many details in Fig. 5-13(b) are still irrelevant to the question. The shape and size of the islands, the width of the river, the lengths of the bridges—none of these things really matter. So, then, what does matter? Surprisingly little. *The only thing that truly matters to the solution of this problem is the relationship between land masses (islands and banks) and bridges: Which land masses are connected to each other and by how many bridges?* This information is captured by the red edges in Fig. 5-13(c). Thus, when we strip the map of all its superfluous information, we end up with the graph model shown in Fig. 5-13(d). The four vertices of the graph represent each of the four land masses; the edges represent the seven bridges. In this graph an *Euler circuit* would represent a stroll around the town that crosses each bridge once and ends back at the starting point; an *Euler path* would represent a stroll that crosses each bridge once but does not return to the starting point.

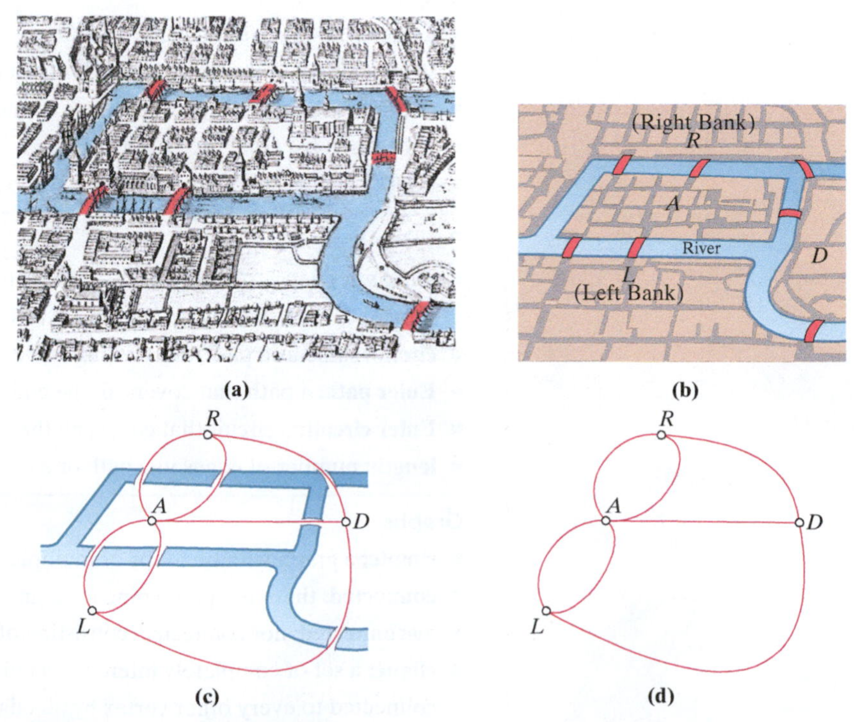

FIGURE 5-13 (a) Königsberg map, (b) a leaner version, (c) an even leaner version, and (d) the graph model.

As big moments go this one may not seem like much, but Euler's idea to turn a puzzle about walking across bridges in a quaint medieval city into an abstract question about graphs was a "eureka" moment in the history of mathematics.

EXAMPLE 5.13 MODELING THE SECURITY GUARD PROBLEM

In Example 5.1 we were introduced to the problem of the security guard who needs to patrol, on foot, the streets of the Sunnyside neighborhood [Fig. 5-14(a)]. The graph in Fig. 5-14(b)—where each edge represents a block of the neighborhood and each vertex an intersection—is a graph model of this problem. The questions raised in Example 5.1 can now be formulated in the language of graphs.

1. Does the graph in Fig. 5-14(b) have an Euler circuit that starts and ends at *S*?
2. What is the fewest number of edges that have to be added to the graph so that there is an Euler circuit?

We will learn how to answer such questions in the next couple of sections.

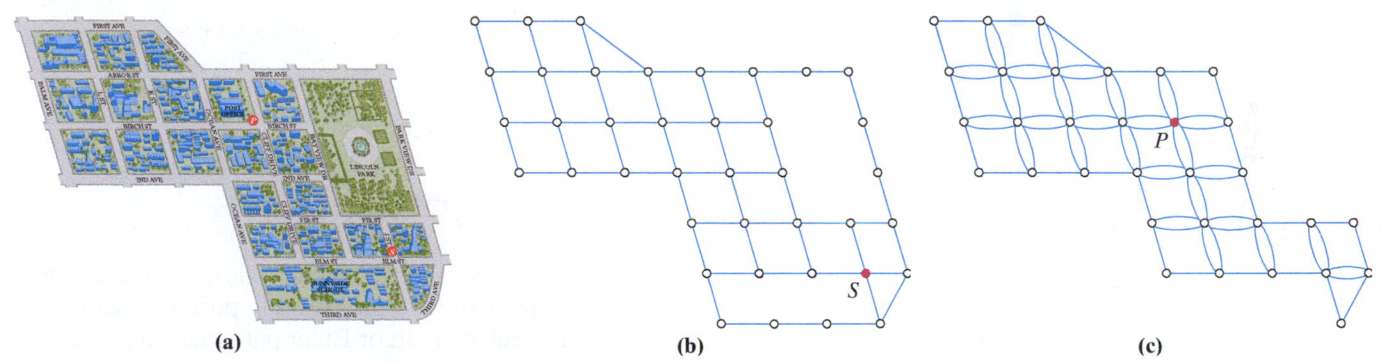

FIGURE 5-14 (a) The Sunnyside neighborhood. (b) A graph model for the security guard. (c) A graph model for the mail carrier.

EXAMPLE 5.14 MODELING THE MAIL CARRIER PROBLEM

Unlike the security guard, the mail carrier in Example 5.2 must make two passes through every block that has homes on both sides of the street (she has to physically place the mail in the mailboxes), must make one pass through blocks that have homes on only one side of the street, and does not have to walk along blocks where there are no houses. In this situation an appropriate graph model requires two edges on the blocks that have homes on both sides of the street, one edge for the blocks that have homes on only one side of the street, and no edges for blocks having no homes on either side of the street. The graph that models this situation is shown in Fig. 5-14(c).

EXAMPLE 5.15 MODELING THE UPS DRIVER PROBLEM

In Example 5.3 we discussed the UPS driver street-routing problem. The circumstances for the UPS driver are slightly different than those of the mail carrier. First, the UPS driver has to cover only those blocks where he has packages to deliver, shown by the red crosses in Fig. 5-15(a). Second, because of other deliveries outside the neighborhood, his route requires that he enter and exit the neighborhood at opposite ends, as shown in Fig. 5-15(a). Third, the driver can deliver packages on both sides of the street in a single pass. Taking all of these factors into account, we use the graph in Fig. 5-15(b) as a model for the UPS driver problem, with the required starting and ending points of the route shown in red.

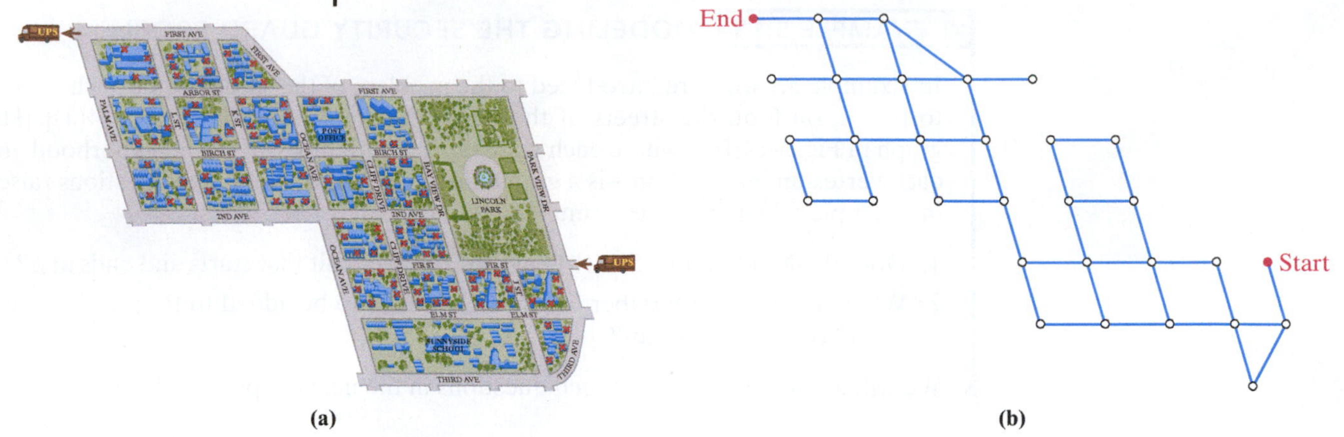

(a) (b)

FIGURE 5-15 (a) The original street map for the UPS driver. (b) A graph model for the problem.

5.3 Euler's Theorems and Fleury's Algorithm

In this section we are going to develop the basic theory that will allow us to determine if a graph has an Euler circuit, an Euler path, or neither. This is important because, as we saw in the previous section, what are Euler circuit or Euler path questions in theory are real-life street-routing questions in practice. The three theorems we are going to see next (all due to Euler) are surprisingly simple and yet tremendously useful.

EULER'S CIRCUIT THEOREM

- If a graph is *connected* and *every vertex is even*, then it has an Euler circuit (at least one, usually more).
- If a graph has *any odd vertices*, then it does not have an Euler circuit.

If we want to know if a graph has an Euler circuit or not, here is how we can use Euler's circuit theorem. First we make sure the graph is connected. (If it isn't, then no matter what else, an Euler circuit is impossible.) If the graph is connected, then we start checking the degrees of the vertices, one by one. As soon as we hit an odd vertex, we know that an Euler circuit is out of the question. If there are no odd vertices, then we know that the answer is yes—the graph does have an Euler circuit! (The theorem doesn't tell us how to find it—that will come soon.) Figure 5-16 illustrates the three possible scenarios. The graph in Fig. 5-16(a) cannot have an Euler circuit for the simple reason that it is disconnected. The graph in Fig. 5-16(b) is connected, but we can quickly spot odd vertices (C is one of them; there are others). This graph has no Euler circuits either. But the graph in Fig. 5-16(c) is connected and all the vertices are even. This graph does have Euler circuits.

The basic idea behind Euler's circuit theorem is that as we travel along an Euler circuit, every time we go through a vertex we "use up" two different edges at that vertex—one to come in and one to go out. We can keep doing this as long as the

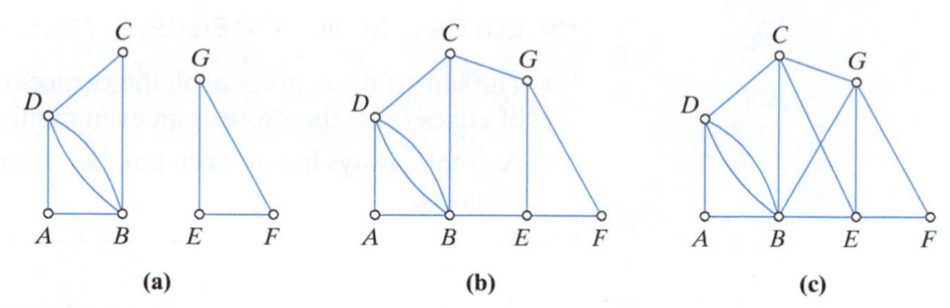

FIGURE 5-16 (a) Not connected; (b) some vertices are odd; (c) all vertices are even.

vertices are even. A single odd vertex means that at some point we are going to come into it and not be able to get out. An analogous theorem will work with Euler paths, but now we do need odd vertices for the starting and ending points of the path. All the other vertices have to be even. Thus, we have the following theorem.

EULER'S PATH THEOREM

- If a graph is *connected* and has *exactly two odd vertices*, then it has an Euler path (at least one, usually more). Any such path must start at one of the odd vertices and end at the other one.
- If a graph has *more than two* odd vertices, then it cannot have an Euler path.

EXAMPLE 5.16 THE BRIDGES OF KÖNIGSBERG PUZZLE SOLVED

Back to the Königsberg bridges problem. In Example 5.12 we saw that the layout of the bridges in the old city can be modeled by the graph in Fig. 5-17(a). This graph has four odd vertices; thus, neither an Euler circuit nor an Euler path can exist. We now have an unequivocal answer to the puzzle: *There is no possible way anyone can walk across all the bridges without having to recross some of them!* How many bridges will need to be recrossed? It depends. If we want to start and end in the same place, we must recross at least two of the bridges. One of the many possible routes is shown in Fig. 5-17(b). In this route the bridge connecting L and D is crossed twice, and so is one of the two bridges connecting A and R. If we are allowed to start and end in different places, we can do it by recrossing just one of the bridges. One possible route starting at A, crossing bridge LD twice, and ending at R is shown in Fig. 5-17(c).

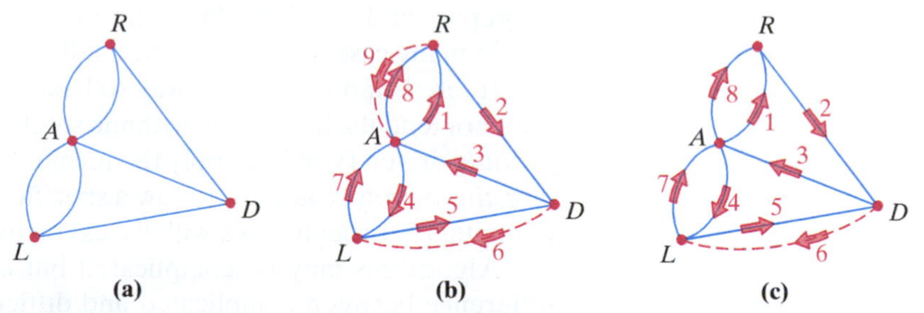

FIGURE 5-17 (a) The original graph with four odd vertices; (b) a walk recrossing bridges DL and RA; (c) a walk recrossing bridge DL only.

Euler's circuit theorem deals with graphs with no odd vertices, whereas Euler's path theorem deals with graphs with two or more odd vertices. The only scenario not covered by the two theorems is that of graphs with just *one* odd vertex. Euler's third theorem rules out this possibility—a graph cannot have just one odd vertex. In fact, Euler's third theorem says much more.

EULER'S SUM OF DEGREES THEOREM

- The sum of the degrees of all the vertices of a graph equals twice the number of edges (and, therefore, is an even number).
- A graph always has an even number of *odd* vertices (i.e., *odd* vertices come in pairs).

Euler's sum of degrees theorem is based on the following basic observation: Take any edge—let's call it XY. The edge contributes once to the degree of vertex X and once to the degree of vertex Y, so, in all, that edge makes a total contribution of 2 to the sum of the degrees. Thus, when the degrees of all the vertices of a graph are added, the total is twice the number of edges. Since the total sum is an even number, it is impossible to have just one odd vertex, or three odd vertices, or five odd vertices, and so on. *The odd vertices of a graph always come in twos.*

Table 5-2 is a summary of Euler's three theorems. It shows the relationship between the number of odd vertices in a connected graph G and the existence of Euler paths or Euler circuits. (The assumption that G is connected is essential—a disconnected graph cannot have Euler paths or circuits regardless of what else is going on.)

Number of odd vertices	Conclusion
0	G has Euler circuit
2	G has Euler path
4, 6, 8, ...	G has neither
1, 3, 5, ...	This is impossible!

TABLE 5-2 ■ Euler's theorems (summary)

Euler's theorems help us answer the following existence question: Does the graph have an Euler circuit, an Euler path, or neither? But when the graph has an Euler circuit or path, how do we find it? For small graphs, simple trial-and-error usually works fine, but real-life applications sometimes involve graphs with hundreds, or even thousands, of vertices. In these cases a trial-and-error approach is out of the question, and what is needed is a systematic strategy that tells us how to create an Euler circuit or path. In other words, we need an *algorithm*.

Fleury's Algorithm

There are many types of problems that can be solved by simply following a set of procedural rules—very specific rules like *when you get to this point, do this, ... after you finish this, do that*, and so on. Given a specific problem X, an **algorithm** for solving X is a set of *procedural rules* that, when followed, always lead to some sort of "solution" to X. X need not be a mathematics problem—algorithms are used, sometimes unwittingly, in all walks of life: directions to find someone's house, the instructions for assembling a new bike, and a recipe for baking an apple pie are all examples of real-life algorithms. A useful analogy is to think of the problem as a *dish* we want to prepare and the algorithm as a *recipe* for preparing that dish.

In many cases, there are several different algorithms for solving the same problem (there is more than one way to bake an apple pie); in other cases, the problem does not lend itself to an algorithmic solution. In mathematics, algorithms are either *formula* driven (you just apply the formula or formulas to the appropriate inputs) or *directive* driven (you must follow a specific set of directives). In this part of the book (Chapters 5 through 8) we will discuss many important algorithms of the latter type.

Algorithms may be complicated but are rarely difficult. (There is a world of difference between complicated and difficult—accounting is complicated, calculus is difficult!) You don't have to be a brilliant and creative thinker to implement most algorithms—you just have to learn how to follow instructions carefully and methodically. For most of the algorithms we will discuss in this and the next three chapters, the key to success is simple: practice, practice, and more practice!

We will now turn our attention to an algorithm that finds an *Euler circuit* or an *Euler path* in a connected graph. Technically speaking, these are two separate algorithms, but in essence they are identical, so they can be described as one. (The algorithm we will give here is attributed to a Frenchman by the name of M. Fleury, who is alleged to have published a description of the algorithm in 1885.

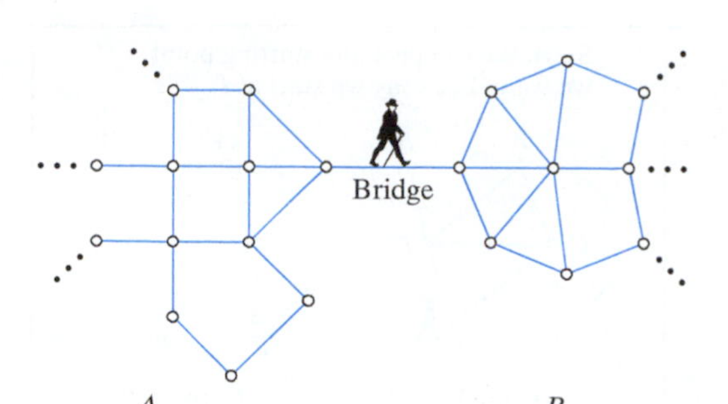

FIGURE 5-18 The bridge separates the two sections. Once you cross from A to B, the only way to get back to A is by recrossing the bridge.

Other than his connection to this algorithm, little else is known about Monsieur Fleury.)

The idea behind Fleury's algorithm can be paraphrased by that old piece of folk wisdom: *Don't burn your bridges behind you.* In graph theory the word *bridge* has a very specific meaning—it is the only edge connecting two separate sections (call them A and B) of a graph, as illustrated in Fig. 5-18. This means that if you are in A, you can only get to B by crossing the bridge. If you do that and then want to get back to A, you will need to recross that same bridge. It follows that if you don't want to recross any edges, you better finish your business at A before you move on to B.

Thus, Fleury's algorithm is based on a simple principle: To find an Euler circuit or an Euler path, *bridges are the last edges you want to cross.* Only do it if you have no other choice! Simple enough, but there is a rub: The graph whose bridges we are supposed to avoid is not necessarily the original graph of the problem. Instead, it is that part of the original graph that has yet to be traveled. The point is this: Once we travel along an edge, we are done with it! We should not cross it again, so from that point on, as far as we are concerned that edge is part of the past. Our only concern now is with the future: How are we going to get around in the *yet-to-be-traveled* part of the graph? Thus, when we talk about bridges that we want to leave as a last resort, we are really referring to *bridges of the to-be-traveled part of the graph.*

FLEURY'S ALGORITHM FOR FINDING AN EULER CIRCUIT (PATH)

- **Preliminaries.** Make sure that the graph is connected and either (1) has no odd vertices (circuit) or (2) has just two odd vertices (path).
- **Start.** Choose a starting vertex. [In case (1) this can be any vertex; in case (2) it must be one of the two *odd* vertices.]
- **Intermediate steps.** At each step, if you have a choice, *don't choose a bridge of the yet-to-be-traveled part* of the graph. However, if you have only one choice, take it.
- **End.** When you can't travel any more, the circuit (path) is complete. [In case (1) you will be back at the starting vertex; in case (2) you will end at the other odd vertex.]

The only complicated aspect of Fleury's algorithm is the bookkeeping. With each new step, the untraveled part of the graph changes and there may be new bridges formed. Thus, in implementing Fleury's algorithm it is critical to separate the *past* (the part of the graph that has already been traveled) from the *future* (the part of the graph that still needs to be traveled). While there are many different ways to accomplish this, a fairly reliable way goes like this: Start with *two* copies of the graph. Copy 1 is to keep track of the "future"; copy 2 is to keep track of the "past." Every time you travel along an edge, *erase* that edge from copy 1, but mark it (say in red) and label it with the appropriate number on copy 2. As you move forward, copy 1 gets smaller and copy 2 gets redder. At the end, copy 1 has disappeared; copy 2 shows the actual Euler circuit or path. [The applet ***Euler Paths and Circuits: Fleury's Algorithm*** (available in ***MyMathLab*** in the *Multimedia Library* or in *Tools for Success* allows you to practice finding Euler circuits (paths) using this strategy.]

It's time to look at a couple of examples.

EXAMPLE 5.17 IMPLEMENTING FLEURY'S ALGORITHM

The graph in Fig. 5-19(a) is a very simple graph—it would be easier to find an Euler circuit just by trial-and-error than by using Fleury's algorithm. Nonetheless, we will do it using Fleury's algorithm. The real purpose of the example is to see the algorithm at work. Each step of the algorithm is explained in Figs. 5-19(b) through (h).

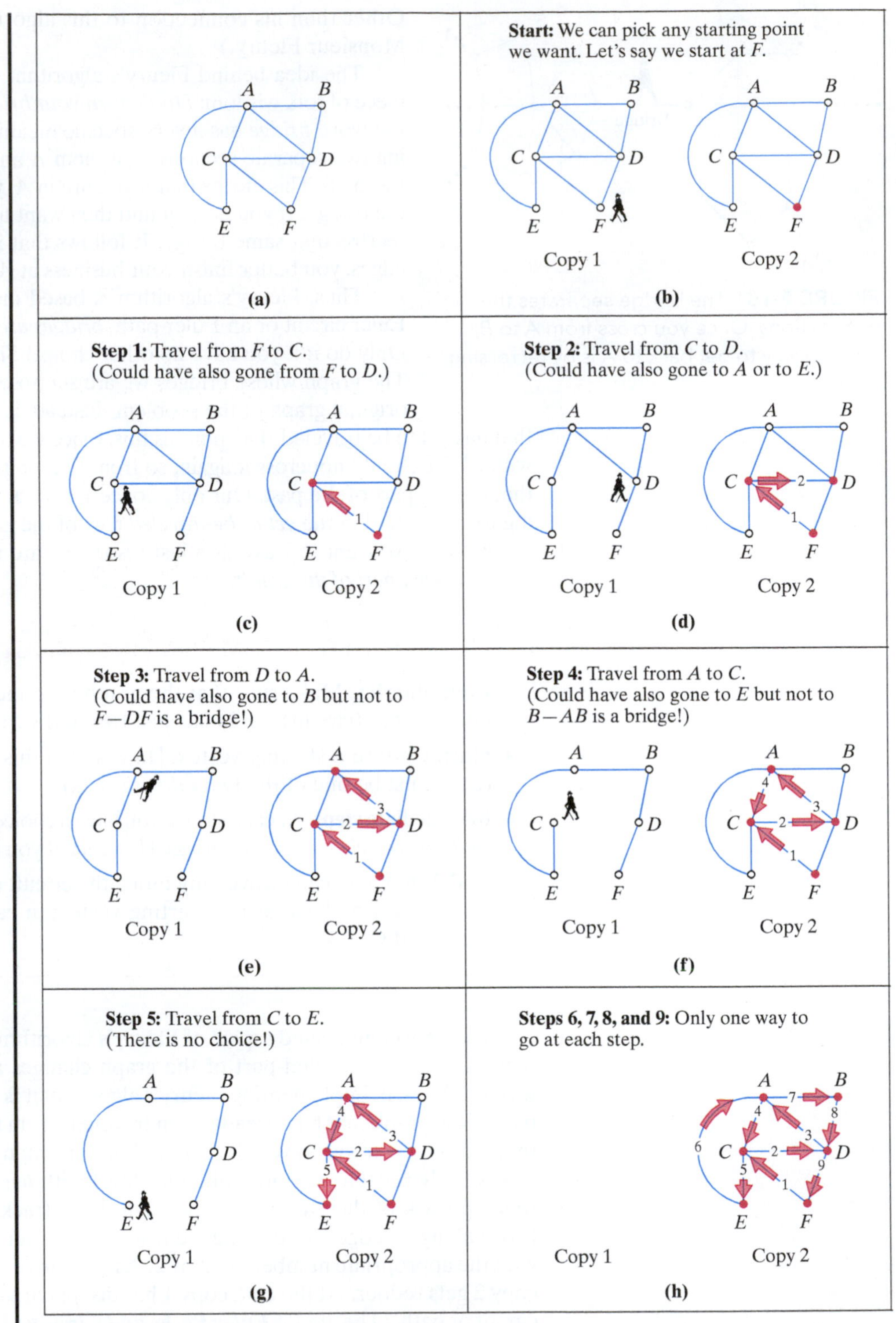

FIGURE 5-19 Fleury's algorithm at work.

EXAMPLE 5.18 FLEURY'S ALGORITHM FOR EULER PATHS

We will apply Fleury's algorithm to the graph in Figure 5-20. Since it would be a little impractical to show each step of the algorithm with a separate picture as we did in Example 5.17, you are going to have to do some of the work. Start by making two copies of the graph. (If you haven't already done so, get some paper, a pencil, and an eraser.)

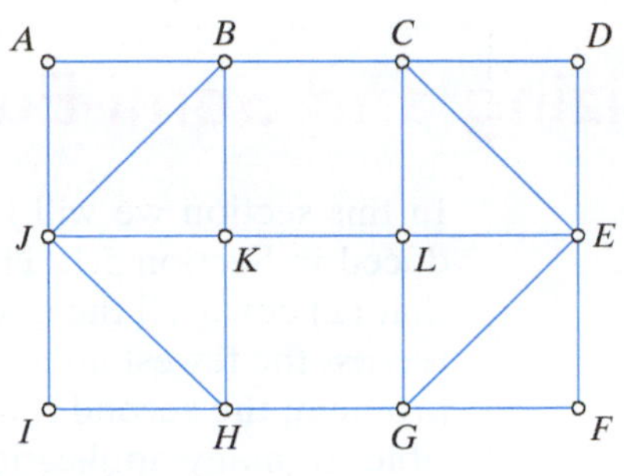

FIGURE 5-20

- **Start.** This graph has two odd vertices, E and J. We can pick either one as the starting vertex. Let's start at J.
- **Step 1.** From J we have five choices, all of which are OK. We'll randomly pick K. (Erase JK on copy 1, and mark and label JK with a 1 on copy 2.)
- **Step 2.** From K we have three choices (B, L, or H). Any of these choices is OK. Say we choose B. (Now erase KB from copy 1 and mark and label KB with a 2 on copy 2.)
- **Step 3.** From B we have three choices (A, C, or J). Any of these choices is OK. Say we choose C. (Now erase BC from copy 1 and mark and label BC with a 3 on copy 2.)
- **Step 4.** From C we have three choices (D, E, or L). Any of these choices is OK. Say we choose L. (EML—that's shorthand for erase, mark, and label.)
- **Step 5.** From L we have three choices (E, G, or K). Any of these choices is OK. Say we choose K. (EML.)
- **Step 6.** From K we have only one choice—to H. Without further ado, we choose H. (EML.)
- **Step 7.** From H we have three choices (G, I, or J). But for the first time, one of the choices is a bad choice. We should not choose G, as HG is a bridge of the yet-to-be-traveled part of the graph [Fig. 5-21(a)]. Either of the other two choices is OK. Say we choose J [Fig. 5-21(b)]. (EML.)

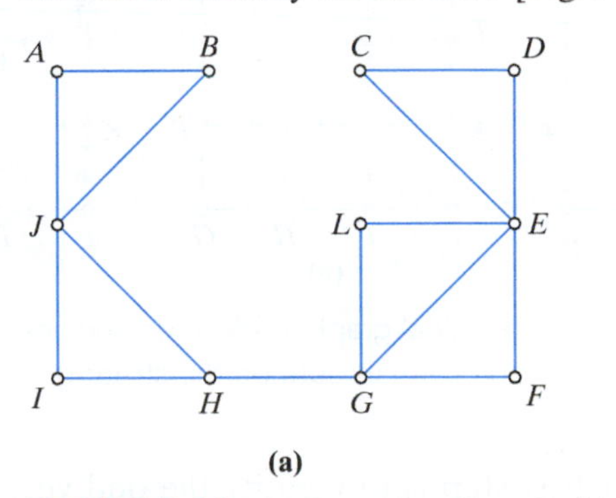

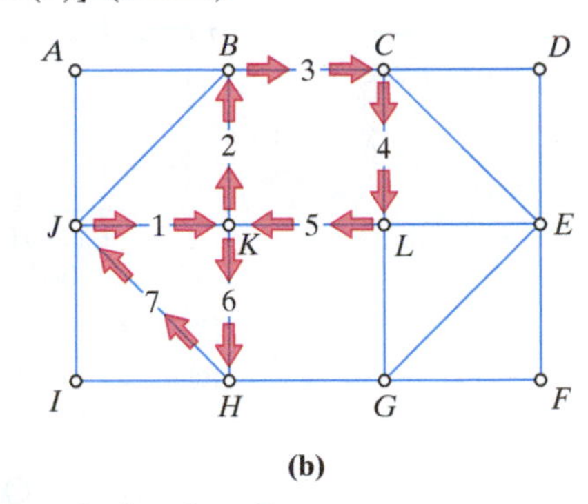

FIGURE 5-21 (a) Copy 1 at Step 7. (b) Copy 2 after Step 7.

- **Step 8.** From J we have three choices (A, B, or I), but we should not choose I, as JI has just become a bridge. Either of the other two choices is OK. Say we choose B. (EML)
- **Steps 9 through 13.** Each time we have only one choice. From B we have to go to A, then to J, I, H, and G.
- **Steps 14 through 21.** Not to belabor the point, let's just cut to the chase. The rest of the path is given by $G, F, E, D, C, E, G, L, E$. There are many possible endings, and you should find a different one by yourself.

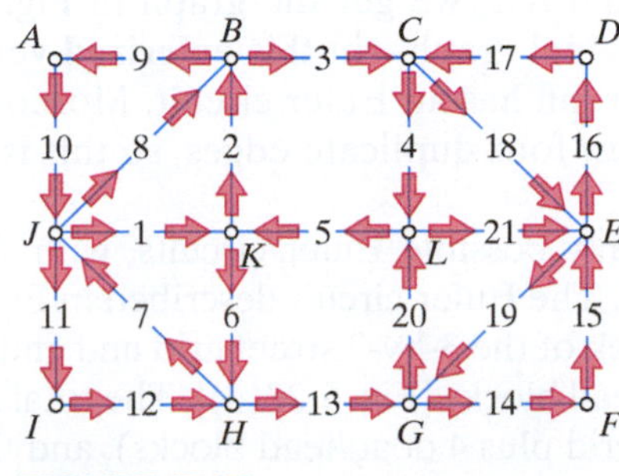

FIGURE 5-22

The completed Euler path (one of hundreds of possible ones) is shown in Fig. 5-22.

5.4 Eulerizing and Semi-Eulerizing Graphs

In this section we will finally answer some of the street-routing problems introduced in Section 5.1. The common thread in all these problems is to find routes that (1) cover all the edges of the graph that models the original problem and (2) recross the fewest number of edges. The first requirement typically comes with the problem; the second requirement comes from the desire to be as efficient as possible. In many applications, each edge represents a unit of cost. The more edges along the route, the higher the cost of the route. In a UPS truck routing problem, for example, the first pass along an edge is a requirement of the job. Any additional passes along that edge represent a wasted expense (these extra passes are often described as *deadhead* travel). Thus, an optimal route is one with the fewest number of deadhead edges. (This is only true under the assumption that each edge equals one unit of cost.)

We are now going to see how the theory developed in the preceding sections will help us design optimal street routes for graphs with many (more than two) odd vertices. The key idea is that we can turn odd vertices into even vertices by adding "duplicate" edges in strategic places.

Eulerizations

EXAMPLE 5.19 COVERING A 3-BY-3 STREET GRID

The graph in Fig. 5-23(a) models a 3-block-by-3-block street grid. The graph has 24 edges, each representing a block of the street grid. How can we find an optimal route that sweeps through the blocks of the street grid and ends back at the starting point?

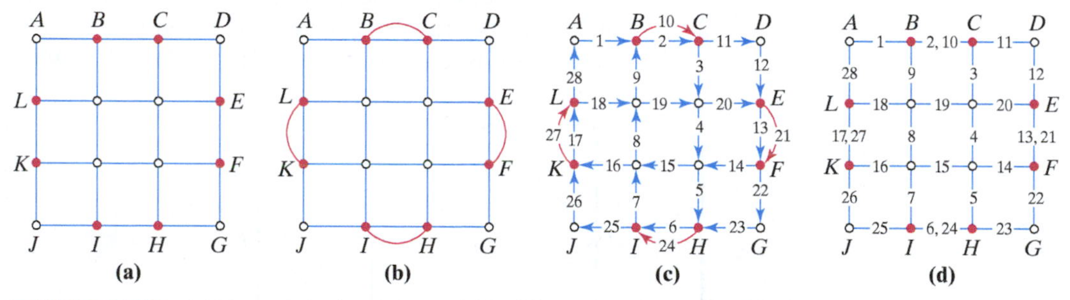

FIGURE 5-23 (a) The original graph model (odd vertices shown in red); (b) an optimal eularization; (c) an Euler circuit; (d) an optimal route on the street grid (just follow the numbers).

Our first step is to identify the odd vertices in the graph model. This graph has eight odd vertices (B, C, E, F, H, I, K, and L), shown in red. When we add a duplicate copy of edges BC, EF, HI, and KL, we get the graph in Fig. 5-23(b). This is called an **eulerization** of the original graph—in this **eulerized** version the vertices are all even, so we know this graph has an Euler circuit. Moreover, with eight odd vertices we need to add *at least* four duplicate edges, so this is the best we can do.

Figure 5-23(c) shows one of the many possible Euler circuits, with the edges numbered in the order they are traveled. The Euler circuit described in Fig. 5-23(c) represents a route that sweeps every block of the 3-by-3 street grid and ends back at the starting point, using only four deadhead blocks [Fig. 5-23(d)]. The total length of this route is 28 blocks (24 blocks in the grid plus 4 deadhead blocks), and this route is optimal—no matter how clever you are or how hard you try, if you want to sweep the grid and start and end at the same point, you will have to pass through a minimum of 28 blocks! (There are many other ways to do it using just 28 blocks, but none with fewer than 28.)

EXAMPLE 5.20 **COVERING A 4-BY-4 STREET GRID**

The graph in Fig. 5-24(a) models a 4-block-by-4-block street grid consisting of 40 blocks. The 12 odd vertices in the graph are shown in red. We want to find a route that sweeps through the 40 blocks of the street grid, ends back at the starting point, and has the fewest number of deadhead blocks. To do this, we first eulerize the graph by adding the fewest possible number of edges. Figure 5-24(b) shows how *not to do it!* This graph violates the cardinal rule of eulerization—you can only duplicate edges that are part of the original graph. Edges *DF* and *NL* are new edges, not duplicates, so Fig. 5-24(b) is out! Figure 5-24(c) shows a legal eulerization, but it is not optimal, as it is obvious that we could have accomplished the same thing by adding fewer duplicate edges. Figure 5-24(d) shows an *optimal eulerization* of the original graph—one of several possible. Once we have an optimal eulerization, we have the blueprint for the optimal route on the street grid. Regardless of the specific details, we now know that the route will travel along 48 blocks—the 40 original blocks in the grid plus 8 deadhead blocks. A route can be found by using Fleury's algorithm on the optimally eulerized graph shown in Fig. 5-24(d). We leave the details to the reader (see Exercise 39).

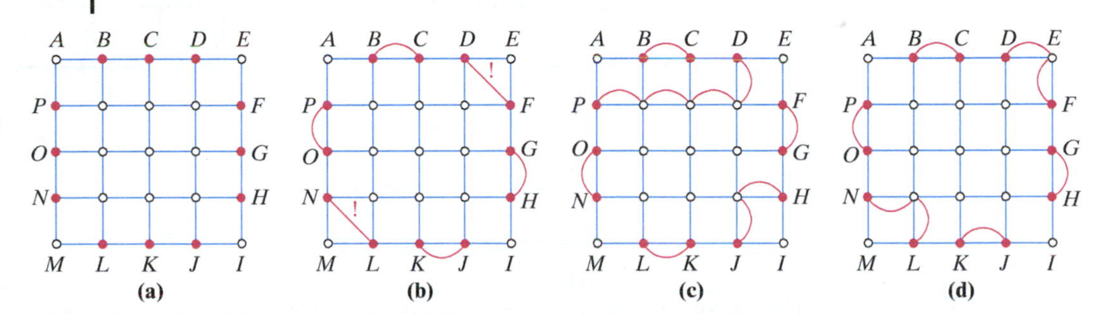

FIGURE 5-24 (a) The original graph model (odd vertices shown in red). (b) Bad move — *DF* and *NL* were not edges of the original graph! (c) An eulerization, but not an optimal one! (d) One of the many possible optimal eulerizations.

EXAMPLE 5.21 **THE SECURITY GUARD PROBLEM SOLVED**

We are now ready to solve the security guard street-routing problem introduced in Example 5.1 and subsequently modeled in Example 5.13. Let's recap the story: The security guard is required to walk each block of the Sunnyside neighborhood at least once and he wants to deadhead as few blocks as possible. He usually parks his car at *S* (there is a donut shop on that corner) and needs to end his route back at the car. Figure 5-25 shows the evolution of a solution: Fig. 5-25(a) shows the original neighborhood that the security guard must cover; Fig. 5-25(b) shows the graph model of the problem (with the 18 odd vertices of the graph highlighted in red); Fig. 5-25(c) shows an eulerization of the graph in (b), with the 9 duplicate edges shown in red. This is the fewest number of edges required to eulerize the graph in Fig. 5-25(b), so the eulerization is optimal. The eulerized graph in Fig. 5-25(c) has all even vertices, and an Euler circuit can be found. Since a circuit can be started at any vertex we will start the circuit at S. Figure 5-25(d) shows one of the many possible optimal routes for the security guard (just follow the numbers). Note that using a different starting vertex will not make the route shorter, so the security guard can continue parking in front of the donut shop—it will not hurt!

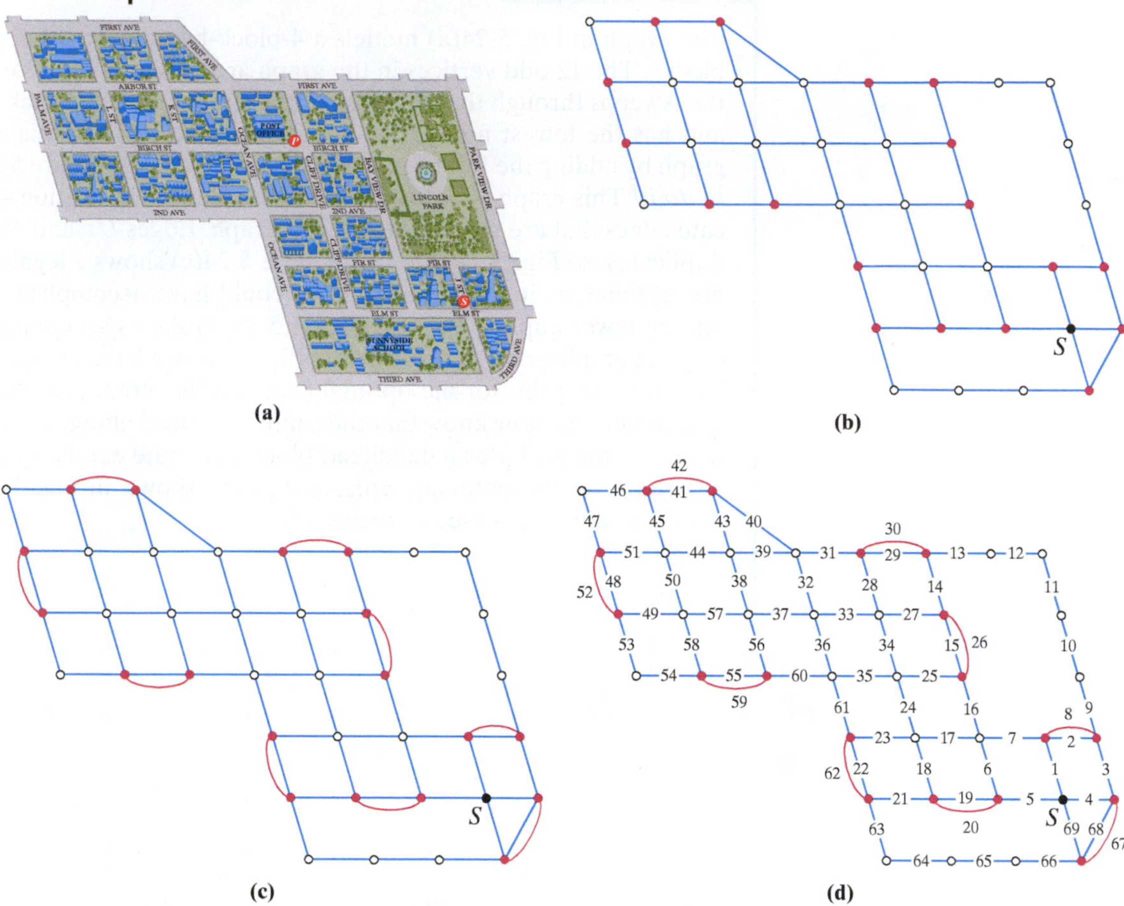

FIGURE 5-25 (a) The Sunnyside neighborhood; (b) graph model for the security guard problem; (c) an optimal eulerization of (b); (d) an optimal route for the security guard.

EXAMPLE 5.22 THE MAIL CARRIER PROBLEM SOLVED

The solution to the mail carrier problem follows essentially the same story line as the one for the security guard. Let's recap this story: The mail carrier needs to start and end her route at the post office (P), cover both sides of the street when there are buildings on both sides, and cover just one side on blocks where there are buildings on only one side, with no need to cover any streets where there are no buildings (like the back side of the park and the school). Fig. 5-26 shows the graph model for the mail carrier (see Example 5.14). The interesting thing about this graph is that every vertex is already even (there are some vertices of degree 2 in the corners, there are a couple of vertices of degree 6, and there are lots of vertices of degrees 4 and 8). This means that the graph does not have to be eulerized, as it has Euler circuits in its present form. An optimal route for the mail carrier can be found by finding an Euler circuit of the graph that starts and ends at P. We now know how to do that using Fleury's algorithm (or just plain trial and error if you prefer) and leave it as an exercise for the reader to do so (Exercise 40).

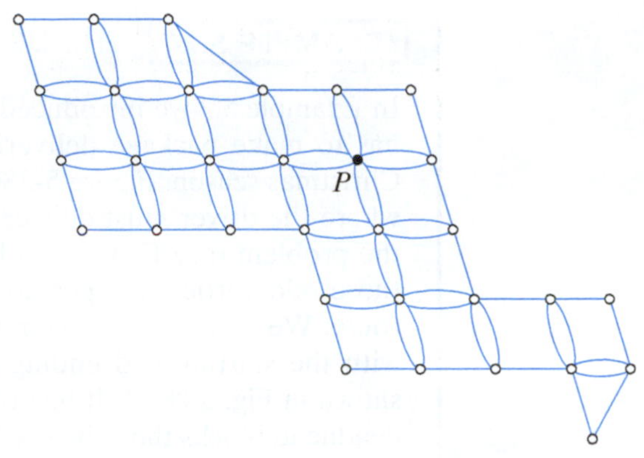

FIGURE 5-26 Graph model for the mail carrier.

Semi-Eulerizations

In cases where a street route is not required to end back where it started (either because we can choose to start and end in different places or because the starting and ending points are required to be different) we are looking for Euler paths, rather than Euler circuits. In these cases we are looking for a graph that has two odd vertices and the rest even. We need the two odd vertices to give us a starting and ending point for our route. The process of adding additional edges to a graph so that all the vertices except two are even is called a **semi-eulerization**, and we say that the graph has been *semi-eulerized.*

EXAMPLE 5.23 THE BRIDGES OF MADISON COUNTY SOLVED

In Example 5.5 we introduced the problem of routing a photographer across all the bridges in Madison County, shown in Fig. 5-27(a). To recap the story: The photographer needs to cross each of the 11 bridges at least once (for her photo shoot). Each crossing of a bridge costs \$25, and the photographer is on a tight budget, so she wants to cover all the bridges once but recross as few bridges as possible. The other relevant fact is that the photographer can start her trip at either bank of the river and end the trip at either bank of the river. Figure 5-27(b) is a graph model of the Madison bridges layout. A la Euler, we let the vertices represent the land masses and the edges represent the bridges. The graph has four odd vertices (R, L, B, and D). The photographer can start the shoot at either bank and end at either bank—say she chooses to start the route at R and end it at L. Figure 5-27(c) shows a semi-eulerization of the graph in (b), with R and L left as odd vertices, and the edge BD crossed a second time just to make vertices B and D even. The numbers in Fig. 5-27(c) show one possible optimal route for the photographer, with a total of 12 bridge crossings and a total cost of \$300.

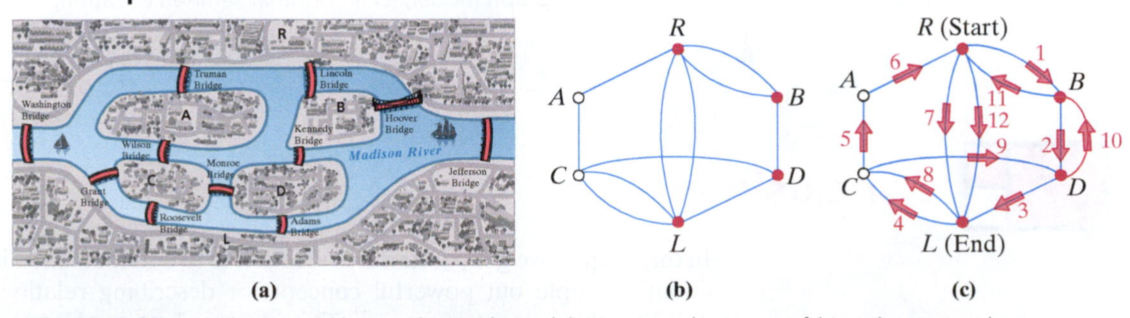

FIGURE 5-27 (a) The original layout; (b) graph model; (c) semi-eulerization of (b) and an optimal route.

EXAMPLE 5.24 THE UPS DRIVER PROBLEM DECONSTRUCTED

In Example 5.3 we introduced the street-routing problem facing a UPS driver who has to make package deliveries around the Sunnyside neighborhood during the Christmas season. Figure 5-28(a) shows all the locations (marked with red crosses) where the driver must deliver packages. Figure 5-28(b) shows the graph model of the problem (see Example 5.15), with the odd vertices highlighted in red. Two of those odd vertices happen to be the designated starting and ending points of his route. We solve this problem by finding an optimal semi-eulerization of the graph with the starting and ending points left alone. One optimal semi-eulerization is shown in Fig. 5-28(c). It has 10 additional edges shown in red. These represent the deadhead blocks the driver will have to cover a second time. The last step is to find an Euler path in Fig. 5-28(c) with the designated starting and ending points—that would give us an optimal route. This can be done most conveniently by taking advantage of the applet ***Euler's Paths and Circuits: Fleury's Algorithm*** (available in *MyMathLab*), and we leave this final detail as an exercise [Exercise 76(a)].

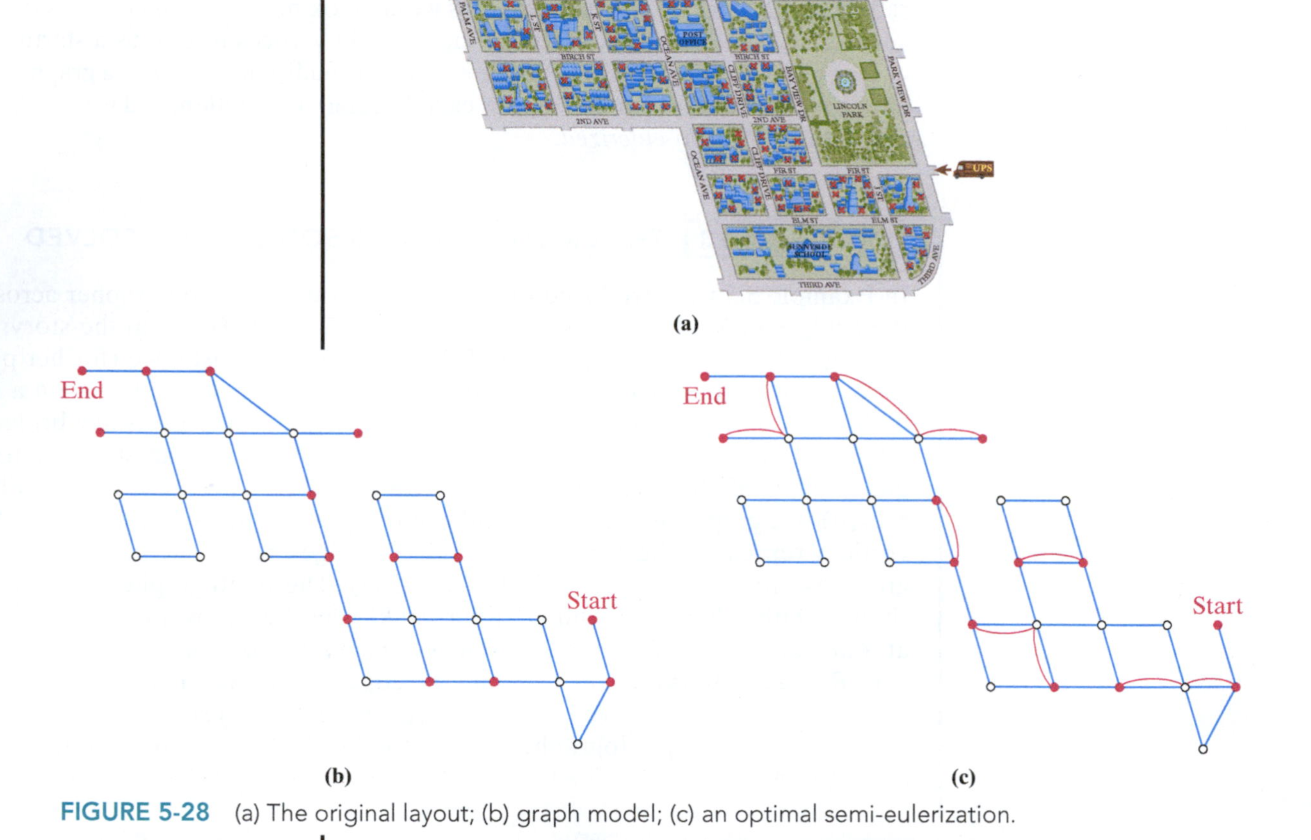

FIGURE 5-28 (a) The original layout; (b) graph model; (c) an optimal semi-eulerization.

Conclusion

In this chapter we got our first introduction to two fundamental ideas. First, we learned about a simple but powerful concept for describing relationships within a set of objects—the concept of a *graph*. This idea can be traced back to Euler, while solving the Königsberg bridge puzzle some 280 years ago. Since then, the study of graphs has grown into one of the most important and useful branches of modern mathematics.

The second important idea of this chapter is the concept of a *graph model*. Every time we take a real-life problem and turn it into a mathematical problem, we are, in effect, modeling. Unwittingly, we have all done some form of mathematical modeling at one time or another—first using arithmetic and later using equations and functions to describe real-life situations. In this chapter we learned about a new type of modeling called graph modeling, in which we use graphs and the mathematical theory of graphs to solve certain types of routing problems.

By necessity, the routing problems that we solved in this chapter were fairly simplistic—crossing a few bridges, patrolling a small neighborhood, routing a UPS driver. We should not be deceived by the simplicity of these examples—larger-scale variations on these themes have significant practical importance. In many big cities, where the efficient routing of municipal services (police patrols, garbage collection, snow removal, etc.) is a major issue, the very theory that we developed in this chapter is being used on a large scale, the only difference being that many of the more tedious details are mechanized and carried out by a computer. (In New York City, for example, garbage collection, curb sweeping, snow removal, and other municipal services have been scheduled and organized using graph models since the 1970s, and the improved efficiency has yielded savings estimated in the tens of millions of dollars a year.)

If you are disappointed that we haven't dug deeper in this chapter into *graphs*, *graph models*, and *algorithms*, don't worry. In the next three chapters we will see many other interesting and important real-life problems that can be solved using these three fundamental ideas of modern mathematics.

KEY CONCEPTS

5.1 Street-Routing Problems

- **routing problems:** problems concerned with routing the delivery of goods or services to a set of destinations, **137**
- **street-routing problems:** problems where a specified set of connections (roads, bridges, edges) must be traveled at least once, **137**

5.2 An Introduction to Graphs

- **graph:** a "picture" consisting of **vertices** (dots) and **edges** (lines) connecting the vertices, **141**
- **vertex set:** the set of vertices of a graph, **142**
- **edge list:** a list of all the edges of a graph, **142**
- **adjacent vertices:** two vertices connected by an edge, **142**
- **loop:** an edge that connects a vertex with itself, **142**
- **multiple edges:** two or more edges connecting the same two vertices, **142**
- **degree:** number of edges meeting at the vertex, **142**
- **odd (even) vertex:** a vertex of odd (even) degree, **142**
- **clique:** a set of vertices with the property that any two are adjacent, **142**
- **simple graph:** a graph with no loops or multiple edges, **142**
- **isolated vertex:** a vertex of degree 0, **143**
- **friendship graph:** a graph where the vertices represent individuals and the edges represent friendships between pairs of individuals, **144**
- **adjacent edges:** two edges with a shared vertex, **145**
- **path:** a sequence of edges each adjacent to the next, with no edge included more than once, and starting and ending at different vertices, **145**
- **circuit:** same as a path but starting and ending at the same vertex, **145**

- **length:** number of edges in a path or a circuit, **145**
- **Euler path:** a path that travels along each edge of a graph once and only once, **146**
- **Euler circuit:** a circuit that travels along each edge of a graph once and only once, **146**
- **connected graph:** a graph such that there is a path going from any vertex to any other vertex, **146**
- **disconnected graph:** a graph that is not connected; it has two or more connected components, **146**
- **components:** the connected "pieces" that make up a graph, **146**
- **bridge:** an edge in a connected graph without which the graph would be disconnected, **147**

5.3 Euler's Theorems and Fleury's Algorithm

- **Euler's Circuit Theorem:** a connected graph has an Euler circuit if and only if all vertices are even, **150**
- **Euler's Path Theorem:** a connected graph has an Euler path if and only it has two odd vertices, **151**
- **Euler's Sum of Degrees Theorem:** the sum of the degrees of all the vertices equals twice the number of edges, **152**
- **algorithm:** a set of procedural rules that when followed produces the answer to some problem, **152**
- **Fleury's algorithm:** an algorithm for finding Euler circuits or Euler paths in a graph; it builds the Euler circuit (path) edge by edge—choosing a bridge of the yet-to-be traveled part of the graph only when there is no other choice, **153**

5.4 Eulerizing and Semi-Eulerizing Graphs

- **eulerization:** the process of duplicating edges in a graph to make it have all even vertices, **156**
- **semi-eulerization:** the process of duplicating edges in a graph to make it have all but two even vertices, **159**

EXERCISES

WALKING

5.1 Street-Routing Problems

No exercises for this section.

5.2 An Introduction to Graphs

1. For the graph shown in Fig. 5-29,
   - **(a)** give the vertex set.
   - **(b)** give the edge list.
   - **(c)** give the degree of each vertex.
   - **(d)** draw a version of the graph without crossing points.

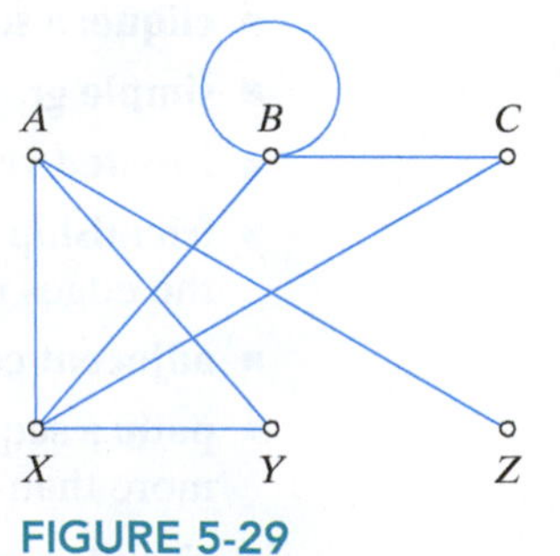

FIGURE 5-29

2. For the graph shown in Fig. 5-30,

(a) give the vertex set.

(b) give the edge list.

(c) give the degree of each vertex.

(d) draw a version of the graph without crossing points.

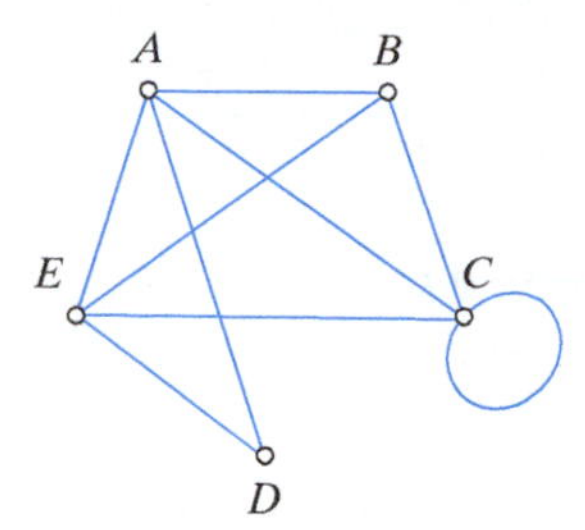

FIGURE 5-30

3. For the graph shown in Fig. 5-31,

(a) give the vertex set.

(b) give the edge list.

(c) give the degree of each vertex.

(d) give the number of components of the graph.

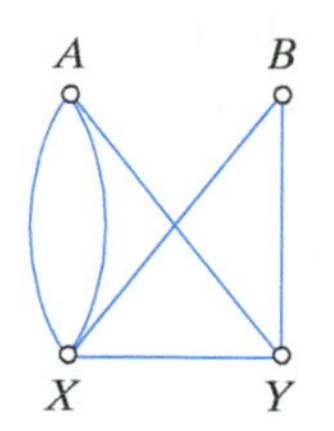

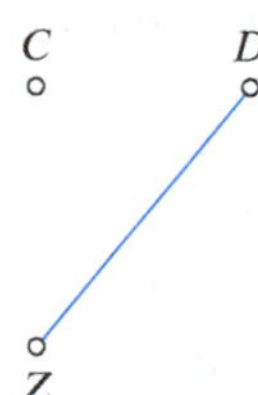

FIGURE 5-31

4. For the graph shown in Fig. 5-32,

(a) give the vertex set.

(b) give the edge list.

(c) give the degree of each vertex.

(d) give the number of components of the graph.

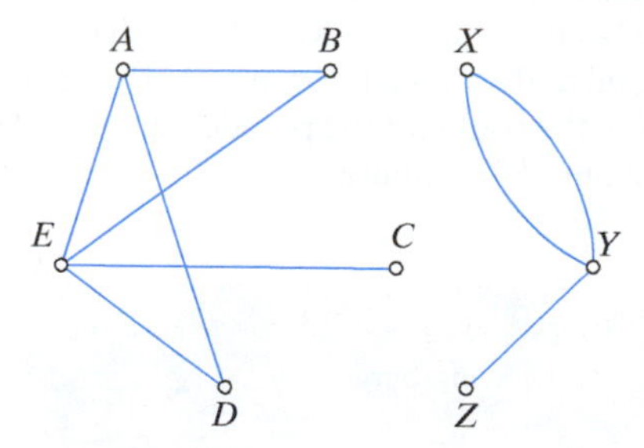

FIGURE 5-32

5. Consider the graph with vertex set $\{K, R, S, T, W\}$ and edge list $RS, RT, TT, TS, SW, WW, WS$. Draw two different pictures of the graph.

6. Consider the graph with vertex set $\{A, B, C, D, E\}$ and edge list $AC, AE, BD, BE, CA, CD, CE, DE$. Draw two different pictures of the graph.

7. Consider the graph with vertex set $\{A, B, C, D, E\}$ and edge list AD, AE, BC, BD, DD, DE. Without drawing a picture of the graph,

(a) list all the vertices adjacent to D.

(b) list all the edges adjacent to BD.

(c) find the degree of D.

(d) find the sum of the degrees of the vertices.

8. Consider the graph with vertex set $\{A, B, C, X, Y, Z\}$ and edge list $AX, AY, AZ, BB, CX, CY, CZ, YY$. Without drawing a picture of the graph,

(a) list all the vertices adjacent to Y.

(b) list all the edges adjacent to AY.

(c) find the degree of Y.

(d) find the sum of the degrees of the vertices.

9. **(a)** Give an example of a connected graph with eight vertices such that each vertex has degree 2.

(b) Give an example of a disconnected graph with eight vertices such that each vertex has degree 2.

(c) Give an example of a graph with eight vertices such that each vertex has degree 1.

10. **(a)** Give an example of a connected graph with eight vertices: six of degree 2 and two of degree 3.

(b) Give an example of a disconnected graph with eight vertices: six of degree 2 and two of degree 3.

(c) Give an example of a graph with eight vertices such that each vertex has degree 4.

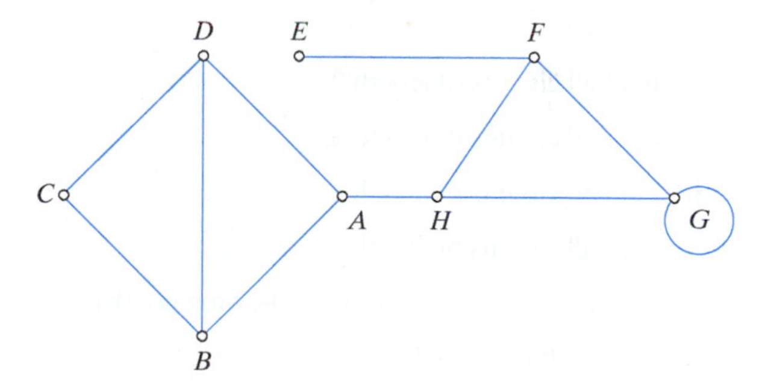

FIGURE 5-33

11. Consider the graph in Fig. 5-33.

(a) Find a path from C to F passing through vertex B but not through vertex D.

(b) Find a path from C to F passing through both vertex B and vertex D.

(c) Find a path of length 4 from C to F.

(d) Find a path of length 7 from C to F.

(e) How many paths are there from C to A?

(f) How many paths are there from H to F?

(g) How many paths are there from C to F?

12. Consider the graph in Fig. 5-33.

(a) Find a path from D to E passing through vertex G only once.

(b) Find a path from D to E passing through vertex G twice.

(c) Find a path of length 4 from D to E.

(d) Find a path of length 8 from D to E.

(e) How many paths are there from D to A?

(f) How many paths are there from H to E?

(g) How many paths are there from D to E?

13. Consider the graph in Fig. 5-33.

(a) Find all circuits of length 1. (*Hint:* Loops are circuits of length 1).

(b) Find all circuits of length 2.

(c) Find all circuits of length 3.

(d) Find all circuits of length 4.

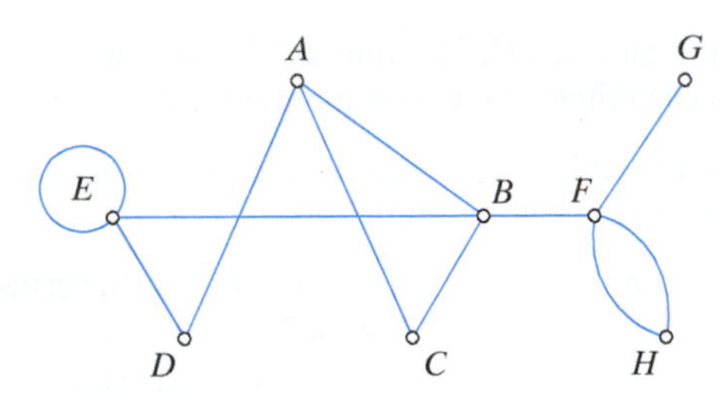

FIGURE 5-34

14. Consider the graph in Fig. 5-34.

(a) Find all circuits of length 1. (*Hint:* Loops are circuits of length 1).

(b) Find all circuits of length 2.

(c) Find all circuits of length 3.

(d) Find all circuits of length 4.

(e) Find all circuits of length 5.

15. List all the bridges in each of the following graphs:

(a) the graph in Fig. 5-33.

(b) the graph with vertex set $\{A, B, C, D, E\}$ and edge list AB, AE, BC, CD, DE.

(c) the graph with vertex set $\{A, B, C, D, E\}$ and edge list AB, BC, BE, CD.

16. List all the bridges in each of the following graphs:

(a) the graph in Fig. 5-34.

(b) the graph with vertex set $\{A, B, C, D, E\}$ and edge list AB, AD, AE, BC, CE, DE.

(c) the graph with vertex set $\{A, B, C, D, E\}$ and edge list AB, BC, CD, DE.

17. Consider the graph in Fig. 5-35.

(a) List all the bridges in this graph.

(b) If you remove *all* the bridges from the graph, how many components will the resulting graph have?

(c) What is the length of the *shortest* path from C to J? Describe such a path.

(d) What is the length of the *longest* path from I to J? Describe such a path.

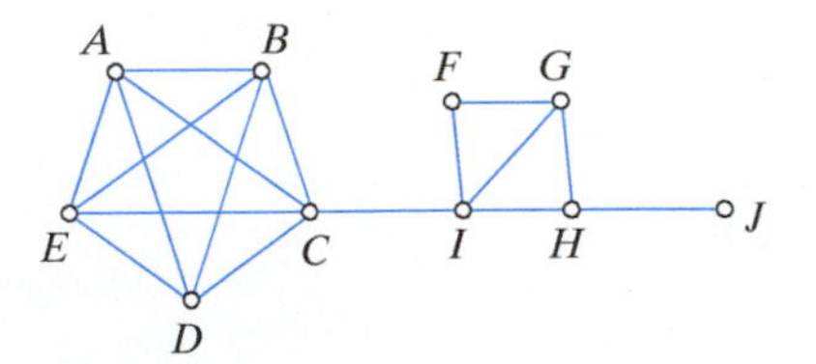

FIGURE 5-35

18. Consider the graph in Fig. 5-36.

(a) List all the bridges in this graph.

(b) If you remove *all* the bridges from the graph, how many components will the resulting graph have?

(c) What is the length of the *shortest* path from E to J? Describe such a path.

(d) What is the length of the *longest* path from I to J? Describe such a path.

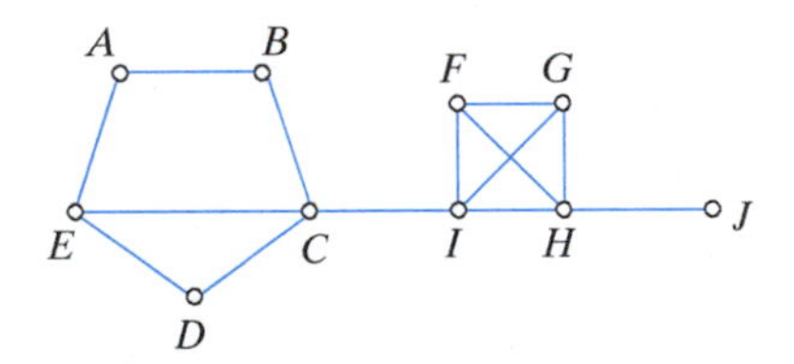

FIGURE 5-36

19. Figure 5-37 shows a map of the downtown area of the picturesque hamlet of Kingsburg, with the Kings River running through the downtown area and the three islands (A, B, and C) connected to each other and both banks by seven bridges. You have been hired by the Kingsburg Chamber of Commerce to organize the annual downtown parade. Part of your job is to plan the route for the parade. Draw a graph that models the layout of Kingsburg.

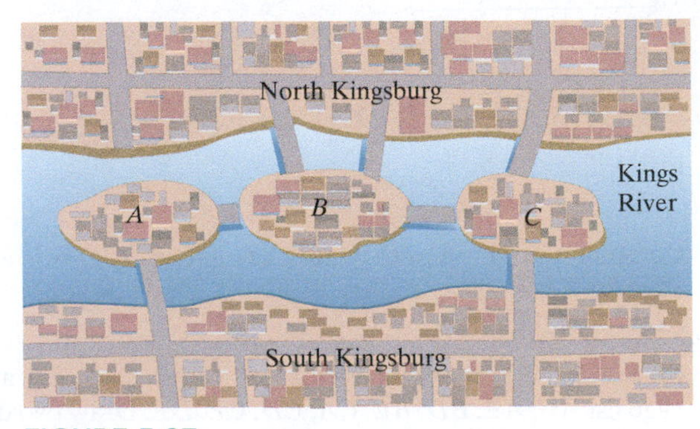

FIGURE 5-37

20. Figure 5-38 is a map of downtown Royalton, showing the Royalton River running through the downtown area and the three islands (A, B, and C) connected to each other and both banks by eight bridges. The Downtown Athletic Club wants to design the route for a marathon through the downtown area. Draw a graph that models the layout of Royalton.

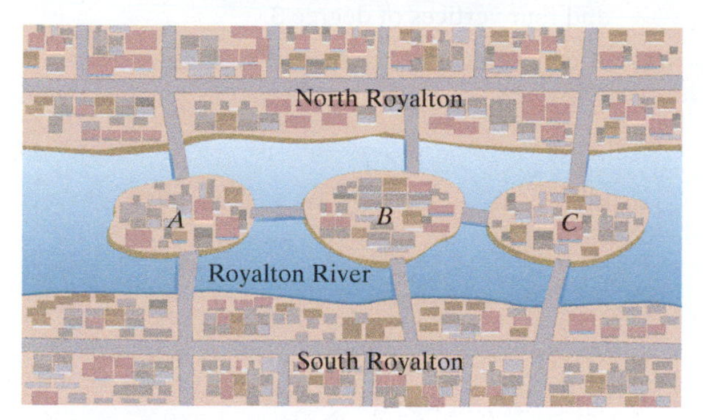

FIGURE 5-38

21. A night watchman must walk the streets of the Green Hills subdivision shown in Fig. 5-39. The night watchman needs to walk only once along each block. Draw a graph that models this street-routing problem.

FIGURE 5-39

22. A mail carrier must deliver mail on foot along the streets of the Green Hills subdivision shown in Fig. 5-39. The mail carrier must make two passes on every block that has houses on both sides of the street (once for each side of the street), but only one pass on blocks that have houses on only one side of the street. Draw a graph that models this street-routing problem.

23. Six teams (A, B, C, D, E, and F) are entered in a softball tournament. The top two seeded teams (A and B) have to play only three games; the other teams have to play four games each. The tournament pairings are A plays against C, E, and F; B plays against C, D, and F; C plays against every team except F; D plays against every team except A; E plays against every team except B; and F plays against every team except C. Draw a graph that models the tournament.

24. The Kangaroo Lodge of Madison County has 10 members (A, B, C, D, E, F, G, H, I, and J). The club has five working committees: the Rules Committee (A, C, D, E, I, and J), the Public Relations Committee (B, C, D, H, I, and J), the Guest Speaker Committee (A, D, E, F, and H), the New Year's Eve Party Committee (D, F, G, H, and I), and the Fund Raising Committee (B, D, F, H, and J).

(a) Suppose we are interested in knowing which pairs of members are on the same committee. Draw a graph that models this problem. (*Hint*: Let the vertices of the graph represent the members.)

(b) Suppose we are interested in knowing which committees have members in common. Draw a graph that models this problem. (*Hint*: Let the vertices of the graph represent the committees.)

25. Table 5-3 summarizes the Facebook friendships between a group of eight individuals [an F indicates that the individuals (row and column) are Facebook friends]. Draw a graph that models the set of friendships in the group. (Use the first letter of the name to label the vertices.)

	Fred	Pat	Mac	Ben	Tom	Hale	Zac	Cher
Fred		F			F	F		
Pat	F				F	F		F
Mac				F			F	
Ben			F				F	
Tom	F	F				F		
Hale	F	F			F			F
Zac			F	F				
Cher		F				F		

TABLE 5-3

26. The Dean of Students' office wants to know how the seven general education courses selected by incoming freshmen are clustered. For each pair of general education courses, if 30 or more incoming freshmen register for both courses, the courses are defined as being "significantly linked." Table 5-4 shows all the significant links between general education courses (indicated by a 1). Draw a graph that models the significant links between the general education courses. (Use the first letter of each course to label the vertices of the graph.)

	Math	Chemistry	Biology	English	Physics	History	Art
Math		1	1	1	1		
Chemistry	1		1				
Biology	1	1		1		1	
English	1		1		1	1	1
Physics	1			1		1	1
History			1	1	1		1
Art				1	1	1	

TABLE 5-4

27. Figure 5-40 shows the downtown area of the small village of Kenton. The village wants to have a Fourth of July parade that passes through all the blocks of the downtown area, except for the 14 blocks highlighted in yellow, which the police department considers unsafe for the parade route. Draw a graph that models this street-routing problem.

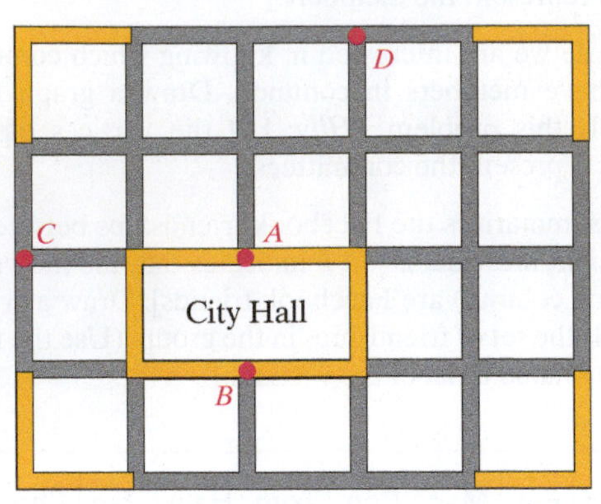

FIGURE 5-40

28. Figure 5-40 shows the downtown area of the small village of Kenton. At regular intervals at night, a police officer must patrol every downtown block at least once, and each of the six blocks along City Hall at least twice. Draw a graph that models this street-routing problem.

5.3 Euler's Theorems and Fleury's Algorithm

In Exercises 29 through 34 choose from one of the following answers and provide a short explanation for your answer using Euler's theorems.
(A) the graph has an Euler circuit.
(B) the graph has an Euler path.
(C) the graph has neither an Euler circuit nor an Euler path.
(D) the graph may or may not have an Euler circuit.
(E) the graph may or may not have an Euler path.
(F) there is no such graph.

29. **(a)** Fig. 5-41(a)

(b) Fig. 5-41(b)

(c) A *connected* graph with eight vertices, all of degree 2

(d) A graph with 8 vertices, all of degree 2. (*Hint:* See Exercise 9).

(e) A connected graph with eight vertices: three vertices of degree 2 and five vertices of degree 3.

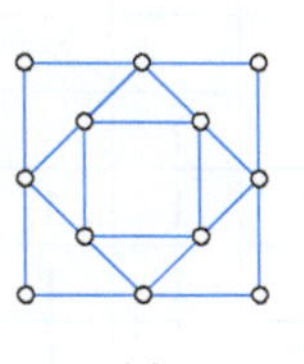
(a)

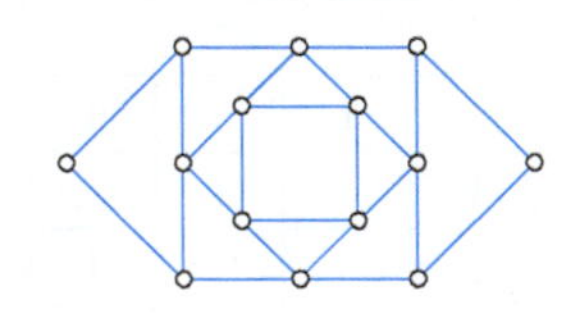
(b)

FIGURE 5-41

30. **(a)** Fig. 5-42(a)

(b) Fig. 5-42(b)

(c) A connected graph with eight vertices: four vertices of degree 2 and four vertices of degree 3

(d) A graph with eight vertices: four vertices of degree 4 and four vertices of degree 3.

(e) A connected graph with nine vertices of degree 3.

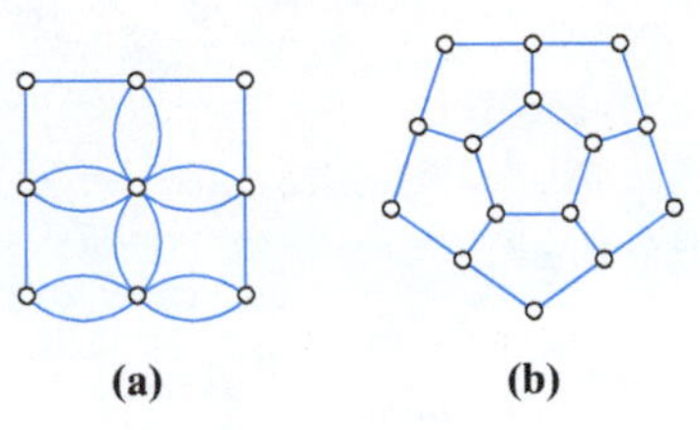
(a) (b)

FIGURE 5-42

31. **(a)** Fig. 5-43(a)

(b) Fig. 5-43(b)

(c) A graph with eleven vertices, all of degree 3

(d) A graph with twelve vertices, all of degree 3.

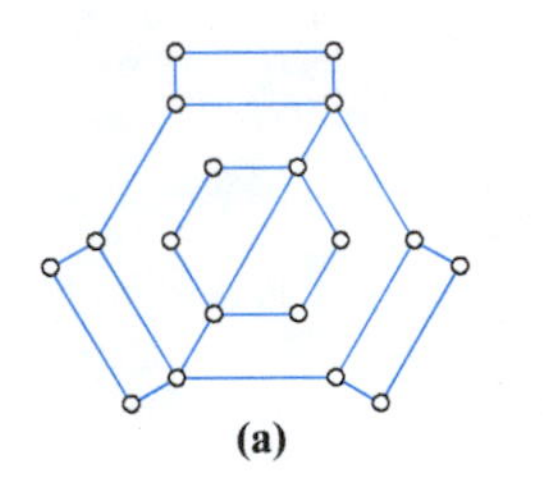
(a)

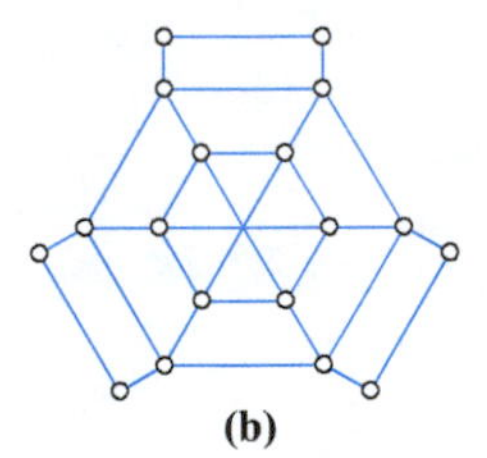
(b)

FIGURE 5-43

32. **(a)** Fig. 5-44(a)

(b) Fig. 5-44(b)

(c) A graph with six vertices of degrees 1, 2, 3, 4, 5, and 6, respectively.

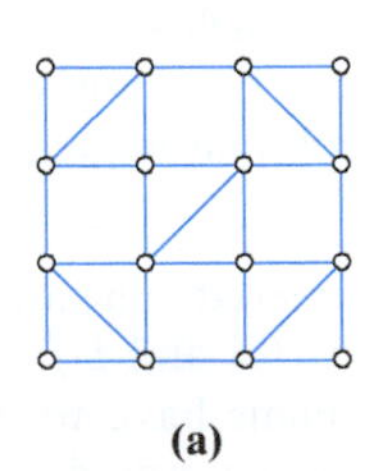
(a)

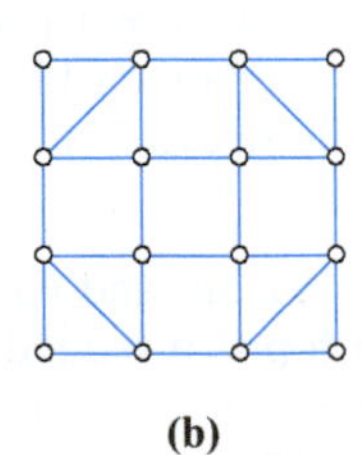
(b)

FIGURE 5-44

33. **(a)** Fig. 5-45(a)

(b) Fig. 5-45(b)

(c) A graph with six vertices: one vertex of degree 0 and five vertices of degree 2.

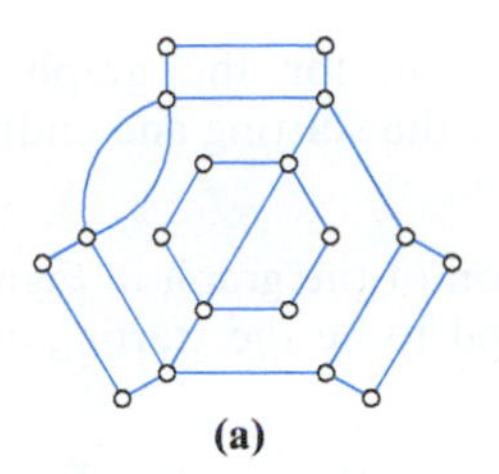

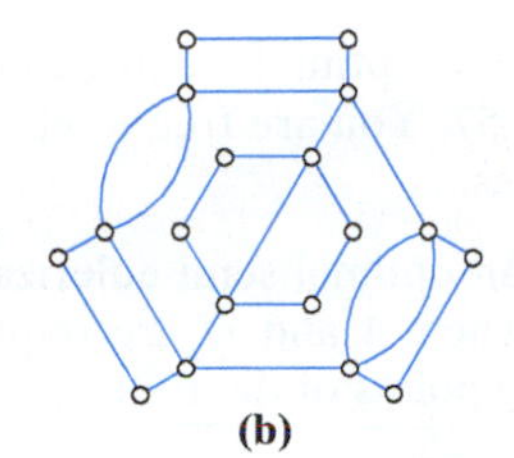

FIGURE 5-45

34. **(a)** Fig. 5-46(a)

(b) Fig. 5-46(b)

(c) A graph with six vertices: two vertices of degree 0, two vertices of degree 2, and two vertices of degree 3.

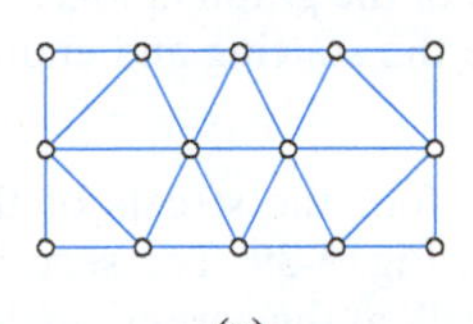

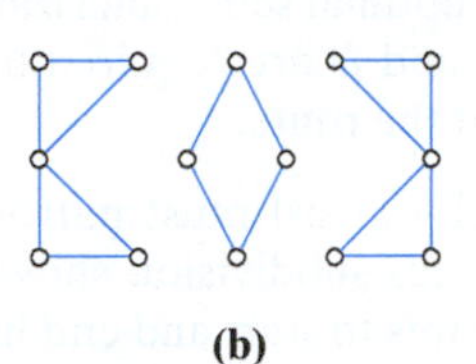

FIGURE 5-46

35. Find an Euler circuit for the graph in Fig. 5-47. Show your answer by labeling the edges 1, 2, 3, and so on in the order in which they are traveled.

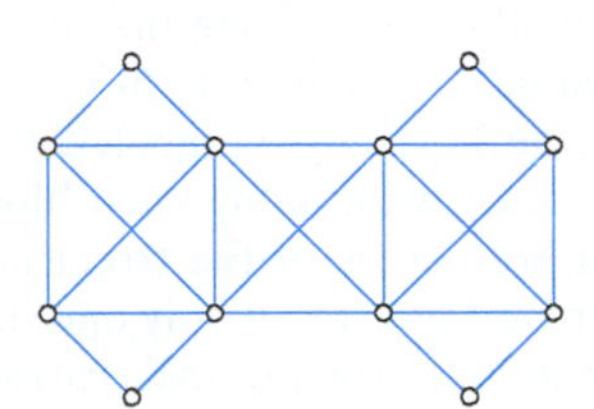

FIGURE 5-47

36. Find an Euler circuit for the graph in Fig. 5-48. Show your answer by labeling the edges 1, 2, 3, and so on in the order in which they can be traveled.

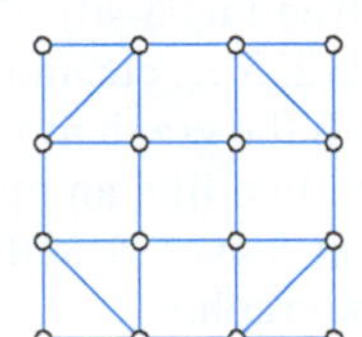

FIGURE 5-48

37. Find an Euler path for the graph in Fig. 5-49. Show your answer by labeling the edges 1, 2, 3, and so on in the order in which they are traveled.

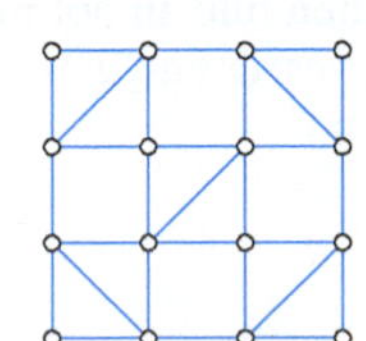

FIGURE 5-49

38. Find an Euler path for the graph in Fig. 5-50. Show your answer by labeling the edges 1, 2, 3, and so on in the order in which they are traveled.

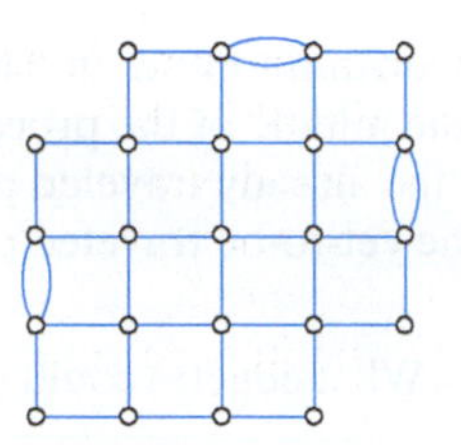

FIGURE 5-50

39. Find an Euler circuit for the graph in Fig. 5-51. Use B as the starting and ending point of the circuit. Show your answer by labeling the edges 1, 2, 3, and so on in the order in which they are traveled.

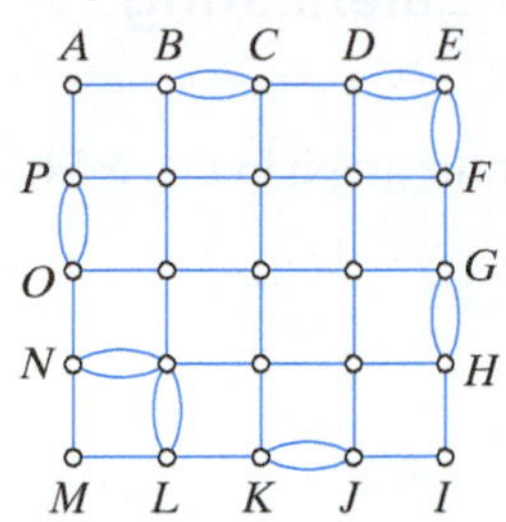

FIGURE 5-51

40. Find an Euler circuit for the graph in Fig. 5-52. Use P as the starting and ending point of the circuit. (This is the route for the mail carrier problem in Example 5-22.) Show your answer by labeling the edges 1, 2, 3, and so on in the order in which they are traveled.

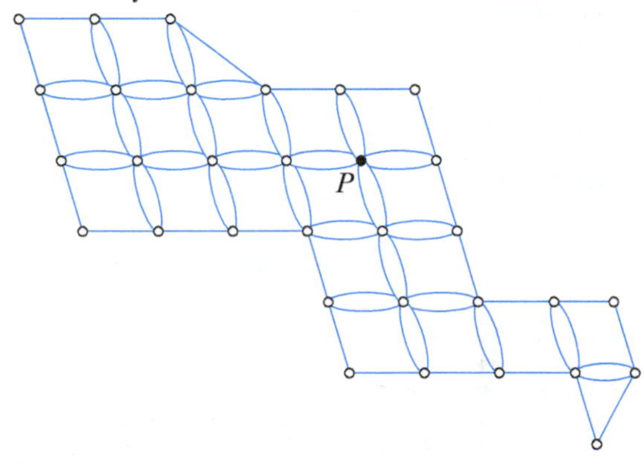

FIGURE 5-52

41. Suppose you are using Fleury's algorithm to find an Euler circuit for a graph and you are in the middle of the process. The graph in Fig. 5-53 shows both the already traveled part of the graph (the red edges) and the yet-to-be traveled part of the graph (the blue edges).

(a) Suppose you are standing at P. What edge(s) could you choose next?

(b) Suppose you are standing at B. What edge should you *not* choose next?

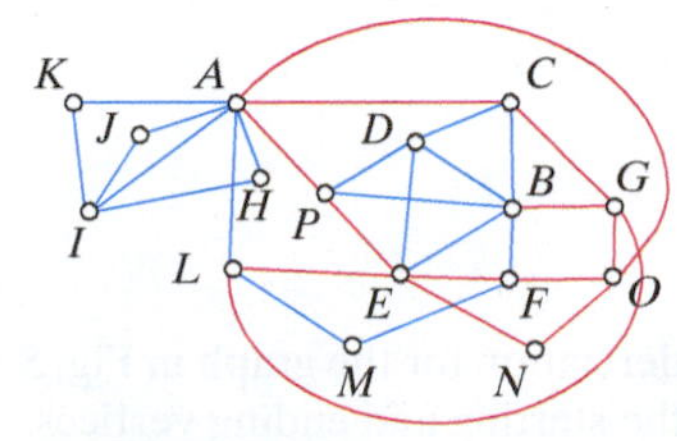

FIGURE 5-53

42. Suppose you are using Fleury's algorithm to find an Euler circuit for a graph and you are in the middle of the process. The graph in Fig. 5-53 shows both the already traveled part of the graph (the red edges) and the yet-to-be traveled part of the graph (the blue edges).

(a) Suppose you are standing at C. What edge(s) could you choose next?

(b) Suppose you are standing at A. What edge should you *not* choose next?

5.4 Eulerizing and Semi-Eulerizating Graphs

43. Find an optimal eulerization for the graph in Fig. 5-54.

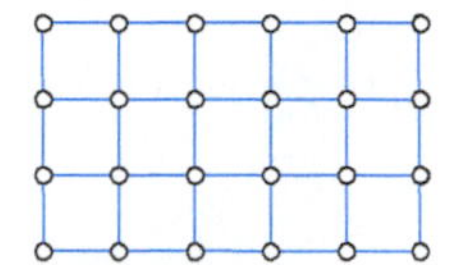

FIGURE 5-54

44. Find an optimal eulerization for the graph in Fig. 5-55.

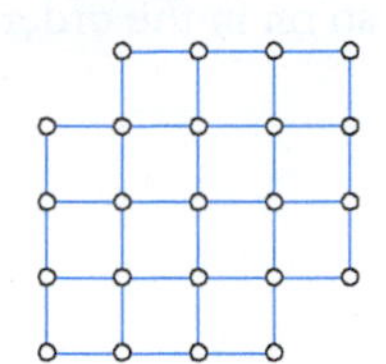

FIGURE 5-55

45. Find an optimal eulerization for the graph in Fig. 5-56.

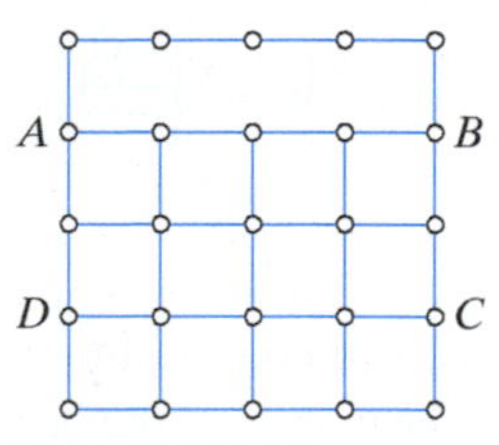

FIGURE 5-56

46. Find an optimal eulerization for the graph in Fig. 5-57.

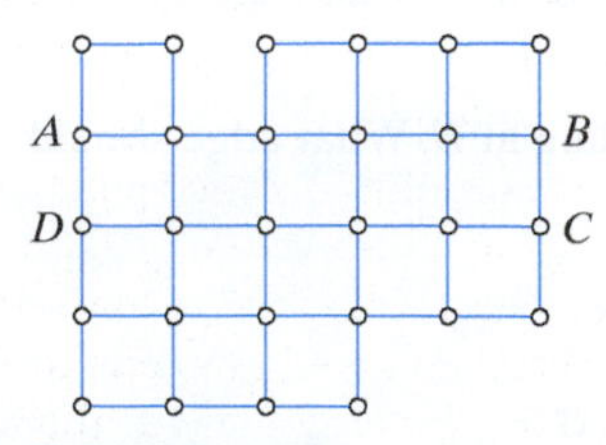

FIGURE 5-57

47. Find an optimal semi-eulerization for the graph in Fig. 5-56. You are free to choose the starting and ending vertices.

48. Find an optimal semi-eulerization for the graph in Fig. 5-57. You are free to choose the starting and ending vertices.

49. Find an optimal semi-eulerization of the graph in Figure 5-56 when A and D are required to be the starting and ending points of the route.

50. Find an optimal semi-eulerization of the graph in Figure 5-57 when A and B are required to be the starting and ending points of the route.

51. Find an optimal semi-eulerization of the graph in Figure 5-56 when B and C are required to be the starting and ending points of the route.

52. Find an optimal semi-eulerization of the graph in Fig. 5-57 when A and D are required to be the starting and ending points of the route.

53. A security guard must patrol on foot the streets of the Green Hills subdivision shown in Fig. 5-39. The security guard wants to start and end his walk at the corner labeled A, and he needs to cover each block of the subdivision at least once. Find an optimal route for the security guard. Describe the route by labeling the edges 1, 2, 3, and so on in the order in which they are traveled. (*Hint*: You should do Exercise 21 first.)

54. A mail carrier must deliver mail on foot along the streets of the Green Hills subdivision shown in Fig. 5-39. His route must start and end at the Post Office, labeled P in the figure. The mail carrier must walk along each block twice if there are houses on both sides of the street and once along blocks where there are houses on only one side of the street. Find an optimal route for the mail carrier. Describe the route by labeling the edges 1, 2, 3, and so on in the order in which they are traveled. (*Hint*: You should do Exercise 22 first.)

55. This exercise refers to the Fourth of July parade problem introduced in Exercise 27. Find an optimal route for the parade that starts at A and ends at B (see Fig. 5-40). Describe the route by labeling the edges 1, 2, 3,... etc. in the order they are traveled. [*Hint*: Start with the graph model for the parade route (see Exercise 27); then find an optimal semi-eulerization of the graph that leaves A and B odd; then find an Euler path in this new graph.]

56. This exercise refers to the Fourth of July parade problem introduced in Exercise 27. Find an optimal route for the parade that starts at C and ends at D (see Fig. 5-40). Describe the route by labeling the edges 1, 2, 3,... etc. in the order they are traveled. [*Hint*: Start with the graph model for the parade route (see Exercise 27); then find an optimal semi-eulerization of the graph that leaves C and D odd; then find an Euler path in this new graph.]

JOGGING

57. Consider the following puzzle: You must trace Fig. 5-58 *without retracing any lines*. We know this is impossible without lifting the pencil, so you are allowed to lift the pencil. What is the minimum number of times that you must lift the pencil in order to trace the entire figure? Explain your answer. (*Hint:* Try Exercise 47 first.)

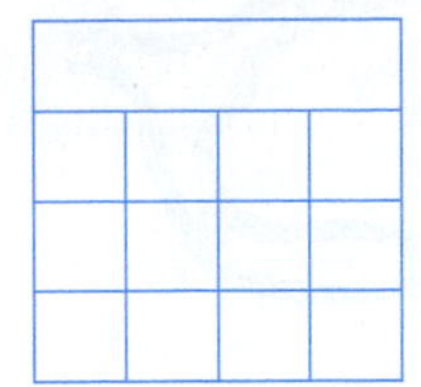

FIGURE 5-58

58. (a) Explain why in every graph the sum of the degrees of all the vertices equals twice the number of edges.

(b) Explain why every graph must have an even number of odd vertices.

59. If G is a connected graph with no bridges, how many vertices of degree 1 can G have? Explain your answer.

60. Regular graphs. A graph is called *regular* if every vertex has the same degree. Let G be a connected regular graph with N vertices.

(a) Explain why if N is odd, then G must have an Euler circuit.

(b) When N is even, then G may or may not have an Euler circuit. Give examples of both situations.

61. Suppose G is a disconnected graph with exactly two odd vertices. Explain why the two odd vertices must be in the same component of the graph.

62. Consider the following game. You are given N vertices and are required to build a graph by adding edges connecting these vertices. Each time you add an edge you must pay $1. You can stop when the graph is connected.

(a) Describe the strategy that will cost you the least amount of money.

(b) What is the minimum amount of money needed to build the graph? (Give your answer in terms of N.)

63. Figure 5-59 shows a map of the downtown area of the picturesque hamlet of Kingsburg. You have been hired by the Kingsburg Chamber of Commerce to organize the annual downtown parade. Part of your job is to plan the route for the parade. An *optimal* parade route is one that keeps the bridge crossings to a minimum and yet crosses each of the seven bridges in the downtown area at least once.

(a) Find an optimal parade route if the parade is supposed to start in North Kingsburg but can end anywhere.

(b) Find an optimal parade route if the parade is supposed to start in North Kingsburg and end in South Kingsburg.

(c) Find an optimal parade route if the parade is supposed to start in North Kingsburg and end on island B.

(d) Find an optimal parade route if the parade is supposed to start in North Kingsburg and end on island A.

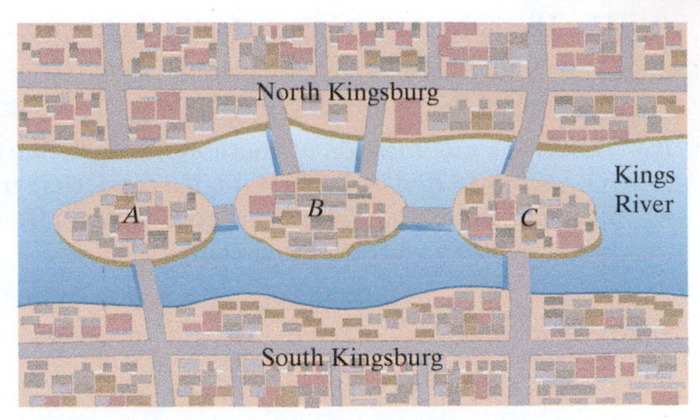

FIGURE 5-59

64. Kissing circuits. When two circuits in a graph have no edges in common but share a common vertex v, they are said to be *kissing at* v.

(a) For the graph shown in Fig. 5-60, find a circuit kissing the circuit A, D, C, A (there is only one), and find two different circuits kissing the circuit A, B, D, A.

(b) Suppose G is a connected graph and every vertex in G is even. Explain why the following statement is true: *If a circuit in G has no kissing circuits, then that circuit must be an Euler circuit.*

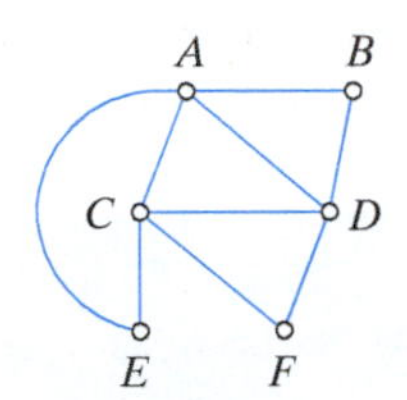

FIGURE 5-60

65. Hierholzer's algorithm. *Hierholzer's algorithm* is another algorithm for finding an Euler circuit in a graph. The basic idea behind Hierholzer's algorithm is to start with an arbitrary circuit and then enlarge it by patching to it a *kissing circuit*, continuing this way and making larger and larger circuits until the circuit cannot be enlarged any farther. (For the definition of kissing circuits, see Exercise 64.) More formally, Hierholzer's algorithm is as follows:

Step 1. Start with an arbitrary circuit C_0.

Step 2. Find a kissing circuit to C_0. If there are no kissing circuits to C_0, then you are finished—C_0 is itself an Euler circuit of the graph [see Exercise 64(b)]. If there is a kissing circuit to C_0, let's call it K_0, and let V denote the vertex at which the two circuits kiss. Go to Step 3.

Step 3. Let C_1 denote the circuit obtained by "patching" K_0 to C_0 at vertex V (i.e., start at V, travel along C_0 back to V, and then travel along K_0 back again to V). Now find a kissing circuit to C_1. (If there are no kissing circuits to C_1, then you are finished—C_1 is your Euler circuit.) If there is a kissing circuit to C_1, let's call it K_1, and let W denote the vertex at which the two circuits kiss. Go to Step 4.

Steps 4, 5, and so on. Continue this way until there are no more kissing circuits available.

(a) Use Hierholzer's algorithm to find an Euler circuit for the graph shown in Fig. 5-61 (this is the graph model for the mail carrier in Example 5.14).

(b) Describe a modification of Hierholzer's algorithm that allows you to find an Euler path in a connected graph having exactly two vertices of odd degree. (*Hint*: A path can also have a kissing circuit.)

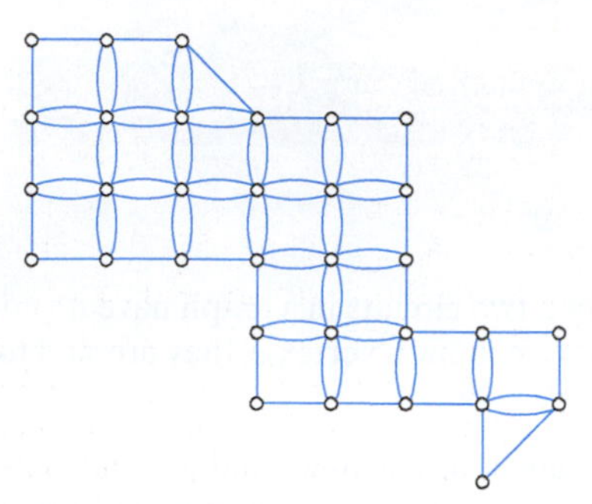

FIGURE 5-61

Exercises 66 through 68 refer to Example 5.23. In this example, the problem is to find an optimal route (i.e., a route with the fewest bridge crossings) for a photographer who needs to cross each of the 11 bridges of Madison County for a photo shoot. The layout of the 11 bridges is shown in Fig. 5-62. You may find it helpful to review Example 5.23 before trying these two exercises.

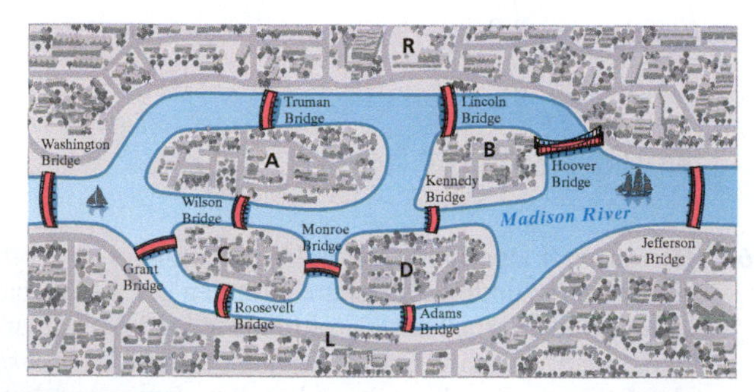

FIGURE 5-62

66. Describe an optimal route for the photographer if the route must start at B and end at L.

67. Describe an optimal route for the photographer if the route must start and end in D and the first bridge crossed must be the Adams Bridge.

68. Describe an optimal route for the photographer if the route must start and end in the same place, the first bridge crossed must be the Adams bridge, and the last bridge crossed must be the Grant Bridge.

69. This exercise comes to you courtesy of Euler himself. Here is the question in Euler's own words, accompanied by the diagram shown in Fig. 5-63.

Let us take an example of two islands with four rivers forming the surrounding water. There are fifteen bridges

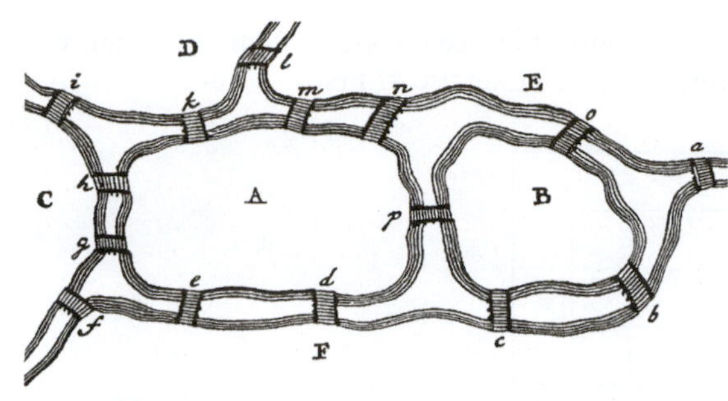

FIGURE 5-63

marked a, b, c, d, etc., across the water around the islands and the adjoining rivers. The question is whether a journey can be arranged that will pass over all the bridges but not over any of them more than once.

What is the answer to Euler's question? If the "journey" is possible, describe it. If it isn't, explain why not.

RUNNING

70. Suppose G is a connected graph with N vertices, all of even degree. Let k denote the number of bridges in G. Find the value(s) of k. Explain your answer.

71. Suppose G is a connected graph with $N-2$ even vertices and two odd vertices. Let k denote the number of bridges in G. Find all the possible values of k. Explain your answer.

72. Complete bipartite graphs. A complete bipartite graph is a graph having the property that the vertices of the graph can be divided into two groups A and B and each vertex in A is adjacent to each vertex in B, as shown in Fig. 5-64. Two vertices in A are never adjacent, and neither are two vertices in B. Let m and n denote the number of vertices in A and B, respectively, and assume $m \leq n$.

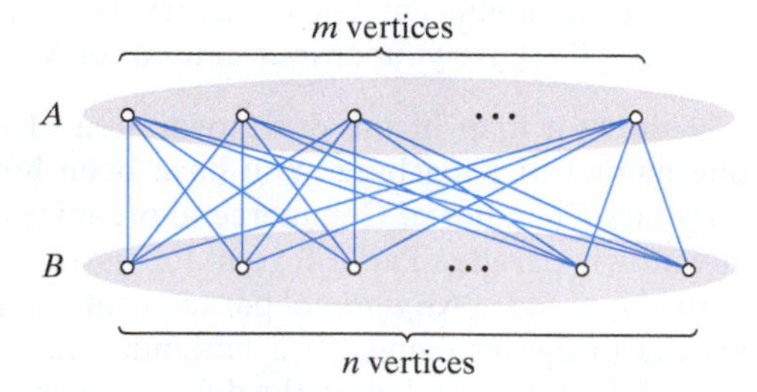

FIGURE 5-64

(a) Describe all the possible values of m and n for which the complete bipartite graph has an Euler circuit. (*Hint*: There are infinitely many values of m and n.)

(b) Describe all the possible values of m and n for which the complete bipartite graph has an Euler path.

73. Suppose G is a simple graph with N vertices ($N \geq 2$). Explain why G must have at least two vertices of the same degree.

APPLET BYTES MyMathLab®

These Applet Bytes are mini-explorations built around the applet* ***Euler Paths and Circuits: Fleury's Algorithm*** *(available in MyMathLab in the Multimedia Library or Tools for Success). The applet allows you to implement Fleury's algorithm on either one of the pre-defined larger graphs available under the Custom menu, or alternatively, on a graph of your own creation.*

74. A policeman has to patrol on foot the streets of the subdivision shown in Fig. 5-65. The policeman needs to start his route at the police station, located at *X*, and end the route at the local coffee shop, located at *Y*. He needs to cover each block of the subdivision at least once, but he wants to make his route as efficient as possible and duplicate the fewest possible number of blocks.

FIGURE 5-65

(a) Open the applet *Euler's Path and Circuits: Fleury's Algorithm* and use the "Create Graph" tab to create a graph model of the subdivision.

(b) While still on the "Create Graph" tab find an optimal semi-eulerization of the graph model that has the vertices corresponding to locations *X* and *Y* as the only two odd vertices. (*Hints:* 1. You will need to add 12 duplicate edges. 2. You can take advantage of the "Show Degrees" box at the bottom of the window to find the odd vertices of the graph.)

(c) Once you found an optimal semi-eulerization of the graph model, go to the "Practice Fleury's Algorithm" tab and find an optimal route for the policeman to cover the subdivision. (Any route will do for this part.)

(d) Suppose that the policeman wants to be able to take a break and stop at the coffee shop *Y* halfway through his route. Find an optimal route that allows him to do so. (*Hints:* 1. You can use the "Reset" button in the "Practice Fleury's Algorithm" tab to start a new route from scratch. 2. There are a total of 81 blocks in the route, so this route should schedule a pass through the coffee shop *Y* at step 40 or 41.)

75. The Security Guard Problem revisited. In Examples 5.1, 5.13, and 5.21 we discussed the problem of finding an optimal route for a security guard patrolling the streets of the Sunnyside neighborhood. In this Applet Byte we revisit the problem and add a twist to it.

(a) Figure 5-25(d) of Example 5.21 shows an optimal route for the security guard. This is one of thousands of possible optimal routes. Since there are so many, you should have no trouble finding a different one using the "Practice Fleury's Algorithm tab of the applet. (*Hint and Good News*: You don't have to build the graph model from scratch—it is already built for you: in the "Create Graph" tab go to the "Custom" menu and choose "Security Guard").

(b) Based on the route you found in (a), record the difference between the two edge labels in the nine double-edges of the graph. These differences indicate the "delays" between the first and second pass through that block. (For example, say a double edge has labels 10 and 19. This means that the security guard covered that block first as the 10th block of the route and then again as the 19th block of the route—a delay of 9 blocks between first and second passes.) We will call a delay of 20 or less a "short" delay, a delay between 21 and 30 a "medium" delay, and a delay of 31 or more a "long" delay. (For security reasons, short delays are bad, medium delays are acceptable and long delays are best.)

(c) Compute the number of short delays, medium delays and long delays of the route you found in (a), and give the route a score (no points for short delays, 1 point for medium delays and 2 points for long delays).

(d) Use the applet to find a route that has as few short delays and as many long delays as possible. Try to maximize the score of your new route.

76. The UPS Driver Problem revisited. In Examples 5.3 and 5.24 we discussed the problem of finding an optimal route for a UPS driver that must deliver packages at multiple locations in the Sunnyside neighborhood. In this Applet Byte we follow up on the unfinished business in Example 5.24.

(a) Figure 5-28(c) of Example 5.24 shows an optimal semi-eulerization of the graph model for the UPS driver problem, but no optimal route is given. In fact, the example ends with "we leave this final detail as an exercise." So, here it is: find an optimal route for the UPS driver. (*Hint*: Choose "UPS Driver" from the "Custom" menu in the "Create Graph" tab and then go to the "Practice Fleury's Algorithm" tab to do the work.)

(b) Count the number of left turns in the route you found in (a). (*Note*: Assume all U-turns are left turns, as this problem is set in the U.S., where we drive on the right side of the road. In England, Ireland, the Virgin Islands, and any other place where they drive on the left side of the road all U-turns are right turns.)

(c) To UPS drivers left turns are bad news (they often have to stop and wait for oncoming traffic to clear), and a good route is one that has as few left turns as possible. Use the applet to find a route with as few left turns as you can muster.

* MyMathLab code required.

August 2015: Curiosity takes a break from drilling to take this selfie at a drilling site called "Big Sky." NASA/JPL-Caltech/MSSS. (For more details, see Example 6.3.)

The Mathematics of Touring

Traveling Salesman Problems

The Mars Science Laboratory *Curiosity* is a six-wheeled rover that looks like a dune buggy on steroids and cost NASA $2.5 billion to build and launch. *Curiosity*, loaded with fancy cameras and all kinds of scientific instruments, left Cape Canaveral, Florida, on November 26, 2011 and landed on the Gale crater region of Mars on August 6, 2012. *Curiosity*'s mission is to explore and drill in the general vicinity of the Gale crater—an area where planetary scientists believe there is a good chance of finding chemical and biological markers that might be evidence of past or present life on Mars. To put it simply, *Curiosity* is on the hunt for tiny Martians, dead or alive.

Curiosity is not built for speed (on a real good *sol*—that's one Earth day in Mars—it might travel 500 to 600 feet) but rather for endurance and the ability to move over and around obstacles that might appear in its path. In other words, *Curiosity* is a rugged *traveler*, moving slowly but steadily through various locations on the rough and uncharted territory that is the surface of Mars. The less time *Curiosity* has to spend moving around, the more time it has to conduct its experiments, so one of the key aspects of *Curiosity*'s mission is to optimize its travels. This is where the mathematics comes in.

The general problem of optimizing the route of a *traveler* that must visit a specified set of *locations* is known as the *traveling salesman problem* (TSP). This name is misleading, and the typical *traveling salesman problem* has nothing to do with a traveling salesperson—the name applies to many important real-life problems, including the routing of a $2.5- billion roving laboratory on the surface of Mars.

This chapter starts with a general description of what constitutes a TSP, followed by several real-life examples of TSPs (Section 6.1). In Section 6.2 we introduce and discuss the key mathematical concepts that are used to model a TSP (*Hamilton circuits, Hamilton paths*, and *complete graphs*). In Sections 6.3, 6.4, and 6.5, we introduce four different algorithms for solving TSPs: the *brute-force*, *nearest-neighbor*, *repetitive nearest-neighbor*, and *cheapest-link* algorithms, and use these algorithms to "solve" the TSPs introduced in Section 6.1. In these sections we also discuss the pros and cons of the various algorithms and introduce the concepts of *inefficient* and *approximate* algorithms.

6.1 What Is a Traveling Salesman Problem?

The term *traveling salesman problem* (**TSP**) is catchy but a bit misleading, since most of the time the problems that fall under this heading have nothing to do with salespeople living out of a suitcase. The expression "traveling salesman" has traditionally been used as a convenient metaphor for many different real-life problems that share a common mathematical structure.

The three elements common to all TSPs (from now on we simply refer to any traveling salesman problem as a TSP) are the following:

- **A traveler.** The traveler could be a person, a vehicle (a bus, a spaceship, an unmanned rover, etc.); it could even be a bee.
- **A set of sites.** These are the places or locations the traveler must visit. We will use N to denote the number of sites.
- **A set of costs.** These are positive numbers associated with the expense of traveling from a site to another site. Here the "cost" variable is not restricted to just *monetary* cost—it can also represent *distance* traveled or *time* spent on travel.

A *solution* to a TSP is a "trip" that starts and ends at one site and visits all the other sites once (but only once). We call such a trip a **tour**. An *optimal solution* (**optimal tour**) is a tour of minimal *total cost*. (Notice that we used *a tour* rather than *the tour*—in general a TSP has more than one optimal solution.)

Let's now look at some examples that should help clarify the types of problems we call TSPs. We will not solve any of these TSPs in this section, but we will discuss solutions in later sections.

EXAMPLE 6.1 A REAL TRAVELING SALESMAN'S TSP

Willy "the Traveler" is a traveling salesman who spends a lot of time on the road calling on customers. He is planning his next business trip, where he will visit customers in five cities we will call A, B, C, D, and E for short $(N = 5)$. Since A is Willy's hometown, he needs to start and end his trip at A.

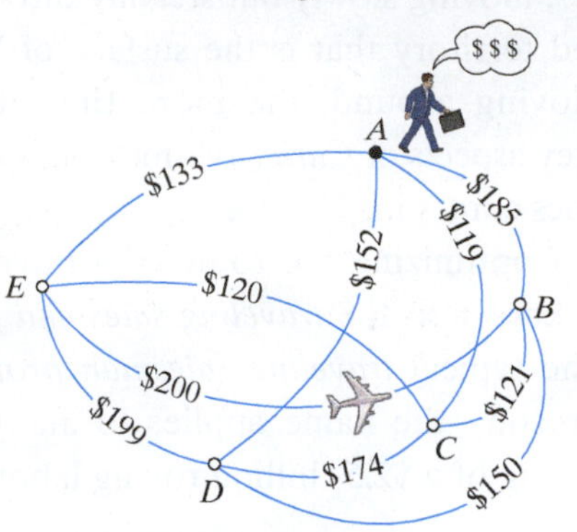

FIGURE 6-1 Cost of travel between cities.

The graph in Fig. 6-1 shows the *cost* (in dollars) of a one-way ticket between any pair of cities (for simplicity we are assuming the cost of a one-way ticket is the same regardless of the direction of travel—something that is not always true in the crazy world of modern airline ticket prices). Like most people, Willy hates to waste money, so among the many possible ways he can organize the sales tour Willy wants to find the *cheapest* (i.e., *optimal*) *tour*. We will see the solution to Willy's TSP in Section 6.3, but if you want to give it a try on your own now, please feel free to do so.

EXAMPLE 6.2 THE INTERPLANETARY MISSION TSP

July, 2016: NASA's Juno spacecraft enters Jupiter's orbit. (NASA/JPL-Caltech.)

It is the year 2050. An unmanned mission to explore the outer planetary moons in our solar system is about to be launched from Earth. The mission is scheduled to visit Callisto, Ganymede, Io, Mimas, and Titan (the first three are moons of Jupiter; the last two of Saturn), collect rock samples at each, and then return to Earth with the loot. The graph in Fig. 6-2 shows the time (in years) required for exploration and travel between any two moons, as well as between Earth and any moon.

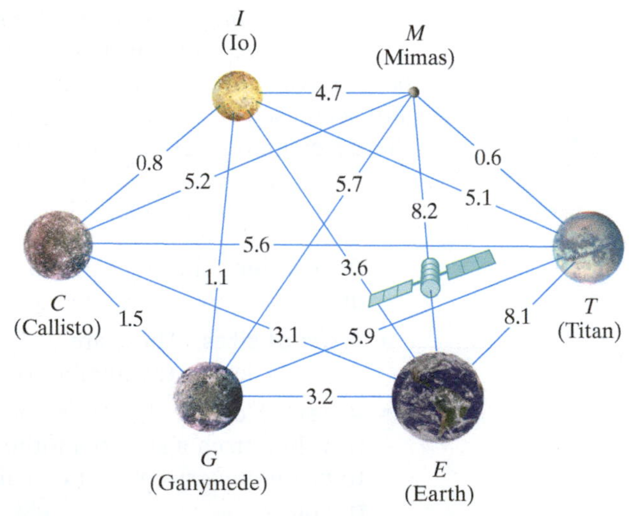

FIGURE 6-2 Travel time between moons (in years).

This is a long mission, and one of the obvious goals of the mission is to complete it in the least time. In this TSP, the *traveler* is the spacecraft, the *sites* are the moons plus the starting and ending site Earth $(N = 6)$, and the *cost variable* is time. As in Example 6.1, the goal is to find an optimal tour. We will solve the inter-planetary moons TSP in Sections 6.3, 6.4, and 6.5.

EXAMPLE 6.3 THE *CURIOSITY* TSP

The story of the Mars Science Laboratory nicknamed *Curiosity* was introduced in the chapter opener. *Curiosity* is a six-wheeled $2.5-billion rover loaded with cameras and instruments that roams around a region of Mars known as the Gale crater, collecting rocks and doing chemical and biological assays of the soil. The primary mission objective is to look for evidence of the existence of life (past or present) on Martian soil. *Curiosity* is a slow traveler (about 30 meters an hour on average), and the Martian terrain is treacherous for a rover, so efficient planning of its travels is a critical part of the mission.

Curiosity (the green star in this image) hits the 10,000 m mark on its odometer on its way to Logan Pass (one of the drilling sites).

Figure 6-3(a) is a graph showing seven locations around the Gale crater (for simplicity we'll call them G_1 through G_7) that *Curiosity* is scheduled to visit, with G_1 being the landing site. The numbers on each edge represent travel distances in meters. [To make some of these distances meaningful, consider the fact that *Curiosity* can cover only about 200 m a day, so the trek from say G_2 to G_5 (8000 m) would take roughly 40 days.] Note that in the graph the edges are not drawn to scale—as in any graph, the positioning of the vertices is irrelevant and the only data that are relevant in this case are the numbers (distances) associated with each edge. In spite of a conscious effort to render the graph as clear and readable as possible, it's hard to get around the fact that there are a lot of numbers in the picture and things get pretty crowded. The distance chart in Fig. 6-3(b) provides exactly the same information as the graph in Fig. 6-3(a) but is a little easier to work with. [For TSPs with more than six sites ($N > 6$), a chart is generally preferable to a graph.]

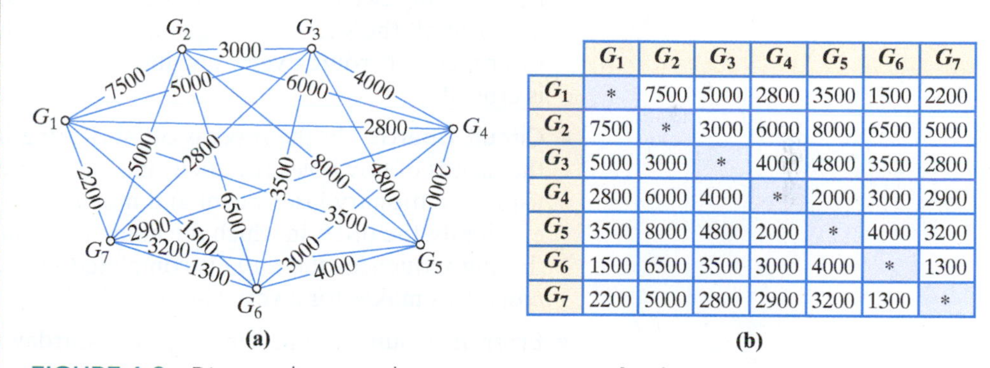

	G_1	G_2	G_3	G_4	G_5	G_6	G_7
G_1	*	7500	5000	2800	3500	1500	2200
G_2	7500	*	3000	6000	8000	6500	5000
G_3	5000	3000	*	4000	4800	3500	2800
G_4	2800	6000	4000	*	2000	3000	2900
G_5	3500	8000	4800	2000	*	4000	3200
G_6	1500	6500	3500	3000	4000	*	1300
G_7	2200	5000	2800	2900	3200	1300	*

FIGURE 6-3 Distance between locations (in meters) for the *Curiosity* TSP.

Given the harsh conditions on Mars, *Curiosity*'s tour will be long and full of risks. As one might imagine, the goal is to find a tour that minimizes distance and thus takes the least amount of travel time. In this TSP, *Curiosity* is the traveler, $N = 7$, and the cost variable is distance. We will solve the *Curiosity* TSP in Section 6.5. ■

EXAMPLE 6.4 THE CONCERT TOUR TSP

The indie rock band Luna Park is planning a concert tour. The 10 cities booked for the tour (including the band's home base A) and the distances between them are given in the mileage chart shown in Fig. 6-4. (Note that in this example we made no attempt to display the distances in a graph—to say the least, it would be messy.)

	A	B	C	D	E	F	G	H	J	K
A	*	185	119	152	133	321	297	277	412	381
B	185	*	121	150	200	404	458	492	379	427
C	119	121	*	174	120	332	439	348	245	443
D	152	150	174	*	199	495	480	500	454	489
E	133	200	120	199	*	315	463	204	396	487
F	321	404	332	495	315	*	356	211	369	222
G	297	458	439	480	463	356	*	471	241	235
H	277	492	348	500	204	211	471	*	283	478
J	412	379	245	454	396	369	241	283	*	304
K	381	427	443	489	487	222	235	478	304	*

FIGURE 6-4 Mileage chart for the concert tour.

The goal is to find the *shortest* (i.e., optimal) concert tour. In this TSP the *traveler* is the rock band (and their entourage), the *sites* are the cities in the concert tour ($N = 10$), and the *cost variable* is distance traveled. We will discuss and partially solve the concert tour TSP in Section 6.4. ■

Beyond the TSPs introduced in Examples 6.1 through 6.4 (we will follow through on these later in the chapter), the following is a short list of other real-world applications where TSPs arise:

- **School buses.** A school bus (the *traveler*) picks up children in the morning and drops them off at the end of the day at designated stops (the *sites*). On a typical rural school bus route there may be 20 to 30 such stops. With school buses, total time on the bus is always the most important variable (students have to get to school on time), and there is a known time of travel (the *cost*) between any two bus stops. Since children must be picked up at every bus stop, a *tour* of all the sites (starting and ending at the school) is required. Since the bus repeats its route every day during the school year, finding an *optimal tour* is crucial.
- **Circuit boards.** In the process of fabricating integrated-circuit boards, tens of thousands of tiny holes (the *sites*) must be drilled in each board. This is done by using a stationary laser beam and moving the board (the *traveler*). To do this efficiently, the order in which the holes are drilled should be such that the entire drilling sequence (the *tour*) is completed in the least amount of time (*optimal cost*). This makes for a very high-tech TSP.
- **Errands around town.** On a typical Saturday morning, an average Joe or Jane (the *traveler*) sets out to run a bunch of errands around town, visiting various sites (grocery store, hair salon, bakery, post office). When gas was cheap, *time* used to be the key *cost* variable, but with the cost of gas these days, people are more likely to be looking for the tour that minimizes the total *distance* traveled (see Exercise 59 for an illustration of this TSP).
- **Bees do it.** In 2010, scientists at the University of London studying the travel patterns of foraging bumblebees as they search for their food source (flower nectar) made a surprising discovery: The bees fly from flower to flower not in the order in which they originally discover the flowers but in the order that gives the shortest overall route. In other words, the bees are routing their foraging trips by solving a TSP. Scientists still don't understand what methods bumblebees use to solve such complex mathematical problems, but the answer is clearly not by doing numerical calculations—the brain of a bee is the size of a grain of sand. Understanding the shortcuts that allow a humble bumblebee to solve a TSP is important, as it may be possible to program the same shortcuts in modern computer algorithms used to solve large-scale TSPs.

Hamilton Paths and Circuits

Sir William Rowan Hamilton (1805–1865)

In 1857, the Irish mathematician Sir William Rowan Hamilton invented a board game the purpose of which was to find a trip along the edges of the graph shown in Fig. 6-5 that visited each of the vertices once and only once, returning at the end to the starting vertex. Hamilton called the game the "Icosian Game"—don't ask why—and tried to market the game to make a little money, but he ended up just selling the rights to a London dealer for 25 pounds. (You can play the Icosian and a few other similar games using the applet ***Hamilton Paths and Circuits*** available in *MyMathLab* in the *Multimedia Library* or in *Tools for Success*.)

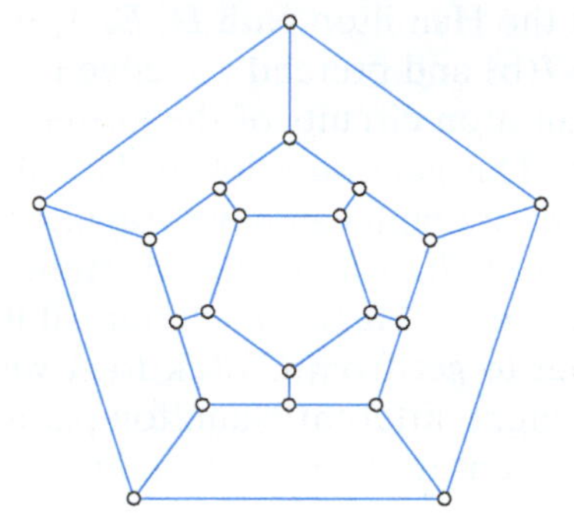

FIGURE 6-5 Hamilton's Icosian game.

The only reason the story of Hamilton's Icosian game is of any interest to us is that it illustrates an important concept in this chapter—that of traveling along the edges of a connected graph with the purpose of visiting each and every one of the *vertices* once (but only once). This leads to the following two definitions:

- **Hamilton path.** A *Hamilton path* in a connected graph is a path that visits each of the *vertices* of the graph once and only once.
- **Hamilton circuit.** A *Hamilton circuit* in a connected graph is a circuit that visits each of the *vertices* of the graph once and only once.

In spite of the similarities in the definitions, Hamilton paths and circuits are very different from Euler paths and circuits and the two should not be confused. With Hamilton the name of the game is to *visit all the vertices* of the graph once; with Euler the name of the game is to *sweep through all the edges* of the graph once.

EXAMPLE 6.5 GRAPHS WITHOUT HAMILTON CIRCUITS

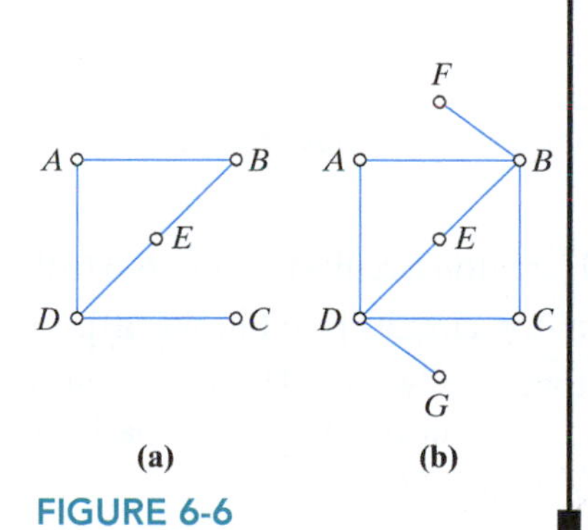

FIGURE 6-6

Figure 6-6 shows two graphs. The graph in Fig. 6-6(a) has no Hamilton circuits because once you visit *C* you are stuck there. On the other hand, the graph has several Hamilton paths. One of them is *C*, *D*, *E*, *B*, *A*; another one is *C*, *D*, *A*, *B*, *E*. We can *reverse* those two paths and get two more: *A*, *B*, *E*, *D*, *C* and *E*, *B*, *A*, *D*, *C*. Clearly, the Hamilton path has to start or end at *C*, so the four Hamilton paths listed above are the only ones possible. Note that the graph has two odd vertices, so from Euler's theorem for paths (p. 151) we know that it has Euler paths as well.

The graph in Fig. 6-6(b) has no Hamilton circuits because once you visit *F* (or *G*) you are stuck there. Nor does the graph have Hamilton paths: The path has to start at either *F* or *G* and end at the other one, and there is no way to visit the other five vertices without going through some vertices more than once. Note that this graph also has two odd vertices, so we know that it has an Euler path.

EXAMPLE 6.6 LISTING HAMILTON CIRCUITS AND PATHS

The graph in Fig. 6-7(a) has no Euler circuits or paths (it has four odd vertices) but has lots of Hamilton circuits. We are going to try to list them all. To organize ourselves we will list the Hamilton circuits using *A* as the starting and ending point of the circuit. We'll start by going clockwise around the outside square and taking the "detour" to visit *E* at different times in the trip. This gives the following four Hamilton circuits: (1) *A*, *B*, *C*, *D*, *E*, *A*; (2) *A*, *B*, *C*, *E*, *D*, *A*; (3) *A*, *B*, *E*, *C*, *D*, *A*; and (4) *A*, *E*, *B*, *C*, *D*, *A*. [Figure 6-7(b) illustrates circuit (1) *A*, *B*, *C*, *D*, *E*, *A*]. We can also go around the outside square counterclockwise and get four more Hamilton circuits: (5) *A*, *D*, *C*, *B*, *E*, *A*; (6) *A*, *D*, *C*, *E*, *B*, *A*; (7) *A*, *D*, *E*, *C*, *B*, *A*; and (8) *A*, *E*, *D*, *C*, *B*, *A*. Notice that these last four are reversals of the first four.

What about Hamilton circuits that start at a different vertex—say for example *B*? Fortunately, we won't have to worry about finding more Hamilton circuits by changing the starting vertex. Any Hamilton circuit that starts and ends at B can be reinterpreted as a Hamilton circuit that starts and ends at A (or any other vertex, for that matter). For example, the circuit B, C, D, E, A, B is just the circuit A, B, C, D, E, A written in a different way [Fig. 6-7(b)]. In other words, a Hamilton circuit is defined by the ordering of the vertices and is independent of which vertex is used as the starting and ending point. This helps a great deal—*once we are sure that we listed all the Hamilton circuits that start and end at A, we know that we have listed all of them!* We will use this observation repeatedly in the next section.

In general, a graph has many more Hamilton paths than circuits. For one thing, each Hamilton circuit can be "broken" into a Hamilton path by deleting one of the edges of the circuit. For example, we can take the Hamilton circuit A, B, C, D, E, A shown in Fig. 6-7(b) and delete EA. This gives the Hamilton path A, B, C, D, E shown in Fig. 6-7(c). We can delete any other edge as well. If we delete CD from the Hamilton circuit A, B, C, D, E, A, we get the Hamilton path D, E, A, B, C [you can see this best if you look again at Fig. 6-7(b) and pretend the edge CD is gone]. You can do this with any of the eight Hamilton circuits of the graph—delete one of its edges and create a Hamilton path. This generates lots of Hamilton paths, so we will not list them all. On top of that, a graph can have Hamilton paths that are not "broken" Hamilton circuits. Figure 6-7(d) shows the Hamilton path A, B, E, D, C. This is not a path we would get from a Hamilton circuit—if it were we would be able to close it into a circuit, but to get from C back to A we would have to go through E once again. There are eight different Hamilton paths that do not come from Hamilton circuits, and we leave it to the reader to find the other seven (Exercise 9).

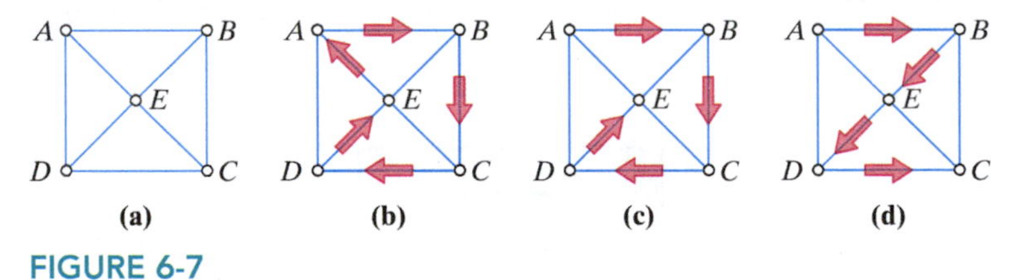

FIGURE 6-7

The following is a recap of the key points we should take away from Examples 6.5 and 6.6:

- Some graphs have neither Hamilton paths nor Hamilton circuits [see Fig. 6-6 (b)].
- Some graphs have Hamilton paths but no Hamilton circuits [see Fig. 6-6(a)].
- Any graph that has Hamilton circuits will automatically have Hamilton paths because you can always "break" a Hamilton circuit into a Hamilton path by deleting one of the edges of the circuit [see Figs. 6-7(b) and (c)].
- A graph can have Hamilton paths that do not come from a "broken" Hamilton circuit [see Fig. 6-7(d)].
- A Hamilton circuit or path can be reversed (i.e., traveled in the opposite direction). This gives a different Hamilton circuit or path. (Reversing the Hamilton circuit A, B, C, D, E, A gives the Hamilton circuit A, E, D, C, B, A.)
- The same Hamilton circuit can be written in many different ways by changing the chosen starting vertex. [A, B, C, D, E, A and B, C, D, E, A, B are two different descriptions of the Hamilton circuit in Fig. 6-7(b).]

Complete Graphs

A simple graph (i.e., no loops or multiple edges) in which the vertices are completely interconnected (every vertex is connected to every other vertex) is called a **complete graph**. Complete graphs are denoted by the symbol K_N, where N is the number of vertices. Figure 6-8 shows the complete graphs for $N = 3, 4, 5$, and 6.

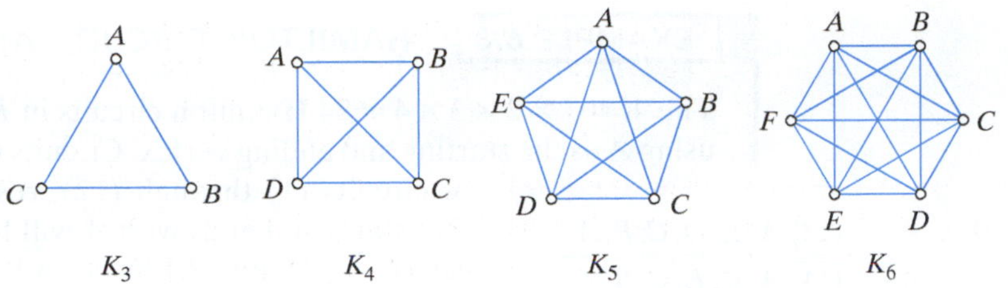

K_3 K_4 K_5 K_6

FIGURE 6-8

Listed below are four key properties of K_N:

1. The *degree* of every vertex in K_N is $N-1$.
2. The number of *edges* in K_N is $\frac{N(N-1)}{2}$.
3. The number of *Hamilton paths* in K_N is $N! = 1 \times 2 \times 3 \times \cdots \times N$.
4. The number of *Hamilton circuits* in K_N is $(N-1)! = 1 \times 2 \times 3 \times \cdots \times (N-1)$.

[*Notes*: (1) follows from the definition of K_N: since every vertex is adjacent to each of the other $N-1$ vertices, the degree of each vertex is $N-1$. (2) follows from (1) and Euler's sum of degrees theorem (p. 152): Since each vertex has degree $N-1$, the sum of all the degrees is $N(N-1)$ and the number of edges is $\frac{N(N-1)}{2}$. Properties (3) and (4) require a little more detailed explanation, so we will start by exploring in some detail the Hamilton circuits and paths in K_4 and K_5.]

EXAMPLE 6.7 HAMILTON CIRCUITS AND PATHS IN K_4

The six Hamilton circuits in K_4 are shown in Fig. 6-9. Listed using A as the starting and ending vertex they are: (a) $A, B, C, D\ A$; (b) $A, D, B, C\ A$; (c) $A, B, D, C\ A$; and their respective reversals (d) $A, D, C, B\ A$; (e) $A, C, B, D\ A$; (f) $A, C, D, B\ A$.

How do we know that Fig. 6-9 shows all possible Hamilton circuits? In a complete graph you can go from any vertex to any other vertex. Starting at A one can go to either B, C, or D; from there to either of the other two, from there to the only one left, and finally come back to A. This gives us $3 \times 2 \times 1 = 6$ different possibilities.

Each of the six Hamilton circuits can be broken into four different Hamilton paths (by deleting one of the edges). For example, the circuit A, B, C, D, A shown in Fig. 6-9(a) gives us the following four Hamilton paths: (1) A, B, C, D; (2) B, C, D, A; (3) C, D, A, B; and (4) D, A, B, C. All in all, we get $6 \times 4 = 24$ Hamilton paths. Since the graph is complete, any Hamilton path can be closed into a Hamilton circuit (just join the starting and ending vertices of the path). It follows that there are no other possible Hamilton paths—just the 24 obtained from Hamilton circuits.

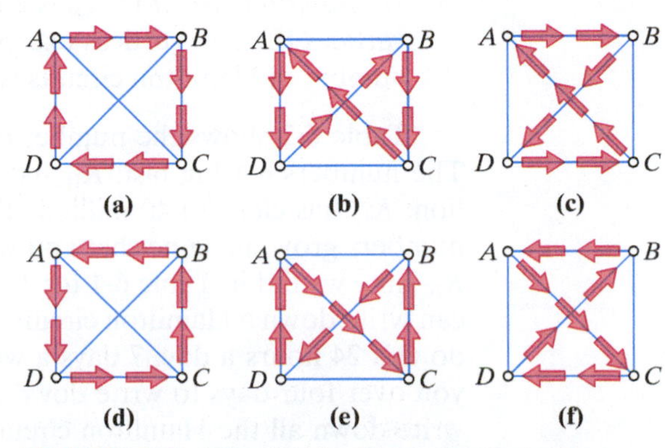

FIGURE 6-9 The six Hamilton circuits in K_4.

EXAMPLE 6.8 HAMILTON CIRCUITS AND PATHS IN K_5

The $4! = 1 \times 2 \times 3 \times 4 = 24$ Hamilton circuits in K_5 are shown in Table 6-1, written using A as the starting and ending vertex. Circuits (13) through (24) are the reversals of circuits (1) through (12), respectively. Any Hamilton circuit that starts and ends with A will have the other four "inside" vertices (B, C, D, and E) listed in between in some order. There are $4 \times 3 \times 2 \times 1 = 24$ ways to list the letters B, C, D, and E in order: 4 choices for the first letter, 3 choices for the second letter, 2 choices for the third letter, and a single choice for the last letter. This explains why Table 6-1 is a complete list of all the possible Hamilton circuits in K_5.

(1) A, B, C, D, E, A	(13) A, E, D, C, B, A
(2) A, B, C, E, D, A	(14) A, D, E, C, B, A
(3) A, B, D, C, E, A	(15) A, E, C, D, B, A
(4) A, B, D, E, C, A	(16) A, C, E, D, B, A
(5) A, B, E, C, D, A	(17) A, D, C, E, B, A
(6) A, B, E, D, C, A	(18) A, C, D, E, B, A
(7) A, C, B, D, E, A	(19) A, E, D, B, C, A
(8) A, C, B, E, D, A	(20) A, D, E, B, C, A
(9) A, C, D, B, E, A	(21) A, E, B, D, C, A
(10) A, C, E, B, D, A	(22) A, D, B, E, C, A
(11) A, D, B, C, E, A	(23) A, E, C, B, D, A
(12) A, D, C, B, E, A	(24) A, E, B, C, D, A

TABLE 6-1 ■ The 24 Hamilton circuits in K_5

There are $24 \times 5 = 120$ Hamilton paths in K_5. For obvious reasons, we are not going to list them, but by now we should know how to generate them: Take any of the 24 Hamilton circuits and break it up by deleting one of its five edges. Figure 6-10 shows three paths chosen randomly out of the 120 possible Hamilton paths: The path in Fig. 6-10(a) is obtained by deleting edge DA from circuit (2) on Table 6-1; the path in, Fig. 6-10(b) is obtained by deleting edge BC from circuit (11). We leave it as an exercise for the reader to figure out the ancestry of the Hamilton path in Fig. 6-10(c).

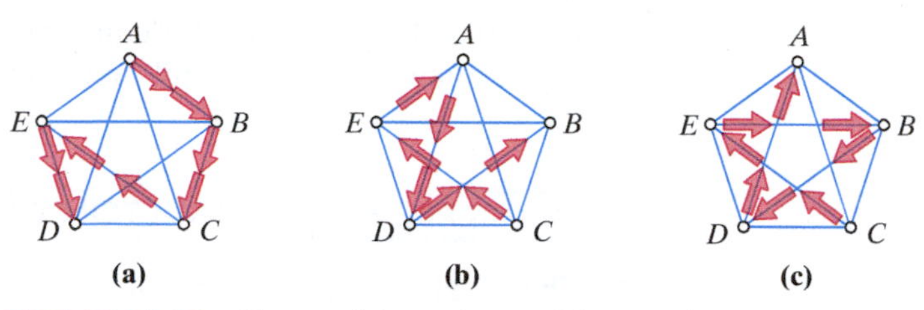

FIGURE 6-10 Three of the 120 possible Hamilton paths in K_5.

Another way to count the number of Hamilton paths in K_5 is to think in terms of *permutations*. (We first introduced permutations and their connection to factorials in Section 2.4. For a quick review of these concepts the reader is encouraged to revisit that section.) Each Hamilton path in K_5 is a permutation of the letters A, B, C, D, and E. It follows that the number of Hamilton paths equals the number of permutations of the five letters. This number is $5! = 120$.

When we generalize the preceding observations to a complete graph with N vertices we get the following improved version of properties (3) and (4):

3. A *Hamilton path* in K_N is equivalent to a *permutation of the N vertices*, and, therefore, the number of Hamilton paths is $N!$
4. A *Hamilton circuit* in K_N is equivalent to a permutation of the $N-1$ "inside" vertices (i.e., all vertices except the starting/ending vertex), and, therefore, the number of Hamilton circuits is $(N-1)!$

Table 6-2 shows the number of Hamilton circuits in K_N for $N = 3$ through 12. The numbers get big fast: K_{10} has 362,800 Hamilton circuits; K_{11} has over 3.6 million; K_{12} has close to 40 million. To understand the implications of how fast these numbers grow, imagine that you want to make a list of all the Hamilton circuits in K_N (like we did in Table 6-1 for K_5). Now suppose that (a) you work really fast and can write down a Hamilton circuit each second and (b) you are superhuman and can do this 24 hours a day, 7 days a week. Even with such superpowers, it would take you over four days to write down all the Hamilton circuits in K_{10}, over a month to write down all the Hamilton circuits in K_{11}, and well over a year to write down all the Hamilton circuits in K_{12}.

N	Hamilton circuits
3	2
4	6
5	24
6	120
7	720
8	5040
9	40,320
10	362,880
11	3,628,800
12	39,916,800

TABLE 6-2

N	SUPERHERO computation time
21	40 minutes
22	14 hours
23	13 days
24	10 months
25	20 years
26	500 years
27	13,000 years
28	350,000 years
29	9.7 million years
30	280 million years

TABLE 6-3

A computer, of course, can do the job a lot faster than even the fastest human. Let's say, for the sake of argument, that you have unlimited access to SUPERHERO, the fastest computer on the planet and that SUPERHERO can generate one *quadrillion* (i.e., a million billion) Hamilton circuits per second. SUPERHERO could crank out the Hamilton circuits of K_{12} in a matter of nanoseconds. Problem solved! Not so fast. Let's up the ante a little and try to use SUPERHERO to generate the Hamilton circuits for K_{25}. Surprise! It would take SUPERHERO 20 years to do it! Table 6-3 shows how long it would take for even the world's fastest supercomputer to generate the Hamilton circuits in K_N for values of N ranging from 21 to 30. The numbers are beyond comprehension, and the implications will become clear in the next section.

The Brute-Force Algorithm

We start this section with the TSP introduced in Example 6.1.

EXAMPLE 6.9 THE REAL TRAVELING SALESMAN'S TSP SOLVED

In Example 6.1 we left Willy the traveling salesman hanging. What Willy would like from us, most of all, is to help him find the optimal (in this case *cheapest*) tour of the five cities in his sales territory. The cost of travel between any two cities is shown in Fig. 6.11 (this is exactly the same Figure as Fig. 6-1). Notice that the underlying graph in Fig. 6-11 is just a fancy version of K_5 (five vertices each adjacent to the other four) with numbers associated with each of the 10 edges. In this TSP the numbers represent the cost of travel (in either direction) along that edge.

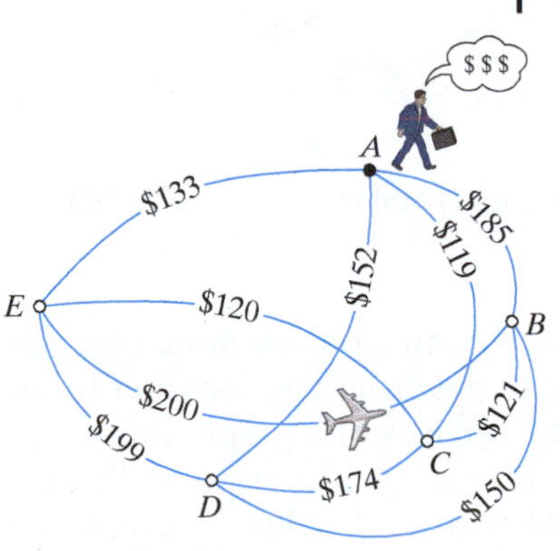

FIGURE 6-11

Table 6-4 lists the 24 possible tours that Willy could potentially take. The first and third columns of Table 6-4 gives the 24 different Hamilton circuits in K_5. This part is an exact copy of Table 6-1. Notice once again that the list is organized so that each tour in the third column is the reversed version of the corresponding tour in the first column. Listing the tours this way saves a lot of work because the cost of a tour is the same regardless of the direction of travel: Once we know the cost of a tour we know the cost of its reversal. (Please note that this is true only because we made the assumption that the cost of travel between two cities is the same regardless of the direction of travel. When the costs vary with the direction of travel then the shortcut won't work.) The middle column shows the total cost of each tour and the calculation that leads to it (for each tour just add the costs of the edges that make the tour).

Tour	Total cost	Tour
(1) A, B, C, D, E, A	$185 + 121 + 174 + 199 + 133 = 812$	(13) A, E, D, C, B, A
(2) A, B, C, E, D, A	$185 + 121 + 120 + 199 + 152 = 777$	(14) A, D, E, C, B, A
(3) A, B, D, C, E, A	$185 + 150 + 174 + 120 + 133 = 762$	(15) A, E, C, D, B, A
(4) A, B, D, E, C, A	$185 + 150 + 199 + 120 + 119 = 773$	(16) A, C, E, D, B, A
(5) A, B, E, C, D, A	$185 + 200 + 120 + 174 + 152 = 831$	(17) A, D, C, E, B, A
(6) A, B, E, D, C, A	$185 + 200 + 199 + 174 + 119 = 877$	(18) A, C, D, E, B, A
(7) A, C, B, D, E, A	$119 + 121 + 150 + 199 + 133 = 722$	(19) A, E, D, B, C, A
(8) A, C, B, E, D, A	$119 + 121 + 200 + 199 + 152 = 791$	(20) A, D, E, B, C, A
(9) A, C, D, B, E, A	$119 + 174 + 150 + 200 + 133 = 776$	(21) A, E, B, D, C, A
(10) A, C, E, B, D, A	$119 + 120 + 200 + 150 + 152 = 741$	(22) A, D, B, E, C, A
(11) A, D, B, C, E, A	$152 + 150 + 121 + 120 + 133 = 676$	(23) A, E, C, B, D, A
(12) A, D, C, B, E, A	$152 + 174 + 121 + 200 + 133 = 780$	(24) A, E, B, C, D, A

TABLE 6-4 ■ The 24 possible tours in Example 6-9 and their costs, with the optimal tour(s) highlighted

Once we have the complete list of tours and their respective costs, we just choose the optimal tour(s). In this case, the optimal tours are (11) and its reversal (23). The cost is \$676. Figure 6-12 shows both optimal tours.

A final note on Willy's sales trip: Was doing all this work worth it? In this case, yes. If Willy just chose the order of the cities at random, the worst-case scenario would be tour (6) or (18), costing \$877. With just a little effort we found an optimal tour costing \$676—a potential savings of \$201.

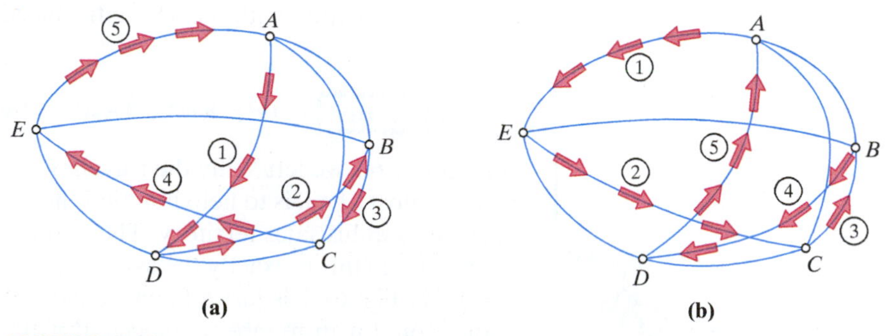

FIGURE 6-12 The two optimal tours for Willy's trip. (Total travel cost = \$676.)

Before moving on to other TSP examples, let's connect a few dots. Our general assumption for a TSP is that there is a way to get from any location to any other location. If we think of the locations as the vertices of a graph, this means that every TSP has an underlying graph that is a *complete graph.* In addition to the underlying graph, there are costs associated with each edge. A graph that has numbers associated with its edges is called a **weighted graph**, and the numbers are called **weights**. A tour in a TSP translates to a Hamilton circuit in the underlying graph, and an optimal tour translates into a Hamilton circuit of least total weight.

In short, the TSP is the concrete, real-life problem and its reformulation in terms of Hamilton circuits in a complete weighted graph is the mathematical model that represents the problem. In this model, *locations are vertices, costs are weights, tours are Hamilton circuits,* and *optimal tours are Hamilton circuits of least total weight.*

We can describe now the approach we used in Example 6.9 to solve Willy's TSP in a somewhat more formal language: We made a list of all possible Hamilton circuits (first and third columns of Table 6-4), calculated the total weight of each (middle column of Table 6-4), and picked the circuits with least total weight. This strategy is formally known as the **brute-force algorithm**.

THE BRUTE-FORCE ALGORITHM

- **Step 1.** Make a list of all the Hamilton circuits of the underlying graph K_N.
- **Step 2.** Calculate the total weight of each Hamilton circuit.
- **Step 3.** Choose a Hamilton circuit with least total weight.

> "In theory, there is no difference between theory and practice. In practice, there is."
>
> – Yogi Berra

The brute-force algorithm is based on a simple idea—when you have a finite number of options, try them all and you will always be able to determine which one is the best. In theory, we should be able to use the brute-force algorithm to solve any TSP. In practice, we can solve only small TSPs this way. It's easy to say "make a list of all the Hamilton circuits . . ."— doing it is something else. Just take a look again at Table 6-3. It would take the world's fastest supercomputer 20 years to do it for K_{25}, and 280 millions years to do it for K_{30}. And keep in mind that in real-world applications a TSP with $N = 30$ sites is considered small. In some cases, a TSP might involve 50 or even 100 sites.

The brute-force algorithm is a classic example of what is formally known as an **inefficient algorithm**—an algorithm for which the computational effort needed to carry out the steps of the algorithm grows disproportionately with the size of the problem. The trouble with inefficient algorithms is that they can only be used to solve small problems and, therefore, have limited practical use. Even the world's fastest computer would be of little use when trying to solve a TSP with $N = 25$ sites using the brute-force algorithm.

Before we conclude this section we tackle one more example of how we might use the brute-force algorithm to solve a TSP.

EXAMPLE 6.10 THE INTERPLANETARY MISSION TSP SOLVED

We are revisiting the TSP introduced in Example 6-2. In this TSP the goal is to find the *fastest* tour for an interplanetary mission to five of the outer moons in our solar system. Moreover, the tour has to start and end at our home planet, Earth. This makes it a TSP with $N = 6$ vertices. Figure 6-13(a) shows the mission time (in years) for travel between any two moons and between Earth and any moon.

Callisto

To use the brute-force algorithm we would start with a list of all the possible Hamilton circuits using Earth as the starting and ending vertex. The problem is that this list has $5! = 120$ different Hamilton circuits. That's more work than we care to do, so a full list is out of the question. Imagine now that we get a hint: The first stop in an optimal tour is Callisto. How much help is that? A lot. It means that the optimal tour must be a Hamilton circuit of the form $E, C, *, *, *, *, E$. The *'s are the letters G, I, M, and T in some order. Since there are only $4! = 24$ possible permutations of these four letters, the brute-force algorithm becomes much more manageable: We make a list of the 24 Hamilton circuits of the form $E, C, *, *, *, *, E$, find their

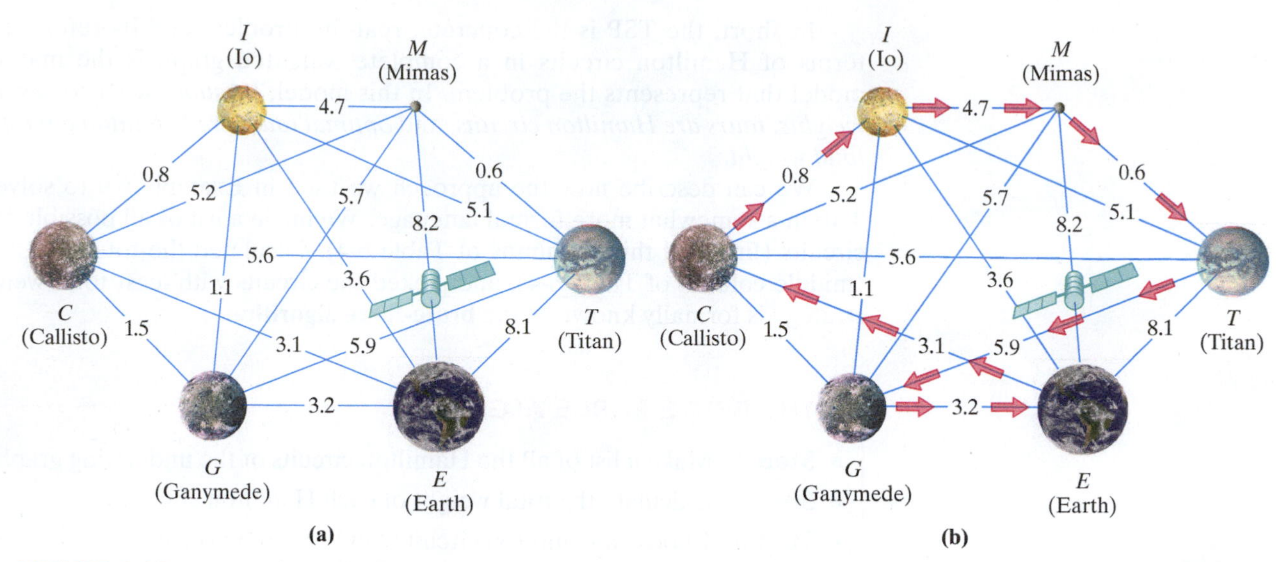

FIGURE 6-13 (a) Graph model for the interplanetary mission TSP. (b) An optimal tour. Total length: 18.3 years.

weights, and pick one (there are several) of least total weight. One of the optimal tours is shown in Fig. 6-13(b). The total length of this tour is 18.3 years. We leave it to the reader to verify the details. [A pretty convenient way to do this is with the *Traveling Salesman* applet available in ***MyMathLab*** (see Exercise 73).] ■

6.4 The Nearest-Neighbor and Repetitive Nearest-Neighbor Algorithms

In this section we introduce a new method for solving TSPs called the *nearest-neighbor algorithm* (NNA). We will illustrate the basic idea of the nearest-neighbor algorithm using the interplanetary mission TSP once again. We already found an optimal interplanetary tour in Fig. 6-13(b), so the point now is to see the nearest-neighbor algorithm at work.

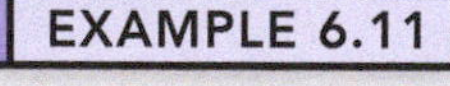

EXAMPLE 6.11 THE INTERPLANETARY MISSION TSP AND THE NEAREST-NEIGHBOR ALGORITHM

Io

Look at the graph in Fig. 6-13(a) once again and imagine planning the mission. Starting from Earth we could choose any of the moons for our first stop. Of all the choices, Callisto makes the most sense because in terms of travel time it is Earth's "nearest neighbor": Among the edges connecting E to the other vertices, EC is the one with the smallest weight. (Technically speaking we should call C the "smallest weight" neighbor, but that sounds a bit strange, so we use *nearest neighbor* as a generic term for the vertex connected by the edge of least weight.)

So we made it to Callisto. Where to next? Following the same logic, we choose to go to Callisto's nearest neighbor. That would be Io. From Io we go to the nearest neighbor we have not yet been to. That would be Ganymede. From Ganymede we go to the nearest unvisited neighbor Mimas, from Mimas we go to the nearest unvisited neighbor Titan (the last moon left), and finally, from Titan the mission comes back to Earth. This tour, called the *nearest-neighbor tour*, is shown in Fig. 6-14. The total length of this tour is 19.4 years, and that's a bit of bad news. In Example 6.10 we found an optimal tour with total length 18.3 years—this tour is more than a year longer.

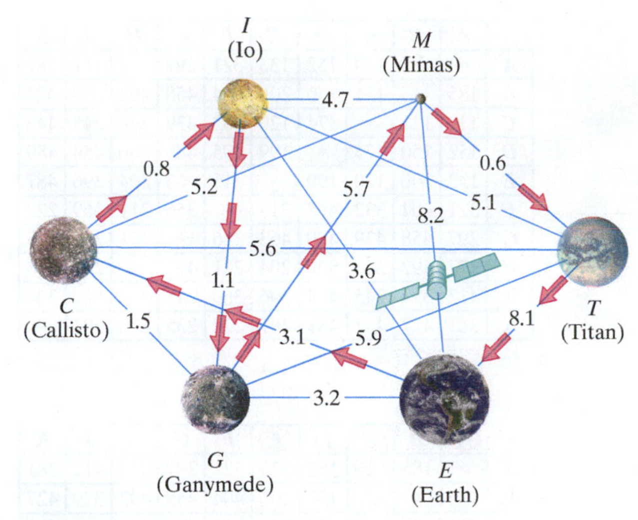

FIGURE 6-14 Nearest-neighbor tour for the interplanetary mission TSP. Total length: 19.4 years.

Example 6.11 illustrates a common situation when trying to find solutions to TSPs. We can come up with very simple and fast methods for finding "good" tours, but these methods cut corners and don't necessarily produce the "best" (optimal) tour.

Formally, we will use the term **approximate algorithm** to describe any algorithm for solving TSPs that produces a tour, but not necessarily an optimal tour. The *nearest-neighbor* algorithm is an example of an *approximate* algorithm. We will discuss a different approximate algorithm called the *cheapest-link* algorithm in Section 6.5, and in Chapter 8 we will see approximate algorithms for solving a different type of problem (not a TSP).

Obviously, approximate algorithms are important and useful. The question is why? Why do we bother with methods that give *approximate* solutions when our objective is to find an *optimal* solution? The answer is that we are making a tradeoff between the solution and the amount of effort it takes to find it. Like in many walks of life, the choice is between perfection and expediency. If it takes 20 years to find the optimal tour in a TSP with $N = 25$ locations but we can find a *suboptimal* tour (i.e., a tour that is close to optimal) in a matter of minutes, shouldn't we consider the second a better option? Approximate algorithms are the mathematical version of the commonly used strategy of cutting corners for the sake of expediency.

To illustrate the usefulness of the nearest-neighbor algorithm we will now tackle a TSP with $N = 10$ sites.

EXAMPLE 6.12 THE CONCERT TOUR TSP AND THE NNA

In Example 6.4 we were introduced to the indie rock band Luna Park, just about to embark on a 10-city concert tour. Figure 6-15(a) shows the distances (in miles) between any two cities. Because the band members all live at A, the tour needs to start and end at A. The goal is to find the optimal (shortest distance) tour.

Notice that we are not looking at a complete weighted graph, but the mileage chart provides exactly the same information as the graph would (and it does it in a much cleaner way). The only algorithm we know that guarantees an optimal tour is the brute-force algorithm, and to use brute force would mean making a list of $9! = 362{,}880$ possible Hamilton circuits. Obviously, we are not about to do that, at least not without help. But, we now know of a quick-and-dirty shortcut—the nearest-neighbor algorithm. Let's try it and see what we get.

To implement the nearest-neighbor algorithm from a chart we use the following strategy: Start with the row labeled A, and look for the smallest number in that row. That number (119) identifies A's nearest neighbor, in this case C [Fig. 6-15(b)]. We

	A	B	C	D	E	F	G	H	J	K
A	*	185	119	152	133	321	297	277	412	381
B	185	*	121	150	200	404	458	492	379	427
C	119	121	*	174	120	332	439	348	245	443
D	152	150	174	*	199	495	480	500	454	489
E	133	200	120	199	*	315	463	204	396	487
F	321	404	332	495	315	*	356	211	369	222
G	297	458	439	480	463	356	*	471	241	235
H	277	492	348	500	204	211	471	*	283	478
J	412	379	245	454	396	369	241	283	*	304
K	381	427	443	489	487	222	235	478	304	*

(a)

	A	B	C	D	E	F	G	H	J	K
A	*	185	119	152	133	321	297	277	412	381
B	185	*	121	150	200	404	458	492	379	427
C	119	121	*	174	120	332	439	348	245	443
D	152	150	174	*	199	495	480	500	454	489
E	133	200	120	199	*	315	463	204	396	487
F	321	404	332	495	315	*	356	211	369	222
G	297	458	439	480	463	356	*	471	241	235
H	277	492	348	500	204	211	471	*	283	478
J	412	379	245	454	396	369	241	283	*	304
K	381	427	443	489	487	222	235	478	304	*

(b)

	A	B	C	D	E	F	G	H	J	K
A	*	185	119	152	133	321	297	277	412	381
B	185	*	121	150	200	404	458	492	379	427
C	119	121	*	174	120	332	439	348	245	443
D	152	150	174	*	199	495	480	500	454	489
E	133	200	120	199	*	315	463	204	396	487
F	321	404	332	495	315	*	356	211	369	222
G	297	458	439	480	463	356	*	471	241	235
H	277	492	348	500	204	211	471	*	283	478
J	412	379	245	454	396	369	241	283	*	304
K	381	427	443	489	487	222	235	478	304	*

(c)

	A	B	C	D	E	F	G	H	J	K
A	*	185	119	152	133	321	297	277	412	381
B	185	*	121	150	200	404	458	492	379	427
C	119	121	*	174	120	332	439	348	245	443
D	152	150	174	*	199	495	480	500	454	489
E	133	200	120	199	*	315	463	204	396	487
F	321	404	332	495	315	*	356	211	369	222
G	297	458	439	480	463	356	*	471	241	235
H	277	492	348	500	204	211	471	*	283	478
J	412	379	245	454	396	369	241	283	*	304
K	381	427	443	489	487	222	235	478	304	*

(d)

FIGURE 6-15 (a) Mileage chart for the concert tour TSP. (b) *A*'s nearest neighbor: *C*. (c) *C*'s nearest neighbor: *E*. (d) *E*'s nearest neighbor: *D*.

now go to row *C* and look for *C*'s nearest neighbor. [Before we do that, it helps to cross out the column for *C* (this helps make sure we don't go back to *C* in the middle of the tour). For similar reasons we also crossed out the *A*-column.] The smallest number available in row *C* is 120, and it identifies *C*'s nearest neighbor *E* [Fig. 6-15(c)]. We now cross out the *E*-column and go to row *E*, looking for a nearest-neighbor. The smallest number available in row *E* is 199, and it identifies *E*'s nearest neighbor *D* [Fig. 6-15(d)]. We cross out the *D*-column and continue in this way: from *D* to its nearest available neighbor *B* (150), from *B* to *J* (379), from *J* to *G* (241), from *G* to *K* (235), from *K* to *F* (222), from *F* to the only city left *H* (211), and we finally end the tour by returning to *A* (277). This is the nearest-neighbor tour: *A*, *C*, *E*, *D*, *B*, *J*, *G*, *K*, *F*, *H*, *A*, and it has a total length of 2153 miles. ■

Example 6.12 clearly illustrates the tradeoff between perfection and expediency. We found a concert tour for the band, and it took only a few minutes, and we did it without the aid of a computer. That's the good news. But how good is the solution we found? More specifically, how much longer is the nearest-neighbor tour found in Example 6.12 than the optimal tour? To answer this question, we would need to know how long the optimal tour is. With the aid of a computer, special software, and a fair amount of effort, I was able to find the answer: The optimal concert tour has a total length of 1914 miles. This information allows us to look back and judge the "goodness" of the solution we found in Example 6.12. We do this by introducing the concept of *relative error*.

- **Relative error of a tour.** Let C denote the total cost of a given tour and Opt denote the total cost of the optimal tour. The *relative error* ε of the tour is given by $\varepsilon = \frac{C - Opt}{Opt}$.

The best way to think of the relative error is as a *percent* (i.e., the amount of error expressed as a percent of the optimal solution).

We are now ready to pass judgment on the nearest-neighbor tour we found in Example 6.12: The relative error of the tour is $\varepsilon = \frac{2153 - 1914}{1914} \approx 0.1249 = 12.49\%$. Is a relative error of 12.49% good or bad? The answer very much depends on the circumstances: When cost is a critical variable—say, for example, the timing of an interplanetary mission—an error of 12.49% is high. When cost is less critical—say, for example, the distance covered by a rock band traveling around in a fancy motor coach—an error of 12.49% might be OK.

The Repetitive Nearest-Neighbor Algorithm

One of the interesting features of the NNA is that the nearest-neighbor tour depends on the choice of starting vertex; change the starting vertex and you *could* end up with a different nearest-neighbor tour. (Note the emphasis on the word "could"; we might change the starting vertex and still get the same tour.) The preceding observation can help us squeeze better tours from the nearest-neighbor algorithm: Try a different starting vertex, and you might get a new and shorter tour; the more vertices you try, the better your chances. This is the key idea behind a refinement of the nearest-neighbor algorithm known as the *repetitive nearest-neighbor* algorithm. Our next example illustrates how this strategy works.

EXAMPLE 6.13 THE CONCERT TOUR TSP REVISITED

In our last example (Example 6.12) we found a concert tour for the Luna Park rock band using the NNA. We used A as the starting and ending vertex because the band lives at A, so in some sense we had no choice. The tour we found had a total length of 2153 miles.

Nearest-Neighbor tour	Total length
(1) $A, C, E, D, B, J, G, K, F, H, A$	2153
(2) $B, C, A, E, D, J, G, K, F, H, B$	2427
(3) $C, A, E, D, B, J, G, K, F, H, C$	2237
(4) $D, B, C, A, E, H, F, K, G, J, D$	2090
(5) $E, C, A, D, B, J, G, K, F, H, E$	2033
(6) $F, H, E, C, A, D, B, J, G, K, F$	2033
(7) $G, K, F, H, E, C, A, D, B, J, G$	2033
(8) $H, E, C, A, D, B, J, G, K, F, H$	2033
(9) $J, G, K, F, H, E, C, A, D, B, J$	2033
(10) $K, F, H, E, C, A, D, B, J, G, K$	2033

TABLE 6-5 ■ Nearest-neighbor tours for every possible starting vertex

We are going to try the same thing again but now will use B as the starting and ending vertex. (Let's disregard for now the fact that the tour really needs to start and end at A. We'll deal with that issue later.) From B we go to its nearest-neighbor C, from C to its nearest-neighbor A, from A to E, and so on. The work is a little tedious, but we can finish the tour in a matter of minutes. We leave the details to the enterprising reader, but the bottom line is that we end up with the tour $B, C, A, E, D, J, G, K, F, H, B$ with a total length of 2427 miles.

We will now repeat this process, using each of the other vertices as the starting/ending vertex and finding the corresponding nearest-neighbor tour. Figuring a couple of minutes per vertex, the process might take about 20 minutes. Table 6-5 summarizes the results. (It would be a good idea to try a couple of these on your own. You might want to use the applet *Traveling Salesman* available in ***MyMathLab*** to do so.) The first column shows the 10 nearest-neighbor tours obtained by running over all possible starting/ending vertices; the second column shows the total length of each tour. We are looking for the best among these. Tours (5) through (10) are all tied for best, each with a total length of 2033 miles. We can take any one of these tours, and get a nice improvement over the original tour we found in Example 6.12.

What about the fact that none of these tours starts and ends at A? That's easy to fix. We can take any of these tours and rewrite them so that they start and end at A. When we do that, regardless of which one we use, we get the tour $A, D, B, J, G, K, F, H, E, C, A$. We call this tour *the repetitive nearest-neighbor tour*. It is not an optimal tour, but an improvement over the nearest-neighbor tour nonetheless. (This tour has a relative error $\varepsilon = \frac{2033 - 1914}{1914} \approx 0.0622 = 6.22\%$.) ■

We conclude this section with a formal description of the *nearest-neighbor* and *repetitive nearest-neighbor* algorithms.

THE NEAREST-NEIGHBOR ALGORITHM

- **Start:** Start at the designated starting vertex. If there is no designated starting vertex pick any vertex.
- **First step:** From the starting vertex go to its *nearest neighbor* (i.e., the vertex for which the corresponding edge has the smallest weight).
- **Middle steps:** From each vertex go to its *nearest neighbor*, choosing only among *the vertices that haven't been yet visited*. (If there is more than one nearest neighbor choose among them at random.) Keep doing this until all the vertices have been visited.
- **Last step:** From the last vertex return to the starting vertex. The tour that we get is called the **nearest-neighbor** tour.

THE REPETITIVE NEAREST-NEIGHBOR ALGORITHM

- Let X be any vertex. Find the nearest-neighbor tour with X as the starting vertex, and calculate the cost of this tour.
- Repeat the process using each of the other vertices of the graph as the starting vertex.
- Of the nearest-neighbor tours thus obtained, choose one with least cost. If necessary, rewrite the tour so that it starts at the designated starting vertex. The tour that we get is called the **repetitive nearest-neighbor** tour.

6.5 The Cheapest-Link Algorithm

Last, but not least, we will introduce a completely different *approximate* algorithm for solving TSPs called the *cheapest-link algorithm* (CLA). (The term "cheapest link" is used here to mean the same thing as "edge of least cost," and the algorithm could just as well be called the "edge of least cost algorithm," but "cheapest link" sounds a little better.) The idea is to piece together a tour by looking at all the possible *links* (i.e., edges) and always choosing the *cheapest* link available (with a couple of restrictions that we will discuss later). Unlike the nearest-neighbor algorithm, the order in which we piece the tour together has nothing to do with the order in which the vertices will be visited. It sounds complicated, but it's not. We will illustrate the cheapest-link algorithm by revisiting some by now familiar examples, starting with the interplanetary mission TSP (Examples 6.2, 6.10, and 6.11). This is an important example in terms of understanding the inner workings of the cheapest-link algorithm—you are encouraged to read it carefully. It is also long, and doodling along will help you keep up with the details. Paper, a pencil, and a red pen are strongly recommended.

EXAMPLE 6.14 THE INTERPLANETARY MISSION TSP AND THE CLA

The graph in Fig. 6-16(a) is a repeat of Fig. 6-2. The weights of the edges represent the cost (in this case time) of travel between any two locations. We will now show how the cheapest-link algorithm handles this TSP.

- **Step 1.** We start by scanning the graph in Fig. 6-16(a) looking for the *cheapest link (edge)*. Here the cheapest link is the one connecting Mimas and Titan

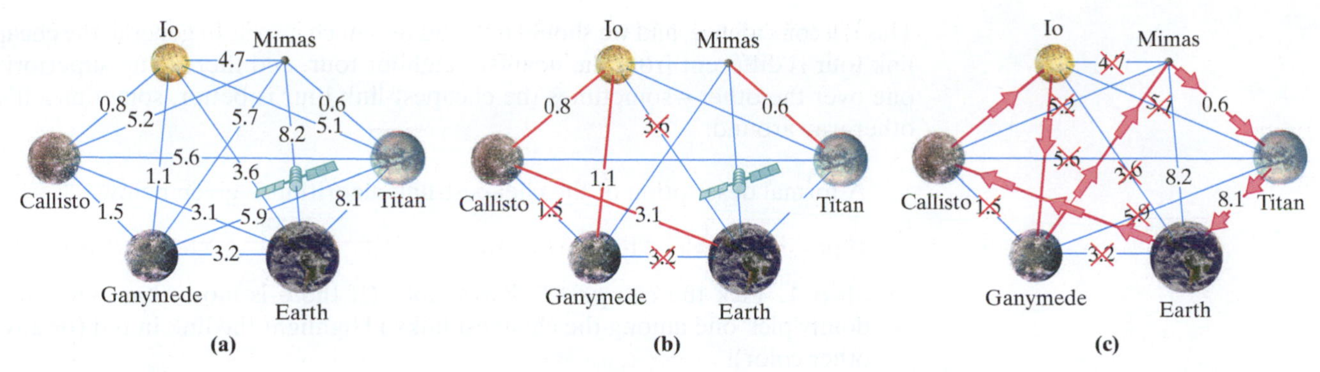

FIGURE 6-16 (a) The interplanetary mission TSP. (b) Partially constructed cheapest-link tour. (c) Cheapest-link tour after completion.

(0.6 years). [For convenience we'll denote it by MT (0.6).] We select that link and indicate that we are doing so by highlighting it in red.

- **Step 2.** We go back to scanning the graph looking for the *next cheapest link*. It is the Callisto-Io link [CI(0.8)]. We proceed to highlight it in red. (Notice that the two red links are not connected, but it doesn't matter at this point.)
- **Step 3.** Once again, we scan the graph looking for the next cheapest link and find IG (1.1). We highlight it in red.
- **Step 4.** The next cheapest link in the graph is CG (1.5). We pick up our red pen and are about to highlight the link when we suddenly realize that this is not a good move. A red CG means that we would be forming a red circuit C, I, G, C. But tours (Hamilton circuits) can't contain partial circuits, so any edge that forms a partial circuit must be ruled out. (For convenience, we call this the **partial-circuit rule.**) So, even though CG is the next available cheapest link, we can't choose it because of the partial-circuit rule. We indicate this fact by ✕-ing out CG (the ✕ is like a little marker saying, "Do not travel along this link"). So, we try again. After CG, the next cheapest link is CE (3.1). No problem here, so we select it and highlight it in red.
- **Step 5.** The next cheapest link in the graph is EG (3.2). If we were to highlight GE in red we would be forming the partial red circuit E, G, I, C, E. Because of the partial-circuit rule, we must ✕-out EG. We scan the graph again and find that the next cheapest link is IE. If we were to highlight IE in red we would have the following problem: *three* red edges (IE, IG, and IC) meeting at one vertex. This is not possible in a tour, since it would require visiting that vertex more than once. (For convenience, we call this the **three-edge rule.**) So we ✕-out IE as well. Note that choosing IE would also violate the *partial-circuit rule*, so there are two different reasons why IE is a bad choice! [If you are doodling along with this narration, your picture at this point should look something like Fig. 6-16(b).] The next four cheapest links in order are IM (4.7), IT (5.1), CM (5.2), and CT (5.6). They all have to be ruled out because of the three-edge rule. This leads us to GM (5.7). This edge works, so we select it and highlight it in red.
- **Step 6.** Since $N = 6$ this should be the last step. The last step is a little easier than the others—there should be only one way to close the circuit. Looking at the graph we see that E and T are the two loose ends that need to be connected. We do that by adding the link ET. Now we are done.

This is the end of the busywork. We found a Hamilton circuit in the graph and from it we get two different tours, depending on the direction we choose to travel. Going clockwise gives the *cheapest-link tour* E, C, I, G, M, T, E with a total length of 19.4 years. Going counterclockwise gives us the reverse tour. ■

Notice that for the interplanetary mission TSP, the cheapest-link tour we just found is exactly the same as the nearest-neighbor tour we found in Example 6.11.

This is a coincidence, and we should not read too much into it. In general, the cheapest-link tour is different from the nearest-neighbor tour, and there is no superiority of one over the other—sometimes the cheapest-link tour is better, sometimes it's the other way around.

A formal description of the cheapest-link algorithm is given below.

THE CHEAPEST-LINK ALGORITHM

- **Step 1.** Pick the *cheapest link* available. (If there is more than one, randomly pick one among the cheapest links.) Highlight the link in red (or any other color).
- **Step 2.** Pick the next cheapest link available and highlight it.
- **Steps 3, 4, . . . , N − 1.** Continue picking and highlighting the cheapest available link that (a) does not violate the *partial-circuit* rule (i.e., does not close a partial circuit) or (b) does not violate the *three-edge* rule (i.e., does not create three edges meeting at the same vertex).
- **Step N.** Connect the two vertices that close the red circuit. Once we have the Hamilton circuit, we can add a direction of travel (clockwise or counterclockwise). Either one gives us a **cheapest-link tour**.

Our next example illustrates how to implement the cheapest-link algorithm when we have to work from a chart rather than a graph. The main difficulty in this situation is that it is not easy to spot violations of the *partial-circuit* and *three-edge* rules when looking at a chart. A simple way around this difficulty is to create an auxiliary picture of the tour as it is being built, one edge at a time.

EXAMPLE 6.15 THE *CURIOSITY* TSP AND THE CLA

We are finally going to take a look at the Mars rover *Curiosity* TSP introduced in the chapter opener and described in Example 6.3. The table shown in Fig. 6-17(a) gives the distances (in meters) between the seven drilling sites that *Curiosity* is scheduled to visit. We will use the cheapest-link algorithm working directly out of the distance chart. All we will need is an additional auxiliary graph that will help us visualize the links as we move through the steps of the algorithm. Figure 6-17(b) shows the auxiliary graph when we start—a blank slate of seven vertices G_1 through G_7 and no edges.

- **Step 1.** We scan the distance chart looking for the smallest possible number. (*Note*: In a distance chart the half below the diagonal is the mirror image of the half above the diagonal, so we have to scan only one of the two halves.) The smallest number is 1300, and it belongs to the edge G_6G_7. This is the cheapest link, and we select it. We indicate this by connecting G_6 and G_7 in the auxiliary graph [Fig. 6-17(c)].
- **Step 2.** We scan the distance chart again looking for the next smallest number. The number is 1500, and it belongs to the edge G_1G_6. We select this edge and indicate that we are doing so by connecting G_1 and G_6 in the auxiliary graph [Fig. 6-17(d)].
- **Step 3.** The next smallest number in the chart is 2000, and it belongs to the edge G_4G_5. We select this edge and connect G_4 to G_5 in the auxiliary graph [Fig. 6-17(e)].
- **Step 4.** The next smallest number in the chart is 2200, and it belongs to the edge G_1G_7. When we go to the auxiliary graph we see that we can't select this

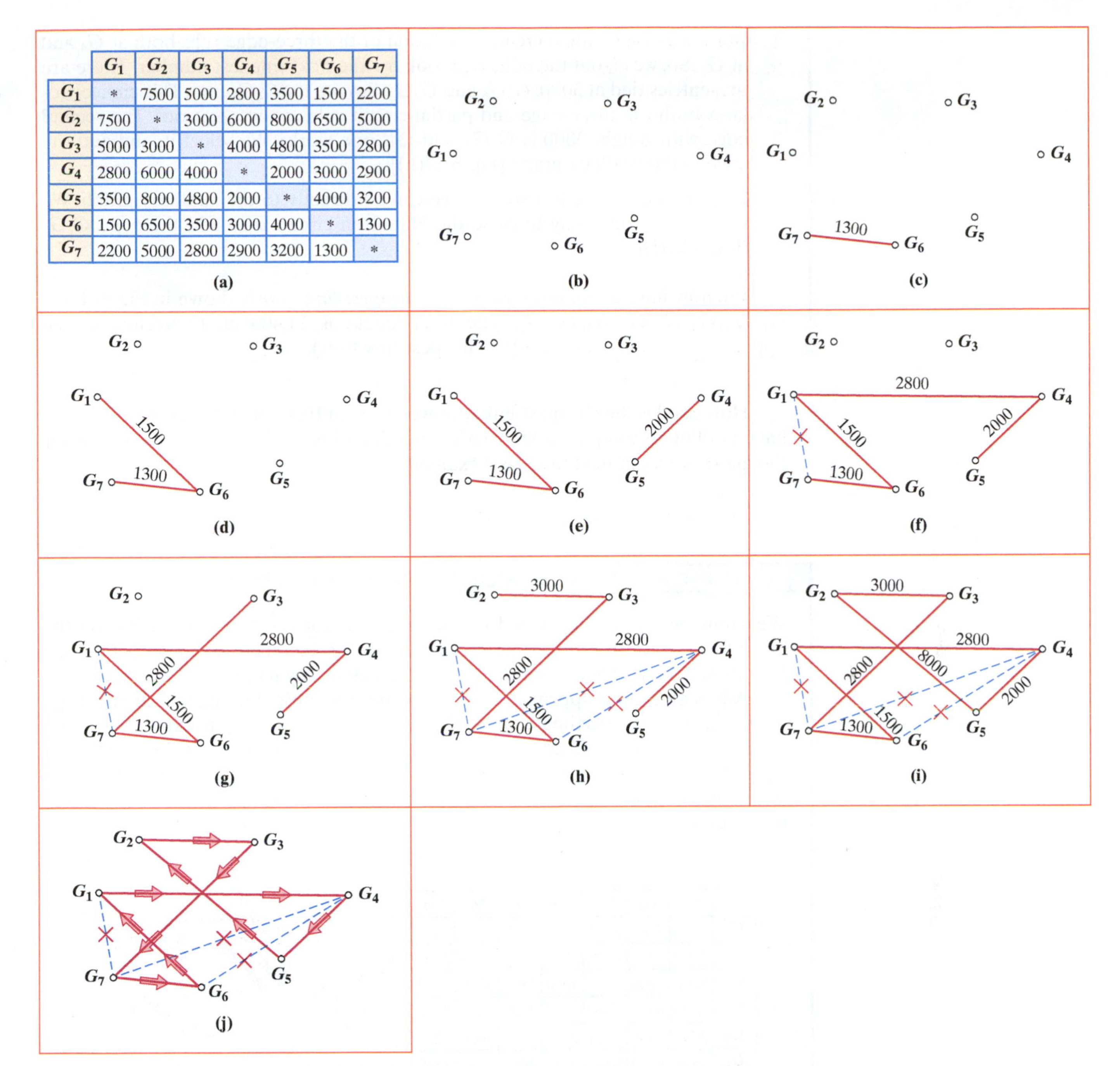

	G_1	G_2	G_3	G_4	G_5	G_6	G_7
G_1	*	7500	5000	2800	3500	1500	2200
G_2	7500	*	3000	6000	8000	6500	5000
G_3	5000	3000	*	4000	4800	3500	2800
G_4	2800	6000	4000	*	2000	3000	2900
G_5	3500	8000	4800	2000	*	4000	3200
G_6	1500	6500	3500	3000	4000	*	1300
G_7	2200	5000	2800	2900	3200	1300	*

FIGURE 6-17 (a) Distance chart for the *Curiosity* TSP, (b) through (i) implementing the steps of the CLA, and (j) cheapest-link tour (21,400 m).

edge because it would create the circuit G_1, G_7, G_6, G_1—a violation of the partial-circuit rule. We X-out G_1G_7 and scan the chart again, looking for the next smallest number after 2200. Here there is a tie with two entries equal to 2800: G_1G_4 and G_3G_7. We choose randomly one of these, say G_1G_4. This one is OK, and we select it by adding that edge to the auxiliary graph [Fig. 6-17(f)].

- **Step 5.** We now try G_3G_7, the other edge with a cost of 2800. It works, so we add the edge G_3G_7 to the auxiliary graph [Fig. 6-17(g)].
- **Step 6.** The next smallest number in the chart is 2900, and it belongs to the edge G_4G_7. When we look at the auxiliary graph, we can see that this edge won't

work because it would create a violation of the three-edge rule, both at G_4 and at G_7. So, we X-out the edge and look for the next smallest number. There are two entries tied at 3000: G_4G_6 and G_2G_3. Say we try G_4G_6 first. This choice violates both the three-edge and partial-circuit rules, so we X-it out. The second edge with weight 3000 is G_2G_3, and this one works. We select it, and add the edge to the auxiliary graph [Fig. 6-17(h)].

- **Step 7.** This is the last step, so we can go directly to the auxiliary graph. We see that the only way to close the Hamilton circuit is to add the edge G_2G_5 [Fig. 6-17(i)].

We now have a Hamilton circuit. A *cheapest-link tour* is shown in Fig. 6-17(j): $G_1, G_4, G_5, G_2, G_3, G_7, G_6, G_1$. The total length is 21,400 m. (Traveling in the opposite direction gives the other cheapest-link tour).

How good is the cheapest-link tour we found in Example 6.15? At this point we have nothing to compare it with, so we just don't know. We will be able to answer the question in our next (and last) example.

EXAMPLE 6.16 THE *CURIOSITY* TSP AND THE NNA

We know only one way to find the optimal tour for the *Curiosity* TSP—use the brute-force algorithm—but this would require us to create a list with $6! = 720$ Hamilton circuits. That's a lot of circuits to check out by hand.

A less ambitious approach is to try the nearest-neighbor algorithm and hope we get a good approximate tour. We leave the details as an exercise for the reader [see Exercise 73(a)], but here it is: the *nearest-neighbor tour* (starting at G_1) is $G_1, G_6, G_7, G_3, G_2, G_4, G_5, G_1$ [Fig. 6-18(b)]. The total length of this tour is 20,100 m. This is a respectable improvement over the cheapest-link tour found in Example 6.15.

	G_1	G_2	G_3	G_4	G_5	G_6	G_7
G_1	*	7500	5000	2800	3500	1500	2200
G_2	7500	*	3000	6000	8000	6500	5000
G_3	5000	3000	*	4000	4800	3500	2800
G_4	2800	6000	4000	*	2000	3000	2900
G_5	3500	8000	4800	2000	*	4000	3200
G_6	1500	6500	3500	3000	4000	*	1300
G_7	2200	5000	2800	2900	3200	1300	*

(a)

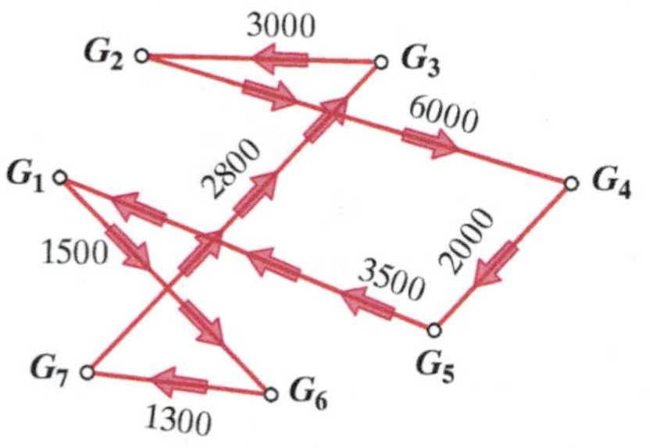

(b)

FIGURE 6-18 (a) Distance chart for the *Curiosity* TSP; (b) nearest-neighbor tour (20,100 m).

Out of curiosity, we decided to find, once and for all, an *optimal tour* for *Curiosity*. Using the *Traveling Salesman* applet in ***MyMathLab*** we were able to check all 720 Hamilton circuits and came up with a surprise: the nearest-neighbor tour shown in Fig. 6-18(b) happens to be an optimal tour as well. In other words, in the case of the *Curiosity* TSP the nearest-neighbor algorithm (in a sense the most basic of all TSP algorithms) produced an optimal solution—a nice (and lucky) turn of events. Too bad we can't count on this happening on a consistent basis. Well, maybe next chapter.

Conclusion

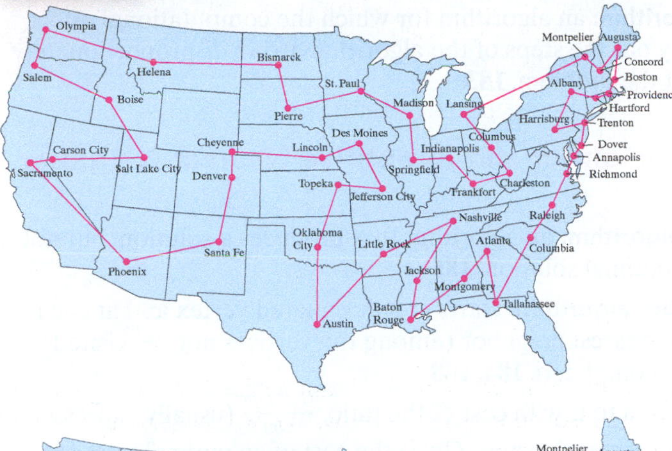

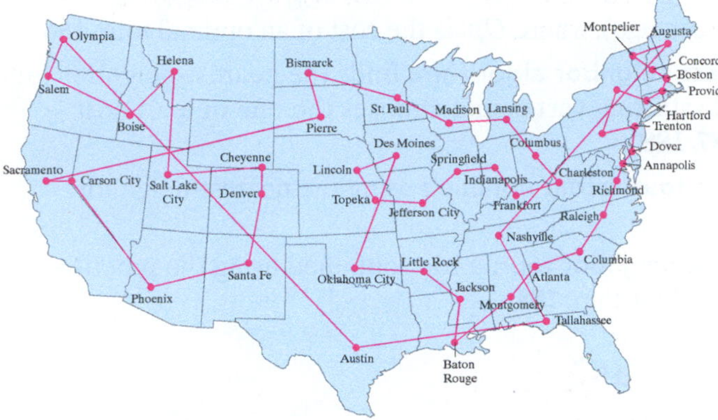

Top: The 48 state capitals in the continental U.S. and the *optimal tour* that visits all of them. Total length approximately 12,000 miles. (Computed using special software and hundreds of hours of supercomputer time.) *Bottom*: A *suboptimal* solution for the same 48 cities. Total length approximately 14,500 miles. [Computed in a matter of minutes using the *nearest-neighbor algorithm* (starting City: Olympia, Washington).]

TSP is the acronym for *traveling salesman problem*. TSPs are some of the most important and perplexing problems in modern mathematics. Important because there is a wide range of real-life applications that can be modeled by TSPs, and perplexing because nobody knows a general algorithm for finding optimal solutions to TSPs that works no matter how large the number of vertices is. It's likely that an *optimal and efficient general algorithm for solving TSPs is a mathematical impossibility* along the lines of Arrow's Impossibility Theorem in Chapter 1, but unlike Arrow, nobody has been able to prove this as a mathematical fact. Great fame and fortune await the first person that can do this.

Alternatively, there are many good *approximate algorithms* that can produce *suboptimal* (i.e., approximate) solutions to TSPs even when the number of vertices is very large. These algorithms represent a departure from the traditional notion that a math problem can have only one answer or that an answer is either right or wrong. In this chapter we discussed several *approximate algorithms* for solving TSPs and learned an important lesson: some math problems can have a perfect solution or an approximate solution. When the perfect solution is beyond the human ability to compute it, a good approximate solution that is easy to compute is not such a bad thing. Regardless of how we choose to tackle *traveling salesman problems*, the acronym TSP should not stand for *Totally Stumped and Perplexed.*

KEY CONCEPTS

6.1 What Is a Traveling Salesman Problem?

- **TSP:** an acronym for *traveling salesman problem*, **173**
- **tour:** a trip that starts and ends at a site and visits all the other sites exactly once, **173**
- **optimal tour:** a tour of minimal total cost, **173**

6.2 Hamilton Paths and Circuits

- **Hamilton path:** a path that visits each vertex of a connected graph once and only once, **177**
- **Hamilton circuit:** a circuit that visits each vertex of a connected graph once and only once, **177**
- **complete graph (K_N):** a graph with no loops or multiple edges such that any two distinct vertices are connected by an edge, **178**

6.3 The Brute-Force Algorithm

- **brute-force algorithm:** an algorithm that checks the cost of every possible Hamilton circuit and chooses the optimal one, **183**
- **inefficient algorithm:** an algorithm for which the computational effort needed to carry out the steps of the algorithm grows disproportionately with the size of the problem, **183**

6.4 The Nearest-Neighbor and Repetitive Nearest-Neighbor Algorithms

- **approximate algorithm:** an algorithm that produces a solution, but not necessarily an optimal solution, **185**
- **nearest-neighbor algorithm:** starts at a designated vertex and at each step it visits the nearest neighbor (among the vertices not yet visited) until the tour is completed, **184, 188**
- **relative error:** for a tour with cost C, the ratio, $\frac{C - Opt}{Opt}$ (usually expressed in the form of a percentage), where Opt is the cost of an optimal tour, **186**
- **repetitive nearest-neighbor algorithm:** finds the nearest-neighbor tour for each possible starting vertex and chooses the one of least cost among them, **187, 188**
- **nearest-neighbor tour:** a tour obtained using the nearest-neighbor algorithm, **188**
- **repetitive nearest-neighbor tour:** a tour obtained using the repetitive nearest-neighbor algorithm, **188**

6.5 The Cheapest-Link Algorithm

- **partial-circuit rule:** a Hamilton circuit (tour) cannot contain any partial circuits, **189**
- **three-edge rule:** a Hamilton circuit (tour) cannot have three edges coming out of a vertex, **189**
- **cheapest-link algorithm:** at each step chooses the cheapest link available that does not violate the partial-circuit rule or the three-edge rule, **190**
- **cheapest-link tour:** a tour obtained using the cheapest-link algorithm, **190**

EXERCISES

WALKING

6.1 What Is a Traveling Salesman Problem?

No exercises for this section.

6.2 Hamilton Paths and Circuits

1. For the graph shown in Fig. 6-19,
   (a) find three different Hamilton circuits.
   (b) find a Hamilton path that starts at A and ends at B.
   (c) find a Hamilton path that starts at D and ends at F.

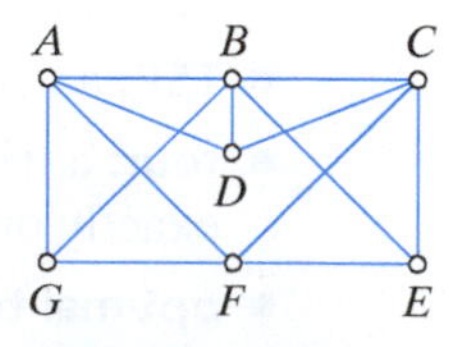

FIGURE 6-19

2. For the graph shown in Fig. 6-20,
   (a) find three different Hamilton circuits.
   (b) find a Hamilton path that starts at A and ends at B.
   (c) find a Hamilton path that starts at F and ends at I.

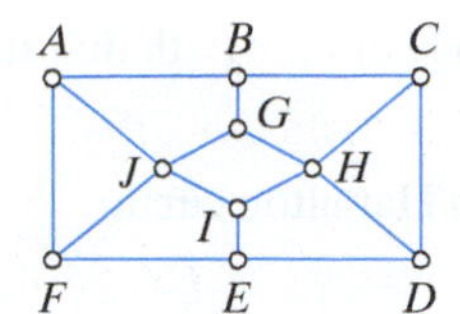

FIGURE 6-20

3. Find all possible Hamilton circuits in the graph in Fig. 6-21. Write your answers using A as the starting/ending vertex.

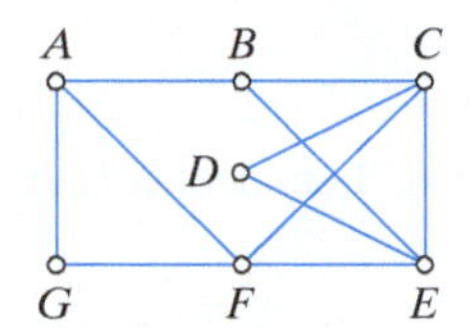

FIGURE 6-21

4. Find all possible Hamilton circuits in the graph in Fig. 6-22. Write your answers using A as the starting/ending vertex.

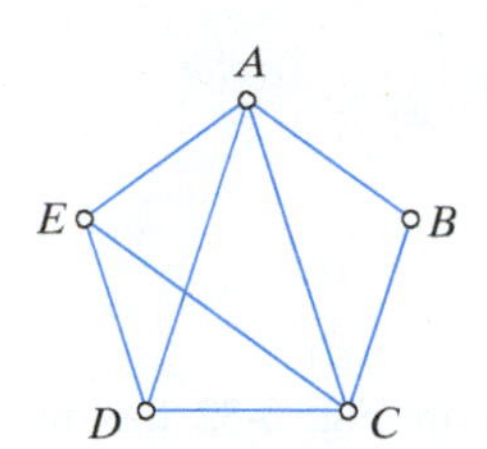

FIGURE 6-22

5. For the graph shown in Fig. 6-23,
   (a) find a Hamilton path that starts at A and ends at E.
   (b) find a Hamilton circuit that starts at A and ends with the edge EA.
   (c) find a Hamilton path that starts at E and ends at A.
   (d) find a Hamilton circuit that starts at E and ends with the edge AE.

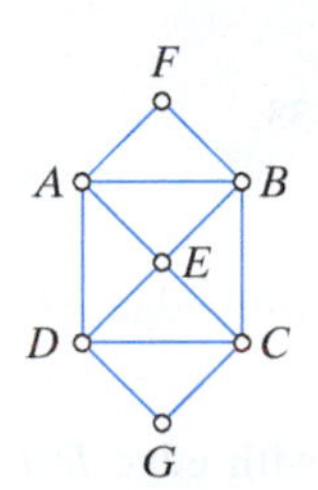

FIGURE 6-23

6. For the graph shown in Fig. 6-24,
   (a) find a Hamilton path that starts at A and ends at E.
   (b) find a Hamilton circuit that starts at A and ends with the edge EA.
   (c) find a Hamilton path that starts at E and ends at A.
   (d) find a Hamilton circuit that starts at E and ends with the edge AE.

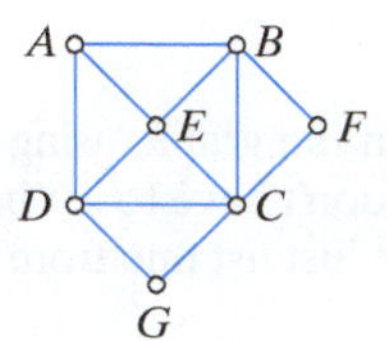

FIGURE 6-24

7. Suppose $D, G, E, A, H, C, B, F, D$ is a Hamilton circuit in a graph.
   (a) Find the number of vertices in the graph.
   (b) Write the Hamilton circuit using A as the starting/ending vertex.
   (c) Find two different Hamilton paths in the graph that start at A.

8. Suppose G, B, D, C, A, F, E, G is a Hamilton circuit in a graph.
   (a) Find the number of vertices in the graph.
   (b) Write the Hamilton circuit using F as the starting/ending vertex.
   (c) Find two different Hamilton paths in the graph that start at F.

9. Consider the graph in Fig. 6-25.
   (a) Find the five Hamilton paths that can be obtained by "breaking" the Hamilton circuit B,A,D,E,C,B (i.e., by deleting just one edge from the circuit).
   (b) Find the eight Hamilton paths that do not come from "broken" Hamilton circuits (i.e., cannot be closed into a Hamilton circuit). (*Hint*: See Example 6.6).

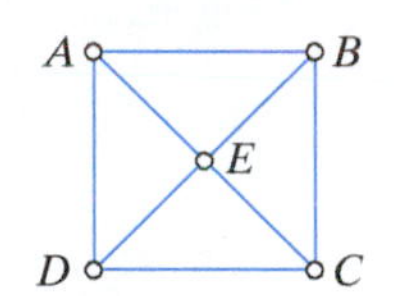

FIGURE 6-25

10. Consider the graph in Fig. 6-26.
   (a) Find all the Hamilton circuits in the graph, using B as the starting/ending vertex. (*Hint*: There are five Hamilton circuits and another five that are reversals of the first five.)
   (b) Find the four Hamilton paths that start at B and do not come from "broken" Hamilton circuits (i.e., cannot be closed into a Hamilton circuit).

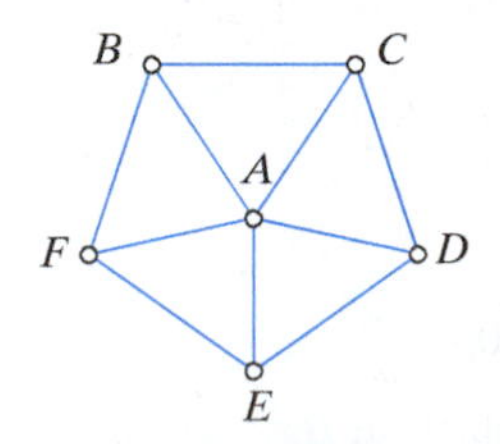

FIGURE 6-26

11. Consider the graph in Fig. 6-27.

(a) Find all the Hamilton circuits in the graph, using A as the starting/ending vertex. You don't have to list both a circuit and its reversal—you can just list one from each pair.

(b) Find all the Hamilton paths that do not come from "broken" Hamilton circuits (i.e., cannot be closed into a Hamilton circuit). You don't have to list both a path and its reversal—you can just list one from each pair.

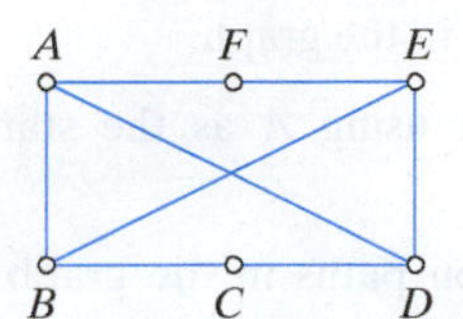

FIGURE 6-27

12. Consider the graph in Fig. 6-28.

(a) Find all the Hamilton circuits in the graph, using A as the starting/ending vertex. You don't have to list both a circuit and its reversal—you can just list one from each pair.

(b) Find all the Hamilton paths that do not come from "broken" Hamilton circuits (i.e., cannot be closed into a Hamilton circuit). You don't have to list both a path and its reversal—you can just list one from each pair. (*Hint*: Such paths must either start or end at C. You can just list all the paths that start at C— the ones that end at C are their reversals.)

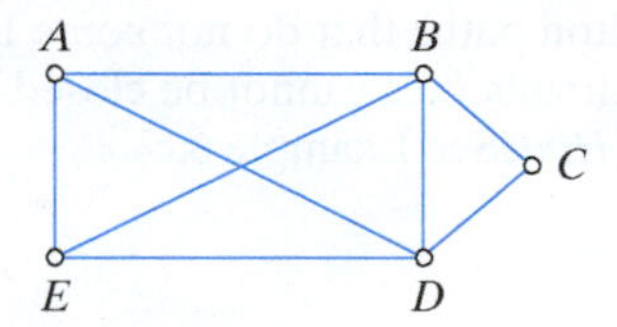

FIGURE 6-28

13. For the graph shown in Fig. 6-29,

(a) find a Hamilton path that starts at A and ends at F.

(b) find a Hamilton path that starts at K and ends at E.

(c) explain why the graph has no Hamilton path that starts at C.

(d) explain why the graph has no Hamilton circuits.

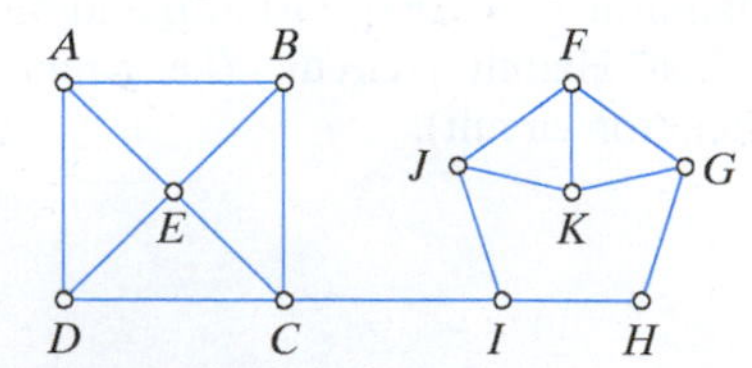

FIGURE 6-29

14. For the graph shown in Fig. 6-30,

(a) find a Hamilton path that starts at B.

(b) find a Hamilton path that starts at E.

(c) explain why the graph has no Hamilton path that starts at A or at C.

(d) explain why the graph has no Hamilton circuit.

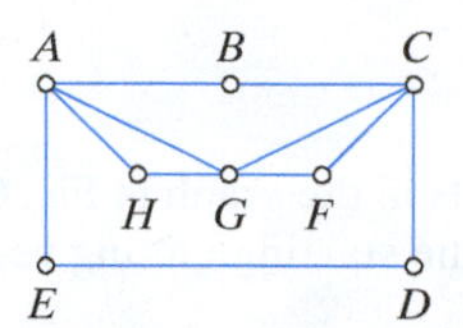

FIGURE 6-30

15. Explain why the graph shown in Fig. 6-31 has neither Hamilton circuits nor Hamilton paths.

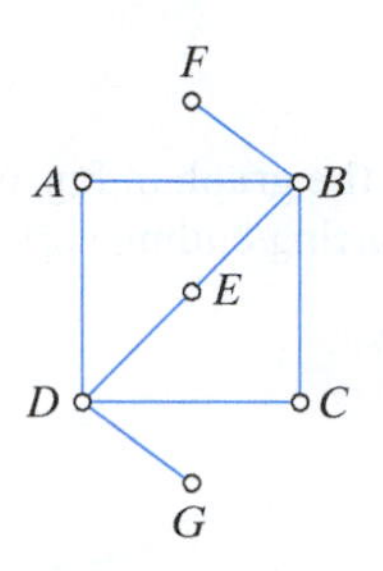

FIGURE 6-31

16. Explain why the graph shown in Fig. 6-32 has neither Hamilton circuits nor Hamilton paths.

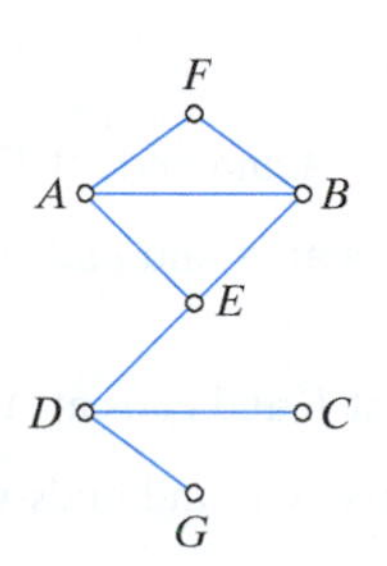

FIGURE 6-32

17. For the weighted graph shown in Fig. 6-33,

(a) find the weight of edge BD.

(b) find a Hamilton circuit that starts with edge BD, and give its weight.

(c) find a Hamilton circuit that ends with edge DB, and give its weight.

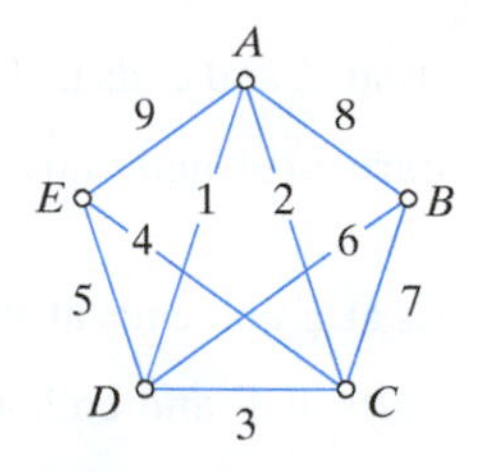

FIGURE 6-33

18. For the weighted graph shown in Fig. 6-34,

(a) find the weight of edge AD.

(b) find a Hamilton circuit that starts with edge AD, and give its weight.

(c) find a Hamilton circuit that ends with edge DA, and give its weight.

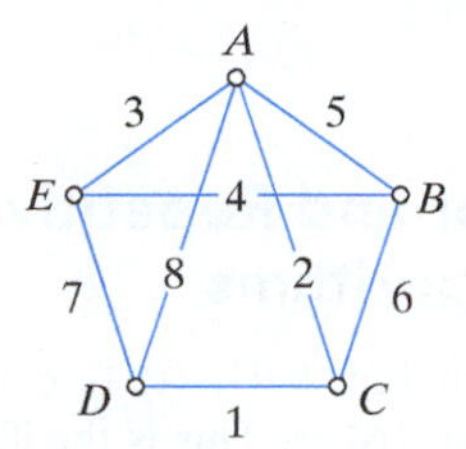

FIGURE 6-34

19. For the weighted graph shown in Fig. 6-35,

(a) find a Hamilton path that starts at A and ends at C, and give its weight.

(b) find a second Hamilton path that starts at A and ends at C, and give its weight.

(c) find the optimal (least weight) Hamilton path that starts at A and ends at C, and give its weight.

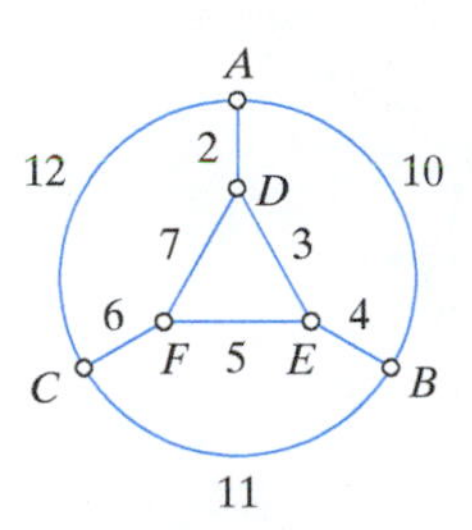

FIGURE 6-35

20. For the weighted graph shown in Fig. 6-36,

(a) find a Hamilton path that starts at B and ends at D, and give its weight.

(b) find a second Hamilton path that starts at B and ends at D, and give its weight.

(c) find the optimal (least weight) Hamilton path that starts at B and ends at D, and give its weight.

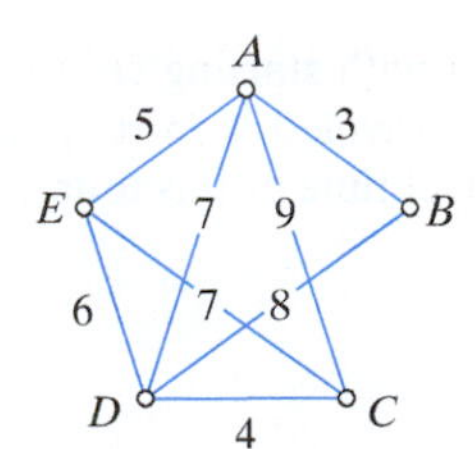

FIGURE 6-36

21. Suppose you have a supercomputer that can generate one *billion* (10^9) Hamilton circuits per second.

(a) Estimate (in years) how long it would take the supercomputer to generate all the Hamilton circuits in K_{21}. (*Hint*: There are about 3.15×10^7 seconds in a year.)

(b) Estimate (in years) how long it would take the supercomputer to generate all the Hamilton circuits in K_{22}. (*Hint*: There are about 3.15×10^7 seconds in a year.)

22. Suppose you have a supercomputer that can generate one *trillion* (10^{12}) Hamilton circuits per second.

(a) Estimate (in years) how long it would take the supercomputer to generate all the Hamilton circuits in K_{26}.

(b) Estimate (in years) how long it would take the supercomputer to generate all the Hamilton circuits in K_{27}.

23. (a) How many edges are there in K_{20}?

(b) How many edges are there in K_{21}?

(c) If the number of edges in K_{50} is x and the number of edges in K_{51} is y, what is the value of $y - x$?

24. (a) How many edges are there in K_{200}?

(b) How many edges are there in K_{201}?

(c) If the number of edges in K_{500} is x and the number of edges in K_{501} is y, what is the value of $y - x$?

25. In each case, find the value of N.

(a) K_N has 120 distinct Hamilton circuits.

(b) K_N has 45 edges.

(c) K_N has 20,100 edges.

26. In each case, find the value of N.

(a) K_N has 720 distinct Hamilton circuits.

(b) K_N has 66 edges.

(c) K_N has 80,200 edges.

6.3 The Brute-Force Algorithm

27. Find an optimal tour for the TSP given in Fig. 6-37, and give its cost.

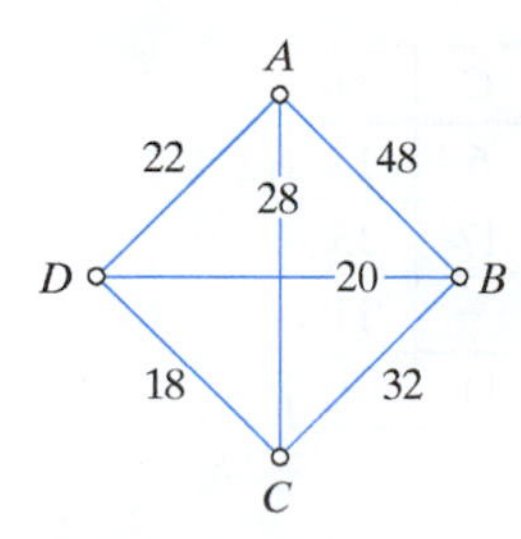

FIGURE 6-37

28. Find an optimal tour for the TSP given in Fig. 6-38, and give its cost.

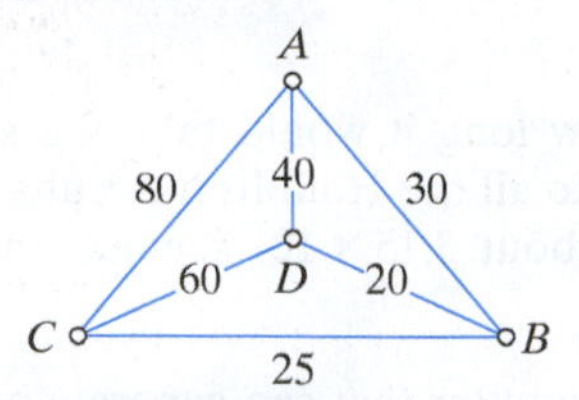

FIGURE 6-38

29. A truck must deliver furniture to stores located in five different cities A, B, C, D, and E. The truck must start and end its route at A. The time (in hours) for travel between the cities is given in Fig. 6-39. Find an optimal tour for this TSP and give its cost in hours. (*Hint*: The edge AD is part of an optimal tour.)

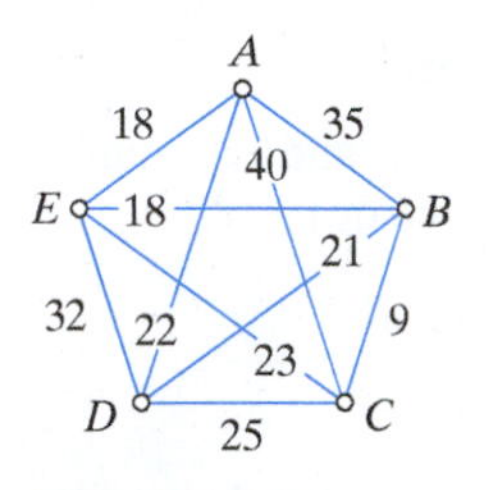

FIGURE 6-39

30. A social worker starts from her home A, must visit clients at B, C, D, and E (in any order), and return home to A at the end of the day. The graph in Fig. 6-40 shows the distance (in miles) between the five locations. Find an optimal tour for this TSP, and give its cost in miles. (*Hint*: The edge AC is part of an optimal tour.)

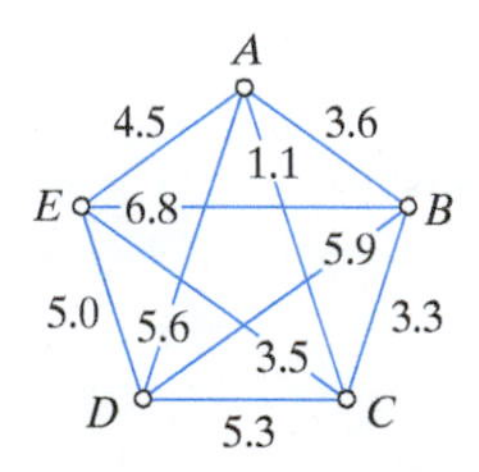

FIGURE 6-40

31. You are planning to visit four cities A, B, C, and D. Table 6-6 shows the time (in hours) that it takes to travel by car between any two cities. Find an optimal tour for this TSP that starts and ends at B.

	A	B	C	D
A	*	12	6	14
B	12	*	17	15
C	6	17	*	11
D	14	15	11	*

TABLE 6-6

32. An unmanned rover must be routed to visit four sites labeled A, B, C, and D on the surface of the moon. Table 6-7 shows the distance (in kilometers) between any two sites. Assuming the rover landed at C, find an optimal tour.

	A	B	C	D
A	*	4	18	16
B	4	*	17	13
C	18	17	*	7
D	16	13	7	*

TABLE 6-7

6.4 The Nearest-Neighbor and Repetitive Nearest-Neighbor Algorithms

33. For the weighted graph shown in Fig. 6-41, (i) find the indicated tour, and (ii) give its cost. (*Note*: This is the TSP introduced in Example 6.1.)

(a) The nearest-neighbor tour with starting vertex B

(b) The nearest-neighbor tour with starting vertex C

(c) The nearest-neighbor tour with starting vertex D

(d) The nearest-neighbor tour with starting vertex E

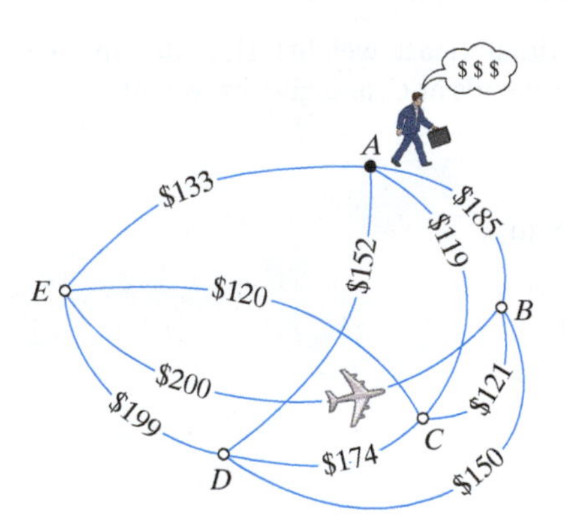

FIGURE 6-41

34. A delivery service must deliver packages at Buckman (B), Chatfield (C), Dayton (D), and Evansville (E) and then return to Arlington (A), the home base. Figure 6-42 shows a graph of the estimated travel times (in minutes) between the cities.

(a) Find the nearest-neighbor tour with starting vertex A. Give the total travel time of this tour.

(b) Find the nearest-neighbor tour with starting vertex D. Write the tour as it would be traveled if starting and ending at A. Give the total travel time of this tour.

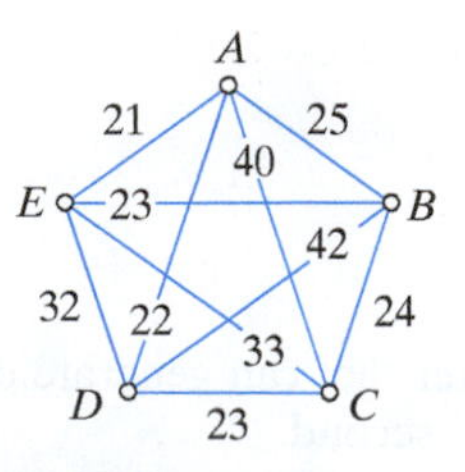

FIGURE 6-42

35. The Brute-Force Bandits is a rock band planning a five-city concert tour. The cities and the distances (in miles) between them are given in the weighted graph shown in Fig. 6-43. The tour must start and end at A. The cost of the chartered bus in which the band is traveling is \$8 per mile.

(a) Find the nearest-neighbor tour with starting vertex A. Give the cost (in \$) of this tour.

(b) Find the nearest-neighbor tour with starting vertex B. Write the tour as it would be traveled by the band, starting and ending at A. Give the cost (in \$) of this tour.

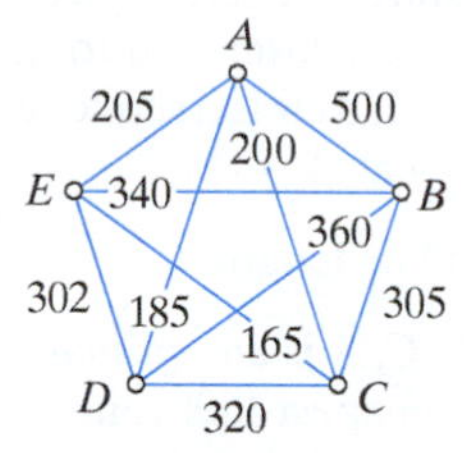

FIGURE 6-43

36. A space mission is scheduled to visit the moons Callisto (C), Ganymede (G), Io (I), Mimas (M), and Titan (T) to collect rock samples at each and then return to Earth (E). The travel times (in years) are given in the weighted graph shown in Fig. 6-44. (*Note*: This is the interplanetary TSP discussed in Example 6.11.)

(a) Find the nearest-neighbor tour with starting vertex E. Give the total travel time of this tour.

(b) Find the nearest-neighbor tour with starting vertex T. Write the tour as it would be traveled by an expedition starting and ending at E. Give the total travel time of this tour.

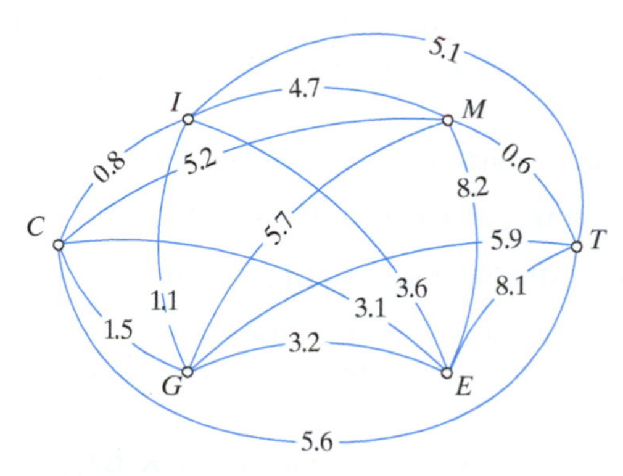

FIGURE 6-44

37. This exercise refers to the furniture truck TSP introduced in Exercise 29 (see Fig. 6-39).

(a) Find the nearest-neighbor tour starting at A.

(b) Find the nearest-neighbor tour starting at B, and give the answer using A as the starting/ending city.

38. This exercise refers to the social worker TSP introduced in Exercise 30 (see Fig. 6-40).

(a) Find the nearest-neighbor tour starting at A.

(b) Find the nearest-neighbor tour starting at C, and give the answer using A as the starting/ending city.

39. Darren is a sales rep whose territory consists of the six cities in the mileage chart shown in Fig. 6-45. Darren wants to visit customers at each of the cities, starting and ending his trip in his home city of Atlanta. His travel costs (gas, insurance, etc.) average \$0.75 per mile.

(a) Find the nearest-neighbor tour with Atlanta as the starting city. What is the total cost of this tour?

(b) Find the nearest-neighbor tour using Kansas City as the starting city. Write the tour as it would be traveled by Darren, who must start and end the trip in Atlanta. What is the total cost of this tour?

Mileage Chart

	Atlanta	Columbus	Kansas City	Minneapolis	Pierre	Tulsa
Atlanta	*	533	798	1068	1361	772
Columbus	533	*	656	713	1071	802
Kansas City	798	656	*	447	592	248
Minneapolis	1068	713	447	*	394	695
Pierre	1361	1071	592	394	*	760
Tulsa	772	802	248	695	760	*

FIGURE 6-45

40. The Platonic Cowboys are a country and western band based in Nashville. The Cowboys are planning a concert tour to the seven cities in the mileage chart shown in Fig. 6-46.

(a) Find the nearest-neighbor tour with Nashville as the starting city. What is the total length of this tour?

(b) Find the nearest-neighbor tour using St. Louis as the starting city. Write the tour as it would be traveled by the band, which must start and end the tour in Nashville. What is the total length of this tour?

Mileage Chart

	Boston	Dallas	Houston	Louisville	Nashville	Pittsburgh	St. Louis
Boston	*	1748	1804	941	1088	561	1141
Dallas	1748	*	243	819	660	1204	630
Houston	1804	243	*	928	769	1313	779
Louisville	941	819	928	*	168	388	263
Nashville	1088	660	769	168	*	553	299
Pittsburgh	561	1204	1313	388	553	*	588
St. Louis	1141	630	779	263	299	588	*

FIGURE 6-46

41. Find the repetitive nearest-neighbor tour (and give its cost) for the furniture truck TSP discussed in Exercises 29 and 37 (see Fig. 6-39).

42. Find the repetitive nearest-neighbor tour for the social worker TSP discussed in Exercises 30 and 38 (see Fig. 6-40).

43. This exercise is a continuation of Darren's sales trip problem (Exercise 39). Find the repetitive nearest-neighbor tour, and give the total cost for this tour. Write the answer using Atlanta as the starting city.

44. This exercise is a continuation of the Platonic Cowboys concert tour (Exercise 40). Find the repetitive nearest-neighbor tour, and give the total mileage for this tour. Write the answer using Nashville as the starting city.

45. Suppose that in solving a TSP you use the nearest-neighbor algorithm and find a nearest-neighbor tour with a total cost of $13,500. Suppose that you later find out that the cost of an optimal tour is $12,000. What was the relative error of your nearest-neighbor tour? Express your answer as a percentage, rounded to the nearest tenth of a percent.

46. Suppose that in solving a TSP you use the nearest-neighbor algorithm and find a nearest-neighbor tour with a total length of 21,400 miles. Suppose that you later find out that the length of an optimal tour is 20,100 miles. What was the relative error of your nearest-neighbor tour? Express your answer as a percentage, rounded to the nearest tenth of a percent.

6.5 Cheapest-Link Algorithm

47. Find the cheapest-link tour (and give its cost) for the furniture truck TSP discussed in Exercise 29 (see Fig. 6-39).

48. Find the cheapest-link tour for the social worker TSP discussed in Exercise 30 (see Fig. 6-40).

49. For the Brute-Force Bandits concert tour discussed in Exercise 35, find the cheapest-link tour, and give the bus cost for this tour (see Fig. 6-43).

50. For the weighted graph shown in Fig. 6-47, find the cheapest-link tour. Write the tour using *B* as the starting vertex.

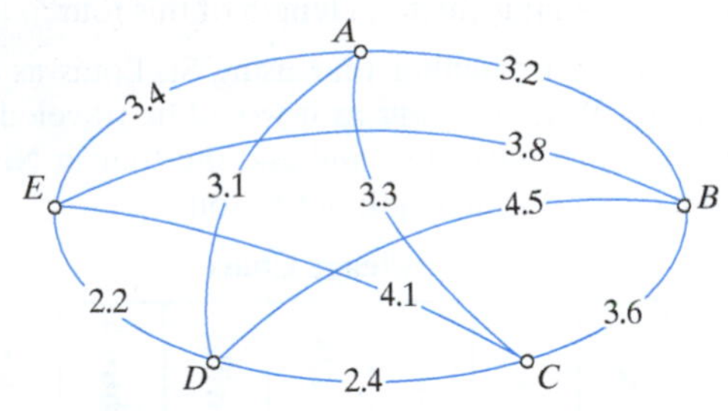

FIGURE 6-47

51. For Darren's sales trip problem discussed in Exercise 39, find the cheapest-link tour, and give the total cost for this tour (see Fig. 6-45).

52. For the Platonic Cowboys concert tour discussed in Exercise 40, find the cheapest-link tour, and give the total mileage for this tour (see Fig. 6-46).

53. A rover on the planet Mercuria has to visit six sites labeled *A* through *F*. Figure 6-48 shows the time (in days) for the rover to travel between any two sites.

(a) Find the cheapest-link tour for these sites and give its length.

(b) Given that the tour *A, B, D, E, C, F, A* is an optimal tour, find the relative error of the cheapest-link tour found in (a).

	A	B	C	D	E	F
A	*	11	19	16	9	10
B	11	*	20	13	17	15
C	19	20	*	21	13	11
D	16	13	21	*	12	16
E	9	17	13	12	*	14
F	10	15	11	16	14	*

FIGURE 6-48

54. A robotic laser must drill holes on five sites (*A, B, C, D*, and *E*) in a microprocessor chip. At the end, the laser must return to its starting position *A* and start all over. Figure 6-49 shows the time (in seconds) it takes the laser arm to move from one site to another. In this TSP, a tour is a sequence of drilling locations starting and ending at *A*.

(a) Find the cheapest-link tour and its length.

(b) Given that the tour *A, D, B, E, C, A* is an optimal tour, find the relative error of the cheapest-link tour found in (a).

	A	B	C	D	E
A	*	1.2	0.7	1.0	1.3
B	1.2	*	0.9	0.8	1.1
C	0.7	0.9	*	1.2	0.8
D	1.0	0.8	1.2	*	0.9
E	1.3	1.1	0.8	0.9	*

FIGURE 6-49

JOGGING

55. Consider a TSP with nine vertices labeled *A* through *I*.

(a) How many tours are of the form *A, G, . . . , A*? (*Hint*: The remaining seven letters can be rearranged in any sequence.)

(b) How many tours are of the form *B, . . . , E, B*?

(c) How many tours are of the form *A, D, . . . , F, A*?

56. Consider a TSP with 11 vertices labeled *A* through *K*.

(a) How many tours are of the form *A, B, . . . , A*? (*Hint*: The remaining nine letters can be rearranged in any sequence.)

(b) How many tours are of the form *C, . . . , K, C*?

(c) How many tours are of the form *D, B, . . . , K, D*?

57. Suppose that in solving a TSP you find an approximate solution with a cost of $1614, and suppose that you later find out that the relative error of your solution was 7.6%. What was the cost of the optimal solution?

58. Suppose that in solving a TSP you find an approximate solution with a cost of $2508, and suppose that you later find out that the relative error of your solution was 4.5%. What was the cost of the optimal solution?

59. You have a busy day ahead of you. You must run the following errands (in no particular order): Go to the post office, deposit a check at the bank, pick up some French bread at the deli, visit a friend at the hospital, and get a haircut at Karl's Beauty Salon. You must start and end at home. Each block on the map shown in Fig. 6-50 is exactly 1 mile.

(a) Draw a weighted graph modeling to this problem.

(b) Find an optimal tour for running all the errands. (Use any algorithm you think is appropriate.)

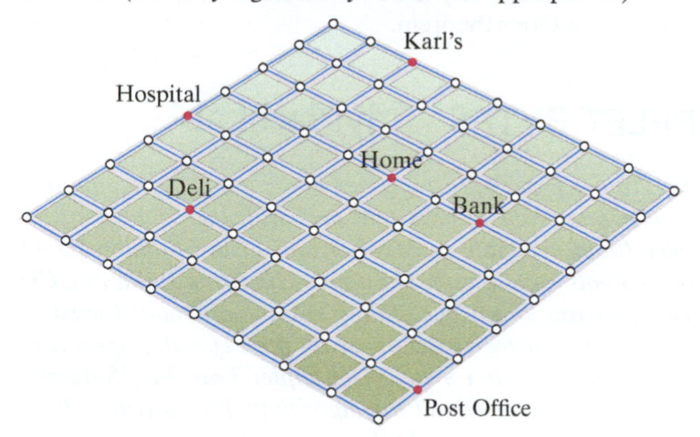

FIGURE 6-50

In Exercises 60 and 61, you are scheduling a dinner party for six people (A, B, C, D, E, and F). The guests are to be seated around a circular table, and you want to arrange the seating so that each guest is seated between two friends (i.e., the guests to the left and to the right are friends of the guest in between). You can assume that all friendships are mutual (when X is a friend of Y, Y is also a friend of X).

60. Suppose that you are told that all possible friendships can be deduced from the following information:

A is friends with B and F; B is friends with A, C, and E; C is friends with B, D, E, and F; E is friends with B, C, D, and F.

(a) Draw a "friendship graph" for the dinner guests.

(b) Find a possible seating arrangement for the party.

(c) Is there a possible seating arrangement in which B and E are seated next to each other? If there is, find it. If there isn't, explain why not.

61. Suppose that you are told that all possible friendships can be deduced from the following information:

A is friends with C, D, E, and F; B is friends with C, D, and E; C is friends with A, B, and E; D is friends with A, B, and E.

Explain why it is impossible to have a seating arrangement in which each guest is seated between friends.

62. If the number of edges in K_{500} is x and the number of edges in K_{502} is y, what is the value of $y - x$?

63. Explain why the cheapest edge in any graph is always part of the Hamilton circuit obtained using the nearest-neighbor algorithm.

64. **(a)** Explain why a graph that has a bridge cannot have a Hamilton circuit.

(b) Give an example of a graph with bridges that has a Hamilton path.

RUNNING

65. Julie is the marketing manager for a small software company based in Boston. She is planning a sales trip to Michigan to visit customers in each of the nine cities shown on the mileage chart in Fig. 6-54. She can fly from Boston to any one of the cities and fly out of any one of the cities back to Boston for the same price (call the arrival city A and the departure city D). Her plan is to pick up a rental car at A, drive to each of the other cities, and drop off the rental car at the last city D. Slightly complicating the situation is that Michigan has two separate peninsulas—an upper peninsula and a lower peninsula—and the only way to get from one to the other is through the Mackinaw Bridge connecting Cheboygan to Sault Ste. Marie. (There is a $3 toll to cross the bridge in either direction.)

Mileage Chart

	Detroit	Lansing	Grand Rapids	Flint	Cheboygan	Sault Ste. Marie	Marquette	Escanaba	Menominee
Detroit	*	90	158	68	280				
Lansing	90	*	68	56	221				
Grand Rapids	158	68	*	114	233				
Flint	68	56	114	*	215				
Cheboygan	280	221	233	215	*	78			
Sault Ste. Marie					78	*	164	174	227
Marquette						164	*	67	120
Escanaba						174	67	*	55
Menominee						227	120	55	*

FIGURE 6-54

(a) Suppose that the rental car company charges 39 cents per mile plus a drop off fee of $250 if A and D are different cities (there is no charge if $A = D$). Find the optimal (cheapest) route and give the total cost.

(b) Suppose that the rental car company charges 49 cents per mile but the car can be returned to any city without a drop off fee. Find the optimal route and give the total cost.

66. m by n grid graphs. An m by n grid graph represents a rectangular street grid that is m blocks by n blocks, as indicated in Fig. 6-51.

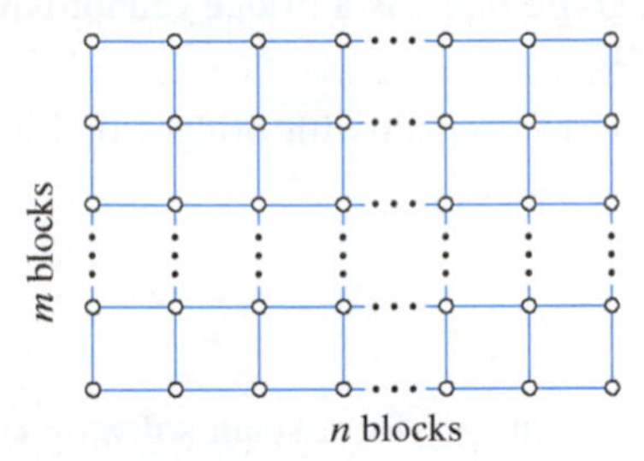

FIGURE 6-51

(a) If m and n are both odd, then the m by n grid graph has a Hamilton circuit. Describe the circuit by drawing it on a generic graph.

(b) If either m or n is even and the other one is odd, then the m by n grid graph has a Hamilton circuit. Describe the circuit by drawing it on a generic graph.

(c) If m and n are both even, then the m by n grid graph does not have a Hamilton circuit. Explain why a Hamilton circuit is impossible.

67. Complete bipartite graphs. A complete bipartite graph is a graph with the property that the vertices can be divided into two sets A and B and each vertex in set A is adjacent to each of the vertices in set B. There are no other edges! If there are m vertices in set A and n vertices in set B, the complete bipartite graph is written as $K_{m,n}$. Figure 6-52 shows a generic bipartite graph.

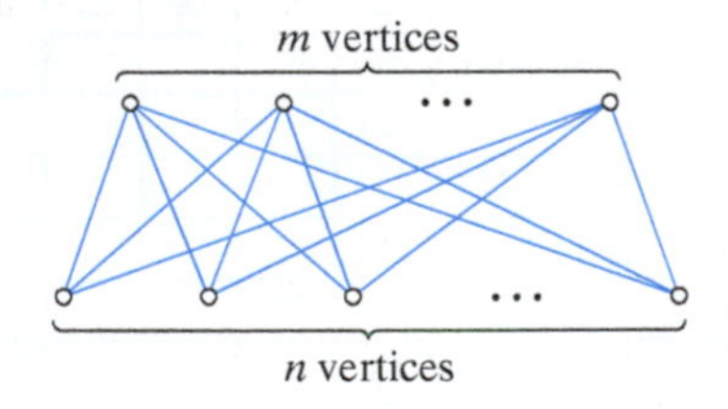

$K_{m,n}$

FIGURE 6-52

(a) For $n > 1$, the complete bipartite graphs of the form $K_{n,n}$ all have Hamilton circuits. Explain why.

(b) If the difference between m and n is exactly 1 (i.e., $|m - n| = 1$), the complete bipartite graph $K_{m,n}$ has a Hamilton path. Explain why.

(c) When the difference between m and n is more than 1, then the complete bipartite graph $K_{m,n}$ has neither a Hamilton circuit nor a Hamilton path. Explain why.

68. Ore's theorem. A connected graph with N ($N > 2$) vertices is said to satisfy *Ore's condition* if $\deg(X) + \deg(Y) \geq N$ for every pair of non-adjacent vertices X and Y of the graph. Ore's theorem states that *if a graph satisfies Ore's condition, then it has a Hamilton circuit.*

(a) Explain why the complete bipartite graph K_n (see Exercise 67) satisfies Ore's condition.

(b) Explain why for $m \neq n$, the complete bipartite graph $K_{m,n}$ (see Exercise 67) does *not* satisfy Ore's condition.

(c) Ore's condition is sufficient to guarantee that a connected graph has a Hamilton circuit but is not a necessary condition. Give an example of a graph that has a Hamilton circuit but does not satisfy Ore's condition.

69. Dirac's theorem. If G is a connected graph with N vertices and $\deg(X) \geq \frac{N}{2}$ for every vertex X, then *G has a Hamilton circuit*. Explain why Dirac's theorem is a direct consequence of Ore's theorem.

APPLET BYTES MyMathLab®

These Applet Bytes are exercises and mini-explorations built around two separate applets that deal with the contents of this chapter (both available in MyMathLab in the Multimedia Library or Tools for Success). The applet* ***Hamilton Paths and Circuits*** *allows you to practice finding Hamilton circuits (or paths) on graphs of your own creation or on pre-defined graphs available under the Custom menu. The applet* ***Traveling Salesman*** *allows you to implement all the algorithms for "solving" TSP's covered in this chapter (for TSP's of up to $N = 8$).*

70. Hamiltonian Games. Open the applet *Hamilton Paths and Circuits* and choose the "Create Graph" tab.

(a) Go to the "Custom" menu at the bottom of the window and choose the "Icosian Game." Go to the "Practice Finding Hamilton Circuits and Paths" tab. Find a Hamilton *circuit* of this graph.

(b) Go to the "Custom" menu at the bottom of the window and choose "Game 2." Go to the "Practice Finding Hamilton Circuits and Paths" tab. Find a Hamilton *circuit* of this graph.

(c) Go to the "Custom" menu at the bottom of the window and choose "Game 3." Go to the "Practice Finding Hamilton Circuits and Paths" tab. Find a Hamilton *path* of this graph.

(d) (If you are looking for a challenge) After completing part (c), click on the "Reset" button and try to find a Hamilton *circuit* in the "Game 3" graph. After a little while (don't spend too much time on it) you will come to the realization that it is impossible—*the graph does not have a Hamilton circuit*! Give an explanation as to why this graph does not have a Hamilton circuit.

71. 2 by 2 Grid. Open the applet *Hamilton Paths and Circuits* and choose the "Create Graph" tab. Go to the "Custom" menu at the bottom of the window and choose "2 by 2 Grid." Go to the "Practice Finding Hamilton Circuits and Paths" tab.

(a) Find a Hamilton *path* of the graph that starts at E and ends at one of the corner vertices.

(b) Go back to the "Create Graph" tab and relabel the vertices using just two labels (say B and W for Black and White), with adjacent vertices having different labels. Rewrite the Hamilton path you found in part (a) as a

* MyMathLab code required.

string of alternating B's and W's. Explain why *any other Hamilton path* would have the same string of B's and W's.

(c) Explain why a Hamilton *circuit* is impossible in the 2 by 2 grid graph. [*Hint*: Think of the colors of the beginning and ending vertices in a string of B's and W's that describes a Hamilton circuit.]

72. Square Grids. A square grid graph is a graph representing a street grid of n blocks by n blocks. (The 2 by 2, 3 by 3, 4 by 4 and 5 by 5 Grids are all available in the "Custom" menu located at the bottom of the "Create Graph" tab of the *Hamilton Path and Circuits* applet.)

(a) Choose the 3 by 3 Grid. Go to the "Practice Finding Hamilton Circuits and Paths and find a Hamilton *circuit* of the graph.

(b) Choose the 5 by 5 Grid. Go to the "Practice Finding Hamilton Circuits and Paths and find a Hamilton *circuit* of the graph.

(c) Using your experience with parts (a) and (b) to describe a Hamilton *circuit* in an n by n grid graph when n is *odd*.

(d) Choose the 4 by 4 Grid. Go to the "Practice Finding Hamilton Circuits and Paths and find a Hamilton *path* of the graph that starts at one of the corner vertices and ends at the center vertex M.

(e) Using your experience with part (d), describe a Hamilton *path* that starts at one of the corner vertices and ends at the "center" vertex of an n by n grid graph when n is *even*. (Note that when n is even there is exactly one center vertex in the grid.)

(f) Explain why the 4 by 4 Grid cannot have a Hamilton *circuit*. [Hint: Do Exercise 71(b) and (c) first. Apply the B and W labeling for alternating vertices idea to the 4 by 4 Grid. What would a Hamilton circuit look like as a sequence of B's and W's?]

(g) Explain why an n by n Grid cannot have a Hamilton *circuit* when n is even.

73. The Interplanetary Mission TSP. [This Applet Byte is a continuation of Example 6.10 (pp. 183–184).] Open the applet *Traveling Salesman* and click on the "Create Graph" tab. From the Complete Graphs menu choose "K6". Relabel the vertices (use E for Earth, C for Callisto, and so on) and change the weights of the edges so that the graph matches the graph model for the interplanetary mission TSP (Fig. 6-13 on p. 184.)

(a) Go to the "Nearest Neighbor" tab. Find the *repetitive nearest-neighbor* tour. (Pick a vertex and find the nearest-neighbor tour that has that vertex as the starting point. Record the tour and its total length. Use the "Reset" button and start all over with a different starting vertex. Repeat until you used all six vertices. Choose the tour of least total length. Rewrite the tour as a tour that starts and ends at E.)

(b) In Example 6.10, Fig. 6-13 (b) shows the tour E, C, I, M, T, G, E as an optimal tour. Use the "Brute Force" tab to find a different optimal tour of the form E, C, *, *, *, *, E.

74. The *Curiosity* TSP. [This Applet Byte is a continuation of Example 6.16 (p. 192).] Open the applet *Traveling Salesman* and click on the "Create Graph" tab. From the Complete Graphs menu choose "K7". Re-label the vertices G1 through G7 and change the weights of the edges so that the graph matches the graph model for the *Curiosity* TSP [Fig. 6-3(a) on p. 175.)

(a) Go to the "Nearest Neighbor" tab. Find the *nearest-neighbor* tour using G1 as the starting vertex.

(b) Go to the "Brute Force" tab. Find an *optimal* tour for the Curiosity rover. [*Hint*: Ordinarily you would have check all 720 possible tours listed in the "Select Tour" menu (that's a lot of checking), but what if someone told you that *there is an optimal tour somewhere between tours 401 and 430*? Well, it's true, so use this hint. You should also time how long it takes you to check the 30 tours and use that to estimate how long it would take you to check the full 720 tours.]

What's Facebook up to these days? If you want to know, you need to understand what's going on under the hood in a social network.

7 The Mathematics of Networks

The Cost of Being Connected

What do Facebook, the Internet, your family, the electrical power grid, the interstate highway system, and the veins and arteries in your body have in common? They all share the same fundamental structure—they are networks.

While the word *network* is used today in many different contents (*social network, computer network, telecommunications network,* etc.), there is a common thread to all of them—a network *connects* things. So, for the purposes of this chapter we will use the term *network* to mean a *connected graph.*

Whether we want to or not, we all belong to many different networks. Some are good (our families); some are not so good (the alumni association calling for a donation every other week); some are useful (the Internet); some are addictive (Facebook). Because of the ubiquitous role networks play in our lives, the study of networks has flourished in the last 30 years. Sociologists, economists, engineers, and urban planners all study different aspects of the theory of networks, but when you peel off all their layers, networks are basically mathematical structures.

The purpose of this chapter is to give a very general introduction to the mathematics of networks and related structures such as trees and spanning trees. Section 7.1 starts with several real-life examples of *networks*. To keeps things manageable, the size of the networks discussed is scaled down, but the general ideas can be applied at much bigger scales. Networks that are minimally connected, which are called *trees*, are also introduced in Section 7.1. Section 7.2 introduces the concept of a *spanning tree* of a network. In the spirit of Chapters 5 and 6, Section 7.2 deals with optimization questions: What is a *minimum spanning tree* (MST) in a *weighted* network? What is a *maximum spanning tree* (MaxST)? *Why do we care*? We conclude the chapter with a discussion of *Kruskal's algorithm* (Sec. 7.3). Kruskal's algorithm is a simple method for finding minimum and maximum spanning trees in weighted networks. The algorithm works efficiently on networks regardless of their size, it always gives an optimal solution, and, most important, it is extremely easy to implement. When it comes to optimization problems one can't ask for better karma.

7.1 Networks and Trees

Networks

A **network** is a just another name for a *connected graph*. (In the context of networks, vertices are often called *nodes* and edges are called *links*.) Most of the networks we will consider in this chapter will be *simple networks* (i.e., without loops or multiple edges), but we do not make this a requirement. In some applications, a network can have loops, multiple edges, or both.

Our first example illustrates how social networks evolve. Social networks—such as Facebook, Twitter, LinkedIn, and Instagram—are networks that connect people through some sort of social relationship—friendship, business, etc. The example is small, but you can imagine the same idea working on a much larger scale.

EXAMPLE 7.1 SOCIAL NETWORKS

Imagine 15 students (named A through O) enrolled in a very popular seminar called *The Mathematics of Social Networks*. One of the goals in a small seminar like this one is to get the students to connect with each other and exchange ideas as much as possible, in other words, to "network." For the purposes of this example we will say that two students in the seminar have *connected* if they have exchanged phone numbers or email addresses (presumably for the purposes of intellectual exchange, but we won't really dwell into their reasons for doing so). We can best visualize the interconnections among students in the seminar by means of a *connections graph*: the vertices (nodes) of the graph are the students, and pairs of students are linked by an edge if they have connected according to our definition of the term.

Figure 7-1(a) shows one possible version of the connections graph. In this scenario the graph is not a network but rather three separate, disconnected networks. Looking at this graph would not make the instructor happy. Figure 7-1(b) shows the new connections graph after two additional connections have been added: *EG* and *CH*. Now the graph becomes a true social network. The instructor is much happier—all students can connect, either directly or through intermediaries.

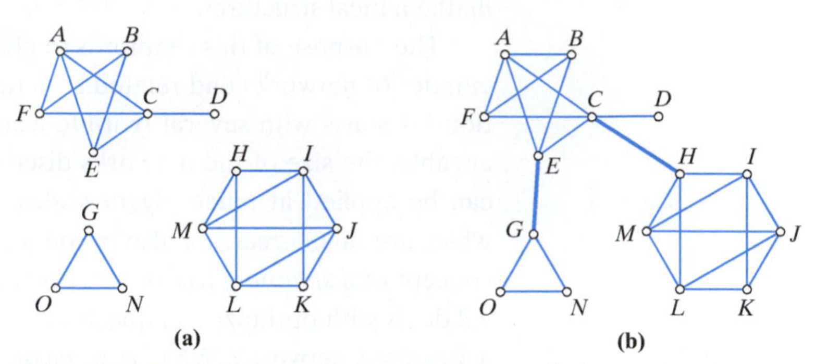

FIGURE 7-1 (a) The graph of connections with three separate components. (b) The graph of connections becomes a network.

Because a network is a connected graph, there are paths going from any vertex to any other vertex—at least one but usually many more. We are usually interested in the *shortest path* connecting a pair of vertices—in other words, a path whose length is as small as possible. We will call the length of a shortest path joining two vertices in a network the **degree of separation** between the two vertices.

EXAMPLE 7.2 SIX DEGREES OF SEPARATION

Consider again the network in Fig. 7-1(b). Let's look at degrees of separation between different pairs of students.

- *A* and *B* are joined by an edge, so we will say there is *one* degree of separation between them.
- *A* and *G* are not directly connected, but there is a path of length 2 connecting them: *A, E, G*. We will say that there are *two* degrees of separation between *A* and *G*.
- *A* and *K* are connected by the path *A, C, H, M, L, K* of length 5, but this is not as short as possible. The path *A, C, H, L, K* has length 4. Can we do even better? No. There is a second path of length 4 (*A, C, H, I, K*), but there are no paths of length 3 or less (to connect *A* and *K* one must go through *H*—that requires at least two edges; to connect *H* and *K* requires at least two more edges.) It follows that the degree of separation between *A* and *K* is *four*.
- *N* and *M* are pretty far apart—the shortest possible path connecting them is *N, G, E, C, H, M*. There you are—five degrees of separation.
- Are there pairs of students with six degrees of separation between them? You bet. Take *N* and *K*. There are two paths of length 6 joining *N* and *K* : *N, G, E, C, H, I, K* and *N, G, E, C, H, L, K*. Moreover, there are no shorter paths joining *N* and *K* (to get from *N* to *K* you have to pass through *H*; it takes at least four edges to get from *N* to *H* and another two to get from *H* to *K*). In addition to *N* and *K*, there are several other pairs of students having six degrees of separation between them.

In Example 7.2 we found many different levels of separation between the students in the seminar. For *A* and *B*, with just one degree of separation, communication is easy; for *N* and *K*, with six degrees of separation, communication is possible but only by going through five separate intermediaries (highly inconvenient).

When it comes to degrees of separation between vertices, the smaller the better, and networks having smaller degrees of separation among the vertices ("tight" networks) are much better at pushing information through the network than networks having vertices that are loosely separated ("loose" networks). One way to measure how tight or loose a network is is by looking at the *largest degree of separation between any pair of vertices* in the network. This number is called the **diameter** of the network.

When we claim that the diameter of a network is a given number D, we have to show two things: (1) that there is at least one pair of vertices having D degrees of separation between them, and (2) that there are no two vertices with degree of separation larger than D.

EXAMPLE 7.3 DIAMETER

Once again let's consider the network shown in Fig. 7-1(b) describing the connections among 15 students in the seminar. We saw in Example 7.2 that there were many different levels of separation among the students—one, two, three, four, five, and even six degrees of separation. We didn't mention pairs of students having seven degrees of separation because there aren't any. Given that six is the largest possible degree of separation in the network, we can restate that fact by saying that the diameter of the network is six. For such a small graph this is a pretty large diameter—in fact, there is an unproven but widely held idea in popular culture that there are never more than six degrees of separation between any two people on the planet.

Social networks such as Facebook, Twitter, and LinkedIn are examples of *organic* networks—networks that evolve and change on their own without any organized or centralized planning. In contrast to these, there are *planned* networks that are designed with a specific purpose in mind. Our next example illustrates on a very small scale one of the most important planned networks in modern life—the electrical power grid.

EXAMPLE 7.4 POWER GRIDS

Figure 7-2(a) shows a map of the main Texas power grid. Figure 7-2(b) shows a small section of the power grid connecting 14 small towns labeled A through N. In this graph the vertices represent the towns and the edges represent the main power lines that carry electricity to the various towns. The graph is connected, so it is indeed a network—any town can draw power from the grid. In addition, there is a weight associated with each edge of the network. In this case the weight of an edge represents the length (in miles) of the power line connecting the two towns. [Figure 7-2(b) is not drawn to exact scale but is close. The reason for this is that the length of the power line is usually close to, but not necessarily the same, as the distance between the towns.] We will return to this example in Section 7.3.

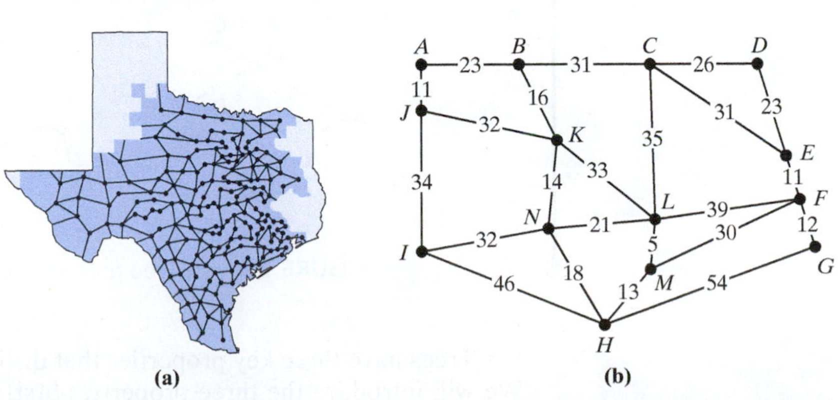

FIGURE 7-2 (a) The main Texas power grid map. (b) Small section of the power grid.

The network in Fig. 7.2(b) is an example of what we call a **weighted network**. Each edge has a *weight* that we will generally think of as representing a *cost* (either money, time, or distance). We will deal with weighted networks later in the chapter. In almost all weighted network problems the weights are positive numbers, but later in this chapter we will make an exception and consider weighted networks where the weights are negative numbers.

Trees

The second important concept in this chapter is that of a *tree*. We all know that trees are important, as they provide shade and help clean up our air, but in the context of this chapter the term **tree** means *a network that has no circuits.*

EXAMPLE 7.5 TREES AND ROOTS

Figure 7-3 shows three networks.

- The network shown in Fig. 7-3(a) has the circuit A, B, G, F, E, D, H, A. It *is not* a tree.
- The network in Fig. 7-3(b) has no circuits. It *is* a tree. (What might look like a circuit in the picture is not—the crossing points of edge HD with edges BG and CG are just crossing points and not vertices.) This tree may not look very tree-like, but we can fix that easily.
- It is obvious that the network in Fig. 7-3(c) has no circuits. It is a tree, and it looks the part. Surprisingly, this tree is the same tree as the one shown in Fig. 7-3(b). We just picked one of the vertices to be the "root" of the tree (in this case B) and built the "branches" of the tree up from the root. We can do this with any tree: Pick any vertex to be the root and build the tree up from there. Figure 7-4 shows two more versions of the same tree—in Fig. 7-4(a) the tree is rooted at A; in Fig. 7-4(b) the tree is rooted at D. (Both trees are shown sideways with the root on the left—just trying to not waste space on the page . . . and save some real trees!)

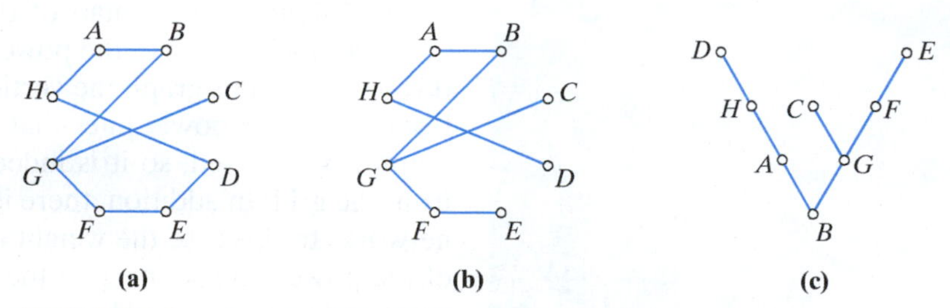

FIGURE 7-3 (a) A network with circuits is not a tree. (b) A tree. (c) The same tree with B as the "root."

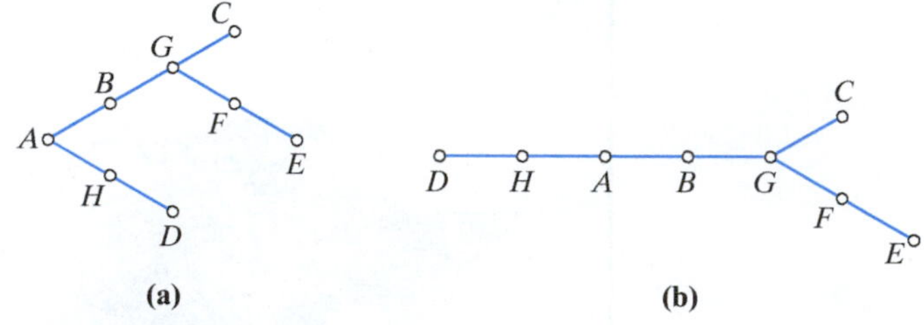

FIGURE 7-4 (a) Tree rooted at A. (b) Same tree rooted at D.

Trees have three key properties that distinguish them from ordinary networks. We will introduce the three properties first, illustrate them with an example, and conclude this section with a more formal version of these properties.

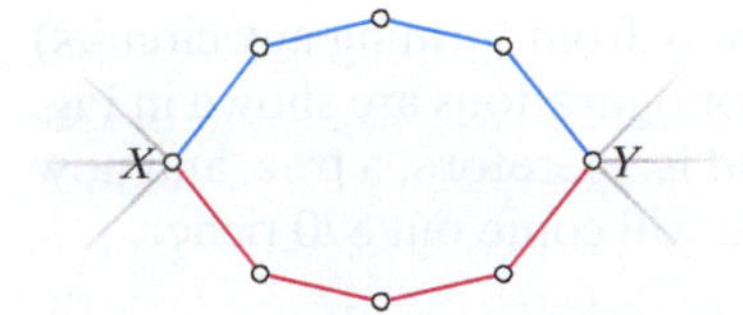

FIGURE 7-5 Two different paths joining X and Y make a circuit.

- **The single-path property.** In a tree, there is *only one path connecting two vertices.* If there were two paths connecting a pair of vertices, those two paths would create a circuit, as illustrated in Fig. 7-5. Conversely, a network that is not a tree must have at least one circuit, and that circuit will always provide alternative paths between its vertices. Look at Fig. 7-5 again: Given a circuit and two vertices (X and Y) in the circuit, there are at least two different paths (red and blue) connecting X and Y.
- **The all-bridges property.** In a tree, *every edge is a bridge.* Essentially this means that a tree has no edges to spare—if we were to delete *any* edge, the tree would become disconnected and would no longer be a network. Figure 7-6 illustrates why every edge must be a bridge: Imagine an edge AB that is *not* a bridge. Then there would have to be an alternate path from A to B (shown as the dashed red curve). But AB together with the alternate path from A to B would form a circuit, and a tree doesn't have circuits. Conversely, if every edge of the network is a bridge then the network must be a tree.

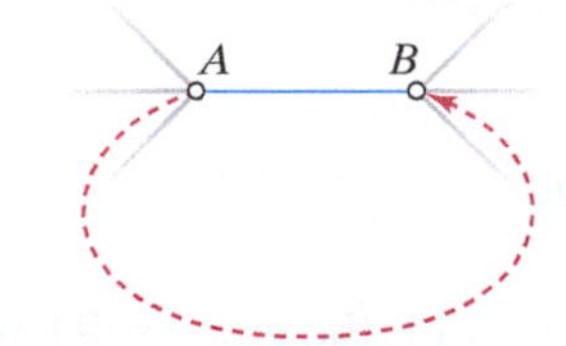

FIGURE 7-6 If AB is not a bridge, then it must be part of a circuit.

- **The $N - 1$ edges property.** A tree with N vertices has $N - 1$ edges. Always. This means that no matter what the shape of the tree is, the number of edges is one less than the number of vertices. The tree in Fig. 7-4 has $N = 8$ vertices. We don't even have to check—the number of edges must be 7. Conversely, a network with N vertices and $N - 1$ edges must be a tree.

From the above properties of trees we inherit the following key property of networks: In a network with N vertices and M edges, $M \geq N - 1$ (i.e., the number of edges is at least $N - 1$). When $M = N - 1$ the network is a tree; when $M > N - 1$ the network has circuits. The difference between the number of edges M and the minimum possible number of edges $N - 1$ is an important number called the **redundancy** of the network.

- **Redundancy of a Network.** In a network with N vertices and M edges, the redundancy R is given by $R = M - (N - 1)$. [$R = 0$ means the network is a tree; $R > 0$ means the network is not a tree.]

EXAMPLE 7.6 CONNECT THE DOTS (AND THEN STOP)

Imagine the following "connect-the-dots" game: Start with eight isolated vertices. The object of the game is to create a network connecting the vertices by adding edges, one at a time. You are free to create any network you want. In this game, bridges are good and circuits are bad. (Imagine, for example, that for each bridge in your network you get a $10 reward, but for each circuit in your network you pay a $10 penalty.)

So grab a marker and start playing. We will let M denote the number of edges you have added at any point in time. In the early stages of the game ($M = 1, 2, \ldots, 6$) the graph is disconnected [Figs. 7-7(a) through (d)].

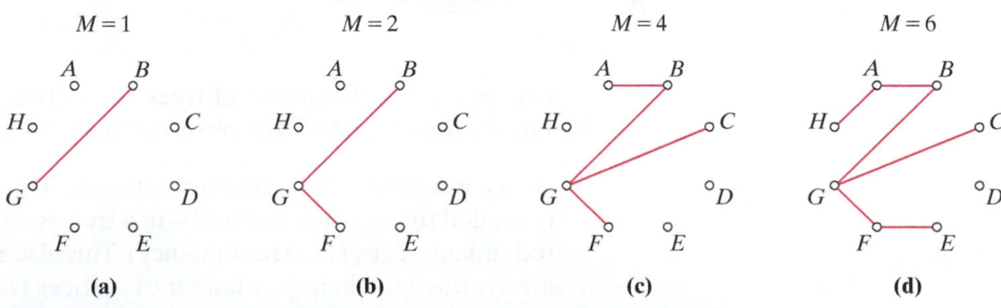

FIGURE 7-7 For small values of M, the graph is disconnected.

As soon as you get to $M = 7$ (and if you stayed away from forming any circuits) the graph becomes connected. Some of the possible configurations are shown in Fig. 7-8. Each of these networks has redundancy $R = 0$ and is, therefore, a tree, and now each of the seven edges is a bridge. Stop here and you will come out \$70 richer.

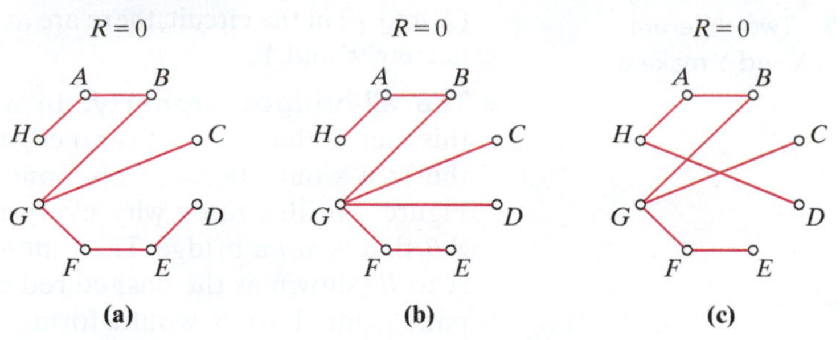

FIGURE 7-8 When $M = 7$, we have a tree ($R = 0$).

Interestingly, this is as good as it will get. When $M = 8$ $(R = 1)$, the graph will have a circuit—it just can't be avoided. In addition, none of the edges in that circuit can be bridges of the graph. As a consequence, the larger the circuit that we create, the fewer the bridges left in the graph. [The graph in Fig. 7-9(a) has a circuit and five bridges, the graph in Fig. 7-9(b) has a circuit and two bridges, and the graph in Fig. 7-9(c) has a circuit and only one bridge.] As the redundancy increases, the number of circuits goes up (very quickly) and the number of bridges goes down [Figs. 7-10(a), (b), and (c)].

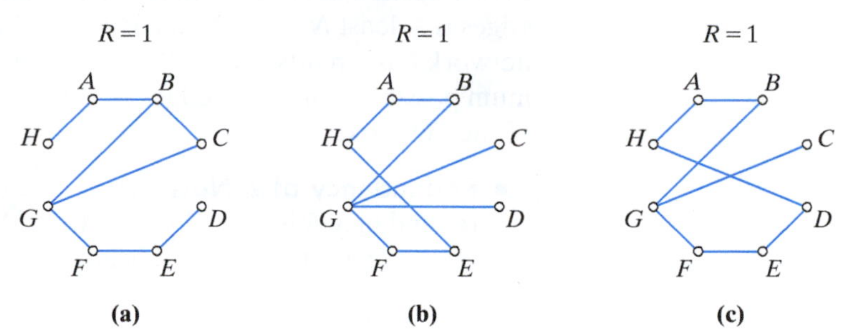

FIGURE 7-9 When $M = 8$ ($R = 1$) we have a network with a circuit.

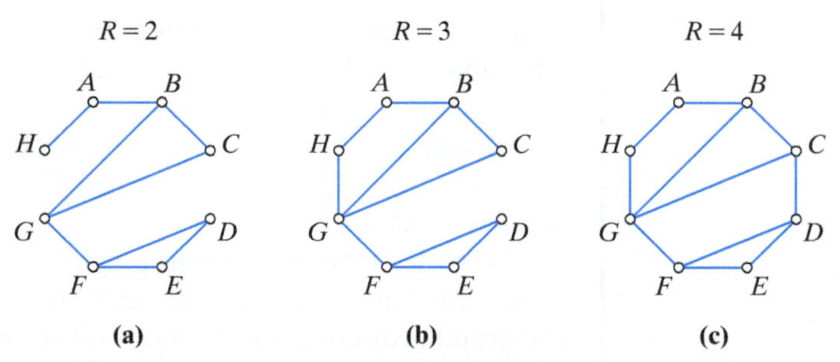

FIGURE 7-10 For larger values of R, we have a network with lots of circuits.

Our preceding discussion of trees and networks can be rephrased and summarized in the following two key observations (and respective conclusions):

- A tree is a *minimally connected* network. This means that every edge of the tree is needed to keep it connected—in a tree every edge is a bridge, and there are no redundant edges (zero redundancy). This also means that the number of edges is always one less than the number of vertices ($R = 0$ implies $M = N - 1$). *Conclusion 1*: *If you want to minimize the number of edges in a network, build a tree.*

- A network that is not a tree must have some redundant edges (positive redundancy). The redundant edges form circuits, and the higher the redundancy the more circuits in the network. Each circuit creates additional paths for connecting vertices in the circuit, so the more circuits the more ways there are to get around in the network. *Conclusion 2: If you want to have lots of alternative routes to get around in a network, increase its redundancy.*

7.2 Spanning Trees, MSTs, and MaxSTs

Spanning Trees

A **spanning tree** in a network is a *subtree* of the network that *spans* all the vertices. The easiest way to explain the meaning of this definition is with a few examples.

EXAMPLE 7.7 SUBTREES AND SPANNING TREES

Figure 7-11(a) shows a small network with 8 vertices and 9 edges. Figure 7-11(b) shows (in red) a *subtree* of the network. The name *subtree* comes from the fact that the red tree has its vertices and edges inside the network. The subtree in Fig. 7-11(b) does not include all the vertices, but the one in Fig. 7-11(c) does. We say that the subtree in Fig. 7-11(c) *spans* the network, and we call such subtrees *spanning trees* of the network. The spanning tree in Fig. 7-11(c) has 7 edges, and any other spanning tree of this network will have 7 edges as well.

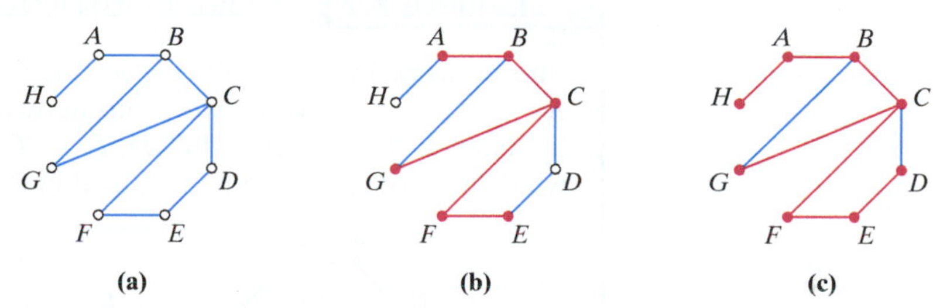

FIGURE 7-11 (a) The original network. (b) A subtree (in red). Vertices H and D are not in the subtree. (c) A spanning subtree.

EXAMPLE 7.8 COUNTING SPANNING TREES

The network in Fig. 7-12(a) has $N = 8$ vertices and $M = 8$ edges. The redundancy of the network is $R = 1$. To find a spanning tree we will have to "discard" one edge. Five of these edges are bridges of the network, and they *will have to be part of any spanning tree.* The other three edges (BC, CG, and GB) form a circuit of length 3, and if we exclude any one of the three edges, then we will have a spanning tree. Thus, the network has three different spanning trees [Figs. 7-12(b), (c), and (d)].

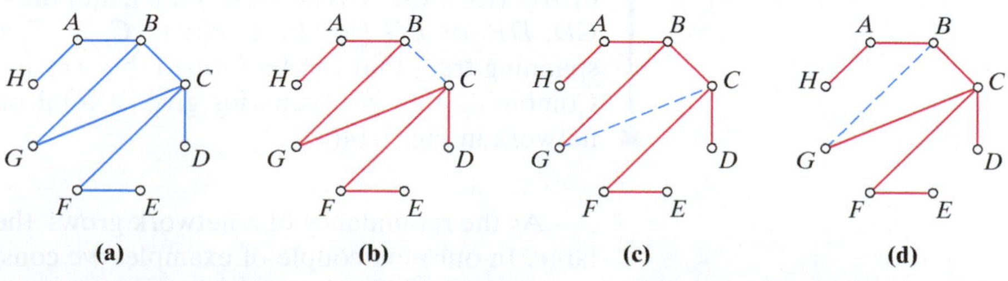

FIGURE 7-12 (a) The original network. (b), (c), and (d) Spanning trees.

The network in Fig. 7-13(a) is the same network as the one in Fig. 7-11(a). The redundancy of the network is $R = 2$, so to find a spanning tree we will have to "discard" two edges. Edges AB and AH are bridges of the network, so they will have to be part of any spanning tree. The other seven edges are split into two separate circuits (B, C, G, B of length 3 and C, D, E, F, C of length 4). A spanning tree can be found by "busting" each of the two circuits. This means excluding any one of the three edges of circuit B, C, G, B and any one of the four edges of circuit C, D, E, F, C. For example, if we exclude BC and CD, we get the spanning tree shown in Fig. 7-13(b). We could also exclude BC and DE and get the spanning tree shown in Fig. 7-13(c), and so on. Given that there are $3 \times 4 = 12$ different ways to choose an edge from the circuit of length 3 and an edge from the circuit of length 4, we will not show all 12 spanning trees. [Figs. 7-13(b) through (e) show some of them.]

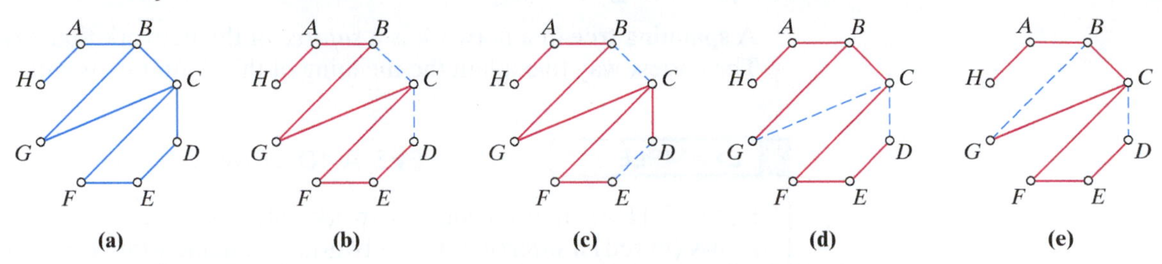

FIGURE 7-13 (a) The original network. (b), (c), (d), and (e) Spanning trees.

EXAMPLE 7.9 MORE COUNTING OF SPANNING TREES

The network in Fig. 7-14(a) is another network with 8 vertices and redundancy $R = 2$. The difference between this network and the one in Fig. 7-13(a) is that here the circuits B, C, G, B and C, D, E, G, C share a common edge CG. Determining which pairs of edges can be excluded in this case is a bit more complicated.

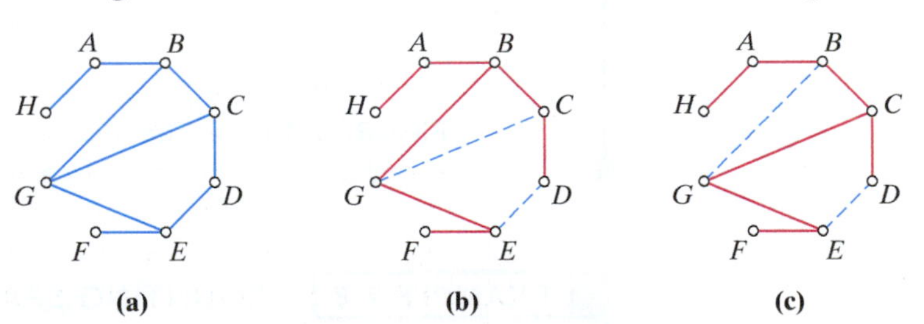

FIGURE 7-14 (a) The original network. (b) and (c) Spanning trees.

If one of the excluded edges is the common edge CG, then the other excluded edge can be any other edge in a circuit. There are five choices (BC, CD, DE, EG, and GB), and each choice will result in a different spanning tree [one of these is shown in Fig. 7-14(b)]. The alternative scenario is to exclude two edges neither of which is the common edge CG. In this case one excluded edge has to be either BC or BG (to "bust" circuit B, C, G, B), and the other excluded edge has to be either CD, DE, or EG (to "bust" circuit C, D, E, G, C). There are $2 \times 3 = 6$ possible spanning trees that can be formed this way [one of these is shown in Fig. 7-14(c)]. Combining the two scenarios gives a total of 11 possible spanning trees for the network in Fig. 7-14(a).

As the redundancy of a network grows, the number of spanning trees gets very large. In our next couple of examples we consider spanning trees in weighted networks of high redundancy.

EXAMPLE 7.10 THE AMAZONIAN CABLE NETWORK

The Amazonia Telephone Company is contracted to provide telephone, cable, and Internet service to the seven small mining towns shown in Fig. 7-15(a). These towns are located deep in the heart of the Amazon jungle, which makes the project particularly difficult and expensive. In this environment the most practical and environmentally friendly option is to create a network of underground fiber-optic cable lines connecting the towns. In addition, it makes sense to bury the underground cable lines along the already existing roads connecting the towns. The existing network of roads is shown in Fig. 7-15(a). Figure 7-15(b) is a network model of all the possible connections between the towns. The vertices of the network represent the towns, the edges represent the existing roads, and the weight of each edge represents the cost (in millions of dollars) of creating a fiber-optic cable connection along that particular edge.

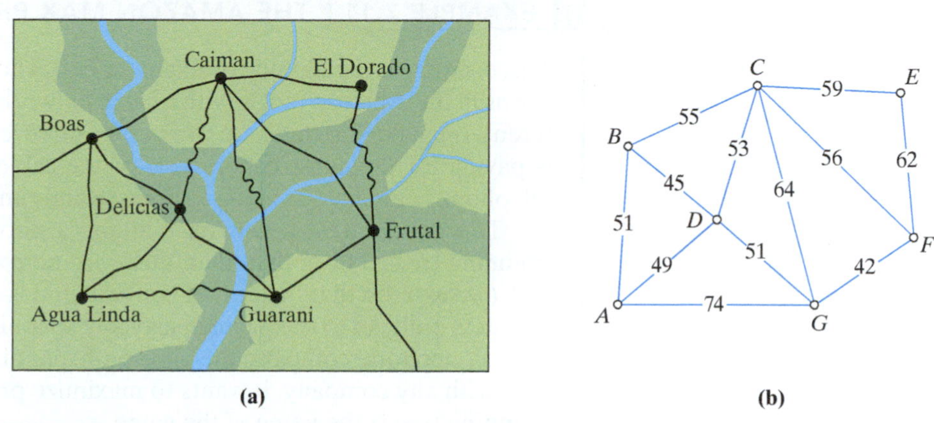

FIGURE 7-15 (a) Network of roads connecting seven towns. (b) Network model showing the cost (in millions) of each connection.

The problem facing the engineers and planners of the Amazonia Telephone Company is how to design a cable network that connects all the towns using the existing network of roads and that costs as little as possible to build—in other words, a *minimum cost* network. The first thing we can say about a minimum cost network is that it must be a *spanning tree* of the original network of roads shown in Fig. 7-15(a). (To connect all the towns means that the cable network must *span* all the vertices; to eliminate redundant connections means that it will have redundancy $R = 0$, and that means a *tree*.) But in this example there are costs involved, and not all spanning trees will have the same cost, so rather than finding any old spanning tree, we need to find the spanning tree with least total weight. We call such a spanning tree a *minimum spanning tree.*

One way to find a minimum spanning tree is to list all possible spanning trees, find the total cost of each, and pick the one with least cost. (This is the same approach we described in Chapter 6 as the *brute-force algorithm.*) The problem is that the network in Fig. 7-15(b) has high redundancy ($R = 6$) and hundreds of possible spanning trees. Sifting through all of them to find the one with least cost is not a good plan. In Section 7.3 we will discuss a better way of finding a minimum spanning tree in the Amazon jungle. ■

Example 7.10 was our introduction to the concept of a *minimum spanning tree.* We now give it a formal definition:

- **Minimum Spanning Tree.** In a weighted network, a **minimum spanning tree (MST)** is *a* spanning tree with least total weight.

Sometimes there is only one MST and we can refer to it as *the* MST, but we can't assume this to be true in general. (For example, if all the weights in the network are the same, then every spanning tree is an MST.)

In most applications, the weights of the network represent *costs*—money, time, or distance. In these cases the goal is to *minimize*. There are some applications, however, where the weights represent *profits*. Profits (or other types of *gains* such as *higher bandwidth* on Internet connections or *increased flows* in pipelines) are things we want more of, rather than less, so instead of minimizing we should be maximizing. This leads to our next definition.

- **Maximum Spanning Tree.** In a weighted network, a **maximum spanning tree (MaxST)** is *a* spanning tree with highest total weight.

Our next example illustrates a MaxST application.

EXAMPLE 7.11 THE AMAZON MAX PROFIT NETWORK

This example is the flip side of Example 7.10. The problem is still to connect the seven towns in the Amazon jungle with a cable network, but the circumstances are quite different. Imagine that there are two parties involved: one party—say the government—is paying for the construction costs of the cable network; the other party—say the telephone company—is going to operate and run the network.

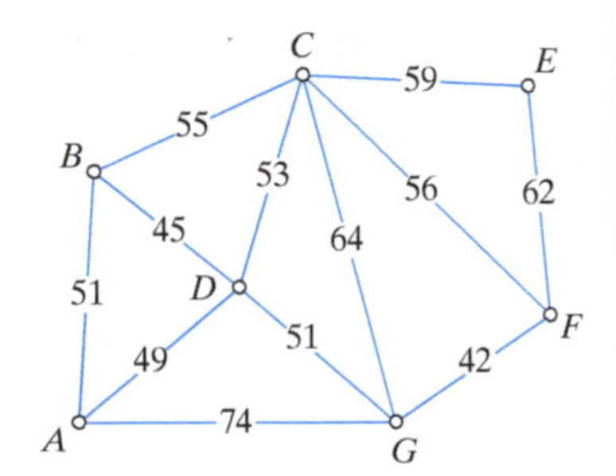

FIGURE 7-16

The government insists on building a network with zero redundancy (i.e., a spanning tree). Other than that, there are no restrictions on the choice of spanning tree (we assume that the cost of construction is the same no matter which spanning tree gets built, so to the paying party any spanning tree will do). On the other hand, to the telephone company the choice of spanning tree is very important—as is the case with any company, it wants to maximize profit, so building the most profitable spanning tree is the name of the game.

Let's assume that the weight of each edge of the network in Fig. 7-16 represents the expected annual *profits* (in millions) to the phone company for operating that segment of the network. (You may have noticed that we are using exactly the same weighted network as that in Example 7.10. As you will see in the next section, this is not a coincidence.) To the phone company, the problem now becomes finding the MaxST of the original network. We will learn how to do this in the next section.

7.3 Kruskal's Algorithm

Unlike the situation with TSPs (see Chapter 6) there are several *efficient* and *optimal* algorithms for finding minimum spanning trees. Moreover, any algorithm that finds minimum spanning trees can be tweaked to find maximum spanning trees as well. In this section we will introduce one such algorithm—a simple algorithm called *Kruskal's algorithm* after American mathematician Joseph Kruskal.

Kruskal's algorithm is a variation of the *cheapest-link algorithm* used for solving TSPs in Chapter 6 (but easier). The minimum spanning tree gets built one edge at a time by choosing at each step the cheapest available edge that *does not create any circuits*. Period. In contrast to the cheapest-link algorithm, with Kruskal's algorithm no other rules need to be enforced—stay away from circuits and you'll be fine. You continue choosing the cheapest available edge that does not create circuits until $N - 1$ edges are chosen. At that point you have your MST.

The following is a formal description of Kruskal's algorithm. (For simplicity, we use "cheapest" to denote "of least weight.")

KRUSKAL'S ALGORITHM

- **Step 1.** Pick the *cheapest edge* available. (In case of a tie, pick one at random.) Mark it (say in red).
- **Step 2.** Pick the next cheapest edge available and mark it.
- **Steps 3, 4, . . . , $N - 1$.** Continue picking and marking the cheapest unmarked edge available that does not create circuits. After step $N - 1$ you are done.

EXAMPLE 7.12 THE AMAZONIAN CABLE NETWORK AND KRUSKAL'S ALGORITHM

In Example 7.10 we raised the following question: What is the optimal fiber-optic cable network connecting the seven towns shown in Fig. 7-17(a)? The weight of each edge is the cost (in millions of dollars) of laying the cable along that segment of the network.

The answer, as we now know, is to find the minimum spanning tree of the network. We will use Kruskal's algorithm to do it. Here are the details:

- **Step 1.** We start by choosing the cheapest edge of the network. In this case we choose GF, and mark it in red (or any other color) as shown in Fig. 7-17(b). (Note that this does not have to be the first link actually built—we are putting the network together on paper only. On the ground, the schedule of construction is a different story, and there are many other factors that need to be considered.)
- **Step 2.** The next cheapest edge available is BD at \$45 million. We choose it for the MST and mark it in red [Fig. 7-17(c)].
- **Step 3.** The next cheapest edge available is AD at \$49 million. Again, we choose it for the MST and mark it in red [Fig. 7-17(d)].

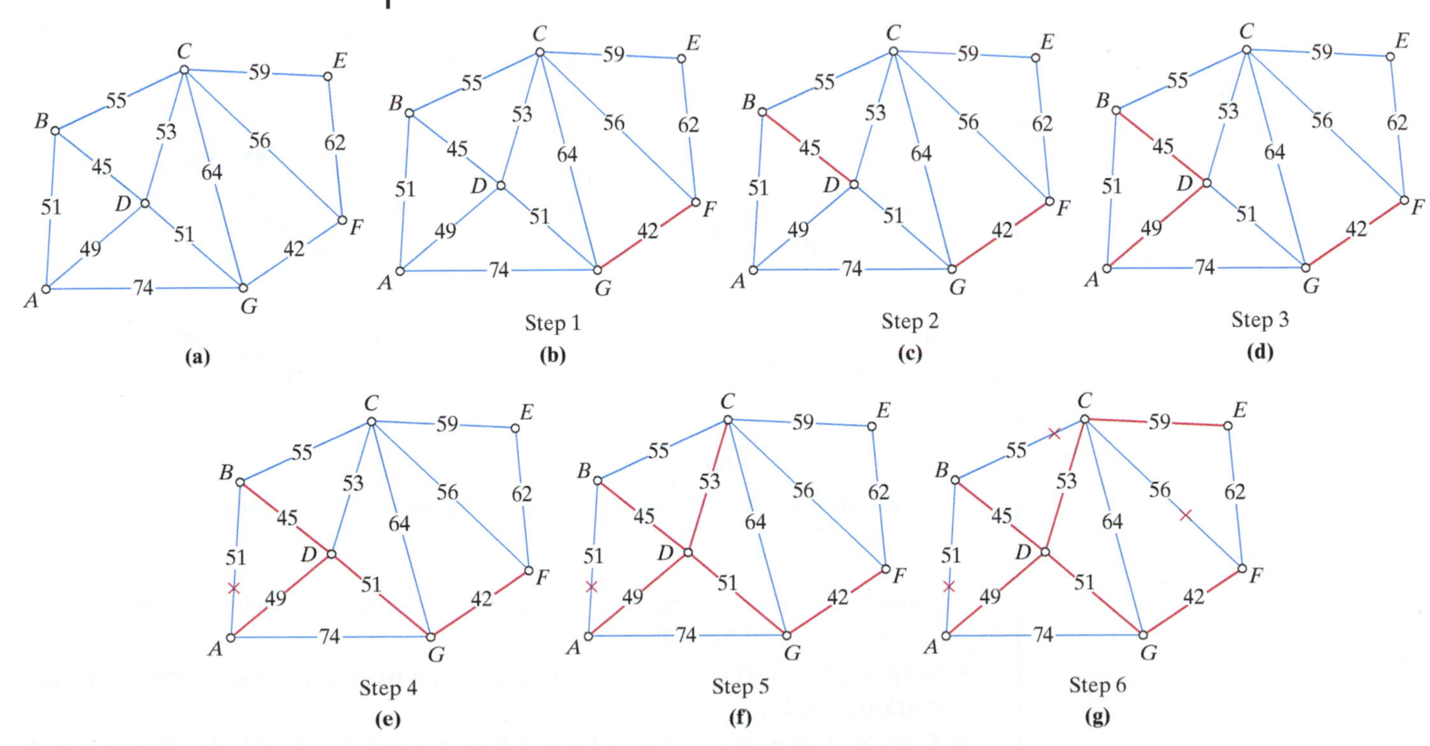

FIGURE 7-17

- **Step 4.** For the next cheapest edge there is a tie between AB and DG, both at $51 million. But we can rule out AB—it would create a circuit in the MST, and we can't have that! (For bookkeeping purposes it is a good idea to erase or cross out the edge.) The edge DG, on the other hand, is just fine, so we mark it in red and make it part of the MST [Fig. 7-17(e)].
- **Step 5.** The next cheapest edge available is CD at $53 million. No problems here, so again, we mark it in red and make it part of the MST [Fig. 7-17(f)].
- **Step 6.** The next cheapest edge available is BC at $55 million, but this edge would create a circuit, so we cross it out. The next possible choice is CF at $56 million, but once again, this choice creates a circuit so we must cross it out. The next possible choice is CE at $59 million, and this is one we do choose. We mark it in red and make it part of the MST [Fig. 7-17(g)].
- **Step. . . .** Wait a second—we are finished! Even without looking at a picture, we can tell we are done—six links is exactly what is needed for an MST on seven vertices.

The total cost of the red MST shown in Fig. 7-17(g) is $299 million.

Any algorithm that can find an MST can also be used to find a MaxST by means of a simple modification: *Change the signs of the weights* in the network. We call the network that we get when we change the signs of all the weights the **negative** of the original network. Switching the signs of the weights switches MaxSTs into MSTs and vice versa. *A MaxST of a network is an MST of the negative network*. This follows from the simple fact that changing the signs of numbers reverses their inequality relationships ($74 > 45$ implies that $-74 < -45$). We will illustrate this idea by returning to the MaxST problem introduced in Example 7.11.

EXAMPLE 7.13 THE AMAZON MAX PROFIT AND KRUSKAL'S ALGORITHM

Figure 7-18(a) shows the *negative* network for the original network in Example 7.11 (Fig. 7-16). We'll use Kruskal's algorithm to find the MST of this negative network. This will be the MaxST that we are looking for.

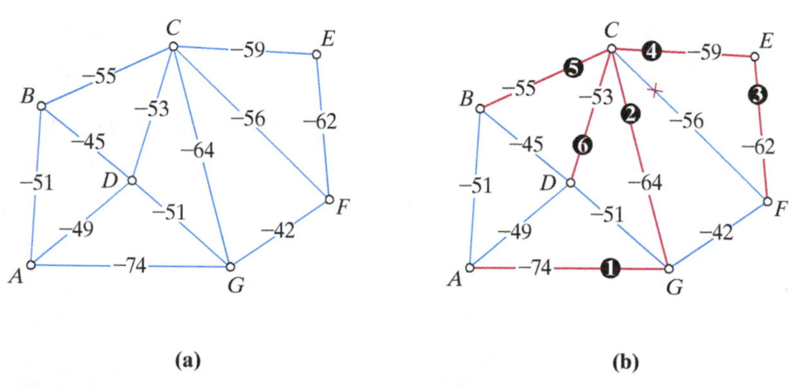

FIGURE 7-18 (a) The *negative* network. (b) The MST of the negative network.

- **Step 1.** The cheapest edge in the negative network is AG, with a weight of -74. We choose it and mark it in red.
- **Step 2.** The next cheapest edge is CG, with a weight of -64. We choose it and mark it in red.
- **Step 3.** The next cheapest edge is EF, with a weight of -62. We choose it and mark it in red.

- **Step 4.** The next cheapest edge is *CE*, with a weight of −59. We choose it and mark it in red.
- **Step 5.** The next cheapest edge is *CF*, with a weight of −56. We choose it and . . . oops! Can't do that. Choosing *CF* would create a red circuit. Spanning trees don't have circuits. We rule out *CF*, and move on to the next cheapest edge, *CB*, with a weight of −55. No circuits with this one, so we choose it for the MST.
- **Step 6.** The next cheapest edge is *CD*, with a weight of −53. No problems here, so we choose it and mark it in red.

At Step 6 we are done! We have our MST for the negative network, shown in Fig. 7-18(b). This network is the MaxST we were looking for in the original network. The total profits that can be expected from this MaxST are given by the sum of the original weights: $74 + 64 + 62 + 59 + 55 + 53 = 367$ million.

EXAMPLE 7.14 FINDING MSTs IN POWER GRIDS USING KRUSKAL'S ALGORITHM

In many rural areas electricity transmission lines are old. Old transmission lines carry lower voltages, leak more power, and are more sensitive to bad weather than modern transmission lines, so a common infrastructure project is to update the older parts of the grid with new transmission lines. The problem is how to choose the newer transmission lines so that they carry power to as many customers as possible while at the same time keeping the cost of the project down. The least costly solution often involves finding an MST.

Figure 7-19(a) shows a section of the electrical power grid connecting 14 rural towns in central Texas (see Example 7.4). The weight of each edge represents the length (in miles) of that particular segment of the grid. When the terrain is flat—as is in central Texas—the cost of replacing a transmission line is proportional to the length of the line (for high-voltage, modern transmission lines it is about $500,000 per mile), so when we minimize length we are also minimizing cost. It follows that the MST of the grid in Fig. 7-19(a) is going to give the cheapest spanning tree of updated power lines for the grid.

We'll find the MST using Kruskal's algorithm. [As a heads-up, we know ahead of time that in this network Kruskal's algorithm will require 13 steps (the network has 14 vertices), so we'll be brief and to the point.]

- **Step 1.** Choose *LM* (5) and mark it in red.
- **Step 2.** Choose *AJ* (11) and mark it in red.
- **Step 3.** Choose *EF* (11) and mark it in red.
- **Step 4.** Choose *FG* (12) and mark it in red.
- **Step 5.** Choose *HM* (13) and mark it in red.
- **Step 6.** Choose *NK* (14) and mark it in red.
- **Step 7.** Choose *BK* (16) and mark it in red.
- **Step 8.** Choose *HN* (18) and mark it in red.
- **Step 9.** Skip *NL* (21) because it closes a circuit. Choose *AB* (23) and mark it in red.
- **Step 10.** Choose *DE* (23) and mark it in red.
- **Step 11.** Choose *CD* (26) and mark it in red.
- **Step 12.** Choose *FM* (30) and mark it in red.
- **Step 13.** Skip *CE* (31), *BC* (31), and *JK* (32) because they all close circuits. Choose *IN* (32) and mark it in red. That's it—we are done.

Figure 7-19(b) shows the MST. The total length of the MST is 234 miles. At an average cost of $500,000 per mile, the total cost of the infrastructure update is $117 million.

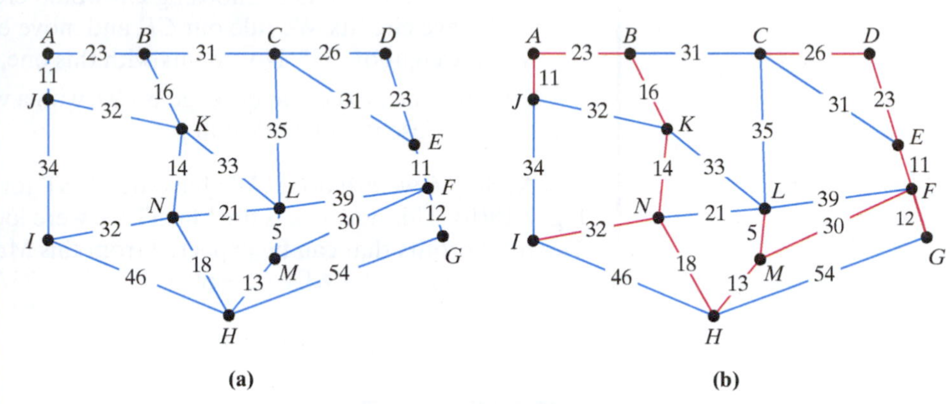

FIGURE 7-19 (a) The power grid. (b) The MST for infrastructure update.

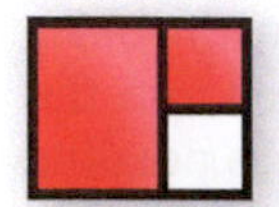

Conclusion

A typical real-world network connects *things* (people, computers, cities, etc.) to each other, making possible the free flow of some *commodity* (communication, information, electricity, etc.) between the different nodes of the network. From the perspective of the people being served by the network, the *more* connected the network is, the better—connections increase the ways the commodity can flow among the nodes, thus improving convenience, reliability, and performance. On the other hand, from the perspective of those planning or servicing the network, the *less* connected the network is, the better—connections can be expensive to build and maintain. As is usually the case, there are tradeoffs that must be made between convenience and cost. When convenience is important, or when costs are low, networks tend to be highly connected and have a lot of redundancy. When costs are high, convenience becomes less of a consideration and networks tend to have little or no redundancy.

In this chapter we discussed networks in general and some of the mathematical concepts associated with building good networks—redundancy, trees, spanning trees, weighted networks, MSTs, and MaxSTs. This was just a small peek into the mathematics of networks, a deep and important topic with obvious applications to our ever more connected lives. If you don't believe that, think about going without electricity or Internet access for a few days or even worse, imagine Facebook or Twitter being down—for just a day.

> "There are three kinds of death in this world. There's heart death, there's brain death, and there's being off the network."
>
> – Guy Almes

KEY CONCEPTS

7.1 Networks and Trees

- **network:** a connected graph, **205**
- **degree of separation:** the length of the shortest path connecting two vertices of a network, **206**
- **diameter:** in a network, the largest degree of separation between pairs of vertices, **207**
- **weighted network:** a network with weights associated with the edges, **208**
- **tree:** a network with no circuits, **208**
- **single-path property:** in a tree there is one and only one path connecting any two vertices, **209**
- **all-bridges property:** in a tree all edges are bridges, **209**
- **$N-1$ edges property:** a tree with N vertices has $N-1$ edges, **209**
- **redundancy (R):** in a network with N vertices and M edges $R = M - (N-1)$, **209**

7.2 Spanning Trees, MSTs, and MaxSTs

- **subtree:** a set of vertices and edges in a network that form a tree, **211**
- **spanning tree:** a subtree whose vertices are *all* the vertices of the network, **211**
- **minimum spanning tree (MST):** in a weighted network, a spanning tree with smallest total weight, **213**
- **maximum spanning tree (MaxST):** in a weighted network, a spanning tree with largest total weight, **214**

7.3 Kruskal's Algorithm

- **Kruskal's algorithm:** finds an MST of a weighted network; at each step, it chooses the cheapest available edge that does not form a circuit, **215**
- **negative network:** the network obtained by changing the signs of the weights in the original network, **216**
- **Kruskal's algorithm for MaxSTs:** applies Kruskal's algorithm for MSTs to the negative network, **216**

EXERCISES

WALKING

7.1 Networks and Trees

1. A computer lab has seven computers labeled A through G. The connections between computers are as follows:
   - A is connected to D and G
   - B is connected to C, E, and F
   - C is connected to B, E, and F
   - D is connected to A and G
   - E is connected to B and C
   - F is connected to B and C
   - G is connected to A and D

   Is the lab set-up a computer network? Explain why or why not.

2. The following is a list of the electrical power lines connecting eight small towns labeled A through H.
   - A power line connecting A and D
   - A power line connecting B and C
   - A power line connecting B and E
   - A power line connecting B and G
   - A power line connecting C and G
   - A power line connecting D and F
   - A power line connecting D and H
   - A power line connecting E and G

   Do the power lines form a network? Explain why or why not.

3. Consider the network shown in Fig. 7-20.
   (a) How many degrees of separation are there between C and E?
   (b) How many degrees of separation are there between A and E?
   (c) How many degrees of separation are there between A and F?

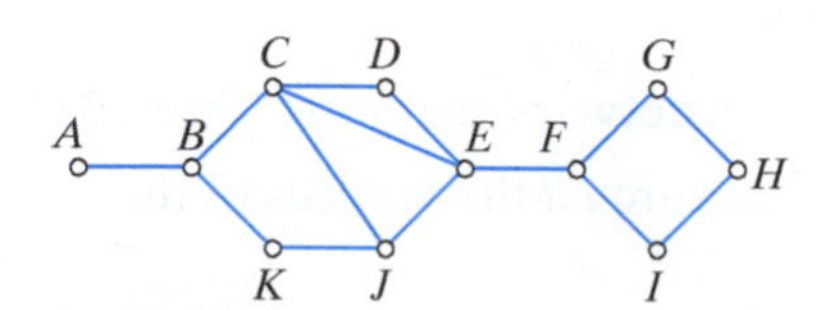

FIGURE 7-20

4. Consider the network shown in Fig. 7-21.
   (a) How many degrees of separation are there between D and A?
   (b) How many degrees of separation are there between A and L?
   (c) How many degrees of separation are there between B and H?

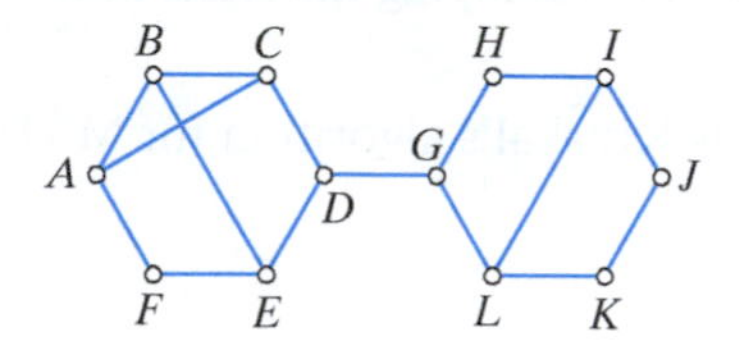

FIGURE 7-21

5. Consider once again the network shown in Fig. 7-20.
   (a) Find two vertices in the network having five degrees of separation between them.
   (b) Find two vertices in the network having six degrees of separation between them.
   (c) If you can, find two vertices in the network having seven degrees of separation between them. If you can't, then explain why you don't think there are any.
   (d) What is the diameter of the network?

6. Consider once again the network shown in Fig. 7-21.
   (a) Find two vertices in the network having five degrees of separation between them.
   (b) Find two vertices in the network having six degrees of separation between them.
   (c) If you can, find two vertices in the network having seven degrees of separation between them. If you can't, then briefly explain why you don't think there are any.
   (d) What is the diameter of the network?

7. Consider the network shown in Fig. 7-22. (This is the network we discussed in Examples 7.1 and 7.2)
   (a) In addition to the pair N and K, there are three other pairs of vertices in the network having six degrees of separation between them. Find them.
   (b) If you can, find two vertices in the network having seven degrees of separation between them. If you can't, then briefly explain why you don't think there are any.
   (c) What is the diameter of the network?

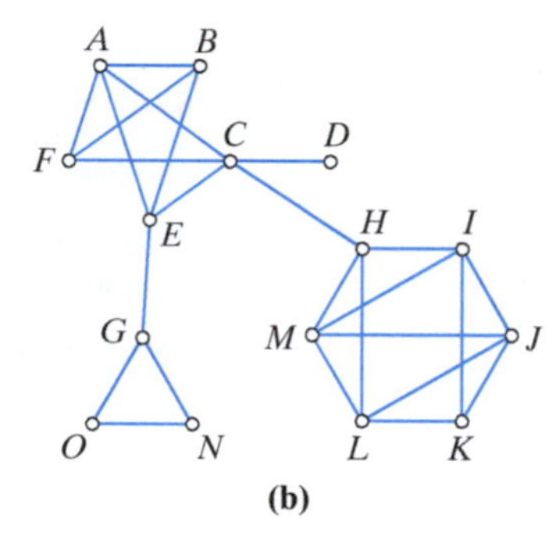

FIGURE 7-22

8. Consider the network shown in Fig. 7-23. (This is the network we discussed in Examples 7.1 and 7.2 with one additional link—the edge GH.)
   (a) Find two vertices in the network having five degrees of separation between them.
   (b) If you can, find two vertices in the graph having six degrees of separation between them. If you can't, then briefly explain why you don't think there are any.
   (c) What is the diameter of the network?

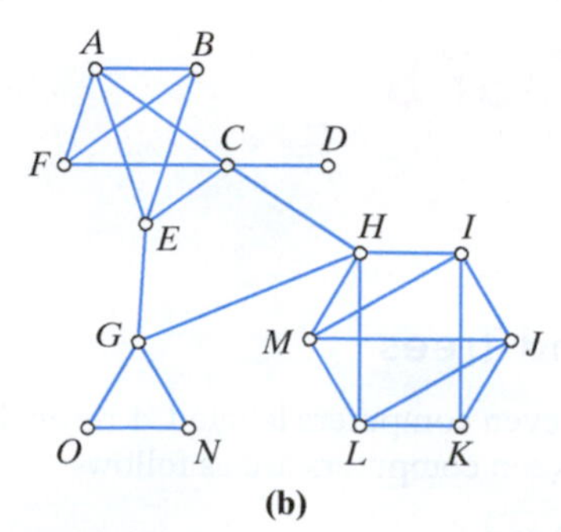

FIGURE 7-23

9. Consider the tree shown in Fig. 7-24.

(a) How many degrees of separation are there between A and J?

(b) How many degrees of separation are there between E and L?

(c) How many degrees of separation are there between M and P?

(d) What is the diameter of the network?

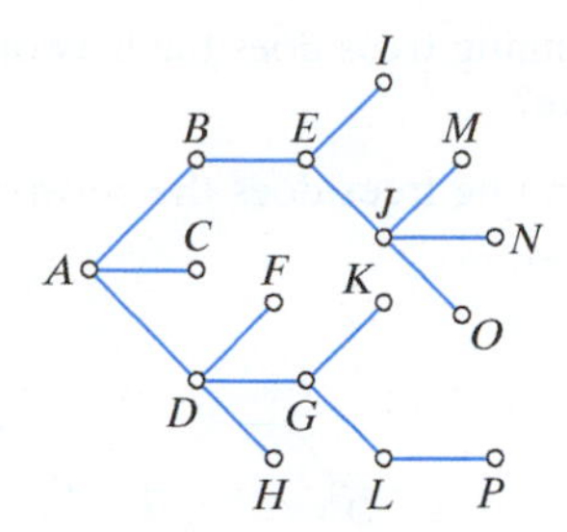

FIGURE 7-24

10. Consider the tree shown in Fig. 7-25.

(a) How many degrees of separation are there between A and P?

(b) How many degrees of separation are there between E and P?

(c) How many degrees of separation are there between L and P?

(d) What is the diameter of the network?

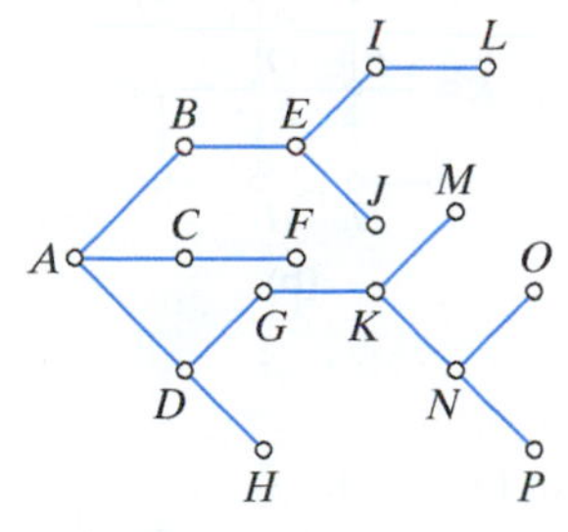

FIGURE 7-25

In Exercises 11 through 24 you are given information about a network. Choose one of the following three options: (A) the network is definitely a tree; (B) the network is definitely not a tree; (C) the network may or may not be a tree (more information is needed). Accompany your answer with a brief explanation for your choice.

11. The network has 15 vertices and 16 edges.

12. The network has 23 vertices and no bridges.

13. The network has 16 vertices and 15 edges.

14. The network has 23 vertices and 22 bridges.

15. The network has redundancy $R = 1$.

16. The network has redundancy $R = 0$.

17. The network has 10 vertices (A through J), and there is only one path connecting A and J.

18. The network has 10 vertices (A through J) and there are two paths connecting C and D.

19. The network has five vertices, no loops, and no multiple edges, and every vertex has degree 4.

20. The network has five vertices, no loops, and no multiple edges, and every vertex has degree 2.

21. The network has five vertices, no loops, and no multiple edges, and has one vertex of degree 4 and four vertices of degree 1.

22. The network has five vertices, no loops, and no multiple edges, and has two vertices of degree 1 and three vertices of degree 2.

23. The network has all vertices of even degree. (*Hint*: You will need to use some concepts from Chapter 5 to answer this question.)

24. The network has two vertices of odd degree and all the other vertices of even degree. (*Hint*: You will need to use some concepts from Chapter 5 to answer this question.)

7.2 Spanning Trees, MSTs, and MaxSTs

25. Consider the network shown in Fig. 7-26.

(a) Find a spanning tree of the network.

(b) Calculate the redundancy of the network.

(c) What is the diameter of the network?

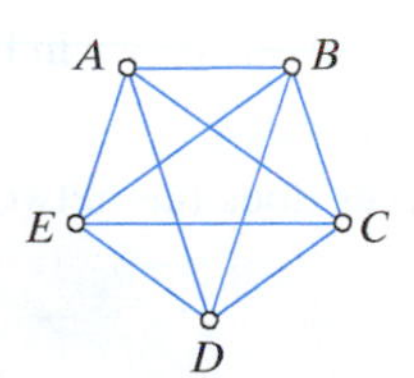

FIGURE 7-26

26. Consider the network shown in Fig. 7-27.

(a) Find a spanning tree of the network.

(b) Calculate the redundancy of the network.

(c) What is the diameter of the network?

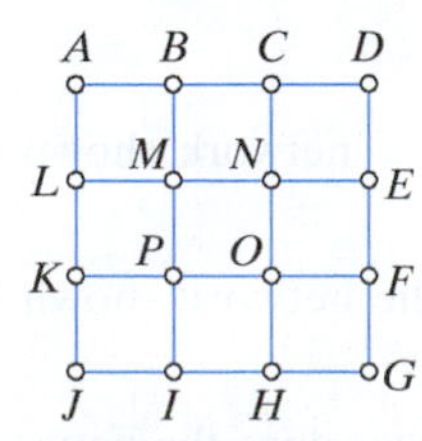

FIGURE 7-27

27. Consider the network shown in Fig. 7-28.

(a) Find a spanning tree of the network.

(b) Calculate the redundancy of the network.

(c) What is the diameter of the network?

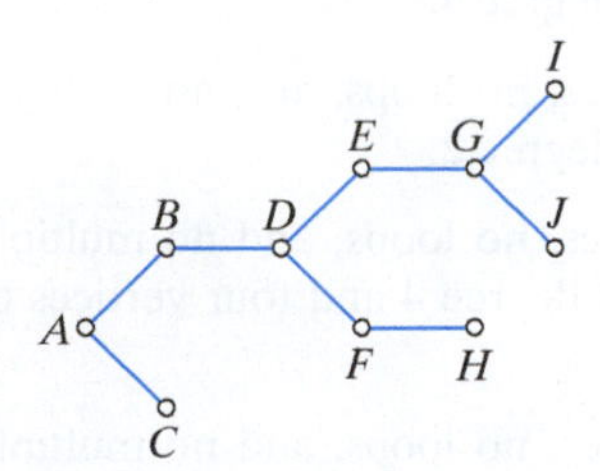

FIGURE 7-28

28. Consider the network shown in Fig. 7-29.

(a) Find a spanning tree of the network.

(b) Calculate the redundancy of the network.

(c) What is the diameter of the network?

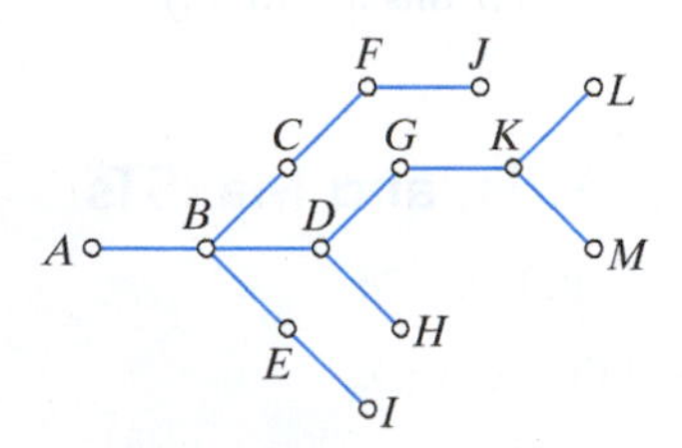

FIGURE 7-29

29. (a) Find all the spanning trees of the network shown in Fig. 7-30(a).

(b) Find all the spanning trees of the network shown in Fig. 7-30(b).

(c) How many different spanning trees does the network shown in Fig. 7-30(c) have?

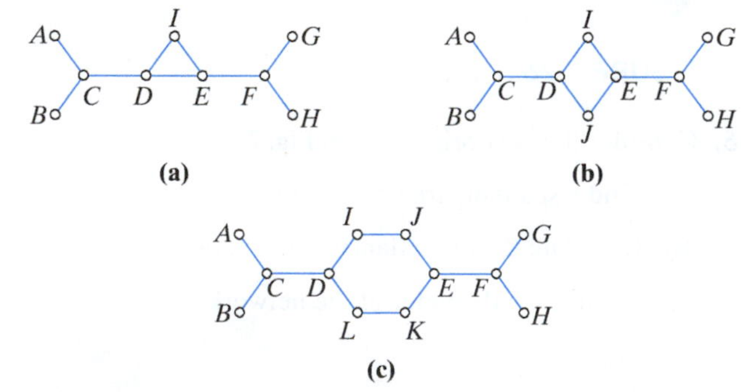

FIGURE 7-30

30. (a) Find all the spanning trees of the network shown in Fig. 7-31(a).

(b) Find all the spanning trees of the network shown in Fig. 7-31(b).

(c) How many different spanning trees does the network shown in Fig. 7-31(c) have?

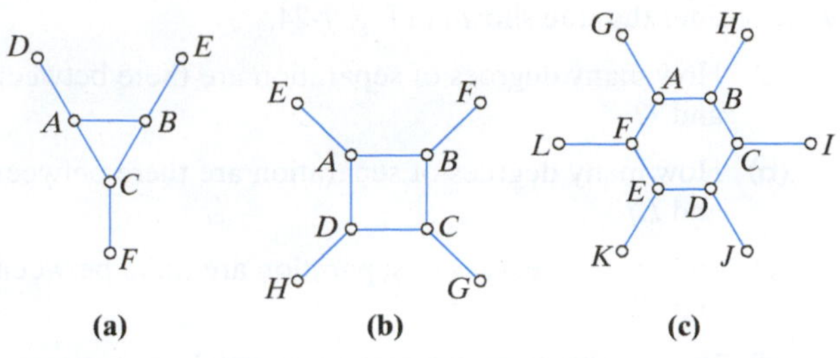

FIGURE 7-31

31. (a) How many different spanning trees does the network shown in Fig. 7-32(a) have?

(b) How many different spanning trees does the network shown in Fig. 7-32(b) have?

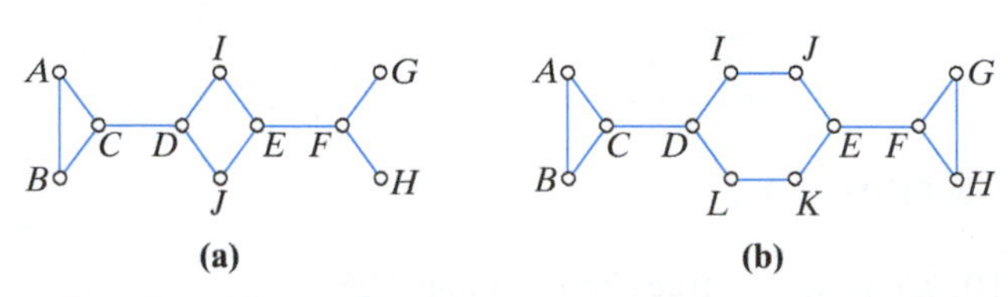

FIGURE 7-32

32. (a) How many different spanning trees does the network shown in Fig. 7-33(a) have?

(b) How many different spanning trees does the network shown in Fig. 7-33(b) have?

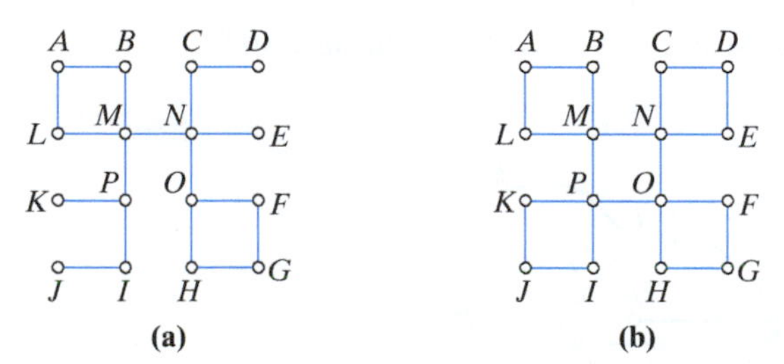

FIGURE 7-33

33. Consider the network shown in Fig. 7-34.

(a) How many different spanning trees does this network have?

(b) Find the spanning tree that has the largest degree of separation between *H* and *G*.

(c) Find a spanning tree that has the smallest degree of separation between *H* and *G*.

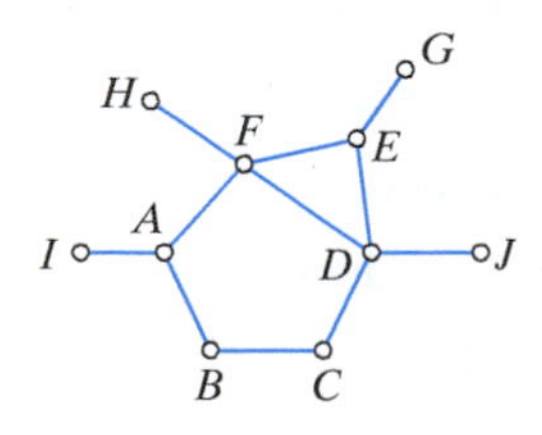

FIGURE 7-34

34. Consider the network shown in Fig. 7-35.

(a) How many different spanning trees does this network have?

(b) Find a spanning tree that has the largest degree of separation between H and J.

(c) Find a spanning tree that has the smallest degree of separation between K and G.

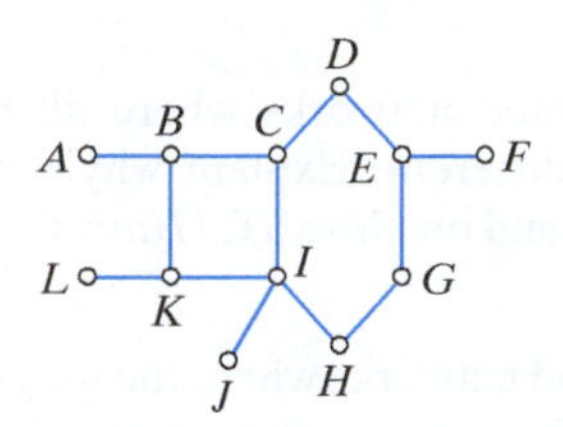

FIGURE 7-35

7.3 Kruskal's Algorithm

35. The 3 by 4 grid shown in Fig. 7-36 represents a network of streets (3 blocks by 4 blocks) in a small subdivision. For landscaping purposes, it is necessary to get water to each of the corners by laying down a system of pipes along the streets. The cost of laying down the pipes is \$40,000 per mile, and each block of the grid is exactly half a mile long. Find the cost of the cheapest network of pipes connecting all the corners of the subdivision. Explain your answer. (*Hint*: First determine the number of blocks in the MST.)

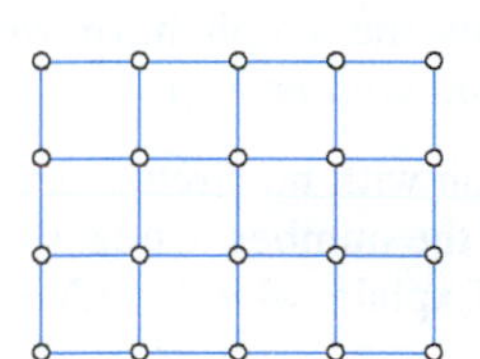

FIGURE 7-36

36. The 4 by 5 grid shown in Fig. 7-37 represents a network of streets (4 blocks by 5 blocks) in a small subdivision. For landscaping purposes, it is necessary to get water to each of the corners by laying down a system of pipes along the streets. The cost of laying down the pipes is \$40,000 per mile, and each block of the grid is exactly half a mile long. Find the cost of the cheapest network of pipes connecting all the corners of the subdivision. Explain your answer. (*Hint*: First determine the number of blocks in the MST.)

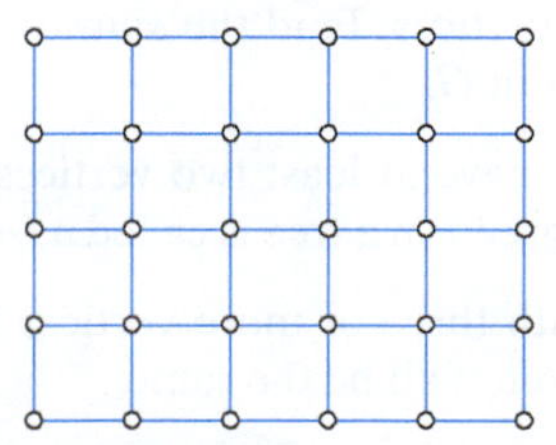

FIGURE 7-37

37. Find the MST of the network shown in Fig. 7-38 using Kruskal's algorithm, and give its weight.

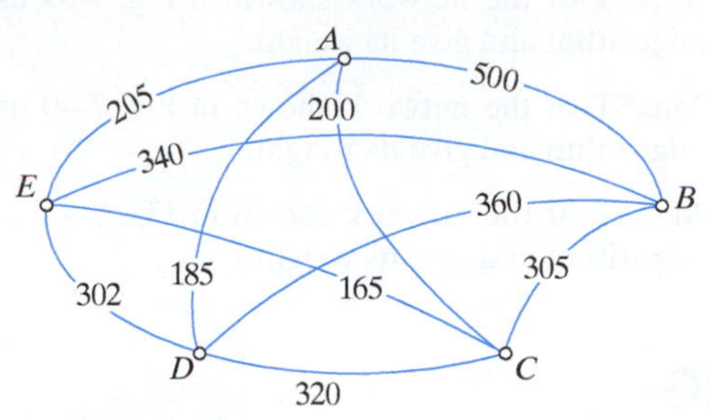

FIGURE 7-38

38. Find the MST of the network shown in Fig. 7-39 using Kruskal's algorithm, and give its weight.

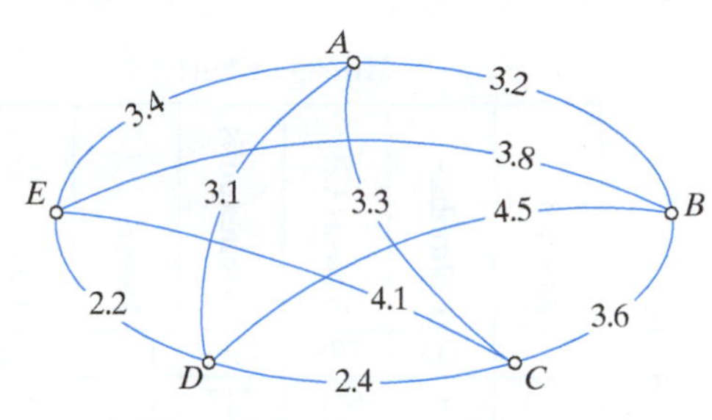

FIGURE 7-39

39. Find the MST of the network shown in Fig. 7-40 using Kruskal's algorithm, and give its weight.

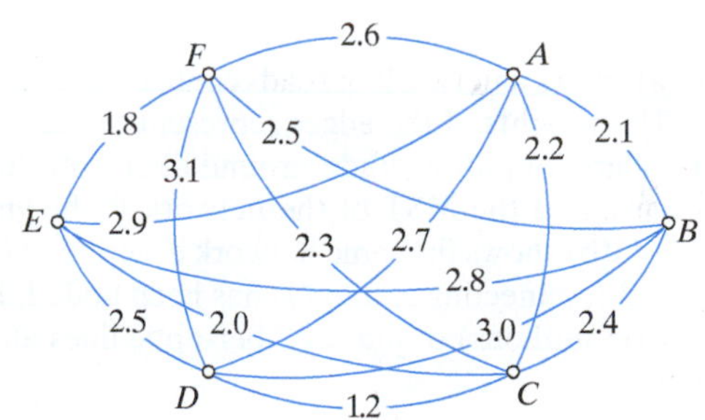

FIGURE 7-40

40. Find the MST of the network shown in Fig. 7-41 using Kruskal's algorithm, and give its weight.

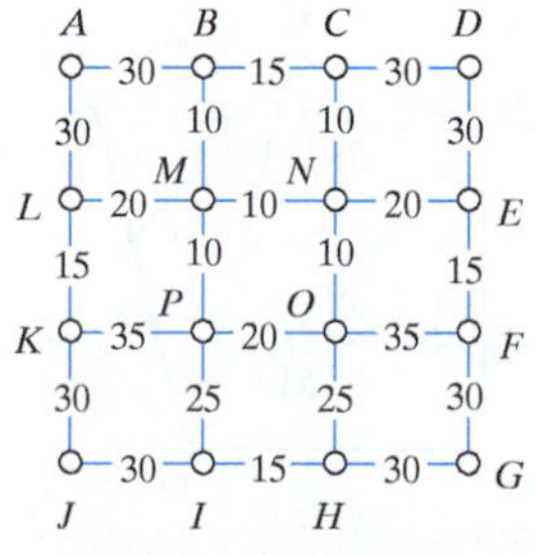

FIGURE 7-41

41. Find the MaxST of the network shown in Fig. 7-38 using Kruskal's algorithm and give its weight.

42. Find the MaxST of the network shown in Fig. 7-39 using Kruskal's algorithm and give its weight.

43. Find the MaxST of the network shown in Fig. 7-40 using Kruskal's algorithm and give its weight.

44. Find the MaxST of the network shown in Fig. 7-41 using Kruskal's algorithm and give its weight.

JOGGING

45. The mileage chart in Fig. 7-42 shows the distances between Atlanta, Columbus, Kansas City, Minneapolis, Pierre, and Tulsa. Working directly from the mileage chart use Kruskal's algorithm to find the MST connecting the six cities. (*Hint*: See Example 6.15 for the use of an auxiliary graph.)

Mileage Chart

	Atlanta	Columbus	Kansas City	Minneapolis	Pierre	Tulsa
Atlanta	*	533	798	1068	1361	772
Columbus	533	*	656	713	1071	802
Kansas City	798	656	*	447	592	248
Minneapolis	1068	713	447	*	394	695
Pierre	1361	1071	592	394	*	760
Tulsa	772	802	248	695	760	*

FIGURE 7-42

46. Figure 7-43(a) shows a network of roads connecting cities A through G. The weights of the edges represent the cost (in millions of dollars) of putting underground fiber-optic lines along the roads, and the MST of the network is shown in red. Figure 7-43(b) shows the same network except that one additional road (connecting E and G) has been added. Let x be the cost (in millions) of putting fiber-optic lines along this new road.

(a) Describe the MST of the network in Fig. 7-43(b) in the case $x > 59$. Explain your answer.

(b) Describe the MST of the network in Fig. 7-43(b) in the case $x < 59$. Explain your answer.

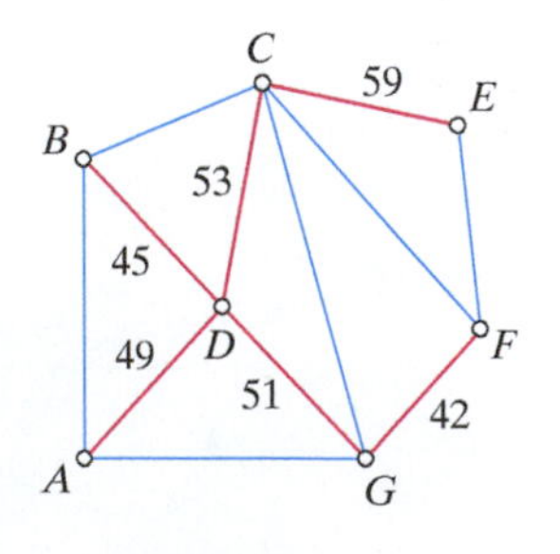

(a)

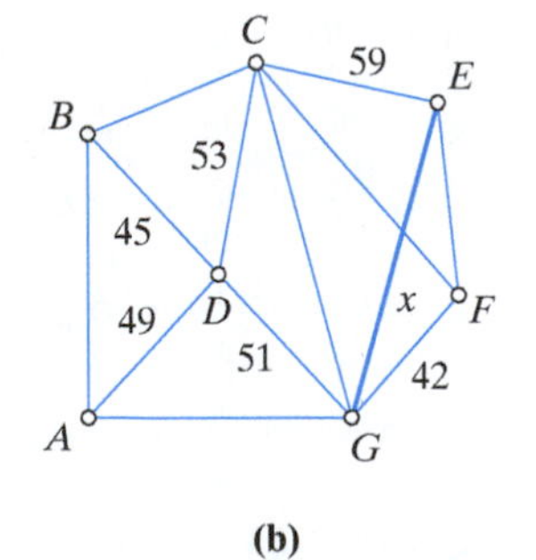

(b)

FIGURE 7-43

47. Consider a network with M edges. Let k denote the number of bridges in the network.

(a) If $M = 5$, list all the possible values of k.

(b) If $M = 123$, describe the set of all possible values of k.

48. Explain why in a network with no loops or multiple edges, the maximum redundancy is given by $R = \frac{(N^2 - 3N + 2)}{2}$. (*Hint*: The maximum redundancy occurs when the network is K_N.)

49. This exercise refers to weighted networks where all the weights in the network are different. Explain why these networks have only one MST and one MaxST. (*Hint*: Think of Kruskal's algorithm.)

50. This exercise refers to weighted networks where the weights in the network are *not* all different (i.e., there are at least two edges with the same weight).

(a) Give an example of a network of this type that has only one MST.

(b) Give an example of a network of this type that has more than one MST.

51. Suppose that in a weighted network there is just one edge (call it XY) with the *smallest* weight. Explain why the edge XY must be in every MST of the network.

52. Suppose G is a disconnected graph with N vertices, M edges, and no circuits.

(a) How many components does the graph have when $N = 9$ and $M = 6$?

(b) How many components does the graph have when $N = 240$ and $M = 236$? Explain your answer.

53. Suppose G is a disconnected graph with no circuits. Let N denote the number of vertices, M the number of edges, and K the number of components. Explain why $M = N - K$. (*Hint*: Try Exercise 52 first.)

54. **Cayley's theorem.** Cayley's theorem says that the number of spanning trees in a complete graph with N vertices is given by N^{N-2}.

(a) List the $4^2 = 16$ spanning trees of K_4.

(b) Which is larger, the number of Hamilton circuits or the number of spanning trees in a complete graph with N vertices? Explain.

RUNNING

55. (a) Let G be a tree with N vertices. Find the sum of the degrees of all the vertices in G.

(b) Explain why a tree must have at least two vertices of degree 1. (A vertex of degree 1 in a tree is called a *leaf*.)

(c) Explain why in a tree with three or more vertices the degrees of the vertices cannot all be the same.

56. Show that if a tree has a vertex of degree K, then there are at least K vertices in the tree of degree 1.

57. A *bipartite graph* is a graph with the property that the vertices of the graph can be divided into two sets A and B so that every edge of the graph joins a vertex from A to a vertex from B (Fig. 7-44). Explain why trees are always bipartite graphs.

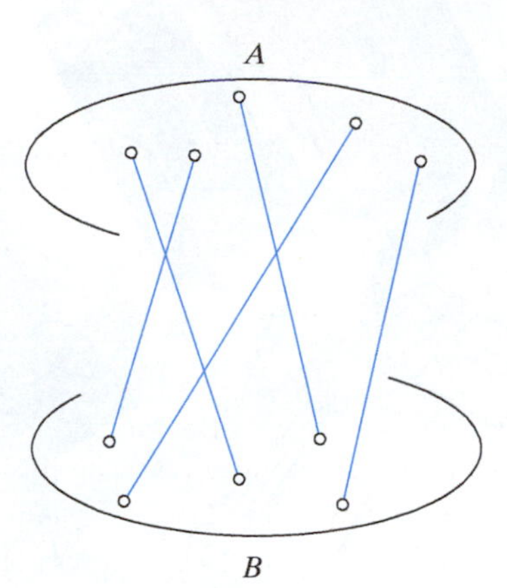

FIGURE 7-44

58. Suppose that in a weighted network there is just one edge (call it XY) with the *largest* weight.

(a) Give an example of a network with more than one MST and such that XY must be in every MST.

(b) Give an example of a network with more than one MST and such that XY is in none of the MSTs.

59. Suppose that there is an edge in a network that must be included in any spanning tree. Give an algorithm for finding the minimum spanning tree that includes a given edge. (*Hint*: Modify Kruskal's algorithm.)

APPLET BYTES MyMathLab®

These Applet Bytes are exercises built around the applet* ***Kruskal's Algorithm*** *(available in MyMathLab in the Multimedia Library or Tools for Success). The applet allows you to create a weighted network and then practice using Kruskal's algorithm to find an MST or MaxST of the network.*

60. Figure 7-45 shows a mileage chart with road distances between Boston (BOS), Dallas (DAL), Houston (HOU), Louisville (LOU), Nashville (NASH), Pittsburgh (PITT), and St. Louis (STL).

(a) Use the "Create Graph" tab to create the weighted network corresponding to the mileage chart. (*Hint*: You don't have to worry about geography—you can place the vertices (cities) anywhere you want or find it convenient.)

(b) Go to the "Practice Kruskal's Algorithm" tab and find the MST of the network.

Mileage Chart

	Boston	Dallas	Houston	Louisville	Nashville	Pittsburgh	St. Louis
Boston	*	1748	1804	941	1088	561	1141
Dallas	1748	*	243	819	660	1204	630
Houston	1804	243	*	928	769	1313	779
Louisville	941	819	928	*	168	388	263
Nashville	1088	660	769	168	*	553	299
Pittsburgh	561	1204	1313	388	553	*	588
St. Louis	1141	630	779	263	299	588	*

FIGURE 7-45

61. Fig. 7-46 shows a portion of the electrical power grid connecting 14 rural towns in central Texas. The weight of each edge represents the length (in miles) of the power line connecting the two vertices.

(a) Use the applet to create the negative of the network in Fig. 7-46.

(b) Find the MaxST of the network.

(c) Assume that power company can sell parts of its power grid to another power company for $1 million per mile of power line. What would be the total price of the MaxST?

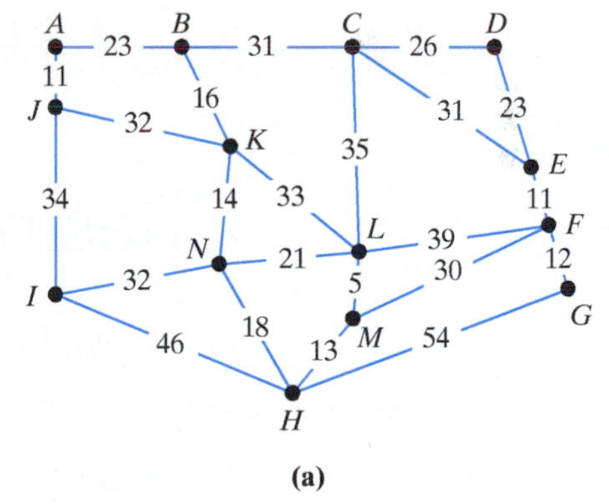

FIGURE 7-46

*MyMathLab code required.

Build a brand new house in less than three hours? Yes, you can. Start with a good set of plans, put a system in place, add just the right number of workers, and follow an *optimal schedule*. For the final result, see the next page.

8 The Mathematics of Scheduling

Chasing the Critical Path

Typically, it takes between four and nine months to build an average American home—rarely less than 120 days. But when a crowd of over 400 workers and volunteers got together the morning of October 1, 2005, in an empty lot in Tyler, Texas, they had a different idea in mind—the crazy thought that they would build a decent-sized, good quality house in less than *three* hours. Two hours and 52 minutes later—a world record—a 2249 square-foot, 3-bedroom, 2-bath house with a two-car garage was standing in the middle of the lot and ready for occupancy.

How was this tremendous feat of human ingenuity accomplished? Obviously, having hundreds of workers was essential to setting the record, but unleashing hordes of workers on a construction site is not enough—as a matter of fact, it is usually counterproductive. As is often said, *too many cooks can spoil the broth.* Most of the credit for the success of the project goes to what was accomplished before the physical work ever started: the creation of a seamless and critically coordinated system of assignments that told every worker what to do and when to do it, so that not a single second would be wasted. In other words, the creation of a master *schedule*.

Building a house in world record time using hundreds of workers is an extreme example of something that we do more often than we realize: When facing a complex project, we break up the project into small tasks, and then plan and execute those individual tasks in a *coordinated and organized* way. We do this when we have to throw a party, repair a car, cook a fancy dinner, or write a book.

> “We knew this house would be built in world record time not by the best human effort, but only if we had a system in place. Clever people would not build it, although we had the most talented people on task; only a foolproof system would.”
>
> – *Brian Conaway (lead organizer of the world-record house-building project)*

This chapter is about the *coordination and organization* part of the story, or, to put it more formally, about *scheduling*. The chapter starts with an introduction to the key elements in any scheduling problem—*processors, tasks, processing times*, and *precedence relations* (Section 8.1). Scheduling problems are best modeled using a special type of graph called a *directed graph* (or digraph for short)—we discuss digraphs and some of their basic properties in Section 8.2. In Section 8.3 we will discuss the general rules for creating schedules—the key concept here is that of a *priority list*. The two most commonly used algorithms for “solving” a scheduling problem are the *decreasing-time algorithm* and the *critical-path algorithm*—we discuss these in Sections 8.4 and 8.5.

8.1 An Introduction to Scheduling

We will now introduce the principal characters in any scheduling story.

- **The processors.** Every job requires workers. We will use the term **processors** to describe the “workers” who carry out the work. While the word *processor* may sound cold and impersonal, it does underscore an important point: processors need not be human beings. In scheduling, a processor could just as well be a robot, a computer, an automated teller machine, and so on. We will use N to represent the number of processors and $P_1, P_2, P_3, \ldots, P_N$ to denote the processors. We will assume throughout the chapter that $N \geq 2$ (scheduling a job with just one processor is trivial and not very interesting).

- **The tasks.** In every complex project there are individual pieces of work, often called "jobs" or "tasks." We will need to be a little more precise than that, however. We will define a **task** as an indivisible unit of work that (either by nature or by choice) cannot be broken up into smaller units. Thus, by definition a task cannot be shared—it is always *carried out by a single processor*. (In general, we will use capital letters $A, B, C, \ldots$, to represent the tasks, although in specific situations it is convenient to use abbreviations, such as WE for "wiring the electrical system" and PL for "plumbing.")

At any particular moment in a project, tasks are in one of the following four states:

- ***Ineligible***: The task cannot be started because some of the prerequisites for the task have not yet been completed ("can't start putting the roof up because the framing hasn't been completed").
- ***Ready***: The task has not been started but could be started at this time ("the framing is done, OK to start putting up the roof").
- ***In execution***: The task is being carried out by one of the processors ("the roof is being put up"), or
- ***Completed*** ("the roof is finished").

- **The processing times.** The **processing time** of a task is the amount of time, without interruption, required by *one processor* to execute that task. When dealing with human processors, there are many variables (ability, attitude, work ethic, etc.) that can affect the processing time of a task, and this adds another layer of complexity to an already complex situation. On the other hand, if we assume a "robotic" interpretation of the processors (either because they are indeed machines or because they are human beings trained to work in a very standardized and uniform way), then scheduling becomes somewhat more manageable.

  To keep things simple we will work under the following three assumptions:

  - ***Versatility***: Any processor can execute any task.
  - ***Uniformity***: The processing time for a task is the same regardless of which processor is executing the task.
  - ***Persistence***: Once a processor starts a task, it will complete it without interruption.

  Under the preceding assumptions, the concept of *processing time* for a task (we will sometimes call it the P-time) makes good sense—and we can conveniently incorporate this information by including it inside parentheses next to the name of the task. Thus, the notation $X(5)$ tells us that the task called X has a processing time of 5 units (be it minutes, hours, days, or any other unit of time) *regardless of which processor is assigned to execute the task.*

- **The precedence relations.** Precedence relations are formal restrictions on the order in which the tasks can be executed, much like those course prerequisites in the school catalog that tell you that you can't take course Y until you have completed course X. In the case of tasks, these prerequisites are called **precedence relations**. A typical precedence relation is of the form *task X precedes task Y* (we also say X is *precedent* to Y), and it means that *task Y cannot be started until task X*

has been completed. A precedence relation can be conveniently abbreviated by writing $X \rightarrow Y$, or described graphically as shown in Fig. 8-1(a). A single scheduling problem can have hundreds or even thousands of precedence relations, each adding another restriction on the scheduler's freedom.

At the same time, it also happens fairly often that there are no restrictions on the order of execution between two tasks in a project. When a pair of tasks X and Y have no precedence requirements between them (neither $X \rightarrow Y$ nor $Y \rightarrow X$), we say that the tasks are **independent**. When two tasks are independent, either one can be started before the other one, or they can both be started at the same time. Graphically, we can tell that two tasks are independent if there are no arrows connecting them [Fig. 8-1(b)].

Two final comments about precedence relations are in order. First, precedence relations are *transitive:* If $X \rightarrow Y$ and $Y \rightarrow Z$, then it must be true that $X \rightarrow Z$. In a sense, the last precedence relation is implied by the first two, and it is really unnecessary to mention it [Fig. 8-1(c)]. Thus, we will make a distinction between two types of precedence relations: *basic* and *implicit*. Basic precedence relations are the ones that come with the problem and that we must follow in the process of creating a schedule. If we do this, the implicit precedence relations will be taken care of automatically.

The second observation is that *precedence relations cannot form a cycle!* Imagine having to schedule the tasks shown in Fig. 8-1(d): X precedes Y, which precedes Z, which precedes W, which in turn precedes X. Clearly, this is logically impossible. From here on, we will always assume that there are no cycles of precedence relations among the tasks.

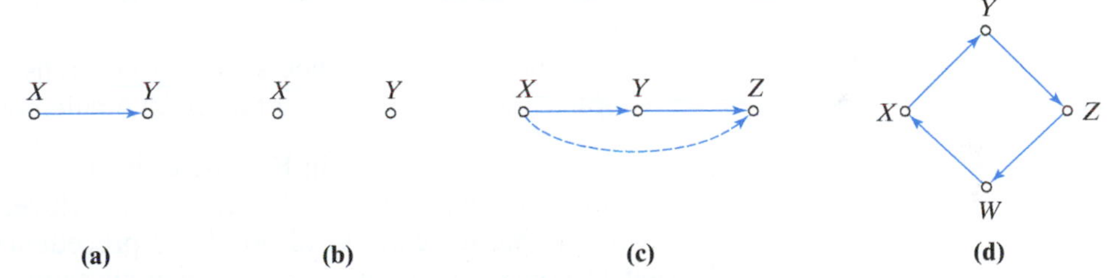

FIGURE 8-1 (a) X is precedent to Y. (b) X and Y are independent tasks. (c) When $X \rightarrow Y$ and $Y \rightarrow Z$ then $X \rightarrow Z$ is implied. (d) These tasks cannot be scheduled because of the cyclical nature of the precedence relations.

Processors, *tasks*, *processing times*, and *precedence relations* are the basic ingredients that make up a scheduling problem. They constitute, in a manner of speaking, the hand that is dealt to us. But how do we play such a hand? To get a small inkling of what's to come, let's look at the following simple example.

EXAMPLE 8.1 REPAIRING A WRECK

Imagine that you just wrecked your car, but thank heavens you are OK, and the insurance company will pick up the tab. You take the car to the best garage in town, operated by the Tappet brothers Click and Clack (we'll just call them P_1 and P_2). The repairs on the car can be broken into four different tasks: (A) exterior body work (4 hours), (B) engine repairs (5 hours), (C) painting and exterior finish work (7 hours), and (D) transmission repair (3 hours). The only precedence relation for this set of tasks is that the painting and exterior finish work cannot be started until the exterior body work has been completed ($A \rightarrow C$). The two brothers always work together on a repair project, but each takes on a different task (so they won't argue with each other). Under these assumptions, how should the different tasks be scheduled? Who should do what and when?

Even in this simple situation, there are many different ways to schedule the repair. Figure 8-2 shows several possible schedules, each one illustrated by means of a timeline. Figure 8-2(a) shows a schedule that is very inefficient. All the short tasks are assigned to one processor (P_1) and all the long tasks to the other processor (P_2)—obviously not a very clever strategy. Under this schedule, the project **finishing time** (the duration of the project from the start of the first task to the completion of the last task) is 12 hours. (We will use *Fin* to denote the project finishing time, so for this project we can write $Fin = 12$ hours.)

Figure 8-2(b) shows what looks like a much better schedule, but it violates the precedence relation $A \rightarrow C$ (as much as we would love to, we cannot start task C until task A is completed). On the other hand, if we force P_2 to be idle for one hour, waiting for the green light to start task C, we get a perfectly good schedule, shown in Fig. 8-2(c). Under this schedule the finishing time of the project is $Fin = 11$ hours.

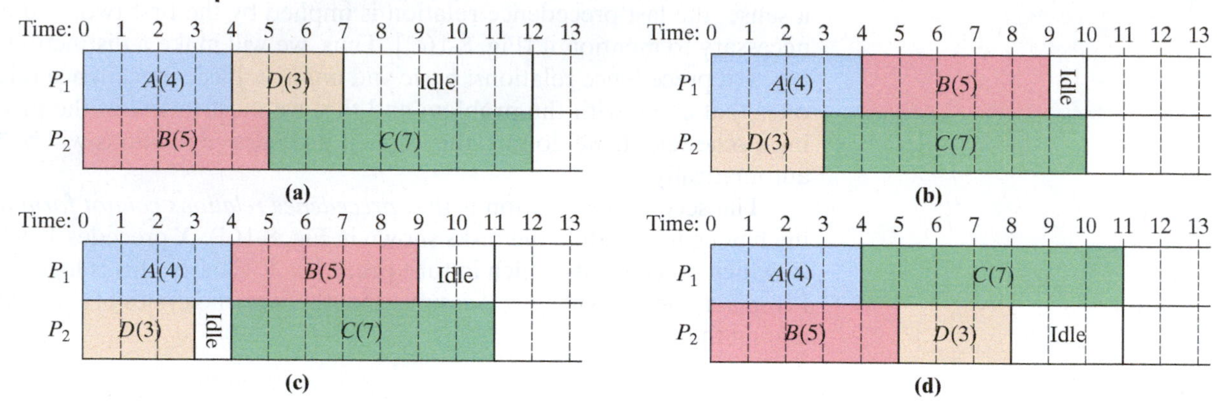

FIGURE 8-2 (a) A legal schedule with $Fin = 12$ hours. (b) An *illegal* schedule (the precedence relation $A \rightarrow C$ is violated). (c) An optimal schedule ($Opt = 11$ hours). (d) A different optimal schedule.

The schedule shown in Fig. 8-2(c) is an improvement over the first schedule. Can we do even better? No! No matter how clever we are and no matter how many processors we have at our disposal, the precedence relation $A(4) \rightarrow C(7)$ implies that 11 hours is a minimum barrier that we cannot break—it takes 4 hours to complete A, 7 hours to complete C, and *we cannot start C until A is completed!* Thus, the schedule shown in Fig. 8-2(c) is an **optimal schedule** and the finishing time of $Fin = 11$ hours is the **optimal finishing time**. (From now on we will use *Opt* instead of *Fin* when we are referring to the optimal finishing time.) Figure 8-2(d) shows a different optimal schedule with finishing time $Opt = 11$ hours. ■

As scheduling problems go, Example 8.1 was a fairly simple one. But even from this simple example, we can draw some useful lessons. First, notice that even though we had only four tasks and two processors, we were able to create several different schedules. The four we looked at were just a sampler—there are other possible schedules that we didn't bother to discuss. Imagine what would happen if we had hundreds of tasks and dozens of processors—the number of possible schedules to consider would be overwhelming. In looking for a good, or even an optimal, schedule, we are going to need a systematic way to sort through the many possibilities. In other words, we are going to need some good *scheduling algorithms*.

The second useful thing we learned in Example 8.1 is that when it comes to the finishing time of a project, there is an *absolute minimum* time that no schedule can break, no matter how good an algorithm we use or how many processors we put to work. In Example 8.1 this absolute minimum was 11 hours, and, as luck would have it, we easily found a schedule [actually two—Figs. 8-2(c) and (d)] with a finishing time to match it. Every project, no matter how simple or complicated, has such an absolute minimum (called the *critical time*) that depends on the processing times and precedence relations for the tasks and not on the number of processors used. We will return to the concept of critical time in Section 8.5.

To set the stage for a more formal discussion of scheduling algorithms, we will introduce the most important example of this chapter. While couched in what seems like science fiction terms, the situation it describes is not totally farfetched—in fact, this example is a simplified version of the types of scheduling problems faced by home builders in general.

EXAMPLE 8.2 BUILDING THAT DREAM HOME (ON MARS)

Mars Base Artist's Conception (NASA)

It is the year 2050, and several human colonies have already been established on Mars. Imagine that you accept a job offer to work in one of these colonies. What will you do about housing?

Like everyone else on Mars, you will be provided with a living pod called a Martian Habitat Unit (MHU). MHUs are shipped to Mars in the form of prefabricated kits that have to be assembled on the spot—an elaborate and unpleasant job if you are going to do it yourself. A better option is to use specialized "construction" robots that can do all the assembly tasks much more efficiently than human beings can. These construction robots can be rented by the hour at the local Rent-a-Robot outlet.

The assembly of an MHU consists of 15 separate tasks, and there are 17 different precedence relations among these tasks that must be followed. The tasks, their respective processing times, and their precedent tasks are all shown in Table 8-1.

Task	Label (*P*-time)	Precedent tasks
Assemble pad	*AP*(7)	
Assemble flooring	*AF*(5)	
Assemble wall units	*AW*(6)	
Assemble dome frame	*AD*(8)	
Install floors	*IF*(5)	*AP, AF*
Install interior walls	*IW*(7)	*IF, AW*
Install dome frame	*ID*(5)	*AD, IW*
Install plumbing	*PL*(4)	*IF*
Install atomic power plant	*IP*(4)	*IW*
Install pressurization unit	*PU*(3)	*IP, ID*
Install heating units	*HU*(4)	*IP*
Install commode	*IC*(1)	*PL, HU*
Complete interior finish work	*FW*(6)	*IC*
Pressurize dome	*PD*(3)	*HU*
Install entertainment unit	*EU*(2)	*PU, HU*

TABLE 8-1 ■ Tasks, *P*-times, and precedence relations for Example 8.2

Here are some of the questions we will address later in the chapter: How can you get your MHU built quickly? How many robots should you rent to do the job? How do you create a suitable work schedule that will get the job done? (A robot will do whatever it is told, but someone has to tell it what to do and when.)

8.2 Directed Graphs

A directed graph, or **digraph** for short, is a graph in which the edges have a direction associated with them, typically indicated by an arrowhead. Digraphs are particularly useful when we want to describe **asymmetric relationships** (X related to Y does not imply that Y must be related to X).

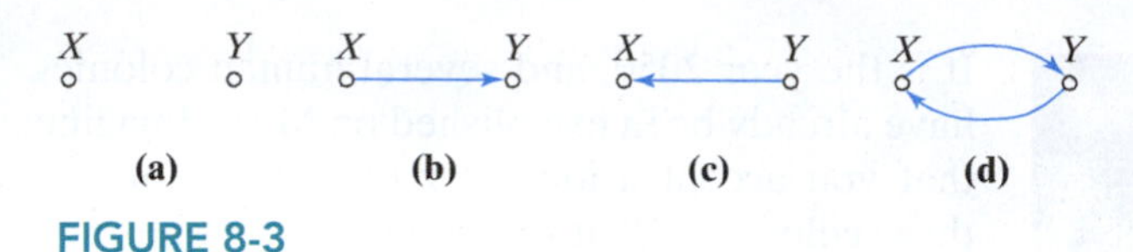

FIGURE 8-3

The classic example of an asymmetric relationship is romantic love: Just because X is in love with Y, there is no guarantee that Y reciprocates that love. Given two individuals X and Y and some asymmetric relationship (say love), we have four possible scenarios: Neither loves the other [Fig. 8-3(a)], X loves Y but Y does not love X [Fig. 8-3(b)], Y loves X but X does not love Y [Fig. 8-3(c)], and they love each other [Fig. 8-3(d)].

To distinguish digraphs from ordinary graphs, we use slightly different terminology.

- In a digraph, instead of talking about edges we talk about **arcs**. Every arc is defined by its *starting vertex* and its *ending vertex*, and we respect that order when we write the arc. Thus, if we write XY, we are describing the arc in Fig. 8-3(b) as opposed to the arc YX shown in Fig. 8-3(c).
- A list of all the arcs in a digraph is called the **arc-set** of the digraph. The digraph in Fig. 8-3(d) has arc-set $\mathcal{A} = \{XY, YX\}$.
- If XY is an arc in the digraph, we say that vertex X is **incident to** vertex Y, or, equivalently, that Y is **incident from** X.
- The arc YZ is said to be **adjacent** to the arc XY if the starting point of YZ is the ending point of XY. (Essentially, this means one can go from X to Z by way of Y.)
- In a digraph, a **path** from vertex X to vertex W $(W \neq X)$ consists of a sequence of arcs $XY, YZ, ZU, \ldots, VW$ such that each arc is adjacent to the one before it and no arc appears more than once in the sequence—it is essentially a trip from X to W along the arcs in the digraph. The best way to describe the path is by listing the vertices in the order of travel: X, Y, Z, and so on. Not surprisingly, the **length** of a path refers to the number of arcs on the path.
- When the path starts and ends at the same vertex, we call it a **cycle** of the digraph. Just like circuits in a regular graph, cycles in digraphs can be written in more than one way—the cycle X, Y, Z, X is the same as the cycles Y, Z, X, Y and Z, X, Y, Z. Just as for paths, the **length** of a cycle is the number of arcs in the cycle.
- In a digraph, the notion of the degree of a vertex is replaced by the concepts of *indegree* and *outdegree*. The **outdegree** of X is the number of arcs that have X as their *starting vertex* (outgoing arcs); the **indegree** of X is the number of arcs that have X as their *ending vertex* (incoming arcs).

The following example illustrates some of the above concepts.

EXAMPLE 8.3 DIGRAPH BASICS

The digraph in Fig. 8-4 has vertex-set $\mathcal{V} = \{A, B, C, D, E\}$ and arc-set $\mathcal{A} = \{AB, AC, BD, CA, CD, CE, EA, ED\}$. In this digraph, A is *incident to* B and C, but not to E. By the same token, A is *incident from* E as well as from C. The indegree of vertex A is 2, and so is the outdegree. The indegree of vertex D is 3, and the outdegree is 0. We leave it to the reader to find the indegrees and outdegrees of each of the other vertices of the graph.

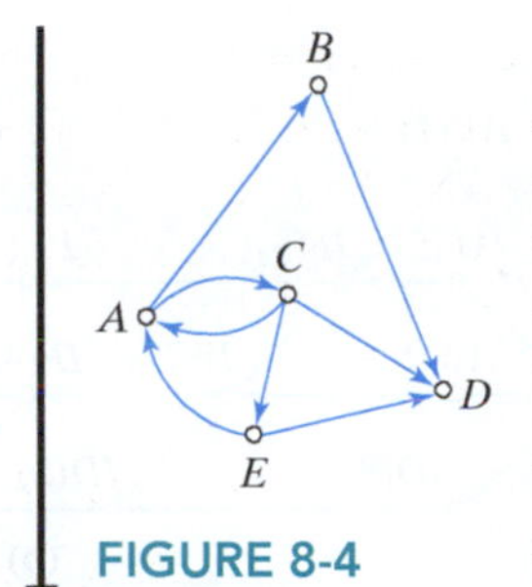

FIGURE 8-4

In this digraph, there are several paths from A to D, such as A, C, D (length 2); A, C, E, D (length 3); A, B, D (length 2); and even A, C, A, B, D (length 4). On the other hand, A, E, D is not a path from A to D (because you can't travel directly from A to E). There are two cycles in this digraph: A, C, E, A of length 3 (which can also be written as C, E, A, C and E, A, C, E) and A, C, A of length 2. Notice that there is no cycle passing through D because D has outdegree 0 and, thus, is a "dead-end," and there is no cycle passing through B because from B you can only go to D, and once there you are stuck.

While love is not to be minimized as a subject of study, there are many other equally important applications of digraphs:

- **The World Wide Web.** The Web is a giant digraph, where the vertices are Web pages and an arc from X to Y indicates that there is a *hyperlink* (informally called a *link*) on Web page X that allows you to jump directly to Web page Y. Web linkages are asymmetric—there may be a link on X that sends you to Y but no link on Y that sends you to X.
- **Traffic flow.** In most cities some streets are one-way streets and others are two-way streets. In this situation, digraphs allow us to visualize the flow of traffic through the city's streets. The vertices are intersections, and the *arcs* represent one-way streets. (To represent a two-way street, we use two arcs, one for each direction.)
- **Telephone traffic.** To track and analyze the traffic of telephone calls through their network, telephone companies use "call digraphs." In these digraphs the vertices are telephone numbers, and an arc from X to Y indicates that a call was initiated from telephone number X to telephone number Y.
- **Tournaments.** Digraphs are frequently used to describe certain types of tournaments, with the vertices representing the teams (or individual players) and the arcs representing the outcomes of the games played in the tournament (the arc XY indicates that X defeated Y). Tournament digraphs can be used in any sport in which the games cannot end in a tie (basketball, tennis, etc.).
- **Organization charts.** In any large organization (a corporation, the military, a university, etc.) it is important to have a well-defined chain of command. The best way to describe the chain of command is by means of a digraph often called an *organization chart*. In this digraph the *vertices* are the individuals in the organization, and an *arc* from X to Y indicates that X is Y's immediate boss (i.e., Y takes orders directly from X).

As you probably guessed by now, digraphs are also used in scheduling. There is no better way to visualize the tasks, processing times, and precedence relations in a project than by means of a digraph in which the vertices represent the tasks (with their processing times indicated in parentheses) and the arcs represent the precedence relations.

EXAMPLE 8.4 PROJECT DIGRAPH FOR THE MARTIAN HABITAT

Let's return to the scheduling problem first discussed in Example 8.2. We can take the tasks and precedence relations given in Table 8-1 and create a digraph like the one shown in Fig. 8-5(a). It is helpful to try to place the vertices of the digraph so that the arcs point from left to right, and this can usually be done with a little trial-and-error. After this is done, it is customary to add two fictitious tasks called START and END, where START indicates the imaginary task of getting the project started (cutting the red ribbon, so to speak) and END indicates the imaginary task of declaring the project complete [Fig. 8-5(b)]. By giving these fictitious tasks zero processing time, we avoid affecting the time calculations for the project.

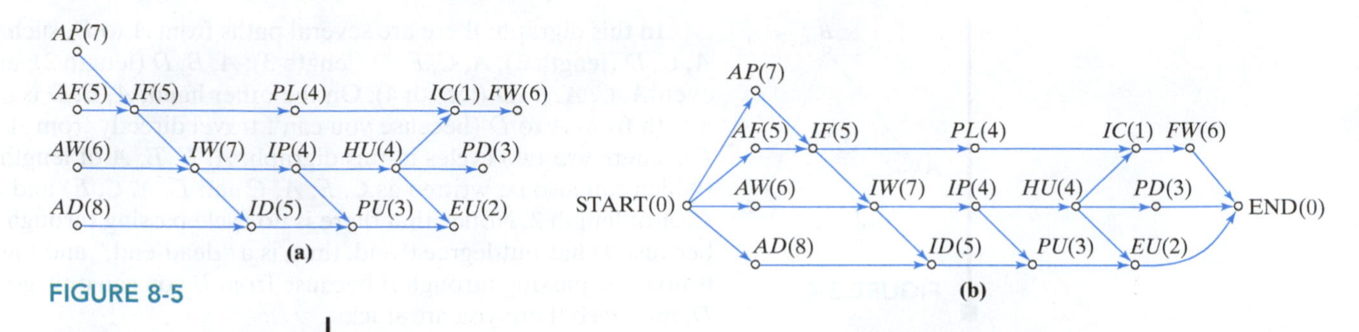

FIGURE 8-5

Every project is best described by its **project digraph**: The vertices of the project digraph are the tasks [including the fictitious tasks START (0) and END (0)], and the arcs are the precedence relations. The project digraph allows us to better visualize the execution of the project as a flow, moving from left to right.

8.3 Priority-List Scheduling

The project digraph is the basic graph model used to package all the information in a scheduling problem, but there is nothing in the project digraph itself that specifically tells us how to create a schedule. We are going to need something else, some set of instructions that indicates the order in which tasks should be executed. We can accomplish this by the simple act of prioritizing the tasks in some specified order, called a *priority list*.

A **priority list** is a list of all the tasks prioritized in the order we prefer to execute them. If task X is ahead of task Y in the priority list, then X gets priority over Y. This means that when it comes to a choice between the two, *X is executed ahead of Y*. However, if X is not yet *ready* for execution, then *we skip over it and move on to the first ready task after X in the priority list*. If there are no ready tasks after X in the priority list, the free processors must sit idle and wait until a task becomes ready.

The process of scheduling tasks using a priority list and following these basic rules is known as the *priority-list model* for scheduling. The priority-list model is a completely general model for scheduling—every priority list produces a schedule, and any schedule can be created from some (usually more than one) "parent" priority list. The trick is going to be to figure out *which* priority lists give us good schedules and which don't. We will come back to this topic in Sections 8.4 and 8.5.

Since each time we change the order of the tasks we get a different priority list, there are as many priority lists as there are ways to order the tasks. For three tasks, there are six possible priority lists; for 4 tasks, there are 24 priority lists; for 10 tasks, there are more than 3 million priority lists; and for 100 tasks, there are more priority lists than there are molecules in the universe.

Clearly, a shortage of priority lists is not going to be our problem. If this sounds familiar, it's because we have seen the idea before—priority lists are nothing more than *permutations of the tasks*. Like sequential coalitions (Chapter 2) and Hamilton circuits (Chapter 6), the number of priority lists is given by a factorial. (For a review of permutations and factorials, see Section 2.4.)

NUMBER OF PRIORITY LISTS

The number of possible priority lists in a project consisting of M tasks is $M! = M \times (M-1) \times \cdots \times 2 \times 1$

Before we proceed, we will illustrate how the priority-list model for scheduling works with a few small but important examples. Even with such small examples, there is a lot to keep track of, and you are well advised to have pencil and paper in front of you as you follow the details.

EXAMPLE 8.5 **PREPARING FOR LAUNCH: PART 1**

Immediately preceding the launch of a satellite into space, last-minute system checks need to be performed by the on-board computers, and it is important to complete these system checks as quickly as possible—for both cost and safety reasons. Suppose that there are five system checks required: $A(6)$, $B(5)$, $C(7)$, $D(2)$, and $E(5)$, with the numbers in parentheses representing the hours it takes one computer to perform that system check. In addition, there are precedence relations: D cannot be started until both A and B have been finished, and E cannot be started until C has been finished. The project digraph is shown in Fig. 8-6.

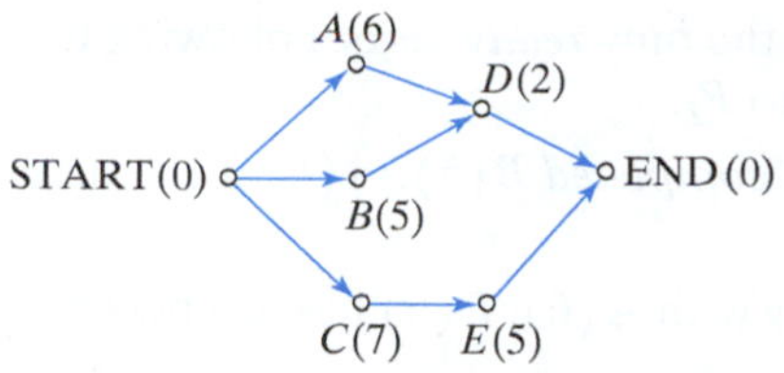

FIGURE 8-6

Let's assume that there are two identical computers on board (P_1 and P_2) that will carry out the individual system checks. How do we use the priority-list model to create a schedule for these two processors?

For starters, we will need a priority list. Suppose that the priority list is given by listing the system checks in alphabetical order, and let's follow the evolution of the project under the priority-list model. We will use T to indicate the elapsed time in hours.

Priority list: $A(6), B(5), C(7), D(2), E(5)$

- **T = 0 (START).** $A(6)$, $B(5)$, and $C(7)$ are the only *ready* tasks. Following the priority list, we assign $A(6)$ to P_1 and $B(5)$ to P_2.
- **T = 5** P_1 is still *busy* with $A(6)$; P_2 has just *completed* $B(5)$. $C(7)$ is the only available *ready* task. We assign $C(7)$ to P_2.
- **T = 6** P_1 has just *completed* $A(6)$; P_2 is *busy* with $C(7)$. $D(2)$ has just become a *ready* task (A and B have been completed). We assign $D(2)$ to P_1.
- **T = 8.** P_1 has just *completed* $D(2)$; P_2 is still *busy* with $C(7)$. There are no *ready* tasks at this time for P_1, so P_1 has to sit *idle*.

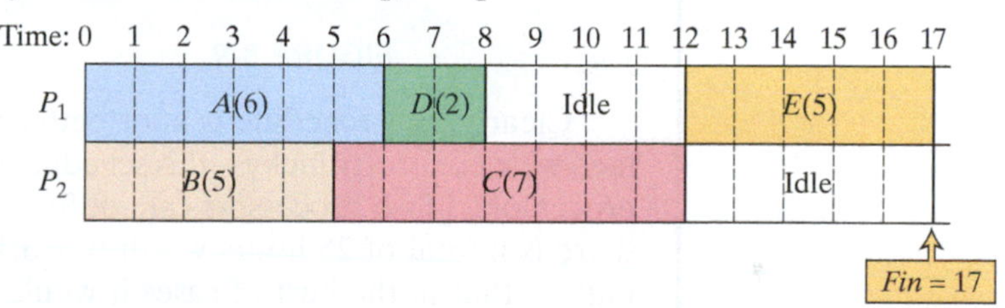

FIGURE 8-7

- **T = 12.** P_1 is *idle*; P_2 has just *completed* $C(7)$. Both processors are *ready* for work. $E(5)$ is the only ready task, so we assign $E(5)$ to P_1, P_2 sits *idle*. (Note that in this situation, we could have just as well assigned $E(5)$ to P_2 and let P_1 sit idle. The processors don't get tired and don't care if they are working or idle, so the choice is random.)
- **T = 17 (END).** P_1 has just *completed* $E(5)$, and, therefore, the project is completed.

The evolution of the entire project together with the project finishing time ($Fin = 17$) are captured in the timeline shown in Fig. 8-7. Is this a good schedule? Given the excessive amount of idle time (a total of 9 hours), one might suspect this is a rather bad schedule. How could we improve it? We might try changing the priority list. ■

EXAMPLE 8.6 **PREPARING FOR LAUNCH: PART 2**

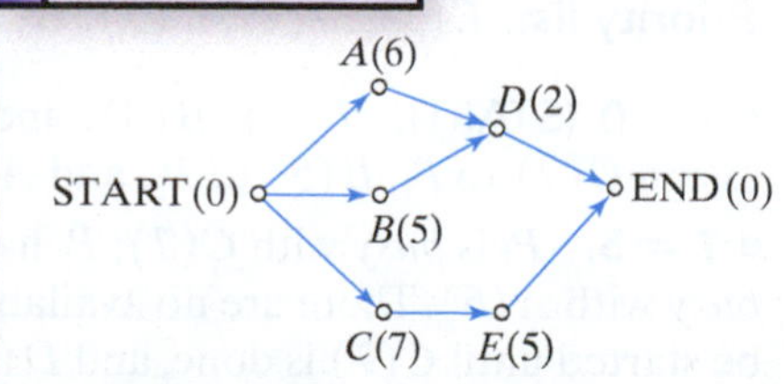

FIGURE 8-8

We are going to schedule the satellite launch system checks with the same two processors but with a different priority list. The project digraph is shown once again in Fig. 8-8. (When scheduling, it's really useful to have the project digraph right in front of you.)

This time let's try a reverse alphabetical order for the priority list. Why? Why not—at this point we are just shooting in the dark! (Don't worry—we will become a lot more enlightened later in the chapter.)

Priority list: $E(5), D(2), C(7), B(5), A(6)$

- **T = 0 (START).** $C(7), B(5)$, and $A(6)$ are the only *ready* tasks. Following the priority list, we assign $C(7)$ to P_1 and $B(5)$ to P_2.
- **T = 5.** P_1 is still *busy* with $C(7)$; P_2 has just *completed* $B(5)$. $A(6)$ is the only available ready task. We assign $A(6)$ to P_2.
- **T = 7.** P_1 has just *completed* $C(7)$; P_2 is *busy* with $A(6)$. $E(5)$ has just become a *ready* task, and we assign it to P_1.
- **T = 11.** P_2 has just *completed* $A(6)$; P_1 is *busy* with $E(5)$. $D(2)$ has just become a *ready* task, and we assign it to P_2.
- **T = 12.** P_1 has just *completed* $E(5)$; P_2 is *busy* with $D(2)$. There are no tasks left, so P_1 sits idle.
- **T = 13. (END).** P_2 has just *completed* the last task, $D(2)$. Project is completed.

The timeline for this schedule is shown in Fig. 8-9. The project finishing time is $Fin = 13$ hours.

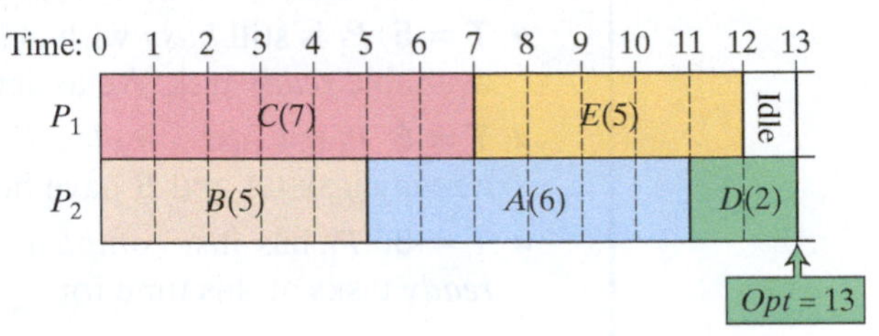

FIGURE 8-9

Clearly, this schedule is a lot better than the one obtained in Example 8.5. In fact, we were pretty lucky—this schedule turns out to be an optimal schedule for two processors. [Two processors cannot finish this project in less than 13 hours because there is a total of 25 hours worth of work (the sum of all processing times), which implies that in the best of cases it would take 12.5 hours to finish the project. But since the processing times are all whole numbers and tasks cannot be split, the finishing time cannot be less than 13 hours! Thus, $Opt = 13$ hours.]

Thirteen hours is still a long time for the computers to go over their system checks. Since that's the best we can do with two computers, the only way to speed things up is to add a third computer to the "workforce." Adding another computer to the satellite can be quite expensive, but perhaps it will speed things up enough to make it worth it. Let's see.

EXAMPLE 8.7 PREPARING FOR LAUNCH: PART 3

We will now schedule the system checks using $N = 3$ computers (P_1, P_2, P_3). For the reader's convenience the project digraph is shown again in Fig. 8-10. We will use the "good" priority list we found in Example 8.6.

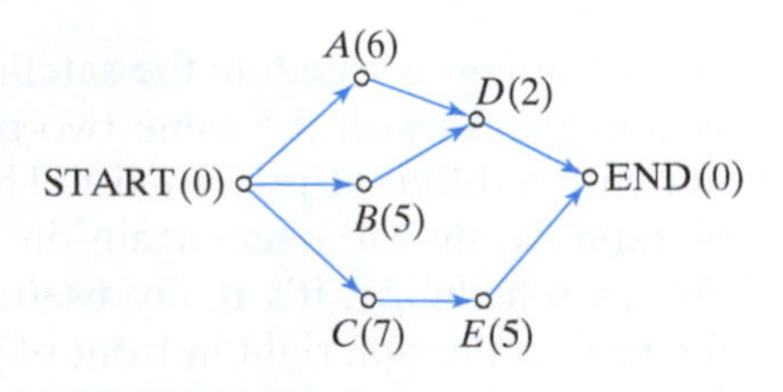

FIGURE 8-10

Priority list: $E(5), D(2), C(7), B(5), A(6)$

- **T = 0 (START).** $C(7)$, $B(5)$, and $A(6)$ are the *ready* tasks. We assign $C(7)$ to P_1, $B(5)$ to P_2, and $A(6)$ to P_3.
- **T = 5.** P_1 is *busy* with $C(7)$; P_2 has just *completed* $B(5)$; and P_3 is *busy* with $A(6)$. There are no available ready tasks for P_2 [$E(5)$ can't be started until $C(7)$ is done, and $D(2)$ can't be started until $A(6)$ is done], so P_2 sits idle until further notice.

- **$T = 6$.** P_3 has just *completed* $A(6)$; P_2 is *idle*; and P_1 is still *busy* with $C(7)$. $D(2)$ has just become a *ready* task. We randomly assign $D(2)$ to P_2 and let P_3 be idle, since there are no other ready tasks. [Note that we could have just as well assigned $D(2)$ to P_3 and let P_2 be idle.]
- **$T = 7$.** P_1 has just *completed* $C(7)$ and $E(5)$ has just become a *ready* task, so we assign it to P_1. There are no other tasks to assign, so P_3 continues to sit idle.
- **$T = 8$.** P_2 has just *completed* $D(2)$. There are no other tasks to assign, so P_2 and P_3 both sit *idle*.
- **$T = 12$ (END).** P_1 has just *completed* the last task, $E(5)$, so the project is completed.

The timeline for this schedule is shown in Fig. 8-11. The project finishing time is *Fin* = 12 hours, a pathetically small improvement over the two-processor schedule found in Example 8.6. The cost of adding a third processor doesn't seem to justify the benefit.

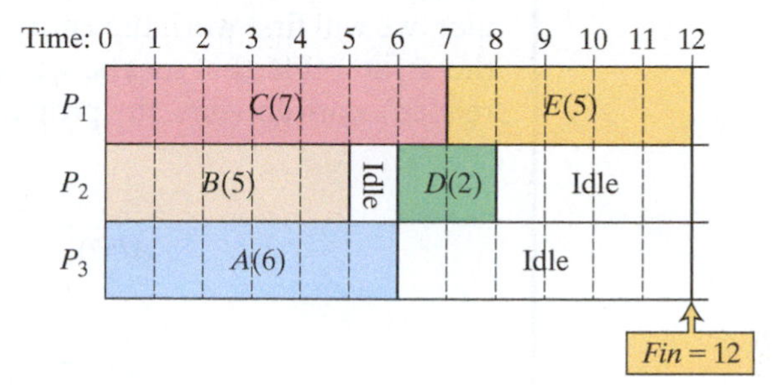

FIGURE 8-11

The Priority-List Model

The previous three examples give us a general sense of how to create a schedule from a project digraph *and* a priority list. We will now formalize the ground rules of the **priority-list model** for scheduling.

At any particular moment in time throughout a project, a processor can be either *busy* or *idle* and a task can be *ineligible, ready, in execution*, or *completed*. Depending on the various combinations of these, there are three different scenarios to consider:

- *All processors are busy.* In this case, there is nothing we can do but wait.
- *One processor is free.* In this case, we scan the priority list from left to right, looking for the first *ready* task in the priority list and assign it to the free processor. (Remember that for a task to be *ready*, all the tasks that are precedent to it must have been completed.) If there are no ready tasks at that moment, the processor stays idle until things change.
- *More than one processor is free.* In this case, the *first ready* task on the priority list is given to one free processor, the second ready task is given to another free processor, and so on. If there are more free processors than ready tasks, some of the processors will remain idle until one or more tasks become ready. Since the processors are identical and tireless, the choice of which free processor is assigned which ready task is completely arbitrary.

It's fair to say that the basic idea behind the priority-list model is not difficult, but there is a lot of bookkeeping involved, and that becomes critical when the number of tasks is large. At each stage of the schedule we need to keep track of the status of each task—which tasks are *ready* for processing, which tasks are *in execution*, which tasks have been *completed*, which tasks are still *ineligible*. One convenient recordkeeping strategy goes like this: On the priority list itself *ready* tasks are circled

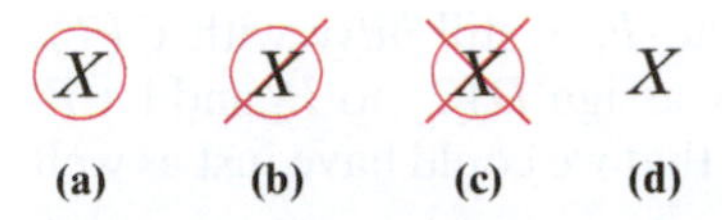

FIGURE 8-12 "Road" signs on a priority list. (a) Task X is *ready.* (b) Task X is *in execution.* (c) Task X is *completed.* (d) Task X is *ineligible.*

in red [Fig. 8-12(a)]. When a ready task is picked up by a processor and goes into *execution*, put a single red slash through the red circle [Fig. 8-12(b)]. When a task that has been in execution is completed, put a second red slash through the circle [Fig. 8-12(c)]. At this point, it is also important to check the project digraph to see if any new tasks have all of a sudden become eligible. Tasks that are *ineligible* remain unmarked [Fig. 8-12(d)].

As they say, the devil is in the details, so a slightly more substantive example will help us put everything together—the project digraph, the priority list model, and the bookkeeping strategy.

EXAMPLE 8.8 ASSEMBLING A MARTIAN HOME WITH A PRIORITY LIST

This is the third act of the Martian Habitat Unit (MHU) building project. We are finally ready to start the project of assembling that MHU, and, like any good scheduler, we will first work the entire schedule out with pencil and paper. Let's start with the assumption that maybe we can get by with just two robots (P_1 and P_2). For the reader's convenience, the project digraph is shown again in Fig. 8-13.

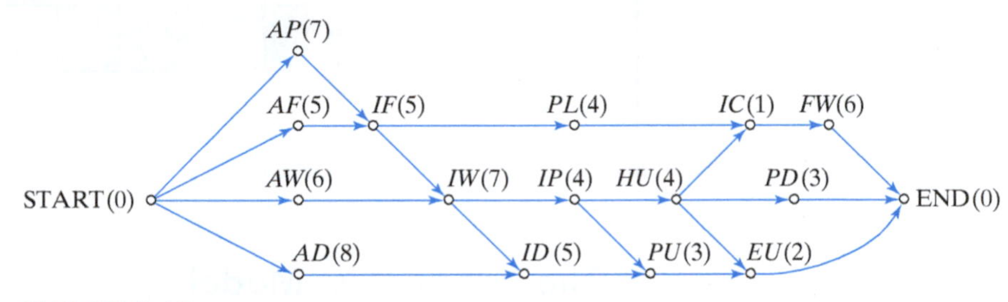

FIGURE 8-13

Let's start with a random priority list.

Priority list: $AD(8)$, $AW(6)$, $AF(5)$, $IF(5)$, $AP(7)$, $IW(7)$, $ID(5)$, $IP(4)$, $PL(4)$, $PU(3)$, $HU(4)$, $IC(1)$, $PD(3)$, $EU(2)$, $FW(6)$. (Ready tasks are circled in red.)

- **T = 0 (START).** Status of processors: P_1 starts AD; P_2 starts AW. We put a single red slash through AD and AW.

  Priority list: AD, AW, AF, IF, AP, IW, ID, IP, PL, PU, HU, IC, PD, EU, FW.
- **T = 6.** P_1 busy (executing AD); P_2 completed AW (put a second slash through AW) and starts AF (put a slash through AF).

  Priority list: AD, AW, AF, IF, AP, IW, ID, IP, PL, PU, HU, IC, PD, EU, FW.
- **T = 8.** P_1 completed AD and starts AP; P_2 is busy (executing AF).

  Priority list: AD, AW, AF, IF, AP, IW, ID, IP, PL, PU, HU, IC, PD, EU, FW.
- **T = 11.** P_1 busy (executing AP); P_2 completed AF, but since there are no ready tasks, it remains idle.

  Priority list: AD, AW, AF, IF, AP, IW, ID, IP, PL, PU, HU, IC, PD, EU, FW.
- **T = 15.** P_1 completed AP. IF becomes a ready task and goes to P_1; P_2 stays idle.

  Priority list: AD, AW, AF, IF, AP, IW, ID, IP, PL, PU, HU, IC, PD, EU, FW.

At this point, we will let you take over and finish the schedule. After a fair amount of work, we obtain the final schedule shown in Fig. 8-14, with project finishing time $Fin = 44$ hours.

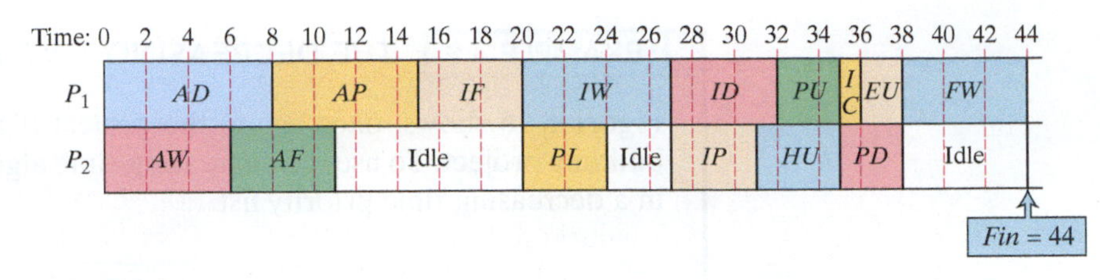

FIGURE 8-14

Scheduling with priority lists is a two-part process: (1) choose a priority list and (2) use the priority list and the ground rules of the priority-list model to come up with a schedule (Fig. 8-15). As we saw in the previous example, the second part is long and tedious, but purely mechanical—it can be done by anyone (or anything) that is able to follow a set of instructions, be it a meticulous student or a computer application. (The applet ***Priority List Scheduling***, available in *MyMathLab* in the *Multimedia Library* or in *Tools for Success* is such an app. Once the user creates the project digraph, selects the number of processors and inputs a priority list, the applet generates a schedule.)

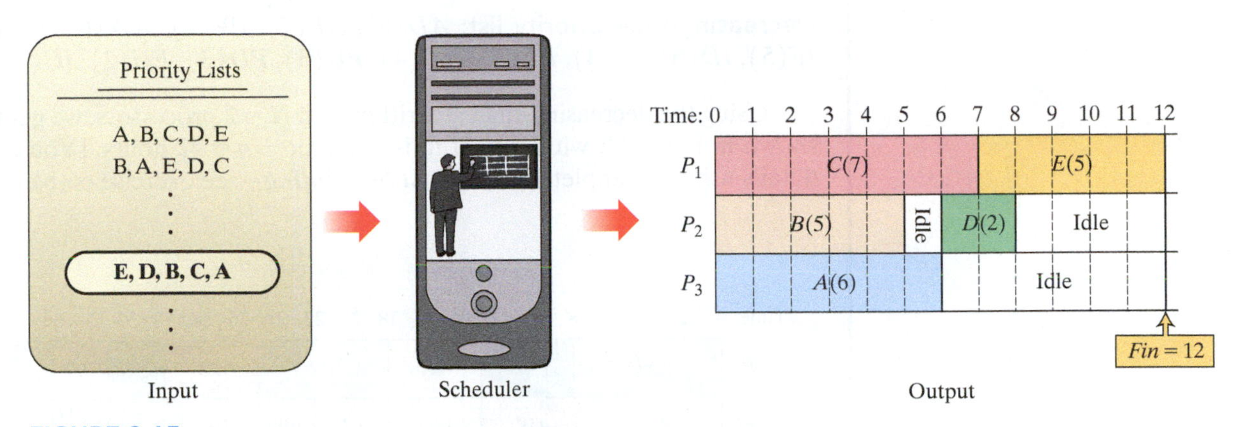

FIGURE 8-15

Ironically, it is the seemingly easiest part of this process—choosing a priority list—that is actually the most interesting. Among all the priority lists there is one (or more) that will give a schedule with the optimal finishing time. We will call these **optimal priority lists**. How do we find an optimal priority list? Short of that, how do we find "good" priority lists, that is, priority lists that give schedules with finishing times reasonably close to the optimal? These are both important questions, and some answers are coming up next.

8.4 The Decreasing-Time Algorithm

Our first attempt to find a good priority list is to formalize what is an intuitive and commonly used strategy: *Prioritize the bigger jobs as the first to complete, and leave the small jobs for last.* In terms of priority lists, this strategy is implemented by creating a priority list in which the tasks are listed in *decreasing* order of processing times—longest first, second longest next, and so on. (When there are two or more tasks with equal processing times, we order them randomly.)

A priority list in which the tasks are listed in decreasing order of processing times is called, not surprisingly, a **decreasing-time priority list**, and the process of creating a schedule using a decreasing-time priority list is called the **decreasing-time algorithm**.

EXAMPLE 8.9 THE DECREASING TIME ALGORITHM GOES TO MARS

Figure 8-16 shows, once again, the project digraph for the Martian Habitat Unit building project. To use the decreasing-time algorithm we first prioritize the 15 tasks in a decreasing-time priority list.

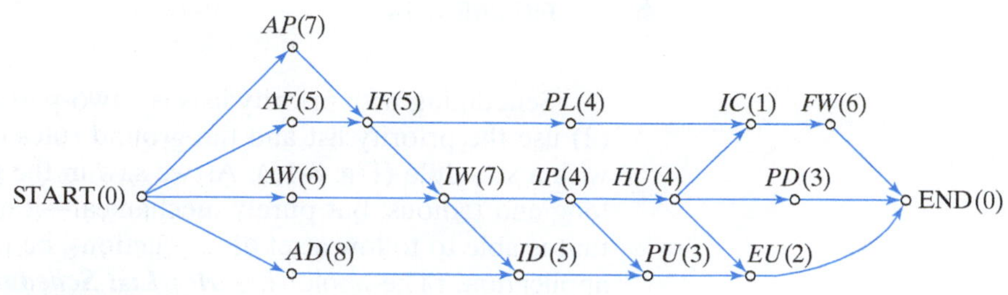

FIGURE 8-16

Decreasing-time priority list: $AD(8)$, $AP(7)$, $IW(7)$, $AW(6)$, $FW(6)$, $AF(5)$, $IF(5)$, $ID(5)$, $IP(4)$, $PL(4)$, $HU(4)$, $PU(3)$, $PD(3)$, $EU(2)$, $IC(1)$

Using the decreasing-time algorithm with $N = 2$ processors, we get the schedule shown in Fig. 8-17, with project finishing time $Fin = 42$ hours. [You can verify the details using the applet ***Priority List Scheduling***—see exercise 66(b).]

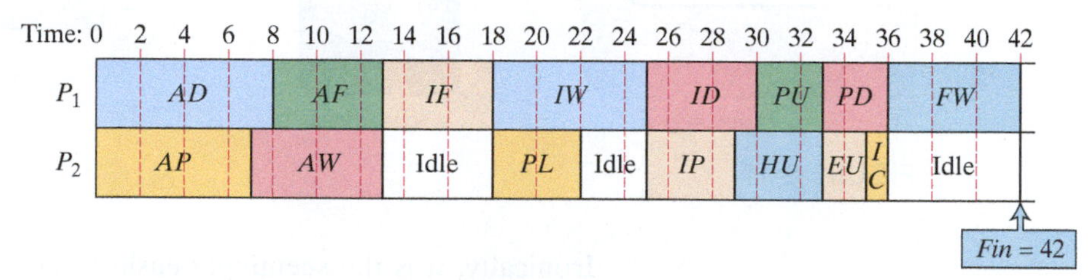

FIGURE 8-17

When looking at the finishing time under the decreasing-time algorithm, one can't help but feel disappointed. The sensible idea of prioritizing the longer jobs ahead of the shorter jobs turned out to be somewhat of a dud—at least in this example! What went wrong? If we work our way backward from the end, we can see that we made a bad choice at $T = 33$ hours. At this point there were three ready tasks [$PD(3)$, $EU(2)$, and $IC(1)$], and both processors were available. Based on the decreasing-time priority list, we chose the two "longer" tasks, $PD(3)$ and $EU(2)$, ahead of the short task, $IC(1)$. This was a bad move! $IC(1)$ is a much more *critical* task than the other two because we can't start task $FW(6)$ until we finish task $IC(1)$. Had we looked ahead at some of the tasks for which PD, EU, and IC are precedent tasks, we might have noticed this.

An even more blatant example of the weakness of the decreasing-time algorithm occurs at the very start of this schedule, when the algorithm fails to take into account that task $AF(5)$ should have a very high priority. Why? $AF(5)$ is one of the two tasks that must be finished before $IF(5)$ can be started, and $IF(5)$ must be finished before $IW(7)$ can be started, which must be finished before $IP(4)$ and $ID(5)$ can be started, and so on down the line.

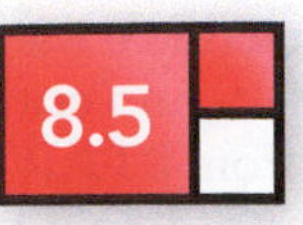

8.5 Critical Paths and the Critical-Path Algorithm

When there is a long path of tasks in the project digraph, it seems clear that the first task along that path should be started as early as possible. This idea leads to the following informal rule: *The greater the total amount of work that lies ahead of a task, the sooner that task should be started.*

To formalize these ideas, we will introduce the concepts of *critical paths* and *critical times.*

- **Critical path (time) for a vertex.** For a given vertex X of a project digraph, the *critical path for X* is the path *from X to END* having the *longest* processing time. (The *processing time* of a path is defined to be the sum of the processing times of all the tasks along the path.) When we add the processing times of all the tasks along the critical path for a vertex X, we get the *critical time for X.* (By definition, the critical time of END is 0.)
- **Critical path (time) for a project.** The path with longest processing time from START to END is called the *critical path for the project*, and the total processing time for this critical path is called the *critical time for the project.*

EXAMPLE 8.10 CRITICAL PATHS IN THE MHU PROJECT DIGRAPH

Figure 8-18 shows the project digraph for assembling the Martian Habitat Unit. We will illustrate the concepts of critical paths and critical times using three different vertices of the project digraph: *HU*, *AD*, and START.

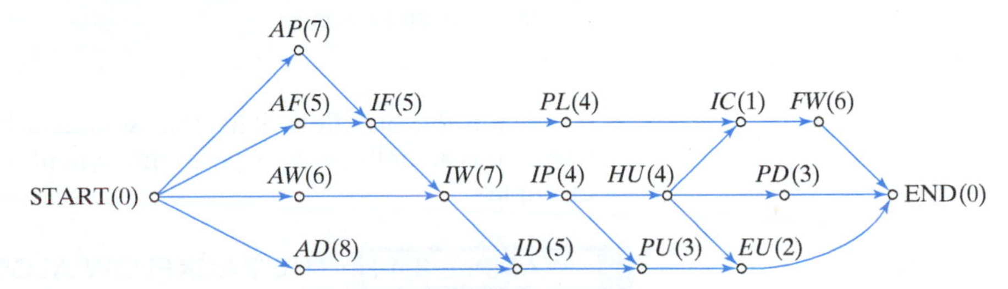

FIGURE 8-18

- Let's start with *HU*—a relatively easy case. A quick look at Fig. 8-18 should convince you that there are only three paths from *HU* to END, namely,
  - *HU*, *IC*, *FW*, END, with processing time $4 + 1 + 6 = 11$ hours,
  - *HU*, *PD*, END, with processing time $4 + 3 = 7$ hours, and
  - *HU*, *EU*, END, with processing time $4 + 2 = 6$ hours.

  Of the three paths, *HU*, *IC*, *FW*, END has the longest processing time, and thus is the *critical path* for vertex *HU*. The *critical time* for *HU* is 11 hours.
- Next, let's find the critical path for *AD*. There is only one path from *AD* to END, namely *AD*, *ID*, *PU*, *EU*, END, which makes the decision especially easy. Since this is the only path, it is automatically the longest path and, therefore, the *critical path* for *AD*. The *critical time* for *AD* is $8 + 5 + 3 + 2 = 18$ hours.
- To find the critical path for the project, we need to find the path from START to END with longest processing time. Since there are dozens of paths from START to END, let's just eyeball the project digraph for a few seconds and take our best guess. . . .

  OK, if you guessed START, *AP*, *IF*, *IW*, *IP*, *HU*, *IC*, *FW*, END, you have good eyes. This is indeed the *critical path*. It follows that the *critical time* for the Martian Habitat Unit project is 34 hours.

We will soon discuss the special role that the critical time and the critical path play in scheduling, but before we do so, let's address the issue of how to find critical paths. In a large project digraph there may be thousands of paths from START to END, and the "eyeballing" approach we used in Example 8.10 is not likely to work. What we need here is an efficient algorithm, and fortunately there is one—it is called the **backflow algorithm**.

THE BACKFLOW ALGORITHM

- **Step 1.** Working backwards (starting at END and working your way back to START), find the critical time of each vertex of the project digraph. [To help with the record keeping, write each critical time in square brackets [ ] to distinguish it from the processing time in parentheses ( ).]

  Here is how you find the critical time of a given vertex $X(p)$:

  - Look at all the vertices that X is *incident to*, and choose among them the one with the largest *critical time*. (Note that since you are working backwards, you should have already found the critical times of all the vertices ahead of X). In Fig. 8-19(a) imagine that the largest number in square brackets (i.e., the largest critical time) is L.
  - Assign to $X(p)$ the critical time $[p + L]$, as shown in Fig. 8-19(b). In other words, *the critical time for a task X equals the processing time of X plus the largest critical time among the vertices that X* is incident to.

- **Step 2.** Once we have the critical time for every vertex in the project digraph, critical paths are found by just following the *path along largest critical times*. In other words, the critical path for any vertex X (and that includes START) is obtained by starting at X and moving to the adjacent vertex with largest critical time, and from there to the adjacent vertex with largest critical time, and so on.

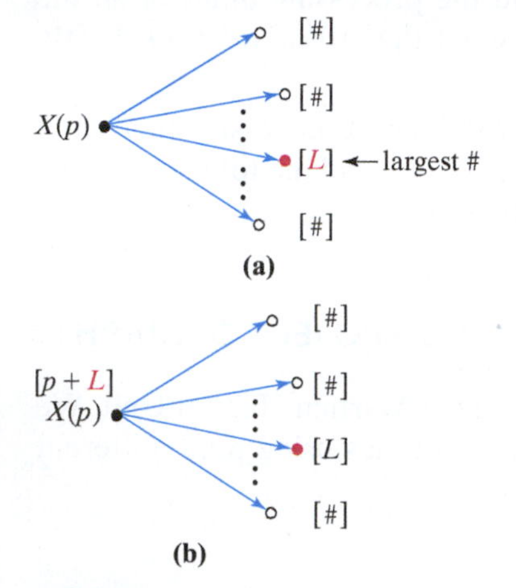

FIGURE 8-19

While the backflow algorithm sounds a little complicated when described in words, it is actually pretty easy to implement in practice, as we will show in the next example.

EXAMPLE 8.11 THE BACKFLOW ALGORITHM AND THE MHU PROJECT DIGRAPH

We are now going to use the backflow algorithm to find the critical time for each of the vertices of the Martian Habitat Unit project digraph (Fig. 8-20).

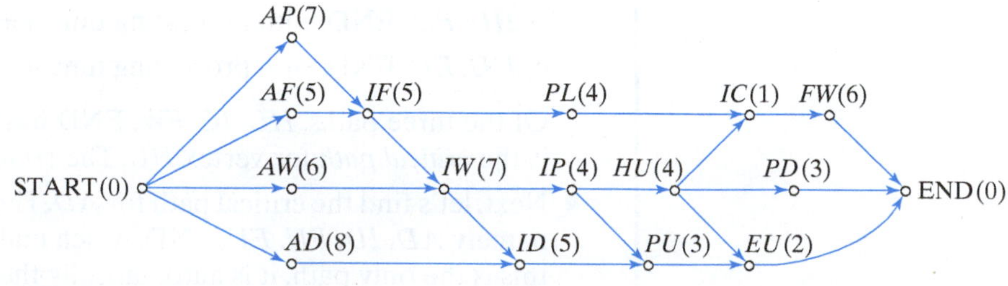

FIGURE 8-20

- **Step 1.**
  - Start at END. The critical time of END is 0, so we add a $[0]$ next to END(0).
  - The backflow now moves to the three vertices that precede END, namely, $FW(6)$, $PD(3)$, and $EU(2)$. In each case the critical time is the processing time plus 0, so the critical times are $FW[6]$, $PD[3]$, and $EU[2]$. We add this information to the project digraph.

- From $FW[6]$, the backflow moves to $IC(1)$. The vertex $IC(1)$ is incident only to $FW[6]$, so the critical time for *IC* is $1+6=7$. We add a $[7]$ next to *IC* in the project digraph.
- The backflow now moves to $HU(4)$, $PL(4)$, and $PU(3)$. There are three vertices $HU(4)$ is incident to ($IC[7]$, $PD[3]$, and $EU[2]$). Of the three, the one with the largest critical time is $IC[7]$. This means that the critical time for *HU* is $4+7=11$. $PL(4)$ is only incident to $IC[7]$, so its critical time is $4+7=11$. $PU(3)$ is only incident to $EU[2]$, so its critical time is $3+2=5$. Add $[11]$, $[11]$, and $[5]$ next to *HU*, *PL*, and *PU*, respectively.
- The backflow now moves to $IP(4)$ and $ID(5)$. $IP(4)$ is incident to $HU[11]$ and $PU[5]$, so the critical time for *IP* is $4+11=15$. $ID(5)$ is only incident to $PU(5)$, so its critical time is $5+5=10$. We add $[15]$ next to *IP*, and $[10]$ next to *ID*.
- The backflow now moves to $IW(7)$. The critical time for *IW* is $7+15=22$. (Please verify that this is correct!)
- The backflow now moves to $IF(5)$. The critical time for *IF* is $5+22=27$. (Ditto.)
- The backflow now moves to $AP(7)$, $AF(5)$, $AW(6)$, and $AD(8)$. Their respective critical times are $7+27=34$, $5+27=32$, $6+22=28$, and $8+10=18$.
- Finally, the backflow reaches START(0). We still follow the same rule—the critical time is $0+34=34$. This is the critical time for the project!

- **Step 2.** The critical time for every vertex of the project digraph is shown in red numbers in Fig. 8-21. We can now find the critical path by following the trail of largest critical times: START, *AP*, *IF*, *IW*, *IP*, *HU*, *IC*, *FW*, END.

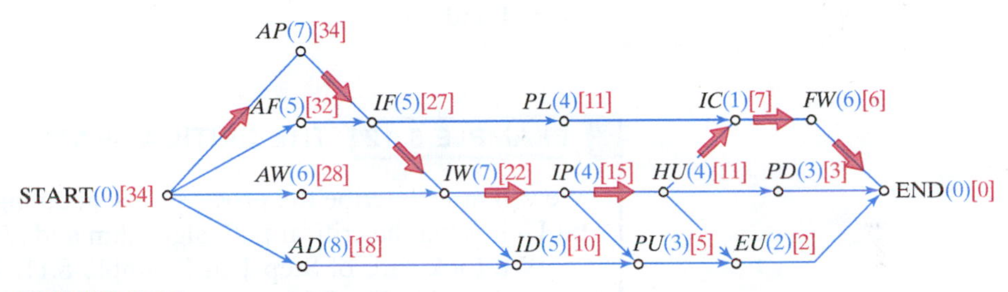

FIGURE 8-21

Why are the critical path and critical time of a project of special significance? We saw earlier in the chapter that for every project there is a theoretical time barrier below which a project cannot be completed, regardless of how clever the scheduler is or how many processors are used. Well, guess what? This theoretical barrier *is the project's critical time.*

If a project is to be completed in the optimal completion time, it is absolutely essential that all the tasks in the critical path be done at the earliest possible time. Any delay in starting up one of the tasks in the critical path will necessarily delay the finishing time of the entire project. (By the way, this is why this path is called *critical.*)

Unfortunately, it is not always possible to schedule the tasks on the critical path one after the other, bang, bang, bang without delay. For one thing, processors are not always free when we need them. (Remember that a processor cannot stop in the middle of one task to start a new task.) Another reason is the problem of uncompleted precedent tasks. We cannot concern ourselves only with tasks along the critical path and disregard other tasks that might affect them through precedence relations. There is a whole web of interrelationships that we need to worry about. Optimal scheduling is extremely complex.

The Critical-Path Algorithm

The concept of critical paths can be used to create very good (although not necessarily optimal) schedules. The idea is to use *critical times* rather than processing times to prioritize the tasks. The priority list we obtain when we write the tasks in decreasing order of *critical times* (with ties broken randomly) is called the **critical-time priority list**, and the process of creating a schedule using the critical-time priority list is called the **critical-path algorithm**.

CRITICAL-PATH ALGORITHM

- **Step 1: Find critical times.** Using the backflow algorithm, find the *critical time* for every task in the project.
- **Step 2: Create priority list.** Using the critical times obtained in Step 1, create a *priority list* with the tasks listed in decreasing order of critical times (i.e., a critical-time priority list).
- **Step 3: Create schedule.** Using the critical-time priority list obtained in Step 2, create the *schedule*.

There are, of course, plenty of small details that need to be attended to when carrying out the critical-path algorithm, especially in Steps 1 and 3. Fortunately, everything that needs to be done we now know how to do. (The applet ***Priority List Scheduling*** makes the implementation of the critical-path algorithm particularly convenient—once you create a project digraph the applet does all the work for Steps 1 and 3.)

EXAMPLE 8.12 THE CRITICAL-PATH ALGORITHM GOES TO MARS

We will now describe the process for scheduling the assembly of the Martian Habitat Unit using the critical-path algorithm and $N = 2$ processors.

We took care of Step 1 in Example 8.11. The critical times for each task are shown in red in Fig. 8-22.

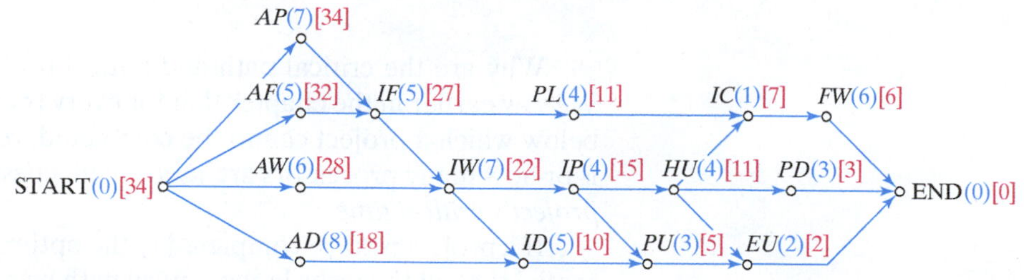

FIGURE 8-22

Step 2 follows directly from Step 1. The critical-time priority list for the project is $AP[34]$, $AF[32]$, $AW[28]$, $IF[27]$, $IW[22]$, $AD[18]$, $IP[15]$, $PL[11]$, $HU[11]$, $ID[10]$, $IC[7]$, $FW[6]$, $PU[5]$, $PD[3]$, $EU[2]$.

Step 3 is a lot of busywork—for the details, see Exercise 69.

The timeline for the resulting schedule is given in Fig. 8-23. The project finishing time is $Fin = 36$ hours. This is a very good schedule, but it is not an optimal schedule. (Figure 8-24 shows the timeline for an *optimal schedule* with finishing time $Opt = 35$ hours.)

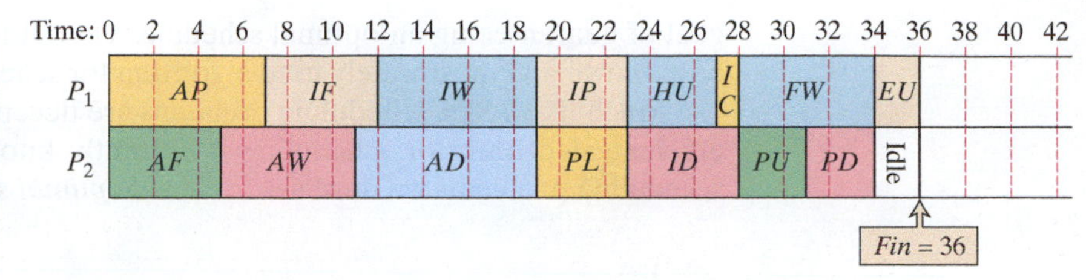

FIGURE 8-23 Timeline for the MHU building project under the critical-path algorithm ($N = 2$).

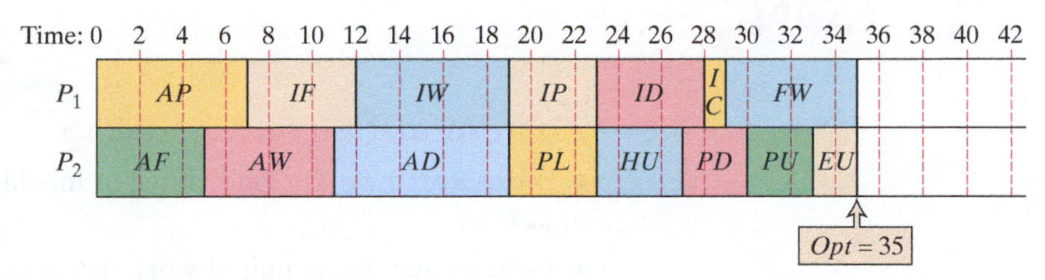

FIGURE 8-24 Timeline for an *optimal schedule* for the MHU building project ($N = 2$).

The critical-path algorithm is an excellent *approximate* algorithm for scheduling a project, but as Example 8.12 shows, it does not always give an optimal schedule. In this regard, scheduling problems are like TSPs (Chapter 6)—there are *efficient approximate* algorithms for scheduling, but no *efficient optimal* algorithm is currently known. Of the standard scheduling algorithms, the critical-path algorithm is by far the most commonly used. Other, more sophisticated algorithms have been developed in the last 40 years, and under specialized circumstances they can outperform the critical-path algorithm, but as an all-purpose algorithm for scheduling, the critical-path algorithm is hard to beat.

Conclusion

In one form or another, the scheduling of human (and nonhuman) activity is a pervasive and fundamental problem of modern life. At its most informal, it is part and parcel of the way we organize our everyday living (so much so that we are often scheduling things without realizing we are doing so). In its more formal incarnation, the systematic scheduling of a set of activities for the purposes of saving either time or money is a critical issue in management science. Business, industry, government, education—wherever there is a big project, there is a schedule behind it.

By now, it should not surprise us that at their very core, scheduling problems are mathematical in nature and that the mathematics of scheduling can range from the simple to the extremely complex. In this chapter we focused on a very specific type of scheduling problem in which we are given a set of *tasks*, a set of *precedence relations* among the tasks, and a set of identical *processors*. The objective is to schedule the tasks by properly assigning tasks to processors so that the *project finishing time* is as small as possible.

To tackle these scheduling problems systematically, we first developed a graph model of the problem, called the *project digraph*, and a general framework by means of which we can create, compare, and analyze schedules, called the *priority-list model*. Within the priority-list model, many strategies can be followed (with each strategy leading to the creation of a specific priority list). In this chapter, we considered two basic strategies for creating schedules. The first was the *decreasing-time algorithm*, a strategy that intuitively makes a lot of sense but that in practice often results in inefficient schedules. The second strategy, called the *critical-path algorithm*, is generally a big improvement over the decreasing-time algorithm, but it falls short of the ideal goal of guaranteeing an optimal schedule. The critical-path algorithm is by far the best known and most widely used algorithm for scheduling in business and industry.

Much like TSPs, scheduling problems are deceptively difficult. No optimal and efficient algorithm for scheduling is currently known, so when facing very large scheduling problems we must settle for *suboptimal* solutions.

KEY CONCEPTS

8.1 An Introduction to Scheduling

- **processor:** a worker (individual or machine) assigned to carry out a task, **227**
- **task:** an indivisible unit of work that is not broken up or shared, **228**
- **processing time (*P*-time):** for a task, the amount of time required by any single processor to fully execute the task, **228**
- **precedence relation:** a prerequisite indicating that a task cannot be started before another task is complete, **228**
- **independent tasks:** a pair of tasks not bound by any precedence relations, **229**
- **finishing time (*Fin*):** the total time elapsed between the start and the end of a project, **230**
- **optimal schedule:** a schedule with the smallest possible finishing time, **230**
- **optimal finishing time (*Opt*):** the finishing time under an optimal schedule, **230**

8.2 Directed Graphs

- **asymmetric relationship:** a relationship between pairs of objects that need not be reciprocal, **232**
- **digraph:** a graph with directed edges (arcs) used to describe an asymmetric relationship between objects, **232**
- **arc:** an arc XY $(X \rightarrow Y)$ indicates that the relationship goes from X to Y, **232**
- **arc-set:** a list of all the arcs in a digraph, **232**
- **incident to (from):** given an arc XY, X is *incident to* Y; Y is *incident from* X, **232**
- **adjacent arcs:** arc YZ is said to be adjacent to arc XY because the starting vertex of YZ is the ending vertex of XY, **232**
- **path:** from X to Z; a sequence of distinct adjacent arcs, starting at X and ending at a different vertex Z, **232**
- **cycle:** a sequence of distinct adjacent arcs that starts and ends at the same vertex, **232**
- **length:** for a path or a cycle, the number of arcs in the path (cycle), **232**
- **outdegree:** for a vertex X, the number of arcs having X as their starting vertex, **232**
- **indegree:** for a vertex Y, the number of arcs having Y as their ending vertex, **232**

- **project digraph:** a digraph having the tasks (including the imaginary tasks START and END) as vertices and the precedence relations as arcs. The processing times are included with the tasks for convenience, **234**

8.3 Priority-List Scheduling

- **priority list:** an ordered list (permutation) of the tasks, **234**
- **priority-list model:** a model for scheduling where *ready* tasks are executed in the order given in the priority list, with *ineligible* tasks skipped over until they become ready, **237**
- **optimal priority list:** a priority list that generates an optimal schedule, **239**

8.4 The Decreasing-Time Algorithm

- **decreasing-time priority list:** a priority list where the tasks are listed in decreasing order of processing times (longest first, shortest last), **239**
- **decreasing-time algorithm:** creates a schedule by choosing a decreasing-time priority list and applying the priority-list model to it, **239**

8.5 Critical Paths and the Critical-Path Algorithm

- **critical path (for a vertex):** for a vertex X, a path from X to END having the longest total processing time, **241**
- **critical time (for a vertex):** the total processing time of the critical path for the vertex, **241**
- **critical path (for a project):** a path from START to END having the longest total processing time, **241**
- **critical time (for a project):** the total processing time of the critical path for the project, **241**
- **backflow algorithm:** an algorithm for finding critical times and critical paths. The algorithm starts at END and works itself backwards toward START, assigning to each vertex a critical time equal to its processing time plus the largest critical time among the vertices ahead of it, **242**
- **critical-time priority list:** a priority list where the tasks are listed in decreasing order of critical times (longest critical time first, shortest last), **244**
- **critical-path algorithm:** creates a schedule by choosing a critical-time priority list and applying the priority-list model to it, **244**

EXERCISES

WALKING

8.1 An Introduction to Scheduling

No exercises for this section.

8.2 Directed Graphs

1. For the digraph shown in Fig. 8-25, find
   - (a) the indegree and outdegree of A.
   - (b) the indegree and outdegree of B.
   - (c) the indegree and outdegree of D.
   - (d) the sum of the indegrees of all the vertices.
   - (e) the sum of the outdegrees of all the vertices.

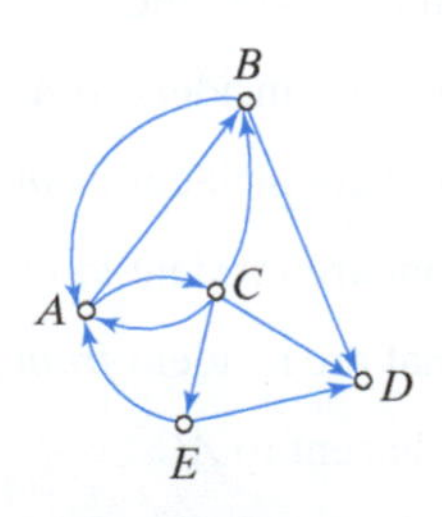

FIGURE 8-25

2. For the digraph shown in Fig. 8-26, find
   (a) the indegree and outdegree of A.
   (b) the indegree and outdegree of C.
   (c) the indegree and outdegree of E.
   (d) the sum of the indegrees of all the vertices.
   (e) the sum of the outdegrees of all the vertices.

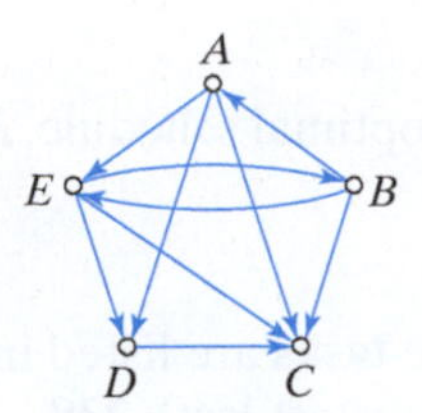

FIGURE 8-26

3. For the digraph in Fig. 8-25, find
   (a) all paths of length 2 from A to D.
   (b) all paths of length 3 from A to D.
   (c) a path of length 4 from A to D.
   (d) a path of length 5 from A to D.
   (e) all paths from D to A.
4. For the digraph in Fig. 8-26, find
   (a) a path of length 1 from B to E.
   (b) a path of length 2 from B to E.
   (c) a path of length 4 from B to E.
   (d) all paths from E to B.
   (e) all paths from D to E.
5. For the digraph in Fig. 8-25, find
   (a) all cycles of length 2.
   (b) all cycles of length 3.
   (c) all cycles of length 4.
   (d) all cycles of length 5.
6. For the digraph in Fig. 8-26, find
   (a) all cycles of length 2.
   (b) all cycles of length 3.
   (c) all cycles of length 4.
7. For the digraph in Fig. 8-25, find
   (a) all vertices that are incident *to* A.
   (b) all vertices that are incident *from* A.
   (c) all vertices that are incident *to* D.
   (d) all vertices that are incident *from* D.
   (e) all the arcs adjacent to AC.
   (f) all the arcs adjacent to CD.
8. For the digraph in Fig. 8-26, find
   (a) all vertices that are incident *to* A.
   (b) all vertices that are incident *from* A.
   (c) all vertices that are incident *to* D.
   (d) all vertices that are incident *from* D.
   (e) all the arcs adjacent to BA.
   (f) all the arcs adjacent to BC.
9. (a) Draw a digraph with vertex-set $\mathcal{V} = \{A, B, C, D\}$ and arc-set $\mathcal{A} = \{AB, AC, AD, BD, DB\}$.
   (b) Draw a digraph with vertex-set $\mathcal{V} = \{A, B, C, D, E\}$ and arc-set $\mathcal{A} = \{AC, AE, BD, BE, CD, DC, ED\}$.
   (c) Draw a digraph with vertex-set $\mathcal{V} = \{W, X, Y, Z\}$ and such that W is incident to X and Y, X is incident to Y and Z, Y is incident to Z and W, and Z is incident to W and X.
10. (a) Draw a digraph with vertex-set $\mathcal{V} = \{A, B, C, D\}$ and arc-set $\mathcal{A} = \{AB, AC, AD, BC, BD, DB, DC\}$.
   (b) Draw a digraph with vertex-set $\mathcal{V} = \{V, W, X, Y, Z\}$ and arc-set $\mathcal{A} = \{VW, VZ, WZ, XV, XY, XZ, YW, ZY, ZW\}$.
   (c) Draw a digraph with vertex-set $\mathcal{V} = \{W, X, Y, Z\}$ and such that every vertex is incident to every other vertex.
11. Consider the digraph with vertex-set $\mathcal{V} = \{A, B, C, D, E\}$ and arc-set $\mathcal{A} = \{AB, AE, CB, CE, DB, EA, EB, EC\}$. Without drawing the digraph, determine
   (a) the outdegree of A.
   (b) the indegree of A.
   (c) the outdegree of D.
   (d) the indegree of D.
12. Consider the digraph with vertex-set $\mathcal{V} = \{V, W, X, Y, Z\}$ and arc-set $\mathcal{A} = \{VW, VZ, WZ, XY, XZ, YW, ZY, ZW\}$. With-out drawing the digraph, determine
   (a) the outdegree of V.
   (b) the indegree of V.
   (c) the outdegree of Z.
   (d) the indegree of Z.
13. Consider the digraph with vertex-set $\mathcal{V} = \{A, B, C, D, E, F\}$ and arc-set $\mathcal{A} = \{AB, BD, CF, DE, EB, EC, EF\}$.
   (a) Find a path from vertex A to vertex F.
   (b) Find a Hamilton path from vertex A to vertex F. (*Note*: A Hamilton path is a path that passes through every vertex of the graph once.)
   (c) Find a cycle in the digraph.
   (d) Explain why vertex F cannot be part of any cycle.
   (e) Explain why vertex A cannot be part of any cycle.
   (f) Find all the cycles in this digraph.

14. Consider the digraph with vertex-set $\mathcal{V} = \{A, B, C, D, E\}$ and arc-set $\mathcal{A} = \{AB, AE, CB, CD, DB, DE, EB, EC\}$.

(a) Find a path from vertex A to vertex D.

(b) Explain why the path you found in (a) is the only possible path from vertex A to vertex D.

(c) Find a cycle in the digraph.

(d) Explain why vertex A cannot be part of a cycle.

(e) Explain why vertex B cannot be part of a cycle.

(f) Find all the cycles in this digraph.

15. The White Pine subdivision is a rectangular area six blocks long and two blocks wide. Streets alternate between one way and two way as shown in Fig. 8-27. Draw a digraph that represents the traffic flow in this neighborhood.

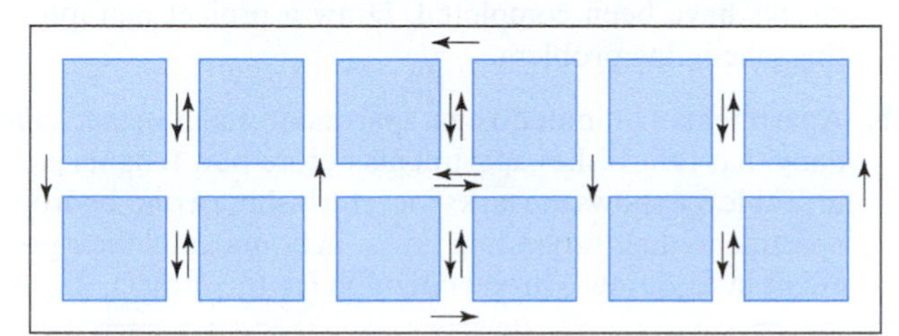

FIGURE 8-27

16. A mathematics textbook consists of 10 chapters. Although many of the chapters are independent of the others, some chapters require that previous chapters be covered first. The following list describes all the chapter dependences: Chapter 1 is a prerequisite to Chapters 3 and 5; Chapters 2 and 9 are both prerequisites to Chapter 10; Chapter 3 is a prerequisite to Chapter 6; and Chapter 4 is a prerequisite to Chapter 7, which in turn is a prerequisite to Chapter 8. Draw a digraph that describes the dependences among the chapters in the book.

17. The digraph in Fig. 8-28 is a *respect* digraph. That is, the vertices of the digraph represent members of a group and an arc XY represents that X respects Y.

(a) If you had to choose one person to be the leader of the group, whom would you pick? Explain.

(b) Who would be the worst choice to be the leader of the group? Explain.

(c) Assume that you know the respect digraph of a group of individuals, and that is the only information available to you. Which individual would be the most reasonable choice for leader of the group?

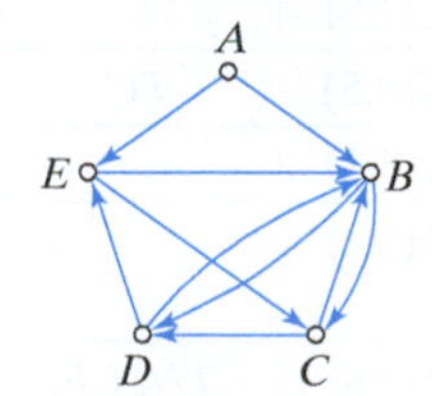

FIGURE 8-28

18. The digraph in Fig. 8-29 is an example of a *tournament* digraph. In this example the vertices of the digraph represent five volleyball teams in a round-robin tournament (i.e., every team plays every other team). An arc XY represents that X defeated Y in the tournament. (*Note*: There are no ties in volleyball.)

(a) Which team won the tournament? Explain.

(b) Which team came in last in the tournament? Explain.

(c) Suppose that you are given the tournament digraph of some tournament. What does the *indegree* of a vertex represent? What does the *outdegree* of a vertex represent?

(d) If T denotes the tournament digraph for a round-robin tournament with N teams, then for any vertex X in T, the indegree of X plus the outdegree of $X = N - 1$. Explain why.

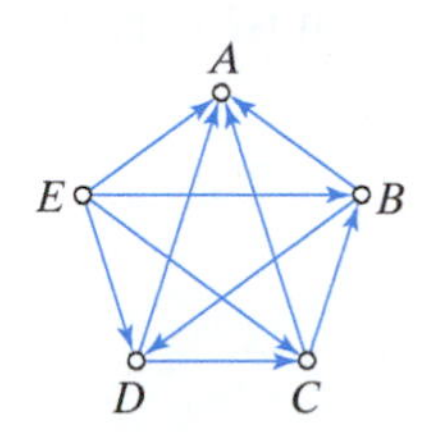

FIGURE 8-29

19. As part of its extended concert tour around the United States, the famous rock band Angelface will be giving a concert at the Smallville Bowl next month. This is a big event for Smallville, and the whole town is buzzing. The digraph in Fig. 8-30 shows the hyperlinks connecting the following six Web sites: (1) www.angelface.com (the rock band's Web site), (2) www.ticketmonster.com (the Web site for TicketMonster, the only ticket agency licensed to sell tickets to the concert), (3) www.knxrock.com (the Web site of Smallville rock station KNXR), (4) www.joetheblogger .com (the personal blog of Joe Fan, a local rock and roll aficionado), and (5) and (6) www.SmallvilleInn.com and www.SmallvilleSuites.com (the Web sites of two local sister hotels owned by the same company and both offering a special rate for out-of-town fans coming to the concert).

(a) Make an educated guess as to which vertex is the most likely to represent (1), the rock band's Web site, and explain the reasoning behind your answer.

(b) Make an educated guess as to which vertex is the most likely to represent (2), the TicketMonster Web site, and explain the reasoning behind your answer.

(c) Make an educated guess as to which vertex is the most likely to represent (3), the Web site of rock station KNXR, and explain the reasoning behind your answer.

(d) Make an educated guess as to which vertex is the most likely to represent (4), Joe Fan's blog, and explain the reasoning behind your answer.

(e) Make an educated guess as to which two vertices are the most likely to represent (5) and (6), the two Smallville sister hotel Web sites.

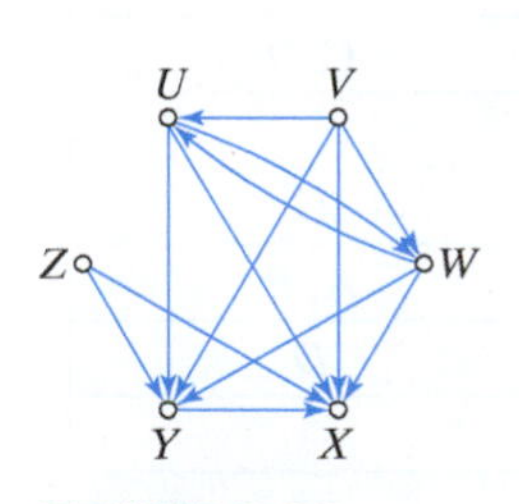

FIGURE 8-30

20. Wobble, a start-up company, is developing a search engine for the Web. Given a particular search word, say Angelface (the name of a rock band), Wobble's strategy is to find all the Web sites containing the term *Angelface,* and then look at Web sites to which those Web sites link, to obtain a ranked list of search results. In other words, if there are lots of links pointing to Web site X from Web sites that mention Angelface, then it is likely that X contains lots of useful information about Angelface. In the digraph shown in Fig. 8-31, the vertices represent Web pages that contain the word *Angelface* and the arcs represent hyperlinks from one Web site to another.

(a) Which Web sites would show up first, second, and third in the search results? Explain your answer.

(b) One of the vertices of the digraph is the official Angelface Web site. Which do you think it is? Explain your answer.

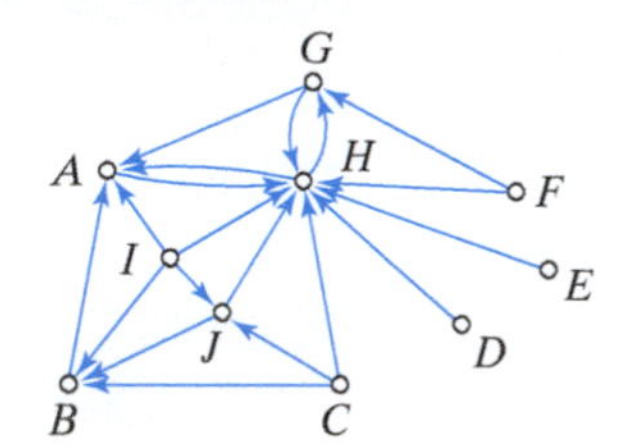

FIGURE 8-31

21. A project consists of eight tasks labeled A through H. The processing time (P-time) and precedence relations for each task are given in Table 8-2. Draw the project digraph.

Task	P-time	Precedent tasks
A	3	
B	10	C, F, G
C	2	A
D	4	G
E	5	C
F	8	A, H
G	7	H
H	5	

TABLE 8-2

22. A project consists of eight tasks labeled A through H. The processing time (P-time) and precedence relations for each task are given in Table 8-3. Draw the project digraph.

Task	P-time	Precedent tasks
A	5	C
B	5	C, D
C	5	
D	2	G
E	15	A, B
F	6	D
G	2	
H	2	G

TABLE 8-3

23. Eight experiments need to be carried out. One of the experiments requires 10 hours to complete, two of the experiments require 7 hours each to complete, two more require 12 hours each to complete, and three of the experiments require 20 hours each to complete. In addition, none of the 20-hour experiments can be started until both of the 7-hour experiments have been completed, and the 10-hour experiment cannot be started until both of the 12-hour experiments have been completed. Draw a project digraph for this scheduling problem.

24. Every fifty thousand miles an aircraft undergoes a complete inspection that involves 10 different diagnostic "checks." Three of the checks require 4 hours each to complete, three more require 7 hours each to complete, and four of the checks require 15 hours each to complete. Moreover, none of the 15-hour checks can be started until all the 4-hour checks have been completed. Draw a project digraph for this scheduling problem.

25. Apartments Unlimited is an apartment maintenance company that refurbishes apartments before new tenants move in. Table 8-4 shows the tasks for refurbishing a one-bedroom apartment, their processing times (in hours), and their precedent tasks. Draw a project digraph for the project.

Tasks	Label (P-time)	Precedent tasks
Bathrooms (clean)	$B(2)$	P
Carpets (shampoo)	$C(1)$	S, W
Filters (replace)	$F(0.5)$	
General cleaning	$G(2)$	B, F, K
Kitchen (clean)	$K(3)$	P
Lights (replace bulbs)	$L(0.5)$	
Paint	$P(6)$	L
Smoke detectors (battery)	$S(0.5)$	G
Windows (wash)	$W(1)$	G

TABLE 8-4

26. A ballroom is to be set up for a large wedding reception. Table 8-5 shows the tasks to be carried out, their processing times (in hours), and their precedent tasks. Draw a project digraph for the project of setting up for the wedding reception.

Tasks	Label (P-time)	Precedent tasks
Set up tables and chairs	$TC(1.5)$	
Set tablecloths and napkins	$TN(0.5)$	TC
Make flower arrangements	$FA(2.2)$	
Unpack crystal, china, and flatware	$CF(1.2)$	
Put place settings on table	$PT(1.8)$	TN, CF
Arrange table decorations	$TD(0.7)$	FA, PT
Set up the sound system	$SS(1.4)$	
Set up the bar	$SB(0.8)$	TC

TABLE 8-5

27. Preparing a banquet for a large number of people requires careful planning and the execution of many individual tasks. Imagine preparing a four-course dinner for a party of 20 friends. Suppose that the project is broken down into 10 individual tasks. The tasks and their processing times (in hours) are as follows: $A(1.5)$, $B(1)$, $C(0.5)$, $D(1.25)$, $E(1.5)$, $F(1)$, $G(3)$, $H(2.5)$, $I(2)$, and $J(1.25)$. The precedence relations are as follows: B cannot be started until A and D have been completed, C cannot be started until B and J have been completed, E cannot be started until D and G have been completed, F cannot be started until E and J have been completed, H cannot be started until F and C have been completed, and J cannot be started until I has been completed. Draw a project digraph for this culinary project.

28. Speedy Landscape Service has a project to landscape the garden of a new model home. The tasks that must be performed and their processing times (in hours) are: collect and deliver rocks, $R(4)$; collect and deliver soil (two loads), $S_1(3)$ and $S_2(3)$; move and position rocks, $RM(4)$; grade soil, $SG(5)$; seed lawn, $SL(3)$; collect and deliver bushes and trees, $B(2)$; plant bushes and trees, $BP(1)$; water the new plantings, $W(1)$; and lay mulch, $M(2)$. The precedence relations between tasks are $R \rightarrow RM$, $RM \rightarrow SG$, $S_1 \rightarrow SG$, $S_2 \rightarrow SG$, $SG \rightarrow SL$, $SG \rightarrow B$, $B \rightarrow BP$, $SL \rightarrow W$, $BP \rightarrow W$, and $BP \rightarrow M$. Draw a project digraph for this landscaping project.

8.3 Priority-List Scheduling

Exercises 29 through 32 refer to a project consisting of 11 tasks (A through K) with the following processing times (in hours): A(10), B(7), C(11), D(8), E(9), F(5), G(3), H(6), I(4), J(7), K(5).

29. **(a)** A schedule with $N = 3$ processors produces finishing time $Fin = 31$ hours. What is the total idle time for all the processors?

(b) Explain why a schedule with $N = 3$ processors must have finishing time $Fin \geq 25$ hours.

30. **(a)** A schedule with $N = 5$ processors has finishing time $Fin = 19$ hours. What is the total idle time for all the processors?

(b) Explain why a schedule with $N = 5$ processors must have finishing time $Fin \geq 15$ hours.

31. Explain why a schedule with $N = 6$ processors must have finishing time $Fin \geq 13$ hours.

32. **(a)** Explain why a schedule with $N = 10$ processors must have finishing time $Fin \geq 11$ hours.

(b) Explain why it doesn't make sense to put more than 10 processors on this project.

Exercises 33 and 34 refer to the Martian Habitat Unit scheduling project with $N = 2$ processors discussed in Example 8.8. The purpose of these exercises is to fill in some of the details left out in Example 8.8.

33. For the priority list in Example 8.8, show the priority-list status at time $T = 26$.

34. For the priority list in Example 8.8, show the priority-list status at time $T = 32$.

35. Consider the project digraph shown in Fig. 8-32(a). (Processing times not relevant to this question have been omitted.) Suppose the project is to be scheduled with two processors, P_1 and P_2, and the priority list is A, B, C, D, E, F, G, H. The schedule starts as shown in Fig. 8-32(b).

(a) Can C be the next task assigned to P_1 at $T = 4$? If not, explain why not.

(b) Can D be the next task assigned to P_1 at $T = 4$? If not, explain why not.

(c) Can H be the next task assigned to P_1 at $T = 4$? If not, explain why not.

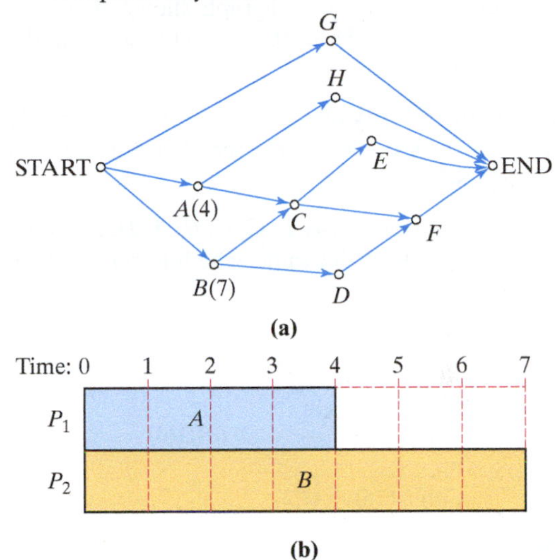

FIGURE 8-32

36. Consider the project digraph shown in Fig. 8-33(a). (Processing times not relevant to this question have been omitted.) Suppose the project is to be scheduled with two processors, P_1 and P_2, and the priority list is $A, B, C, D, E, F, G, H, I$. The schedule starts as shown in Fig. 8-33(b).

(a) Can D be the next task assigned to P_1 at $T = 6$? If not, explain why not.

(b) Can E be the next task assigned to P_1 at $T = 6$? If not, explain why not.

(c) Can G be the next task assigned to P_1 at $T = 6$? If not, explain why not.

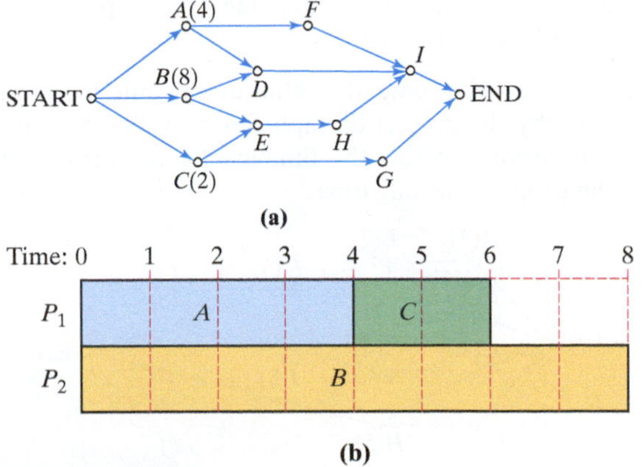

FIGURE 8-33

37. Using the priority list D, C, A, E, B, G, F, schedule the project described by the project digraph shown in Fig. 8-34 using $N = 2$ processors. Show the project timeline and give its finishing time.

38. Using the priority list G, F, E, D, C, B, A, schedule the project described by the project digraph shown in Fig. 8-34 using $N = 2$ processors. Show the project timeline and give its finishing time.

39. Using the priority list D, C, A, E, B, G, F, schedule the project described by the project digraph shown in Fig. 8-34 using $N = 3$ processors. Show the project timeline and give its finishing time.

40. Using the priority list G, F, E, D, C, B, A, schedule the project described by the project digraph shown in Fig. 8-34 using $N = 3$ processors. Show the project timeline and give its finishing time.

41. Explain why the priority lists A, B, D, F, C, E, G and F, D, A, B, G, E, C produce the same schedule for the project shown in Fig. 8-34.

42. Explain why the priority lists E, G, C, B, A, D, F and G, C, E, F, D, B, A produce the same schedule for the project shown in Fig. 8-34.

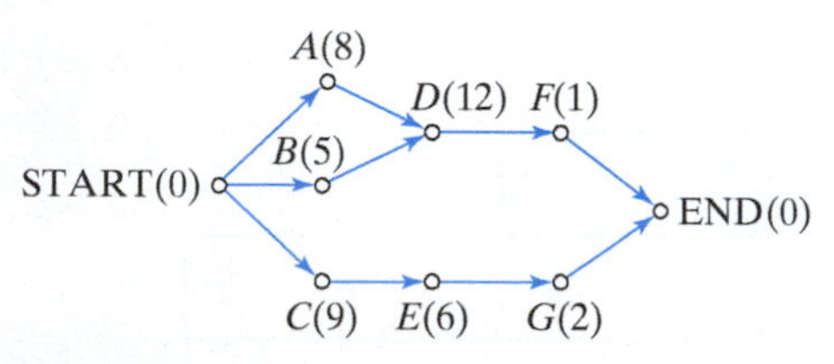

FIGURE 8-34

8.4 The Decreasing-Time Algorithm

43. Use the decreasing-time algorithm to schedule the project described by the project digraph shown in Fig. 8-34 using $N = 2$ processors. Show the timeline for the project, and give the project finishing time.

44. Use the decreasing-time algorithm to schedule the project described by the project digraph shown in Fig. 8-34 using $N = 3$ processors. Show the timeline for the project, and give the project finishing time.

45. Use the decreasing-time algorithm to schedule the project described by the project digraph shown in Fig. 8-35 using $N = 3$ processors. Show the timeline for the project, and give the project finishing time.

46. Use the decreasing-time algorithm to schedule the project described by the project digraph shown in Fig. 8-35 using $N = 2$ processors. Show the timeline for the project, and give the project finishing time.

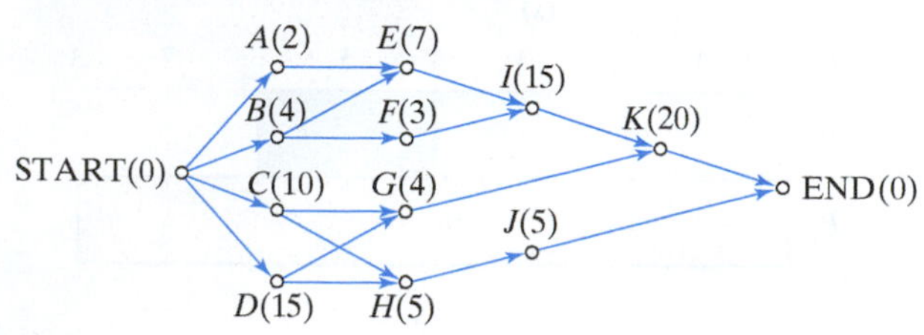

FIGURE 8-35

47. Consider the project described by the project digraph shown in Fig. 8-36, and assume that you are to schedule this project using $N = 2$ processors.

(a) Use the decreasing-time algorithm to schedule the project. Show the timeline for the project and the finishing time *Fin*.

(b) Find an optimal schedule and the optimal finishing time *Opt*.

(c) Use the relative error formula $\varepsilon = \frac{Fin - Opt}{Opt}$ to find the relative error of the schedule found in (a), and express your answer as a percent.

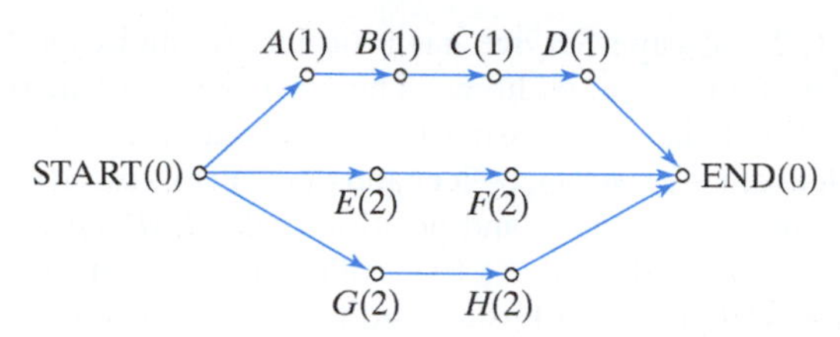

FIGURE 8-36

48. Consider the project described by the project digraph shown in Fig. 8-37, and assume that you are to schedule this project using $N = 2$ processors.

(a) Use the decreasing-time algorithm to schedule the project. Show the timeline for the project and the finishing time *Fin*.

(b) Find an optimal schedule and the optimal finishing time *Opt*.

(c) Use the relative error formula $\varepsilon = \frac{Fin - Opt}{Opt}$ to find the relative error of the schedule found in (a), and express your answer as a percent.

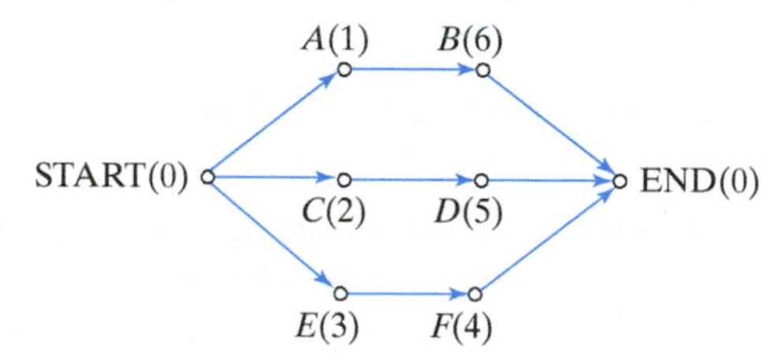

FIGURE 8-37

49. Consider the project described by the project digraph shown in Fig. 8-38, and assume that you are to schedule this project using $N = 3$ processors.

(a) Use the decreasing-time algorithm to schedule the project. Show the timeline for the project and the finishing time *Fin*.

(b) Find an optimal schedule and the optimal finishing time *Opt*.

(c) Use the relative error formula $\varepsilon = \frac{Fin - Opt}{Opt}$ to find the relative error of the schedule found in (a), and express your answer as a percent.

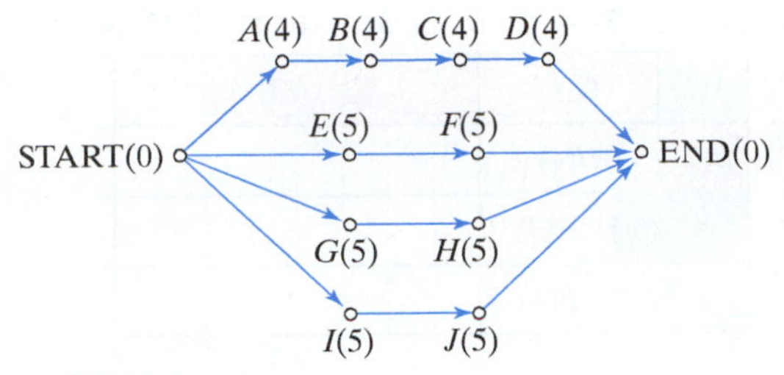

FIGURE 8-38

50. Consider the project described by the project digraph shown in Fig. 8-39, and assume that you are to schedule this project using $N = 4$ processors.

(a) Use the decreasing-time algorithm to schedule the project. Show the timeline for the project and the finishing time *Fin*.

(b) Find an optimal schedule. (*Hint*: *Opt* = 17.)

(c) Use the relative error formula $\varepsilon = \frac{Fin - Opt}{Opt}$ to find the relative error of the schedule found in (a), and express your answer as a percent.

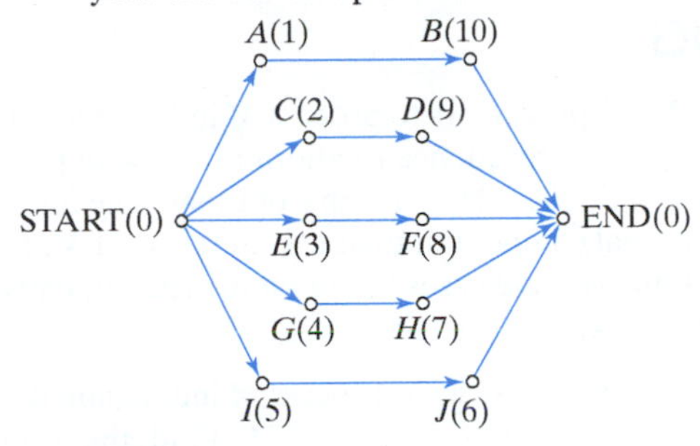

FIGURE 8-39

8.5 Critical Paths and the Critical-Path Algorithm

51. Consider the project digraph shown in Fig. 8-40.

(a) Use the backflow algorithm to find the critical time for each vertex.

(b) Find the critical path for the project.

(c) Schedule the project with $N = 2$ processors using the critical-path algorithm. Show the timeline, and give the project finishing time.

(d) Explain why the schedule obtained in (c) is optimal.

52. Consider the project digraph shown in Fig. 8-40.

(a) Schedule the project with $N = 3$ processors using the critical-path algorithm. Show the timeline, and give the project finishing time.

(b) Explain why the schedule obtained in (a) is optimal.

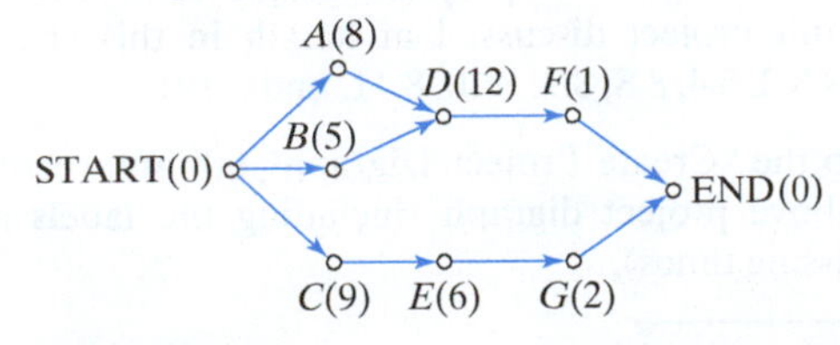

FIGURE 8-40

53. Consider the project digraph shown in Fig. 8-41.

(a) Find the critical path for the project.

(b) Schedule the project with $N = 3$ processors using the critical-path algorithm. Show the timeline, and give the project finishing time.

54. Consider the project digraph shown in Fig. 8-41.

(a) Use the backflow algorithm to find the critical time for each vertex.

(b) Schedule the project with $N = 2$ processors using the critical-path algorithm. Show the timeline, and give the project finishing time.

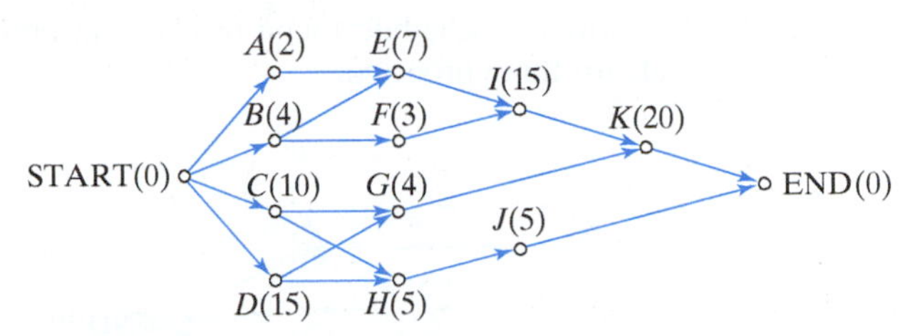

FIGURE 8-41

55. Schedule the Apartments Unlimited project given in Exercise 25 (Table 8-4) with $N = 2$ processors using the critical-path algorithm. Show the timeline, and give the project finishing time. (Note that the project digraph was done in Exercise 25.)

56. Schedule the project given in Exercise 26, Table 8-5 (setting up a ballroom for a wedding) with $N = 3$ processors using the critical-path algorithm. Show the timeline, and give the project finishing time. (Note that the project digraph was done in Exercise 26.)

57. Consider the project described by the project digraph shown in Fig. 8-42.

(a) Find the critical path and critical time for the project.

(b) Find the critical-time priority list.

(c) Schedule the project with $N = 2$ processors using the critical-path algorithm. Show the timeline and the project finishing time.

(d) Find an optimal schedule for $N = 2$ processors.

(e) Use the relative error formula $\varepsilon = \frac{Fin - Opt}{Opt}$ to find the relative error of the schedule found in (c).

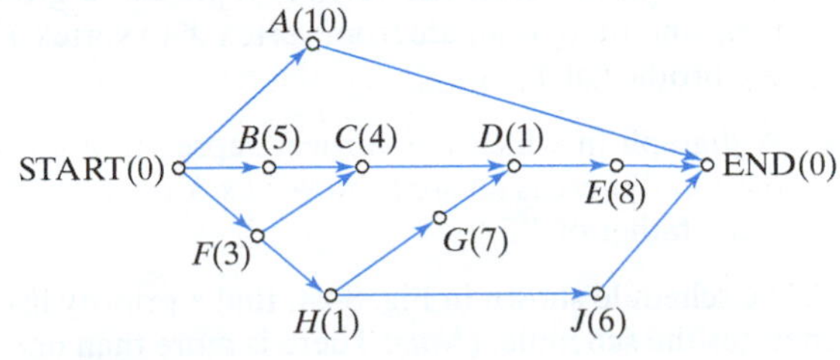

FIGURE 8-42

58. Consider the project digraph shown in Fig. 8-43, with the processing times given in hours.

(a) Find the critical path and critical time for the project.

(b) Find the critical-time priority list.

(c) Schedule the project with $N = 2$ processors using the critical-path algorithm. Show the timeline and the project finishing time.

(d) Explain why any schedule for $N = 2$ processors with finishing time $Fin = 22$ must be an optimal schedule.

(e) Schedule the project with $N = 3$ processors using the critical-path algorithm. Show the timeline and the project finishing time.

(f) Explain why the schedule found in (e) is an optimal schedule for three processors.

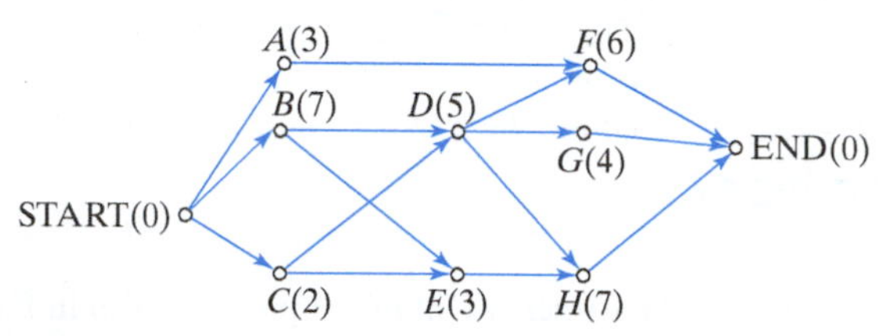

FIGURE 8-43

JOGGING

59. Explain why, in any digraph, the sum of all the indegrees must equal the sum of all the outdegrees.

60. Symmetric and totally asymmetric digraphs. A digraph is called *symmetric* if, whenever there is an arc from vertex X to vertex Y, there is *also* an arc from vertex Y to vertex X. A digraph is called *totally asymmetric* if, whenever there is an arc from vertex X to vertex Y, there *is not* an arc from vertex Y to vertex X. For each of the following, state whether the digraph is symmetric, totally asymmetric, or neither.

(a) A digraph representing the streets of a town in which all streets are one-way streets.

(b) A digraph representing the streets of a town in which all streets are two-way streets.

(c) A digraph representing the streets of a town in which there are both one-way and two-way streets.

(d) A digraph in which the vertices represent a group of men, and there is an arc from vertex X to vertex Y if X is a brother of Y.

(e) A digraph in which the vertices represent a group of men, and there is an arc from vertex X to vertex Y if X is the father of Y.

61. For the schedule shown in Fig. 8-44, find a priority list that generates the schedule. (*Note*: There is more than one possible answer.)

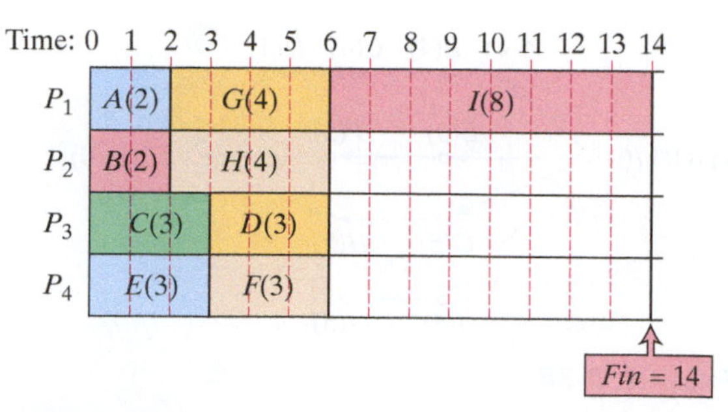

FIGURE 8-44

62. Let W represent the sum of the processing times of all the tasks, N be the number of processors, and Fin be the finishing time for a project.

(a) Explain the meaning of the inequality $Fin \geq W/N$ and why it is true for any schedule.

(b) Under what circumstances is $Fin = W/N$?

(c) What does the value $N \times Fin - W$ represent?

RUNNING

63. You have $N = 2$ processors to process M independent tasks (i.e., there are no precedence relations at all) with processing times $1, 2, 3, \ldots, M$. Find the optimal schedule, and give the optimal completion time Opt in terms of M. (*Hint*: Consider four separate cases based on the remainder when N is divided by 4.)

64. You have $N = 3$ processors to process M independent tasks with processing times $1, 2, 3, \ldots, M$. Find the optimal schedule, and give the optimal completion time Opt in terms of M. (*Hint*: Consider six separate cases based on the remainder when N is divided by 6.)

65. You have $N = 2$ processors to process $M + 1$ independent tasks with processing times $1, 2, 4, 8, 16, \ldots, 2^M$. Find the optimal schedule, and give the optimal completion time Opt in terms of M.

APPLET BYTES MyMathLab®

These Applet Bytes are exercises and mini-explorations built around the applet* ***Priority List Scheduling*** *(available in MyMathLab in the Multimedia Library or Tools for Success). The applet allows you to create a project digraph, enter a priority list and schedule the project using up to $N = 5$ processors. The applet also finds critical times and critical paths for the project.*

66. The main purpose of this Applet Byte is to familiarize you with the *Priority List Scheduling* applet. (In other words, it's a way to get you to play a little with the applet.) Figure 8-45 shows, once again, the project digraph for the Martian Habitat Unit project discussed at length in this chapter (Examples 8.2, 8.4, 8.8, 8.9, 8.10, 8.11, and 8.12).

(a) Go to the "Create Project Digraph" tab and recreate the above project digraph (including the labels and processing times).

* MyMathLab code required.

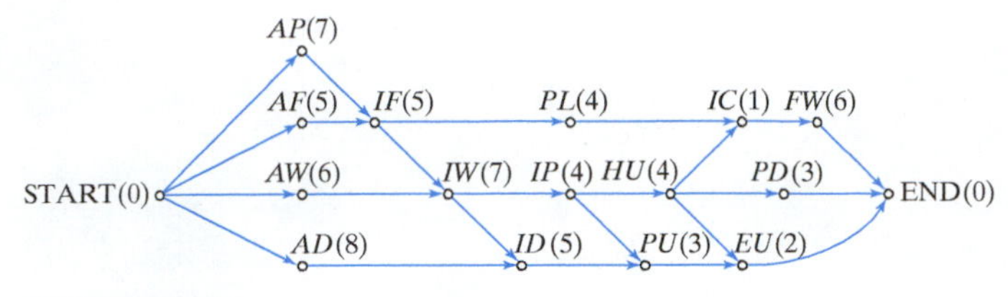

FIGURE 8-45

(b) Go to the "Practice Scheduling" tab, enter the decreasing-time priority list and schedule the project using $N = 2$ processors. Compare the results with Fig. 8-17.

(c) Still using the decreasing-time priority list, schedule the project using $N = 3$ processors.

(d) While still in the "Practice Scheduling" tab, click the "Show critical times" box. From the critical times shown in the graph, enter the critical-time priority list and schedule the project using $N = 2$ processors.

(e) Using the critical-time priority list, schedule the project using $N = 3$ processors. Explain why, in spite of the fact that the schedule shows a lot of idle time for processors 2 and 3, this schedule is optimal.

(f) Edit the critical-time priority list and change it to *AP, AF, AW, IF, AD, IW, IP, PL, ID, HU, PD, IC, FW, PU, EU.* Now schedule the project using $N = 2$ processors. Explain why this schedule is optimal.

67. One of the most paradoxical aspects of scheduling with priority lists is the fact that sometimes *you can speed up the completion of a project by eliminating a processor.* How can this be? This applet byte explores how this can happen. Figure 8-46 shows the project digraph.

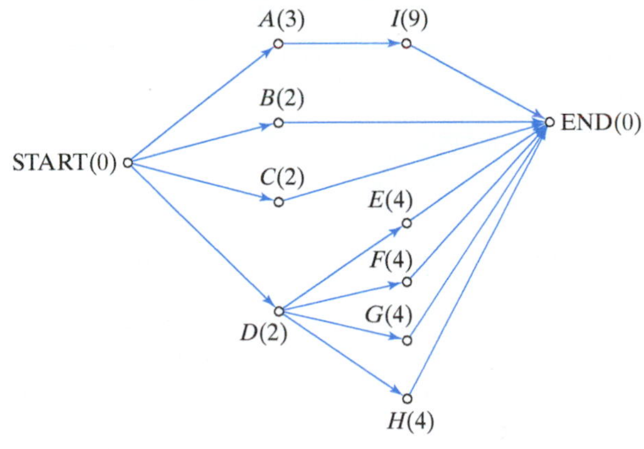

FIGURE 8-46

(a) Go to the "Create Project Digraph" tab and recreate the above project digraph.

(b) Go to the "Practice Scheduling" tab, enter the priority list *A, B, C, D, E, F, G, H, I,* and schedule the project using $N = 4$ processors. Explain why this schedule is not optimal.

(c) Using the same priority list as in (b) schedule the project using $N = 3$ processors. Explain why this schedule is optimal.

(d) Explain why the completion time for the project improved when you reduced the number of processors from 4 to 3. (*Hint:* It has to do with the assignment of task *I*.)

(e) Use the critical path algorithm to schedule the project with $N = 4$ processors and explain why the schedule obtained is optimal.

68. In this Applet Byte we will use once again the project digraph in Fig. 8-46 to explore a second paradoxical aspect of scheduling with priority lists—the fact that you can *slow down the completion of a project by speeding up the processors.*

(a) Recreate the project digraph shown in Fig. 8-46, and find the schedule for the priority list *A, B, C, D, E, F, G, H, I* using $N = 3$ processors. [If you did Exercise 67(c) you know that this schedule is optimal.]

(b) Go back to the "Create Project Digraph" and change the processing times of all tasks to be one unit less (change 2's to 1's, 3's to 2's, . . . , 9 to 8). (Imagine that the processing times shown are in days, and that using better software you can speed up the processors so that all tasks can be completed in one less day.) Schedule this project still using the priority list *A, B, C, D, E, F, G, H, I,* and $N = 3$ processors.

(c) Compare the schedules you obtained in parts (a) and (b) and explain what happened.

(d) Find an optimal priority list for the project in part (b) using $N = 3$ processors.

PART 3

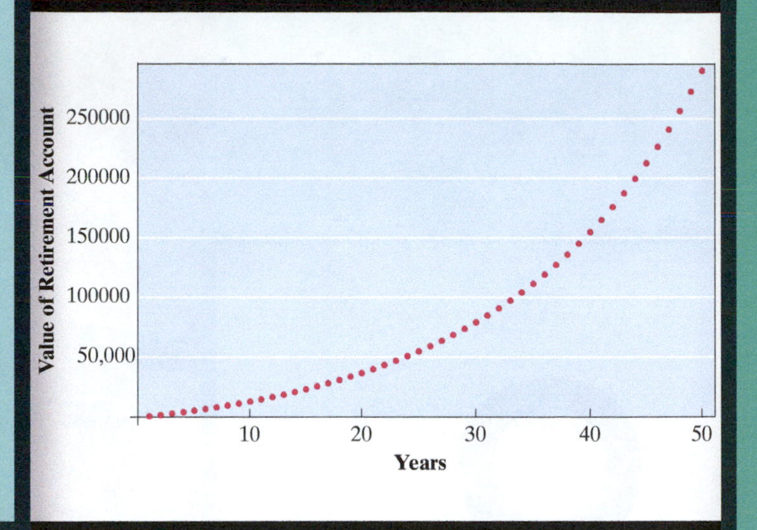

Growth

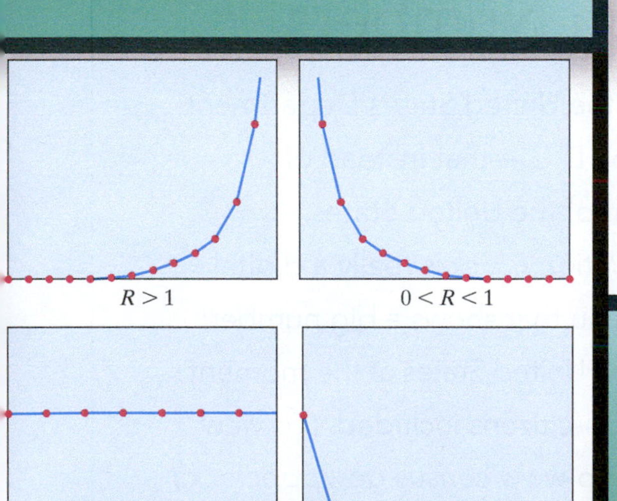

Compounding	T	F	APY
annually	1	1.06	6.000%
semi-annually	2	$\left(1+\frac{0.06}{2}\right)^2 = 1.0609$	6.090%
quarterly	4	$\left(1+\frac{0.06}{4}\right)^4 \approx 1.061364$	6.1364%
monthly	12	$\left(1+\frac{0.06}{12}\right)^{12} \approx 1.061678$	6.1678%
continuously	∞	$e^{0.06} \approx 1.061837$	6.1837%

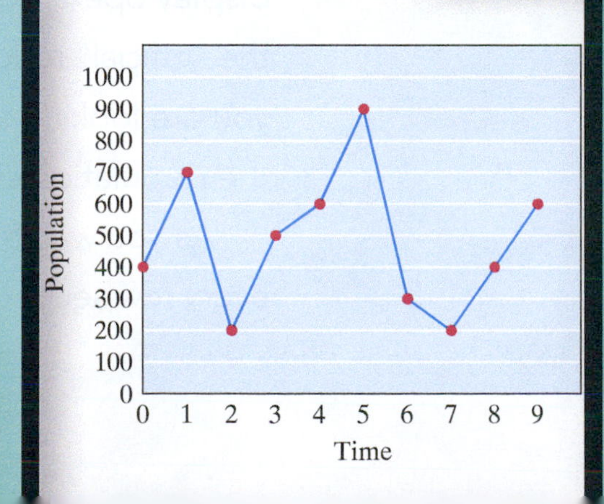

9

October 17, 2006: U.S. population hits 300 million (maybe). (For more on keeping track of the U.S. population, see Example 9.8.)

Population Growth Models

There Is Strength in Numbers

There is a big clock—located inside the United States Department of Commerce Building in Washington D.C.—that instead of keeping time keeps the official count of the United States population. The *United States Population Clock* is really a digital display operated by the Census Bureau that shows a big number: the "official" resident population of the United States at the moment you are looking at it—citizens and non-citizens included. (To view the population clock in real time, go to *www.census.gov/popclock*.)

As you watch the U.S. population clock in action you see that every few seconds the count increases by 1. What happened? Did some hospital call the Census Bureau with news that another baby had just been born? No. The ticking of the U.S. population clock is based on a mathematical model for the growth of the U.S. population. Like any other model, this model is not intended to be a perfect representation of reality but just an approximation—that number on the clock is almost certain to be off.

So, if the population figure in the clock is wrong, what is the clock good for? You'd be surprised. This is the remarkable thing about population models—they are incredibly useful when it comes to the "big picture" even though they are typically wrong in the details (the U. S. population clock is almost certain to be accurate in the millions digits, but after that all bets are off). In spite of their lack of accuracy, population models are successfully used by economists to make economic forecasts, by epidemiologists to project (and prepare for) epidemics, by urban planners to plan the future of cities, by demographers to develop population pyramids, by biologists to study animal populations, and so much more.

In this chapter we will look at three different mathematical models of population growth. (Be careful how you interpret the meaning of the word *growth* here. In its mathematical usage growth can be positive or negative and "negative growth" is just a fancy way of saying "loss" or "decline.") In Section 9.1 we discuss the basic ideas behind *mathematical models* of population growth and introduce *sequences* in general and *population sequences* in particular. In Section 9.2 we introduce the *linear growth model* and the concept of an *arithmetic sequence*, we learn about an important formula called the *arithmetic sum formula*, and discuss several real-life applications of linear growth (including the Census Bureau model for the growth of the U.S. population). In Section 9.3 we introduce the *exponential growth model* and the concept of a *geometric sequence*, and we introduce an extremely important and useful formula called the *geometric sum formula* (much more on exponential growth and the geometric sum formula will come in Chapter 10). Finally, in Section 9.4 we introduce the *logistic growth model* and the *logistic equation*, and discuss examples of logistic growth in animal populations.

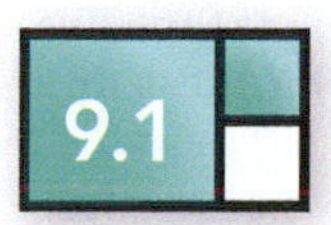

> "Remember that all models are wrong; the practical question is how wrong do they have to be to not be useful."
>
> – George E. P. Box

9.1 Sequences and Population Sequences

Sequences

In mathematics the word *sequence* has a very specific meaning: A **sequence** is an *infinite, ordered* list of numbers. In principle, the numbers can be any type of number—positive, negative, zero, rational, or irrational. The individual numbers in a sequence are called the **terms** of the sequence.

The simplest way to describe a sequence is using a list format—start writing the terms of the sequence, in order, separated by commas. The list, however, is infinite, so at some point one has to stop writing. At that point, a ". . ." is added as a symbolic way of saying "and so on." For lack of a better term, we will call this the **infinite list** description of the sequence.

How many terms should we write at the front end before we appeal to the ". . ."? This is a subjective decision, but the idea is to write enough terms so that a reasonable third party looking at the sequence can figure out how the sequence continues. No matter what we do, the "..." is always a leap of faith, and we should strive to make that leap as small as possible. Some sequences become clear with four or five terms, others take more.

EXAMPLE 9.1 HOW MANY TERMS ARE ENOUGH?

- Consider the sequence that starts with $1, 2, 4, 8, 16, 32, \ldots$. We could have continued writing down terms, but it seems reasonable to assume that at this point most people would agree that the sequence continues with $64, 128, \ldots$. The leap of faith here is small.
- Consider the sequence that starts with $3, 5, 7, \ldots$. Are there enough terms here so that we can figure out what comes next? A good guess is that the sequence continues with $9, 11, 13, \ldots$ but this is not the only reasonable guess. Perhaps the sequence is intending to describe the odd prime numbers, and in that case the next three terms of the sequence would be $11, 13, 17, \ldots$. We can conclude that $3, 5, 7, \ldots$ does not provide enough terms to draw a clear conclusion about the sequence and that a decent description should have included a few additional terms.
- Consider the sequence that starts with 1, 11, 21, 1211, 111221, 312211, 13112221, 1113213211, 31131211131221, ... These are the first nine terms of the sequence—can you guess what the tenth term is? Even after nine terms, what comes next is far from obvious. (This "mystery" sequence represents an interesting and challenging puzzle, and we leave it to the reader to try to find the next term and decipher the puzzle. Give it a try! The answer is given on page 289.)

The main lesson to be drawn from Example 9.1 is that describing a sequence using an infinite list is simple and convenient, but it doesn't work all that well with the more exotic sequences. Are there other ways? Yes. Before we get to them, we introduce some useful notation for sequences.

- **Sequence notation.** A generic sequence can be written in infinite list form as

$$A_1, A_2, A_3, A_4, A_5, \ldots$$

The A is a variable representing a symbolic name for the sequence. Each term of the sequence is described by the sequence name and a numerical subscript that represents the position of the term in the sequence. You may think of the subscript as the "address" of the term. This notation makes it possible to conveniently describe any term by its position in the sequence: A_{10} represents the 10th term, A_{100} represents the 100th term, and A_N represents a term in a generic position N in the sequence.

The aforementioned notation makes it possible to describe some sequences by just giving an **explicit formula** for the generic Nth term of the sequence. That formula then is used with $N = 1$ for the first term, $N = 2$ for the second term, and so on.

EXAMPLE 9.2 SEQUENCES DESCRIBED BY A NICE FORMULA

Consider the sequence defined by the formula $A_N = 2^N + 1$. The first four terms of this sequence are

$$A_1 = 2^1 + 1 = 3,\ A_2 = 2^2 + 1 = 5,\ A_3 = 2^3 + 1 = 9, \text{ and } A_4 = 2^4 + 1 = 17.$$

If we are interested in the 10th term, we just plug $N = 10$ into the formula and get $A_{10} = 2^{10} + 1 = 1025$. If we are interested in the 100th term of the sequence we apply the formula once again and get $A_{100} = 2^{100} + 1$. Oops! What now? This is a huge number, and it may or may not be worth spelling it out in full. It is often the case that leaving the answer as $A_{100} = 2^{100} + 1$ makes more sense than writing down a 31-digit number, but if you really must have it, then here it is:

$$A_{100} = 1{,}267{,}650{,}600{,}228{,}229{,}401{,}496{,}703{,}205{,}377.$$

Example 9.2 worked out well because the formula defining the sequence was fairly simple. (Yes, the terms get very large, but that is a separate issue.) Unfortunately, as our next example shows, sometimes we will have to deal with sequences defined by some pretty complicated explicit formulas.

EXAMPLE 9.3 SEQUENCES DESCRIBED BY A NOT SO NICE FORMULA

A very important and much-studied sequence is called the *Fibonacci sequence*, named after the Italian mathematician Leonardo Pisano, better known as Fibonacci (circa 1175–1250).

Leonardo Pisano, aka Fibonacci (circa 1175–1250).

There are several alternative formulas that can be used to define the Fibonacci sequence, and here is one of them (using F to represent the terms of the sequence):

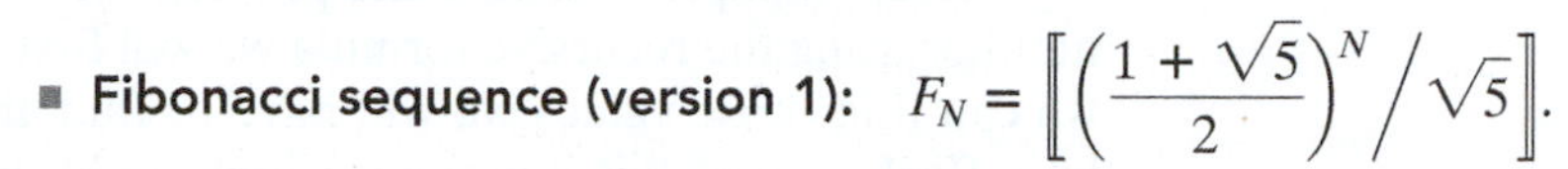

- **Fibonacci sequence (version 1):** $F_N = \left[\!\left[\left(\frac{1+\sqrt{5}}{2}\right)^N \Big/ \sqrt{5} \right]\!\right].$

($[\![\ ]\!]$ means that the number inside the brackets is rounded to the nearest integer).

To calculate the terms of this sequence using such a nasty formula is going to require at the very least a good calculator. You might want to confirm this with your own calculator or computer, but the first seven terms generated by the formula are as follows: $F_1 = [\![0.72\ldots]\!] = 1$, $F_2 = [\![1.17\ldots]\!] = 1$, $F_3 = [\![1.89\ldots]\!] = 2$, $F_4 = [\![3.06\ldots]\!] = 3$, $F_5 = [\![4.95\ldots]\!] = 5$, $F_6 = [\![8.02\ldots]\!] = 8$, and $F_7 = [\![12.98\ldots]\!] = 13$.

With this formula we don't have to compute the terms in sequential order—if we want to (and with the right equipment), we can dive directly into deeper waters: $F_{12} = [\![144.0013\ldots]\!] = 144$; $F_{25} = [\![75024.999\ldots]\!] = 75025$, and so on.

An alternative way to describe a sequence is to use a *recursive* formula. In contrast to an explicit formula, a **recursive formula** defines a term of a sequence using previous terms of the sequence. A recursive formula is not always possible, but when it is, it can provide a much nicer definition for a sequence than an explicit formula. A case in point is the Fibonacci sequence.

EXAMPLE 9.4 RECURSIVE FORMULA FOR THE FIBONACCI SEQUENCE

The recursive formula for the Fibonacci sequence is surprisingly simple, especially when compared to the explicit formula given in Example 9.3. It is amazing that the two definitions describe the same sequence.

- **Fibonacci sequence (version 2):** $F_N = F_{N-1} + F_{N-2}$, and $F_1 = 1$, $F_2 = 1$.

The key part of the above formula is the recursive rule $F_N = F_{N-1} + F_{N-2}$. Stated in plain English, the rule says that *each term of the sequence is obtained by adding its two preceding terms*. Since this rule cannot be applied to the first or second term (neither has two preceding terms to work with), the first two terms must be given separately.

Applying the recursive formula requires us to find the terms of the sequence in order. The first two terms, $F_1 = 1$, $F_2 = 1$, are given in the definition. The next term comes from the recursive formula: $F_3 = F_2 + F_1 = 1 + 1 = 2$. Another turn of the crank gives $F_4 = F_3 + F_2 = 2 + 1 = 3$. Do it again, and we get $F_5 = F_4 + F_3 = 3 + 2 = 5$. We can continue this way generating, in order, the terms of the sequence for as long as we want, and the beauty of it is that all we have to do to find each term is add the previous two numbers.

Here is the Fibonacci sequence written in infinite list form: 1, 1, 2, 3, 5, 8, 13, 21, 34, 55, 89, 144, 233, ...

Examples 9.3 and 9.4 illustrate the fact that the same sequence can be defined in two completely different ways. In the case of the Fibonacci sequence, we saw first a definition based on a nasty looking *explicit* formula (version 1) and then a different definition based on a beautifully simple *recursive* formula (version 2). It is tempting to declare the latter version clearly superior to the first, but before we do so, there is one important detail to consider—with the recursive formula we cannot find a term of the sequence unless we know the terms that come before it. Think of computing a value of F_N as analogous to getting to the peak of a mountain. The recursive formula is the mathematical version of climbing up to the peak one step at a time; the explicit formula is the mathematical version of being dropped at the peak by a helicopter.

As an example, consider the problem of finding the value of F_{100}. To find this number using the recursive formula we will first have to find F_{99} and F_{98}, and before we can find those values we will have to find and F_{97} and F_{96}, and so on down the line. Each step of this journey is simple (add two numbers), but there are a lot of steps. In contrast, if we use the explicit formula we have to carry out one calculation—it is not an easy one, but it can be done with the right tools (a good scientific calculator or a computer program). Which is better? You decide.

Population Sequences

For the rest of this chapter we will focus on special types of sequences called *population sequences*. For starters, let's clarify the meaning of the word **population.** In its original meaning, the word refers to human populations (the Latin root of the word is *populus*, which means "people") but over time the scope of the word has been expanded to apply to many other "things"—animals, bacteria, viruses, Web sites, plastic bags, money, etc. The main characteristic shared by all these "things" is that their quantities change over time, and to track the ebb and flow of these changes we use a *population sequence*.

A **population sequence** describes the size of a population as it changes over time, measured in discrete time intervals. A population sequence starts with an *initial* population (you have to start somewhere), and it is customary to think of the start as time zero. The size of the population at time zero is the first term of the population sequence. After some time goes by (it may be years, hours, seconds, or even nanoseconds), there is a "change" in the population—up, down, or it may even stay unchanged. We call this change a *transition*, and the population after the first transition is the *first generation*. The size of the *first* generation is the *second* term of the population sequence. After another transition the population is in its second generation, and the size of the *second* generation is given by the *third* term of the population sequence. The population sequence continues this way, each term describing the size of the population at a particular generation.

Notice that in the above description there is a slight mismatch between the generations and the terms of the population sequence: The first generation is represented by the second term of the sequence, the second generation by the third term, and so on. This is a little annoyance that can be fixed by starting the subscripts of the population sequence at $N = 0$. This means that for population sequences we will adopt a slightly different notation than the one we use with ordinary sequences.

- **Population sequence notation.** A generic population sequence is described by

$$P_0, P_1, P_2, P_3, P_4, \ldots$$

where P_0 is the size of the initial population, P_1 is the size of the population in the first generation, P_2 is the size of the population in the second generation, and so on (Fig. 9-1).

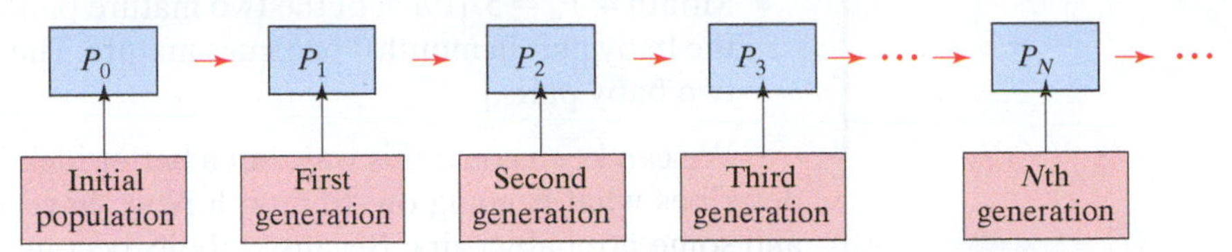

FIGURE 9-1 A generic population sequence. P_N is the size of the population in the Nth generation.

A convenient way to visualize a population sequence is with a **time-series graph**. In a typical two-dimensional, time-series graph, the horizontal axis is used to represent time and the vertical axis is used to represent the size of the population. The terms of the sequence are represented by either isolated points, as in Fig. 9-2(a), or by points connected with lines, as in Fig. 9-2(b). The former is called a **scatter plot**, the latter a **line graph**.

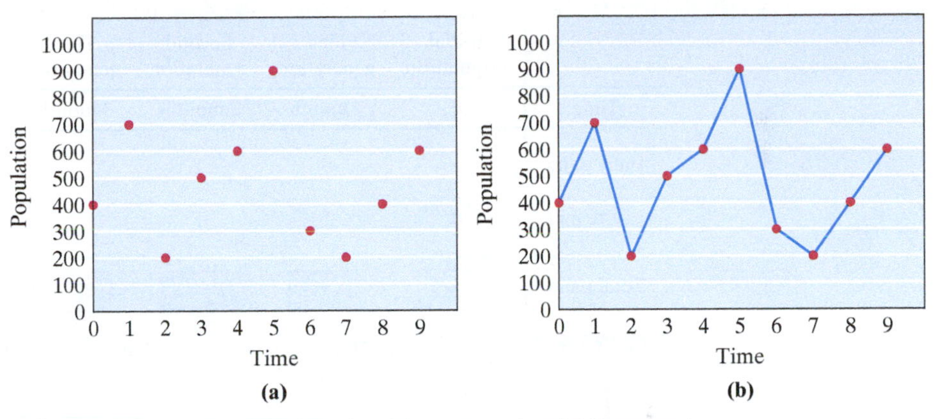

FIGURE 9-2 Time-series graphs: (a) scatter plot; (b) line graph.

EXAMPLE 9.5 FIBONACCI'S RABBITS

More than eight centuries ago, in his seminal book *Liber Abaci* published in 1202, Fibonacci proposed the following population growth problem:

> *A man puts one pair of rabbits in a certain place entirely surrounded by a wall. How many pairs of rabbits can be produced from that pair in a year if the nature of these rabbits is such that every month each pair bears a new pair which from the second month on becomes productive?*

Fibonacci's description of the problem can be restated into the following three facts: (1) each month, each pair (male/female) of *mature* (he called them productive) rabbits produces a pair of baby rabbits, (2) it takes one month for a pair of baby rabbits to become mature, and (3) the initial population consists of one pair (male/female) of baby rabbits.

We can use a population sequence to track the evolution of the rabbit population over time. (Since everything is stated in terms of male/female pairs, we will let the terms of the sequence represent pairs of rabbits.)

- Start: $P_0 = 1$. [The initial population has one pair (male/female) of baby rabbits.]
- Month 1: $P_1 = 1$. [The starting pair becomes mature.]
- Month 2: $P_2 = 2$. [The mature pair produces a baby pair. There are now two pairs—the original pair and their baby pair.]

- Month 3: $P_3 = 3$. [The mature pair in month 2 produces another baby pair; the baby pair in month 2 becomes mature. There are now two mature pairs and one baby pair.]
- Month 4: $P_4 = 5$. [Each of the two mature pairs in month 3 produces a baby pair; the baby pair in month 3 becomes mature. There are now three mature pairs and two baby pairs.]

We can keep going this way, but a better idea is to try to find a general rule that describes what is going on. In month N we have P_N pairs—some are mature pairs and some are baby pairs. Because rabbits become mature the month after they are born, the number of mature pairs in month N equals all pairs from month $N-1$. This number is P_{N-1}. Because every mature pair produces a baby pair the following month, the number of baby pairs in month N equals the *number of mature* pairs in month $N-1$, which in turn equals all pairs from month $N-2$. This number is P_{N-2}.

Figure 9-3 illustrates the generational changes for the first six generations. The blue arrows in the figure represent the fact that the number of baby pairs each month equals the number of mature pairs the previous month; the red arrows represent the fact that every pair of rabbits (baby or mature) is a mature pair one month later.

	Initial population	First generation	Second generation	Third generation	Fourth generation	Fifth generation	Sixth generation
Time	0	1 month	2 months	3 months	4 months	5 months	6 months
Baby pairs	1	0	1	1	2	3	5
Mature pairs	0	1	1	2	3	5	8
Total pairs	$P_0 = 1$	$P_1 = 1$	$P_2 = 2$	$P_3 = 3$	$P_4 = 5$	$P_5 = 8$	$P_6 = 13$

FIGURE 9-3 Fibonacci's rabbits: P_N is the number of male/female pairs in the Nth generation.

We now have a nice recursive formula for the rabbit population in month N: $P_N = P_{N-1} + P_{N-2}$, and starting terms $P_0 = 1$, $P_1 = 1$. We can use this formula to track the growth of the rabbit population over a 12-month period and answer Fibonacci's original question ("how many pairs of rabbits . . . in a year?"):

$$P_0 = 1, P_1 = 1, P_2 = 2, P_3 = 3, P_4 = 5, P_5 = 8, P_6 = 13,$$

$$P_7 = 21, P_8 = 34, P_9 = 55, P_{10} = 89, P_{11} = 144, P_{12} = 233.$$

There is an obvious connection between the population sequence describing the growth of the rabbit population in Example 9.5 and the Fibonacci sequence discussed in Example 9.4. Other than a change in notation (the F's changed to P's and the subscripts started at 0 instead of 1), they are the same sequence.

Fibonacci's description of how his rabbit population would grow was an oversimplified *mathematical model* of how real-life rabbit populations live and breed. In real life, things are quite a bit more complicated: real rabbits are not monogamous, produce litters of varying sizes, die, and so on. We can't capture all the variables that affect a rabbit population in a simple equation, and Fibonacci never intended to do that.

In practical terms, then, is there any value to simplified mathematical models that attempt to describe the complex behavior of populations? The answer is Yes. We can make very good predictions about the ebb and flow of a population over time even when we don't have a completely realistic set of rules describing the population's behavior. Mathematical models that describe population growth are based on a "big-picture" principle: Capture the variables that are really influential in determining how the population changes over time, put them into an equation (or several equations in the case of more complicated models) describing how these variables interact, and forget about the small details. In the next three sections we will put this idea into practice and explore three classic models of population "growth." [A cautionary note on the terminology: In discussions of populations, the word *growth* takes on a very general meaning—growth can be *positive* (the numbers go up), *negative* (the numbers go down), or *zero* (the numbers stay the same).]

9.2 The Linear Growth Model

A population grows according to a **linear growth** model if in each generation the population changes by a fixed amount d. When a population grows according to a linear growth model, that population grows *linearly*, and the population sequence is called an arithmetic sequence. Linear growth and arithmetic sequences go hand in hand, but they are not synonymous. *Linear growth* is a term we use to describe a special type of population growth, while an *arithmetic sequence* is an abstract concept that describes a special type of number sequence.

Linear growth models occur mostly when studying *inanimate* populations—that is, populations of things that are not alive and, therefore, do not reproduce. Our first two examples are made-up, but describe typical real-life situations involving linear growth.

EXAMPLE 9.6 HOW MUCH GARBAGE CAN WE TAKE?

The city of Cleansburg is considering a new law that would restrict the monthly amount of garbage allowed to be dumped in the local landfill to a maximum of 120 tons a month. There is a concern among local officials that unless this restriction on dumping is imposed, the landfill will reach its maximum capacity of 20,000 tons in a few years. Currently, there are 8000 tons of garbage in the landfill. Suppose the law is passed right now, and the landfill collects the maximum allowed (120 tons) of garbage each month from here on. (a) How much garbage will there be in the landfill five years from now? (b) How long would it take the landfill to reach its maximum capacity of 20,000 tons?

We can answer these questions by modeling the amount of garbage in the landfill as a population that grows according to a linear growth model. A very simple way to think of the growth of the garbage population is the following: Start with an initial population of $P_0 = 8000$ tons and *each month add $d = 120$ tons to whatever the garbage population was in the previous month*. This formulation gives the *recursive* formula $P_N = P_{N-1} + 120$, with $P_0 = 8000$ to get things started. Figure 9-4 illustrates the first few terms of the population sequence based on the recursive formula. For the purposes of answering the questions posed at the start of this example, the recursive formula is not particularly convenient. Five years, for example, equals 60 months, and we would prefer to find the value of P_{60} without having to compute the first 59 terms in the sequence.

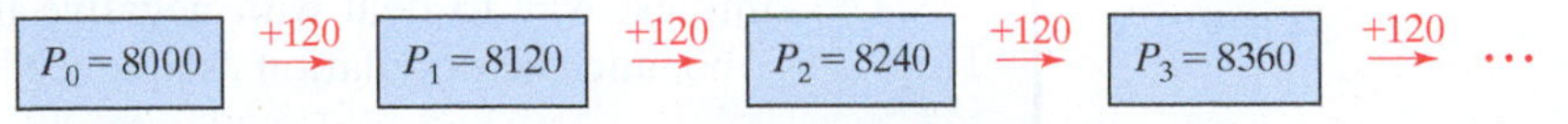

FIGURE 9-4 Population sequence for Example 9.6 ($P_N = P_{N-1} + 120$, $P_0 = 8000$).

We can get a nice explicit formula for the growth of the garbage population using a slightly different interpretation: *In any given month N, the amount of garbage in the landfill equals the original 8000 tons plus 120 tons for each month that has passed.* This formulation gives the *explicit* formula $P_N = 8000 + 120 \times N$. Figure 9-5 illustrates the growth of the population viewed in terms of the explicit formula.

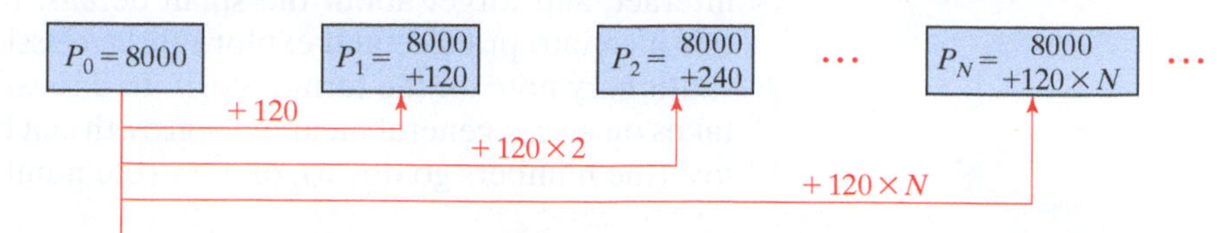

FIGURE 9-5 Population sequence for Example 9.6 ($P_N = 8000 + 120 \times N$).

The explicit formula $P_N = 8000 + 120 \times N$ will allow us to quickly answer the two questions raised at the start of the example. (a) After five years (60 months), the garbage population in the landfill is given by $P_{60} = 8000 + 120 \times 60 = 8000 + 7200 = 15{,}200$. (b) If X represents the month the landfill reaches its maximum capacity of 20,000, then $20{,}000 = 8000 + 120X$. Solving for X gives $X = 100$ months. The landfill will be maxed out 8 years and 4 months from now.

The line graph in Fig. 9-6 shows the projections for the garbage population in the landfill until the landfill reaches its maximum capacity. Not surprisingly, the line graph forms a straight line. This is always true in a linear growth model (and the reason for the name *linear*)—the growth of the population follows a straight line.

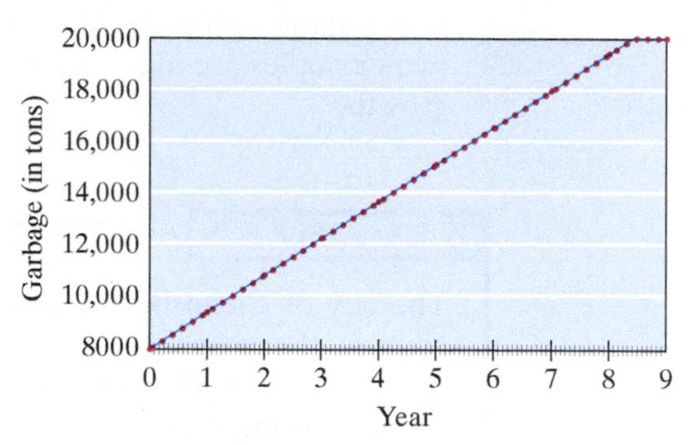

FIGURE 9-6 Line graph for Example 9.6.

EXAMPLE 9.7 NEGATIVE LINEAR GROWTH AND DONUTS

Crunchy Donuts is a chain of donut stores. At the height of their popularity, Crunchy Donuts were the rage—people stood in line for hours (OK, maybe minutes) for a chance to bite into a warm Crunchy. But times have changed. Bad management and changes in people's eating habits have forced Crunchy Donuts Corporation to start closing stores. On January 1, 2016 Crunchy Donuts had 1290 stores open for business, but since then it has been closing stores at the rate of 60 stores a month. If they continue closing stores at this rate, how long will they stay in business? When will that very last Crunchy be consumed?

We will model the number of Crunchy Donut stores still open as a population with the generations changing once a month—not exactly how things happen but a good approximation of reality. What makes this example different from Example 9.6 is that here we have *negative* linear growth: The population is *declining* at a constant rate of 60 stores a month.

The simplest way to deal with negative linear growth is to let d be negative. Here $d = -60$, and the population "sequence" written in list form is

$$1290, 1230, 1170, 1110, \ldots, ???$$

Here is the reason for the strange "???" at the end and for the word *sequence* being inside quotation marks: In this situation the list is not infinite—at some point the numbers will become negative, but a negative number of donut stores doesn't make sense. In fact, what we really have here is a finite list that starts as an arithmetic sequence but stops right before the terms become negative numbers.

We can find the number of generations it takes before the terms become negative by setting the explicit formula for the Nth term of the sequence equal to 0 and then solving the equation for N. When N is not an integer (as is often the case) we round it *down* to the nearest integer. Here are the steps applied to our model:

- Explicit formula for the Nth term set equal to 0: $P_N = 1290 - 60N = 0$.
- Solve above equation for N: $1290 = 60N \Rightarrow N = \frac{1290}{60} = 21.5$.
- The last positive term of the population sequence is $P_{21} = 1290 - 60 \times 21 = 30$.

$P_{21} = 30$ means that on October 1, 2017 (21 months after January 1, 2016) the model predicts that Crunchy Donuts will be down to 30 stores and sometime around the middle of October someone will be eating the last Crunchy ever. This is only a model, but it's still sad to see them go.

We will now generalize the ideas introduced in Examples 9.6 and 9.7 and introduce some useful terminology.

- A population grows *linearly* (i.e., according to a linear growth model) if in each generation the population changes by a constant amount d. The constant d is called the **common difference** and the population sequence is called an **arithmetic sequence**.
- When d is positive we have *positive growth* (i.e., the population is increasing); when d is negative we have *negative growth* (i.e., the population is decreasing); when $d = 0$ we have *zero growth* (i.e., the population stays constant).
- A population that grows linearly, with initial population P_0 and common difference d, is described by the population sequence

$$P_0, P_0 + d, P_0 + 2d, P_0 + 3d, P_0 + 4d, \ldots$$

Since a population sequence cannot have negative terms, in the case of negative growth the "sequence" ends at the last positive term.
- An *explicit* formula for the Nth term of the population sequence is $P_N = P_0 + Nd$; a *recursive* formula for the Nth term of the population sequence $(N \neq 0)$ is $P_N = P_{N-1} + d$.
- The line graph describing a population that grows linearly is always a straight line. When $d > 0$ the line graph has positive slope [Fig. 9-7(a)]; when $d < 0$ the line graph has negative slope [Fig. 9-7(b)]; when $d = 0$ the line graph is horizontal [Fig. 9-7(c)].

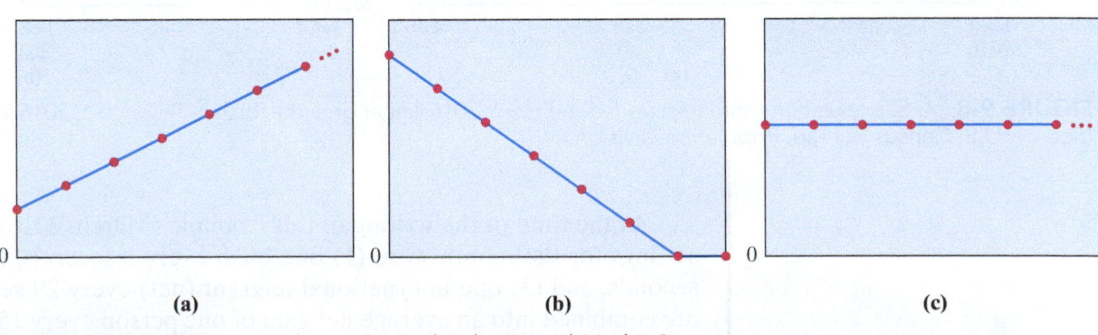

FIGURE 9-7 Linear growth. (a) $d > 0$, (b) $d < 0$, and (c) $d = 0$.

Our next example illustrates an important model of population growth. In this model the population grows on a straight line for a while, then the model is recalibrated and the population grows on a different straight line, and so on. This type of growth is called *piecewise linear*.

EXAMPLE 9.8 A SHORT-TERM MODEL OF THE U.S. POPULATION

The U.S. Census Bureau runs a virtual population clock that shows the official United States population at any moment in time. If you go right now to *www.census.gov/popclock* you can find out what the official United States population is at this moment. If you go back a minute later the numbers will have changed. Why and how did they change?

The ticking of the population clock is based on a mathematical model for the growth of the U.S. population developed by the Census Bureau. The Census Bureau model is based on a combination of just three variables (the Census Bureau refers to these as the *component settings* for the model): (1) the average frequency of births in the United States, (2) the average frequency of deaths in the United States, and (3) the average frequency of international *net migration* into the United States. These component settings are recalibrated each month to account for seasonal changes, but throughout each monthly period they remain constant. This means that the short-term (monthly) model of the U.S. population growth is a linear growth model, and a line graph showing the population growth throughout one month is a line segment. Figure 9-8(a) shows the line graph for the U.S. population from January 1 to February 1, 2016. It is a straight line. If we look at two consecutive months, the line graph need not be a straight line. It might instead be two separate line segments (with slightly different slopes). Figure 9-8(b) shows the line graph for the U.S. population from January 1 to March 1, 2016. It looks almost like a straight line, but it isn't—between February 1 and March 1, the line graph gets just a tad steeper.

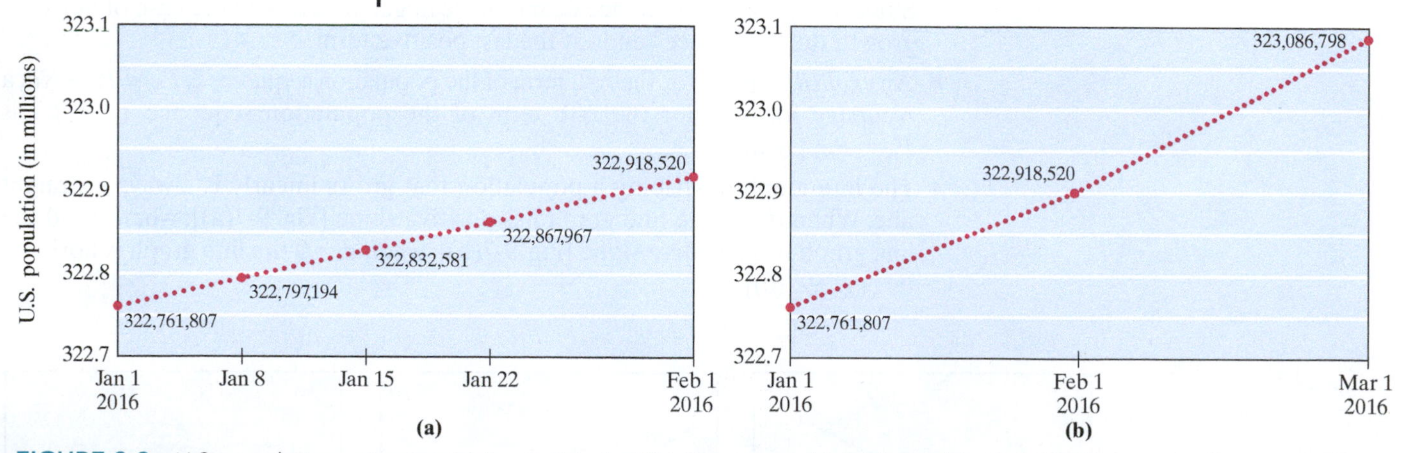

FIGURE 9-8 U.S. population estimates: (a) Jan. 1–Feb. 1, 2016 (linear growth); (b) Jan. 1–March 1, 2016 (piecewise linear growth). (*Source:* U.S. Census Bureau, Population Division.)

At the time of the writing of this example (March, 2016), the three component settings for the month were (1) one birth every 8 seconds, (2) one death every 11 seconds, and (3) one international migrant (net) every 29 seconds. These numbers are combined into an average net gain of one person every 15 seconds. [Here is how you get the 15 seconds (if you have the stomach to deal with parts of a person):

(1) one birth every 8 seconds is equivalent to 1/8 (one eighth) of a person born every second, (2) one death every 11 seconds is equivalent to 1/11 (one eleventh) of a person dying every second, and (3) one net migrant every 29 seconds is equivalent to 1/29 (one twenty-ninth) of a person migrating into the United States every second. The net result is that the U.S. population grows by $\frac{1}{8} - \frac{1}{11} + \frac{1}{29} = \frac{175}{2552} \approx 0.0686$ parts of a person per second. This is equivalent to one person every $\frac{1}{0.0686} \approx 15$ seconds.] This number stays constant throughout the month: One person gained every 15 seconds means 240 persons added every hour, 5760 persons added each day, and a grand total gain of 178,560 persons for the month of March, 2016.

EXAMPLE 9.9 REDUCING SINGLE-USE PLASTIC BAG CONSUMPTION

According to the Environmental Protection Agency, about 380 billion plastic bags are consumed each year in the United States. Of these, only about 1% are recycled—the remaining 99% end up in landfills or polluting waterways, rivers, and oceans. (*Source:* U.S. Environmental Protection Agency, *www.epa.gov.*)

Of the 380 billion plastic bags consumed, an estimated 100 billion are *single-use* plastic bags of the type used at the grocery store for bagging groceries. The typical fate of a single-use plastic bag is to be neither reused nor recycled. As consumers, we can do something about this—take reusable cloth bags with us when we go grocery shopping. But how much of a dent can this minor change in grocery shopping habits make on the overall problem?

In the United States, on average, 450 single-use plastic grocery bags are consumed each year by each shopping adult in the population. Giving it a positive spin, for each shopping adult who stops using single-use plastic bags at the grocery store, 450 single-use plastic bags can be subtracted from the 100 billion consumed each year. That's not much. But suppose that instead of thinking in terms of individuals we thought in terms of population percentages. There are roughly 210 million shopping-age adults in the U.S. population. For each 1% of this population that stops using single-use plastic bags, 0.945 billion bags (i.e., 945 million) can be subtracted from the 100 billion consumed each year.

This gives us a realistic linear growth model for annual consumption of single-use plastic bags: $P_N = 100 - (0.945)N$, in which P_N represents the annual number of single-use plastic bags consumed (in billions) and N represents the percent of the adult U.S. population that switches to the use of reusable cloth bags.

Figure 9.9 shows a line graph for this model. It shows that it is indeed possible to make a dent in the single-use plastic bag problem.

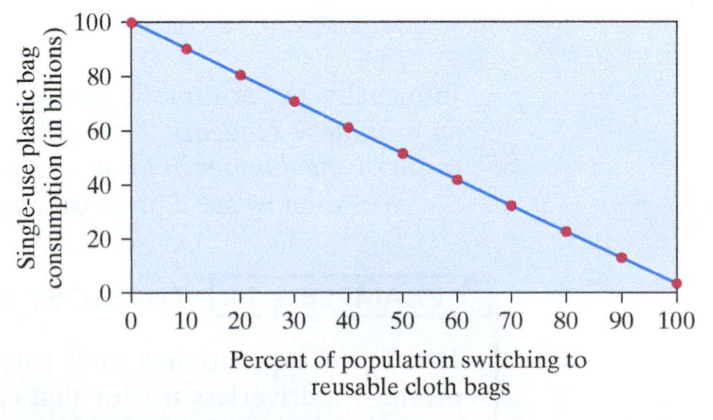

FIGURE 9-9 Line graph for Example 9.9.

Arithmetic Sums and the Arithmetic Sum Formula

Suppose you are given an arithmetic sequence—say, 5, 8, 11, 14, 17, . . . —and you are asked to add the first N terms of that sequence. How would you do it? When N is small, you would probably just add them the old-fashioned way. (Take, for example, the sum of the first five terms: $5 + 8 + 11 + 14 + 17$. Nothing to it—you just do it and come up with 55.) But what if you were asked to add lots of terms—say the first 500 terms—of the sequence? Adding 500 numbers, even with a calculator, does not seem like a very enticing idea. Fortunately, there is a nice trick that allows us to easily add any number of consecutive terms in any arithmetic sequence. Before giving the general formula, let's see the trick in action.

EXAMPLE 9.10 ADDING THE FIRST 500 TERMS OF 5, 8, 11, 14, . . .

Since a population sequence starts with P_0, the first 500 terms are $P_0, P_1, P_2, \ldots, P_{499}$. In the particular case of the sequence 5, 8, 11, 14, 17, . . . we have $P_0 = 5$, $d = 3$, and $P_{499} = 5 + 3 \times 499 = 1502$. The sum we want to find is

$$S = 5 + 8 + 11 + \cdots + 1496 + 1499 + 1502.$$

Now here comes the trick: (1) write the sum in the normal way, (2) below the first sum, write the sum again but do it backwards (making sure you line up the plus signs), and (3) add the columns, term by term. In our case, we get

$$\begin{aligned}
&(1) \quad S = 5 + 8 + 11 + \cdots + 1496 + 1499 + 1502.\\
&(2) \quad S = 1502 + 1499 + 1496 + \cdots + 11 + 8 + 5.\\
&(3) \quad 2S = 1507 + 1507 + 1507 + \cdots + 1507 + 1507 + 1507.
\end{aligned}$$

The key is that what happened in (3) is no coincidence. In each column we get the same number: $1507 = 5 + 1502 =$ (starting term) + (ending term). Rewriting (3) as $2S = 500 \times 1507$ and solving for S gives the sum we want: $S = \frac{1507 \times 500}{2} = 376{,}750$.

We will now generalize the trick we used in the preceding example. In the solution $S = \frac{1507 \times 500}{2}$ the 1507 represents the sum of the starting and ending terms, the 500 represents the number of terms being added, and the 2 is just a 2. The generalization of this observation gives a very useful formula we will call the **arithmetic sum formula**.

ARITHMETIC SUM FORMULA

If $P_0, P_1, P_2, \ldots$ are the terms of an arithmetic sequence, then

$$P_0 + P_1 + P_2 + \cdots + P_{N-1} = \frac{(P_0 + P_{N-1})N}{2}.$$

Informally, the arithmetic sum formula says *to find the sum of consecutive terms of an arithmetic sequence, first add the first and the last terms of the sum, multiply the result by the number of terms being added, and divide by two.*

We will now see a practical application of the arithmetic sum formula.

EXAMPLE 9.11 THE COST OF BUILDING UP INVENTORY

Jane Doe Corporation, a small tractor manufacturer, has developed a radically new product: a driverless tractor that can be controlled remotely with a joystick by an operator sitting in front of a console. The company can manufacture a maximum of

30 tractors per week and wants to build up its inventory to meet the projected demand when the new tractor is unveiled. The plan is to go into an accelerated production schedule ahead of the product launch date. The company will produce 30 tractors each week for a period of 72 weeks and place the tractors in storage at a cost of $10 per tractor per week. How much should the company budget for storage costs over the 72-week production period?

Table 9-1 shows the weekly storage cost starting at the end of week 1 (the start of storage) and ending at the end of week 72. The storage costs represent the first 72 terms of an arithmetic sequence that starts with $300 and ends with $21,600. The arithmetic sum formula gives the total storage costs over the 72-week production period: $\frac{(\$300 + \$21{,}600) \times 72}{2} = \$788{,}400$.

Week	0	1	2	3	...	70	71	72
Production	0	30	30	30	...	30	30	30
Storage cost	0	$300	$600	$900	...	$21,000	$21,300	$21,600

TABLE 9-1 ■ Weekly storage costs for new tractors

Sometimes the use of the arithmetic sum formula requires us to do a little detective work, as in the case where we are given just the first term and last term of the sum and the common difference d of the arithmetic sequence.

EXAMPLE 9.12 $(4 + 13 + 22 + \cdots + 922) = ?$

Suppose you are asked to add the terms of an arithmetic sequence with common difference $d = 9$. The first term in the sum is 4; the last term is 922. To use the arithmetic sum formula we need to know how many terms we are adding. Call that number N. Then $922 = P_{N-1}$ (remember, we start at P_0). On the other hand, the explicit formula for linear growth gives $P_{N-1} = 4 + 9(N - 1)$. Combining the two gives $922 = 4 + 9(N - 1) = 9N - 5 \Rightarrow 9N = 927 \Rightarrow N = 103$. We now have what we need to use the arithmetic sum formula:

$$4 + 13 + 22 + \cdots + 922 = \frac{(4 + 922) \times 103}{2} = 47{,}689.$$

An interesting application of the arithmetic sum formula is in finding the sum of consecutive *odd* (or *even*) numbers. We'll show how it works for odd numbers in the next example and leave the application to even numbers as exercises (see Exercises 34 and 66).

EXAMPLE 9.13 $(1 + 3 + 5 + 7 + \cdots + 999) = ?$

First notice that the odd numbers form an arithmetic sequence with common difference $d = 2$. The only tricky part is finding the number of terms in the sum $1 + 3 + 5 + 7 + \cdots + 999$. Writing

$$999 = 1 + 2(N - 1) = 1 + 2N - 2 = 2N - 1,$$

and solving for N, gives $N = 500$. Now we are ready to use the arithmetic sum formula:

$$1 + 3 + 5 + \cdots + 999 = [(1 + 999) \times 500]/2 = 250{,}000.$$

For the sweet general formula for the sum of the first N odd numbers, check out Exercise 67.

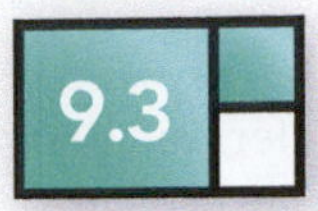

9.3 The Exponential Growth Model

Imagine you are presented with the following choice: (1) You can have \$100,000, to be paid to you one month from today, or (2) you can have 1 penny today, 2 pennies tomorrow, 4 pennies the day after tomorrow, and so on—each day doubling your payoff from the day before—for a full month (say 31 days). Which of the two options would you choose? We'll come back to this question later in the section, but you should make your choice now, and no cheating—once you make your choice you have to stay with it.

Before we start a full discussion of exponential growth, we need to spend a little time explaining the mathematical meaning of the term *growth rate*. In this chapter we will focus on growth rates as they apply to population models, but the concept applies to many other situations besides populations. In Chapter 10, for example, we will discuss growth rates again, but in the context of money and finance.

When the size of a population "grows" from some value X to some new value Y, we want to describe the growth in relative terms, so that the growth in going from $X = 2$ to $Y = 4$ is the same as the growth in going from $X = 50$ to $Y = 100$.

- **Growth rate.** The *growth rate* r of a population as it changes from an initial value X (the *baseline*) to a new value Y (the *end-value*) is given by the ratio $r = \frac{Y - X}{X}$. (*Note*: It is customary to express growth rates in terms of percentages, so, as a final step, r is converted to a percent.)

One important thing to keep in mind about the definition of growth rate is that it is not symmetric—the growth rate when the baseline is X and the end-value is Y is very different from the growth rate when the baseline is Y and the end-value is X.

EXAMPLE 9.14 GROWTH RATES, END-VALUES, AND BASELINES

	Baseline X	End-value Y	Growth rate $r = \frac{(Y - X)}{X}$
(1)	2	4	$\frac{(4-2)}{2} = 1 = 100\%$
(2)	50	100	$\frac{(100-50)}{50} = 1 = 100\%$
(3)	100	50	$\frac{(50-100)}{100} = -\frac{1}{2} = -50\%$
(4)	10	12	$\frac{(12-10)}{10} = \frac{2}{10} = 20\%$
(5)	10	7.5	$\frac{(7.5-10)}{10} = -\frac{2.5}{10} = -25\%$
(6)	1321	1472	$\frac{(1472-1321)}{1321} \approx 11.43\%$

TABLE 9-2 ■ Computation of growth rates

Table 9-2 shows the growth rates for several different baseline/end-value combinations. The first column of Table 9-2 shows the baseline population X, the second column shows the value for the new population Y, and the third column shows the computation of the growth rate r. All the computations except for (6) can and should be done without the use of a calculator.

Sometimes we are given the baseline X and the growth rate r and need to compute the end-value Y. If we solve the equation $r = \frac{Y - X}{X}$ for Y, we get $Y = rX + X = (r + 1)X$. In this formula r must be in fractional or decimal form. Table 9-3 shows several examples of the computation of the end-value Y given the baseline X and the growth rate r.

The last variation of this theme is to compute the baseline X given the end-value Y and the growth rate r. The formula for X in terms of Y and r is $X = \frac{Y}{r+1}$. It is just a twisted version of the formula $Y = (r + 1)X$.

	Baseline X	Growth rate r	End-value Y = (r + 1)X
(1)	50	$100\% = 1$	$2 \times 50 = 100$
(2)	50	$-100\% = -1$	$0 \times 50 = 0$
(3)	50	$50\% = 0.5$	$1.5 \times 50 = 75$
(4)	100	$-20\% = -0.2$	$0.8 \times 100 = 80$
(5)	37,314	$5.4\% = 0.054$	$1.054 \times 37{,}314$
(6)	37,314	$-5.4\% = -0.054$	$0.946 \times 37{,}314$

TABLE 9-3 ■ Computation of end-values

We are finally ready to define *exponential growth.*

- A population grows **exponentially** if in each generation the population "grows" by the same constant *factor* R called the **common ratio**. In this case the population sequence takes the form

$$P_0,\ P_1 = RP_0,\ P_2 = R^2P_0,\ P_3 = R^3P_0,\ P_4 = R^4P_0,\ \ldots \qquad (1)$$

(Note the difference between linear and exponential growth: In linear growth we *add* a fixed constant, in exponential growth we *multiply* by a fixed constant.)

- An *explicit* formula for the Nth term of the population sequence given in equation (1) is $P_N = R^N P_0$; a *recursive* formula for the Nth term of the population sequence ($N \neq 0$) is $P_N = RP_{N-1}$.
- Any numerical sequence in which a term is obtained by multiplying the preceding term by the same constant R is called a **geometric sequence**. While geometric sequences in general can have negative terms ($1, -2, 4, -8, \ldots$ is a geometric sequence with $R = -2$), population sequences cannot have any negative terms. This implies that for population sequences we must include the assumption that $R \geq 0$.
- If a population sequence grows exponentially, the *growth rate* from one generation to the next is constant. If we call this constant growth rate r, we can describe the population sequence in a slightly different form:

$$P_0,\ P_1 = (1+r)P_0,\ P_2 = (1+r)^2P_0,\ P_3 = (1+r)^3P_0,\ P_4 = (1+r)^4P_0,\ \ldots \qquad (2)$$

Equations (1) and (2) are two different versions of the exponential growth model, connected to each other by the relation between the common ratio R and the growth rate r:

$$R = (1+r), \text{ or equivalently, } r = (R-1)$$

- The relation $R = (1+r)$ combined with the restriction $R \geq 0$ imposes a restriction on the values of the growth rate: $r \geq -1$ (i.e., a population growth rate cannot go below $-1 = -100\%$). When $R > 1$ the growth rate $r = (R-1)$ is positive and the line graph of the population sequence is increasing [Fig. 9-10(a)]; when $0 < R < 1$ the growth rate r is negative and the line graph of the population sequence is decreasing [Fig. 9-10(b)]; when $R = 1$ the growth rate r is zero and the line graph of the population sequence is horizontal [Fig. 9-10(c)]; and finally, $R = 0$ means $r = -1$ and a total collapse of the population—it becomes extinct after one generation [Fig. 9-10(d)]. (This last scenario is so unusual as to be practically impossible, but in the next section we will see a model that makes the possibility of extinction much more realistic.)

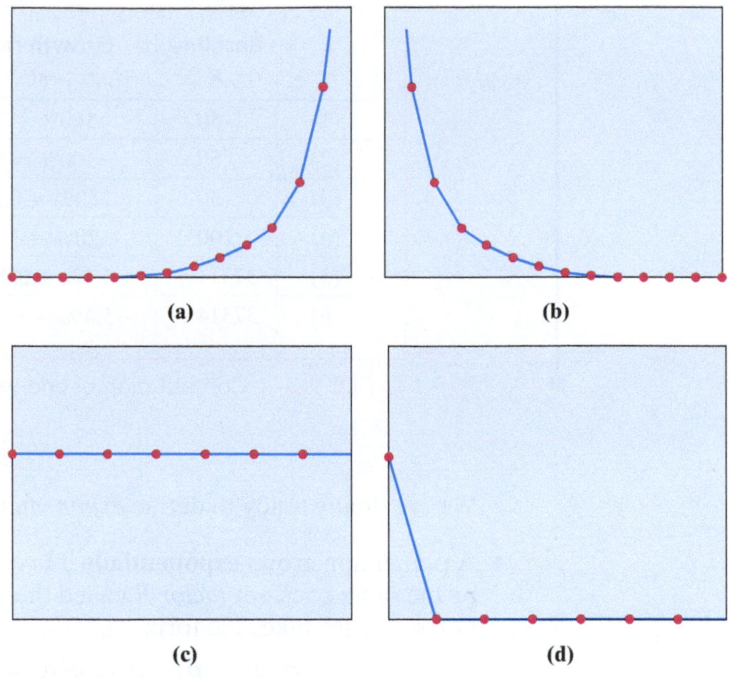

FIGURE 9-10 Exponential growth. (a) $R > 1 \, (r > 0)$, (b) $0 < R < 1 \, (-1 < r < 0)$, (c) $R = 1 \, (r = 0)$, (d) $R = 0 \, (r = -1)$.

Exponential growth models are useful to describe the growth of living organisms such as bacteria and viruses, the growth of human or animal populations under unrestricted breeding conditions, and even the growth of intellectual products such as e-mails, Web pages, and data sets. In all real-life applications, exponential growth is assumed to occur only for a limited amount of time—there is always a point at which exponential growth stops because the conditions that support the model can no longer be sustained. Our next example illustrates this point.

EXAMPLE 9.15 THE SPREAD OF AN EPIDEMIC

In their early stages, infectious diseases such as HIV or the swine flu spread following an exponential growth model—each infected individual infects roughly the same number of healthy individuals over a given period of time. Formally, this translates into the recursive formula $P_N = (1 + r)P_{N-1}$ where P_N denotes the number of infected individuals in the population at time N and r denotes the growth rate of the infection.

Every epidemic starts with an original group of infected individuals called "population zero." Let's consider an epidemic in which "population zero" consists of just one infected individual ($P_0 = 1$) and such that, on average, each infected individual transmits the disease to one healthy individual each month. This means that every month the number of infected individuals doubles and the growth rate is $r = 1 = 100\%$. Under this model the number of infected individuals N months after the start of the epidemic is given by $P_N = 2^N$.

Figure 9-11 shows the growth of the epidemic during its first year. By the end of the first year the number of infected individuals is $P_{12} = 2^{12} = 4096$. That's not too bad. But suppose that no vaccines are found to slow down the epidemic and the growth rate for infected individuals continues at 100% per month. At the end of the second year the number of infected individuals would equal $P_{24} = 2^{24} = 16{,}777{,}216$. Eight months after that the number of infected individuals would equal $P_{32} = 2^{32} \approx 4.3$ billion (more than half of the world's population); one month later every person on the planet would be infected.

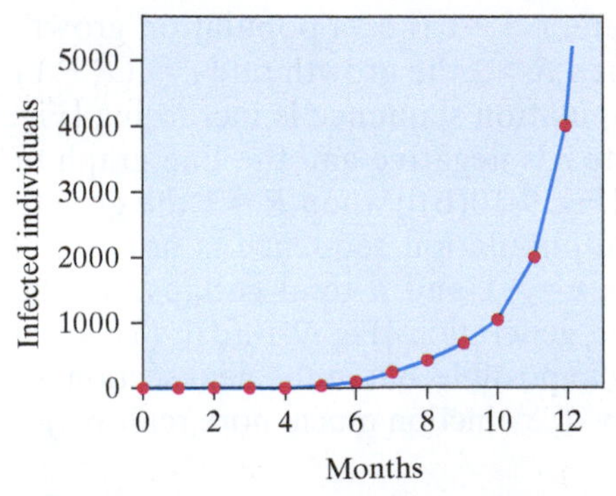

FIGURE 9-11 The first year of an epidemic.

Example 9.15 illustrates what happens when exponential growth continues unchecked, and why, when modeling epidemics, exponential growth is a realistic model for a while, but there must be a point in time where the rate of infection has to level off and the model must change. Otherwise, the human race would have been wiped out many times over.

Our next example shows the flip side of an epidemic—how vaccines and good public health policy help to eradicate a disease. Smallpox and polio are two examples of diseases that have been practically eradicated, but to keep things simple we will use a fictitious example.

EXAMPLE 9.16 ERADICATING THE GAMMA VIRUS INFECTION

Thanks to a new vaccine and good public health policy, the number of reported cases of the Gamma virus infection has been *dropping* by 70% a year since 2015, when there were 1 million reported cases of the infection. If the present rate continues, how many reported cases of Gamma virus infection can we predict by the year 2022? How long will it take to eradicate the virus?

In this example we are dealing with negative exponential growth. The growth rate is $r = -70\% = -0.7$, and the common ratio is $R = 1 - 0.7 = 0.3$. The initial population is $P_0 = 1{,}000{,}000$.

According to the model, the number of cases of the Gamma virus in 2022 should be $P_7 = 1{,}000{,}000 \times (0.3)^7 \approx 219$. By 2023 this number will drop to about 66 cases, by 2024 to about 20 cases, by 2025 to 6 cases, and by 2026 to 2 cases. Then, by 2027, we can expect that the Gamma virus will have been eradicated.

The negative growth of Gamma virus cases is illustrated in Figure 9-12.

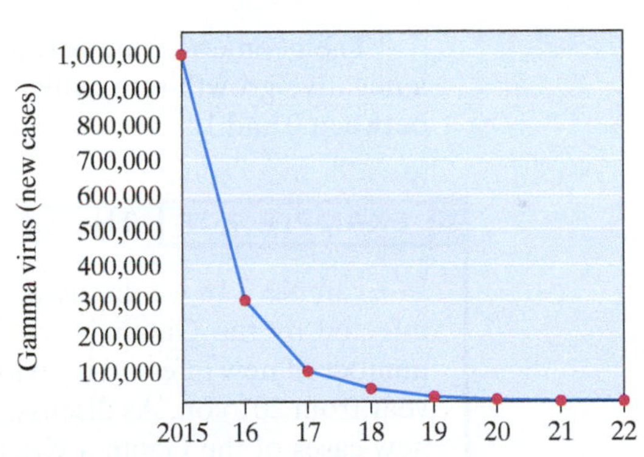

FIGURE 9-12 Negative exponential growth ($r = -0.7$, $R = 0.3$).

The Geometric Sum Formula

Suppose you want to find the sum of the first 100 (or 1000) terms of a geometric sequence. Just as with arithmetic sequences, there is a handy formula that will allow us to do it without having to actually add the terms. We will call this formula the **geometric sum formula**.

GEOMETRIC SUM FORMULA

$$P_0 + RP_0 + R^2P_0 + \cdots + R^{N-1}P_0 = \frac{(R^N - 1)P_0}{R - 1}$$

Before we move on, a couple of comments about the geometric sum formula are in order.

- The left-hand side of the formula represents the sum of the first N terms of a geometric sequence with initial term P_0 and common ratio R. (As tempting as it is to think that the number of terms is $N-1$, we have to remind ourselves that the count starts at 0 and not at 1.) It follows that on the right-hand side of the formula, the exponent of R is the number of terms being added.
- The geometric sum formula works for all values of R with one important exception: When $R=1$ the denominator on the right-hand side equals 0, and the formula does not work. When $R=1$, however, no fancy formula is needed: Each term of the sum equals P and the sum equals NP_0.

The next set of examples illustrate the application of the geometric sum formula. (For a guided derivation of the formula itself, see Exercise 70).

EXAMPLE 9.17 ADDING THE FIRST 20 TERMS OF 5, 15, 45, 135, . . .

The sequence 5, 15, 45, 135, . . . is a geometric sequence with $R=3$. Say we need to calculate the sum of the first 20 terms of this sequence, namely

$$5+15+45+135+\cdots+3^{19}\times 5.$$

Without the geometric sum formula this would be quite a chore; with the geometric formula it's a breeze (but you still need a calculator—we are dealing with surprisingly large numbers):

$$5+15+45+135+\cdots+3^{19}\times 5=\frac{(3^{20}-1)\times 5}{(3-1)}=8{,}716{,}961{,}000$$

The geometric sum formula can also be applied to population sequences with a negative growth rate, which, as we now know, means that the common ratio R is between 0 and 1.

EXAMPLE 9.18 THE COST OF ERADICATING THE GAMMA VIRUS

In Example 9.16 we discussed the negative growth of the population of individuals infected by the Gamma virus. Thanks to the development of a new vaccine, the number of new infected individuals is expected to drop by roughly 70% a year every year from 2015 on. As discussed in detail in Example 9.16, the annual population of new cases of the Gamma virus is described by a geometric sequence with common ratio $R=0.3$ and initial population $P_0=1{,}000{,}000$ (Table 9-4). According to the model, the Gamma virus will have been eradicated by the year 2027, when the number of infected individuals drops to essentially 0.

Year	2015	2016	2017	2018	. . .	2025	2026
Number of cases	1,000,000	300,000	90,000	27,000	. . .	6	2

TABLE 9-4

The most important aspect of the strategy for eradicating an infectious disease by way of a vaccine is an effective inoculation campaign. The ideal is to inoculate all infected individuals with the vaccine, and one of the key issues in doing so is to figure out how many doses of the vaccine need to be manufactured. In the case of the Gamma virus vaccine, that number can be found by adding all the entries in the second row of Table 9-4 (we are assuming that one shot is all it takes to kill the virus). This calculation can best be done using the geometric sum formula. There are twelve terms in the sum (one for each year from 2015 to 2026), the initial population is 1,000,000, and the common ratio is 0.3. The total is given by

$$\frac{1{,}000{,}000 \times (0.3^{12} - 1)}{(0.3 - 1)} \approx 1{,}428{,}571.$$

That total is an approximation based on the mathematical model for the eradication plan. These numbers are good but not perfect. Just to be safe, let's say that 1.5 million doses of the vaccine would do the job.

It is worth mentioning that in the foregoing calculation, both numerator and denominator are negative. If you prefer not having to deal with negative numbers, here is a little trick: Flip the $(0.3^{12} - 1)$ to $(1 - 0.3^{12})$ *and* flip the $(0.3 - 1)$ to $(1 - 0.3)$. This changes the sign of both numerator and denominator but leaves the answer unchanged. It's a convenient trick when using the geometric sum formula with $R < 1$.

EXAMPLE 9.19 A MONTH'S WORTH OF DOUBLING PENNIES

We started this section with a unique proposition: (1) Take a lump payment of \$100,000 one month from now, or (2) take 1 cent today, 2 cents tomorrow, 4 cents the next day, and so on for 31 days. Which one did you think was the better offer? If you chose (2) you were wise. After 31 days, you would have a grand total of

$$1 + 2 + 2^2 + 2^3 + \cdots + 2^{30} \text{ cents.}$$

This is a geometric sum with $R = 2$ and $P_0 = 1$. Applying the geometric sum formula gives $1 + 2 + 2^2 + 2^3 + \cdots + 2^{30} = \frac{2^{31} - 1}{2 - 1} = 2^{31} - 1 = 2{,}147{,}483{,}647$ cents. That's \$21,474,836 plus some spare change! (Yes, that's right—\$21 million plus!)

The geometric sum formula has many important and interesting applications to finance (amortizing loans, estimating the value of a retirement plan, calculating the value of an annuity, etc.), and we will discuss some of these applications in Chapter 10.

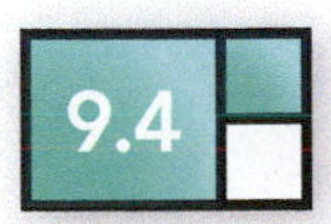

9.4 The Logistic Growth Model

One of the key tenets of population biology is the idea that there is an inverse relation between the growth rate of a population and its density. Small populations have plenty of room to spread out and grow, and thus their growth rates tend to be high. As the population density increases, however, there is less room to grow and there is more competition for resources—the growth rate tends to taper off. Sometimes the population density is so high that resources become scarce or depleted, leading to negative population growth or even to extinction.

The effects of population density on growth rates were studied in the 1950s by behavioral psychologist John B. Calhoun. Calhoun's now classic studies showed that when rats were placed in a closed environment, their behavior and growth rates were normal as long as the rats were not too crowded. When their environment became too crowded, the rats started to exhibit abnormal behaviors, such as infertility and cannibalism, which effectively put a brake on the rats' growth rate. In extreme cases, the entire rat population became extinct.

Calhoun's experiments with rats are but one classic illustration of the general principle that a *population's growth rate is negatively impacted by the population's density*. This principle is particularly important in cases in which the population is confined to a limited environment. Population biologists call such an environment the **habitat**. The habitat might be a cage (as in Calhoun's rat experiments), a lake (for a population of fish), a garden (for a population of snails), and, of course, Earth itself (everyone's habitat).

In 1838, the Belgian mathematician Pierre François Verhulst proposed a mathematical model of population growth for species living within a fixed habitat. Verhulst called his model the **logistic growth model**.

The logistic growth model is based on two principles:

1. Every biological population living in a confined habitat has a natural intergenerational growth rate that we call the **growth parameter** of that population. The growth parameter of a population depends on the kind of species that makes up the population and the nature of its habitat—a population of beetles in a garden has a different growth parameter than a population of gorillas in the rainforest, and a population of gorillas in the rainforest has a different growth parameter than a population of gorillas in a zoo. Given a specific species and a specific habitat for that species, we will assume the growth parameter is a constant we will denote by r.
2. The actual growth rate of a specific population living in a specific habitat depends not just on the growth parameter r (otherwise we would have an exponential growth model) but also on the amount of "elbow room" available for the population to grow (a variable that changes from generation to generation). When the population is small (relative to the size of the habitat) and there is plenty of elbow room for the population to grow, the growth rate is roughly equal to the growth parameter r and the population grows more or less exponentially [Fig. 9-13(a)]. As the population gets bigger and there is less space for the population to grow, the growth rate gets proportionally smaller [Fig. 9-13(b)]. Sometimes there is a switch to negative growth, and the population starts decreasing for a few generations to get back to a more sustainable level [Fig. 9-13(c)].

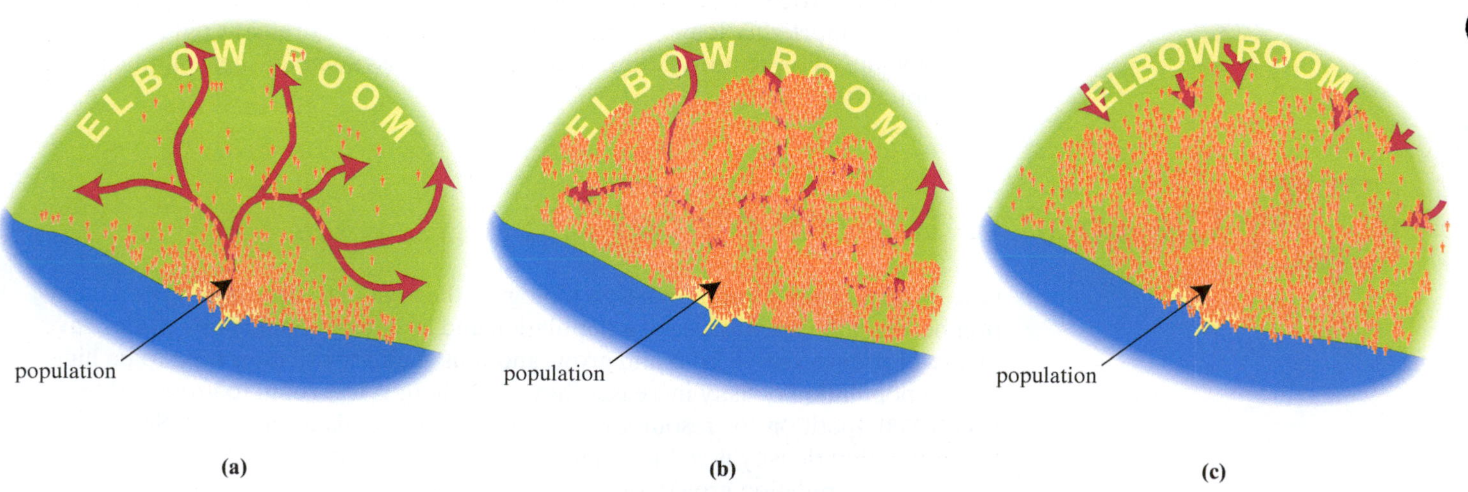

FIGURE 9-13 (a) Growth rate approximately r, (b) growth rate gets smaller, and (c) negative growth rate.

We will now discuss how the two loosely stated principles above can be formalized into a mathematical model. The first step is to quantify the concept of "elbow room" in a habitat. This can be done by introducing a related concept: the *carrying capacity*. For a given species and a given habitat, the **carrying capacity** C of a habitat is the *maximum* number of individuals that the habitat can carry. Once we accept the idea that each habitat has a carrying capacity C for a particular population, we can quantify the amount of "elbow room" for that population as the difference $(C - P_N)$ where P_N is the size of the population in the Nth generation.

The simplest way to put populations, habitats, carrying capacity, and elbow room into a mathematical model is to combine the population size P_N and the carrying capacity C into a single fraction $p_N = \frac{P_N}{C}$, the ***p*-value** of the population. The p-value of a population represents the fraction (or percentage) of the carrying capacity that is occupied by that population and is analogous to the occupancy rate at a hotel. Using p-values makes things a lot easier. We don't really worry about what the carrying capacity C of the habitat is—the p-value has already taken that into account: A p-value of 0.6 means that the population is occupying 60% of the carrying capacity of the habitat; a p-value of 1 means 100% occupancy of the habitat. Using p-values we can also express the relative elbow room for a population in its habitat: The relative elbow room for the Nth generation is $(1 - p_N)$. [Think of $(1 - p_N)$ as the difference between 100% occupancy and the current occupancy rate at the Habitat Hotel.]

We are finally ready to put all of the aforementioned together. Everything will be expressed in terms of population p-values.

- In the Nth generation, the p-value of the population is p_N and the amount of elbow room for that population is $(1 - p_N)$.
- The growth rate of the population in going from the Nth generation to the next generation is proportional to the growth parameter r and the amount of elbow room $(1 - p_N)$, i.e.,

$$\text{growth rate for the } N\text{th generation} = r(1 - p_N).$$

- In any given generation, the growth rate times the size of the population equals the size of the population in the next generation, i.e.,

$$p_{N+1} = r(1 - p_N)p_N.$$

LOGISTIC EQUATION

$p_{N+1} = r(1 - p_N)p_N$. The equation gives a recursive formula for the growth of a population under the logistic growth model.

Using the logistic equation, we can analyze the behavior of any population living in a fixed habitat. In the next few examples, we will explore the remarkable population patterns that can emerge from the logistic equation. All we will need for our "ecological experiments" is the value p_0 of the starting population (this number is called the **seed**), the value of the growth parameter r, and a good calculator, or better yet, a spreadsheet. [A note of warning: all p-values are expressed as decimals between 0 (the population is extinct) and 1 (the population has completely filled up the habitat). The calculations were done in a computer and carried to 16 decimal places before being rounded off to 3 or 4 decimal places. You are encouraged to follow along with your own calculator, but don't be surprised if the numbers don't match exactly. Round-off errors are a way of life with the logistic equation.]

EXAMPLE 9.20 A STABLE EQUILIBRIUM

Fish farming is big business these days, so you decide to give it a try. You have access to a large, natural pond in which you plan to set up a rainbow trout hatchery. The carrying capacity of the pond is $C = 10{,}000$ fish, and the growth parameter of this type of rainbow trout is $r = 2.5$. We will use the logistic equation to model the growth of the fish population in your pond.

You start by seeding the pond with an initial population of 2000 rainbow trout (i.e., 20% of the pond's carrying capacity, or $p_0 = 0.2$). After the first year (trout have an annual hatching season) the population is given by

$$p_1 = r(1 - p_0)p_0 = 2.5 \times (1 - 0.2) \times (0.2) = 0.4$$

The population of the pond has doubled, and things are looking good! Unfortunately, most of the fish are small fry and not ready to be sent to market. After the second year the population of the pond is given by

$$p_2 = 2.5 \times (1 - 0.4) \times (0.4) = 0.6$$

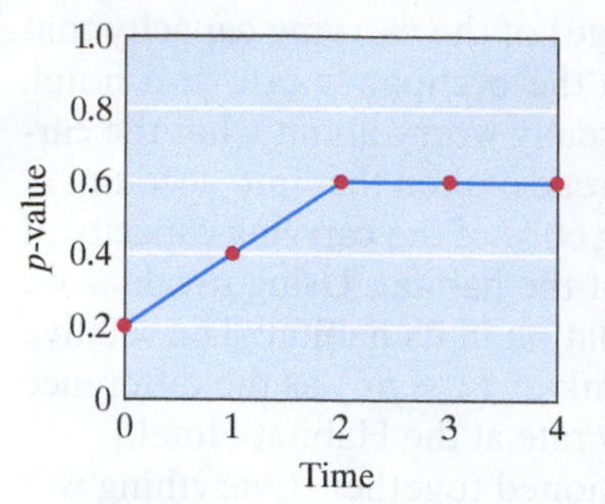

FIGURE 9-14 $r = 2.5$, $p_0 = 0.2$.

The population is no longer doubling, but the hatchery is still doing well. You are looking forward to even better yields after the third year. But on the third year you get a big surprise:

$$p_3 = 2.5 \times (1 - 0.6) \times (0.6) = 0.6$$

Stubbornly, you wait for better luck the next year, but

$$p_4 = 2.5 \times (1 - 0.6) \times (0.6) = 0.6$$

From the second year on, the hatchery is stuck at 60% of the pond capacity—nothing is going to change unless external forces come into play. We describe this situation as one in which the population is at a *stable equilibrium*. Figure 9-14 shows a line graph of the pond's fish population for the first four years.

EXAMPLE 9.21 AN ATTRACTING POINT

Consider the same setting as in Example 9.20 (same pond and the same variety of rainbow trout with $r = 2.5$), but suppose you initially seed the pond with 3000 rainbow trout (30% of the pond's carrying capacity). How will the fish population grow if we start with $p_0 = 0.3$?

The first six years of population growth are as follows:

$$p_1 = 2.5 \times (1 - 0.3) \times (0.3) = 0.525$$

$$p_2 = 2.5 \times (1 - 0.525) \times (0.525) \approx 0.6234$$

$$p_3 = 2.5 \times (1 - 0.6234) \times (0.6234) \approx 0.5869$$

$$p_4 = 2.5 \times (1 - 0.5869) \times (0.5869) \approx 0.6061$$

$$p_5 = 2.5 \times (1 - 0.6061) \times (0.6061) \approx 0.5968$$

$$p_6 = 2.5 \times (1 - 0.5968) \times (0.5968) \approx 0.6016$$

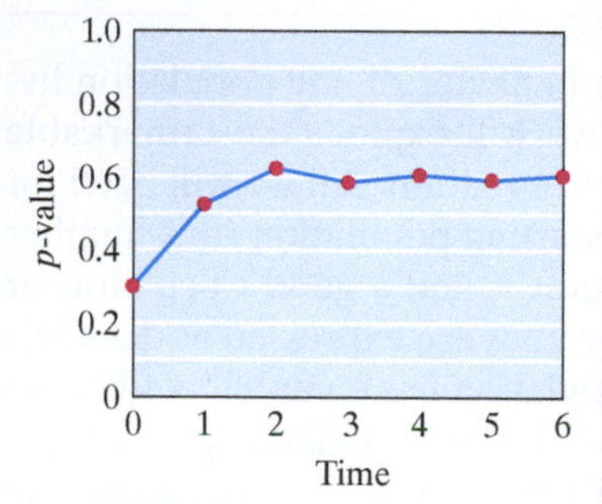

FIGURE 9-15 $r = 2.5$, $p_0 = 0.3$.

Clearly, something different is happening here. The trout population appears to be fluctuating—up, down, up again, back down—but always hovering near the value of 0.6. We leave it to the reader to verify that as one continues with the population sequence, the p-values inch closer and closer to 0.6 in an oscillating (up, down, up, down, ...) manner. The value 0.6 is called an *attracting point* of the population sequence. Figure 9-15 shows a line graph of the pond's fish population for the first six years.

EXAMPLE 9.22 COMPLEMENTARY SEEDS

In Example 9.21 we seeded the pond at 30% of its carrying capacity ($p_0 = 0.3$). If we seed the pond with the complementary seed $p_0 = 1 - 0.3 = 0.7$, we end up with the same populations:

$$p_1 = 2.5 \times 0.3 \times 0.7 = 2.5 \times 0.7 \times 0.3 = 0.525$$

$$p_2 = 2.5 \times (1 - 0.525) \times (0.525) \approx 0.6234$$

and so on.

Example 9.22 points to a simple but useful general rule about logistic growth—the seeds p_0 and $(1 - p_0)$ always produce the same population sequence. This follows because in the expression $p_1 = r(1 - p_0)p_0$, p_0 and $(1 - p_0)$ play interchangeable roles—if you change p_0 to $(1 - p_0)$, then you are also changing $(1 - p_0)$ to p_0. Nothing gained, nothing lost! Once the values match in the first generation, the rest of the p-values follow suit. The moral of this observation is that you should never seed your pond at higher than 50% of its carrying capacity.

EXAMPLE 9.23 A TWO-CYCLE PATTERN

You decided that farming rainbow trout is too difficult. You are moving on to raising something easier—goldfish. The particular variety of goldfish you will grow has growth parameter $r = 3.1$.

Suppose you start by seeding a tank at 20% of its carrying capacity $(p_0 = 0.2)$. The first 16 p-values of the goldfish population (for brevity, the details are left to the reader) are

$$\begin{array}{llll} p_0 = 0.2, & p_1 = 0.496, & p_2 \approx 0.775, & p_3 \approx 0.541, \\ p_4 \approx 0.770, & p_5 \approx 0.549, & p_6 \approx 0.767, & p_7 \approx 0.553, \\ p_8 \approx 0.766, & p_9 \approx 0.555, & p_{10} \approx 0.766, & p_{11} \approx 0.556, \\ p_{12} \approx 0.765, & p_{13} \approx 0.557, & p_{14} \approx 0.765, & p_{15} \approx 0.557, \quad \ldots \end{array}$$

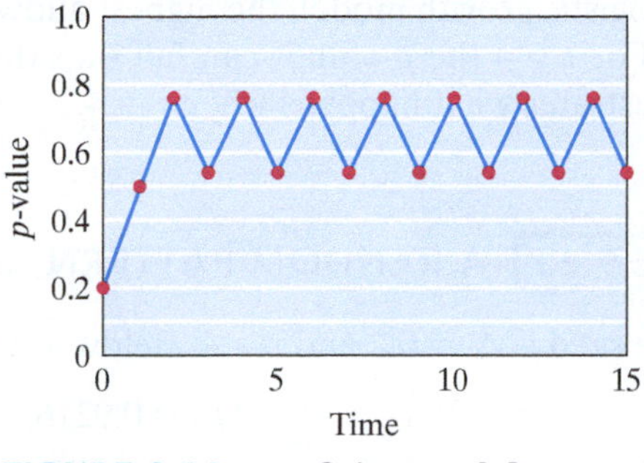

FIGURE 9-16 $r = 3.1$, $p_0 = 0.2$.

An interesting pattern emerges here. After a few breeding seasons, the population settles into a two-cycle pattern, alternating between a high-population period at 0.765 and a low-population period at 0.557. Figure 9-16 convincingly illustrates the oscillating nature of the population sequence.

Example 9.23 describes a situation not unusual in population biology—animal populations that alternate cyclically between two different levels of population density. Even more complex cyclical patterns are possible when we increase the growth parameter just a little.

EXAMPLE 9.24 A FOUR-CYCLE PATTERN

You are now out of the fish-farming business and have acquired an interest in entomology—the study of insects. Let's apply the logistic growth model to study the population growth of a type of flour beetle with growth parameter $r = 3.5$. The seed will be $p_0 = 0.44$. (There is no particular significance to the choice of the seed—you can change the seed and you will still get an interesting population sequence.)

Following are a few specially selected p-values. We leave it to the reader to verify these numbers and fill in the missing details.

$$\begin{array}{llll} p_0 = 0.440, & p_1 \approx 0.862, & p_2 \approx 0.415, & p_3 \approx 0.850, \\ p_4 \approx 0.446, & p_5 \approx 0.865, & \ldots & p_{20} \approx 0.497, \\ p_{21} \approx 0.875, & p_{22} \approx 0.383, & p_{23} \approx 0.827, & p_{24} \approx 0.501, \\ p_{25} \approx 0.875, & \ldots & & \end{array}$$

It took a while, but we can now see a pattern: Since $p_{25} = p_{21}$, the population will repeat itself in a four-period cycle ($p_{26} = p_{22}$, $p_{27} = p_{23}$, $p_{28} = p_{24}$, $p_{29} = p_{25} = p_{21}$, etc.), an interesting and surprising turn of events. Figure 9-17 shows the line graph of the first 26 p-values.

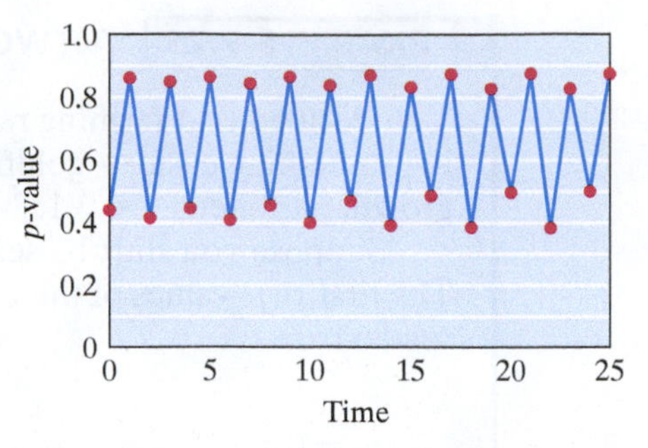

FIGURE 9-17 $r = 3.5, p_0 = 0.44$.

The cyclical behavior exhibited in Example 9.24 is not unusual, and many insect populations follow cyclical patterns of various lengths—7-year locusts, 17-year cicadas, and so on.

In the logistic growth model, the highest allowed value of the growth parameter r is $r = 4$. (For $r > 4$ the p-values can fall outside the permissible range.) Our last example illustrates what happens when $r = 4$.

EXAMPLE 9.25 A RANDOM PATTERN

Suppose the seed is $p_0 = 0.2$ and $r = 4$. Below are the p-values p_0 through p_{20}.

$p_0 = 0.2$, $p_1 = 0.64$, $p_2 = 0.9216$, $p_3 \approx 0.289$,
$p_4 \approx 0.8219$, $p_5 \approx 0.5854$, $p_6 \approx 0.9708$, $p_7 \approx 0.1133$,
$p_8 \approx 0.402$, $p_9 \approx 0.9616$, $p_{10} \approx 0.1478$, $p_{11} \approx 0.5039$,
$p_{12} \approx 0.9999$, $p_{13} \approx 0.0004$, $p_{14} \approx 0.001$, $p_{15} \approx 0.0039$,
$p_{16} \approx 0.0157$, $p_{17} \approx 0.0617$, $p_{18} \approx 0.2317$, $p_{19} \approx 0.7121$, $p_{20} = 0.82$.

Figure 9-18 is a line graph plotting these p-values.

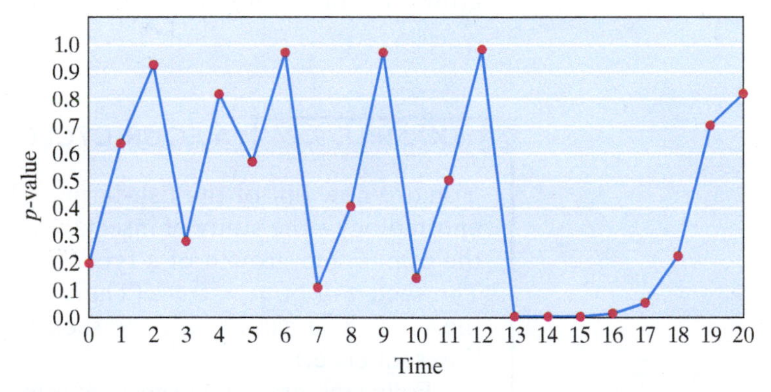

FIGURE 9-18 $r = 4.0, p_0 = 0.2$.

The surprise here is the absence of any predictable pattern. In fact, no matter how much further we continue computing p-values, we will find no pattern—to an outside observer the p-values for this population sequence appear to be quite erratic and seemingly random. Of course, we know better—they are all coming from the logistic equation. ■

The logistic growth model exhibits many interesting surprises. In addition to Exercises 57 through 62 you are encouraged to experiment on your own much like we did in the preceding examples: Choose a seed p_0 between 0 and 0.5, choose an r between 3 and 4, and let her (the logistic equation) rip!

Conclusion

In this chapter we discussed three classic models of population growth. In the *linear growth model*, the population is described by an *arithmetic sequence* of the form P_0, $P_0 + d$, $P_0 + 2d$, $P_0 + 3d, \ldots$. In each transition period the population grows by the addition of a fixed amount d called the *common difference*. Linear growth is most common in situations in which there is no "breeding" such as populations of inanimate objects—commodities, resources, garbage, and so on.

In the *exponential growth model*, the population is described by a *geometric sequence* of the form P_0, RP_0, R^2P_0, $R^3P_0, \ldots$. In each transition period the population grows by multiplication by a positive constant R called the *common ratio*. Exponential growth is typical of situations in which there is some form of "breeding" in the population and the amount of breeding is directly proportional to the size of the population.

In the *logistic growth model* populations are described in relative terms by the logistic equation $p_{N+1} = r(1 - p_N)p_N$. This model is used to describe the growth of biological populations living in a fixed habitat and whose growth rates are in direct proportion to the amount of "elbow room" in the habitat. When confined to a single-species habitat, many animal populations grow according to the logistic growth model.

KEY CONCEPTS

9.1 Sequences and Population Sequences

- **sequence:** an ordered, infinite list of numbers, **259**
- **term:** the individual numbers in a sequence, **259**
- **explicit formula:** for a sequence, a formula that gives the Nth term explicitly, without the use of other terms, **260**
- **recursive formula:** for a sequence, a formula that gives the Nth term of the sequence as a function of other terms of the sequence, **261**
- **population sequence:** a sequence of the form $P_0, P_1, P_2, P_3, \ldots$, where P_N represents the size of the population in the Nth generation, **262**
- **time-series graph:** a graph of a population sequence where time is measured on the horizontal axis and the size of the population is measured on the vertical axis, **263**
- **scatter-plot:** a times-series graph with isolated points representing the values of a population sequence, **263**
- **line graph:** a times-series graph where the values (points) are connected with lines, **263**

9.2 The Linear Growth Model

- **linear growth:** a model of population growth based on the recursive rule $P_N = P_{N-1} + d$, where d is a constant, **265**
- **arithmetic sequence:** a sequence of the form P_0, $P_0 + d$, $P_0 + 2d$, $P_0 + 3d, \ldots$, **267**
- **common difference:** the constant difference d between successive terms of an arithmetic sequence $(d = P_N - P_{N-1})$, **267**
- **arithmetic sum formula:** gives the sum of N consecutive terms of an arithmetic sequence $\left[(\text{first term} + \text{last term}) \times \frac{N}{2}\right]$, **270**

9.3 The Exponential Growth Model

- **growth rate:** the ratio $\frac{(Y - X)}{X}$ as a variable changes from a baseline value X to an end-value Y, **272**

- **exponential growth:** a model of population growth based on the recursive rule $P_N = RP_{N-1}$, where R is a non-negative constant, **273**
- **common ratio:** the constant ratio R between successive terms of a geometric sequence, $(R = \frac{P_N}{P_{N-1}})$, **273**
- **geometric sequence:** a sequence of the form $P_0, RP_0, R^2P_0, R^3P_0, \ldots$, **273**
- **geometric sum formula:** gives the sum of N consecutive terms of a geometric sequence with common ratio R, $\left[(\text{first term})\frac{(R^N - 1)}{(R - 1)}\right]$, **275**

9.4 The Logistic Growth Model

- **habitat:** a confined geographical area inhabited by a population, **277**
- **logistic growth model:** a model of population growth based on the logistic equation, **278**
- **growth parameter:** the natural, unrestricted growth rate of a species in a specific habitat, **278**
- **carrying capacity:** the maximum number of individuals of a given species that a habitat can carry, **278**
- **p-value:** the ratio $\frac{P}{C}$ where P is the size of the population and C is the carrying capacity of the habitat; it represents the fraction or percent of the habitat occupied by the population, **278**
- **logistic equation:** the recursive equation $p_{N+1} = r(1 - p_N)p_N$, where p_N is the p-value of the population in the Nth generation and r is the growth parameter of the species, **279**
- **seed:** for a population growing under the logistic growth model, the p-value of the starting population, **279**

EXERCISES

WALKING

9.1 Sequences and Population Sequences

1. Consider the sequence defined by the explicit formula $A_N = N^2 + 1$.

(a) Find A_1.

(b) Find A_{100}.

(c) Suppose $A_N = 10$. Find N.

2. Consider the sequence defined by the explicit formula $A_N = 3^N - 2$.

(a) Find A_3.

(b) Use a calculator to find A_{15}.

(c) Suppose $A_N = 79$. Find N.

3. Consider the sequence defined by the explicit formula $A_N = \frac{4N}{N+3}$.

(a) Find A_1.

(b) Find A_9.

(c) Suppose $A_N = \frac{5}{2}$. Find N.

4. Consider the sequence defined by the explicit formula $A_N = \frac{2N + 3}{3N - 1}$.

(a) Find A_1.

(b) Find A_{100}.

(c) Suppose $A_N = 1$. Find N.

5. Consider the sequence defined by the explicit formula $A_N = (-1)^{N+1}$.

(a) Find A_1.

(b) Find A_{100}.

(c) Find all values of N for which $A_N = 1$.

6. Consider the sequence defined by the explicit formula $A_N = (-\frac{1}{N})^{N-1}$.

(a) Find A_1.

(b) Find A_4.

(c) Find all values of N for which A_N is positive.

7. Consider the sequence defined by the recursive formula $A_N = 2A_{N-1} + A_{N-2}$ and starting with $A_1 = 1$, $A_2 = 1$.

(a) List the next four terms of the sequence.

(b) Find A_8.

8. Consider the sequence defined by the recursive formula $A_N = A_{N-1} + 2A_{N-2}$ and starting with $A_1 = 1$, $A_2 = 1$.

   (a) List the next four terms of the sequence.

   (b) Find A_8.

9. Consider the sequence defined by the recursive formula $A_N = A_{N-1} - 2A_{N-2}$ and starting with $A_1 = 1$, $A_2 = -1$.

   (a) List the next four terms of the sequence.

   (b) Find A_8.

10. Consider the sequence defined by the recursive formula $A_N = 2A_{N-1} - 3A_{N-2}$ and starting with $A_1 = -1$, $A_2 = 1$.

    (a) List the next four terms of the sequence.

    (b) Find A_8.

11. Consider the sequence 1, 4, 9, 16, 25, ...

    (a) List the next two terms of the sequence.

    (b) Assuming the sequence is denoted by $A_1, A_2, A_3, \ldots$, give an explicit formula for A_N.

    (c) Assuming the sequence is denoted by $P_0, P_1, P_2, \ldots$, give an explicit formula for P_N.

12. Consider the sequence 1, 2, 6, 24, 120, ...

    (a) List the next two terms of the sequence.

    (b) Assuming the sequence is denoted by $A_1, A_2, A_3, \ldots$, give an explicit formula for A_N.

    (c) Assuming the sequence is denoted by $P_0, P_1, P_2, \ldots$, give an explicit formula for P_N.

13. Consider the sequence 0, 1, 3, 6, 10, 15, 21, ...

    (a) List the next two terms of the sequence.

    (b) Assuming the sequence is denoted by $A_1, A_2, A_3, \ldots$, give an explicit formula for A_N.

    (c) Assuming the sequence is denoted by $P_0, P_1, P_2, \ldots$, give an explicit formula for P_N.

14. Consider the sequence 2, 3, 5, 9, 17, 33, ...

    (a) List the next two terms of the sequence.

    (b) Assuming the sequence is denoted by $A_1, A_2, A_3, \ldots$, give an explicit formula for A_N.

    (c) Assuming the sequence is denoted by $P_0, P_1, P_2, \ldots$, give an explicit formula for P_N.

15. Consider the sequence $1, \frac{8}{5}, 2, \frac{16}{7}, \frac{20}{8}, \ldots$.

    (a) List the next two terms of the sequence.

    (b) If the notation for the sequence is $A_1, A_2, A_3, \ldots$, give an explicit formula for A_N.

16. Consider the sequence $3, 2, \frac{5}{4}, \frac{6}{8}, \frac{7}{16}, \ldots$.

    (a) List the next two terms of the sequence.

    (b) If the notation for the sequence is $A_1, A_2, A_3, \ldots$, give an explicit formula for A_N.

17. Airlines would like to board passengers in the order of decreasing seat numbers (largest seat number first, second largest next, and so on), but passengers don't like this policy and refuse to go along. If two passengers randomly board a plane the probability that they board in order of decreasing seat numbers is $\frac{1}{2}$; if three passengers randomly board a plane the probability that they board in order of decreasing seat numbers is $\frac{1}{6}$; if four passengers randomly board a plane the probability that they board in order of decreasing seat numbers is $\frac{1}{24}$; and if five passengers randomly board a plane, the probability that they board in order of decreasing seat numbers is $\frac{1}{120}$. Using the sequence $\frac{1}{2}, \frac{1}{6}, \frac{1}{24}, \frac{1}{120}, \ldots$ as your guide,

    (a) determine the probability that if six passengers randomly board a plane they board in order of decreasing seat numbers.

    (b) determine the probability that if 12 passengers randomly board a plane they board in order of decreasing seat numbers.

18. When two fair coins are tossed the probability of tossing two heads is $\frac{1}{4}$; when three fair coins are tossed the probability of tossing two heads and one tail is $\frac{3}{8}$; when four fair coins are tossed the probability of tossing two heads and two tails is $\frac{6}{16}$; when five fair coins are tossed the probability of tossing two heads and three tails is $\frac{10}{32}$. Using the sequence $\frac{1}{4}, \frac{3}{8}, \frac{6}{16}, \frac{10}{32}, \ldots$ as your guide,

    (a) determine the probability of tossing two heads and four tails when six fair coins are tossed.

    (b) determine the probability of tossing two heads and 10 tails when 12 fair coins are tossed. (*Hint:* Find an explicit formula first.)

9.2 The Linear Growth Model

19. Consider a population that grows linearly following the recursive formula $P_N = P_{N-1} + 125$, with initial population $P_0 = 80$.

    (a) Find P_1, P_2, and P_3.

    (b) Give an explicit formula for P_N.

    (c) Find P_{100}.

20. Consider a population that grows linearly following the recursive formula $P_N = P_{N-1} + 23$, with initial population $P_0 = 57$.

    (a) Find P_1, P_2, and P_3.

    (b) Give an explicit formula for P_N.

    (c) Find P_{200}.

21. Consider a population that grows linearly following the recursive formula $P_N = P_{N-1} - 25$, with initial population $P_0 = 578$.

    (a) Find P_1, P_2, and P_3.

    (b) Give an explicit formula for P_N.

    (c) Find P_{23}.

22. Consider a population that grows linearly following the recursive formula $P_N = P_{N-1} - 111$, with initial population $P_0 = 11{,}111$.

    (a) Find P_1, P_2, and P_3.

    (b) Give an explicit formula for P_N.

    (c) Find P_{100}.

23. Consider a population that grows linearly, with $P_0 = 8$ and $P_{10} = 38$.

(a) Give an explicit formula for P_N. **(b)** Find P_{50}.

24. Consider a population that grows linearly, with $P_5 = 37$ and $P_7 = 47$.

(a) Find P_0.

(b) Give an explicit formula for P_N.

(c) Find P_{100}.

25. Official unemployment rates for the U.S. population are reported on a monthly basis by the Bureau of Labor Statistics. For the period October, 2011, through January, 2012, the official unemployment rates were 8.9% (Oct.), 8.7% (Nov.), 8.5% (Dec.), and 8.3% (Jan.). (*Source:* U.S. Bureau of Labor Statistics, *www.bls.gov*.) If the unemployment rates were to continue to decrease following a linear model,

(a) predict the unemployment rate on January, 2013.

(b) predict when the United States would reach a zero unemployment rate.

26. The world population reached 6 billion people in 1999 and 7 billion in 2012. (*Source:* Negative Population Growth, *www.npg.org*.) Assuming a linear growth model for the world population,

(a) predict the year when the world population would reach 8 billion.

(b) predict the world population in 2020.

27. The Social Security Administration uses a linear growth model to estimate life expectancy in the United States. The model uses the explicit formula $L_N = 66.17 + 0.96N$ where L_N is the life expectancy of a person born in the year $1995 + N$ (i.e., $N = 0$ corresponds to 1995 as the year of birth, $N = 1$ corresponds to 1996 as the year of birth, and so on). (*Source:* Social Security Administration, *www.socialsecurity.gov*.)

(a) Assuming the model continues to work indefinitely, estimate the life expectancy of a person born in 2012.

(b) Assuming the model continues to work indefinitely, what year will you have to be born so that your life expectancy is 90?

28. While the number of smokers for the general adult population is decreasing, it is not decreasing equally across all subpopulations (and for some groups it is actually increasing). For the 18-to-24 age group, the number of smokers was 8 million in 1965 and 6.3 million in 2009. (*Source:* "Trends in Tobacco Use," American Lung Association, 2011.) Assuming the number of smokers in the 18-to-24 age group continues decreasing according to a negative linear growth model,

(a) predict the number of smokers in the 18-to-24 age group in 2015 (round your answer to the nearest thousand).

(b) predict in what year the number of smokers in the 18-to-24 age group will reach 5 million.

29. Use the arithmetic sum formula to find the sum $\underbrace{2 + 7 + 12 + \cdots + 497}_{100 \text{ terms}}$.

30. Use the arithmetic sum formula to find the sum $\underbrace{21 + 28 + 35 + \cdots + 413}_{57 \text{ terms}}$.

31. An arithmetic sequence has first term $P_0 = 12$ and common difference $d = 3$.

(a) The number 309 is which term of the arithmetic sequence?

(b) Find the sum $12 + 15 + 18 + \cdots + 309$.

32. An arithmetic sequence has first term $P_0 = 1$ and common difference $d = 9$.

(a) The number 2701 is which term of the arithmetic sequence?

(b) Find $1 + 10 + 19 + \cdots + 2701$.

33. Find the sum

(a) $1 + 3 + 5 + 7 + \cdots + 149$. (*Hint:* See Example 9.13.)

(b) $\underbrace{1 + 3 + 5 + \cdots}_{100 \text{ terms}}$.

34. Find the sum

(a) $2 + 4 + 6 + \cdots + 98$.

(b) $\underbrace{2 + 4 + 6 + \cdots}_{75 \text{ terms}}$.

35. The city of Lightsville currently has 137 streetlights. As part of an urban renewal program, the city council has decided to install and have operational 2 additional streetlights at the end of each week for the next 52 weeks. Each streetlight costs $1 to operate for 1 week.

(a) How many streetlights will the city have at the end of 38 weeks?

(b) How many streetlights will the city have at the end of N weeks? (Assume $N \leq 52$.)

(c) What is the cost of operating the original 137 lights for 52 weeks?

(d) What is the additional cost for operating the newly installed lights for the 52-week period during which they are being installed?

36. A manufacturer currently has on hand 387 widgets. During the next 2 years, the manufacturer will be increasing his inventory by 37 widgets per week. (Assume that there are exactly 52 weeks in one year.) Each widget costs 10 cents a week to store.

(a) How many widgets will the manufacturer have on hand after 20 weeks?

(b) How many widgets will the manufacturer have on hand after N weeks? (Assume $N \leq 104$.)

(c) What is the cost of storing the original 387 widgets for 2 years (104 weeks)?

(d) What is the additional cost of storing the increased inventory of widgets for the next 2 years?

9.3 The Exponential Growth Model

37. A population grows according to an exponential growth model. The initial population is $P_0 = 11$ and the common ratio is $R = 1.25$.

(a) Find P_1.

(b) Find P_9.

(c) Give an explicit formula for P_N.

38. A population grows according to an exponential growth model, with $P_0 = 8$ and $P_1 = 12$.

(a) Find the common ratio R.

(b) Find P_9.

(c) Give an explicit formula for P_N.

39. A population grows according to the recursive rule $P_N = 4P_{N-1}$, with initial population $P_0 = 5$.

(a) Find P_1, P_2, and P_3.

(b) Give an explicit formula for P_N.

(c) How many generations will it take for the population to reach 1 million?

40. A population *decays* according to an exponential growth model, with $P_0 = 3072$ and common ratio $R = 0.75$.

(a) Find P_5.

(b) Give an explicit formula for P_N.

(c) How many generations will it take for the population to fall below 200?

41. Crime in Happyville is on the rise. Each year the number of crimes committed increases by 50%. Assume that there were 200 crimes committed in 2010, and let P_N denote the number of crimes committed in the year $2010 + N$.

(a) Give a recursive description of P_N.

(b) Give an explicit description of P_N.

(c) If the trend continues, approximately how many crimes will be committed in Happyville in the year 2020?

42. Since 2010, when 100,000 cases were reported, each year the number of new cases of equine flu has decreased by 20%. Let P_N denote the number of new cases of equine flu in the year $2010 + N$.

(a) Give a recursive description of P_N.

(b) Give an explicit description of P_N.

(c) If the trend continues, approximately how many new cases of equine flu will be reported in the year 2025?

43. In 2010 the number of mathematics majors at Bright State University was 425; in 2011 the number of mathematics majors was 463. Find the growth rate (expressed as a percent) of mathematics majors from 2010 to 2011.

44. Avian influenza A(H5N1) is a particularly virulent strain of the bird flu. In 2008 there were 44 cases of avian influenza A(H5N1) confirmed worldwide; in 2009 the number of confirmed cases worldwide was 73. (*Source:* World Health Organization, *www.who.int.*) Find the growth rate in the number of confirmed cases worldwide of avian influenza A(H5N1) from 2008 to 2009. Express your answer as a percent.

45. In 2010 the undergraduate enrollment at Bright State University was 19,753; in 2011 the undergraduate enrollment was 17,389. Find the "growth" rate in the undergraduate enrollment from 2010 to 2011. Give your answer as a percent.

46. In 2009 there were 73 cases of avian influenza A(H5N1) confirmed worldwide; in 2010 the number of confirmed cases worldwide was 48. (*Source:* World Health Organization, *www.who.int.*) Find the "growth" rate in the number of confirmed cases worldwide of avian influenza A(H5N1). Give your answer as a percent.

47. Consider the geometric sequence $P_0 = 2, P_1 = 6, P_2 = 18, \ldots$.

(a) Find the common ratio R.

(b) Use the geometric sum formula to find the sum $P_0 + P_1 + \cdots + P_{20}$.

48. Consider the geometric sequence $P_0 = 4, P_1 = 6, P_2 = 9, \ldots$

(a) Find the common ratio R.

(b) Use the geometric sum formula to find the sum $P_0 + P_1 + \cdots + P_{24}$.

49. Consider the geometric sequence $P_0 = 4, P_1 = 2, P_2 = 1, \ldots$

(a) Find the common ratio R.

(b) Use the geometric sum formula to find the sum $P_0 + P_1 + \cdots + P_{11}$.

50. Consider the geometric sequence $P_0 = 10, P_1 = 2, P_2 = 0.4, \ldots$.

(a) Find the common ratio R.

(b) Use the geometric sum formula to find the sum $P_0 + P_1 + \cdots + P_{24}$.

51. Find the sum

(a) $1 + 2 + 2^2 + 2^3 + \cdots + 2^{15}$.

(b) $1 + 2 + 2^2 + 2^3 + \cdots + 2^{N-1}$. (*Hint:* The answer is an expression in N.)

52. Find the sum

(a) $1 + 3 + 3^2 + 3^3 + \cdots + 3^{10}$.

(b) $1 + 3 + 3^2 + 3^3 + \cdots + 3^{N-1}$. (*Hint:* The answer is an expression in N.)

9.4 The Logistic Growth Model

53. A population grows according to the logistic growth model, with growth parameter $r = 0.8$. Starting with an initial population given by $p_0 = 0.3$,

(a) find p_1.

(b) find p_2.

(c) determine what percent of the habitat's carrying capacity is taken up by the third generation.

54. A population grows according to the logistic growth model, with growth parameter $r = 0.6$. Starting with an initial population given by $p_0 = 0.7$,

(a) find p_1.

(b) find p_2.

(c) determine what percent of the habitat's carrying capacity is taken up by the third generation.

55. For the population discussed in Exercise 53 ($r = 0.8$, $p_0 = 0.3$),

(a) find the values of p_1 through p_{10}.

(b) what does the logistic growth model predict in the long term for this population?

56. For the population discussed in Exercise 54 ($r = 0.6$, $p_0 = 0.7$),

(a) find the values of p_1 through p_{10}.

(b) what does the logistic growth model predict in the long term for this population?

57. A population grows according to the logistic growth model, with growth parameter $r = 1.8$. Starting with an initial population given by $p_0 = 0.4$,

(a) find the values of p_1 through p_{10}.

(b) what does the logistic growth model predict in the long term for this population?

58. A population grows according to the logistic growth model, with growth parameter $r = 1.5$. Starting with an initial population given by $p_0 = 0.8$,

(a) find the values of p_1 through p_{10}.

(b) what does the logistic growth model predict in the long term for this population?

59. A population grows according to the logistic growth model, with growth parameter $r = 2.8$. Starting with an initial population given by $p_0 = 0.15$,

(a) find the values of p_1 through p_{10}.

(b) what does the logistic growth model predict in the long term for this population?

60. A population grows according to the logistic growth model, with growth parameter $r = 2.5$. Starting with an initial population given by $p_0 = 0.2$,

(a) find the values of p_1 through p_{10}.

(b) what does the logistic growth model predict in the long term for this population?

61. A population grows according to the logistic growth model, with growth parameter $r = 3.25$. Starting with an initial population given by $p_0 = 0.2$,

(a) find the values of p_1 through p_{10}.

(b) what does the logistic growth model predict in the long term for this population?

62. A population grows according to the logistic growth model, with growth parameter $r = 3.51$. Starting with an initial population given by $p_0 = 0.4$,

(a) find the values of p_1 through p_{10}.

(b) what does the logistic growth model predict in the long term for this population?

JOGGING

63. Each of the following sequences follows a linear, an exponential, or a logistic growth model. For each sequence, determine which model applies (if more than one applies, then indicate all the ones that apply).

(a) 2, 4, 8, 16, 32, ...

(b) 2, 4, 6, 8, 10, ...

(c) 0.8, 0.4, 0.6, 0.6, 0.6, ...

(d) 0.81, 0.27, 0.09, 0.03, 0.01, ...

(e) 0.49512, 0.81242, 0.49528, 0.81243, 0.49528, ...

(f) 0.9, 0.75, 0.6, 0.45, 0.3, ...

(g) 0.7, 0.7, 0.7, 0.7, 0.7, ...

64. Each of the line graphs shown in Figs. 9-19 through 9-24 describes a population that grows according to a linear, an exponential, or a logistic model. For each line graph, determine which model applies.

(a)

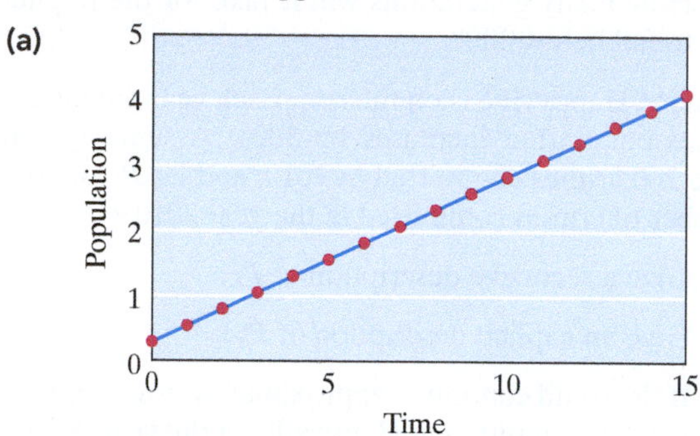

FIGURE 9-19

(b)

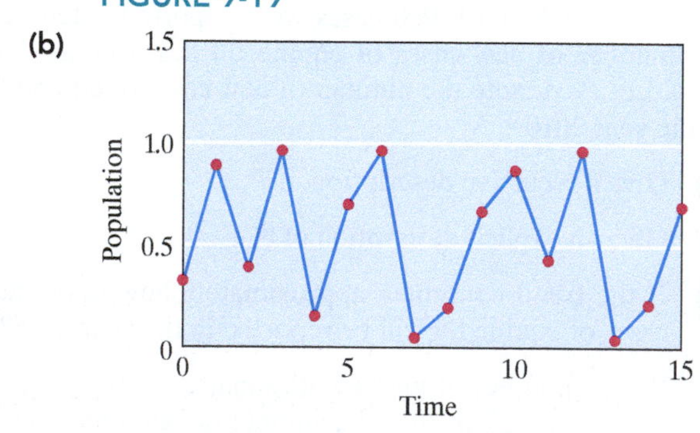

FIGURE 9-20

(c)

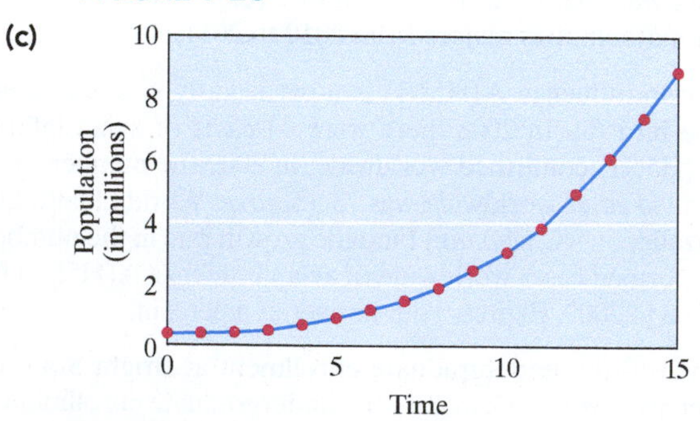

FIGURE 9-21

(d)

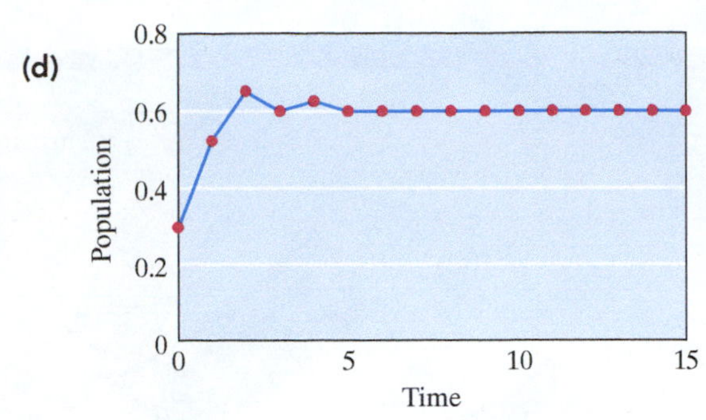

FIGURE 9-22

(e)

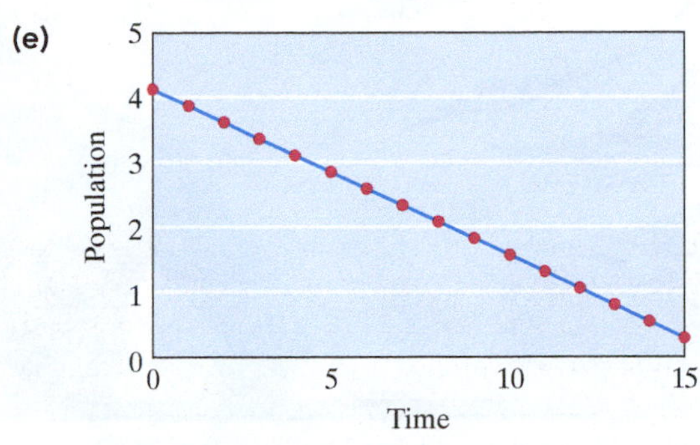

FIGURE 9-23

(f)

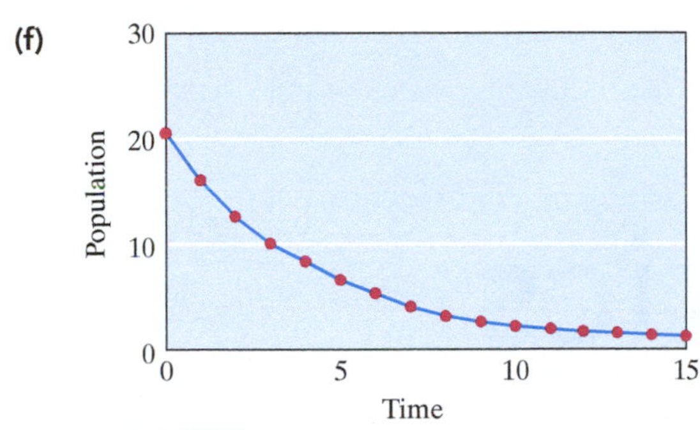

FIGURE 9-24

65. Show that the sum of the first N terms of an arithmetic sequence with first term P_0 and common difference d is

$$\frac{N}{2}\left[2P_0 + (N-1)d\right].$$

66. Find a formula for the sum of the first N even numbers. Show all the steps of your derivation. (*Hint*: Try Exercise 34 first.)

67. Find a formula for the sum of the first N odd numbers. Show all the steps of your derivation. (*Hint*: Try Exercise 33 first.)

68. **(a)** Find a right triangle whose sides are consecutive terms of an arithmetic sequence with common difference $d = 2$.

(b) Find a right triangle whose sides are consecutive terms of a geometric sequence with common ratio R.

69. Give an example of a geometric sequence in which P_0, P_1, P_2, and P_3 are integers, and all the terms from P_4 on are fractions.

RUNNING

70. **Derivation of the geometric sum formula.** This exercise guides you through a step-by-step derivation of the geometric sum formula.

Step 1: Start by setting up the equation $S = P_0 + RP_0 + R^2P_0 + \cdots + R^{N-1}P_0$. (In other words, we use S to denote the left-hand side of the geometric sum formula. The plan is to show that S also equals the right-hand side of the geometric sum formula.)

Step 2: Multiply both sides of the equation in Step 1 by R. This gives an equation for RS.

Step 3: Using the equations in Step 2 and Step 1, find an equation for $RS - S$. Simplify.

Step 4: Solve the equation in Step 3 for S. Show that you end up with the right-hand side of the geometric sum formula.

71. Suppose that you are in charge of stocking a lake with a certain type of alligator with a growth parameter $r = 0.8$. Assuming that the population of alligators grows according to the logistic growth model, is it possible for you to stock the lake so that the alligator population is constant? Explain.

72. Consider a population that grows according to the logistic growth model with initial population given by $p_0 = 0.7$. What growth parameter r would keep the population constant?

73. Consider a population that grows according to the logistic growth model with growth parameter $r\,(r > 1)$. Find p_0 in terms of r so that the population is constant.

74. Suppose $r > 3$. Using the logistic growth model, find a population p_0 such that $p_0 = p_2 = p_4 \ldots$, but $p_0 \neq p_1$.

75. The purpose of this exercise is to understand why we assume that, under the logistic growth model, the growth parameter r is between 0 and 4.

(a) What does the logistic equation give for p_{N+1} if $p_N = 0.5$ and $r > 4$? Is this a problem?

(b) What does the logistic equation predict for future generations if $p_N = 0.5$ and $r = 4$?

(c) If $0 \leq p \leq 1$, what is the largest possible value of $(1-p)p$?

(d) Explain why, if $0 < p_0 < 1$ and $0 < r < 4$, then $0 < p_N < 1$, for every positive integer N.

76. Show that if $P_0, P_1, P_2, \ldots$ is an arithmetic sequence, then $2^{P_0}, 2^{P_1}, 2^{P_2}, \ldots$ must be a geometric sequence.

Solution to the "Mystery" Sequence in Example 9.1:

The sequence 1, 11, 21, 1211, 111221, 312211, 13112221, 1113213211, 31131211131221, . . . is known as the "look-and-say" sequence. Each term of the sequence after the first 1 is obtained by calling out loud the entries in the previous term, as in "one 1" leads to 11, "two 1's" leads to 21; "one 2, one 1" leads to 1211, and so on. Calling out the last term shown above "one 3, two 1's, one 3, one 1, one 2, three 1's, one 3, one 1, two 2's, one 1" gives the next term: 13211311123113112211.

"That money talks, I'll not deny, I heard it once: It said 'Goodbye'."
Richard Armour

10 Financial Mathematics

Money Matters

Sixty percent of U.S. adults neither have a budget nor keep track of their spending;[1] 64% of U.S. adults do not understand how compound interest works;[2] 33% of U.S. adults carry some credit card debt from month to month (11% carry $2500 or more), and 6% admit to having been late making a credit card payment;[3] 45% of working-age U.S. households have no retirement savings at all; the median retirement savings account balance among all working-age households is $3000;[4] and 75% of U.S. adults agree that they could benefit from advice and answers to everyday financial questions.[5]

[1,3,5] *The 2015 Consumer Financial Literacy Survey*, National Foundation for Credit Counseling (NFCC), 2015.
[2] *TNS Survey*, 2008.
[4] *Retirement Security 2015: Roadmap for Policy Makers*, National Institute on Retirement Security (NIRS), 2015.

Every recent survey analyzing how much Americans know about financial matters, and how well they deal with their personal finances, points to the same answers: not very much and not very well.

When it comes to money, you can run but you can't hide. Willingly or not, you have to make financial decisions every day. Some are minor ("Should I get gas at the station on the right or make a U-turn and go to the station across the highway, where gas is 9 cents a gallon cheaper?); some are a little more challenging ("If I buy that new car, should I take the $2000 dealer rebate or the special 1.2% financing offer for 60 months?"); and some have long-term consequences ("Should I invest in real estate or start a retirement savings account?").

The primary purpose of this chapter is to provide you with a few tools needed for a basic understanding of financial mathematics. Reading this chapter is not going to turn you into a Warren Buffett, but it should help you better understand your financial options and stay away from common traps. Think of this chapter as a basic primer for survival in that jungle called the world of finance.

Percentages play an important role in financial mathematics, and this chapter starts with a thorough review of the meaning, use, and abuse of percentages (Section 10.1). A second fundamental concept in financial mathematics is *interest*—the price one pays for using someone else's money. In Section 10.2 we discuss *simple* interest. Our discussion in this section includes investing in government and corporate *bonds* (sometimes a good idea) and borrowing money from *payday loan* companies (always a bad idea). In Section 10.3 we discuss *compound* interest. Our discussion in this section includes investments such as certificates of deposit and savings accounts. In Section 10.4 we discuss retirement savings: how the right savings strategy, combined with the magic of compound interest, can produce a surprisingly large nest-egg for your retirement days. (Here is where I will try to nudge you to act and start your retirement savings plan as soon as possible!) In Section 10.5 we focus on *credit cards* and *installment loans*—the two main generators of *consumer debt* in a modern economy.

> “Bread, cash, dosh, dough, loot, lucre, moolah, readies, the wherewithal: call it what you like, money matters.”
>
> – *Niall Ferguson, The Ascent of Money*

One final note: Although this chapter has its share of formulas, the purpose of the chapter is not to bury you with formulas. Formulas are unavoidable when you are doing financial mathematics, but not the end-all. In fact, using the applet *Financial Calculator* (available in *MyMathLab* in the *Multimedia Library* or in *Tools for Success*), you can carry out any of the calculations discussed in this chapter. The formulas are built into the applet—all you have to know is how to ask the right question. The main goal of the chapter is to provide you with a few useful tools that might get you started on the road to becoming a good money manager and an informed consumer. And, if it helps you become rich some day, so much the better.

10.1 Percentages

A general truism is that people don't like dealing with fractions. There are exceptions, of course, but most people would rather avoid fractions whenever possible. The most likely culprit for "fraction phobia" is the difficulty of dealing with fractions with different denominators. One way to get around this difficulty is to express

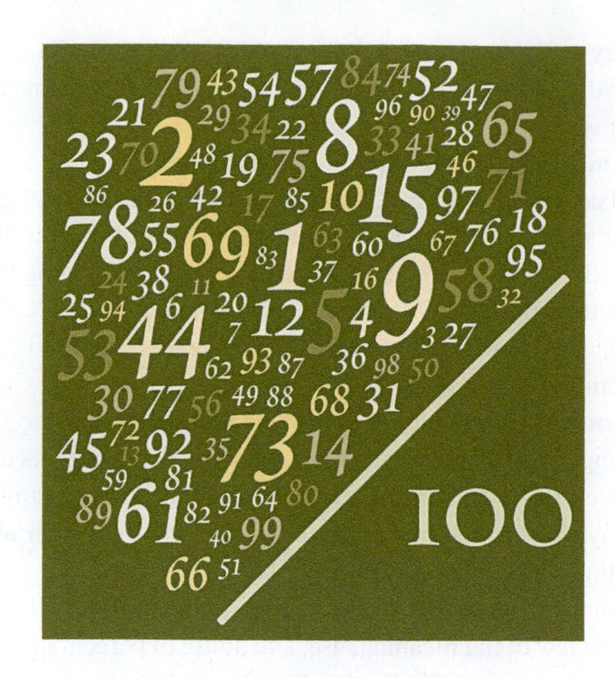

fractions using a common, standard denominator, and that denominator is 100. A "fraction" with denominator 100 can be interpreted as a **percentage**, and the percentage symbol (%) is used to indicate the presence of the hidden denominator 100. Thus,

$$x\% = \frac{x}{100}$$

Percentages are used in several ways. (1) They serve as a common yardstick for comparing different ratios and proportions; (2) they provide a useful way of dealing with fees, taxes, and tips; and (3) they help us better understand how things increase or decrease relative to some given baseline.

Our first example illustrates the use of percentages when comparing ratios.

EXAMPLE 10.1 COMPARING TEST SCORES

Suppose that in your English Lit class you scored 19 out of 25 on the quiz, 49.2 out of 60 on the midterm, and 124.8 out of 150 on the final exam. Without reading further, can you guess which one was your best score? Not easy, right?

The numbers 19, 49.2, and 124.8 are called *raw scores*. Since each raw score is based on a different total, it is hard to compare them directly, but we can do it easily once we express each score as a percentage of the total number of points possible.

- Quiz score = 19/25: Here we can do the arithmetic in our heads. If we just multiply both numerator and denominator by 4, we get 19/25 = 76/100 = 76%.
- Midterm score = 49.2/60: Here the arithmetic is a little harder, so one might want to use a calculator. 49.2 ÷ 60 = 0.82 = 82%. (Since 0.82 = 82/100, it follows that 0.82 = 82%.) This score is a definite improvement over the quiz score.
- Final exam = 124.8/150: Once again, we use a calculator and get 124.8 ÷ 150 = 0.832 = 83.2%. This score is the best one. Well done!

Example 10.1 illustrates the simple but important relation between decimals and percentages: decimals can be converted to percentages through multiplication by 100, as in

$$0.76 = 76\%,\ 1.325 = 132.5\%, \text{ and } 0.005 = 0.5\%.$$

Conversely, percentages can be converted to decimals through division by 100, as in

$$100\% = 1.0,\ 83.2\% = 0.832,\ 7\tfrac{1}{2}\% = 0.075.$$

A note of caution: Later in the chapter we deal with very small percentages such as 0.035%, $\frac{1}{5}\%$, and even $\frac{1}{100}\%$. In these cases there is a natural tendency to "lose" some zeroes along the way. Please be extra careful when converting very small percentages to decimals: $0.035\% = 0.00035$, $\frac{1}{5}\% = 0.2\% = 0.002$, and $\frac{1}{100}\% = 0.01\% = 0.0001$.

Our next two examples illustrate how we use percentages to deal with tips and taxes.

EXAMPLE 10.2 THE (3/20)th RESTAURANT TIP

Imagine you take an old friend out to dinner at a nice restaurant for her birthday. The final bill comes to \$56.80. Your friend suggests that since the service was good, you should tip (3/20)th of the bill. What kind of tip is that?

After a moment's thought, you realize that your friend, who can be a bit annoying at times, is simply suggesting you should tip the standard 15%. After all, $3/20 = 15/100 = 15\%$.

Although 3/20 and 15% are mathematically equivalent, the latter is a much more convenient and familiar way to express the amount of the tip. To compute the actual tip, you simply multiply the amount of the bill by 0.15. In this case we get $0.15 \times \$56.80 = \$8.52$. (People use various tricks to do this calculation in their heads. For an exact calculation you can first take 10% of the bill—in this case \$5.68—and then add the other 5% by computing half of the 10%—in this case \$2.84. If you don't want to worry about nickels and dimes you can just round the \$5.68 to \$6, add another \$3 for the 5%, and make the tip \$9.)

EXAMPLE 10.3 SHOPPING FOR AN iPOD

Imagine you have a little discretionary money saved up and you decide to buy yourself the latest iPod. After a little research you find the following options:

- Option 1: You can buy the iPod at Optimal Buy, a local electronics store. The price is \$299. There is an additional 6.75% sales tax. Your total cost out the door is

$$\begin{aligned} \$299 + (0.0675)\$299 &= \$299 + \$20.1825 \\ &= \$319.18 \quad \text{(rounded to the nearest penny)} \end{aligned}$$

  The above calculation can be shortened by observing that the original price (100%) plus the sales tax (6.75%) can be combined for a total of 106.75% of the original price. Thus, the entire calculation can be carried out by a single multiplication:

$$(1.0675)\$299 = \$319.18$$

- Option 2: At Hamiltonian Circuits, another local electronic store, the sales price is \$312.99, but you happen to have a 5% off coupon good for all electronic products. Using the 5% off coupon, makes the price 95% of \$312.99, or

$$(0.95)\$312.99 = \$297.34$$

  There is still the 6.75% sales tax that needs to be added, and as we saw in Option 1, the quick way to do so is to multiply by 1.0675. Thus, the final price out the door is

$$(1.0675)\$297.34 = \$317.41$$

  For efficiency we can combine the two separate steps (take the discount and add the sales tax) into a single calculation:

$$(1.0675)(0.95)\$312.99 = \$317.41$$

- Option 3: You found an online merchant in Portland, Oregon, that sells the iPod for \$325, and the price includes a 5% shipping/processing charge that you don't have to pay if you pick up the iPod at the store in Portland (there is no sales tax in Oregon). The \$325 is higher than the price at either local store, but you are in luck: your best friend from Portland is coming to visit and can pick up the iPod for you and save you the 5% shipping/processing charge. What would your cost be then?

  Unlike option 2, in this situation we do *not* take a 5% discount on the final price of \$325. Here the 5% was added to the iPod's baseline price to come up with the final cost of \$325, that is, 105% of the baseline price equals \$325. Using B for the unknown baseline price, we have

$$(1.05)B = \$325, \quad \text{or} \quad B = \frac{\$325}{1.05} = \$309.52$$

Although option 3 is your cheapest option, you decide it's not worth imposing on your friend just to save a few bucks and head to Hamiltonian Circuits with your 5% off coupon.

Our next example illustrates the use of percentages to describe relative increases or decreases. Percentage increases and decreases are particularly useful when tracking the ebb and flow of some quantity over time such as following prices in the stock market.

EXAMPLE 10.4 THE DOW JONES INDUSTRIAL AVERAGE

The Dow Jones Industrial Average (DJIA) is one of the most commonly used indicators of the overall state of the stock market in the United States. (As of the writing of this example the DJIA hovered around 18,000.) We are going to illustrate the ups and downs of the DJIA with fictitious numbers.

- Day 1: On a particular day, the DJIA closed at 17,975.84.
- Day 2: The stock market has a good day and the DJIA closes at 18. 191.55. This is an *increase* of 215.71 from the previous day. To express the increase as a percentage, we ask, 215.71 is what percent of 17,975.84 (the day 1 value that serves as our baseline)? The answer is obtained by simply dividing 215.71 by 17,975.84 (and then rewriting it as a percentage). Thus, the percentage increase from day 1 to day 2 is

$$215.71 \div 17{,}975.84 \approx 0.012 = 1.2\%$$

  Here is a little shortcut for the same computation, particularly convenient when you use a calculator (all it takes is one division):

$$18{,}191.55 \div 17{,}975.84 \approx 1.012$$

  All we have to do now is to mentally subtract 1 from the above number. This gives us once again $0.012 = 1.2\%$.

- Day 3: The stock market has a pretty bad day and the DJIA closes at 17,463.89. This represents a decrease from the *previous day* of 727.66. A better way to think of a decrease is as a negative increase, so we say that the "increase" from day 2 to day 3 was $17{,}463.89 - 18{,}191.55 = -727.66$. Once again, this increase can be expressed as a percentage of the base 18,191.55 by dividing:

$$-727.66 \div 18{,}191.55 = -0.04 = -4\%$$

The following list summarizes the basic rules for working with percentages introduced in Examples 10.1 through 10.4.

1. To express the ratio A/B as a percentage, write the ratio as a decimal (i.e., carry out the division $A \div B$), and multiply by 100.
2. To express $P\%$ as a decimal, divide P by 100. When P is itself given as a decimal, move the decimal point two places to the left.
3. To find $P\%$ of B, multiply p times B (where p is $P\%$ *expressed in decimal form*).
4. If you start with a baseline quantity B and add $P\%$ to it, you end up with a final value F that is $(100 + P)\%$ of B. Using p again for the decimal form of $P\%$, this can be expressed by the formula

$$F = (1 + p)B$$

This formula is valid for both positive and negative values of p.

5. Solving the formula $F = (1 + p)B$ for B, gives the formula for finding the baseline B when you know the final value F and the percentage increase p: $B = \frac{F}{(1+p)}$.
6. Solving the formula $F = (1 + p)B$ for p, gives the formula for finding the percentage increase when you know the baseline B and the final value F: $p = \left(\frac{F}{B}\right) - 1 = \frac{F-B}{B}$. The percentage "increase" p can be positive, negative, or zero. When p is positive we have a real increase $(F > B)$; when p is negative we have a decrease $(F < B)$; when p is zero we have no change $(F = B)$.

Note: Please pay attention to the fact that the formulas given in 5 and 6 directly follow from solving the main formula $F = (1 + p)B$ for B and p, respectively. You really only need to make one deposit to the memory bank.

Percentage Traps

Although percentages are unquestionably useful, they can also be misinterpreted, misused, and manipulated. We will discuss a few of the misinterpretations, misuses, and manipulations next.

One of the most common ways people misinterpret percentages is by adding them. If you get a 15% raise one year and a 10% raise the next, your two raises combined come out to more than 25%. If your stock portfolio goes up 25% one day and goes down 20% the next day I have news for you—you didn't make any money. If you go to a department store and a pair of shoes is marked down 20% and then there is an additional markdown of 30%, you are not getting the shoes at half-price. Our next example shows how these things happen.

EXAMPLE 10.5 COMBINING PERCENTAGES

- A 15% raise followed by a 10% raise equals a 26.5% raise. Here is the explanation: If your starting salary is \$$B$, after a 15% raise your salary will be $(1.15)\$B$. After a second raise of 10%, your salary will be $(1.10)(1.15)\$B = (1.265)\$B$. This represents a 26.5% raise of your starting salary.
- If the value of your stock portfolio is \$$B$, after a 25% increase it will be worth $(1.25)\$B$. When you follow that with a 20% drop you have $(0.8)(1.25)\$B$. But $(0.8)(1.25) = 1$, so when all is said and done you are breaking even.
- If the original price of a pair of shoes is \$$B$, after a 20% markdown the price will be $(0.8)\$B$; after an additional 30% markdown the price will be $(0.7)(0.8)\$B = (0.56)\$B$. This represents a combined markdown of 44%.

The second trap with percentages involves the use and abuse of very large numbers. While percentages above 100% are not impossible, one should always be on the alert when large percentages (2000%, 300%, even a measly 110%) are being thrown around. When the coach says that he is happy because his players "gave 110%" he is just using a figure of speech. Nobody can give more than 100% effort, and even that is subject to question. 100% compared to what? Where is the baseline?

A very common misuse of large percentages occurs when stores overstate the discounts they are offering. This is illustrated in our next example.

EXAMPLE 10.6 THE 200% DISCOUNT TRAP

A big electronics store is having a going-out-of-business sale and is advertising a 200% discount on a 55-inch flat-screen TV. That sounds like quite a deal, so you decide to look into it. The TV was originally priced at $3600, and was now marked down to $1200. The salesman claims that reducing the price from $3600 to $1200, represents a 200% discount because $3600 is a 200% increase over $1200. The argument, of course, is completely backwards. The baseline for the decrease is $3600, the final value is $1200, and the percentage change is given by $ $\frac{(1200 - 3600)}{3600} \approx -0.6667$. The correct discount is 66.67%.

Switching the roles of the baseline B and the final value F is a common trick, but here is something to keep in mind: if B and F are both positive, discounts max out at 100%. A discount of 100% means that the item is being given away for free, and a discount higher than 100% means that the store should be paying you to take the item.

The final percentage trap we will briefly discuss involves using percentages to compare percentages.

EXAMPLE 10.7 THE 1% INCREASE TRAP

You just started a job as a sales rep with a major corporation. Your benefits package includes health insurance, and your contribution towards the health insurance premiums is 4% of your salary. Shortly after you start on the job you receive a memo from corporate headquarters announcing that due to rising health insurance premiums the employee contribution will have to increase, but that the increase is very small, just a 1% increase. Next month, when you look at your deductions, you notice that 5% of your salary is being deducted for your health insurance premium.

But jumping from 4% to 5% is not a 1% increase—it is really a 25% increase (never mind those % symbols—just consider the percentage increase when going from 4 to 5).

The point of Example 10.7 is to illustrate the distinction between a 1% increase and an increase of *one percentage point*: 5% is one percentage point higher than 4%, but is 25% larger! Of course, the confusion between percentage points and percent increases can only come up when the quantities being discussed are percentages themselves.

10.2 Simple Interest

We begin this section with a general discussion of *interest*—one of the fundamental concepts in finance.

Interest

Interest is the price someone pays for the use of someone else's money. You can think of interest as a *rental cost*, like the cost of renting a car or an apartment, but applied to money. If you borrow money from a bank you must pay interest to the bank because *you are renting their money*; if you deposit money in a savings account the bank must pay interest to you because *they are renting your money* (so that they can turn around and rent it to someone else).

A more formal description of the concept of interest is as follows: If L (the lender) lends $\$P$ to B (the borrower), then eventually L expects to get more than $\$P$ back from B. [*Note*: this rule applies to business transactions—it does not (or should not) apply when L and B are friends or relatives.] If $\$F$ is the sum B eventually repays L, then the interest is the difference between the principal and the repayment: $I = F - P$. That's the easy version of the story. The devil is in the details, in particular, what are the rules that determine the value of I? This is the main topic we will cover in this and the next section.

The key variables involved in the computation of the interest on a loan are the *principal* P, the *interest rate* r, the *term* of the loan t, the *repayment schedule* and the *type of interest* used, in particular, whether the interest is *simple* or *compounded*. We will define these concepts next.

- **Principal (*P*).** This is the word used to describe the sum of money that the lender L lends the borrower B. (We will assume from here on that all sums of money are given in dollars, so the $ in front of a variable representing money will no longer be needed. When describing a specific sum of money, say $100, it is still important to use the $.)
- **Interest rate (*r*).** This is the rate paid by the borrower B to the lender L for the use of the principal P for a specific unit of time—usually a year. When the interest rate is expressed as a percentage of the principal, and the unit of time is a year, the rate is called an **annual percentage rate (APR)**. The APR is the standard way that financial institutions describe the interest rates on their loans.
- **Term (*t*).** The *term* (or *life*) of a loan is the length of time the borrower is borrowing the money. Typically, the term of a loan is measured in either years or months. (Short-term loans with terms of days or weeks are less common, but they do exist. We will see an example of such a loan later in this section.)
- **Repayment Schedule.** This is a schedule—agreed upon by lender and borrower—for the repayment of the loan. In *single payment* loans the borrower repays the loan in a single lump sum payment at the end of the term. In *installment loans* the borrower repays the loan by making equal monthly payments over the term of the loan. In *credit card loans* the repayment schedule is (up to a point) at the discretion of the borrower. We will discuss how different types of repayment schedules impact the interest on the loan in Section 10.5.
- **Simple interest.** When a loan is based on *simple* interest, the interest rate is applied only to the principal P. In this situation the interest paid on the loan per unit of time is constant (think of a rental where the rental cost is the same each time period). Since an early partial payment on a simple interest loan does not impact the total interest paid, the typical repayment schedule for a simple interest loan is a single payment at the end of the loan.
- **Compound interest.** When a loan is based on *compound* interest, the interest accumulates—first you pay interest on the principal, next you pay interest on the principal and the interest, and so on (think of a rental where the rental cost goes up each time period). For compound interest loans, making partial payments during the term of the loan reduces the total interest paid, and in many cases such as car loans and mortgage loans, monthly payments called *installment payments* are required by the lender. We will discuss auto loans and mortgage loans in Section 10.5.

For the rest of this section we will focus on loans and investments made under simple interest.

The Simple Interest Formula

In simple interest, the cost of the loan is constant over time—the borrower pays the same amount of interest on each time period over the term of the loan. For simplicity let's assume the most common situation: the interest rate r is an APR expressed in decimal form. If P is the principal amount borrowed, the interest on the loan is rP *per year* over the term of the loan, so that if the term of the loan is t years, the total interest paid on the loan is $I = trP$. We call this the **simple interest formula**.

SIMPLE INTEREST FORMULA

$I = Prt$, where I is the interest paid, P is the principal, r is the APR expressed as a decimal, and t is the term of the loan in years. [An alternative version of the simple interest formula gives the *future payoff* F (principal plus interest): $F = P + Prt = P(1 + rt)$.]

EXAMPLE 10.8 BORROWING UNDER SIMPLE INTEREST

Suppose you find a credit union that will give you a student loan under simple interest (this is quite unusual, as most lenders would much rather lend money under compound interest). The APR on the loan is 6%, $(r = 0.06)$, and the term of the loan is three years. Under these terms the credit union will lend you up to a maximum of $6000.

Let's suppose that you are thinking about borrowing $P = \$5000$. At the end of the three-year term $(t = 3)$ you will have to repay the credit union the original principal of $5000 plus interest equal to $I = \$5000 \times 0.06 \times 3 = \$900$.

Now suppose that you decide you can afford to borrow a little more—just enough so that your interest at the end of the three-year term is $1000. How much should you borrow? Here you know the interest I, and the principal P is the unknown. The simple interest formula gives $\$1000 = P \times 0.06 \times 3$. Solving for P gives $P = \$5555.56$.

A more interesting situation arises if you want to pay off the loan early. Suppose that two and a half years after you take out the $5000 loan you come into some money and decide you want to pay the loan back. Can you compute the interest by plugging $t = 2.5$ into the simple interest formula? The answer depends on the fine print in your loan agreement. If the loan agreement allows you to pay off the loan before the end of the original term, then yes, the interest you will owe after 2.5 years is given by $I = \$5000 \times 0.06 \times 2.5 = \$750$, and your payoff will be $F = \$5750$. A more likely scenario, however, is that your loan term is fixed at three years and whether you pay off the loan early or not the credit union will demand a payoff of $5900.

One of the lessons of Example 10.8 is that it is always a good idea to know all the details (not just the APR) of any loan agreement you sign. Fortunately, you don't have to pore over the fine print on your own—the Consumer Credit Protection Act requires lenders to fully disclose, and in clear language, the terms of any loan agreement into which they enter. And if you have any questions, you should always ask.

Bonds

Simple interest is also used on certain types of investment instruments, such as government and corporate *bonds*. Government agencies, including the United States government, states, cities, and municipalities issue bonds as a way to raise money. Private corporations do the same. When you buy a **bond**, you are essentially lending your money to the agency issuing the bond. Bonds can be *government*, *muni*, or *corporate* bonds, depending on whether the issuing agency is a government (state or federal), a municipality, or a corporation. The typical bond is defined by the *term* of the bond (3 years, 5 years, etc.), the APR, and the *face value* F of the bond. The **face value** (also called the **par value**), of a bond is the amount you get back from the issuing agency when the bond reaches *maturity* (i.e., at the end of the term).

EXAMPLE 10.9 BUYING MUNICIPAL BONDS

Suppose you have the opportunity to buy a five-year municipal bond with a face value of $1000 and an APR of 3.25% $(r = 0.0325)$. What is the original price of the bond?

We can answer this question using the simple interest formula, but we have to be a little careful: the face value is the payoff F, so the formula that applies now is $F = P + Prt$. The unknown in this case is the principal P, and solving for P gives $P = \frac{F}{(1+rt)}$. In our example, $F = \$1000$, $r = 0.0325$, and $t = 5$, so $P = \frac{1000}{1 + 0.0325 \times 5} = \frac{1000}{1.1625} = \$860.22$ (rounded to the nearest penny).

In the formula $F = P + Prt = (1 + rt)P$, the face value and the principal are in direct proportion: If the principal on a bond with $1000 face value is P, the principal on the same bond with a different face value is proportional to P—the principal on a $5000 bond is $5P$, the principal on a $2500 bond is $(2.5)P$, and so on.

Payday Loans

> "If you are considering taking out a payday loan, I'd like to tell you about a great alternative. It's called 'Anything Else'."
>
> – *Sarah Silverman*

A **payday loan** is a short-term loan (the typical term is a few days or weeks) where one can borrow a small sum (usually a maximum of $1000 or $1500) using a future paycheck or a personal check as collateral. Typically, the loan has to be repaid within a couple of weeks. The payday loan (also known as *cash advance*) business is big business, and in many neighborhoods (especially poor neighborhoods) finding a payday loan company is easier than finding a grocery store.

There is sound logic behind the payday loan business: sometimes people are short on cash or have an emergency where they need quick access to money and don't have the time or patience to go to a bank and apply for a conventional loan. Helping people out of temporary jams is a good thing to do, and in this sense, payday loan companies provide a good service. And, since payday loan companies operate with higher risk factors, it is reasonable and expected that they will charge higher interest rates than conventional banks. But how high is too high? Example 10.10 answers this question using data from a typical payday loan company.

EXAMPLE 10.10 THE STEEP PRICE OF QUICK CASH

Your best friend has put together a fantastic road trip for spring break. All you need is $400 for your share of the gas, food, lodging, and miscellaneous expenses. You really want to go, but you are totally broke, you have no savings, your credit card is maxed out, and your last semester grades were not so great, so asking your parents is out of the question.

You always wondered about that *QuickCash&Go* branch in your neighborhood. What would it take to get a quick $400 loan from them? You have that tax refund check coming soon that you can use as collateral, so you decide to check it out.

QuickCash&Go charges $15 for every $100 lent. The term of the loan is 14 days—you must pay back the principal plus interest in two weeks! The total interest on a $400 loan is $I = 4 \times \$15 = \$60$. An interest charge of $60 on a $400 loan may not seem exorbitant, but let's check out the APR on this loan. Solving the simple interest formula $I = Prt$ for r gives $r = \frac{I}{Pt}$. In our example, $I = \$60$ and $P = \$400$. What about t? Since the interest rate is annual, we need to express the term of the loan in years: 14 days $= \frac{14}{365}$ years means that $t = \frac{14}{365}$. Putting it all together gives $r = \frac{60}{400} \times \frac{365}{14} \approx 3.9107 \approx 391.07\%$. Yikes!

For legal reasons, we did not use a real payday loan company in Example 10.10, but *QuickCash&Go's* APR of 391.07% on a 14-day loan is not unusual in the payday loan industry, where APRs can run as high as 400%. There has been a backlash against the types of interest rates charged by payday loan companies: 17 states have banned payday loan companies from doing business; other states have passed "usury" laws that put a cap on the APRs that can be charged by lenders in that state. Meanwhile, if you are ever in an emergency and need quick cash, do whatever you can to find a better source of quick money than a payday loan company.

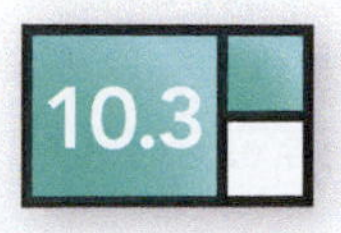

Compound Interest

Under *simple interest* the interest rate is applied only to the principal. Under **compound interest**, not only does the principal generate interest, so does the previously accumulated interest. All other things being equal, money invested under compound interest grows a lot faster than money invested under simple interest, and this difference gets magnified over time. If you are investing for the long haul (a college trust fund, a retirement account, etc.), always look for compound interest.

Annual Compounding

EXAMPLE 10.11 YOUR TRUST FUND FOUND!

Imagine that you have just discovered the following bit of startling news: On the day you were born, your Uncle Nick deposited \$5000 in your name in a trust fund account. The trust fund account pays a 6% APR and the interest is compounded once a year at the end of the year. One of the provisions of the trust fund was that you couldn't touch the money until you turned 18. You are now 18 years, 10 months old and you are wondering, How much money is in the trust fund account now? How much money would there be in the trust fund account if I wait until my next birthday when I turn 19? What is the *future value* of the trust fund (i.e. how much money would there be in the account) if I left the money in for retirement and waited until I turned 60?

Here is an abbreviated timeline of the money in your trust fund account, starting with the day you were born:

- Day you were born: Uncle Nick opens the account with a \$5000 deposit.
- First birthday: 6% interest is added to the account. Balance in account is $(1.06)\$5000$. (We'll leave the expression alone and do the arithmetic later.)
- Second birthday: 6% interest is added to the previous balance. Balance in account is $(1.06)(1.06)\$5000 = (1.06)^2\,\$5000$.
- Third birthday: 6% interest is added to the previous balance. Balance in account is $(1.06)(1.06)^2\,\$5000 = (1.06)^3\,\$5000$.

At this point you might have noticed that the exponent of (1.06) in the calculation of the account balance goes up by 1 on each birthday and, in fact, matches the birthday. Thus,

- Eighteenth birthday: The balance in the account is $(1.06)^{18}\,\$5000$. It is finally time to pull out a calculator and do the computation:

$$(1.06)^{18}\,\$5000 = \$14{,}271.70 \quad \text{(rounded to the nearest penny)}$$

- Today: Remember—right now you are supposed to be 18 years and 10 months old. Since the bank only credits interest to your trust fund account once a year and you haven't turned 19 yet, the balance in the account is still \$14,271.70.

- Nineteenth birthday: The future value of the account on your nineteenth birthday will be

$$(1.06)^{19}\ \$5000 = \$15{,}128 \quad \text{(rounded to the nearest penny)}$$

Moving further along into the future, we have

- Sixtieth birthday: The future value of the trust fund account is

$$(1.06)^{60}\ \$5000 = \$164{,}938.45$$

which is an amazing return for a $5000 investment (if you are willing to wait, of course)!

Figure 10-1(a) plots the growth of the money in the account for the first 18 years. Figure 10-1(b) plots the growth of the money in the account for 60 years.

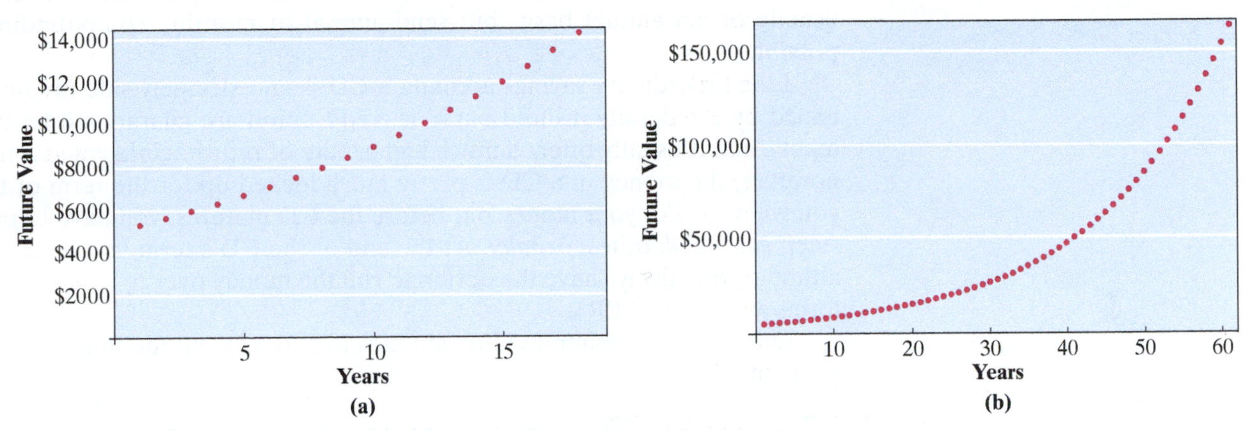

FIGURE 10-1 Cumulative growth of $5000 at 6% interest compounded annually.

EXAMPLE 10.12 YOUR TRUST FUND FOUND! (TIMES TWO)

Let's revisit the story of the trust fund set up for you by your Uncle Nick. Same story, but now we will double down on some of the numbers.

First, let's see how the numbers would unfold if Uncle Nick had initially deposited $10,000 instead of $5000 into the trust account. Without going into the same amount of detail as we did in Example 10.11, but following the same pattern, we have

- Eighteenth birthday: Balance on the trust fund is $(1.06)^{18}\ \$10{,}000 = \$28{,}543.39$.
- Sixtieth birthday: Balance on the trust fund is $(1.06)^{60}\ \$10{,}000 = \$329{,}876.91$.

The key thing to notice is that the numbers above are double the corresponding numbers in Example 10.11 (there is a difference of a penny due to rounding errors). No big surprises here—when you double the initial investment, you also double the balance on the account (at each particular moment in time).

Next, let's see what happens if we double the APR (from 6% to 12%) but leave the initial deposit at $5000.

- Eighteenth birthday: Balance on the trust fund is $(1.12)^{18}\ \$5000 = \$38{,}449.83$.
- Sixtieth birthday: Balance on the trust fund is $(1.12)^{60}\ \$5000 = \$4{,}487{,}984.67$.

When we compare these numbers with the numbers in Example 10.11, we see that doubling the interest more than doubles the balance, and the effect gets larger with time. (After 60 years, the balance on a 12% APR is more than 27 times the balance on a 6% APR for the same initial deposit!)

To get the formula for annual compounding we generalize the computation used in Example 10.11 to any principal P, any APR r, and any term t given in whole number of years (when the term is not a whole number we need to round it down, since a fraction of a year won't generate any interest).

ANNUAL COMPOUNDING FORMULA

The future value F of P dollars *compounded annually* for t years with an APR r (expressed in decimal form) is $F = P(1 + r)^t$.

Certificates of Deposit

A **certificate of deposit (CD)** is a type of investment that is a hybrid between an ordinary savings account and a bond. Much like a bond, when you buy a CD you are lending your money—in this case to a bank or credit union—for a fixed term and at a fixed APR. Unlike a bond, however, the interest on a CD is *compounded*, usually on an annual basis, but semi-annual or monthly compounding are also possible.

Like an ordinary savings account, a CD is an extremely safe investment (CDs issued by a federally insured bank or credit union are guaranteed by the government), and it usually offers a much higher rate of return. Unlike a savings account, however, the money in a CD is pretty much locked up for the term of the CD—if you want to take your money out before the CD matures, you have to pay a pretty steep *early withdrawal penalty*. At the end of the CD's term you have to cash out, although you always have the option to roll the money over to a new CD (with new terms and a new APR).

Our next four examples illustrate the effect that compounding has on the future value of a CD.

EXAMPLE 10.13 FUTURE VALUE OF A CD UNDER ANNUAL COMPOUNDING

Suppose you invest $1500 on a CD with an APR of 4.8% compounded annually. Let t denote the length (in years) of the CD's term. Using the annual compounding formula with $P = \$1500$ and $r = 0.048$ gives the future value F of the CD when it matures in t years: $F = \$1500(1.048)^t$. Table 10-1 shows the values of F corresponding to 1-, 2-, 5-, 10-, and 15-year CDs.

t	$F = \$1500(1.048)^t$
1	$1572.00
2	$1647.46
5	$1896.26
10	$2397.20
15	$3030.47

TABLE 10-1 ■ Annual compounding ($P = \$1500$, APR = 4.8%, F rounded to the nearest penny)

An interesting fact to observe in Table 10-1 is that the 15-year CD roughly doubles the original $1500 investment. There is a general rule of thumb, called the *rule of 72*, that estimates how many years it takes to double an original investment under annual compounding. The rule of 72 is independent of the principal and only depends on the APR.

RULE OF 72

To estimate how long it takes (in years) to double an investment under annual compounding, take the number 72 and *divide it by the APR* (here you use the percentage value of the APR).

If we apply the rule of 72 to an APR of 4.8% we get a doubling time of $72 \div 4.8 = 15$ years. Compare this result with the last row of Table 10-1. That's amazingly accurate! (OK, time to fess up—using an APR of 4.8% is stacking the deck in favor of the rule of 72 because, unlike most APRs, 4.8 divides evenly into 72.)

For a more typical use of the rule of 72 consider a CD compounded annually with a 7% APR. The rule of 72 gives an estimate of $72 \div 7 = 10.2857\ldots$ years for the doubling time. We interpret this to mean that 10 years is not quite long enough to double our money and 11 years is plenty long.

General Compounding

Every so often, one can find a bank or credit union that offers CDs compounded more than once a year—twice a year (*semi-annually*), four times a year (*quarterly*), even 12 times a year (*monthly*). In the case of savings accounts, there is *daily* compounding and even *continuous* compounding. The more frequent the compounding the better it is for the investor as more and more interest accumulates on the principal, but as we will soon see, the frequency of compounding has less of an impact than one would expect.

EXAMPLE 10.14 FUTURE VALUE OF A CD UNDER QUARTERLY COMPOUNDING

For comparison purposes we will use the same principal ($P = \$1500$) and the same APR (4.8%) as in Example 10.13. The only difference is that we are now looking at a CD that compounds quarterly (four times a year).

We will start by computing the future value of the CD after one year (four quarters). To do so we need to find the balance on the account at the beginning of each quarter—that is the amount that generates the interest for that quarter. The computation is analogous to the one we used in Example 10.13 with one important difference: the interest rate for the quarter is not 4.8% (that's the APR!) but rather 1.2% (the APR divided by four!).

The timeline showing the balance on the account at the start of each quarter is given below. (All sums are left as uncalculated numerical expressions—the time for number crunching is at the very end.)

- First quarter: Starting balance is $P = \$1500$.
- Second quarter: Starting balance: $\$1500 \times (1.012)$.
- Third quarter: Starting balance: $\$1500 \times (1.012)^2$.
- Fourth quarter: Starting balance: $\$1500 \times (1.012)^3$.
- Balance at the end of one year: $\$1500 \times (1.012)^4 = \$1573.31$.

t	T	$F = \$1500 \times (1.012)^T$
1	4	\$1573.31
2	8	\$1650.20
5	20	\$1904.15
10	40	\$2417.20
15	60	\$3068.47

TABLE 10-2 ■ Quarterly compounding ($P = \$1500$, APR = 4.8%)

We can use the same basic idea to find the future value of the CD after any number of years. The only number that will change is the exponent of (1.012). Table 10-2 shows the future value of the CD after $t = 1, 2, 5, 10$, and 15 years under quarterly compounding. Notice the introduction of a new variable T representing the total number of times the interest was compounded.

EXAMPLE 10.15 FUTURE VALUE OF A CD UNDER MONTHLY COMPOUNDING

t	T	$F = \$1500(1.004)^T$
1	12	\$1573.61
2	24	\$1650.82
5	60	\$1905.96
10	120	\$2421.79
15	180	\$3077.23

TABLE 10-3 ■ Monthly compounding ($P = \$1500$, APR = 4.8%)

We now consider the future value of a CD under monthly compounding (12 times a year). For comparison purposes we will still use the principal $P = \$1500$ and the APR of 4.8%, as in Examples 10.13 and 10.14.

Now the interest rate in each compounding period is $\frac{4.8\%}{12} = 0.4\% = 0.004$, and there are 12 compounding periods per year. The future value of the CD after one year is given by $F = \$1500 \times (1.004)^{12} = \$1573.61$. Table 10-3 shows the future value of the CD after $t = 1, 2, 5, 10$, and 15 years under monthly compounding. The total number of times the interest was compounded is given by T.

The idea behind Examples 10.14 and 10.15 can be generalized into what is known as the *general compounding formula.*

GENERAL COMPOUNDING FORMULA

$F = P(1 + p)^T$, where F is the future value, P is the principal, p is the periodic interest rate, and T is the number of times the interest is compounded.

The general compounding formula $F = P(1 + p)^T$ is an extension of the annual compounding formula $F = P(1 + r)^t$, but with two changes: The APR r is replaced by the **periodic interest rate** p (obtained by dividing the APR by the number of compounding periods in one year), and the number of years t in the exponent is replaced by T (the total number of times the interest is compounded over the life of the investment).

Table 10-4 shows the details of the general compounding formula when applied to a 15-year CD $(t = 15)$ with a principal $P = \$1500$ and an APR of 4.8% using annual, semi-annual, quarterly, monthly, and daily compounding.

Compounding	p	T	$F = \$1500(1 + p)^T$
annually	0.048	15	\$3030.47
semi-annually	0.024	30	\$3055.55
quarterly	0.012	60	\$3068.47
monthly	0.004	180	\$3077.23
daily*	$\frac{0.048}{365}$	$365 \times 15 = 5475$	\$3081.50

*In financial calculations, every year is assumed to have 365 days.

TABLE 10-4 ■ Future values of a 15-year-term \$1500 CD with APR of 4.8% (values rounded to the nearest penny)

The surprising thing about the future values in Table 10-4 is how little they change as we crank up the frequency of compounding. You would think that over 15 years the difference between daily compounding and monthly compounding should be huge—in reality the difference buys you a cup of coffee and a bagel (if you are lucky)!

A remarkable property of the general compounding formula is that there is a point at which cranking up the frequency of compounding makes very little difference to the future value. If you were to add one more row to Table 10-4 for compounding on an hourly basis (24 hours a day, 365 days a year), you would see an increase in the future value (over daily compounding) of 14 cents.

Continuous Compounding

The ultimate form of compounding is **continuous compounding**—interest compounded infinitely often over infinitely small compounding periods. The idea sounds great, but how do you calculate it? There is a surprisingly nice formula for computing future values under continuous compounding called the *continuous compounding formula*. The surprise is that the formula involves an unexpected guest—the irrational number $e = 2.71828...$ (You may remember seeing e in the context of logarithms, but e plays an important role in financial mathematics as well.) To understand why e shows up in the continuous compounding formula requires some calculus, so we are just going to give the formula and omit any justification for it.

CONTINUOUS COMPOUNDING FORMULA

$F = Pe^{rt}$, where F is the future value, P is the principal, r is the APR (in decimal form), and t is the number of years of continuous compounding.

- To use the continuous compounding formula you will need a calculator with an e^x key. Since e is an irrational number with an infinite, nonrepeating decimal expansion, a calculator can only give an approximate value for e^{rt} and consequently for F itself. You can minimize the error by using a good scientific calculator. There are many good online scientific calculators available for free.
- The continuous compounding formula works even when the term t is a fractional part of a year. For example, for a term of 5 years and 3 months you can use $t = 5\frac{1}{4} = 5.25$; for a term of 500 days you can use $t = \frac{500}{365} = 1\frac{135}{365} = 1.36986301$, and so on.

EXAMPLE 10.16 FUTURE VALUE OF A CD UNDER CONTINUOUS COMPOUNDING

For the last time, let's consider a CD with principal $P = \$1500$ and APR of 4.8%, as in Examples 10.13 through 10.15, but this time the compounding is continuous. Table 10-5 shows the future value of the CD for terms of 1, 2, 5, 10, and 15 years.

Compare the numbers in Table 10-5 with those in Tables 10-2 and 10-3, and you will understand why banks and credit unions don't mind offering their customers continuous compounding on savings accounts and CDs. It's better business to entice customers with continuous compounding than to offer higher APRs. (For example, a CD with 4.8% APR compounded quarterly is a much better offer than a CD with 4.7% APR compounded continuously, even though the latter sounds a lot more impressive.)

t	$F = \$1500e^{(0.048)t}$
1	\$1573.76
2	\$1651.14
5	\$1906.87
10	\$2424.11
15	\$3081.65

TABLE 10-5 ▪ Continuous compounding ($P = \$1500$, APR = 4.8%)

Before we move on, here are two useful observations that apply to *all* the compounding formulas we have discussed so far in this section (annual, general, and continuous):

1. There is a *linear* relationship between the future value F and the principal P. Essentially this means that if we leave the other variables (APR, compounding frequency, and time) fixed, then P and F change proportionally: double P and you double F, triple P and you triple F, cut P in half and you cut F in half. To illustrate this point, take a \$1500 CD compounded continuously for 15 years with a 4.8% APR. We saw in Example 10.16 (Table 10-5) that the future value is $F = \$1500e^{(0.048)15} = \$3081.65$. Now, if we double the principal to \$3000 and leave the other variables the same, the future value is given by $F = \$3000e^{(0.048)15} = \$6163.30$, exactly twice the amount of the future value under the \$1500 principal.

2. The relationship between F and the other variables (APR, compounding frequency, and time) is *not* linear. To illustrate this point, take once again the $1500 CD compounded continuously for 15 years with a 4.8% APR. The future value is $F = \$1500e^{(0.048)15} = \$3081.65$. Now, if we double the APR to 9.6% and leave the other variables the same, the future value is given by $F = \$1500e^{(0.096)15} = \$6331.04$, more than double the future value under the 4.8% APR. And if we double instead the number of years to 30, we get a future value of $F = \$1500e^{(0.048)30} = \$6331.04$.

Savings Accounts

Other than putting your money under the mattress, a savings account is the most conservative type of investment you can make. Savings accounts are extremely safe, and extremely *liquid*—in general you can take part or all of the money out at any time without a penalty (sometimes there are a few restrictions that apply for an initial time period).

Unlike a CD, a savings account has no term (you can leave your money in the savings account for as long as you want) and the APR is not locked—it may go up or down depending on interest rates in general. In a typical savings account the interest is compounded continuously. So what's not to like? There is just one problem with savings accounts—they have very low APRs, and, therefore, you don't get very much interest. That's the price you have to pay if you want a secure, safe, and liquid investment.

In the old days (20 or more years ago), when you opened a savings account you would get a passbook that would allow you to keep a record of your deposits and withdrawals. These days most people use an app on their smart phone to do the same. Even if you never plan to open a savings account or have a nice phone app that does the thinking for you, you should go over the next example carefully. The basic ideas behind the calculations in this example also apply to the calculation of credit card balances (Example 10.24), and that is something that concerns practically everyone.

EXAMPLE 10.17 CONTINUOUS COMPOUNDING IN SAVINGS ACCOUNTS

Imagine you have a savings account with an APR of 1.8% compounded continuously. Your passbook shows a starting balance of $1500 on January 1, a withdrawal of $250 on March 27, a deposit of $400 on July 10, and a withdrawal of $300 on September 2. Let's track the balances on your account for one calendar year (Jan. 1–Dec. 31). [You can conveniently follow all of the calculations using the *Finance Calculator* applet.]

The deposit and withdrawals break up the calendar year into periods of various lengths. Table 10-6 (column 3) shows the length of each period and how the continuous compounding formula is applied to compute the balance in the account (column 5). There are 86 days in the period from January 1 to March 27—measured in years gives $t = \frac{86}{365}$. The balance on March 27 is given by $F = \$1500e^{(0.018)\left(\frac{86}{365}\right)} = \$1506.38$. After a $250 withdrawal on March 27, the principal for the next period of 105 days (March 28–July 10) equals $1256.38. The balance on July 10 is $(\$1256.38)e^{(0.018)\left(\frac{105}{365}\right)} = \$1262.90$. After a $400 deposit on July 10, the principal for the next period of 54 days (July 11–Sept. 2) equals $1662.90. The balance on Sept. 2 is $(\$1662.90)e^{(0.018)\left(\frac{54}{365}\right)} = \$1667.33$. After a $300 withdrawal on Sept. 2, the principal equals $1367.33. Since there are no other deposits or withdrawals, the last period is Sept. 3 to Dec. 31 (120 days) and the balance at the end of the calendar year is $(\$1367.33)e^{(0.018)\left(\frac{120}{365}\right)} = \$1375.45$.

Date	P	Number of days in period	t	Balance $F = Pe^{rt}$	Deposit (Withdrawal)	Ending balance
Jan. 1	$0	0	0	$0	$1500	$1500
March 27	$1500	86	$\frac{86}{365}$	$1506.38	($250)	$1256.38
July 10	$1256.38	105	$\frac{105}{365}$	$1262.90	$400	$1662.90
Sept. 2	$1662.90	54	$\frac{54}{365}$	$1667.33	($300)	$1367.33
Dec. 31	$1367.33	120	$\frac{120}{365}$	$1375.45	$0	$1375.45

TABLE 10-6 ▪ Savings account balances for Example 10.17 ($r = 0.018$)

Annual Percentage Yield

The **annual percentage yield (APY)** (sometimes called the *effective rate*) is the annual percentage increase on the value of an investment (or loan) when both the APR and the frequency of compounding are taken into account. When comparing investments with different APRs and different compounding frequencies, the APY is the right tool to use. Which is a better investment: a 4.8% APR compounded quarterly or a 4.75% APR compounded continuously? Which one would you choose? We'll answer the question in our next example.

Given the APR, the simplest way to find the APY is to use the appropriate compounding formula to compute the value of $1 at the end of one year. Depending on the type of compounding, we use either the general or the continuous compounding formula.

EXAMPLE 10.18 COMPARING INVESTMENTS

Which is a better investment: a 4.8% APR compounded quarterly or a 4.75% APR compounded continuously? We can answer this question by calculating the APY of each investment.

- 4.8% APR compounded quarterly $\left(p = \frac{0.048}{4} = 0.012\right)$: In one year, $P = \$1$ becomes $F = \$1 \times (1.012)^4 \approx \$1.04887$ (here we used the general compounding formula with $T = 4$). The percentage increase can be read directly from the decimal part of F—it is 4.887%. The APY, therefore, is 4.887%.
- 4.75% APR compounded continuously: In one year, $P = \$1$ becomes $F = \$1 \times e^{0.0475} = \$1.048646$ (here we used the continuous compounding formula with $t = 1$). The APY, therefore, is 4.8646%.

Comparing the two APYs, we can see that an APR of 4.8% compounded quarterly is just a tad better than an APR of 4.75% compounded continuously. (A note of caution: This is one place where you do *not* want to round off your calculations to the nearest penny. Had we rounded the values of F to the nearest penny, we would have ended up with $F = \$1.05$ and an incorrect APY of 5% in both cases. By going six decimal places deep in our calculations, we were able to pick up the difference between the two options.)

ANNUAL PERCENTAGE YIELD

The APY is given by:

- $\text{APY} = \left(1 + \frac{r}{T}\right)^T - 1$ when the compounding takes place T times a year.
- $\text{APY} = e^r - 1$ when the compounding is continuous.

(In both cases, r is the APR expressed as a decimal.)

Rather than memorize two more formulas, remember that what we are doing is finding the value of one dollar at the end of one year (using either the general or the continuous compounding formula), subtracting 1, and expressing the result as a percentage. In all cases, we need to run our calculations to at least five or six decimal places of precision. The next example illustrates the point.

EXAMPLE 10.19 COMPUTATION OF APYS

Table 10-7 shows the computation of APYs for a 6% APR under various compounding frequencies.

Compounding	T	F	APY
annually	1	1.06	6.000%
semi-annually	2	$\left(1+\frac{0.06}{2}\right)^2 = 1.0609$	6.090%
quarterly	4	$\left(1+\frac{0.06}{4}\right)^4 \approx 1.061364$	6.1364%
monthly	12	$\left(1+\frac{0.06}{12}\right)^{12} \approx 1.061678$	6.1678%
continuously	∞	$e^{0.06} \approx 1.061837$	6.1837%

TABLE 10-7 ■ Computation of APYs for a 6% APR

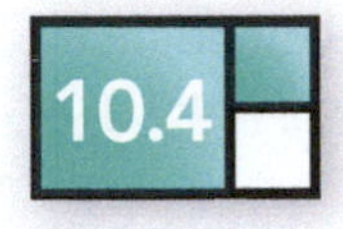

10.4 Retirement Savings

Saving money for retirement? Pleeease—that's so tomorrow! There are so many pressing demands on our hard-earned money. It's much easier to spend now and worry about the future later.

Well, if you think that way, you have plenty of company. A 2015 study conducted by the National Institute on Retirement Security (NIRS) found that the median retirement account balance among all American working households is $3000, and 45% of these households have no retirement savings at all.

Little or no retirement savings, increased longevity, expanding health care costs, and decreasing Social Security benefits—these are the makings of a perfect storm, and unfortunately, many working Americans will be caught in this storm. Surprisingly, there is a good recipe that will enable you to avoid the storm altogether, and the basic ingredients are a healthy dose of self-discipline and a little dash of mathematics.

We start this section by asking the following question: What happens when you set aside a fixed sum of money on a regular basis, leave it alone, and let it compound at a fixed interest rate over a long period of time? To keep things simple, we will assume that the money is *deposited once a year at the end of the year* and that it is *compounded once a year*. In Example 10.20 we will look at what happens when you set aside $1000 a year for 44 years with an annual return of 6% (we'll call this a "conservative strategy"). In Example 10.21 we will double the amount set aside each year to $2000 but cut in half the amount of time over which you do it, so that the time becomes 22 years (we'll call this a "delayed strategy"). Before we go into the details, try guessing which of these two strategies will generate more money (or maybe it's the same in both). You might even take a stab at estimating roughly how much money you will have for retirement in each case.

EXAMPLE 10.20 RETIREMENT SAVINGS I: A CONSERVATIVE STRATEGY

According to a 2015 Gallup poll, the typical American expects to retire at the age of 65. So let's say that you plan to retire at 65, and you start working at 21. You will work for a full 44 years, and you plan to start a retirement savings account right away, at the age of 21. You will set aside $1000 each year (that sounds like a lot, but when you think about it, it's less than the cost of one latte a day) and will get a fixed return of 6% each year. How much money will you have in your retirement savings account at the end of 44 years? Will there be enough for you to retire comfortably?

We will break this problem down into small bites, starting from the beginning and keeping in mind that the money is deposited into the retirement savings account at the end of each year.

- **First $1000 deposit:** You deposit the first $1000 *at the end* of Year 1. At the end of Year 2 this $1000, plus the 6% interest, becomes $1060. For convenience we will write this in the form $\$1000(1.06)$. At the end of Year 3 the original $1000 becomes $\$1000(1.06)^2$, at the end of Year 4 it becomes $\$1000(1.06)^3$, and so on. At the end of Year 44 the first $1000 deposit has grown to $\$1000(1.06)^{43}$. We will leave this figure uncalculated for now.
- **Second $1000 deposit:** You make the second $1000 deposit *at the end* of Year 2. At the end of Year 3 this $1000 grows to $\$1000(1.06)$, at the end of Year 4 to $\$1000(1.06)^2, \ldots$, and at the end of Year 44 to $\$1000(1.06)^{42}$.
- **Third $1000 deposit:** You make the third $1000 deposit *at the end* of Year 3. By now the pattern should be clear—at the end of Year 44 this $1000 has grown to $\$1000(1.06)^{41}$.

⋮

- **Forty-Third $1000 deposit:** You make a $1000 deposit *at the end* of Year 43. At the end of Year 44 this money has been in the account for only one year and has grown to $\$1000(1.06)$.
- **Last $1000 deposit:** Should you deposit $1000 into the retirement savings account *at the end* of Year 44? On the one hand it's pointless, since you are about to retire and close the retirement savings account right away. On the other hand, it makes the calculations a little "nicer," and there is really no harm in doing it. So for no other reason than convenience, we will assume your last $1000 deposit goes in as $1000 and comes back out as $1000.

We can now piece the whole story together: The money in your retirement savings account at the end of 44 years (we'll call it V) under the assumptions we made ($1000 invested every year at the end of the year at 6% annual interest) is given by

$$V = \$1000 + \$1000(1.06) + \cdots + \$1000(1.06)^{41} + \$1000(1.06)^{42} + \$1000(1.06)^{43}$$

It so happens that the above sum is a geometric sum consisting of $N = 44$ terms, initial term $P_0 = \$1000$, and common ratio $R = 1.06$. What a perfect opportunity to bring back the *geometric sum formula* from Chapter 9!

$$V = \$1000\left[\frac{(1.06)^{44} - 1}{1.06 - 1}\right] = \$199{,}758.03$$

Congratulations! With just a little sacrifice and discipline, you will have generated a retirement savings account worth nearly $200,000. ■

Before we go into our next example, it will be useful to generalize the retirement savings calculations in Example 10.20 into what we will call, for ease of reference, the *retirement savings formula for annual contributions*.

RETIREMENT SAVINGS FORMULA (ANNUAL CONTRIBUTIONS)

$$V = P\left[\frac{(1+r)^Y - 1}{r}\right]$$

Here P represents the annual contribution made *at the end of each year*, Y represents the number of years of retirement savings, and r represents the annual interest rate expressed in decimal form. [*Note:* The lonely r in the denominator is what's left after the original denominator $(1 + r) - 1$ is simplified.]

Three relevant observations about the retirement savings formula are in order. First, the retirement savings formula is just a special application of the geometric sum formula introduced in Section 9.3. Second, the formula assumes that payments into the retirement savings account are made *at the end* of each year and don't start generating interest until the end of the following year. Third, the last payment made at the end of year Y generates no interest, and we throw it in only to simplify the calculations. (If that last payment is *not* made, then the total value of the retirement savings account becomes $P\left[\frac{(1+r)^Y - 1}{r}\right] - P$.)

EXAMPLE 10.21 RETIREMENT SAVINGS II: A DELAYED STRATEGY

When workers without a retirement savings account are asked why they don't have one, the most common answer, especially among young workers, is that there is always time to create such an account later and catch up. In other words, save more once you start, but wait until later to get started.

Let's imagine that your situation is just like that in Example 10.20 (you start working at 21 and plan to retire at 65), but your strategy toward retirement savings is different: You will postpone funding your retirement savings account for 22 years, but then you will double down and contribute \$2000 a year to the account in each of the last 22 years. (The APR hasn't changed—it is still 6%.) You are thinking that this will enable you to catch up and save about the same amount as in Example 10.20. Let's see.

Without any further ado we apply the retirement savings formula, with $P =$ \$2000, $Y = 22$, and $r = 0.06$, and get

$$V = \$2000\left[\frac{(1.06)^{22} - 1}{0.06}\right] = \$86{,}784.58$$

Your retirement savings account would now have slightly under \$87,000, compared to the roughly \$200,000 you would have accumulated by saving *half as much for twice as long*. OK, what if you decided to up the ante even more and contribute \$4000 a year for the last 22 years? Now the retirement savings formula would give

$$V = \$4000\left[\frac{(1.06)^{22} - 1}{0.06}\right] = \$173{,}569.16$$

Surprise—even a \$4000 annual contribution over 22 years generates less in retirement savings than a \$1000 annual contribution over 44 years!

Before we move on, a couple of comments about the retirement savings formula are in order.

1. There is a *linear* relationship between the value V of the retirement savings account and the annual contributions P (haven't we heard this line before?). For example, we saw in Example 10.21 that a \$2000 annual contribution for 22 years with a 6% APR results in a retirement savings account with a value of $V =$ \$86,784.58, while a \$4000 annual contribution under the same conditions results in a retirement savings account with a value of $V =$ \$173,569.16—exactly twice as much! We leave it as an exercise for the reader to explain why this is not a coincidence (Exercise 75).
2. The relationship between the value of the retirement savings account and the number of years of retirement savings is far from linear—in fact, it is *exponential*. Table 10.8 illustrates this point. The table shows the value V of a retirement savings account with an annual contribution of \$1000 and an APR of 6% as the number of years Y increases from 10 to 50 (in increments of 5 years); Fig. 10-2 is a scatterplot showing a more detailed picture of how V grows in relation to Y. (Notice the similarity to Fig. 10-1, except that here the growth is even more pronounced.)

	Y = 10	Y = 15	Y = 20	Y = 25	Y = 30	Y = 35	Y = 40	Y = 45	Y = 50
V	\$13,181	\$23,276	\$36,786	\$54,865	\$79,058	\$111,435	\$154,762	\$212,744	\$290,336

TABLE 10-8 ■ Retirement savings at the end of Y years per \$1000 of annual savings with a 6% APR.

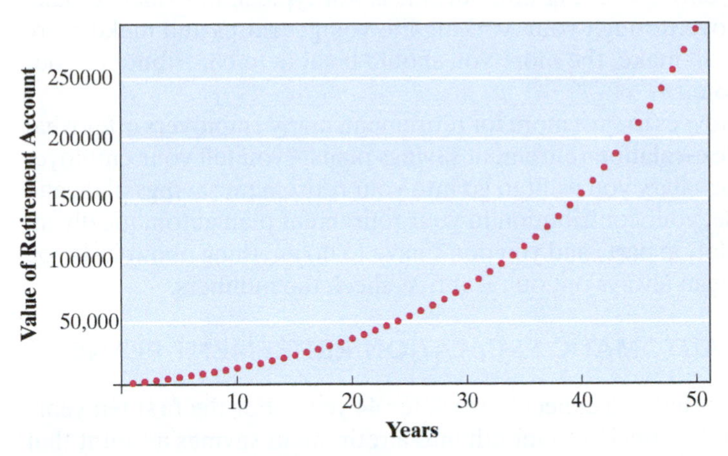

FIGURE 10-2 Retirement savings at the end of Y years per \$1000 of annual savings with a 6% APR.

If there is a moral to Table 10-8 and Fig. 10-2, it's simply this: *If you are serious about setting aside money for a comfortable retirement, start saving early!*

In Examples 10.20 and 10.21 we discussed retirement savings plans based on making *annual* deposits into the account; that was the easiest way to get us started. In real life, however, the more common type of retirement savings plan is one based on *monthly* deposits into the account. In fact, most employer-sponsored retirement plans (*IRA*'s, *401k*'s, *403b*'s, etc.) use an automatic monthly contribution system. That is, you tell your employer how much to take out of your paycheck each month, and the money goes directly into a retirement fund of your choosing. If you are lucky and have a great employer, your employer matches your contribution.

EXAMPLE 10.22 TAX-DEFERRED EMPLOYER-SPONSORED RETIREMENT PLANS

Let's suppose that your employer offers a tax-deferred retirement plan (the contributions are not taxed until you take the money out of the account). In the typical employer-sponsored retirement account, the money is taken directly out of your paycheck, and you don't have to think too much about it. You just tell your employer how much you want taken out each month and where you want the money to go.

So that we can compare apples with apples, we will once again use 44 years of savings, an annual interest rate of 6% *compounded monthly*, and a *monthly* contribution of \$83.33 ($\$83.33 \times 12 = \$999.96$, and that's the closest we can come to matching a \$1000 annual contribution). As before, contributions are made at the end of the period (in this case at the end of the month), and the last contribution is made at the end of the last month even though it doesn't generate any interest.

To cut to the chase, here is the calculation that gives us the value of the retirement savings account at the end of 44 years.

$$V = \$83.33\left[\frac{(1.005)^{528} - 1}{0.005}\right] = \$215{,}346.82$$

In this case the \$83.33 is the monthly contribution, the $0.005 = 0.06/12$ is the *monthly interest rate*, and the 528 is the number of monthly contributions made over the full 44-year period $(528 = 12 \times 44)$.

The clear conclusion we can draw from comparing the results in Example 10.22 with the results in Example 10.20 is that it is definitely worth contributing to your retirement savings account on a monthly basis, rather than annually. All other variables essentially being equal, the value of your retirement savings jumped from \$199,758.03 to \$215,346.82.

The monthly contribution version of the retirement savings formula is

RETIREMENT SAVINGS FORMULA (MONTHLY CONTRIBUTION)

$$V = M\frac{[1 + (r/12)]^{T} - 1}{(r/12)}$$

(continued)

Here M represents the monthly contribution made *at the end of each month*, T represents the *total number of monthly contributions*, and r represents the annual interest rate expressed in decimal form.

So far, we have assumed that you make the same contribution toward retirement throughout your entire working life, but this is not typical, nor does it make much sense. As you move through your working life you get raises and make more money, and the more you make, the more you should be able to contribute to your retirement savings account.

To encourage employees to save more for retirement, many employers offer what are known as **automatic escalation** retirement savings plans—you tell your employer what percentage of your salary you want to go into your retirement savings plan, and that's it. With each raise, your contribution to your retirement plan automatically increases (this is the escalation part) and you don't have to do anything about it. If you change your mind, you can always opt out. But first check the numbers.

EXAMPLE 10.23 AUTOMATIC ESCALATION RETIREMENT PLANS

Once again, let's assume that you expect to work for 44 years. For the first ten years of employment, you contribute \$150 a month into a retirement savings account that pays 6% annual interest compounded monthly. After ten years you get a major promotion and salary raise, so you decide that you can contribute \$225 a month to your retirement savings account (a \$75 bump from your previous contributions). Then after another ten years you increase your contribution to \$350 a month (an additional \$125 bump), and you contribute at that level for the next ten years. Finally, for the last 14 years you contribute \$500 a month (an additional \$150 bump) to your retirement savings account (see Table 10.9).

For computational purposes, the easiest way to think about your escalating retirement savings plan is to break it down into four separate fixed-contribution plans (this is just a mental experiment—there really is just one plan). The breakdown goes like this: plan A, where you contribute \$150 a month for 44 years (528 months); plan B, where you contribute \$75 for 34 years (408 months); plan C, where you contribute \$125 for 24 years (288 months); and plan D, where you contribute \$150 for 14 years (168 months). Once again, you can best see this by looking at Table 10.9.

We can now use the retirement savings formula (monthly version) with each of the four separate "plans":

- Plan A savings: $\$150\left[\frac{(1.005)^{528}-1}{0.005}\right] = \$387{,}639.78$
- Plan B savings: $\$75\left[\frac{(1.005)^{408}-1}{0.005}\right] = \$99{,}774.25$
- Plan C savings: $\$125\left[\frac{(1.005)^{288}-1}{0.005}\right] = \$80{,}139.47$
- Plan D savings: $\$150\left[\frac{(1.005)^{168}-1}{0.005}\right] = \$39{,}345.71$

Total savings in your retirement savings account: \$606,899.21. Nice work!

	Contribution	Years	Number of Months	Value
Plan A	\$150	1 to 44	528	\$387,639.78
Plan B	\$75	11 to 44	408	\$99,774.25
Plan C	\$125	21 to 44	288	\$80,139.47
Plan D	\$150	31 to 44	168	\$39,345.71
Total				**\$606,899.21**

TABLE 10-9 ■ Automatic escalation with bumps of \$75, \$125, and \$150

A final comment: Example 10.23 is a very simplified version of how an automatic escalation plan works. In the real world, and in normal economic times, an employee gets a salary raise on a regular basis, and every time an employee gets a raise that employee has an opportunity to bump up the contribution to her or his retirement savings account. Over a hypothetical 44 years of employment we would expect many such opportunities. In our example we used only four bumps to keep the calculations simple, but the general idea is the same even when the details are a bit more complicated. And, of course, the more often you escalate, the larger your retirement savings stash will be.

In conclusion, *start your retirement savings early*, *contribute as much as possible*, and *escalate your contribution as often as you can*. It's not rocket science—just taking advantage of a cool little bit of math will get you there.

10.5 Consumer Debt

In this section we will focus on the two most common generators of consumer debt in a modern economy—*credit cards* and *installment loans*. According to the Federal Reserve, total consumer debt in the United States in 2015 was roughly \$3.53 trillion. Credit card debt accounted for over one fourth of it (936 billion), and much of the rest was some form of installment loan debt such as student loans, car loans, and mortgages.[1]

Credit Card Debt

A credit card is one of the most convenient and useful financial instruments in a modern economy. Merchants love them because consumers spend more than in a cash-only economy, and consumers love them because it allows them to purchase goods or services and pay for them at a later date. When properly used credit cards are a great deal, but credit cards also have a dark side to them—they generate massive amounts of consumer debt and financial stress.

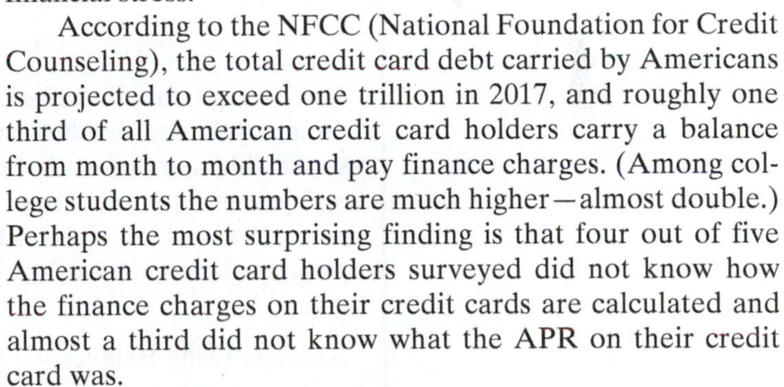

According to the NFCC (National Foundation for Credit Counseling), the total credit card debt carried by Americans is projected to exceed one trillion in 2017, and roughly one third of all American credit card holders carry a balance from month to month and pay finance charges. (Among college students the numbers are much higher—almost double.) Perhaps the most surprising finding is that four out of five American credit card holders surveyed did not know how the finance charges on their credit cards are calculated and almost a third did not know what the APR on their credit card was.

The method used to calculate finance charges can vary from one credit card company to another, but they all follow the same basic script: As long as you pay the *balance due* in full before the *payment due date* you get a free ride and everything you purchase on the credit card is interest free (no finance charges); if you carry a *revolving balance* (i.e., you don't pay the *balance due* in full) you incur *finance charges* on the balance and on any new purchases on the credit card made during the next *billing cycle*. Your previous balance minus your payment plus the finance charges plus any new charges will show up on your next statement as your new balance. One more thing: if you make your payment after the due date you incur a *late payment penalty*.

Our next example illustrates in detail how finance charges are calculated on a typical credit card.

[1] *www.federalreserve.gov/releases/g19/current/default.htm*

EXAMPLE 10.24 DEMYSTIFYING YOUR CREDIT CARD STATEMENT

A typical credit card statement covers a monthly billing cycle that always starts on the same day of the month (randomly chosen by the credit card company). Depending on the month, the number of days in the billing cycle is either 30 or 31 except for February, when it is 28 (or 29 in a leap year).

Let's suppose that you have a MasterCard with an APR of 17.24% on standard purchases (the APR on cash advances is quite a bit higher, but we won't worry about that in this example). The billing cycle starts on the 25th of each month and ends on the 24th of the next month. You are usually very good about paying the full balance due on the card each month, but in early April you went to Mexico with some friends on spring break vacation and charged more than usual to your credit card. For the April 25–May 24 billing cycle you were unable to pay the full balance on your card and had to carry over a $500 balance to your next cycle.

Calculation of Interest Charges. At the end of each billing period we calculate your interest charges using the Average Daily Balance method: (a) we first compute your Average Daily Balance(1), and (b) we then multiply your Average Daily Balance by the periodic interest rate(2).
(1) For each day in the billing period, your daily balance equals the beginning balance that day plus any new transactions minus any payments and credits posted that day. To find the Average Daily Balance we add the daily balances for each day in the billing period and divide by the number of days in the billing period.
(2) The periodic interest rate equals the APR times the number of days in the billing period divided by 365.

Typical fine print on a credit card statement

When your May 25–June 24 credit card bill arrives, these are the charges:

- May 25: Previous balance ($500)
- May 28: Phone bill payment ($85.75)
- June 5: Gasoline purchase ($45)
- June 16: Groceries ($73.60)

Since you didn't pay your previous balance in full, there will also be a finance charge in your bill. We will next show how the *finance charge* is computed. On most credit cards, finance charges are computed using simple interest applied to the **average daily balance** for the billing cycle, so there are two things that you need to calculate—the *periodic interest rate* and the *average daily balance.*

The periodic interest rate is found by taking the number of days in the billing cycle, dividing by 365, and multiplying by the APR r. In this example, the APR on your MasterCard is $r = 0.1724$, and the number of days in the May 25–June 24 billing cycle is 31 (May has 31 days). This gives,

$$\text{periodic interest rate} = (31 \div 365) \times 0.1724 \approx 0.01464.$$

The *average daily balance* is found by adding the daily balance for each day in the billing cycle and dividing the total by the number of days in the billing cycle. This sounds like a lot of work, but the calculation can be simplified by breaking the billing cycle into periods where the daily balance is constant. A table such as Table 10-10 helps. We see from Table 10-10 that over the billing cycle there were three days with a daily balance of $500.00, eight days with a daily balance of $585.75, and so on. This gives

$$\text{Average Daily Balance} = \$\frac{(3 \times 500) + (8 \times 585.75) + (11 \times 630.75) + (9 \times 704.35)}{31} = \$627.85.$$

Period	Daily balance	Number of days
May 25–May 27	$500.00	3
May 28–June 4	$500 + $85.75 = $585.75	8
June 5–June 15	$585.75 + $45.00 = $630.75	11
June 16–June 24	$630.75 + $73.60 = $704.35	9

TABLE 10-10 ■ Daily balances for the May 25–June 24 billing cycle (31 days)

The finance charges for the billing cycle can now be calculated. Since credit card interest is simple interest, the finance charge is found by multiplying the average daily balance by the periodic interest rate. In this case we get,

$$\text{finance charge} = \$627.85 \times 0.01464 = \$9.19.$$

Finally, here is the new balance that will show up on your credit card bill on the next billing cycle:

$$\text{New balance} = \text{previous balance} + \text{purchases} + \text{finance charge} =$$
$$\$500 + \$204.35 + \$9.19 = \$713.54.$$

If you pay off this balance in full on your next bill, you are back on the credit card gravy train where all your purchases are interest free. If you don't, interest charges will continue to accumulate on all your future credit card purchases. ■

When properly used, credit cards offer the savvy consumer flexibility, safety, and free short-term credit. Here are a few basic tips that will help you make the best use of your credit card(s).

- *Pay your balances in full each month.* In and of itself, the $9.19 finance charge in Example 10.24 may not seem like much, but when credit card debt is carried over a long period of time, finance charges will add up to some serious money. The best investment you can make with your money is to pay off your credit card balances in full. Every once in a while, an emergency may arise where you can't pay off the full balance—make a real effort to pay it off as soon as you can. (If you can borrow money at a lower interest rate than the APR on your card it makes sense to do it just to pay off your outstanding credit card balance.)
- *Know the APR as well as other fees on your credit card.* Pay attention to *teaser rates*—introductory rates that are very low (sometimes 0%) for a few months but then turn into a much higher rate later. If you pay your balances in full each month then the APR on your card is not nearly as important as if you don't. Some credit cards have annual fees, some don't. If your credit card has an annual fee you may want to consider whether paying that annual fee is worth it.
- *Try not to use your credit card for cash advances.* The APR on cash advances is much higher than the APR on purchases, and unlike standard purchases, finance charges on cash advances are computed from the moment you take the cash.
- *Pay attention to the Due Date on your statement and always make your payments on time.* There is a steep late payment penalty for paying after the statement due date. Even if you are completely broke, pay at least the minimum payment due each month (and get help!).
- *It's not only plastic!* There is a basic psychological principle underlying credit card use—buying with a credit card is a lot easier than buying with cash. Study after study has shown that all other things being equal, people spend a lot more when they have a credit card than when they don't. The attitude captured by the saying "It's only plastic" is something we all need to be careful of and resist. It may be plastic for a few days or weeks, but when it's time to pay the bill it becomes real money.

Installment Loans

Credit card debt is rarely intentional—consumers charge more than they can afford on their credit cards, are unable to pay the full balance when the bill is due, and start to accumulate credit card debt. Installment loan debt, on the other hand, is purposeful and organized: The borrower pays the loan plus interest back to the lender in equal installments paid at regular intervals over the life of the loan. When the last installment is paid, the loan is completely paid off. The process of paying off a loan by making regular installment payments is called **amortization**, and we often say that a loan is being *amortized* over a certain length of time.

Installment loans provide a very important function in a modern economy—they allow consumers to finance big-ticket purchases such as cars or homes, items that they would otherwise be unable to afford. Installment loans typically have much lower APRs than other types of consumer loans because the item financed by the loan (house, car, etc.) serves as collateral.

The Amortization Formula In the United States, almost all installment loans are based on monthly payments and the interest on the loan is compounded on a monthly basis. From here on, we will assume this is the case.

The following formula, known as the amortization formula, gives the monthly payment M on a loan with principal P, APR r, and paid over T monthly installments.

AMORTIZATION FORMULA

$$M = P\left[\frac{p(1+p)^T}{(1+p)^T - 1}\right], \text{ where } p = \frac{r}{12} \text{ is the monthly interest rate.}$$

The amortization formula is based on a combination of the monthly compounding formula (Section 10.3), and the geometric sum formula (Section 9.3). To keep things simple, we will not show the derivation of the formula here.

If you do an Internet search for "amortization calculator," you will find plenty of online amortization calculators (or *mortgage* calculators) that allow you to find the monthly payments given the loan principal P, the APR r, and the number of payments T, but you must be careful—not all calculators are based on the same assumptions. (The formula given above is based on the assumption that the first installment is paid one month after the origination of the loan, which is the standard for most installment loans.) Some calculators also allow you to calculate the principal P you can borrow when you know the monthly payment M, the APR r, and the number of payments T. The applet *Finance Calculator* enables you to easily do all of these calculations (and more).

EXAMPLE 10.25 FINANCING A NEW CAR

You have just landed a really good job and are in the market for a new car. You found a nice energy-efficient car you really like and negotiate a good price with the dealer: \$23,995 including taxes and license fees. In addition, the car manufacturer is offering a fantastic special promotion—you can choose between a \$2000 manufacturer's rebate or special financing at 1.2% APR for 60 months. If you choose the manufacturer's rebate then you have to finance the balance at the dealer's going rate of 6.48% APR for a 60-month loan. Which one should you choose?

If you take the \$2000 manufacturer's rebate and finance the balance through the dealer's financing you will have a principal of $P = \$21{,}995$, an APR $r = 0.0648$, a monthly interest rate $p = \frac{0.0648}{12} = 0.0054$, and a total of $T = 60$ monthly payments. The amortization formula gives monthly payments of

$$M = \$21{,}995 \times \frac{0.0054 \times (1.0054)^{60}}{(1.0054)^{60} - 1} = \$430.15.$$

If instead of the rebate you choose to go with the special financing offer you will have a principal of $P = \$23{,}995$, an APR $r = 0.012$, a monthly interest rate $p = \frac{0.012}{12} = 0.001$, and a total of $T = 60$ monthly payments. The amortization formula gives monthly payments of

$$M = \$23{,}995 \times \frac{0.001 \times (1.001)^{60}}{(1.001)^{60} - 1} = \$412.23.$$

Clearly, the special financing is a better deal than the manufacturer's rebate—you save almost \$18 per monthly payment and over \$1000 over the life of the loan. Not a bad thing to know!

Our final example for this chapter is about refinancing a home mortgage loan, a decision almost all homeowners face at one time or another: Should I take that "great" deal being offered to refinance my home at a lower APR? Is it worth it if I have to pay refinancing costs? If you own a home, you probably have received some of these refinancing offers. If you don't, you no doubt know someone who has. In any case, once you face one of these situations, you are trapped—no matter what you do, you will be making a decision and a big one at that! If you ignore the offer you are making the decision to stay with your current loan; if you take the offer you are making the decision to abandon your current loan. Clearly, since a choice is unavoidable, you should have the tools to make the proper choice.

EXAMPLE 10.26 TO REFINANCE OR NOT TO REFINANCE

Imagine that you are a homeowner and have just received an offer in the mail to refinance your home loan. The offer is for a 3.9% APR on a 30-year mortgage. (As with all mortgage loans, the interest is compounded monthly.) In addition, there are loan origination costs: $1500 closing costs plus 1 *point* (1% of the amount of the loan). You are trying to decide if this offer is worth pursuing by comparing it with your current mortgage—a 30-year mortgage for $180,000 with a 4.8% APR. Obviously, a 3.9% APR is a lot better than a 4.8% APR, but do the savings justify your up-front expenses for taking out the new loan? Besides, you have made 30 monthly payments on your current loan already. Will all these payments be wasted?

For a fair comparison between the two options (take out a new loan or keep the current loan), we will compare the monthly payments on your current loan with the monthly payments you would be making if you took out the new loan *for the balance of what you owe on your current loan.* (This way we are truly comparing apples with apples.) We can then determine if the monthly savings on your payments justify the up-front loan origination costs of the new loan. The computation will involve several steps, and each step is based on an application of the amortization formula. You can follow these steps easily using the *Finance Calculator* applet.

- **Step 1.** Compute the monthly payment M on a $180,000, 30-year mortgage with 4.8% APR. Here $P = \$180{,}000$, $r = 0.048$, $p = \frac{0.048}{12} = 0.004$, and $T = 360$. The amortization formula gives monthly payments of

$$M = \$180{,}000 \times \frac{0.004 \times (1.004)^{360}}{(1.004)^{360} - 1} = \$944.40.$$

- **Step 2.** Compute the balance on your current mortgage after the 30 monthly payments of $944.40 that you have already made. Here we use the amortization formula to solve for P given $M = \$944.40$, $r = 0.048$, $p = \frac{0.048}{12} = 0.004$, and $T = 330$. The unknown is P. The amortization formula gives

$$\$944.40 = P \times \frac{0.004 \times (1.004)^{330}}{(1.004)^{330} - 1}$$

Solving for P gives

$$P = \$944.40 \times \frac{(1.004)^{330} - 1}{0.004 \times (1.004)^{330}} = \$172{,}863.07.$$

- **Step 3.** Compute the monthly payments on the new mortgage offer. Now the principal is $P = \$172{,}863.07$, the APR is $r = 0.039$, the periodic interest rate is $p = 0.039/12 = 0.00325$, and you have to make a total of $T = 330$ monthly payments. The new monthly payments (we'll call them M') are given by

$$M' = \$172{,}863.07 \times \frac{0.00325 \times (1.00325)^{330}}{(1.00325)^{330} - 1} = \$854.78.$$

If you refinance, you will save \$89.62 in monthly payments over your current mortgage. Your up-front cost to take out the new mortgage is \$1500 + \$1728.63 = \$3228.63. At \$89.62 per month in savings, it will take about 36 months to recover this cost (\$3228.63/\$89.62 = 36.025 . . .). If you think you will be staying in your home for more than three years, it makes good sense to take the refinancing offer. If you plan to sell your home in the next couple of years or feel that interest rates will be coming down and you will find an even lower APR, you are better off staying put.

Conclusion

Whether it's finding a good deal on an iPod, borrowing money to finance a vacation, deciding how much to save for retirement, or figuring out the financing of a new car, some of the most confounding decisions we make as independent adults are about money: how to spend, borrow, invest, and manage it. Good money management involves many intangible things—luck, timing, common sense, self-discipline, and so forth—and one very tangible thing—understanding the basics of financial mathematics.

> "Waste neither time nor money, but make the best use of both."
>
> – Ben Franklin

This chapter started with a discussion of percentages. In everyday life percentages offer a convenient way to understand ratios and proportions (as in "I got a 76% on the quiz") as well as increases and decreases (as in "We are offering a 30% discount on all Rolling Stone CDs"). In financial mathematics percentages are the standard way to describe profits and losses as well as *interest rates* paid on investments or charged on loans.

Given the introductory nature of this chapter, we focused on the most basic types of financial transactions: investments such as bonds, CDs, and retirement savings accounts, and consumer loans such as credit cards, auto loans, and mortgages. Whether an investment or a loan, the basic principle behind the transaction is the same: one party (the investor/lender) gives *money* to another party (the investee/borrower), with the hope that over *time* the investment/loan turns a *profit*. The three key words in the previous sentence (money, time, and profit) represent the three key ingredients in all the formulas we discussed in this chapter: (1) the amount of money invested, (2) the length of the investment, and (3) the interest expected by the lender or investor as a reward for the willingness to temporarily part with money.

The material covered in this chapter represents a good start toward helping you become a good manager of your money and make educated financial decisions. Although there is always going to be an element of uncertainty in most financial decisions, the mathematics you learned here should allow you to understand the nature of this uncertainty so that you can make the most informed possible choice. This, and a little luck, might indeed make you rich.

KEY CONCEPTS

10.1 Percentages

- **percentage:** a rate or proportion per hundred ($x\% = \frac{x}{100}$), **292**

10.2 Simple Interest

- **interest:** the price a borrower pays for the use of a lender's money, **296**
- **principal (*P*):** the sum of money lent by a lender to a borrower, **297**
- **interest rate (*r*):** the rate paid by the borrower to the lender for the use of the principal for a specific unit of time or time period, **297**

- **annual percentage rate (APR):** an annual interest rate expressed as a percentage of the principal, **297**
- **term (*t*):** the duration or *life* of a loan (usually measured in years), **297**
- **repayment schedule:** a schedule agreed upon by lender and borrower for the repayment of a loan, **297**
- **simple interest:** in each time period the interest is applied only to the principal, **297**
- **simple interest formula:** $I = Prt$ [or, equivalently, $F = P + Prt = P(1 + rt)$], **298**
- **bond:** a type of investment in which an individual (the bond holder) lends money to a government agency or corporation for a fixed period of time, **298**
- **face value:** for a bond, the amount of money the individual gets paid back when the bond reaches maturity (also called *par value*), **299**
- **payday loan:** a short-term loan (usually 14 days) in which a person can borrow a small sum using a future paycheck as collateral, **299**

10.3 Compound Interest

- **compound interest:** in each time period the interest is applied to both the principal and the previously accumulated interest, **300**
- **annual compounding formula:** $F = P(1 + r)^t$, where r is the APR expressed as a decimal and t is the whole number of years in the term of the loan (when the term is not a whole number it must be rounded down), **302**
- **certificate of deposit (CD):** a type of investment in which an individual lends a fixed sum of money to a bank or credit union for a specific term and with a fixed APR, **302**
- **future value:** for a CD (or other investments), the amount the investor collects at the end of the term of the investment, **302**
- **rule of 72:** a basic rule of thumb for estimating the length of time it takes an investment to double under a fixed APR compounded annually (72/APR), **303**
- **periodic interest rate:** $\frac{r}{k}$, where k is the number of times the interest compounds in one year and r is the APR expressed as a decimal, **304**
- **general compounding formula:** $F = P(1 + p)^T$, where p is the periodic interest rate and T is the total number of times the interest is compounded over the life of the loan, **304**
- **continuous compounding:** interest compounded infinitely often over infinitely small compounding periods, **305**
- **continuous compounding formula:** $F = Pe^{rt}$, where $e = 2.71828\ldots$ is the base of the natural logarithms, **305**
- **annual percentage yield (APY):** annual percentage increase on the value of an investment, taking into account the APR and the compounding frequency, **307**

10.4 Retirement Savings

- **retirement savings formula (annual):** $V = P\left[\frac{(1 + r)^Y - 1}{r}\right]$, where P is the annual contribution to the retirement savings account (made at the end of the year), r is the APR expressed in decimal form, and Y is the number of annual contributions made, **309**

- **retirement savings formula (monthly):** $V = M\frac{[1 + (r/12)]^T - 1}{(r/12)}$, where M is the monthly contribution to the retirement savings account (made at the end of each month), r is the APR expressed in decimal form, and T is the total number of monthly contributions made, **311**
- **automatic escalation plan:** a retirement savings plan where an employee's contributions automatically increase as the employee's salary increases, **312**

10.5 Consumer Debt

- **average daily balance:** for a credit card, the sum of the daily balances for each day in the billing cycle divided by the number of days in the billing cycle, **314**
- **amortization:** the process of paying off a loan by making regular installment payments, **316**
- **amortization formula:** $M = P\left[\frac{p(1+p)^T}{(1+p)^T - 1}\right]$, ($P$ is the loan principal, M is the monthly payment, T is the number of payments over the life of the loan, and $p = \frac{r}{12}$ is the monthly interest rate), **316**

EXERCISES

WALKING

10.1 Percentages

1. Express each of the following percentages as a decimal.

(a) 6.25%

(b) $3\frac{3}{4}\%$

(c) $\frac{7}{10}\%$

2. Express each of the following percentages as a decimal.

(a) 8.75%

(b) $4\frac{1}{8}\%$

(c) $\frac{4}{5}\%$

3. Express each of the following percentages as a decimal.

(a) 0.82%

(b) 0.05%

4. Express each of the following percentages as a decimal.

(a) 0.25%

(b) 0.003%

5. Suppose that your lab scores in a biology class were 61 out of 75 points in Lab 1, 17 out of 20 points in Lab 2, and 118 out of 150 in Lab 3. Compare your lab scores and rank them in order, from best to last.

6. There were four different sections of *Financial Mathematics 101* offered last semester. In section A, 31 out of 38 students passed the class; in section B, 47 out of 56 students passed the class; in section C, 34 out of 45 students passed the class; and in section D, 45 out of 52 students passed the class. Compare the passing rates in each section and rank them from best to worst.

7. A 250-piece puzzle is missing 14% of its pieces from its box. How many pieces are in the box?

8. Jefferson Elementary School has 750 students. The Friday before spring break 22% of the student body was absent. How many students were in school that day?

9. At the Happyville Mall, you buy a pair of earrings that are marked \$6.95. After sales tax, the bill was \$7.61. What is the tax rate in Happyville (to the nearest tenth of a percent)?

10. Arvin's tuition bill for last semester was \$5760. If he paid \$6048 in tuition this semester, what was the percentage increase on his tuition?

11. For three consecutive years the tuition at Tasmania State University increased by 10%, 15%, and 10%, respectively. What was the overall percentage increase of tuition during the three-year period?

12. For three consecutive years the cost of gasoline increased by 8%, 15%, and 20%, respectively. What was the overall percentage increase of the cost of gasoline during the three-year period?

13. A shoe store marks up the price of its shoes at 120% over cost. A pair of shoes goes on sale for 20% off and then goes on the clearance rack for an additional 30% off. A customer walks in with a 10% off coupon good on all clearance items and buys the shoes. Express the store's profit on these shoes as a percentage of the original cost.

14. Home values in Middletown have declined 7% per year for each of the past four years. What was the total percentage decrease in home values during the four-year period? Round your answer to the nearest tenth of a percentage point.

15. Over a period of one week, the Dow Jones Industrial Average (DJIA) did the following: On Monday the DJIA went up by 2.5%, on Tuesday it went up by 12.1%, on Wednesday it went down by 4.7%, on Thursday it went up by 0.8%, and on Friday it went down by 5.4%. What was the percentage increase/decrease of the DJIA over the week? Round your answer to the nearest tenth of a percentage point.

16. Over a period of one week, the Dow Jones Industrial Average (DJIA) did the following: On Monday the DJIA went down by 3.2%, on Tuesday it went up by 11.2%, on Wednesday it went up by 0.7%, on Thursday it went down by 4.8%, and on Friday it went down by 9.4%. What was the percentage increase/decrease of the DJIA over the week? Round your answer to the nearest tenth of a percentage point.

10.2 Simple Interest

17. Suppose you borrow $875 for a term of four years at simple interest and 4.28% APR. How much is the total (principal plus interest) you must pay back on the loan?

18. Suppose you borrow $1250 for a term of three years at simple interest and 5.1% APR. How much is the total (principal plus interest) you must pay back on the loan?

19. Suppose you purchase a four-year bond with an APR of 5.75%. The face value of the bond is $4920. Find the purchase price of the bond.

20. Suppose you purchase a 15-year U.S. savings bond with an APR of 4%. The face value of the bond is $8000. Find the purchase price of the bond.

21. Suppose you purchase an eight-year bond for $5400. The face value of the bond when it matures is $8316. Find the APR.

22. Suppose you purchase a six-year muni bond for $6000. The face value of the bond when it matures is $7620. Find the APR.

23. Find the APR of a bond that doubles its value in 12 years. Round your answer to the nearest hundredth of a percent.

24. Find the APR of a bond that doubles its value in 20 years. Round your answer to the nearest hundredth of a percent.

25. Advance America is a payday loan company that offers quick, short-term loans using the borrower's future paychecks as collateral. Advance America charges $17 for each $100 loaned for a term of 14 days. Find the APR charged by Advance America.

26. CashNet USA is a payday loan company that offers quick, short-term loans using the borrower's future paychecks as collateral. CashNet USA charges $25 for each $100 loaned for a term of 14 days. Find the APR charged by CashNet USA.

10.3 Compound Interest

For all answers involving money, round the answer to the nearest penny.

27. Find the future value of an investment of $P = \$3250$ compounded annually with a 9% APR for a term of

(a) four years. **(b)** five and a half years.

28. Find the future value of an investment of $P = \$1237.50$ compounded annually with a 8.25% APR for a term of

(a) three years.

(b) four and a half years.

29. Between 1990 and 2010 the average annual inflation rate was 3.5%. Find the salary in 2010 dollars that would be equivalent to a $25,000 salary in 1990.

30. Between 2000 and 2011 the average annual inflation rate was 3%. Find the salary in 2000 dollars that would be equivalent to a $50,000 salary in 2011.

31. Consider a CD paying a 3% APR compounded monthly.

(a) Find the periodic interest rate.

(b) Find the future value of the CD if you invest $1580 for a term of three years.

32. Consider a CD paying a 3.6% APR compounded monthly.

(a) Find the periodic interest rate.

(b) Find the future value of the CD if you invest $3250 for a term of four years.

33. Consider a CD paying a 3.65% APR compounded daily.

(a) Find the periodic interest rate.

(b) Find the future value of the CD if you invest $1580 for a term of three years.

34. Consider a CD paying a 4.38% APR compounded daily.

(a) Find the periodic interest rate.

(b) Find the future value of the CD if you invest $3250 for a term of four years.

35. Consider a CD paying a 3% APR compounded continuously. Find the future value of the CD if you invest $1580 for a term of three years.

36. Consider a CD paying a 3.6% APR compounded continuously. Find the future value of the CD if you invest $3250 for a term of four years.

37. Consider a CD paying a 3.65% APR compounded continuously. Find the future value of the CD if you invest $1580 for a term of 500 days. Round your answer to the nearest dollar.

38. Consider a CD paying a 3.6% APR compounded continuously. Find the future value of the CD if you invest $3250 for a term of 1000 days. Round your answer to the nearest dollar.

39. Suppose you invest $$P$ on a CD paying 2.75% interest compounded continuously for a term of three years. At the end of the term you get $868.80 from the bank. Find the value of the original principal P.

40. Suppose you invest $$P$ on a CD paying 1.85% interest compounded continuously for a term of five years. At the end of the term you get $1645.37 from the bank. Find the value of the original principal P.

41. Find the APY for an APR of 6% compounded

(a) yearly. **(b)** semi-annually.

(c) monthly. **(d)** continuously.

42. Find the APY for an APR of 3.6% compounded

(a) yearly.

(b) semi-annually.

(c) monthly.

(d) continuously.

43. For an investment having an APY of 6%, estimate the number of years needed to double the principal.

44. For an investment having an APY of 7.2%, estimate the number of years needed to double the principal.

10.4 Retirement Savings

45. Find the value of a retirement savings account paying an APR of 6.6% after 45 years (contributions made at the end of each year, including the last year) when the annual contribution is

(a) $1500 (b) $750 (c) $2250

46. Find the value of a retirement savings account paying an APR of 5.4% after 40 years (contributions made at the end of each year, including the last year) when the annual contribution is

(a) $1200

(b) $600

(c) $1800

47. Find the value of a retirement savings account with an annual contribution of $1500 (contributions made at the end of each year, including the last year), and an APR of 6.6% after

(a) 40 years.

(b) 50 years.

(c) 52 years.

[*Note*: You may want to compare your answers to each other and to your answer to Exercise 45(a).]

48. Find the value of a retirement savings account with an annual contribution of $1200 (contributions made at the end of each year, including the last year), and an APR of 5.4% after

(a) 35 years.

(b) 45 years.

(c) 50 years.

[*Note*: You may want to compare your answers to each other and to your answer to Exercise 46(a).]

49. Find the value of a retirement savings account paying an APR of 6.6% (compounded monthly) after 45 years of monthly contributions (contributions made at the end of each month, including the last month) when the monthly contribution is

(a) $125

(b) $62.50

(c) $187.50

50. Find the value of a retirement savings account paying an APR of 5.4% (compounded monthly) after 40 years of monthly contributions (contributions made at the end of each month, including the last month) when the monthly contribution is

(a) $100 (b) $50 (c) $150

51. Find the value of a retirement savings account with a monthly contribution of $125 (contributions made at the end of each month, including the last month) and an APR of 6.6% (compounded monthly) after

(a) 40 years. (b) 50 years. (c) 52 ½ years.

[*Note*: You may want to compare your answers to each other and to your answer to Exercise 49(a).]

52. Find the value of a retirement savings account with a monthly contribution of $100 (contributions made at the end of each month, including the last month) and an APR of 5.4% (compounded monthly) after

(a) 35 years.

(b) 45 years.

(c) 49 ½ years.

[*Note*: You may want to compare your answers to each other and to your answer to Exercise 50(a).]

53. What should your monthly contribution be if your goal is to have $500,000 in your retirement savings account after 40 years? Assume the APR is 6.6% compounded monthly and that contributions are made at the end of each month, including the last month.

54. What should your monthly contribution be if your goal is to have $400,000 in your retirement savings account after 50 years? Assume the APR is 5.4% compounded monthly and that contributions are made at the end of each month, including the last month.

55. Consider a retirement savings account where the monthly contribution is $125 for the first 20 years, is increased to $225 for the next 15 years, and then is increased once again to $400 for the last 5 years. The APR is always 6.6% compounded monthly. What is the value of the account at the end of 40 years?

56. Consider a retirement savings account where the monthly contribution is $100 for the first 20 years, is increased to $225 for the next 15 years, and then is increased once again to $425 for the last 10 years. The APR is always 5.4% compounded monthly. What is the value of the account at the end of 45 years?

10.5 Consumer Debt

57. Suppose you purchase a car and you are going to finance $14,500 for 36 months at an APR of 6% compounded monthly. Find the monthly payments on the loan.

58. Suppose you purchase a car and you are going to finance $18,700 for 60 months at an APR of 4.8% compounded monthly. Find the monthly payments on the loan.

59. Suppose you want to buy a car. The dealer offers a financing package consisting of a 3.6% APR compounded monthly for a term of four years. Suppose that you want your monthly payments to be at most $400. What is the maximum amount that you should finance? (Round your answer to the nearest dollar.)

60. Suppose you want to buy a car. The dealer offers a financing package consisting of a 6% APR compounded monthly for a term of five years. Suppose that you want your monthly payments to be at most $320. What is the maximum amount that you should finance? Give your answer to the nearest dollar.

61. The Simpsons are planning to purchase a new home. To do so, they will need to take out a 30-year home mortgage loan of $160,000 through Middletown Bank. Annual interest rates for 30-year mortgages at Middletown Bank are 5.75% compounded monthly.

(a) Compute the Simpsons' monthly mortgage payment under this loan.

(b) How much interest will the Simpsons pay over the life of the loan?

62. The Smiths are refinancing their home mortgage to a 15-year loan at 5.25% annual interest compounded monthly. Their outstanding balance on the loan is $95,000.

(a) Under their current loan, the Smiths' monthly mortgage payment is $1104. How much will the Smiths be saving in their monthly mortgage payments by refinancing? (Round your answer to the nearest dollar.)

(b) How much interest will the Smiths pay over the life of the new loan?

63. Ken just bought a house. He made a $25,000 down payment and financed the balance with a 20-year home mortgage loan with an interest rate of 5.5% compounded monthly. His monthly mortgage payment is $950. What was the selling price of the house?

64. Cari just bought a house. She made a $35,000 down payment and financed the balance with a 30-year home mortgage loan with an interest rate of 5.75% compounded monthly. Her monthly mortgage payment is $877. What was the selling price of the house?

JOGGING

65. Elizabeth went on a fabulous vacation in May and racked up a lot of charges on her credit card. When it came time to pay her June credit card bill, she left a balance of $1200. Elizabeth's credit card billing cycle runs from the nineteenth of each month to the eighteenth of the next month, and her interest rate is 19.5%. She started the billing cycle June 19–July 18 with a previous balance of $1200. In addition, she made three purchases, with the dates and amounts shown in Table 10-11. On July 15 she made an online payment of $500.00 that was credited to her balance the same day.

(a) Find the average daily balance on the credit card account for the billing cycle June 19–July 18.

(b) Compute the interest charged for the billing cycle June 19–July 18.

(c) Find the new balance on the account at the end of the June 19–July 18 billing cycle.

Date	Amount of purchase/payment
6/21	$179.58
6/30	$40.00
7/5	$98.35
7/15	Payment $500.00

TABLE 10-11

66. Reid's credit card cycle ends on the twenty-fifth of every month. The interest rate on Reid's Visa card is 21.99%, and the billing cycle runs from the twenty-sixth of a month to the twenty-fifth of the following month. At the end of the July 26–Aug. 25 billing cycle, Reid's balance was $5000. During the next billing cycle (Aug. 26–Sept. 25) Reid made three purchases, with the dates and amounts shown in Table 10-12. On September 22 Reid made an online payment of $200 that was credited towards his balance the same day.

(a) Find the average daily balance on the credit card account for the billing cycle Aug. 26–Sept. 25.

(b) Find the interest charged for the billing cycle Aug. 26–Sept. 25.

(c) Find the new balance on the account at the end of the Aug. 26–Sept. 25 billing cycle.

Date of purchase	Amount of purchase
8/31	$148.55
9/12	$30.00
9/19	$103.99

TABLE 10-12

67. How much should a retailer mark up her goods so that when she has a 25% off sale, the resulting prices will still reflect a 50% markup (on her cost)?

68. Joe, a math major, calculates that in the last three years tuition at Tasmania State University has increased a total of exactly 12.4864%. He also knows that tuition increased by the same percentage each year. Determine this percentage.

69. You have a coupon worth x% off any item (including sale items) in a store. The particular item you want is on sale at y% off the marked price of $\$P$. (Assume that both x and y are positive integers smaller than 100.)

(a) Give an expression for the price of the item assuming that you first got the y% off sale price and then had the additional x% taken off using your coupon.

(b) Give an expression for the price of the item assuming that you first got the x% off the original price using your coupon and then had the y% taken off from the sale.

(c) Explain why it makes no difference in which order you have the discounts taken.

70. You buy a \$500 certificate of deposit (CD), and at the end of two years you cash it for \$561.80. What is the annual yield of this investment?

71. You are purchasing a home for \$120,000 and are shopping for a loan. You have a total of \$31,000 to put down, including the closing costs of \$1000 and any loan fee that might be charged. Bank A offers a 10% APR amortized over 30 years with 360 equal monthly payments. There is no loan fee. Bank B offers a 9.5% APR amortized over 30 years with 360 equal monthly payments. There is a 3% loan fee (i.e., a one-time up-front charge of 3% of the loan). Which loan is better?

72. You want to purchase a new car. The price of the car is \$24,035. The dealer is currently offering a special promotion: (1) You can choose a \$1500 rebate up front and finance the balance at 6% for 60 months, or (2) 0% financing for the first 36 months and 6% financing for the remaining 24 months of your loan. Which is the better deal? Justify your answer by computing your monthly payments over 60 months under each of the two options.

73. According to a *Philadelphia Inquirer* finance column, "Borrow \$100,000 with a 6% fixed-rate mortgage and you'll pay nearly \$116,000 in interest over 30 years. Put an extra \$100 a month into principal payments and you'd pay just \$76,000—and be done with mortgage payments nine years earlier." ("Are Extra Mortgage Payments Good Idea?" in Philadelphia Inquirer. Copyright © 2012 by Philadelphia Inquirer. Used by permission of Philadelphia Inquirer.)

(a) Verify that the increase in the monthly payment that is needed to pay off the mortgage in 21 years is indeed close to \$100 and that roughly \$40,000 will be saved in interest.

(b) How much should the monthly payment be increased so as to pay off the mortgage in 15 years? How much interest is saved in doing so?

(c) How much should the monthly payment be increased so as to pay off the mortgage in t years ($t < 30$)? Express the answer in terms of t.

74. Investor A invests P for t years in a savings account with an APR r compounded continuously. Investor B invests the same principal P for twice as long ($2t$ years) in a savings account with an APR that is half as much ($r/2$), also compounded continuously. When all is said and done, which of the two investors ends up with more money? Explain your answer.

75. Linear relationship between V and P in the retirement savings formula. [In all the retirement savings accounts, assume that the frequency of contributions (annual or monthly), the term, and the APR all remain the same.]

(a) Explain why, in the retirement savings formula (be it the annual or the monthly version), the value V of the account is *proportional* to the contribution P. In other words, explain why if a contribution of P generates an account with value V, then a contribution of cP (where c is any positive constant) generates an account with value cV. (*Hint*: What happens in the retirement savings formula when you replace P by cP?)

(a) Explain why if a contribution of P generates a retirement savings account with value V, and a contribution of Q generates a retirement savings account with value W, then a contribution of $(P + Q)$ generates a retirement savings account with value $(V + W)$.

76. Linear relationship between P and M in the amortization formula. [In all the installment loans, assume that the term of the loan and the APR remain the same.]

(a) Explain why, in the amortization formula, the monthly payment M is *proportional* to the principal P. In other words, explain why if the monthly payment on a loan with principal P is M, then the monthly payment on a loan with principal cP (where c is any positive constant) is cM. (*Hint*: What happens in the amortization formula when you replace P by cP?)

(b) Explain why if the monthly payment on a loan with principal P is M, and the monthly payment on a second loan with principal Q is N, then the monthly payment on a loan with principal $(P + Q)$ is $(M + N)$.

PART 4

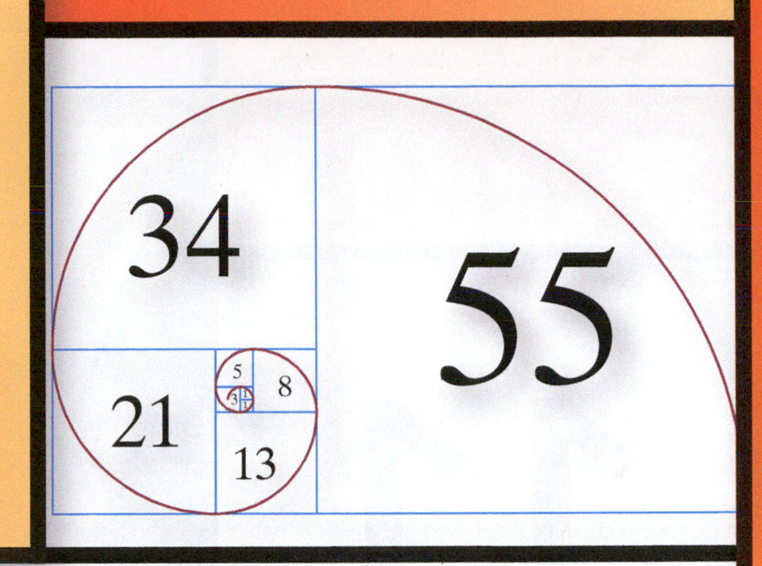

Shape and Form

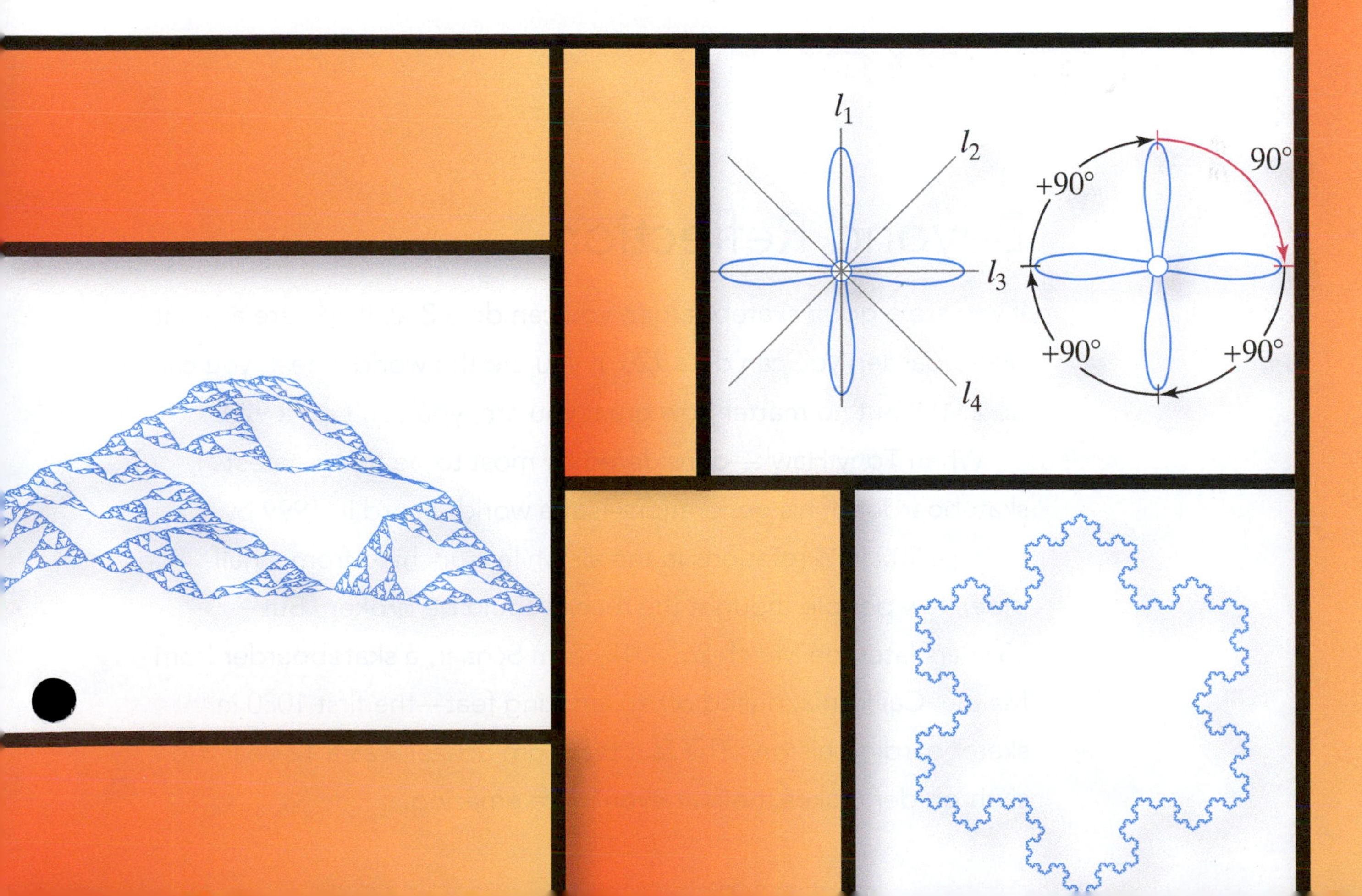

11

March 16, 2012: Skateboarder Tom Schaar lands first-ever 1080.

The Mathematics of Symmetry

Beyond Reflection

If you are a good skateboarder, you can do a 360. If you are a great skateboarder, you can do a 720. If you are the world's best, you can do a 900. But no matter how great you are, you can't do a 920.

When Tony Hawk—considered by most to be the greatest skateboarder of his generation—set a world record in 1999 by doing a 900 (2 ½ rotations in the air while launching from a half-pipe), few people thought the record could be broken. But 13 years later, on March 26, 2012, Tom Schaar, a skateboarder from Malibu, California, pulled off an amazing feat—the first 1080 in skateboarding history. The fact that Tom Schaar was a 12-year-old sixth-grader makes the feat even more amazing.

The shape of a skateboard is such that in order not to crash-land, a skateboarder can only rotate in the air a whole number of revolutions or a whole number plus one-half. After the 900, the next possible move is the 1080—there is nothing in between.

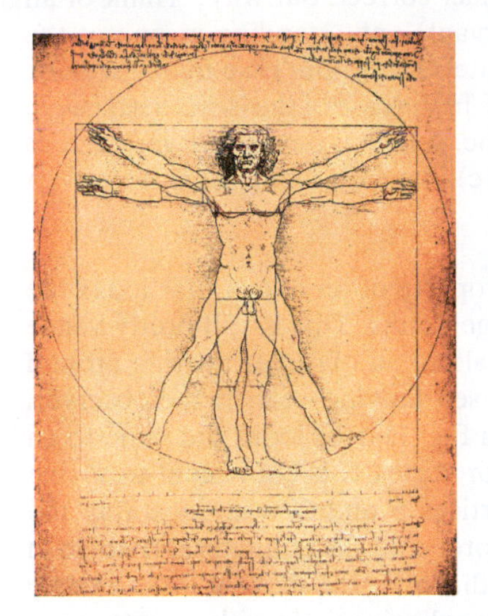

A skateboard is a fairly simple object—it has an oblong deck and four wheels (a pair in the front and a pair in the back). Its original function was to roll forward or backward with equal ease—the jumping, spinning, and assorted tricks came much later, and they impress and amaze precisely because they are acts that defy the skateboard's design. In contrast to a skateboard, a human body has a very complex design, a design that matches the complexity of what the human body is meant to do. There are many things that distinguish the design of a skateboard from the design of a human body, and one of the most significant is the nature of their *symmetries.*

In this chapter we will discuss the symmetry of things. By "things" we mean *physical objects*—things you can touch and feel such as a flower, a dollar bill, etc.—and *shapes*—things you can draw or imagine such as a square, the letter Z, a wallpaper design, etc. We will use the word *object* to describe any of these things. For technical reasons only, we will focus on the *symmetries of two-dimensional objects*—a discussion of symmetry for three-dimensional objects is quite a bit more complicated and beyond the scope of this text.

This chapter starts with a discussion of *rigid motions* (Section 11.1)—a key concept in the study of symmetry. There are only four *basic rigid motions* for two-dimensional objects in the plane, and we discuss each of these separately: *reflections* (Section 11.2), *rotations* (Section 11.3), *translations* (Section 11.4), and *glide reflections* (Section 11.5). In Section 11.6 we use rigid motions to formally define the concepts of *symmetry* and the *symmetry type* of an object. In Section 11.7 we discuss *border* and *wallpaper patterns* and their classification in terms of symmetry types.

11.1 Rigid Motions

As are many other core concepts, symmetry is rather hard to define, and we will not even attempt a proper definition until Section 11.6. We will start our discussion with just an informal stab at a geometric interpretation of symmetry.

EXAMPLE 11.1 SYMMETRIES OF A TRIANGLE

Figure 11-1 shows three triangles: (a) a scalene triangle (all three sides are different), (b) an isosceles triangle, and (c) an equilateral triangle. In terms of symmetry, how do these triangles differ? Which one is the most symmetric? Least symmetric? (What do you think?)

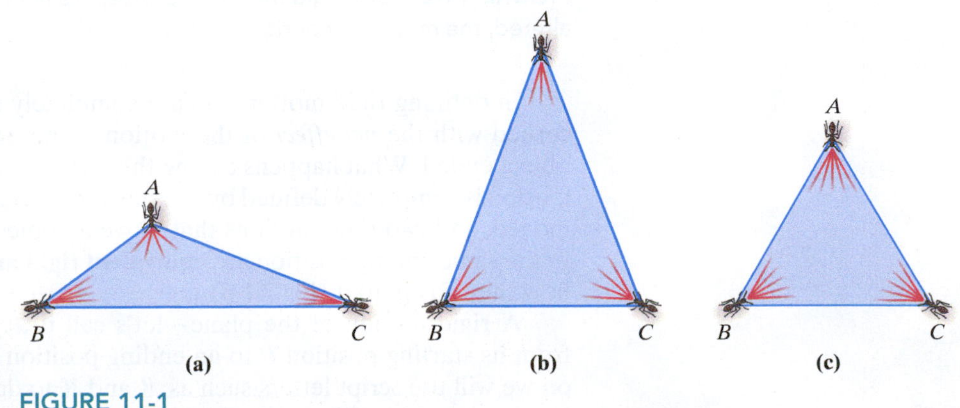

FIGURE 11-1

Even without a formal understanding of what symmetry is, most people would answer that the equilateral triangle in (c) is the most symmetric and the scalene triangle in (a) is the least symmetric. This is in fact correct, but why? Think of an imaginary observer—say a tiny (but very observant) ant—standing at one of the vertices of a triangle, looking toward the opposite side. In the case of the scalene triangle (a), the view from each vertex is different. In the case of the isosceles triangle (b), the view from vertices B and C is the same, but the view from vertex A is different. In the case of the equilateral triangle (c), the view is the same from each of the three vertices.

Let's say, for starters, that *symmetry* is a property of an object that looks the same to an observer standing at different vantage points. This is still pretty vague but a start nonetheless. Now instead of talking about an observer moving around to different vantage points think of the object itself moving—forget the observer. For example, saying that the isosceles triangle in Fig. 11-1(b) looks the same to an observer whether he or she stands at vertex B or vertex C is equivalent to saying that the triangle itself can be moved so that vertices B and C swap locations and the triangle as a whole looks exactly as it did before. And the equilateral triangle in Fig. 11-1(c) can be moved in even more ways so that after the move it looks exactly as it did before. Thus, we might think of symmetry as having to do with ways to move an object so that when all the moving is done, the object looks exactly as it did before.

Given the preceding observations, it is not surprising that to understand symmetry we need to understand the various ways in which we can "move" an object. This is our lead-in to what is a key concept in this chapter—the notion of a *rigid motion*.

Rigid Motions

The act of taking an object and moving it from some starting position to some ending position *without altering its shape or size* is called a **rigid motion** (and sometimes an *isometry*). If, in the process of moving the object, we stretch it, tear it, or generally alter its shape or size, the motion is *not* a rigid motion. Since in a rigid motion the size and shape of an object are not altered, distances between points are preserved: *The distance between any two points X and Y in the starting position is the same as the distance between the same two points in the ending position.* Figure 11-2 illustrates the difference between a motion that is rigid and one that is not. In (a), the motion does not change the shape of the object; only its position in space has changed. In (b), both position *and* shape have changed.

FIGURE 11-2 (a) A rigid motion preserves distances between points. (b) If the shape is altered, the motion is not rigid.

In defining rigid motions we are completely result oriented. We are only concerned with the *net effect* of the motion—where the object started and where the object ended. What happens during the "trip" is irrelevant. This implies that a rigid motion is completely defined by the starting and ending positions of the object being moved, and two rigid motions that move an object from the same starting position to the same ending position are **equivalent rigid motions**—never mind the details of how they go about it (Fig. 11-3).

A rigid motion of the plane—let's call it $\mathcal{M}$—moves each point in the plane from its starting position P to an ending position P', also in the plane. (From here on we will use script letters such as $\mathcal{M}$ and $\mathcal{N}$ to denote rigid motions, which should eliminate any possible confusion between the point M and the rigid motion $\mathcal{M}$.) We

(a) (b)

FIGURE 11-3 Equivalent rigid motions (a) and (b) move an object from the same starting position to the same ending position.

will call the point P' the **image** of the point P under the rigid motion $\mathcal{M}$ and describe this informally by saying that $\mathcal{M}$ *moves P to P'*. (We will also stick to the convention that the image point has the same label as the original point but with a prime symbol added.) It may happen that a point P is moved back to itself under $\mathcal{M}$, in which case we call P a **fixed point** of the rigid motion $\mathcal{M}$.

Even though there are infinitely many things we can do when we move an object, the *types* of motions we can use to do all those things are surprisingly limited. In fact, for two-dimensional objects moving in a plane, there are just *four* categories of rigid motion: (1) *reflection,* (2) *rotation,* (3) *translation,* and (4) *glide reflection.* There is no other way to rigidly move in two dimensions. The above four are called the **basic rigid motions** in the plane. Essentially, this means that when you move a two-dimensional object in the plane, no matter how complicated the motion might be, the net result is equivalent to one, and only one, of these four basic types. (If you are curious, in the case of three-dimensional objects moving in space, there are six basic types of rigid motions: *reflections*, *rotations*, *translations*, *glide reflections,* and two new types called *rotary reflections* and *screw displacements*.)

We will now discuss each of the four basic rigid motions of the plane in a little more detail.

11.2 Reflections

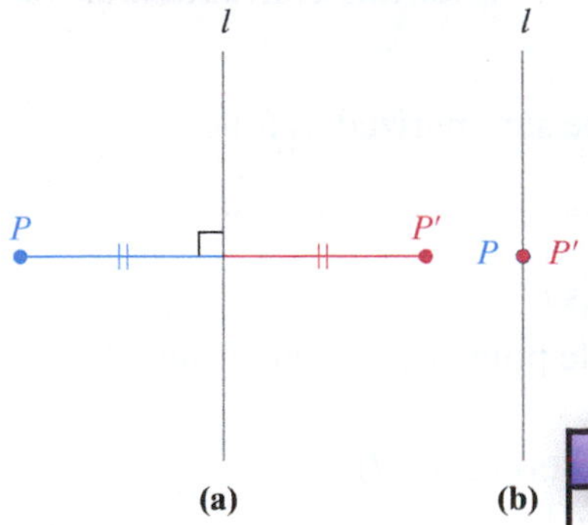

A **reflection** in the plane is a rigid motion that moves an object into a new position that is a mirror image of the starting position. In two dimensions, the "mirror" is a line called the **axis** of reflection.

From a purely geometric point of view, a reflection can be defined by showing how it moves a generic point P in the plane. This is shown in Fig. 11-4: The image of any point P is found by drawing a line through P perpendicular to the axis of reflection l and finding the point P' on the opposite side of l at the same distance as P from l. Points on the axis itself are *fixed points* of the reflection [Fig. 11-4(b)].

(a) (b)

FIGURE 11-4

EXAMPLE 11.2 REFLECTIONS OF A TRIANGLE

Figure 11-5 shows three cases of reflection of a triangle ABC. In all cases the original triangle ABC is shaded in blue and the reflected triangle $A'B'C'$ is shaded in red. In (a), the axis of reflection l does not intersect the triangle ABC. In (b), the axis of

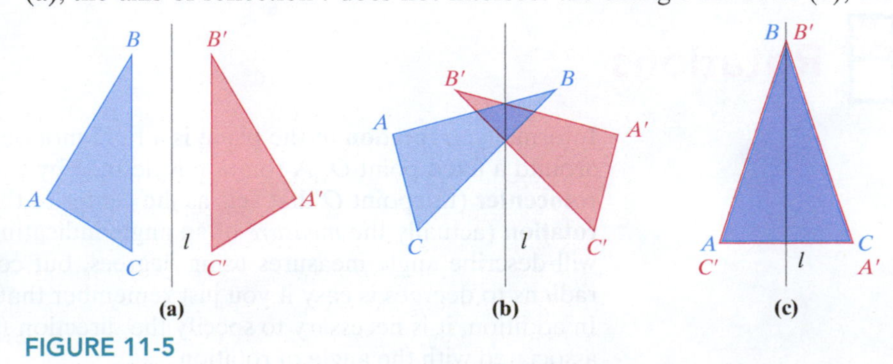

FIGURE 11-5

reflection l cuts through the triangle ABC—here the points where l intersects the triangle are fixed points of the triangle. In (c), the reflected triangle $A'B'C'$ falls on top of the original triangle ABC. The vertex B is a fixed point of the triangle, but the vertices A and C swap positions under the reflection. ■

The following are simple but useful properties of a reflection.

1. *A reflection is completely determined by its axis l.* If we know the axis of reflection, we can find the image of any point P under the reflection (just drop a perpendicular to the axis through P and find the point on the other side of the axis that is at an equal distance).
2. *A reflection is completely determined by two points (a point and its image).* If we know a point P and its image P' under the reflection (and assuming P' is different from P), we can find the axis l of the reflection (it is the perpendicular bisector of the segment PP'). Once we have the axis l of the reflection, we can find the image of any other point (property 1).
3. *A reflection has infinitely many fixed points.* The *fixed points* of a reflection are all the points on the axis l. Points not on the axis of reflection are never fixed.
4. *Reflections are improper rigid motions.* An **improper rigid motion** is one that switches the *left-right* and *clockwise-counterclockwise* orientations of objects: the reflected image of a left hand is a right hand, and in the reflected image of a clock the hands move counterclockwise (Fig. 11-6).
5. *Under a reflection, the image of the image is the original object.* For any point P, if P' is its image then the image of P' is P. This means that when we apply the same reflection twice, every point ends up in its original position and the rigid motion is equivalent to *not having moved the object at all*.

(a)

(b)

FIGURE 11-6 Reflections switch the *left-right* and *clockwise-counterclockwise* orientations of objects.

A rigid motion that is equivalent to not moving the object at all is called the **identity** motion. At first blush it may seem somewhat silly to call the identity motion a motion (after all, nothing moves), but there are very good mathematical reasons to do so, and we will soon see how helpful this convention is for studying and classifying symmetries.

The five properties of reflections listed above can be summarized as follows.

PROPERTIES OF REFLECTIONS

- A reflection is completely determined by its axis l.
- A reflection is completely determined by a single point-image pair P and P' (as long as $P' \neq P$).
- A reflection has infinitely many fixed points (all points on l).
- A reflection is an *improper* rigid motion.
- When the same reflection is applied twice, we get the *identity* motion.

11.3 Rotations

Informally, a **rotation** in the plane is a rigid motion that pivots or swings an object around a fixed point O. A rotation is defined by two pieces of information: (1) the **rotocenter** (the point O that acts as the center of the rotation) and (2) the **angle of rotation** (actually the *measure* of an angle indicating the amount of rotation). (We will describe angle measures using degrees, but converting degrees to radians or radians to degrees is easy if you just remember that 180 degrees equals π radians.) In addition, it is necessary to specify the direction (clockwise or counterclockwise) associated with the angle of rotation.

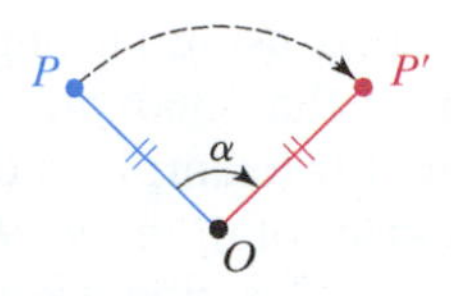

FIGURE 11-7

Figure 11-7 illustrates geometrically how a *clockwise* rotation with rotocenter O and angle of rotation α moves a point P to the point P'.

EXAMPLE 11.3 ROTATIONS OF A TRIANGLE

Figure 11-8 illustrates three cases of rotation of a triangle ABC. In all cases the original triangle ABC is shaded in blue and the reflected triangle $A'B'C'$ is shaded in red. In (a), the rotocenter O lies outside the triangle ABC. The 90° *clockwise* rotation moved the triangle from the "12 o'clock position" to the "3 o'clock position." (Note that a 90° *counterclockwise* rotation would have moved the triangle ABC to the "9 o'clock position.") In (b), the rotocenter O is at the center of the triangle ABC. The 180° rotation turns the triangle "upside down." For obvious reasons, a 180° rotation is often called a *half-turn.* (With half-turns the result is the same whether we rotate clockwise or counterclockwise, so it is unnecessary to specify a direction.) In (c), the 360° rotation moves every point back to its original position—from the rigid motion point of view it's as if the triangle had not moved.

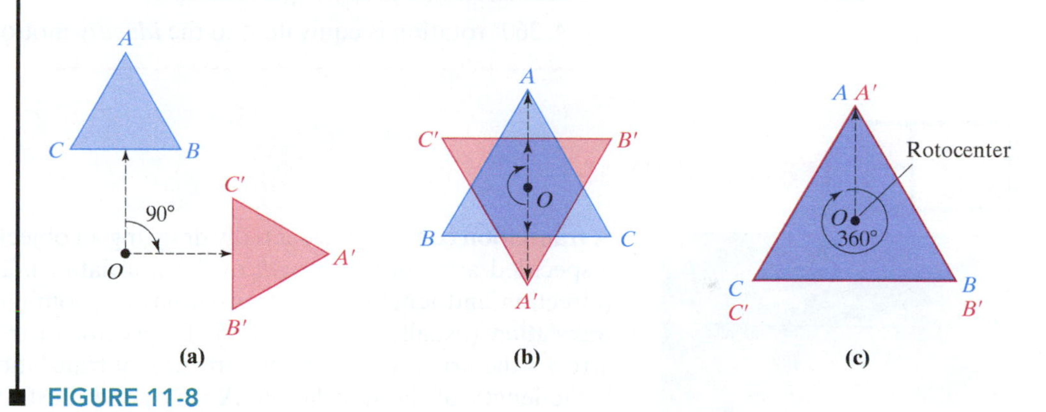

FIGURE 11-8

The following are some important properties of a rotation.

1. *A rotation is completely determined by four points (two points and their respective images).* Unlike a reflection, a rotation *cannot* be determined by a single point-image pair P and P'; it takes a second point-image pair Q and Q' to nail down the rotation. The reason is that infinitely many rotations can move P to P': Any point located on the perpendicular bisector of the segment PP' can be a rotocenter for such a rotation, as shown in Fig. 11-9(a). Given a second pair of points Q and Q' we can identify the rotocenter O as the point where the perpendicular bisectors of PP' and QQ' meet, as shown in Fig. 11-9(b). Once we have identified the rotocenter O, the angle of rotation α is given by the measure of angle POP' (or for that matter QOQ'—they are the same). [*Note*: In the special case where PP' and QQ' happen to have the same perpendicular bisector, as in Fig. 11-9(c), the rotocenter O is the intersection of PQ and $P'Q'$.]
2. *The rotocenter is the only fixed point of a rotation.* This is true for all rotations except for the identity (where all points are fixed points).
3. *A rotation is a proper rigid motion.* A **proper rigid motion** is one where the left-right and clockwise-counterclockwise orientations are preserved. A rotated left hand remains a left hand, and the hands of a rotated clock still move in the clockwise direction.

   A common misconception is to confuse a 180° rotation with a reflection, but they are definitely not the same—the rotation is proper, and the reflection is improper (Fig. 11-10).
4. *A 360° rotation is the identity motion.* A 360° rotation is equivalent to a 0° rotation, and a 0° rotation is just the identity motion. (The expression "going around full circle" is the well-known colloquial version of this property.)

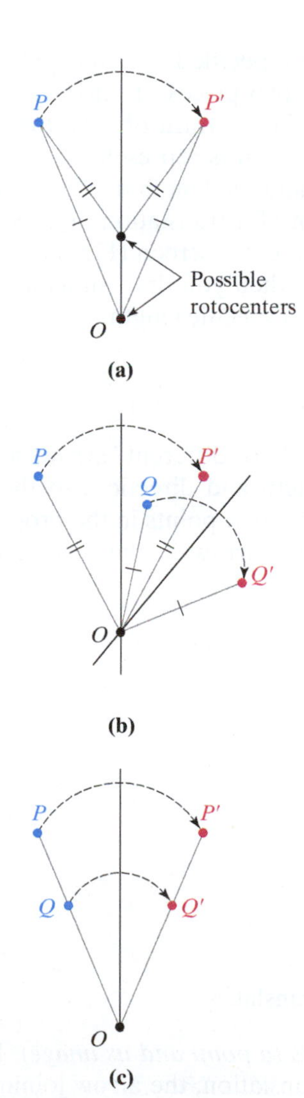

FIGURE 11-9

FIGURE 11-10 (a) 180° rotation, (b) reflection.

FIGURE 11-10 (a) 180° rotation, (b) reflection.

(a)

FIGURE 11-10 (a) 180° rotation, (b) reflection.

(b)

It follows that all rotations can be described using an angle of rotation between 0° and 360°. For angles larger than 360° we divide the angle by 360° and just use the remainder (for example, a clockwise rotation by 759° is equivalent to a clockwise rotation by 39°; the remaining 720° count as 0°). In addition, we can describe a rotation using clockwise or counterclockwise orientations (for example, a clockwise rotation by 39° is equivalent to a counterclockwise rotation by 321°).

The four properties of rotations listed above can be summarized as follows.

PROPERTIES OF ROTATIONS

- A rotation is completely determined by *two* point-image pairs P, P' and Q, Q'.
- A rotation that is not the identity motion has only *one* fixed point, its rotocenter.
- A rotation is a *proper* rigid motion.
- A 360° rotation is equivalent to the *identity* motion.

11.4 Translations

FIGURE 11-11

A **translation** consists of essentially dragging an object in a specified *direction* and by a specified amount (the *length* of the translation). The two pieces of information (direction and length of the translation) are combined in the form of a **vector of translation** (usually denoted by v). The vector of translation is represented by an arrow—the arrow points in the direction of translation and the length of the arrow is the length of the translation. A very good illustration of a translation in a two-dimensional plane is the dragging of the cursor on a computer screen (Fig. 11-11). Regardless of what happens in between, the net result when you drag an icon on your screen is a translation in a specific direction and by a specific length.

EXAMPLE 11.4 TRANSLATION OF A TRIANGLE

Figure 11-12 illustrates the translation of a triangle ABC. Two "different" arrows are shown in the figure, but they both have the same length and direction, so they describe the same vector of translation v. As long as the arrow points in the proper direction and has the right length, the placement of the arrow in the picture is immaterial.

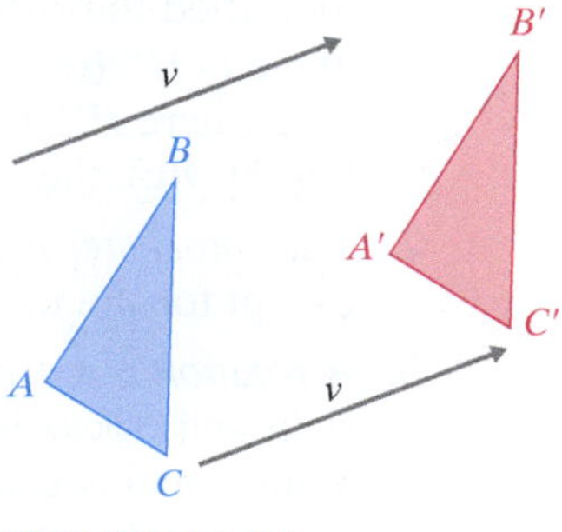

FIGURE 11-12

The following are some important properties of a translation.

1. *A translation is completely determined by two points (a point and its image).* If we are given a point P and its image P' under a translation, the arrow joining

P to P' gives the vector of the translation. Once we know the vector of the translation, we know where the translation moves any other point. Thus, a single point-image pair P and P' is all we need to completely determine the translation.

2. *A translation has no fixed points.* In a translation, every point gets moved some distance and in some direction, so there can't be any fixed points.
3. *A translation is a proper rigid motion.* When an object is translated, left-right and clockwise-counterclockwise orientations are preserved: A translated left hand is still a left hand, and the hands of a translated clock still move in the clockwise direction.
4. *A translation followed by the same translation in the opposite direction is the identity.* The effect of a translation with vector v can be undone by a translation of the same length but in the opposite direction (Fig. 11-13). The vector for this opposite translation can be conveniently described as $-v$. Thus, a translation with vector v followed with a translation with vector $-v$ is equivalent to the identity motion.

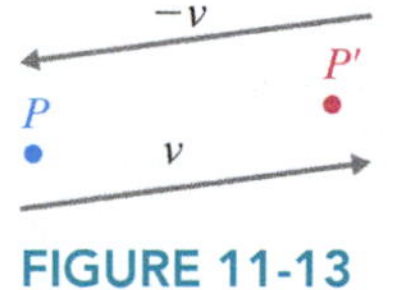

FIGURE 11-13

The four properties of translations listed above can be summarized as follows.

PROPERTIES OF TRANSLATIONS

- A translation is completely determined by a single point-image pair P and P'.
- A translation has *no* fixed points.
- A translation is a *proper* rigid motion.
- A translation with vector v followed by the translation with vector $-v$ is equivalent to the *identity* motion.

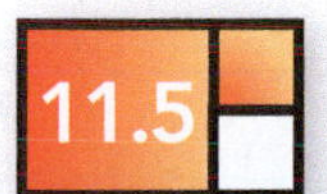

11.5 Glide Reflections

FIGURE 11-14

A **glide reflection** is a rigid motion obtained by combining a translation (the glide) with a reflection. Moreover, the axis of reflection *must* be parallel to the direction of translation. Thus, a glide reflection is described by two things: the vector of the translation v and the axis of the reflection l, and these two *must* be parallel. The footprints left behind by someone walking on soft sand (Fig. 11-14) are a classic example of a glide reflection: right and left footprints are images of each other under the combined effects of a reflection (left foot–right foot) and a translation (the step).

EXAMPLE 11.5 GLIDE REFLECTION OF A TRIANGLE

Figure 11-15 illustrates the result of applying the glide reflection with vector v and axis l to the triangle ABC. We can do this in two different ways, but the final result will be the same. In Fig. 11-15(a), the translation is applied first, moving triangle ABC to the intermediate position $A^*B^*C^*$. The reflection is then applied to $A^*B^*C^*$, giving the final position $A'B'C'$. If we apply the reflection first, the triangle ABC gets moved to the intermediate position $A^*B^*C^*$ [Fig. 11-15(b)] and then translated to the final position $A'B'C'$.

Notice that any point and its image under the glide reflection [for example, A and A' in Fig. 11-15(a)] are on opposite sides but equidistant from the axis l. This implies that the midpoint of the segment joining a point and its image under a glide reflection *must* fall on the axis l [point M in Fig. 11-15(a)].

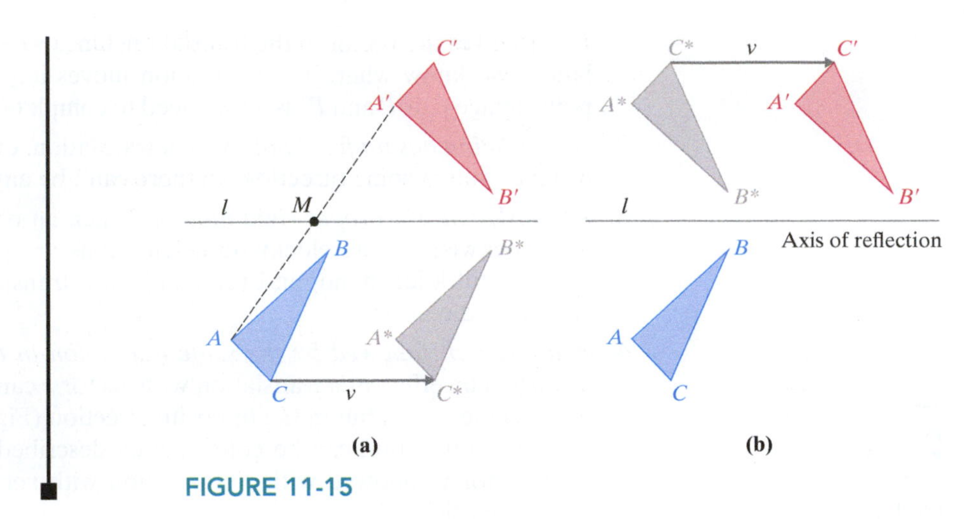

FIGURE 11-15

The following are a few basic properties of a glide reflection.

1. *A glide reflection is completely determined by four points (two points and their respective images).* Given a point-image pair P and P' under a glide reflection, we do not have enough information to determine the glide reflection, but we do know that the axis l must pass through the midpoint of the line segment PP'. Given a second point-image pair Q and Q', we can determine the axis of the reflection: It is the line passing through the points M (midpoint of the line segment PP') and N (midpoint of the line segment QQ'), as shown in Fig. 11-16(a). Once we find the axis of reflection l, we can find the image of one of the points—say P'—under the reflection. This gives the intermediate point P^*, and the vector that moves P to P^* is the vector of translation v, as shown in Fig. 11-16(b). [In the event that the midpoints of PP' and QQ' are the same point M, as shown in Fig. 11-16(c), we can still find the axis l by drawing a line perpendicular to the line PQ passing through the common midpoint M.]

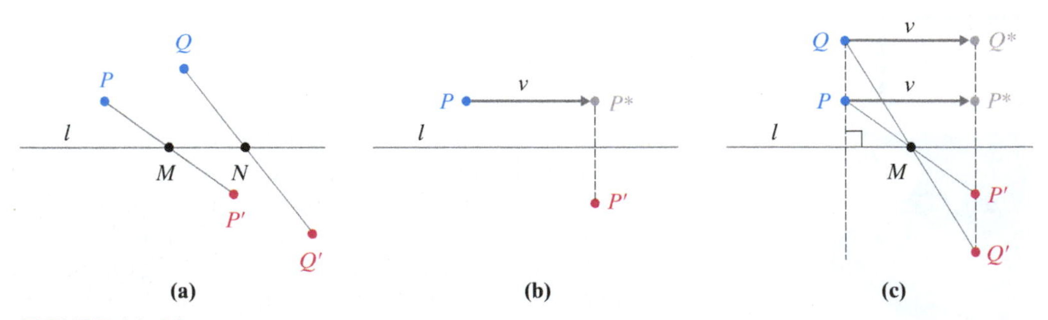

FIGURE 11-16

2. *A glide reflection has no fixed points.* A fixed point of a glide reflection would have to be a point that ends up exactly where it started after it is first translated and then reflected. This cannot happen because the translation moves every point and the reflection cannot undo the action of the translation.
3. *A glide reflection is an improper rigid motion.* A glide reflection is a combination of a proper rigid motion (the translation) and an improper rigid motion (the reflection). Since the translation preserves left-right and clockwise-counterclockwise orientations but the reflection reverses them, the net result under a glide reflection is that orientations are reversed.
4. *A glide reflection followed by the same glide reflection in the opposite direction is the identity.* To be more precise, if we move an object under a glide reflection with vector of translation v and axis of reflection l and then follow it with another glide reflection with vector of translation $-v$ and axis of reflection still l, we get the identity motion. It is as if the object was not moved at all.

The four properties of glide reflections listed above can be summarized as follows.

PROPERTIES OF GLIDE REFLECTIONS

- A glide reflection is completely determined by *two* point-image pairs P, P' and Q, Q'.
- A glide reflection has *no* fixed points.
- A glide reflection is an *improper* rigid motion.
- When a glide reflection with vector v and axis of reflection l is followed with a glide reflection with vector $-v$ and the same axis of reflection l we get the *identity* motion.

11.6 Symmetries and Symmetry Types

With an understanding of the four basic rigid motions and their properties, we can now look at the concept of symmetry in a much more precise way. Here, finally, is a good definition of *symmetry*, one that probably would not have made much sense at the start of this chapter: A **symmetry** of an object is a rigid motion that moves the object back onto itself.

One useful way to think of a symmetry is this: You observe the position of an object, and then, while you are not looking, the object is moved. If you can't tell that the object was moved, the rigid motion is a symmetry. It is important to note that this does not necessarily force the rigid motion to be the identity motion. Individual points may be moved to different positions, even though the whole object is moved back into itself. And, of course, the identity motion is itself a symmetry, one possessed by every object and that from now on we will simply call the *identity*.

Since there are only four basic kinds of rigid motions of two-dimensional objects in two-dimensional space, there are also only four possible types of symmetries: *reflection symmetries, rotation symmetries, translation symmetries*, and *glide reflection symmetries.*

EXAMPLE 11.6 THE SYMMETRIES OF A SQUARE

What are the possible rigid motions that move the square in Fig. 11-17(a) back onto itself? First, there are *reflection symmetries*. For example, if we use the line l_1 in Fig. 11-17(b) as the axis of reflection, the square falls back into itself with points A and B interchanging places and C and D interchanging places. It is not hard to think of three other reflection symmetries, with axes l_2, l_3, and l_4 as shown in Fig. 11-17(b). Are there any other types of symmetries? Yes—the square has *rotation symmetries* as well, shown in Fig. 11-17(c). Using the center of the square O as the rotocenter,

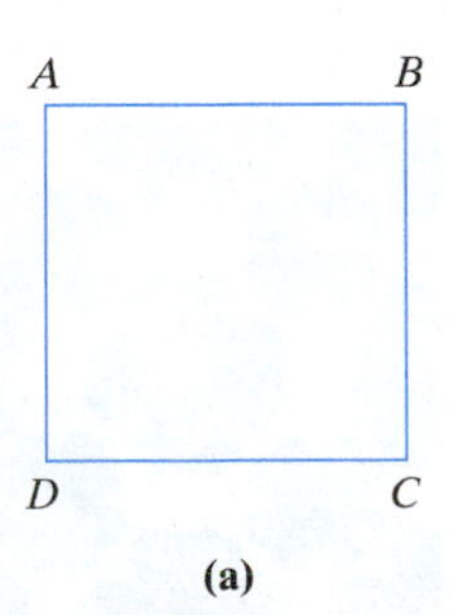

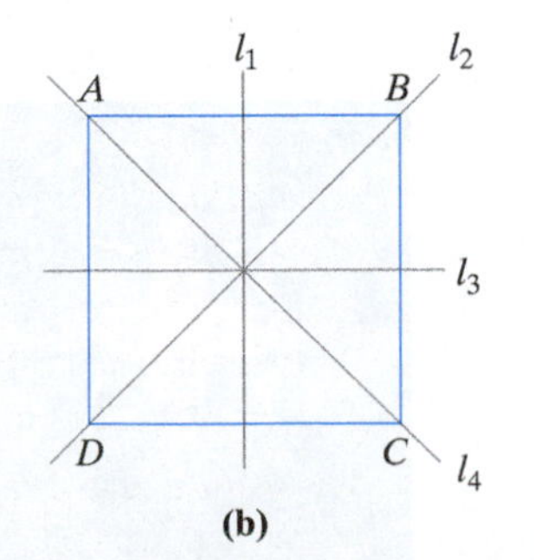

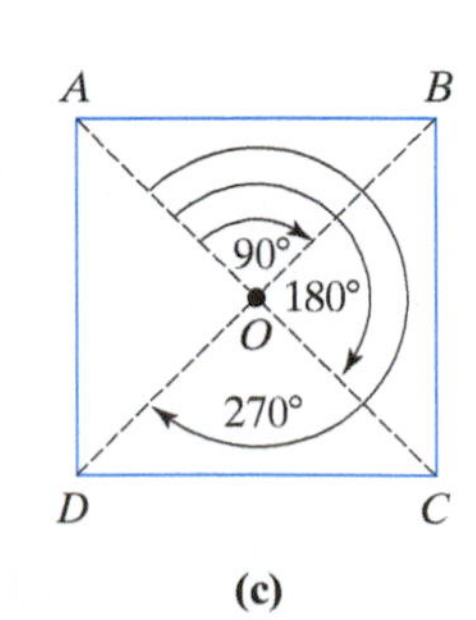

FIGURE 11-17 (a) The original square, (b) reflection symmetries (axes l_1, l_2, l_3, and l_4), and (c) rotation symmetries (rotocenter O and angles of 90°, 180°, 270°, and 360°, respectively).

we can rotate the square by an angle of 90°. This moves the upper-left corner A to the upper-right corner B, B to the lower-right corner C, C to the lower-left corner D, and D to the upper-right corner A. Likewise, rotations with rotocenter O and angles of 180°, 270°, and 360°, respectively, are also symmetries of the square. Notice that the 360° rotation is just the identity.

All in all, we have easily found eight symmetries for a square: Four of them are reflections, and the other four are rotations. Could there be more? What if we combined one of the reflections with one of the rotations? (A symmetry combined with another symmetry, after all, has to be itself a symmetry.) It turns out that the eight symmetries we listed are all there are—no matter how we combine them we always end up with one of the eight (see Exercise 74).

EXAMPLE 11.7 THE SYMMETRIES OF A PROPELLER

Let's now consider the symmetries of the shape shown in Fig. 11-18(a)—a two-dimensional version of a boat propeller (or a ceiling fan if you prefer) with four blades. Once again, we have a shape with four reflection symmetries [the axes of reflection are l_1, l_2, l_3, and l_4 as shown in Fig. 11-18(b)] and four rotation symmetries [with rotocenter located at the center of the propeller and angles of 90°, 180°, 270°, and 360°, respectively, as shown in Fig. 11-18(c)]. And, just as with the square, there are no other possible symmetries.

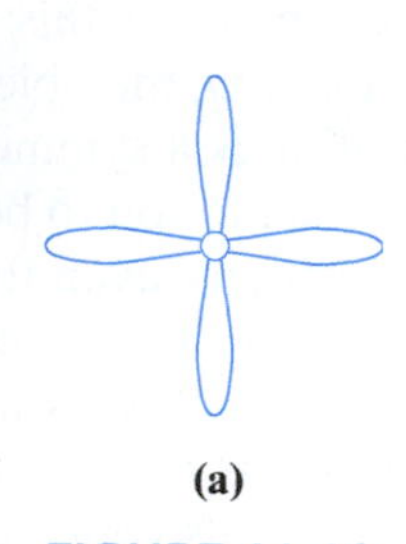

(a)

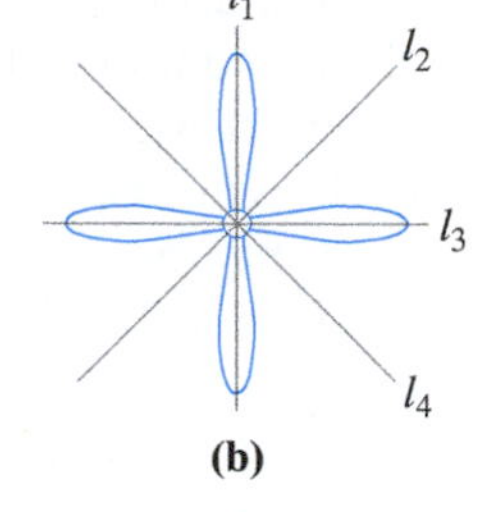

(b)

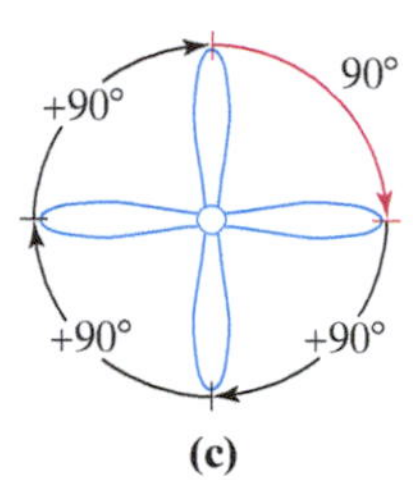

(c)

FIGURE 11-18

An important lesson lurks behind Examples 11.6 and 11.7: *Two different-looking objects can have exactly the same set of symmetries.* A good way to think about this is that the square and the propeller, while certainly different objects, are members of the same "symmetry family" and carry exactly the same symmetry genes. Formally, we will say that two objects or shapes are of the same **symmetry type** if they have *exactly* the same set of symmetries. The propeller in Fig. 11-18 and a square both have the same symmetry type called D_4 (shorthand for four reflections plus four rotations). The four-point star, four-leaf clover, and decorative tile shown in Fig. 11-19 also have four reflection and four rotation symmetries and, therefore, have symmetry type D_4.

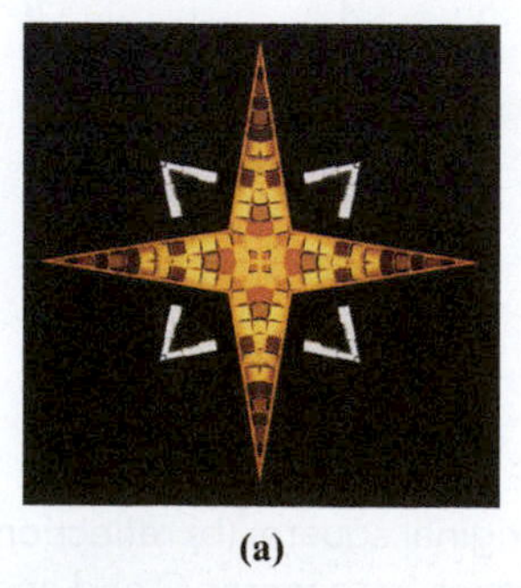

(a)

(b)

(c)

FIGURE 11-19 Objects with symmetry type D_4.

EXAMPLE 11.8 THE SYMMETRY TYPE Z_4

Let's consider now the propeller shown in Fig. 11-20(a). This propeller is only slightly different from the one in Example 11.7, but from the symmetry point of view the difference is significant—here we still have the four rotation symmetries (90°, 180°, 270°, and 360°), but there are no reflection symmetries! [This makes sense because the individual blades of the propeller have no reflection symmetry. As can be seen in Fig. 11-20(c), a vertical reflection is not a symmetry, and neither are any of the other reflections.] This object belongs to a new symmetry family called Z_4 (shorthand for the symmetry type of objects having four rotations only).

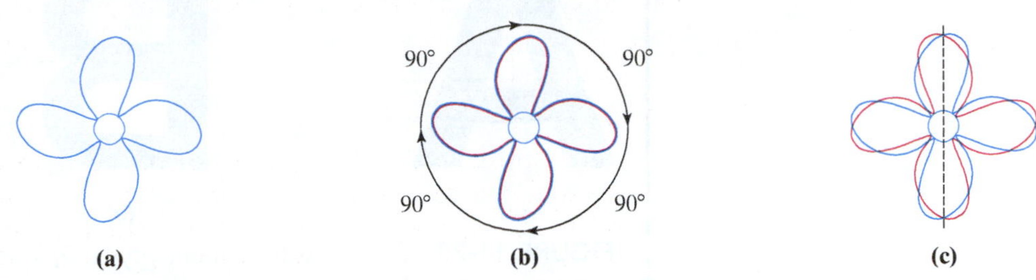

FIGURE 11-20 A propeller with symmetry type Z_4 (four rotation symmetries, no reflection symmetries).

EXAMPLE 11.9 THE SYMMETRY TYPE Z_2

Here is one last propeller example. Every once in a while a propeller looks like the one in Fig. 11-21(a), which is kind of a cross between Figs. 11-20(a) and 11-19(a)—only opposite blades are the same. This figure has no reflection symmetries, and a 90° rotation won't work either [Fig. 11-20(b)]. The only symmetries of this shape are a half-turn (i.e., 180° rotation) and the identity (i.e., a 360° rotation), as shown in Fig. 11-20(c).

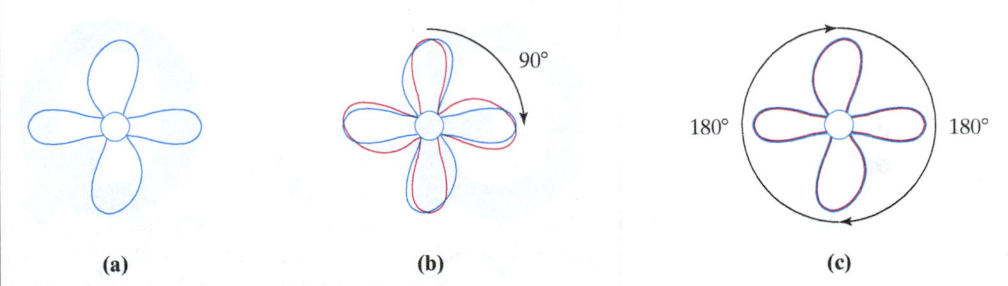

FIGURE 11-21 A propeller with symmetry type Z_2 (two rotation symmetries, no reflection symmetries).

An object having only two rotation symmetries (the identity and a half-turn) is said to be of symmetry type Z_2. Figure 11-22 shows a few additional examples of shapes and objects with symmetry type Z_2.

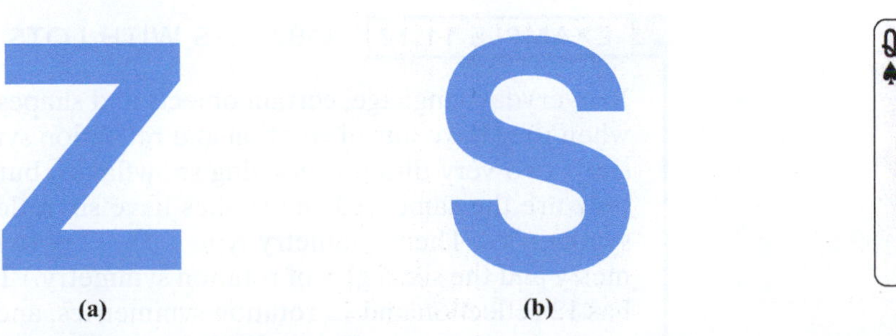

FIGURE 11-22 Objects with symmetry type Z_2. (a) The letter Z, (b) the letter S (in some fonts but not in others), and (c) the queen of spades (and many other cards in the deck).

EXAMPLE 11.10 THE SYMMETRY TYPE D_1

One of the most common symmetry types occurring in nature is that of objects having a single reflection symmetry plus a single rotation symmetry (the identity). This symmetry type is called D_1. Figure 11-23 shows several examples of shapes and objects having symmetry type D_1. Notice that it doesn't matter if the axis of reflection is vertical, horizontal, or slanted.

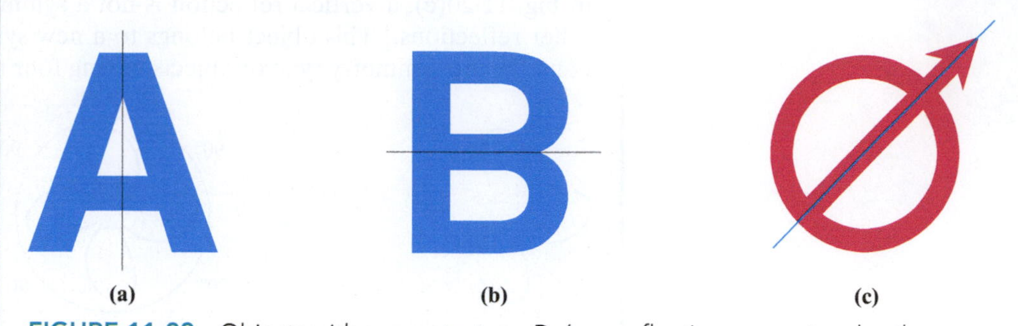

FIGURE 11-23 Objects with symmetry type D_1 (one reflection symmetry plus the identity symmetry).

EXAMPLE 11.11 THE SYMMETRY TYPE Z_1

Many objects and shapes are informally considered to have no symmetry at all, but this is a little misleading, since *every object has at least the identity symmetry*. Objects whose only symmetry is the identity are said to have symmetry type Z_1. Figure 11-24 shows a few examples of objects of symmetry type Z_1—there are plenty of such objects around.

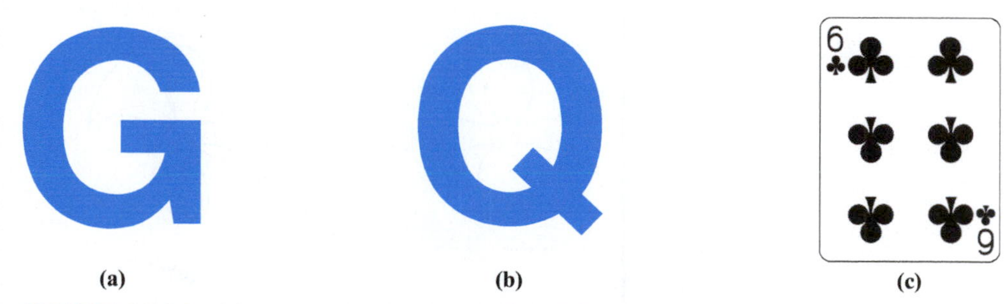

FIGURE 11-24 Objects with symmetry type Z_1 (only symmetry is the identity symmetry). Why doesn't the six of clubs have a half-turn symmetry? (The answer is given by the two middle clubs.)

EXAMPLE 11.12 OBJECTS WITH LOTS OF SYMMETRY

In everyday language, certain objects and shapes are said to be "highly symmetric" when they have lots of rotation and reflection symmetries. Figures 11-25(a) and (b) show two very different looking snowflakes, but from the symmetry point of view they are the same: All snowflakes have six reflection symmetries and six rotation symmetries. Their symmetry type is D_6. (Try to find the six axes of reflection symmetry and the six angles of rotation symmetry.) The mandala shown in Fig. 11-25(c) has 12 reflection and 12 rotation symmetries, and its symmetry type is D_{12}. The spiral design shown in Fig. 11-25(d) has 38 rotation and 38 reflection symmetries. Its symmetry type is D_{38}.

(a)

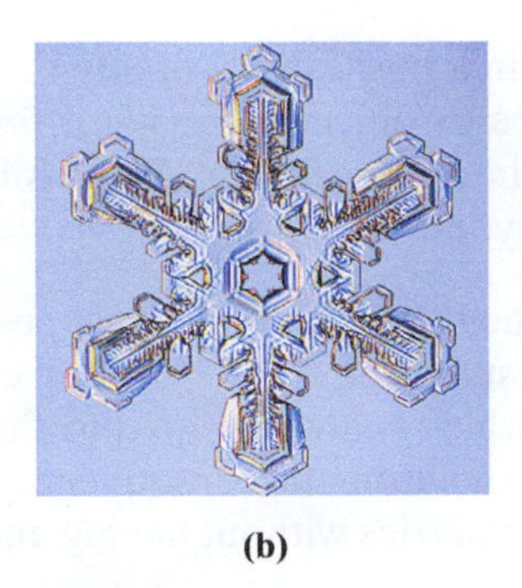
(b)

(c)

(d)

FIGURE 11-25 (a) and (b): Snowflakes (D_6). (c): Mandala (D_{12}). (d): Spiral design (D_{38}).

In each of the objects in Example 11.12, the number of reflections matches the number of rotations. This was also true in Examples 11.6, 11.7, and 11.10. Coincidence? Not at all. When a finite object has *both* reflection and rotation symmetries, the number of rotation symmetries (which includes the identity) has to match the number of reflection symmetries! Any finite object or shape with exactly N reflection symmetries and N rotation symmetries is said to have symmetry type D_N. The D comes from *dihedral symmetry*—the technical mathematical term used to describe this group of symmetry types.

EXAMPLE 11.13 THE SYMMETRY TYPE D_∞

Are there two-dimensional objects with infinitely many symmetries? The answer is yes—circles and circular washers. A circle has infinitely many reflection symmetries (any line passing through the center of the circle can serve as an axis) as well as infinitely many rotation symmetries (use the center of the circle as a rotocenter and any angle of rotation will work). We call the symmetry type of the circle D_∞ (the ∞ is the mathematical symbol for "infinity").

We now know that if a finite two-dimensional shape has rotations *and* reflections, it must have exactly the same number of each. In this case, the shape belongs to the D family of symmetries, specifically, it has symmetry type D_N. However, we also saw in Examples 11.8, 11.9, and 11.11 shapes that have rotations, *but no* reflections. In this case, we used the notation Z_N to describe the symmetry type, with the subscript N indicating the actual number of rotations.

EXAMPLE 11.14 SHAPES WITH ROTATIONS, BUT NO REFLECTIONS

Figure 11-26(a) shows a fan with five blades. The fan has five rotation symmetries but no reflection symmetries, since the blades themselves have no reflection symmetry. The symmetry type of this fan is Z_5. The rotor shown in Fig. 11-26(b) has 13 rotation symmetries but no reflection symmetries—its symmetry type is Z_{13}. The airplane turbine engine shown in Fig. 11-26(c) has 45 equally spaced asymmetric blades—its symmetry type is Z_{45}.

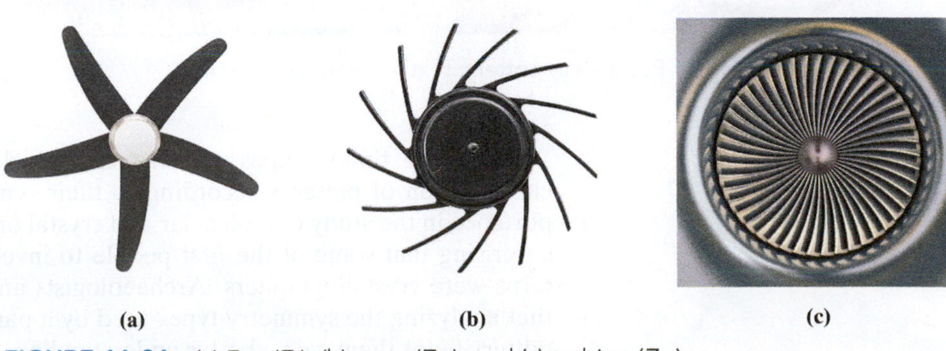
(a) (b) (c)

FIGURE 11-26 (a) Fan (Z_5), (b) rotor (Z_{13}), and (c) turbine (Z_{45}).

We are now in a position to classify the possible symmetries of any *finite* two-dimensional shape or object. (The word *finite* is in there for a reason, which will become clear in the next section.) The possibilities boil down to a surprisingly short list of symmetry types:

- D_N. This is the symmetry type of shapes with both rotation and reflection symmetries. The subscript N ($N = 1, 2, 3$, etc.) denotes the number of reflection symmetries, which is always equal to the number of rotation symmetries. (The rotations are an automatic consequence of the reflections—an object can't have reflection symmetries without having an equal number of rotation symmetries.)
- Z_N. This is the symmetry type of shapes with rotation symmetries only. The subscript N ($N = 1, 2, 3$, etc.) denotes the number of rotation symmetries.
- D_∞. This is the symmetry type of a circle and of circular objects such as rings and washers, the only possible two-dimensional shapes or objects with an infinite number of rotations and reflections.

11.7 Patterns

Well, we've come a long way, but we have yet to see examples of shapes having translation or glide reflection symmetry. If we think of objects as being finite, translation symmetry is impossible: A slide will always move the object to a new position different from its original position! But if we broaden our horizons and consider infinite "objects," translation and glide reflection symmetries are indeed possible.

We will formally define a **pattern** as an infinite "object" consisting of an infinitely repeating basic design called the **motif** of the pattern. The reason we have "object" in quotation marks is that a pattern is really an abstraction—in the real world there are no infinite objects as such, although the idea of an infinitely repeating motif is familiar to us from such everyday objects as pottery, textiles, and tile designs (Fig. 11-27).

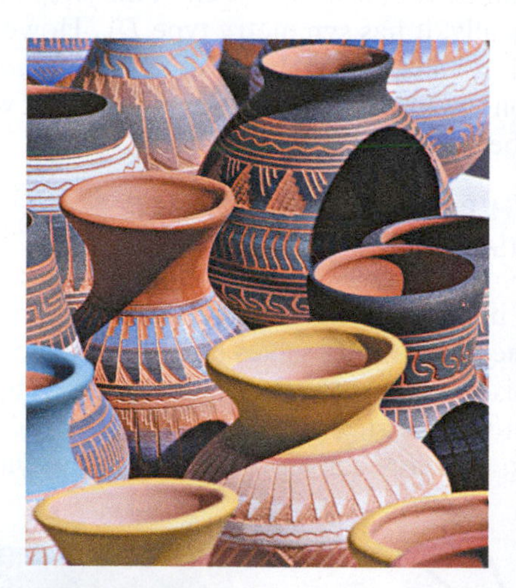

FIGURE 11-27 Patterns in pottery, textiles, and tile designs.

Just like finite shapes, *patterns* can be classified by their symmetries. The classification of patterns according to their symmetry type is of fundamental importance in the study of molecular and crystal organization in chemistry, so it is not surprising that some of the first people to investigate the symmetry types of patterns were crystallographers. Archaeologists and anthropologists have also found that analyzing the symmetry types used by a particular culture in their textiles and pottery helps them gain a better understanding of that culture.

We will briefly discuss the symmetry types of *border* and *wallpaper* patterns. A comprehensive study of patterns is beyond the scope of this book, so we will not go into as much detail as we did with finite shapes.

Border Patterns

Border patterns (also called *linear* patterns) are patterns in which a basic *motif* repeats itself indefinitely in a single direction (or creates the illusion of doing so), as in a decorative fabric, a ribbon, or a ceramic pot (Fig. 11-28).

The most common direction in a border pattern (what we will call the *direction of the pattern*) is horizontal, but in general a border pattern can be in any direction (vertical, slanted 45°, etc.). (For typesetting in a book, it is much more convenient to display a border pattern horizontally, so you will only see examples of horizontal border patterns.)

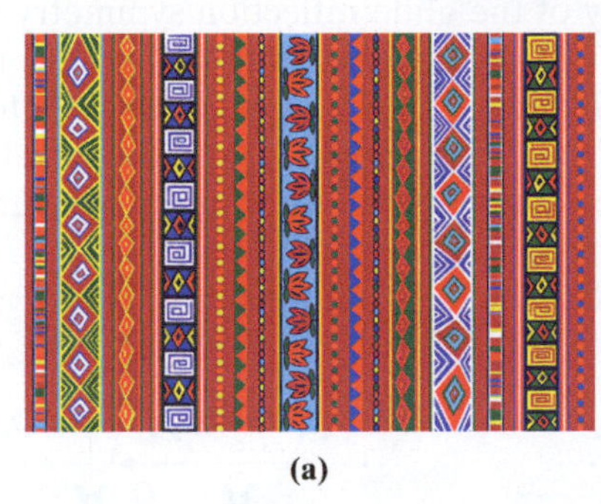
(a)

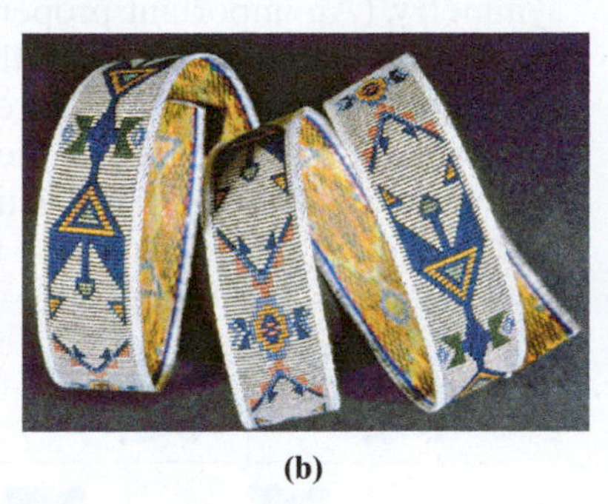
(b)

(c)

FIGURE 11-28 (a) African textile patterns, (b) ribbon with Navajo pattern, and (c) Hopi Pueblo pot.

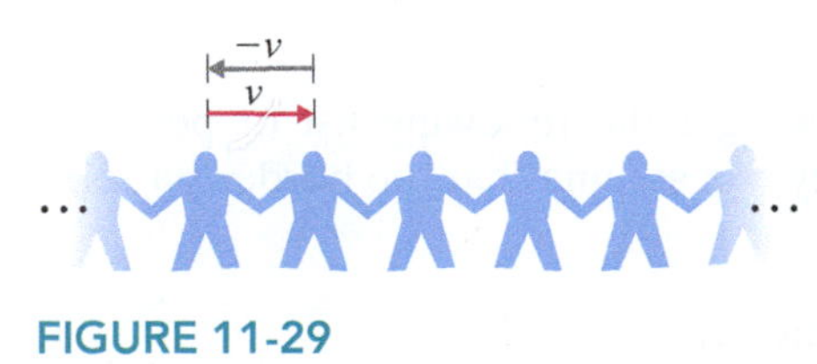

FIGURE 11-29

We will now discuss what kinds of symmetries are possible in a border pattern. Fortunately, the number of possibilities is quite small.

- **Translations.** A border pattern always has *translation symmetries*—they come with the territory. There is a *basic* translation symmetry v (v moves each copy of the motif one unit to the right), the opposite translation $-v$, and any multiple of v or $-v$ (Fig. 11-29).
- **Reflections.** A border pattern can have (a) no reflection symmetry [Fig. 11-30(a)], (b) *horizontal* reflection symmetry only [Fig. 11-30(b)], (c) *vertical* reflection symmetries only [Fig. 11-30(c)], or (d) both *horizontal* and *vertical* reflection symmetries [Fig. 11-30(d)]. In this last case the border pattern automatically picks up a half-turn symmetry as well. In terms of reflection symmetries, Fig. 11-30 illustrates the only four possibilities for reflection symmetry in a border pattern.

··· J J J J J J ···
(a)

··· B B B B B B ···
(b)

··· A A A A A A ···
(c)

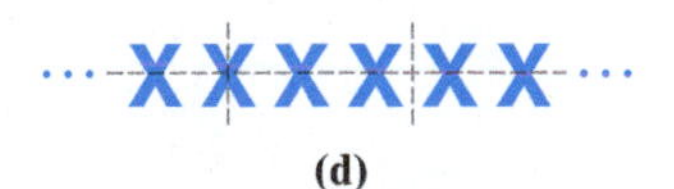
(d)

FIGURE 11-30

- **Rotations.** Like with any other object, the identity is a rotation symmetry of every border pattern, so every border pattern has at least one rotation symmetry. The only other possible rotation symmetry of a border pattern is a 180° rotation. Clearly, no other angle of rotation can take a horizontal pattern and move it back onto itself. Thus, in terms of rotation symmetry there are two kinds of border patterns: those whose only rotation symmetry is the identity and those having a half-turn (180° rotation) symmetry in addition to the identity. The former are border patterns that have a right side "up" [Fig. 11-31(a)]; the latter are border patterns that have no "up" or "down"—they look the same either way [Fig. 11-31(b)].

···AAAAAA···

(a)

···ZZZZZ···

(b)

FIGURE 11-31

- **Glide reflections.** A border pattern can have a *glide reflection symmetry*, but there is only one way this can happen: The axis of reflection *has* to be a line along the center of the pattern, and the *reflection part of the glide reflection is not by itself a symmetry of the pattern.* This means that a border pattern having horizontal reflection symmetry such as the one shown in Fig. 11-32(a) is not considered to have glide reflection symmetry. On the other hand, the border pattern shown in Fig. 11-32(b) does not have horizontal reflection symmetry (the footprints do not get reflected into footprints), but a glide by the vector w combined with a reflection along the axis l results in an honest-to-goodness glide reflection symmetry. (An important property of the glide reflection symmetry is that the vector w is always half the length of the basic translation symmetry v. This implies that two consecutive glide reflection symmetries are equivalent to the basic translation symmetry.) The border pattern in Fig. 11-32(c) has vertical reflection symmetry as well as glide reflection symmetry. In these cases half-turn symmetry (rotocenter O) comes free in the bargain.

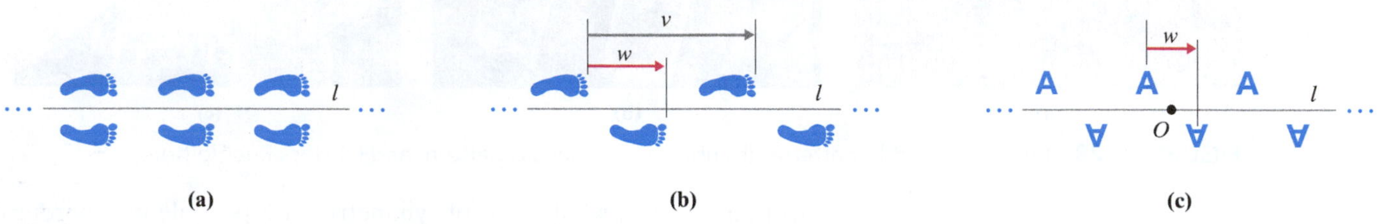

FIGURE 11-32

Combining the preceding observations, we get the following list of possible symmetries of a border pattern. (For simplicity, we assume that the border pattern is in a horizontal direction.)

1. The *identity:* Everything has identity symmetry.
2. *Translation:* By definition all border patterns have translation symmetry (Fig. 11-29).
3. *Horizontal reflection:* Some border patterns have it, some don't. There is only one possible horizontal axis of reflection, and it must run through the middle of the pattern [Fig. 11-30(b)].
4. *Vertical reflection:* Some border patterns have it, some don't. Vertical axes of reflection (i.e., axes perpendicular to the direction of the pattern) can run through the middle of a motif or between two motifs [Fig. 11-30(c)].
5. *Half-turn:* Some border patterns have it, some don't. Rotocenters can be located at the center of a motif or between two motifs [Fig. 11-31(b)].
6. *Glide reflection:* Some border patterns have it, some don't. Neither the reflection nor the glide can be a symmetry on its own. The length of the glide is half that of the basic translation. The axis of the reflection runs through the middle of the pattern [Figs. 11-32(b) and (c)].

Based on the various possible combinations of these symmetries, border patterns can be classified into just *seven different symmetry types*, which we list next. (Since all border patterns have identity symmetry and translation symmetry, we will only make reference to the additional symmetries.)

- **Type 11.** This symmetry type represents border patterns that have no symmetries beyond the identity and translation symmetry. The pattern of J's in Fig. 11-30(a) is an example of this symmetry type.

- **Type 1m.** This symmetry type represents border patterns with just horizontal reflection symmetry. The pattern of B's in Fig. 11-30(b) is an example of this symmetry type.
- **Type m1.** This symmetry type represents border patterns with *just vertical reflection symmetry*. The pattern of A's in Fig. 11-30(c) is an example of this symmetry type.
- **Type mm.** This symmetry type represents border patterns with *both a horizontal and a vertical reflection symmetry*. When both of these symmetries are present, there is also *half-turn symmetry*. The pattern of X's in Fig. 11-30(d) is an example of this symmetry type.
- **Type 12.** This symmetry type represents border patterns with *only half-turn symmetry*. The pattern of Z's in Fig. 11-31(b) is an example of this symmetry type.
- **Type 1g.** This symmetry type represents border patterns with *only glide reflection symmetry*. The pattern of footprints in Fig. 11-32(b) is an example of this symmetry type.
- **Type mg.** This symmetry type represents border patterns with *vertical reflection and glide reflection symmetry*. When both of these symmetries are present, there is also *half-turn symmetry*. The pattern in Fig. 11-32(c) is an example of this symmetry type.

Table 11-1 summarizes the symmetries for each of the seven border pattern symmetry types.

	Translation	Horizontal reflection	Vertical reflection	Half-turn	Glide reflection	Example
11	Yes	No	No	No	No	Fig. 11-30(a)
1m	Yes	Yes	No	No	No	Fig. 11-30(b)
m1	Yes	No	Yes	No	No	Fig. 11-30(c)
mm	Yes	Yes	Yes	Yes	No	Fig. 11-30(d)
12	Yes	No	No	Yes	No	Fig. 11-31(b)
1g	Yes	No	No	No	Yes	Fig. 11-32(b)
mg	Yes	No	Yes	Yes	Yes	Fig. 11-32(c)

TABLE 11-1 ■ The Seven Symmetry Types of Border Patterns

Wallpaper Patterns

Wallpaper patterns are patterns that cover the plane by repeating a *motif* indefinitely along *two or more* nonparallel directions. Typical examples of such patterns can be found in wallpaper (of course), carpets, and textiles.

With wallpaper patterns things get a bit more complicated, so we will skip the details. The possible symmetries of a wallpaper pattern are as follows:

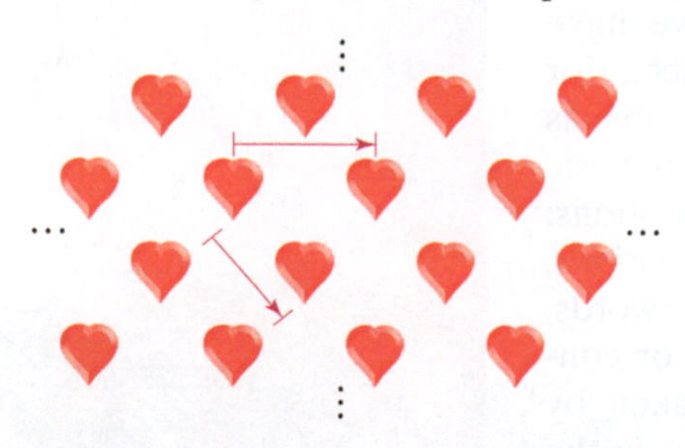

FIGURE 11-33

- **Translations.** Every wallpaper pattern has translation symmetry in at least two different (nonparallel) directions (Fig. 11-33).
- **Reflections.** A wallpaper pattern can have (a) no reflections, (b) reflections in only one direction, (c) reflections in two nonparallel directions, (d) reflections in three nonparallel directions, (e) reflections in four nonparallel directions, and (f) reflections in six nonparallel directions. There are no other possibilities. (Note that it is not possible for a wallpaper pattern to have reflections in exactly five different directions.)

- **Rotations.** In terms of rotation symmetries, a wallpaper pattern can have (a) the identity only, (b) two rotations (identity and 180°), (c) three rotations (identity, 120°, and 240°), (d) four rotations (identity, 90°, 180°, and 270°), and (e) six rotations (identity, 60°, 120°, 180°, 240°, and 300°). There are no other possibilities. (Once again, note that a wallpaper pattern cannot have just five different rotations.)
- **Glide reflections.** A wallpaper pattern can have (a) no glide reflections, (b) glide reflections in only one direction, (c) glide reflections in two nonparallel directions, (d) glide reflections in three nonparallel directions, (e) glide reflections in four nonparallel directions, and (f) glide reflections in six nonparallel directions. There are no other possibilities.

In the early 1900s, it was shown mathematically that there are *only 17 possible symmetry types for wallpaper patterns.* This is quite a surprising fact—it means that the hundreds and thousands of wallpaper patterns one can find at a decorating store all fall into just 17 different symmetry families.

Conclusion

FIGURE 11-34

There is no doubt that a *Z* and a queen of spades are two very different things (Fig. 11-34)—they look different and they serve completely different purposes. In a very fundamental way, however, they share a very important characteristic: They have exactly the same set of symmetries (identity and half-turn). Understanding the symmetries of an object and being able to sort and classify the objects in our physical world by their symmetries is a rare but useful skill, and artists, designers, architects, archaeologists, chemists, and mathematicians often rely on it. In this chapter we explored some of the basic ideas behind the mathematical study of symmetry.

Although we live in a three-dimensional world inhabited by mostly three-dimensional objects and shapes, in this chapter we restricted ourselves to studying the symmetries of two-dimensional objects and shapes for the simple reason that it is a lot easier.

A first important step in understanding symmetries is to understand the different kinds of *rigid motions*—motions that move an object while preserving its original shape. It is somewhat surprising that despite the infinite freedom we have to move an object, for two-dimensional objects there are just four basic types of rigid motions: *reflections*, *rotations*, *translations*, and *glide reflections*. In other words, no matter how complicated or convoluted the actual "trip" taken by the object might appear to be, in the

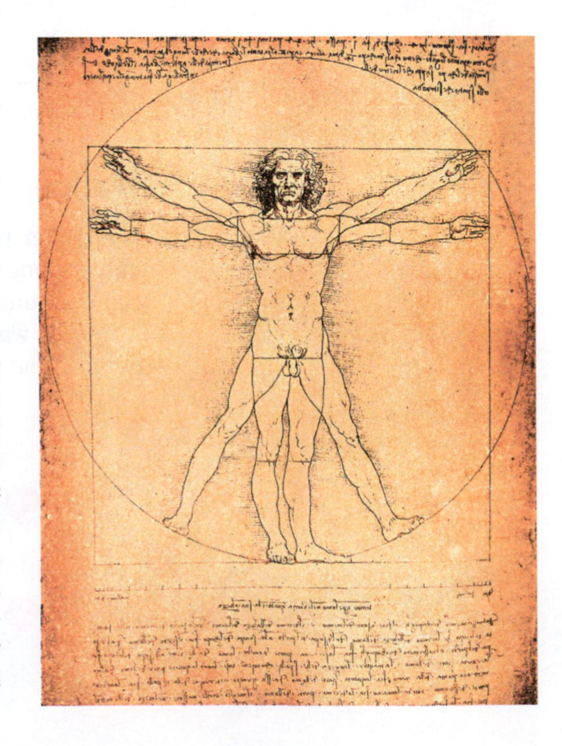

> "Symmetry is a vast subject, significant in art and nature. Mathematics lies at its root, and it would be hard to find a better one on which to demonstrate the working of the mathematical intellect."
>
> – Hermann Weyl

final analysis the motion is equivalent to a single reflection, a single rotation, a single translation, or a single glide reflection.

From the four different kinds of rigid motions in the plane we inherit the four different possible kinds of symmetry in the plane—*reflection, rotation, translation,* and *glide reflection*. Finite objects can only have rotations, or reflections *and* rotations (in an equal number). Patterns can have a little more flexibility in their combinations of the four symmetries but not as much as one would think—only seven symmetry types are possible in the case of border patterns and only 17 symmetry types are possible in the case of wallpaper patterns.

KEY CONCEPTS

11.1 Rigid Motions

- **rigid motion:** a motion that moves an object without altering its shape or size, **328**
- **equivalent rigid motions:** two rigid motions that move an object from the same starting position to the same ending position, regardless of what happens in between, **328**
- **image:** the ending position of a point or object after a rigid motion is applied, **329**
- **fixed point:** any point that is not moved by the rigid motion, **329**
- **basic rigid motions (planar):** reflection, rotation, translation, glide reflection, **329**

11.2 Reflections

- **reflection:** a rigid motion that moves a point to the opposite side and at the same distance from the axis of reflection, **329**
- **improper rigid motion:** a rigid motion that changes the left-right and clockwise-counterclockwise orientations of objects, **330**
- **identity rigid motion:** the rigid "motion" that does not move an object at all, **330**

11.3 Rotations

- **rotation:** a rigid motion that swings or pivots an object around a fixed point, **330**
- **rotocenter:** the fixed point or center of the rotation, **330**
- **proper rigid motion:** a rigid motion that preserves the left-right and clockwise-counterclockwise orientations of objects, **331**

11.4 Translations

- **translation:** a rigid motion that moves an object in a specified direction and by a specified length, **332**
- **vector of translation:** an arrow showing the direction and length of the translation, **332**

11.5 Glide Reflections

- **glide reflection:** a rigid motion that combines a translation and a reflection (the axis of the reflection must be parallel to the vector of the translation), **333**

11.6 Symmetries and Symmetry Types

- **symmetry:** a rigid motion that moves an object back onto itself, **335**
- **symmetry type:** a classification of objects based on their symmetries (objects that have the same set of symmetries have the same symmetry type), **336**
- D_N**:** the symmetry type of objects having N reflection symmetries and N rotation symmetries, **340**
- Z_N**:** the symmetry type of objects having N rotation symmetries (and *no* reflection symmetries), **340**

11.7 Patterns

- **pattern:** an infinite "object" consisting of an infinitely repeating basic motif, **340**
- **motif:** the infinitely repeating basic building block of a pattern, **340**
- **border pattern:** a pattern where the motif repeats itself in a single direction, **341**
- **wallpaper pattern:** a pattern that fills the plane by repeating a motif in two or more nonparallel directions, **343**

EXERCISES

WALKING

11.1 Rigid Motions

No exercises for this section.

11.2 Reflections

1. In Fig. 11-35, indicate which point is the image of P under
   - **(a)** the reflection with axis l_1.
   - **(b)** the reflection with axis l_2.
   - **(c)** the reflection with axis l_3.
   - **(d)** the reflection with axis l_4.

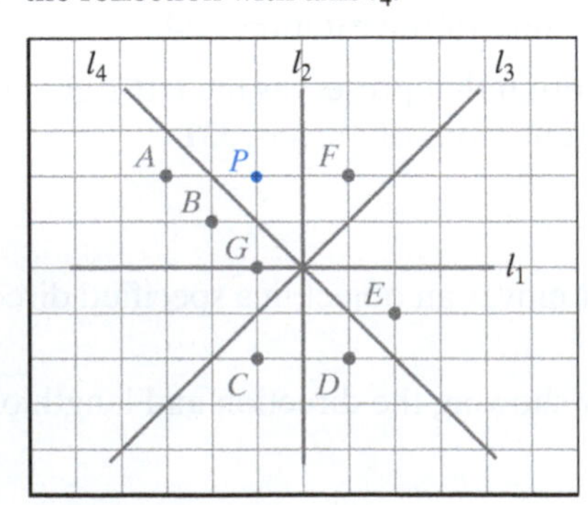

FIGURE 11-35

2. In Fig. 11-36, indicate which point is the image of P under
   - **(a)** the reflection with axis l_1.
   - **(b)** the reflection with axis l_2.
   - **(c)** the reflection with axis l_3.
   - **(d)** the reflection with axis l_4.

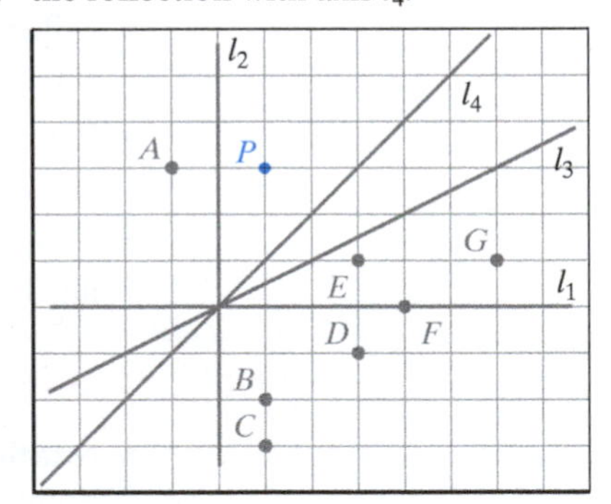

FIGURE 11-36

3. In Fig. 11-37, A' is the image of A under a reflection.
   - **(a)** Find the axis of the reflection.
   - **(b)** Find the image of A' under the reflection.

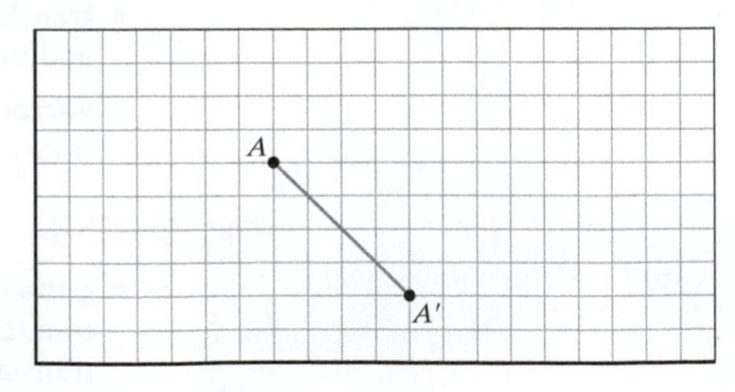

FIGURE 11-37

4. In Fig. 11-38, P' is the image of P under a reflection.
   (a) Find the axis of the reflection.
   (b) Find the image of P' under the reflection.

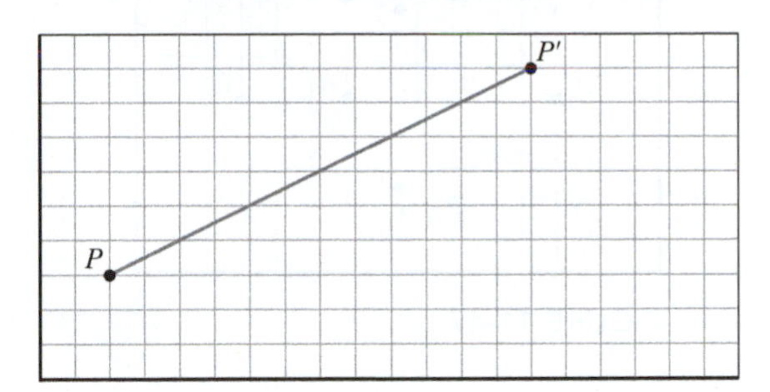

FIGURE 11-38

5. In Fig. 11-39, l is the axis of reflection.
   (a) Find the image of S under the reflection.
   (b) Find the image of quadrilateral $PQRS$ under the reflection.

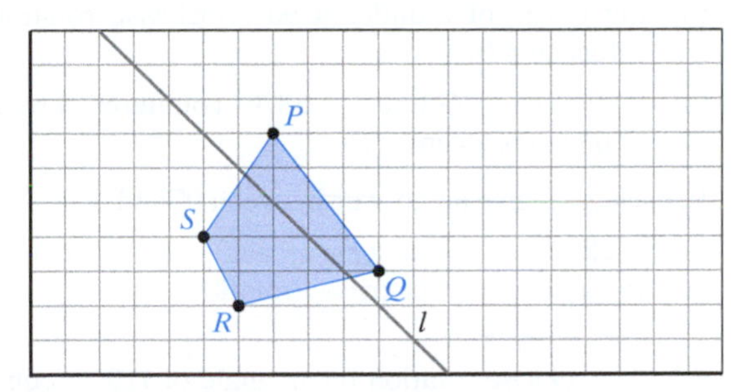

FIGURE 11-39

6. In Fig. 11-40, l is the axis of reflection.
   (a) Find the image of P under the reflection.
   (b) Find the image of the parallelogram $PQRS$ under the reflection.

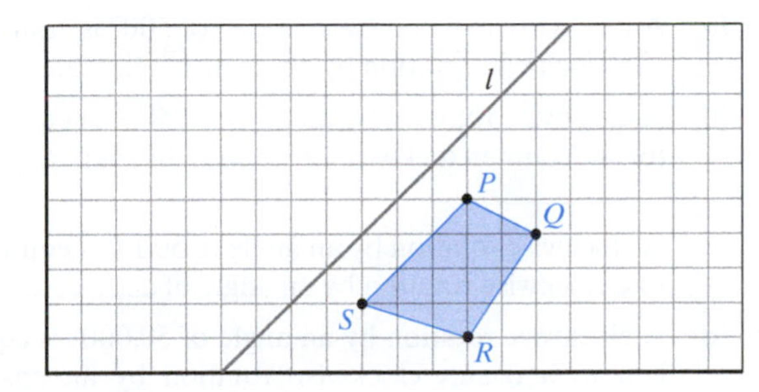

FIGURE 11-40

7. In Fig. 11-41, P' is the image of P under a reflection.
   (a) Find the axis of the reflection.
   (b) Find the image of S under the reflection.
   (c) Find the image of the quadrilateral $PQRS$ under the reflection.
   (d) Find a point on the quadrilateral $PQRS$ that is a fixed point of the reflection.

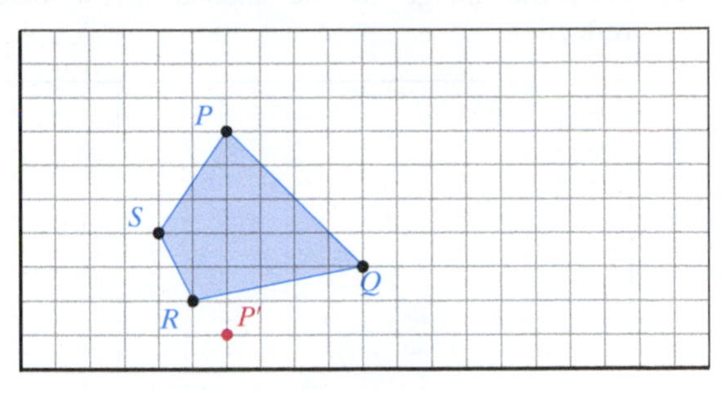

FIGURE 11-41

8. In Fig. 11-42, P' is the image of P under a reflection.
   (a) Find the axis of the reflection.
   (b) Find the image of S under the reflection.
   (c) Find the image of the quadrilateral $PQRS$ under the reflection.
   (d) Find a point on the quadrilateral $PQRS$ that is a fixed point of the reflection.

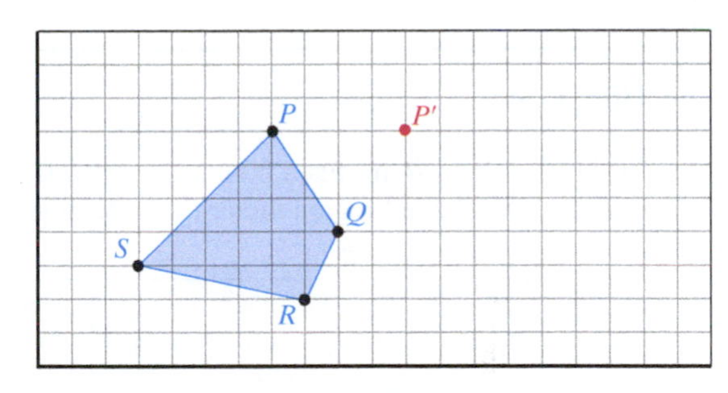

FIGURE 11-42

9. In Fig. 11-43, P' is the image of P under a reflection.
   (a) Find the axis of the reflection.
   (b) Find the image of the shaded arrow under the reflection.

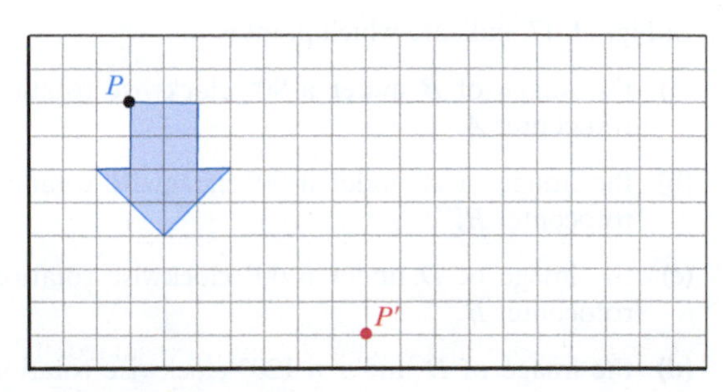

FIGURE 11-43

10. In Fig. 11-44, R' is the image of R under a reflection.
   (a) Find the axis of reflection.
   (b) Find the image of the shaded arrow under the reflection.

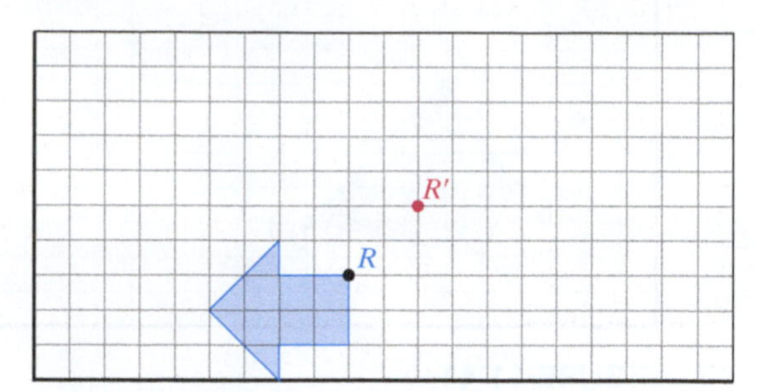

FIGURE 11-44

11. In Fig. 11-45, A and B are fixed points of a reflection. Find the image of the shaded region under the reflection.

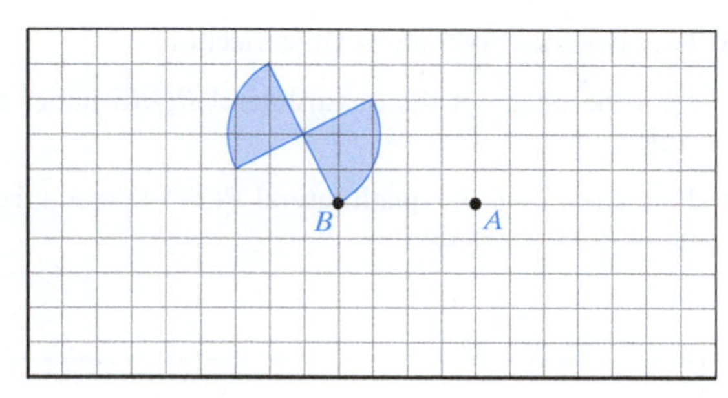

FIGURE 11-45

12. In Fig. 11-46, A and B are fixed points of a reflection. Find the image of the shaded region under the reflection.

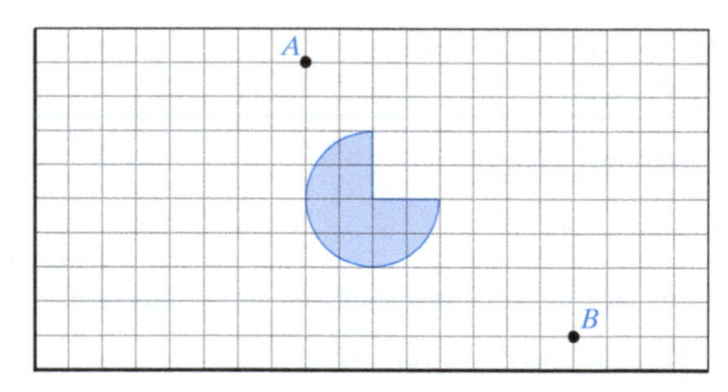

FIGURE 11-46

11.3 Rotations

13. In Fig. 11-47, indicate which point is
   (a) the image of B under a 90° clockwise rotation with rotocenter A.
   (b) the image of A under a 90° clockwise rotation with rotocenter B.
   (c) the image of D under a 60° clockwise rotation with rotocenter B.
   (d) the image of D under a 120° clockwise rotation with rotocenter B.
   (e) the image of I under a 3690° clockwise rotation with rotocenter A.

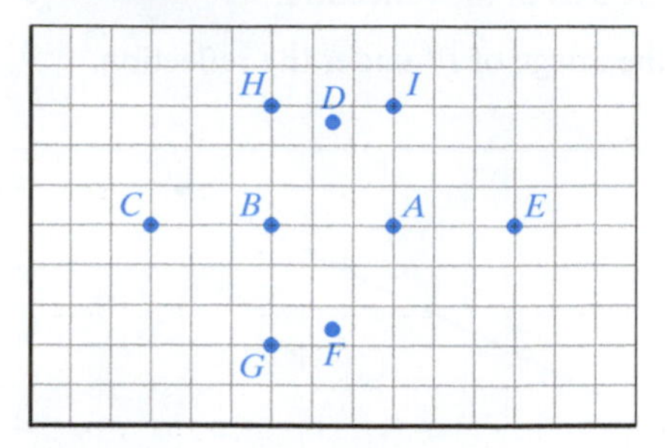

FIGURE 11-47

14. In Fig. 11-47, indicate which point is
   (a) the image of C under a 90° counterclockwise rotation with rotocenter B.
   (b) the image of F under a 60° clockwise rotation with rotocenter A.
   (c) the image of F under a 120° clockwise rotation with rotocenter B.
   (d) the image of I under a 90° clockwise rotation with rotocenter H.
   (e) the image of G under a 3870° counterclockwise rotation with rotocenter B.

15. In each case, give an answer between 0° and 360°.
   (a) A clockwise rotation by an angle of 710° is equivalent to a counterclockwise rotation by an angle of ________.
   (b) A clockwise rotation by an angle of 710° is equivalent to a clockwise rotation by an angle of ________.
   (c) A counterclockwise rotation by an angle of 7100° is equivalent to a clockwise rotation by an angle of ________.
   (d) A clockwise rotation by an angle of 71,000° is equivalent to a clockwise rotation by an angle of ________.

16. In each case, give an answer between 0° and 360°.
   (a) A clockwise rotation by an angle of 500° is equivalent to a clockwise rotation by an angle of ________.
   (b) A clockwise rotation by an angle of 500° is equivalent to a counterclockwise rotation by an angle of ________.
   (c) A clockwise rotation by an angle of 5000° is equivalent to a clockwise rotation by an angle of ________.
   (d) A clockwise rotation by an angle of 50,000° is equivalent to a counterclockwise rotation by an angle of ________.

17. In Fig. 11-48, a rotation moves B to B' and C to C'.
   (a) Find the rotocenter.
   (b) Find the image of triangle ABC under the rotation.

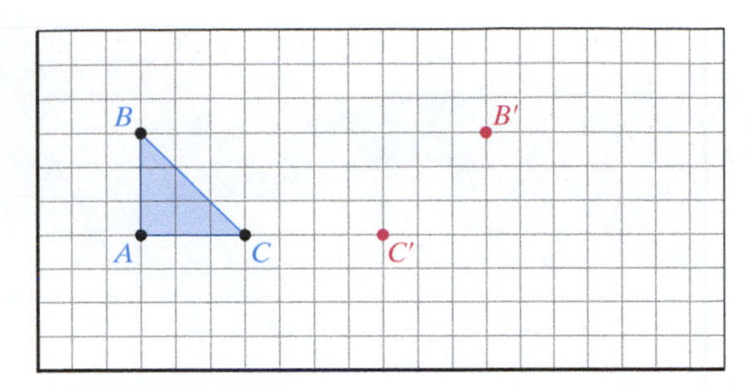

FIGURE 11-48

18. In Fig. 11-49, a rotation moves A to A' and B to B'.

(a) Find the rotocenter O.

(b) Find the image of the shaded arrow under the rotation.

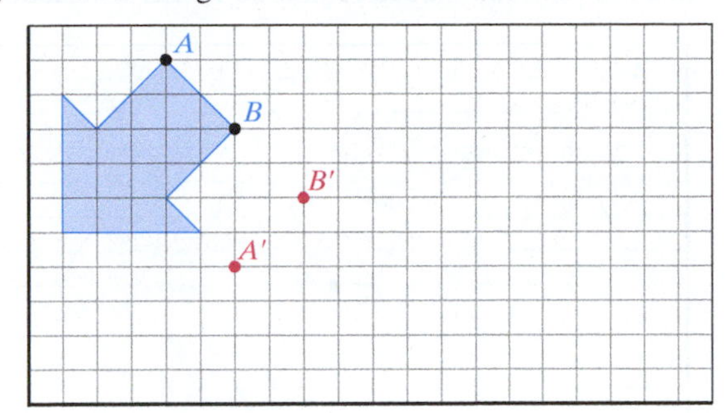

FIGURE 11-49

19. In Fig. 11-50, a rotation moves A to A' and B to B'.

(a) Find the rotocenter O.

(b) Find the image of the shaded arrow under the rotation.

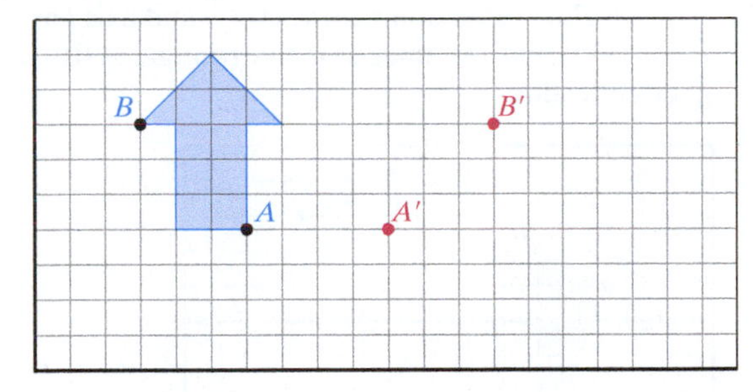

FIGURE 11-50

20. In Fig. 11-51, a rotation moves Q to Q' and R to R'.

(a) Find the rotocenter. **(b)** Find the angle of rotation.

(c) Find the image of quadrilateral $PQRS$ under the rotation.

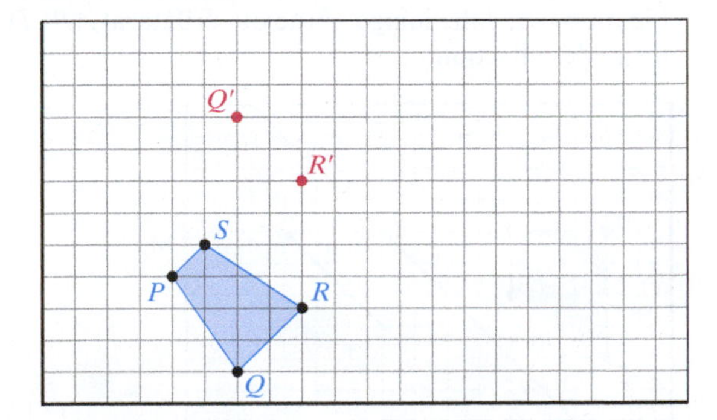

FIGURE 11-51

21. In Fig. 11-52, find the image of triangle ABC under a 60° clockwise rotation with rotocenter O.

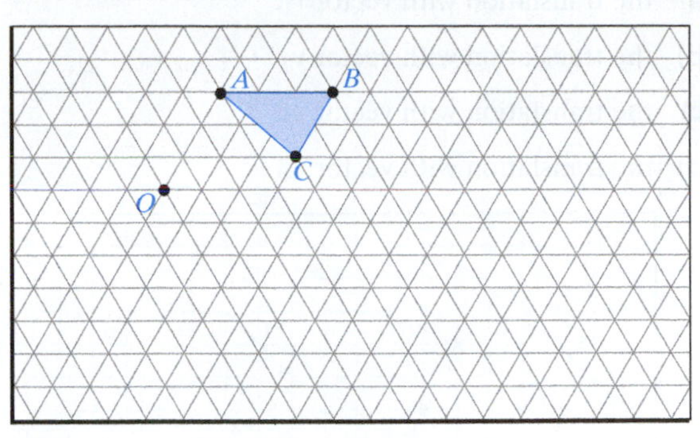

FIGURE 11-52

22. In Fig. 11-53, find the image of $ABCD$ under a 60° counterclockwise rotation with rotocenter O.

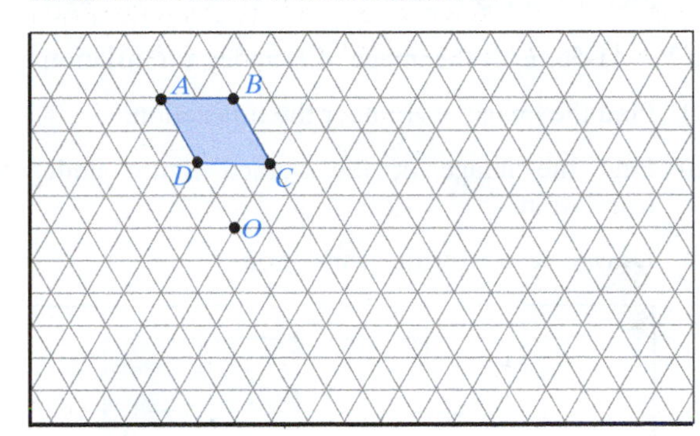

FIGURE 11-53

11.4 Translations

23. In Fig. 11-54, indicate which point is the image of P under

(a) the translation with vector v_1.

(b) the translation with vector v_2.

(c) the translation with vector v_3.

(d) the translation with vector v_4.

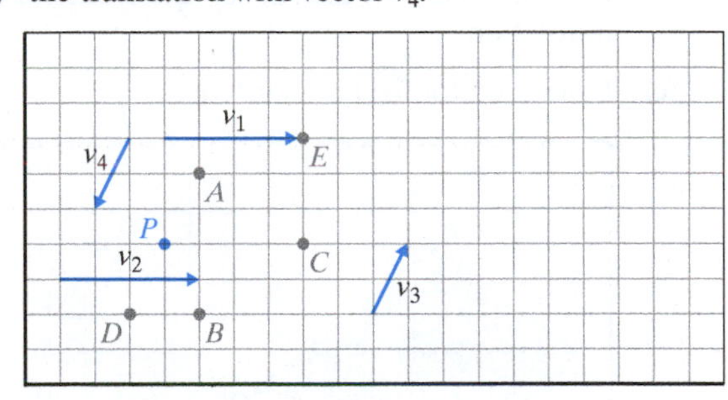

FIGURE 11-54

24. In Fig. 11-55, indicate which point is the image of P under

(a) the translation with vector v_1.

(b) the translation with vector v_2.

(c) the translation with vector v_3.

(d) the translation with vector v_4.

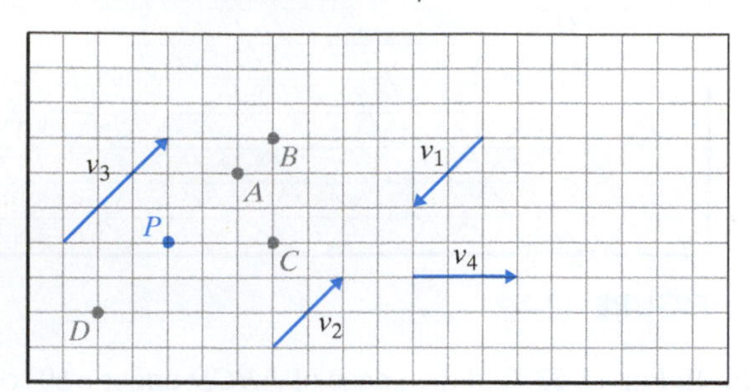

FIGURE 11-55

25. In Fig. 11-56, E' is the image of E under a translation.

(a) Find the image of A under the translation.

(b) Find the image of the shaded figure under the translation.

(c) Draw a vector for the translation.

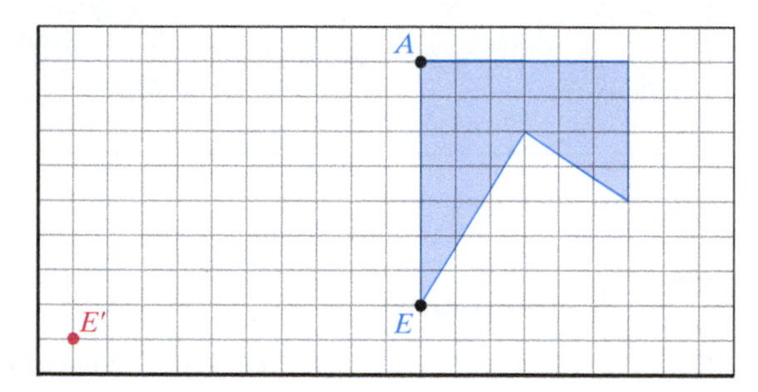

FIGURE 11-56

26. In Fig. 11-57, Q' is the image of Q under a translation.

(a) Find the image of P under the translation.

(b) Find the image of the shaded quadrilateral under the translation.

(c) Draw a vector for the translation.

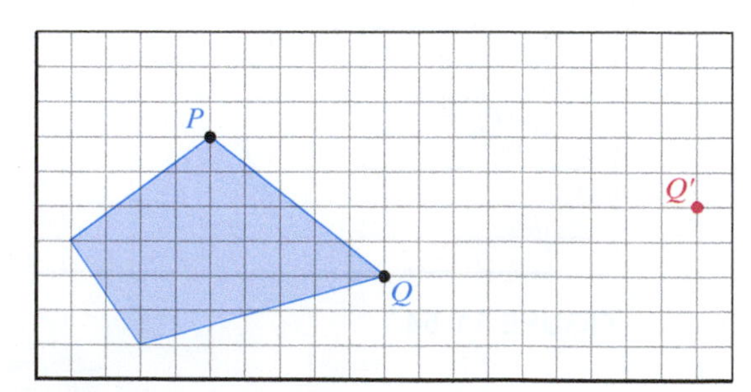

FIGURE 11-57

27. In Fig. 11-58, D' is the image of D under a translation. Find the image of the shaded trapezoid under the translation.

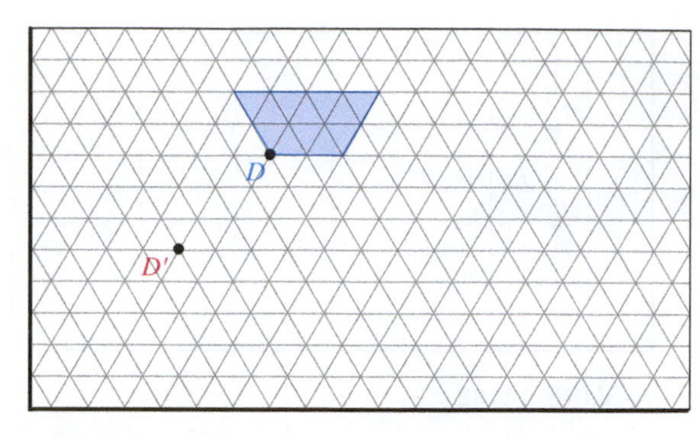

FIGURE 11-58

28. In Fig. 11-59, $Q'$is the image of Q under a translation. Find the image of the shaded region under the translation.

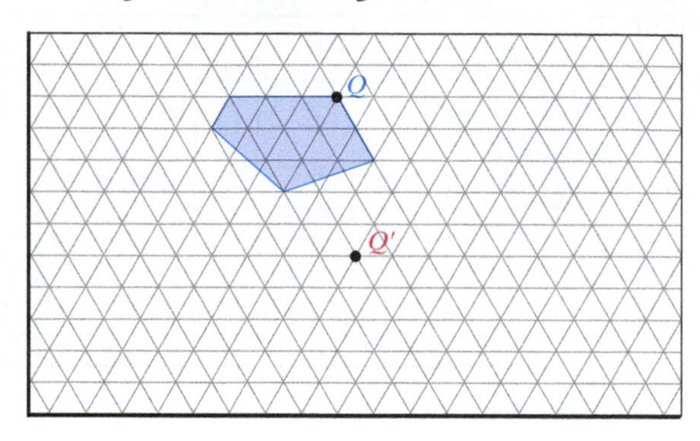

FIGURE 11-59

11.5 Glide Reflections

29. Given a glide reflection with vector v and axis l as shown in Fig. 11-60, find the image of the triangle ABC under the glide reflection.

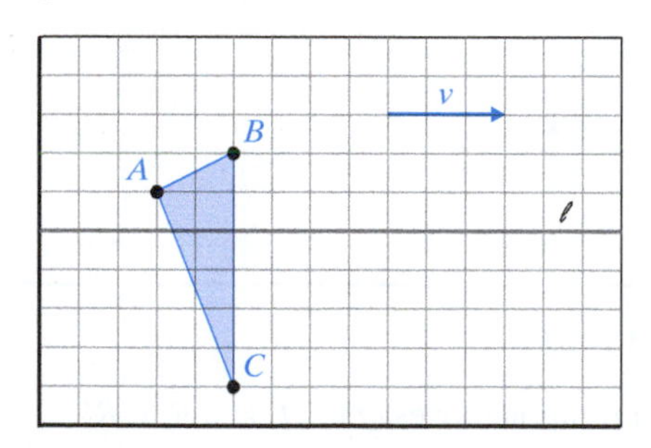

FIGURE 11-60

30. Given a glide reflection with vector v and axis l as shown in Fig. 11-61, find the image of the quadrilateral $ABCD$ under the glide reflection.

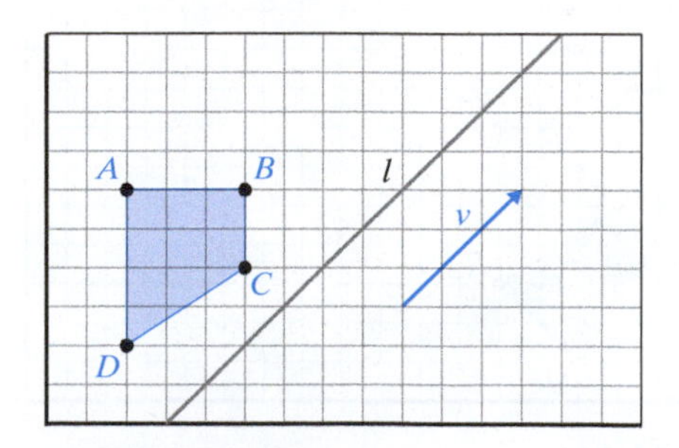

FIGURE 11-61

31. In Fig. 11-62, D' is the image of D under a glide reflection having axis l. Find the image of the polygon $ABCDE$ under the glide reflection.

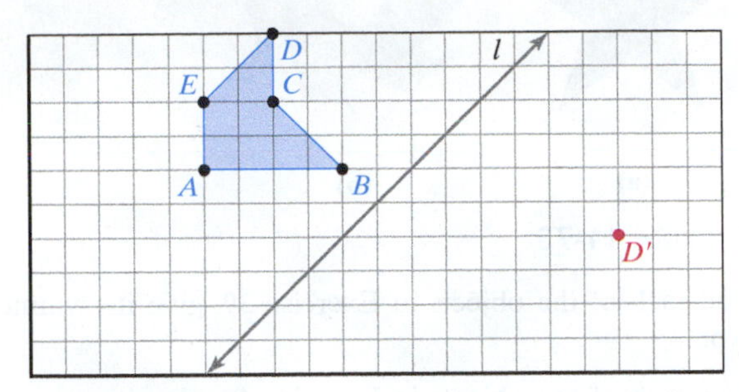

FIGURE 11-62

32. In Fig. 11-63, P' is the image of P under a glide reflection having axis l. Find the image of the quadrilateral $PQRS$ under the glide reflection.

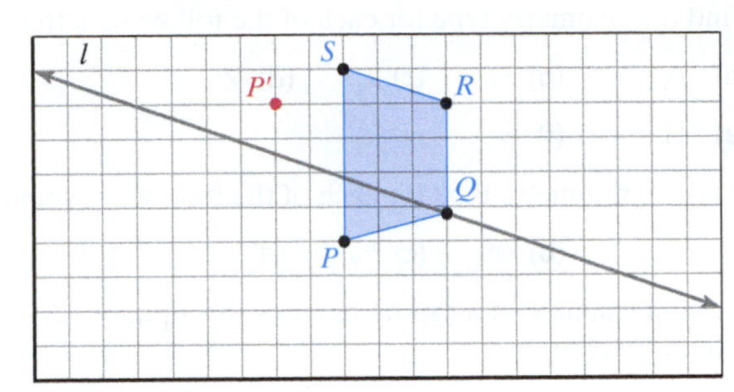

FIGURE 11-63

33. In Fig. 11-64, B' is the image of B and D' is the image of D under a glide reflection.

(a) Find the axis of reflection.

(b) Find the image of A under the glide reflection.

(c) Find the image of the shaded figure under the glide reflection.

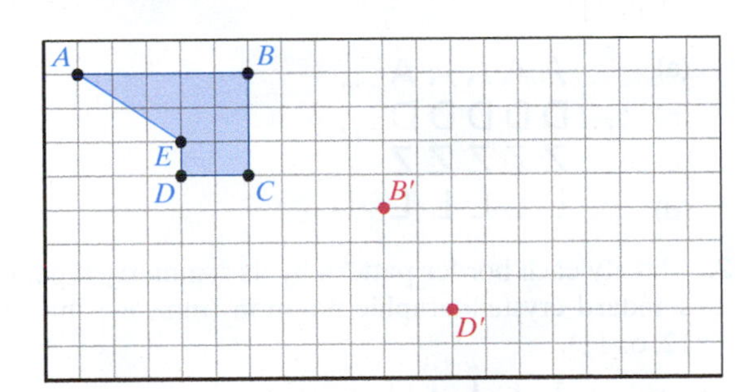

FIGURE 11-64

34. In Fig. 11-65, A' is the image of A and C' is the image of C under a glide reflection.

(a) Find the axis of reflection.

(b) Find the image of B under the glide reflection.

(c) Find the image of the shaded figure under the glide reflection.

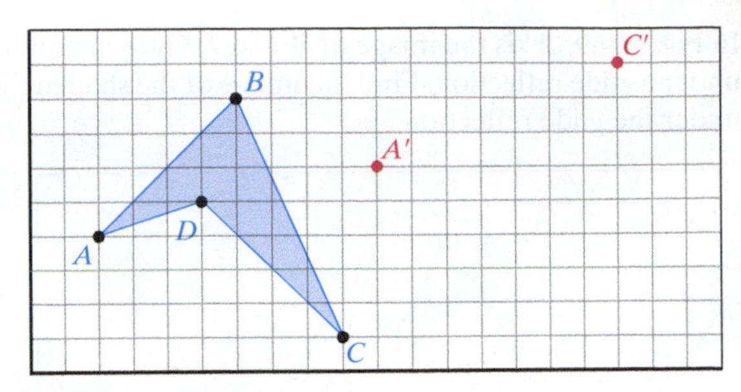

FIGURE 11-65

35. In Fig. 11-66, P' is the image of P and Q' is the image of Q under a glide reflection. Find the image of the shaded figure under the glide reflection.

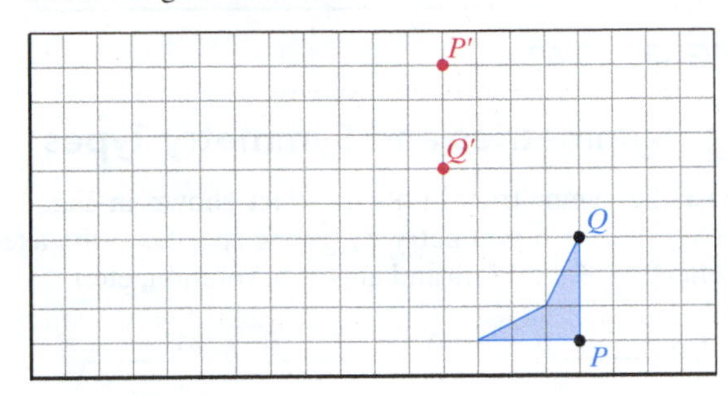

FIGURE 11-66

36. In Fig. 11-67, P' is the image of P and Q' is the image of Q under a glide reflection. Find the image of the shaded figure under the glide reflection.

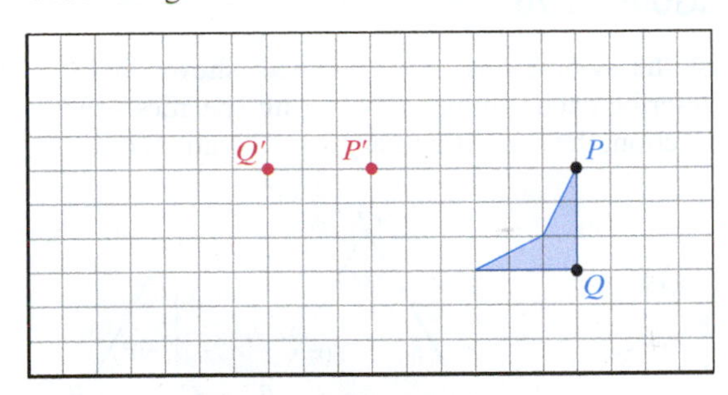

FIGURE 11-67

37. In Fig. 11-68, D' is the image of D and C' is the image of C under a glide reflection. Find the image of the shaded figure under the glide reflection.

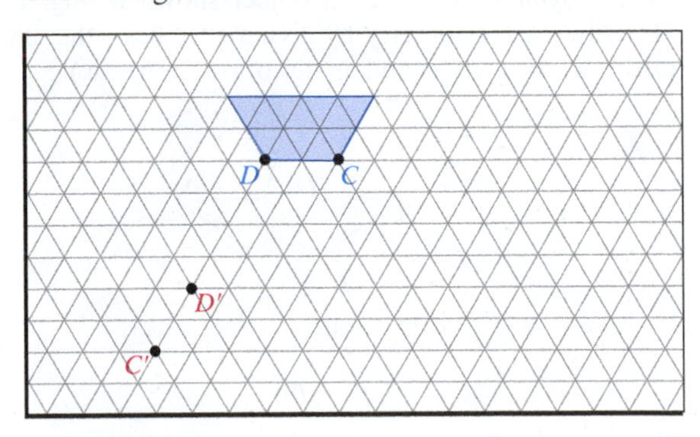

FIGURE 11-68

38. In Fig. 11-69, A' is the image of A and D' is the image of D under a glide reflection. Find the image of the shaded figure under the glide reflection.

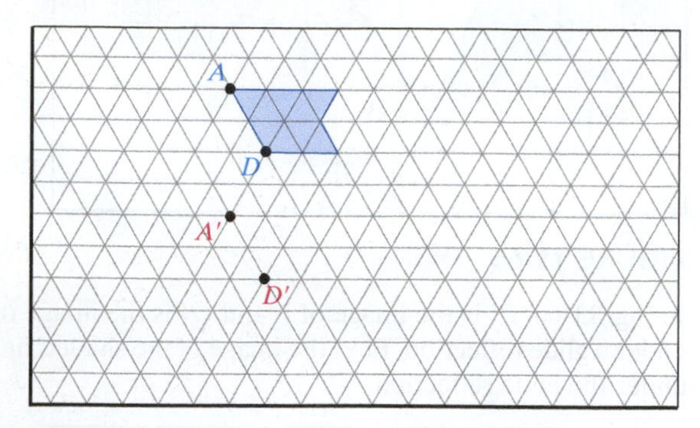

FIGURE 11-69

11.6 Symmetries and Symmetry Types

39. List the symmetries of each object shown in Fig. 11-70. (Describe each symmetry by giving specifics—the axes of reflection, the centers and angles of rotation, etc.)

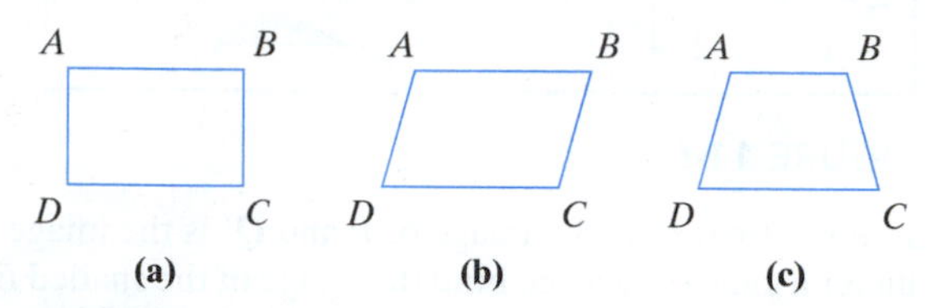

FIGURE 11-70

40. List the symmetries of each object shown in Fig. 11-71. (Describe each symmetry by giving specifics—the axes of reflection, the centers and angles of rotation, etc.)

FIGURE 11-71

41. List the symmetries of each object shown in Fig. 11-72. (Describe each symmetry by giving specifics—the axes of reflection, the centers and angles of rotation, etc.)

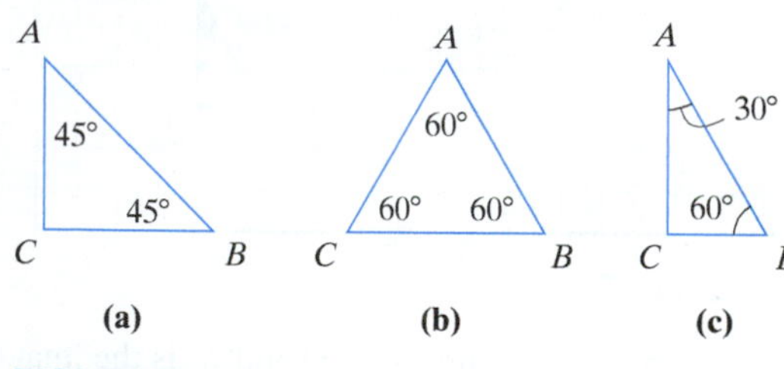
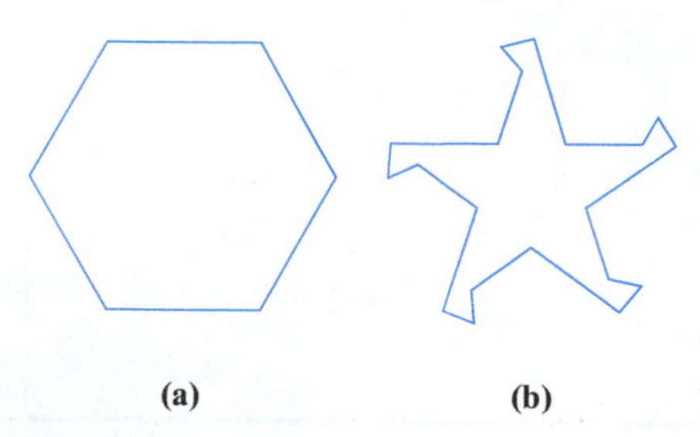

(a) (b)

FIGURE 11-72

42. List the symmetries of each object shown in Fig. 11-73. (Describe each symmetry by giving specifics—the axes of reflection, the centers and angles of rotation, etc.)

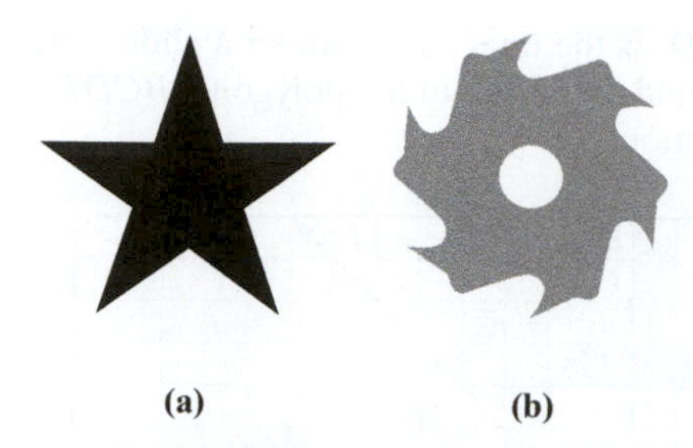

(a) (b)

FIGURE 11-73

43. For each of the objects in Exercise 39, give its symmetry type.

44. For each of the objects in Exercise 40, give its symmetry type.

45. For each of the objects in Exercise 41, give its symmetry type.

46. For each of the objects in Exercise 42, give its symmetry type.

47. Find the symmetry type for each of the following letters.

(a) A **(b)** D **(c)** L **(d)** Z
(e) H **(f)** N

48. Find the symmetry type for each of the following symbols.

(a) $ **(b)** @ **(c)** % **(d)** × **(e)** &

49. Give an example of a capital letter of the alphabet that has symmetry type

(a) Z_1. **(b)** D_1. **(c)** Z_2. **(d)** D_2.

50. Give an example of a one- or two-digit number that has symmetry type

(a) Z_1. **(b)** D_1. **(c)** Z_2. **(d)** D_2.

11.7 Patterns

51. Classify each border pattern by its symmetry type. Use the standard crystallographic notation (mm, mg, m1, 1m, 1g, 12, or 11).

(a) . . . A A A A A . . .
(b) . . . D D D D D . . .
(c) . . . Z Z Z Z Z . . .
(d) . . . L L L L L . . .

52. Classify each border pattern by its symmetry type. Use the standard crystallographic notation (mm, mg, m1, 1m, 1g, 12, or 11).

(a) . . . J J J J J . . .
(b) . . . H H H H H . . .
(c) . . . W W W W W . . .
(d) . . . N N N N N . . .

53. Classify each border pattern by its symmetry type. Use the standard crystallographic notation (mm, mg, m1, 1m, 1g, 12, or 11).

(a) . . . qpqpqpqp . . .
(b) . . . pdpdpdpd . . .
(c) . . . pbpbpbpb . . .
(d) . . . pqbdpqbd . . .

54. Classify each border pattern by its symmetry type. Use the standard crystallographic notation (mm, mg, m1, 1m, 1g, 12, or 11).

(a) ...qbqbqbqb...
(b) ...qdqdqdqd...
(c) ...dbdbdbdb...
(d) ...qpdbqpdb...

55. Find the symmetry type of a border pattern that consists of repeating a motif with symmetry type Z_2.

56. Find the symmetry type of a border pattern that consists of repeating a motif with symmetry type Z_1.

57. Consider a border pattern in a horizontal direction with a repeating motif that has horizontal reflection symmetry. If the motif has symmetry type D_4, what is the symmetry type of the border pattern?

58. Consider a border pattern in a horizontal direction with a repeating motif that has horizontal reflection symmetry. If the motif has symmetry type D_2, what is the symmetry type of the border pattern?

JOGGING

59. The minute hand of a clock is pointing at the number 9, and it is then wound clockwise 7080 degrees.

(a) How many full hours has the hour hand moved?

(b) At what number on the clock does the minute hand point at the end?

60. Name the rigid motion (translation, reflection, glide reflection, or rotation) that moves footprint 1 in Fig. 11-74 onto footprint

(a) 2 (b) 3 (c) 4 (d) 5

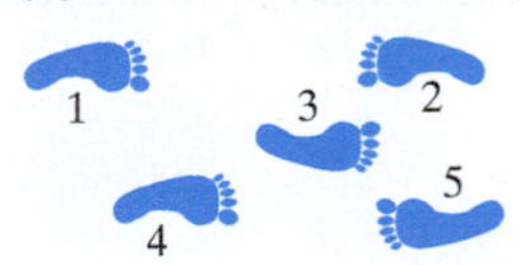

FIGURE 11-74

61. In each case, determine whether the rigid motion is a reflection, rotation, translation, or glide reflection or the identity motion.

(a) The rigid motion is proper and has exactly one fixed point.

(b) The rigid motion is proper and has infinitely many fixed points.

(c) The rigid motion is improper and has infinitely many fixed points.

(d) The rigid motion is improper and has no fixed points.

Exercises 62 through 69 deal with combining rigid motions. Given two rigid motions $\mathcal{M}$ and $\mathcal{N}$, we can combine the two rigid motions by first applying $\mathcal{M}$ and then applying $\mathcal{N}$ to the result. The rigid motion defined by combining $\mathcal{M}$ and $\mathcal{N}$ ($\mathcal{M}$ goes first, $\mathcal{N}$ goes second) is called the ***product*** *of $\mathcal{M}$ and $\mathcal{N}$.*

62. In Fig. 11-75, l_1, l_2, l_3, and l_4 are axes of reflection. In each case, indicate which point is the image of P under

(a) the product of the reflection with axis l_1 and the reflection with axis l_2.

(b) the product of the reflection with axis l_2 and the reflection with axis l_1.

(c) the product of the reflection with axis l_2 and the reflection with axis l_3.

(d) the product of the reflection with axis l_3 and the reflection with axis l_2.

(e) the product of the reflection with axis l_1 and the reflection with axis l_4.

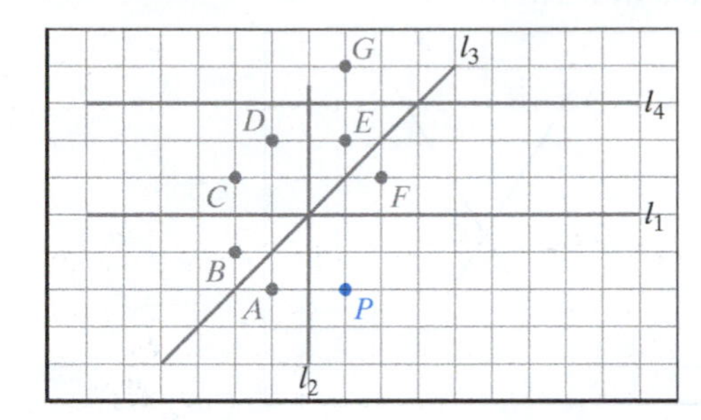

FIGURE 11-75

63. In Fig. 11-76, indicate which point is the image of P under

(a) the product of the reflection with axis l and the 90° clockwise rotation with rotocenter A.

(b) the product of the 90° clockwise rotation with rotocenter A and the reflection with axis l.

(c) the product of the reflection with axis l and the 180° rotation with rotocenter A.

(d) the product of the 180° rotation with rotocenter A and the reflection with axis l.

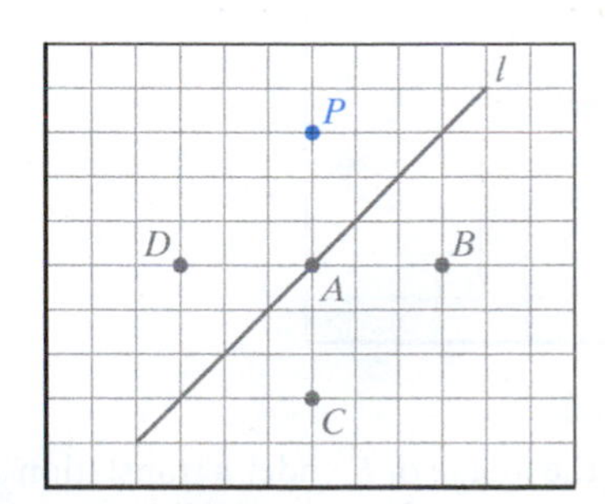

FIGURE 11-76

64. In each case, state whether the rigid motion $\mathcal{M}$ is proper or improper.

(a) $\mathcal{M}$ is the product of a proper rigid motion and an improper rigid motion.

(b) $\mathcal{M}$ is the product of an improper rigid motion and an improper rigid motion.

(c) $\mathcal{M}$ is the product of a reflection and a rotation.

(d) $\mathcal{M}$ is the product of two reflections.

65. Suppose that a rigid motion $\mathcal{M}$ is the product of a reflection with axis l_1 and a reflection with axis l_2, where l_1 and l_2 intersect at a point C. Explain why $\mathcal{M}$ must be a rotation with center C.

66. Suppose that the rigid motion $\mathcal{M}$ is the product of the reflection with axis l_1 and the reflection with axis l_3, where l_1 and l_3 are parallel. Explain why $\mathcal{M}$ must be a translation.

67. In Fig. 11-77, l_1 and l_2 intersect at C, and the angle between them is α.

(a) Give the rotocenter, angle, and direction of rotation of the product of the reflection with axis l_1 and the reflection with axis l_2.

(b) Give the rotocenter, angle, and direction of rotation of the product of the reflection with axis l_2 and the reflection with axis l_1.

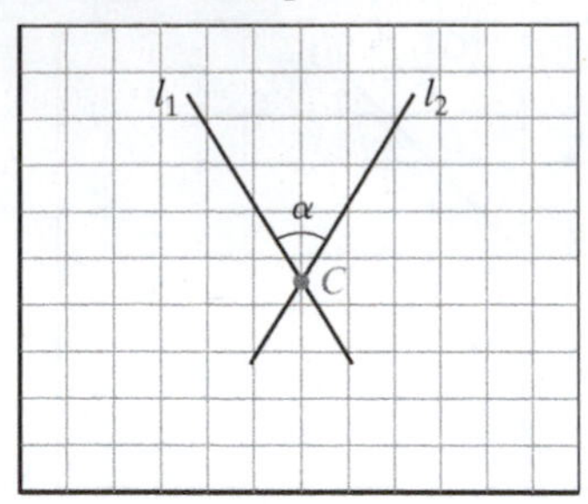

FIGURE 11-77

68. In Fig. 11-78, l_1 and l_3 are parallel and the distance between them is d.

(a) Give the length and direction of the vector of the translation that is the product of the reflection with axis l_1 and the reflection with axis l_3.

(b) Give the length and direction of the vector of the translation that is the product of the reflection with axis l_3 and the reflection with axis l_1.

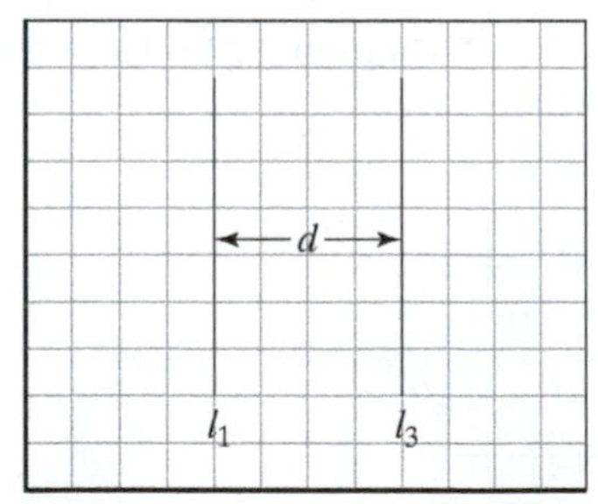

FIGURE 11-78

69. In Fig. 11-79, P' is the image of P under a translation $\mathcal{M}$ and Q' is the image of Q under a translation $\mathcal{N}$.

(a) Find the images of P and Q under the product of $\mathcal{M}$ and $\mathcal{N}$.

(b) Show that the product of $\mathcal{M}$ and $\mathcal{N}$ is a translation. Give a geometric description of the vector of the translation.

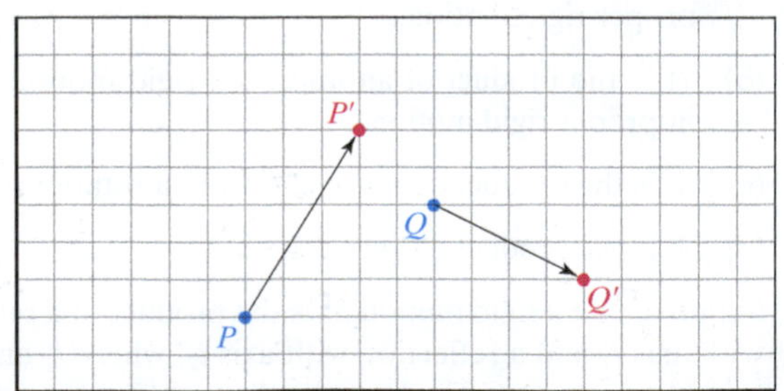

FIGURE 11-79

70. (a) Explain why a border pattern cannot have a reflection symmetry along an axis forming a 45° angle with the direction of the pattern.

(b) Explain why a border pattern can have only horizontal and/or vertical reflection symmetry.

71. A rigid motion $\mathcal{M}$ moves the triangle PQR into the triangle $P'Q'R'$ as shown in Fig. 11-80. Explain why the rigid motion $\mathcal{M}$ must be a glide reflection.

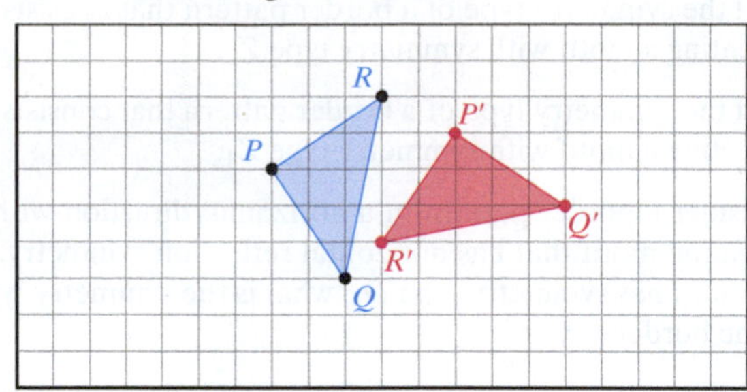

FIGURE 11-80

RUNNING

72. Construct border patterns for each of the seven symmetry types using copies of the symbol ♥ (and rotated versions of it).

73. Let the six symmetries of the equilateral triangle ABC shown in Fig. 11-81 be denoted as follows: r_1 denotes the reflection with axis l_1; r_2 denotes the reflection with axis l_2; r_3 denotes the reflection with axis l_3; R_1 denotes the 120° clockwise rotation with rotocenter O; R_2 denotes the 240° clockwise rotation with rotocenter O; I denotes the identity symmetry. Complete the symmetry "multiplication table" by entering, in each row and column of the table, the product of the row and the column (i.e., the result of applying first the symmetry in the row and then the symmetry in the column). For example, the entry in row r_1 and column r_2 is R_1 because the product of the reflection r_1 and the reflection r_2 is the rotation R_1.

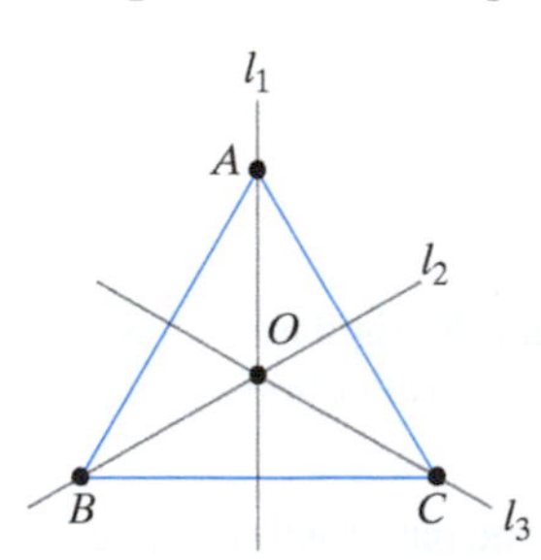

FIGURE 11-81

	r_1	r_2	r_3	R_1	R_2	I
r_1		R_1				
r_2						
r_3						
R_1						
R_2						
I						

74. Let the eight symmetries of the square $ABCD$ shown in Fig. 11-82 be denoted as follows: r_1 denotes the reflection with axis l_1; r_2 denotes the reflection with axis l_2; r_3 denotes the reflection with axis l_3; r_4 denotes the reflection with axis l_4; R_1 denotes the 90° clockwise rotation with rotocenter O; R_2 denotes the 180° clockwise rotation with rotocenter O; R_3 denotes the 270° clockwise rotation with rotocenter O; I denotes the identity symmetry.

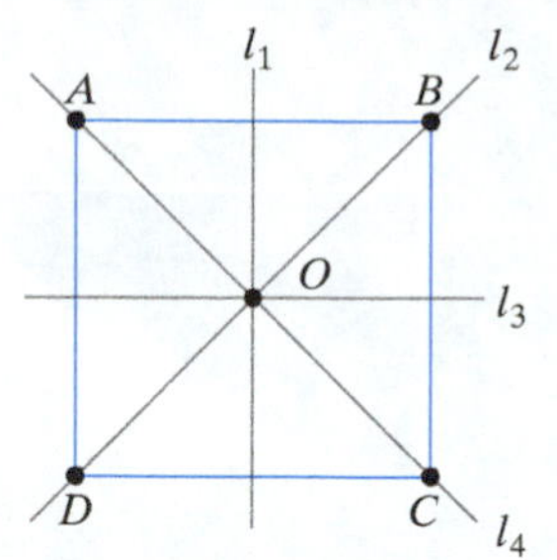

FIGURE 11-82

Complete the symmetry multiplication table below (*Hint*: Try Exercise 73 first.)

	r_1	r_2	r_3	r_4	R_1	R_2	R_3	I
r_1								
r_2								
r_3								
r_4								
R_1								
R_2								
R_3								
I								

75. List all the symmetries of the wallpaper pattern shown in Fig. 11-83.

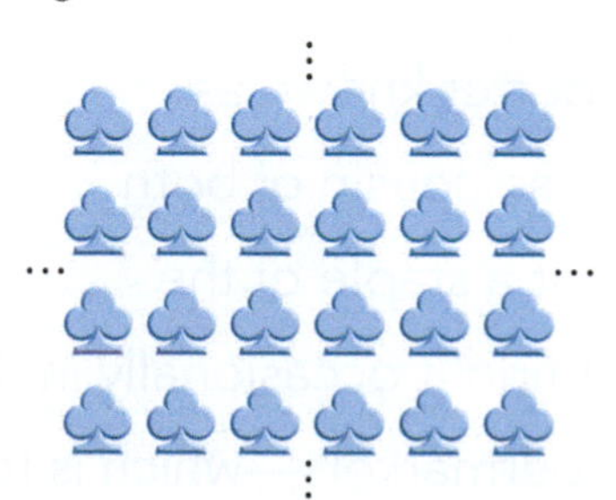

FIGURE 11-83

76. List all the symmetries of the wallpaper pattern shown in Fig. 11-84.

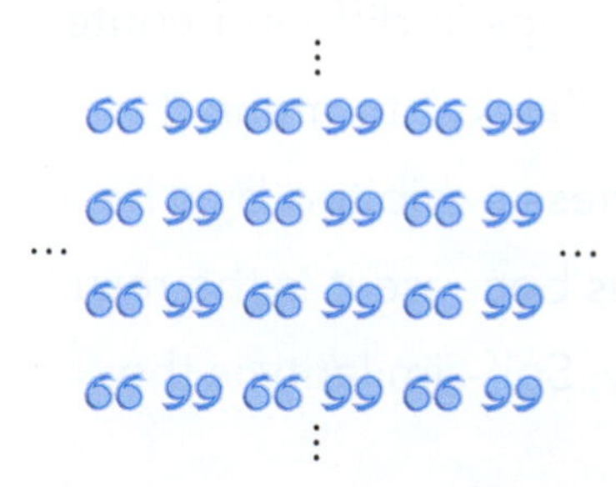

FIGURE 11-84

77. List all the symmetries of the wallpaper pattern shown in Fig. 11-85.

FIGURE 11-85

78. List all the symmetries of the wallpaper pattern shown in Fig. 11-86.

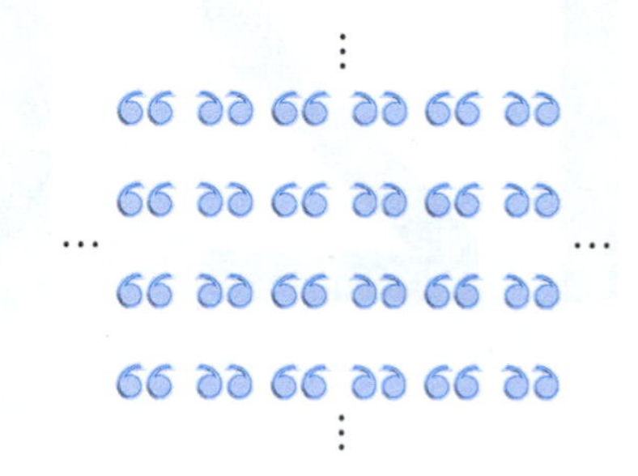

FIGURE 11-86

79. List all the symmetries of the wallpaper pattern shown in Fig. 11-87.

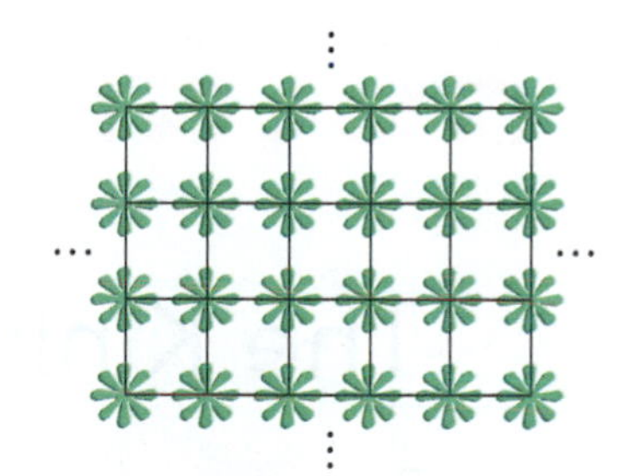

FIGURE 11-87

80. List all the symmetries of the wallpaper pattern shown in Fig. 11-88.

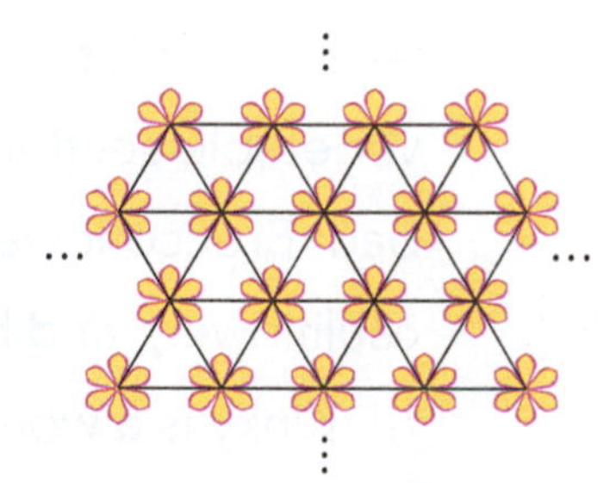

FIGURE 11-88

12

Broccoflower—there is more to this exotic veggie than just good, healthy eats. (For more on its self-similar structure, see pages 361–362.)

Fractal Geometry

The Kinky Nature of Nature

Romanesco broccoli (*Brassica oleracea*)—better known as *broccoflower*—is an edible flower and a close cousin of both broccoli and cauliflower. Broccoflower is not a staple of the American diet—if you are lucky, you might find it occasionally in the vegetable section of one of your better supermarkets—which is too bad: Broccoflower is nutritious, delicious (milder and nuttier than cauliflower), and beautiful to behold. It is also kinky. Very, very kinky.

Kinky is a word that means different things in different contexts, and in the context of this chapter we use "kinky" to mean the opposite of "smooth." The kind of kinkiness exhibited by a head of broccoflower is of particular interest to us because it is the result of a special property called *self-similarity*. Self-similarity is the characteristic of an object that looks like parts of itself, and these parts in turn look like even smaller parts of themselves, and so on (at least for a while). Self-similarity is an extremely useful property, and many things in nature have some form of it—trees, clouds, mountains, rivers, our vascular system of veins and arteries, and so on.

Whatever nature creates, man wants to imitate. By and large, this is a good idea (nature usually knows what it is doing), but imitating nature is easier said than done. How do you imitate the kinky structure of a head of broccoflower, or an oak tree, or a mountain? The first step in this direction is the development of a new type of geometry called *fractal geometry* and the creation of super-kinky geometric shapes called *fractals*. These remarkable shapes are made possible only by incorporating self-similarity into the structure of their geometry. This chapter is an introduction to the basic ideas of fractal geometry.

The chapter is organized around a series of classic examples of geometric fractals— the *Koch snowflake* (followed by a general discussion of *self-similarity*) in Section 12.1, the *Sierpinski gasket* (followed by the *chaos game*) in Section 12.2, the *twisted Sierpinski gasket* in Section 12.3, and the *Mandelbrot set* in Section 12.4. As we discuss each of these examples, we will explore the connection between fractal geometry and the self-organizing kinkiness exhibited by many objects in nature.

12.1 The Koch Snowflake and Self-Similarity

Our first example of a *geometric fractal* is a shape known as the *Koch snowflake* [named after the Swedish mathematician Helge von Koch (1870–1954)]. Like other geometric fractals we will discuss in this chapter, the Koch snowflake is constructed by means of a *recursive process*, a process in which the same set of rules is applied repeatedly in an infinite feedback loop—the output at one stage becomes the input at the next stage. (We discussed recursively defined number sequences in Chapter 9. In contrast, the recursive process considered here will involve geometric shapes.)

The construction of the **Koch snowflake** proceeds as follows:

- **Start.** Start with a shaded *equilateral* triangle [Fig. 12-1(a)]. We will refer to this starting triangle as the *seed* of the Koch snowflake. The size of the seed triangle is irrelevant, so for simplicity we will assume that the sides are of length 1.
- **Step 1.** To the *middle third* of each of the sides of the seed triangle, add a smaller equilateral triangle with sides of length $\frac{1}{3}$, as shown in Fig. 12-1(b). The result is the 12-sided "snowflake" shown in Fig. 12-1(c)—a shape just a tad kinkier than the seed triangle.
- **Step 2.** To the *middle third* of each of the 12 sides of the "snowflake" in Step 1, add an equilateral triangle with sides of length one-third the length of that side. The result is a "snowflake" with 48 sides (12×4), each of length $(\frac{1}{3})^2 = \frac{1}{9}$, as shown in Fig. 12-2(a). (The snowflake got kinkier.)

For ease of reference, we will describe the process of adding an equilateral triangle to the *middle third* of each side of a given "snowflake" as *procedure KS*. This will make the rest of the construction a lot easier to describe. Notice that

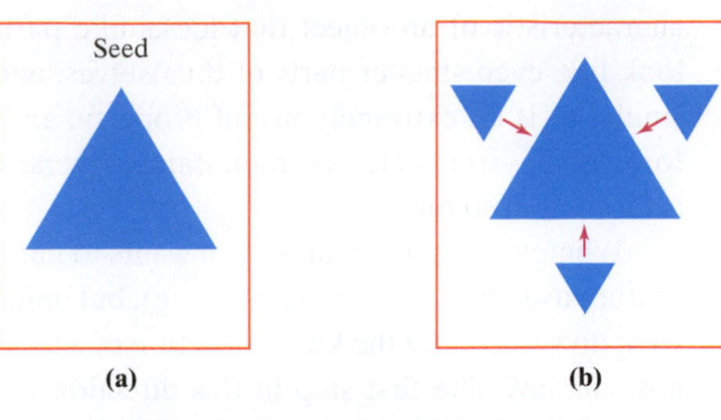

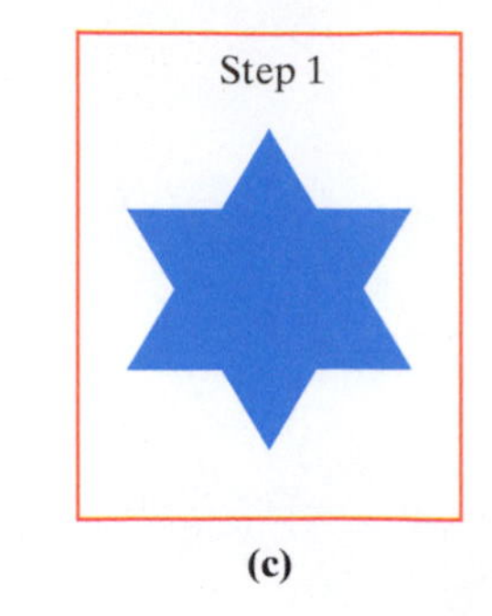

FIGURE 12-1 Koch snowflake: Seed and Step 1.

every time we apply *procedure KS* the snowflake gets kinkier—each side "crinkles" into four new sides and the sides get shorter by a factor of one-third.

- **Step 3.** Apply *procedure KS* to the "snowflake" in Step 2. This gives the more elaborate "snowflake" shown in Fig. 12-2(b), with 192 sides (48×4), each of length $(\frac{1}{3})^3 = \frac{1}{27}$.
- **Step 4.** Apply *procedure KS* to the "snowflake" in Step 3. This gives the "snowflake" shown in Fig. 12-2(c). You definitely don't want to do this by hand— 192 tiny little equilateral triangles are being added! (You can, however, walk your way through the early steps of this construction without sweating the work by using the applet ***Geometric Fractals***, available in *MyMathLab* in the *Multimedia Library* or in *Tools for Success*.)

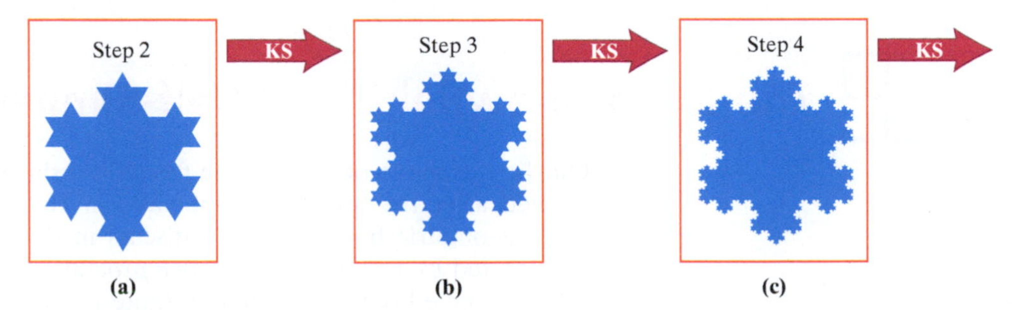

FIGURE 12-2 Koch snowflake: Steps 2, 3, and 4.

- **Steps 5, 6, etc.** At each step apply *procedure KS* to the "snowflake" obtained in the previous step.

At every step of this recursive process the rules that define procedure KS generate a new "snowflake," but after a while it's hard to tell that there are any changes. There is already a very small difference between the snowflakes in Figs. 12-2(b) and (c), and after a few more steps the images become *visually stable*—to the naked eye there is no difference between one snowflake and the next. For all practical purposes what we are seeing is a rendering of a *Koch snowflake* [Fig. 12-3(a)].

A note of caution: Because the Koch snowflake is constructed in an infinite sequence of steps, a perfect rendering of it is impossible—Fig. 12-3(a) is only an approximation. This is no reason to be concerned or upset—the Koch snowflake is an abstract shape that we can study and explore by looking at imperfect pictures of it (much like we do in high school geometry when we study circles even though it is impossible to draw a perfect circle).

The most "happening" part of the Koch snowflake is its boundary (that's where all the kinkiness occurs), so it is often more convenient to only look at the boundary and forget about the interior. If we do that, we get the **Koch curve** (sometimes

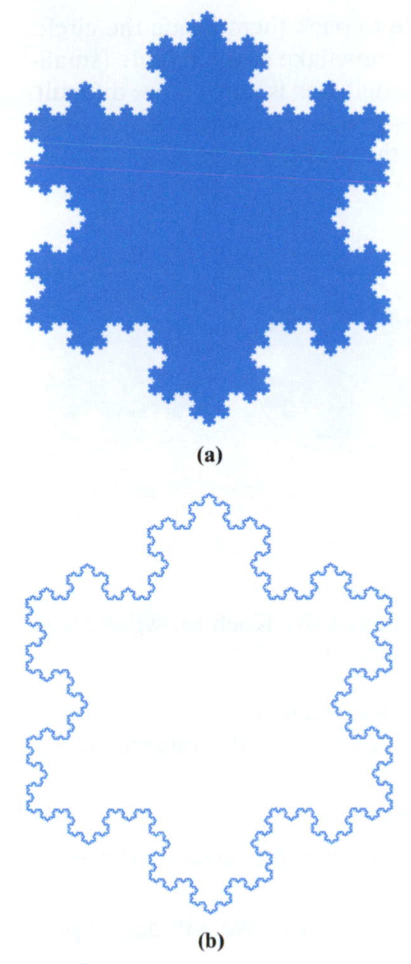

FIGURE 12-3 (a) Koch snowflake. (b) Koch curve.

called the *snowflake curve*) shown in Fig. 12-3(b). From now on, our discussion will alternate between the *Koch snowflake* (a two-dimensional region, with an *area*) and the *Koch curve* (a closed curve, with a *length*).

One advantage of recursive processes is that they allow for very simple and efficient definitions, even when the objects being defined are quite complicated. The Koch curve, for example, is a fairly complicated shape, but we can define it in two lines using a form of shorthand we will call a **replacement rule**—a scale-independent rule that specifies how to substitute one piece for another.

KOCH CURVE REPLACEMENT RULE

- **Start:** Start with an equilateral triangle.
- **Replacement rule:** In each step, replace any individual line segment ——— with the "crinkled" version _/_ (with the point always facing to the outside of the snowflake).

One of the most remarkable facts about the Koch snowflake is that it occupies a relatively small area and yet its boundary is infinitely long—a notion that seems to defy common sense.

- **Boundary.** To compute the length of the boundary let's first compute the perimeter of the "snowflakes" obtained in the first few steps of the construction (Fig. 12-4). The seed triangle has perimeter $P_0 = 3$. In Step 1, each side of length 1 is replaced by four sides of length $\frac{1}{3}$: The perimeter is $P_1 = 3 \times (\frac{4}{3}) = 4$. In Step 2, each side of length $\frac{1}{3}$ is replaced by four sides of length $\frac{1}{9}$: The perimeter is $P_2 = 3 \times (\frac{4}{3})^2 = \frac{48}{9}$. As the replacement process continues, at each step we replace a side by four sides that are $\frac{1}{3}$ as long. Thus, at any given step, the perimeter of the snowflake at that step is $\frac{4}{3}$ times the perimeter of the snowflake at the preceding step. This implies that the perimeters keep growing with each step, and growing very fast indeed. *After infinitely many steps the perimeter is infinite.*

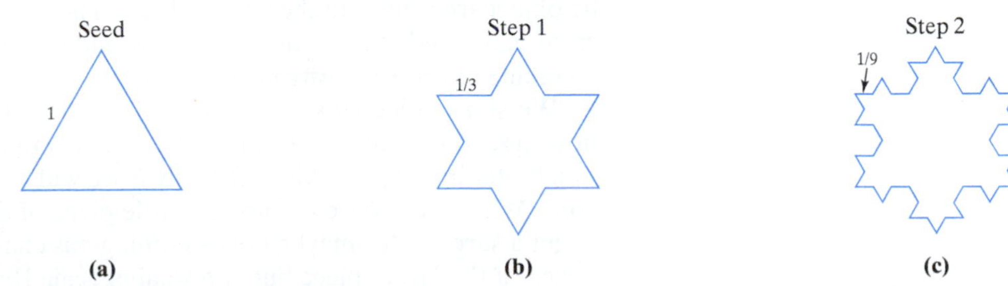

FIGURE 12-4 (a) Three sides of length 1 ($P_0 = 3$). (b) Twelve sides of length $\frac{1}{3}$ ($P_1 = 4$). (c) 48 sides of length $\frac{1}{9}$ ($P_2 = \frac{48}{9} = 5\frac{1}{3}$).

- **Area.** Computing the exact area of the Koch snowflake is considerably more difficult, but it is not hard to convince oneself that the Koch snowflake has a finite area. The key observation is that at every step of the construction the snowflakes stay within the boundary of a given circle—the circle that circumscribes the seed triangle (Fig. 12-5). Essentially, what is happening here is that even though we are adding infinitely many triangles, at each step the triangles get smaller and smaller and we are always able to pack them inside the circle. The easy conclusion is that the area of the Koch snowflake must be finite (smaller than the area of the circle). To calculate the actual area is much more difficult, so we leave out the details, but the bottom line is this: The area of the Koch snowflake is exactly $\frac{8}{5}$ (or 1.6) times the area of the seed triangle.

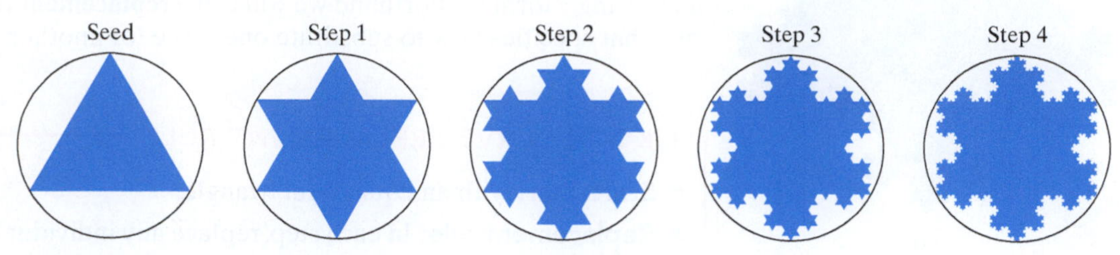

FIGURE 12-5 At every step the snowflakes remain inside the circle that circumscribes the seed triangle.

The following is a summary of the key properties of the Koch snowflake and the Koch curve.

- **Replacement rule.** In each step, replace any line segment ——— with the "crinkled" version _/_ (with the point always facing to the outside of the snowflake).
- **Boundary (Koch curve).** Infinitely long boundary.
- **Area.** Finite area [equal to (1.6)A, where A denotes the area of the seed triangle].
- **Self-similarity.** *Exact*, *infinite*, and *universal* self-similarity. We will define precisely what this means next.

Self-Similarity

The most important characteristic shared by the shapes and objects, which we will see in this chapter, is that they are all *self-similar*. In broad terms, we say that an object is **self-similar** when it is similar to some part of itself. In other words, parts of the object are similar to the whole object. The self-similarity can occur at many different scales and in many different ways, and we will discuss the many variations on the theme of self-similarity next.

We start with a closer look at our new friend, *the Koch curve*. Figure 12-6(a) shows a section of the Koch curve. We do know that the curve is pretty kinky, but to get a better idea of how kinky it really is we will look at it with an imaginary microscope. When we choose to look at a little piece of the original curve and magnify it, we get a surprise (or maybe not)—nothing has changed! The little piece is an exact replica of the larger piece but at a smaller scale [Fig. 12-6(b)]. And, if we crank up the magnification and choose to look at an even smaller piece it happens again—we get a copy of the original piece at an even smaller scale [Fig. 12-6(c)]. Frustrated, we continue cranking the magnification and looking at smaller and smaller bits of the curve, but nothing changes—more of the same at smaller and smaller scales. Boring, you say? Not once you understand how useful this can be.

The kind of self-similarity exhibited by the Koch curve has three key properties—it is *exact, infinite,* and *universal.*

- **Exact self-similarity.** An object or shape has *exact self-similarity* if different parts of the object look *exactly* the same at different scales. (Using the microscope analogy, this means that as you crank up the microscope's magnification you continue seeing parts that are identical in shape.)

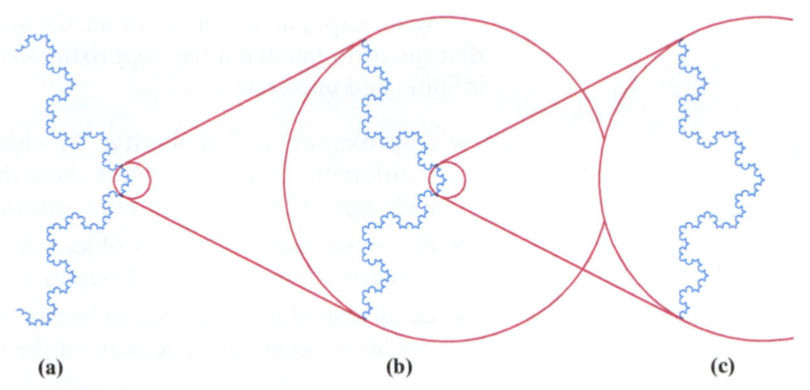

FIGURE 12-6 (a) A small piece of the Koch curve. (b) The piece in (a) magnified ×9.(c) The same piece magnified ×81.

- **Infinite self-similarity.** An object or shape has *infinite self-similarity* if the self-similarity occurs at infinitely many different scales. (This means that if you were able to continue cranking up your microscope's magnification indefinitely, the self-similar parts would continue showing up indefinitely.)
- **Universal self-similarity.** An object or shape has *universal self-similarity* if the self-similarity occurs in every part of the object. (Again using the microscope analogy, you don't have to point the microscope in a specific direction—no matter where you zoom you will eventually find self-similar pieces.)

For a different perspective on self-similarity, let's take another look at our friend the broccoflower. We start with a full head of broccoflower [Fig. 12-7(a)]—for convenience, we will call it the "mother." The mother is made up of a series of buds of various sizes (the "first generation"). Figure 12-7(b) is a close-up of some of the first generation buds at the center of the original picture. These buds all look very much like the mother but they are not identical to it. Let's say that they are approximately (but not exactly) self-similar. Figure 12-7(c) is a close-up of some of the first-generation buds. We can clearly see now the "second-generation" buds, looking very much like the mother and the first-generation buds but not identical to any. The self-similar replication continues but only for another three generations. Figure 12-7(d) shows the "fifth generation," and we can see that the self-similar replication is beginning to break down. These buds no longer look like the mother or like earlier generations. It is also worth noting that the buds of a broccoflower head are spatially arranged in a very special way—the first-generation buds form a spiral arrangement inside the mother, each second-generation bud is part of a spiral inside its parent bud, and so on. Because of this spiral arrangement, the self-similarity is not universal—it occurs only along certain directions.

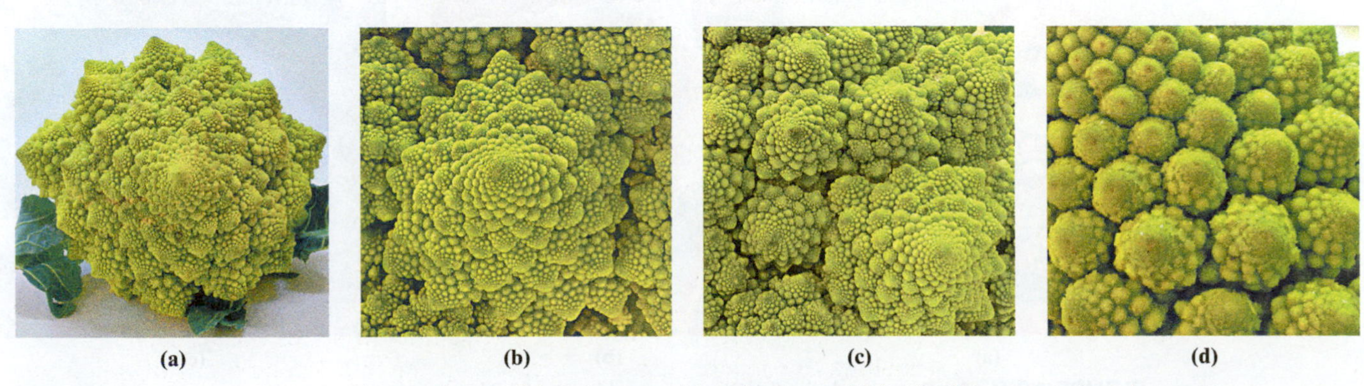

FIGURE 12-7 (a) Full head of broccoflower. (b) A close-up. (c) A close-up of (b). (d) Fifth generation close-up.

What happened with the broccoflower head is the typical kind of self-similarity that can be found in nature—*approximate*, *finite*, and *local* (the alternatives to exact, infinite, and universal).

- **Approximate self-similarity.** An object or shape has *approximate self-similarity* if different parts of the object share the same structure as the whole object and look *approximately* (but not exactly) the same at different scales.
- **Finite self-similarity.** An object or shape has *finite self-similarity* if the self-similarity occurs at multiple scales but eventually stops.
- **Local self-similarity.** An object or shape has *local self-similarity* if the self-similarity occurs in some parts of the object but not everywhere.

We can find self-similarity in many shapes in nature: trees, rivers, seashells our circulatory system, and so on. Figure 12-8 illustrates a few examples of nature's affinity for self-similar shapes. In all cases, the self-similarity is *approximate*, *finite*, and *local*, and (for some unexplained reason) quite beautiful.

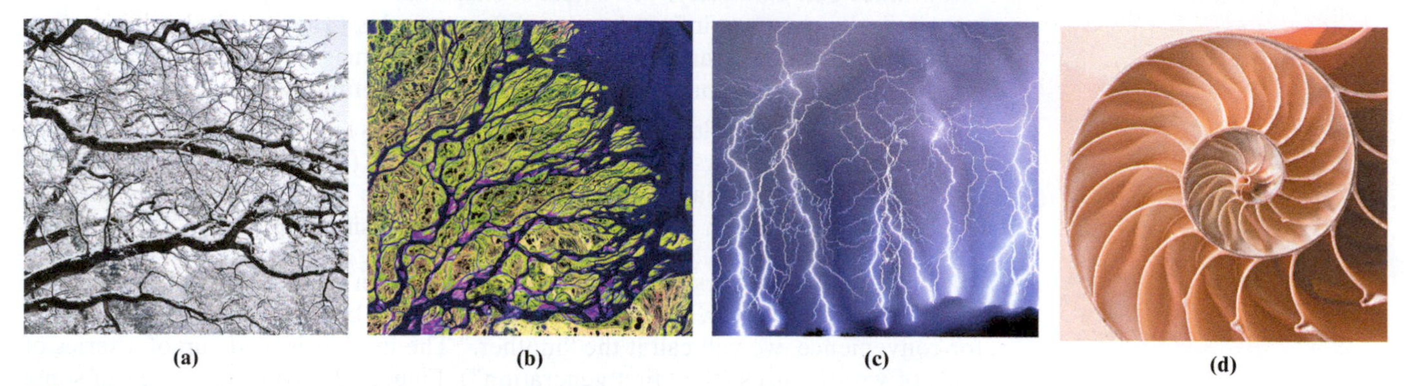

(a) (b) (c) (d)

FIGURE 12-8 (a) Oak tree. (b) River basin as seen from space. (c) Lightning. (d) Interior of a chambered nautilus.

The appeal of self-similarity transcends natural shapes, and self-similarity is often used in man-made objects and images: Russian dolls, cake decorations, abstract art, and so on (Fig. 12-9). Typically, the implied message that these man-made creations try to convey is that the self-similar replication is infinite and that the images keep repeating forever at smaller and smaller scales, but obviously this is an illusion—the self-similarity is *finite* and has to stop after a few iterations (it is impossible for the human hand to draw or physically create an infinite repetition).

For infinite self-similarity we need to think in abstract, mathematical terms and create virtual objects and shapes (such as the Koch snowflake) that can only be seen in the mind's eye. We will study several of these in the remainder of this chapter.

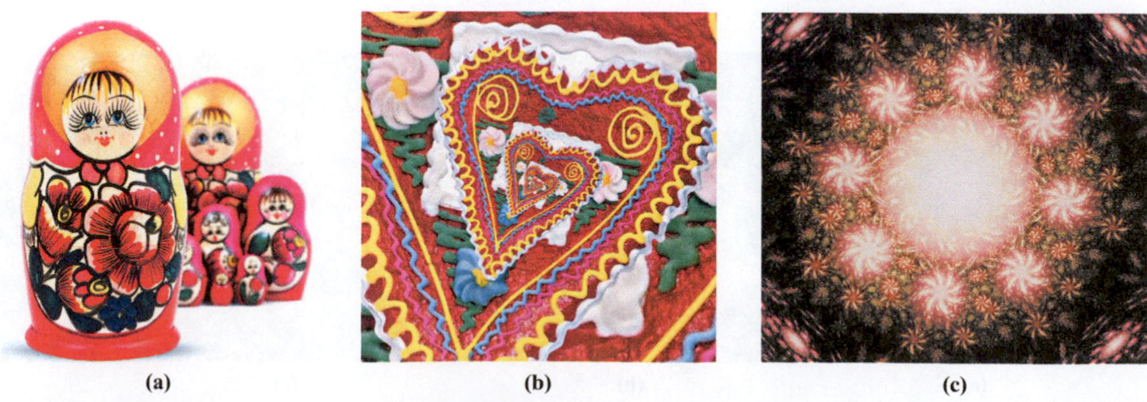

(a) (b) (c)

FIGURE 12-9 (a) Russian dolls. (b) Gingerbread heart. (c) Abstract art.

12.2 The Sierpinski Gasket and the Chaos Game

Waclaw Sierpinski (1882–1969)

With the insight gained by our study of the Koch snowflake, we will now look at another well-known geometric fractal called the *Sierpinski gasket*, named after the Polish mathematician Waclaw Sierpinski.

Just like with the Koch snowflake, the construction of the **Sierpinski gasket** starts with a solid triangle, but this time, instead of repeatedly *adding* smaller and smaller versions of the original triangle, we will *remove* them according to the following procedure:

- **Start.** Start with a seed—the solid triangle *ABC* [Fig. 12-10(a)]. Note that in this case the triangle does not have to be equilateral. Any triangle will do.
- **Step 1.** Remove the triangle connecting the midpoints of the sides of the seed triangle. This gives the shape shown in Fig. 12-10(b)—consisting of three solid triangles, each a half-scale version of the seed and a hole where the middle triangle used to be.

  For convenience, we will call this process of hollowing out the middle triangle of a solid triangle *procedure SG*.
- **Step 2.** Apply *procedure SG* to each of the three solid triangles in Fig. 12-10(b). The result is the "gasket" shown in Fig. 12-10(c) consisting of nine solid triangles, each at one-fourth the scale of the seed triangle, plus three small holes of the same size and one larger hole in the middle.
- **Step 3.** Apply *procedure SG* to each of the nine solid triangles in Fig. 12-10(c). The result is the "gasket" shown in Fig. 12-10(d) consisting of 27 solid triangles, each at one-eighth the scale of the original triangle, nine small holes of the same size, three medium-sized holes, and one large hole in the middle.
- **Steps 4, 5, etc.** Apply *procedure SG* to each solid triangle in the "gasket" obtained in the previous step.

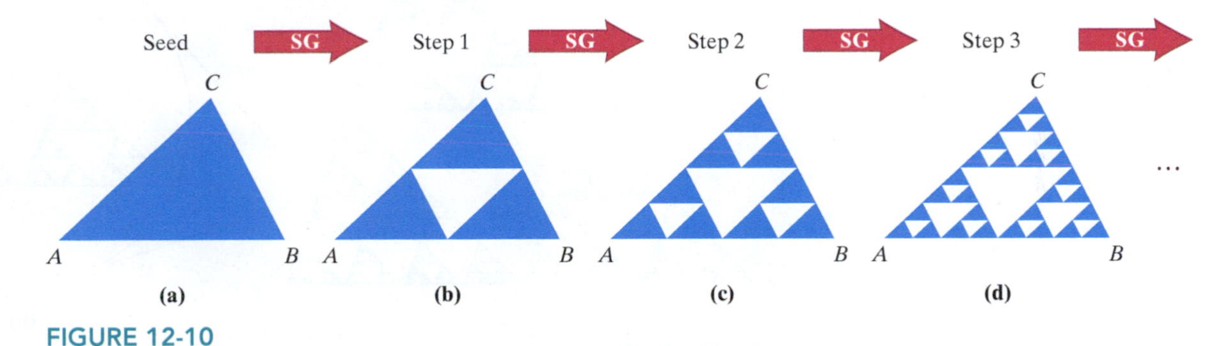

FIGURE 12-10

After a few more steps the figure becomes visually stable—the naked eye cannot tell the difference between the gasket obtained at one step and the gasket obtained at the next step. At this point we have a good rendering of the Sierpinski gasket itself. In your mind's eye you can think of Fig. 12-11 as a picture of the Sierpinski gasket. [Full disclosure—It is just a late step approximation. (You can use the applet ***Geometric Fractals*** available in *MyMathLab* to walk your way through the first six steps of the construction.)

FIGURE 12-11

The Sierpinski gasket is clearly a fairly complicated geometric shape, and yet it can be defined in two lines using the following recursive *replacement rule*.

SIERPINSKI GASKET REPLACEMENT RULE

- **Start:** Start with a shaded seed triangle.
- **Replacement rule:** In each step, replace any individual solid triangle with the "hollowed" version.

Looking at Fig. 12-11, we might think that the Sierpinski gasket is made of a huge number of tiny triangles, but this is just an optical illusion—there are no solid triangles in the Sierpinski gasket, just specks of the original triangle surrounded by a sea of white triangular holes. If we were to magnify any small part of the Sierpinski gasket [Fig. 12-12(a)], we would see more of the same—specks surrounded by white triangles [Fig. 12-12(b)]. This, of course, is another example of *self-similarity*. As with the Koch curve, the self-similarity of the Sierpinski gasket is *exact*, *infinite*, and *universal*.

As a geometric object existing in the plane, the Sierpinski gasket should have an area, but it turns out that its area is infinitesimally small, smaller than any positive quantity. Paradoxical as it may first seem, the mathematical formulation of this fact is that the Sierpinski gasket has *zero area*. At the same time, the boundary of the "gaskets" obtained at each step of the construction grows without bound, which implies that the Sierpinski gasket has an infinitely long boundary (see Exercises 21 and 22).

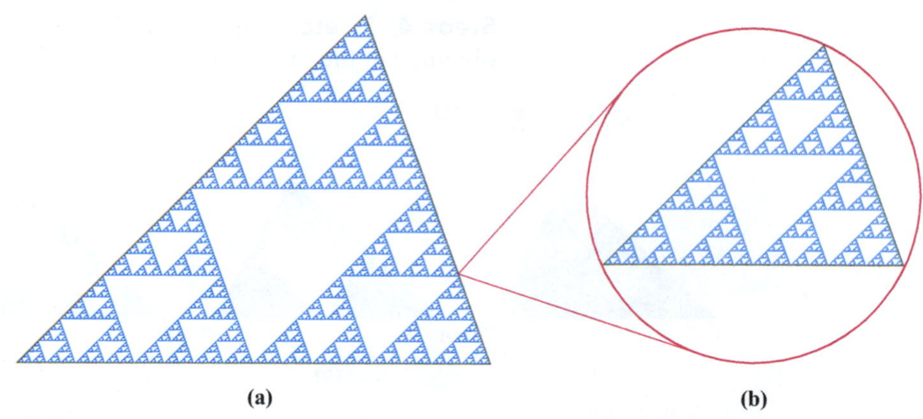

FIGURE 12-12 (a) Sierpinski gasket. (b) Sierpinski gasket detail (magnification ×256).

The following is a summary of the key properties of the Sierpinski gasket:

- **Replacement rule.** In each step, replace any individual solid triangle with the "hollowed" version.
- **Boundary.** Infinitely long boundary.
- **Area.** Infinitely small (zero area).
- **Self-similarity.** Exact, infinite, and universal self-similarity.

The Chaos Game

For a change of pace, we will now play a game of chance called the *chaos game*. All we need to play the chaos game is a single six-sided die and an arbitrary triangle with vertices labeled A, B, and C [Fig. 12-13(a)].

The purpose of the die is to randomly choose one of the three vertices of the triangle with equal probability. We can do this in many different ways. For example, let's say that a roll of 1 or 2 maps to vertex A; a roll of 3 or 4 maps to vertex B; and a roll of 5 or 6 maps to vertex C. Here now are the rules for the **chaos game**:

- **Start.** Roll the die and mark the vertex that corresponds to that roll. Say we roll a 5. This means our starting point is C [Fig. 12-13(b)].
- **Step 1.** Roll the die again. Say we roll a 2, which corresponds to vertex A. We now *move to position M_1 halfway between the previous position C and the winning vertex A*. This is our new point M_1 [Fig. 12-13(c)].
- **Step 2.** Roll the die again, and *move to the point halfway between the last position M_1 and the new winning vertex*. [Say we roll a 3—the move then is to M_2 halfway between M_1 and B as shown in Fig. 12-13(d).] Mark the new position M_2.
- **Steps 3, 4, etc.** Continue rolling the die, each time *marking the point halfway between the last position and the winning vertex.*

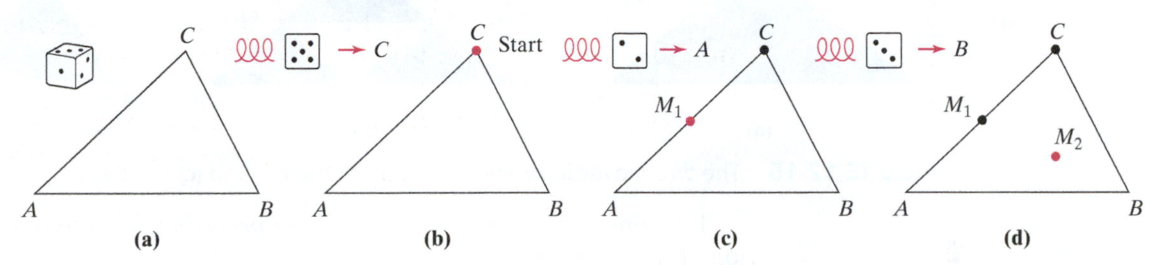

FIGURE 12-13 The chaos game: Always move from the previous position toward the chosen vertex and stop halfway.

What happens after you roll the die more than just a few times? Figure 12-14(a) shows the pattern of points after 50 rolls of the die—just a bunch of scattered dots—but when you continue rolling the die, a pattern emerges [Fig. 12-14(b), (c), and (d)]: the unmistakable tracks of a Sierpinski gasket! The longer we play the chaos game, the more the track of points left behind looks like a Sierpinski gasket. After 100,000 rolls of the die, it would be impossible to tell the difference between the two. (You can convince yourself by playing the chaos game with the applet ***Geometric Fractals*** available in *MyMathLab*. Using the applet, you can see—almost instantaneously—the track of points left behind by thousands of rolls of a virtual die.)

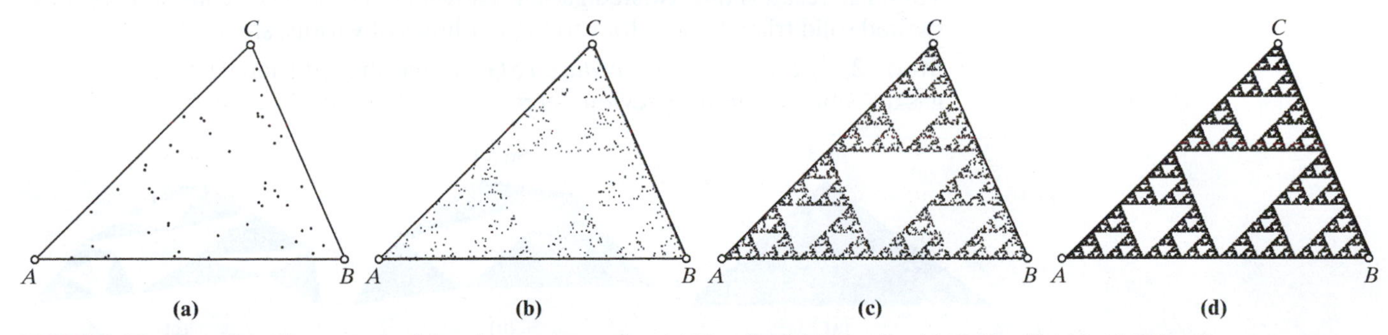

FIGURE 12-14 The "footprint" of the chaos game after (a) 50 rolls, (b) 500 rolls, (c) 5000 rolls, (d) 10,000 rolls.

This is a truly surprising turn of events. The chaos game is ruled by the laws of chance; thus, we would expect that essentially a random pattern of points would be generated. Instead, we get an approximation to the Sierpinski gasket, and the longer we play the chaos game, the better the approximation gets. An important implication of this is that it is possible to generate geometric fractals using simple rules based on the laws of chance.

12.3 The Twisted Sierpinski Gasket

Our next construction is a variation of the original Sierpinski gasket. For lack of a better name, we will call it the *twisted Sierpinski gasket*.

The construction starts out exactly like the one for the regular Sierpinski gasket, with a solid seed triangle [Fig. 12-15(a)] from which we cut out the middle triangle, [whose vertices we will call M, N, and L as shown in Fig. 12-15(b)]. The next move (which we will call the "twist") is new: Each of the points M, N, and L is moved a small amount in a random direction—as if jolted by an earthquake—to new positions M', N', and L'. The resulting shape—after the twist—is shown in Fig. 12-15(c).

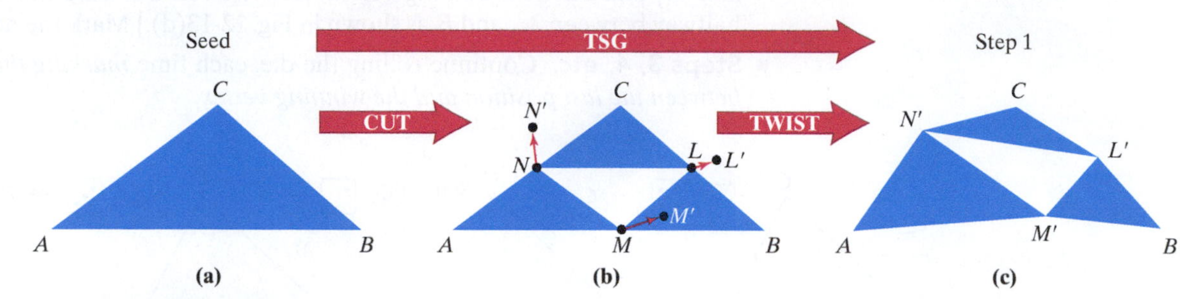

FIGURE 12-15 The two moves in *procedure TSG*: (b) the *cut* and (c) the *twist*.

For convenience, we will use the term *procedure TSG* to describe the combination of the following two moves:

- **Cut.** Remove the "middle" triangle [Fig. 12-15(b)].
- **Twist.** Translate each of the vertices of the removed triangle by a small random amount and in a random direction [Fig. 12-15(c)].

When we repeat *procedure TSG* in an infinite recursive process, we get the **twisted Sierpinski gasket**:

- **Start.** Start with a solid seed triangle [Fig. 12-16(a)].
- **Step 1.** Apply *procedure TSG* to the seed triangle. This gives the "twisted gasket" shown in Fig. 12-16(b), with three twisted solid triangles and a triangular hole in the middle.
- **Step 2.** To each of the three solid triangles in Fig. 12-16(b) apply *procedure TSG*. The result is the "twisted gasket" shown in Fig. 12-16(c), consisting of nine twisted solid triangles and four triangular holes of various sizes.
- **Steps 3, 4, etc.** Apply *procedure TSG* to each shaded triangle in the "twisted gasket" obtained in the previous step.

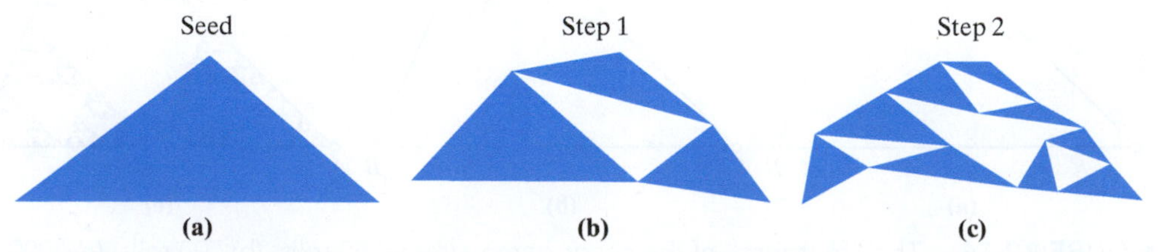

FIGURE 12-16 First two steps in the construction of the twisted Sierpinski gasket.

Figure 12-17(a) shows a twisted Sierpinski gasket by the time you get to Step 7 of the construction. Past this point the figure will remain unchanged to the naked eye. Notice the striking resemblance to a snow-covered mountain. With a few of the standard tools of computer graphics—color, lighting, and shading—a twisted Sierpinski gasket can be morphed to give a very realistic-looking mountain. [Figures 12-17(b) and (c) show two different mountain scenes: one of them is a photo of a real mountain, the other a computer-generated image created using a three-dimensional version of the twisted Sierpinski gasket construction. Can you tell which one is which? (If you are dying to find out, check the photo credits at the back of the book.)]

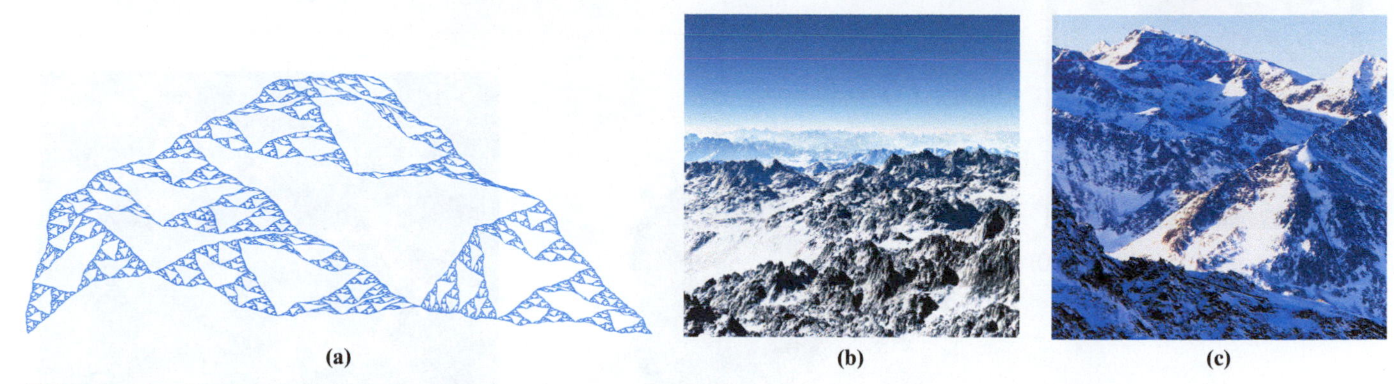

FIGURE 12-17 (a) Twisted Sierpinski gasket. (b), (c) Real mountain, fake mountain: which is which?

We find approximate self-similarity most commonly in nature (trees, rivers, mountains, etc.), but the twisted Sierpinski gasket is the rare example of a man-made geometric shape having approximate self-similarity—when we magnify any part of the gasket we see similar, but not identical, images (Fig. 12-18).

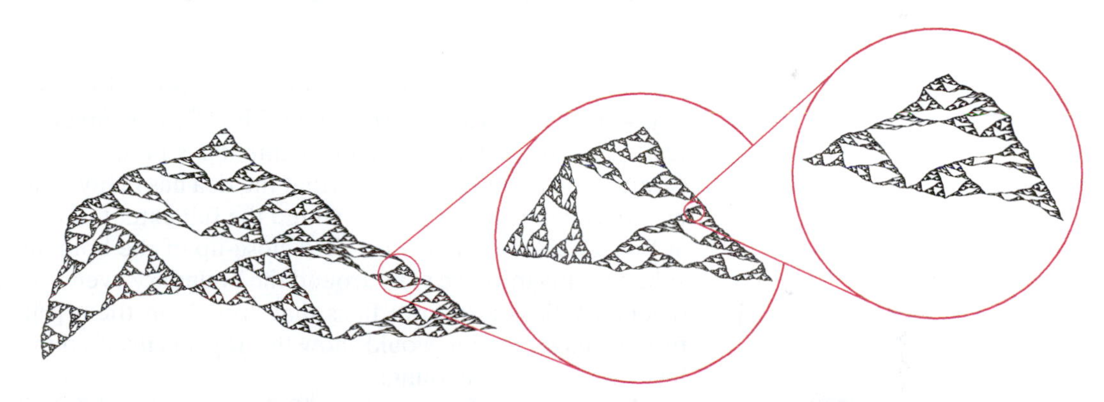

FIGURE 12-18 Approximate self-similarity in the twisted Sierpinski gasket.

The following is a summary of the key properties of the twisted Sierpinski gasket:

- **Self-similarity.** Approximate, infinite, and universal self-similarity.
- **Boundary.** Infinitely long boundary.
- **Replacement rule.** Whenever you have a solid triangle, apply *procedure TSG* to it (i.e., first "cut" it, then "twist" it).
- **Area.** Infinitely small.

12.4 The Mandelbrot Set

In this section we will introduce one of the most interesting and beautiful geometric fractals ever created, an object called the *Mandelbrot set* after the Polish-American mathematician Benoit Mandelbrot. Some of the mathematics behind the Mandelbrot set goes a bit beyond the level of this book, so we will describe the overall idea and try not to get bogged down in the details.

Benoit Mandelbrot (1924–2010)

We will start this section with a brief visual tour. The **Mandelbrot set** is the shape shown in Fig. 12-19(a), a strange-looking blob of black. Using a strategy we will describe later, the different regions outside the Mandelbrot set can be colored according to their mathematical properties, and when this is done [Fig. 12-19(b)], the Mandelbrot set comes to life like a switched-on neon sign.

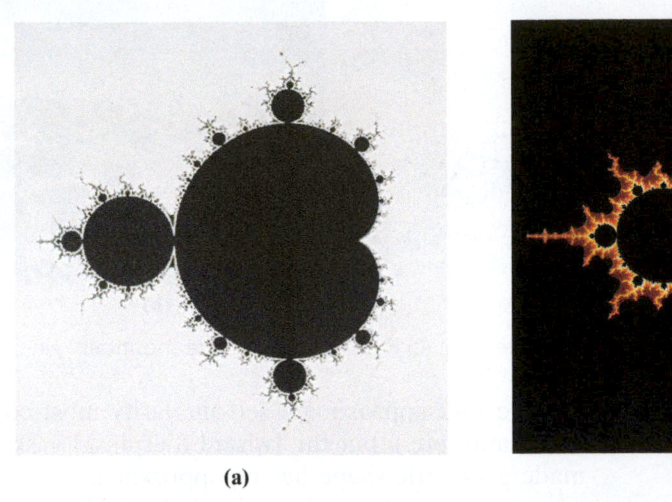

FIGURE 12-19 (a) The Mandelbrot set (black) on a white background. (b) The Mandelbrot set with a mathematically lighted background.

> "So Natr'alists observe, A Flea Hath smaller Fleas that on him prey and these have smaller Fleas to bite'em and so proceed, ad infinitum."
> – *Jonathan Swift*

In the wild imagination of some, the Mandelbrot set looks like some sort of bug—an exotic extraterrestrial flea. The "flea" is made up of a heart-shaped body (called a *cardioid*), a head, and an antenna coming out of the middle of the head. A careful look at Fig. 12-19 shows that the flea has many "smaller fleas that prey on it," but we can only begin to understand the full extent of the infestation when we look at Fig. 12-20(a)—a finely detailed close-up of the boundary of the Mandelbrot set. When we magnify the view around the boundary even further, we can see that these secondary fleas also have fleas that "prey" on them [Figs. 12-20(b) and (c)], and further magnification would show this repeats itself ad infinitum. Clearly, Jonathan Swift was on to something!

Looking carefully at Figs. 12-20(a), (b), and (c), it appears that the Mandelbrot set has some strange form of *infinite* and *approximate* self-similarity—at infinitely many levels of magnification we see the same theme—similar but never identical fleas surrounded by similar smaller fleas. At the same time, we can see that there is tremendous variation in the regions surrounding the individual fleas. The images we see are a peek into a psychedelic coral reef—a world of strange "urchins" and "seahorse tails" in Fig. 12-20(b), "anemone" and "starfish" in Fig. 12-20(c). Further magnification shows an even more exotic and beautiful landscape. Figure 12-21(a) is a close-up of one of the seahorse tails in Fig. 12-20(b). A further close-up of a section of Fig. 12-21(a) is shown in Fig. 12-21(b), and an even further magnification of it is seen in Fig. 12-21(c), revealing a tiny copy of the Mandelbrot set surrounded by a beautiful arrangement of swirls, spirals, and seahorse tails. [The magnification for Fig. 12-21(c) is approximately 10,000 times the original.]

What we see in these pictures is a truly amazing form of approximate self-similarity—anywhere we choose to look we will find (if we crank up the magnification enough) copies of the original Mandelbrot set, always surrounded by an infinitely changing, but always stunning, background. The infinite, approximate self-similarity of the Mandelbrot set manages to blend infinite repetition and infinite variety, creating a landscape as consistently exotic and diverse as nature itself.

FIGURE 12-20 (a) Detailed close-up of a small region on the boundary. (b) An even tighter close-up of the boundary. (c) A close-up of one of the secondary "fleas." (d) A close-up of a tiny tendril near the boundary of the secondary "flea" in (c). (e) A close-up of a small section of the "seahorse tail" in (d). (f) Upon further magnification (10,000 times the original image) four seahorse tails surround a tiny version of the Mandelbrot set.

Complex Numbers and Mandelbrot Sequences

How does this magnificent mix of beauty and complexity called the Mandelbrot set come about? Incredibly, the Mandelbrot set itself can be described mathematically by a recursive process involving simple computations with *complex numbers*.

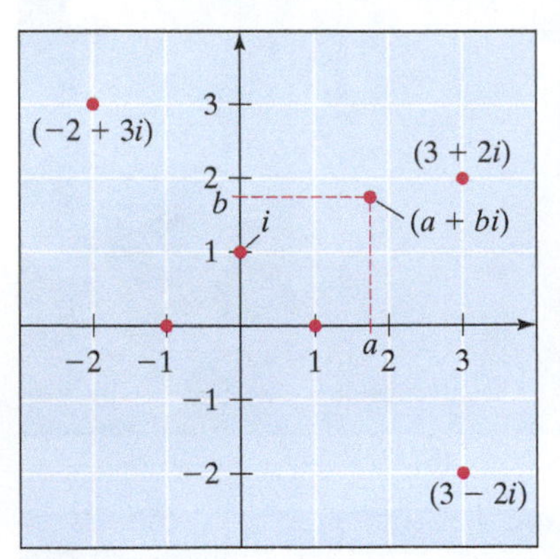

FIGURE 12-21 Every complex number is a point in the Cartesian plane; every point in the Cartesian plane is a complex number.

You may recall having seen complex numbers in high school algebra. Among other things, complex numbers allow us to take square roots of negative numbers and solve quadratic equations of any kind. The basic building block for complex numbers is the number $i = \sqrt{-1}$. Starting with i we can build all other complex numbers using the general form $a + bi$, where a and b are real numbers. For example, $3 + 2i$, $-0.125 + 0.75i$, and even $1 + 0i = 1$ or $-0.75 + 0i = -0.75$. Just like real numbers, complex numbers can be added, multiplied, divided, and squared. (For a quick review of the basic operations with complex numbers, see Exercises 35 through 38.)

For our purposes, the most important fact about complex numbers is that they have a geometric interpretation: The complex number $(a + bi)$ can be identified with the point (a, b) in a Cartesian coordinate system, as shown in Fig. 12-21. This identification means that every complex number can be thought of as a point in the plane and that operations with complex numbers have geometric interpretations (see Exercises 39 and 40).

The key concept in the construction of the Mandelbrot set is that of a *Mandelbrot sequence*. A *Mandelbrot sequence* is an infinite sequence of complex numbers that starts with an arbitrary complex number s we call the *seed*, and then each successive term in the sequence is obtained recursively by *adding the seed s to the square of the previous term*. Figure 12-22 shows a schematic illustration of how a generic Mandelbrot sequence is generated.

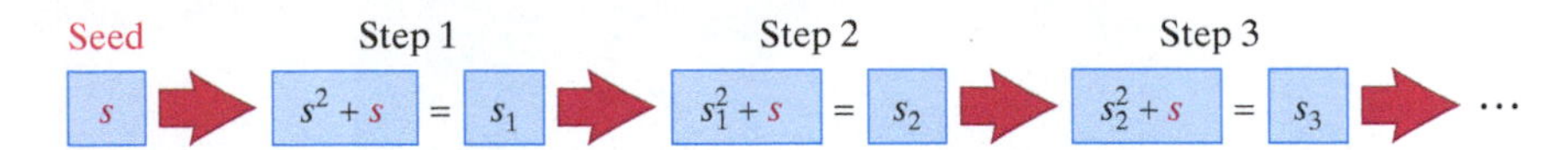

FIGURE 12-22 A Mandelbrot sequence with seed s. The red arrow is shorthand for "square the number and add the seed."

Much like a Koch snowflake and a Sierpinski gasket, a **Mandelbrot sequence** can be defined by means of a recursive rule.

MANDELBROT SEQUENCE RECURSIVE RULE

- **Start:** Choose an arbitrary complex number s. We will call s the seed of the Mandelbrot sequence. Set the seed s to be the initial term of the sequence $(s_0 = s)$.
- **Recursive Rule:** To find the next term in the sequence, square the preceding term and add the seed $[s_{N+1} = (s_N)^2 + s]$.

The following set of examples will illustrate the different patterns of growth exhibited by Mandelbrot sequences as we vary the seeds. These different patterns of growth are going to tell us how to generate the Mandelbrot set itself and the incredible images that we saw in our visual tour. The idea goes like this: Each point in the Cartesian plane is a complex number and thus the seed of some Mandelbrot sequence. The pattern of growth of that Mandelbrot sequence determines whether the seed is inside the Mandelbrot set (black point) or outside (nonblack point). In the case of nonblack points, the color assigned to the point is also determined by the pattern of growth of the corresponding Mandelbrot sequence. Let's try it.

EXAMPLE 12.1 ESCAPING MANDELBROT SEQUENCES

Figure 12-23 shows the first few terms of the Mandelbrot sequence with seed $s = 1$. (Since integers and decimals are also complex numbers, they make perfectly acceptable seeds.)

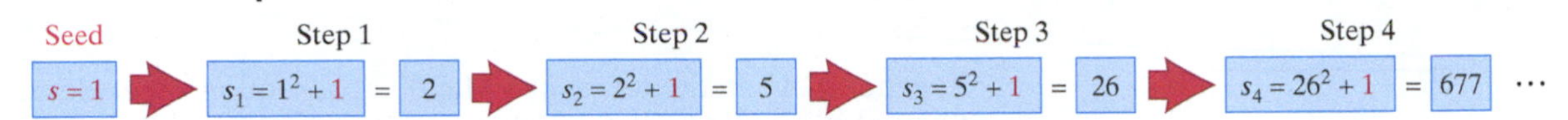

FIGURE 12-23 Mandelbrot sequence with seed $s = 1$ (*escaping very quickly*).

The pattern of growth of this Mandelbrot sequence is clear—the terms are getting larger and larger, and they are doing so very quickly. Geometrically, it means that the points in the Cartesian plane that represent the numbers in this sequence are getting farther and farther away from the origin. For this reason, we call this sequence an *escaping* Mandelbrot sequence.

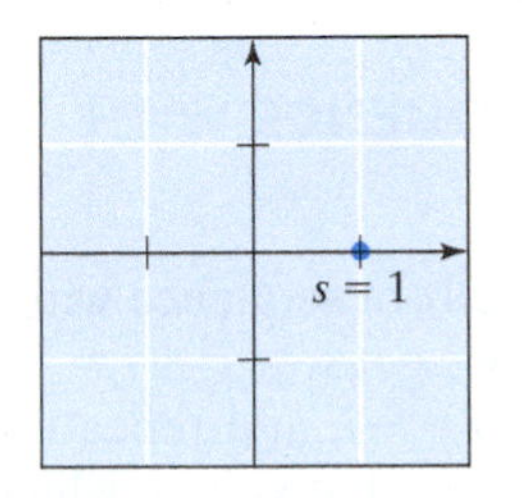

FIGURE 12-24 The seed $s = 1$ is assigned a "cool" color (blue).

In general, when the points that represent the terms of a Mandelbrot sequence move farther and farther away from the origin, we will say that the Mandelbrot sequence is **escaping**. The basic rule that defines the Mandelbrot set is that seeds of escaping Mandelbrot sequences are *not* in the Mandelbrot set and must be assigned some color other than black. While there is no specific rule that tells us what color should be assigned, the overall color palette is based on how fast the sequence is escaping. The typical approach is to use "hot" colors such as reds, yellows, and oranges for seeds that escape slowly and "cool" colors such as blues and purples for seeds that escape quickly. The seed $s = 1$, for example, escapes very quickly, and the corresponding point in the Cartesian plane is painted blue (Fig. 12-24). Note that there is nothing special about blue—we arbitrarily chose a "cool" color.

EXAMPLE 12.2 PERIODIC MANDELBROT SEQUENCES

Figure 12-25 shows the first few terms of the Mandelbrot sequence with seed $s = -1$.

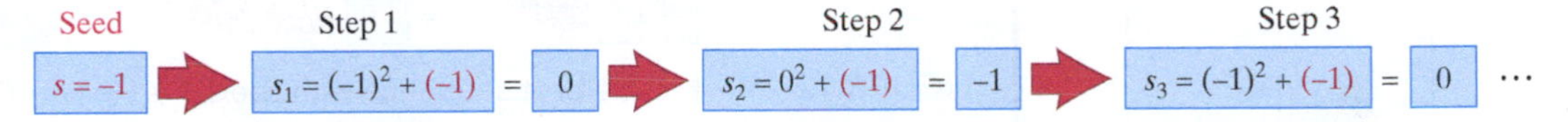

FIGURE 12-25 Mandelbrot sequence with seed $s = -1$ (*periodic*).

The pattern that emerges here is also clear—the numbers in the sequence alternate between 0 and −1. In this case, we say that the Mandelbrot sequence is *periodic*.

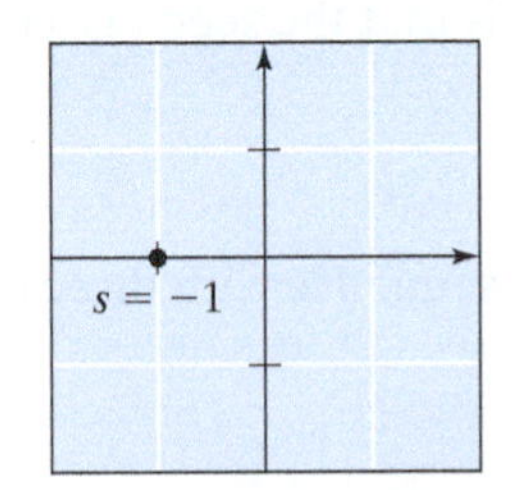

FIGURE 12-26 The seed $s = -1$ is in the Mandelbrot set (black point).

In general, a Mandelbrot sequence is said to be **periodic** if at some point the numbers in the sequence start repeating themselves in a cycle. When the Mandelbrot sequence is periodic, the seed is part of the Mandelbrot set and, thus, is assigned the color black (Fig. 12-26).

EXAMPLE 12.3 ATTRACTED MANDELBROT SEQUENCES

Figure 12-27 shows the first few terms in the Mandelbrot sequence with seed $s = -0.75$.

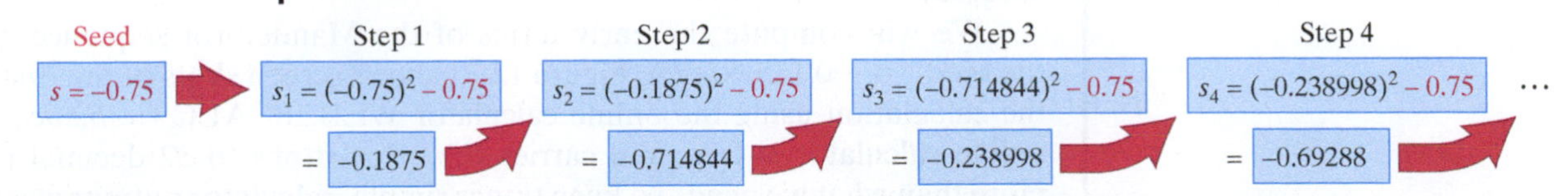

FIGURE 12-27 Mandelbrot sequence with seed $s = -0.75$ (*attracted*).

Here the growth pattern is not obvious, and additional terms of the sequence are needed. Further computation (a calculator will definitely come in handy) shows that as we go farther and farther out in this sequence, the terms get closer and closer to the value −0.5 (see Exercise 51). In this case, we will say that the sequence is *attracted* to the value −0.5.

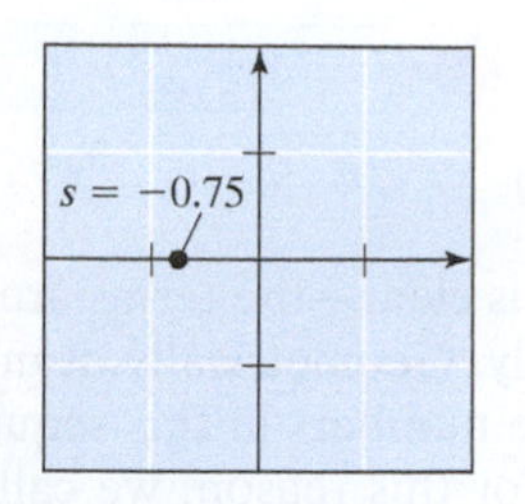

FIGURE 12-28 The seed $s = -0.75$ is in the Mandelbrot set (black point).

In general, when the terms in a Mandelbrot sequence get closer and closer to a fixed complex number a, we say that a is an **attractor** for the sequence or, equivalently, that the sequence is **attracted** to a. Just as with periodic sequences, when a Mandelbrot sequence is attracted, the seed s is part of the Mandelbrot set and is colored black (Fig. 12-28).

So far, all our examples have been based on rational number seeds (mostly to keep things simple), but the truly interesting cases occur when the seeds are complex numbers. The next two examples deal with complex number seeds.

EXAMPLE 12.4 A PERIODIC MANDELBROT SEQUENCE WITH COMPLEX TERMS

In this example we will examine the growth of the Mandelbrot sequence with seed $s = i$. Starting with $s = i$ (and using the fact that $i^2 = -1$), we get $s_1 = i^2 + i = -1 + i$. If we now square s_1 and add i, we get $s_2 = (-1+i)^2 + i = -i$, and repeating the process gives $s_3 = (-i)^2 + i = -1 + i$. At this point we notice that $s_3 = s_1$, which implies $s_4 = s_2$, $s_5 = s_1$, and so on (Fig. 12-29). This, of course, means that this Mandelbrot sequence is *periodic*, with its terms alternating between the complex numbers $-1 + i$ (odd terms) and $-i$ (even terms).

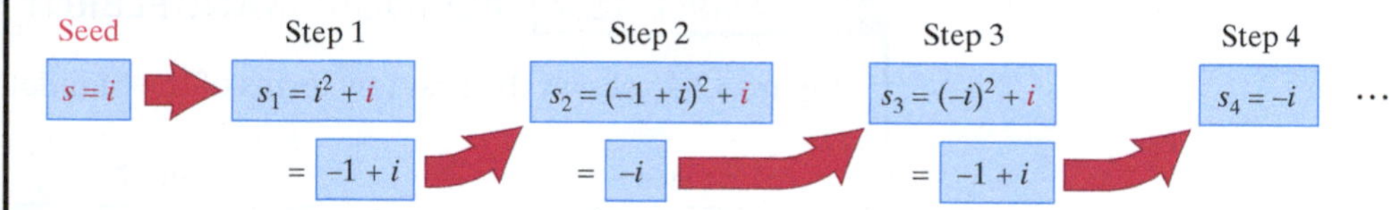

FIGURE 12-29 Mandelbrot sequence with seed $s = i$ (*periodic*).

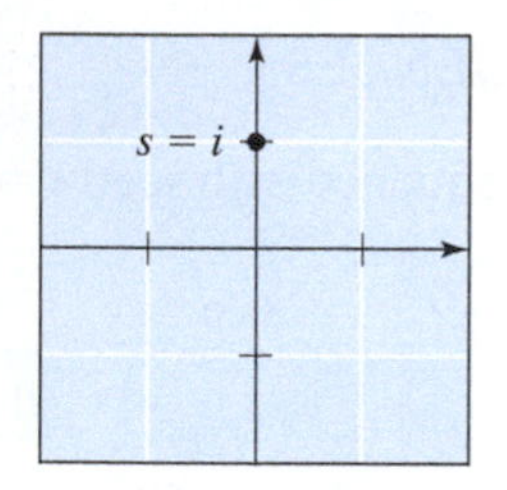

FIGURE 12-30 The seed $s = i$ is in the Mandelbrot set (black point).

The key conclusion from the preceding computations is that the seed i is a black point inside the Mandelbrot set (Fig. 12-30).

The next example illustrates the case of a Mandelbrot sequence with three complex attractors.

EXAMPLE 12.5 A MANDELBROT SEQUENCE WITH THREE COMPLEX ATTRACTORS

For this example you will need a good scientific calculator that handles complex number arithmetic and can carry the calculations to at least eight decimal places of accuracy.

We will compute the early terms of the Mandelbrot sequence generated by the seed $s = -0.125 + 0.75i$. Figure 12-31 shows screen shots of the first few steps of the calculation using the online calculator WEB2.0CALC (*web2.0calc.com*). This online calculator is free, and carries the calculations to 32 decimal places —way more than what we need. To keep things simple, calculator outputs were rounded to eight decimal places when needed [as when going from Fig. 12-31(d) to (e).]

(-0.125+0.75i)^2+(-0.125+0.75i)

(a)

-0.671875+0.5625i

(b)

(-0.671875+0.5625i)^2+(-0.125+0.75i)

(c)

0.010009765625-0.005859375i

(d)

(0.01000977-0.00585938i)^2+(-0.125+0.75i)

(e)

-0.1249341368385315+0.7498826979077148i

(f)

FIGURE 12-31 (a) s_1 (input) (b) s_1 (output) (c) s_2 (input) (d) s_2 (output) (e) s_3 (input rounded to 8 decimal places) (f) s_3 (output).

The first eight terms of the sequence are shown below. (We leave it to the reader to verify the values of s_4 through s_8.)

$s_0 = -0.125 + 0.75i$ $\quad s_1 = -0.671875 + 0.5625i$ $\quad s_2 = 0.01000977 - 0.00585938i$

$s_3 = -0.12493414 + 0.74988270i$ $\quad s_4 = -0.67171552 + 0.56262810i$ $\quad s_5 = 0.00965136 - 0.00585206i$

$s_6 = -0.12494110 + 0.74988704i$ $\quad s_7 = -0.67172029 + 0.56261658i$ $\quad s_8 = 0.00967073 - 0.00584194i$

We can see that the terms in this Mandelbrot sequence are complex numbers that essentially cycle around in sets of three and are approaching three different attractors. Since this Mandelbrot sequence is attracted, the seed $s = -0.125 + 0.75i$ represents another point in the Mandelbrot set. ■

The Mandelbrot Set

Given all the previous examples and discussion, a formal definition of the **Mandelbrot set** using seeds of Mandelbrot sequences sounds incredibly simple: If the Mandelbrot sequence is *periodic* or *attracted*, the seed is part of the Mandelbrot set and is assigned the color black; if the Mandelbrot sequence is *escaping*, the seed is outside the Mandelbrot set and is assigned a color that depends on the speed with which the sequence is escaping (hot colors for slowly escaping sequences, cool colors for quickly escaping sequences). There are a few technical details that we omitted, but essentially these are the key ideas behind the amazing pictures that we saw in Fig. 12-20. In addition, of course, a computer is needed to carry out the computations and generate the images.

Because the Mandelbrot set provides a bounty of aesthetic returns for a relatively small mathematical investment, it has become one of the most popular mathematical playthings of our time. Hundreds of software programs that allow one to explore the beautiful landscapes surrounding the Mandelbrot set are available, and many of these programs are freeware. You can find plenty of these by entering "Mandelbrot set" into your search engine.

Conclusion

The study of fractals and their geometry has become a hot mathematical topic. It is a part of mathematics that combines complex and interesting theories, beautiful graphics, and acute relevance to the real world. In this chapter we only scratched the surface of this deep and rich topic.

The word **fractal** (from the Latin *fractus*, meaning "broken up or fragmented") was coined by Benoit Mandelbrot in the mid-1970s to describe objects as diverse as the Koch curve, the Sierpinski gasket, the twisted Sierpinski gasket, and the Mandelbrot set, as well as many shapes in nature, such as clouds, mountains, trees, rivers, and the broccoflower. These objects share one key characteristic: They all have some form of self-similarity. It is the self-similarity that gives these shapes a strikingly different geometry from the one we find in the traditional geometry of human-made shapes: In the former we get *kinky geometry*, in the latter we get *smooth geometry*. The difference is apparent when we compare Figs. 12-32 and 12-33.

Smooth (i.e., old school) geometry was developed by the Greeks about 2000 years ago and passed on to us essentially unchanged. It was (and still is) a great triumph of the human mind, and it has allowed us to develop much of our technology, engineering, architecture, and so on. The problem is that smooth geometry is not a particularly good tool for describing kinky natural shapes or for studying chaotic natural phenomena. Fractal geometry, by contrast, is. Today fractal geometry is used to study the patterns of clouds and how they affect the weather, to diagnose the pattern of contractions of a human heart, to design more efficient cell phone antennas, and to create the truly realistic forgeries of nature that animate many of the latest video games and science fiction movies, as illustrated in the fractal images shown in Fig. 12-34.

FIGURE 12-32 The smooth geometry of human-made objects.

FIGURE 12-33 The kinky geometry of nature.

FIGURE 12-34 Fractal geometry: computer-generated forgeries of nature.

KEY CONCEPTS

12.1 The Koch Snowflake and Self-Similarity

- **Koch snowflake:** a geometric shape defined by a seed equilateral triangle and a recursive rule where, at each step, smaller equilateral triangles are added to the boundary of the snowflake generated in the previous step, **357**
- **Koch curve:** the curve formed by the boundary of a Koch snowflake, **358**
- **replacement rule:** a scale-independent rule that specifies how to replace a part with another part, **359**
- **self-similarity:** a property of certain objects and shapes having parts that are similar to the whole object, **360**

- **exact self-similarity:** a type of self-similarity where different parts of the object look exactly the same but at different scales, **360**
- **infinite self-similarity:** a type of self-similarity that occurs at infinitely many scales, **361**
- **universal self-similarity:** a type of self-similarity that occurs everywhere in an object, **361**
- **approximate self-similarity:** a type of self-similarity where parts of the object share the same structure as the whole object and look approximately (but not exactly) the same at different scales, **362**
- **finite self-similarity:** a type of self-similarity that occurs at multiple scales but eventually stops, **362**
- **local self-similarity:** a type of self-similarity that occurs in parts of the object but not everywhere, **362**

12.2 The Sierpinski Gasket and the Chaos Game

- **Sierpinski gasket:** a geometric shape defined by an arbitrary solid seed triangle and a replacement rule where smaller and smaller triangles are recursively removed from the middle of the remaining solid triangles, **363**
- **chaos game:** a game in which points are recursively chosen inside a given triangle, each point halfway between the previous point and a randomly chosen vertex of the triangle, **365**

12.3 The Twisted Sierpinski Gasket

- **twisted Sierpinski gasket:** a geometric shape defined by adding a "twist" at each step of the Sierpinski gasket construction, **366**

12.4 The Mandelbrot Set

- **Mandelbrot set:** a geometric shape defined by points in the Cartesian plane representing complex numbers with the property that the Mandelbrot sequences they generate are either periodic or attracted, **368, 373**
- **Mandelbrot sequence:** a sequence of complex numbers that starts with a seed s and recursively generates each term of the sequence by adding s to the square of the preceding term $[s_{N+1} = (s_N)^2 + s]$, **370**
- **escaping Mandelbrot sequence:** a Mandelbrot sequence with the property that the terms of the sequence get larger and larger (i.e., the points they represent are getting farther and farther from the origin of the Cartesian plane), **371**
- **periodic Mandelbrot sequence:** a Mandelbrot sequence with the property that at some point the terms of the sequence start repeating themselves in a cyclical pattern, **372**
- **attracted Mandelbrot sequence:** a Mandelbrot sequence with the property that at some point the terms of the sequence start to get closer and closer to one or several attractors, **372**

EXERCISES

WALKING

12.1 The Koch Snowflake and Self-Similarity

1. Consider the construction of a Koch snowflake starting with a seed triangle having sides of length 81 cm. Let M denote the number of sides, L the length of each side, and P the perimeter of the "snowflake" obtained at the indicated step of the construction. Complete the missing entries in Table 12-1.

	M	L	P
Start	3	81 cm	243 cm
Step 1	12	27 cm	324 cm
Step 2			
Step 3			
Step 4			
Step 5			

TABLE 12-1

2. Consider the construction of a Koch snowflake starting with a seed triangle having sides of length 18 cm. Let M denote the number of sides, L the length of each side, and P the perimeter of the "snowflake" obtained at the indicated step of the construction. Complete the missing entries in Table 12-2.

	M	L	P
Start	3	18 cm	54 cm
Step 1	12	6 cm	72 cm
Step 2			
Step 3			
Step 4			
Step 5			

TABLE 12-2

3. Consider the construction of a Koch snowflake starting with a seed triangle having area $A = 81$. Let R denote the number of triangles added at a particular step, S the area of each added triangle, T the total new area added, and Q the area of the "snowflake" obtained at a particular step of the construction. Complete the missing entries in Table 12-3.

	R	S	T	Q
Start	0	0	0	81
Step 1	3	9	27	108
Step 2	12	1	12	120
Step 3				
Step 4				
Step 5				

TABLE 12-3

4. Consider the construction of a Koch snowflake starting with a seed triangle having area $A = 729$. Let R denote the number of triangles added at a particular step, S denote the area of each added triangle, T the total new area added, and Q the area of the "snowflake" obtained at a particular step of the construction. Complete the missing entries in Table 12-4.

	R	S	T	Q
Start	0	0	0	729
Step 1	3	81	243	972
Step 2	12	9	108	1080
Step 3				
Step 4				
Step 5				

TABLE 12-4

Exercises 5 through 8 refer to a variation of the Koch snowflake called the ***quadratic Koch fractal****. The construction of the quadratic Koch fractal is similar to that of the Koch snowflake, but it uses squares instead of equilateral triangles as the shape's building blocks. The following recursive construction rule defines the quadratic Koch fractal:*

Quadratic Koch Fractal

- **Start.** *Start with a solid seed square [Fig. 12-35(a)].*
- **Step 1.** *Attach a smaller square (sides one-third the length of the sides of the seed square) to the middle third of each side [Fig. 12-35(b)].*
- **Step 2.** *Attach a smaller square (sides one-third the length of the sides of the previous side to the middle third of each side [Fig. 12-35(c)]. (Call this procedure QKF.)*
- **Steps 3, 4, etc.** *At each step, apply procedure QKF to the figure obtained in the preceding step.*

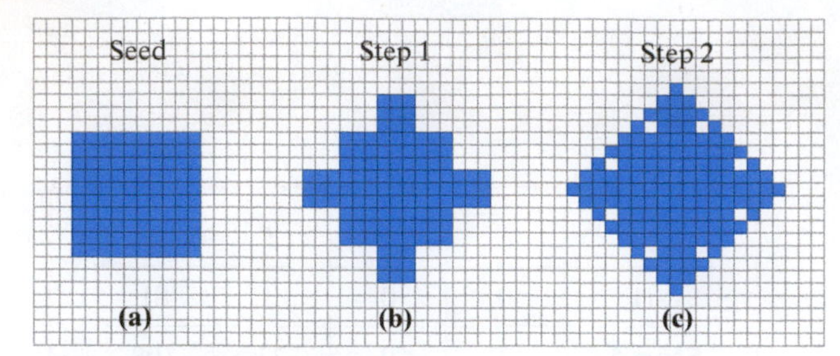

FIGURE 12-35

5. Assume that the seed square of the *quadratic Koch fractal* has sides of length 81 cm. Let M denote the number of sides, L the length of each side, and P the perimeter of the shape obtained at the indicated step of the construction. Complete the missing entries in Table 12-5.

	M	L	P
Start	4	81 cm	324 cm
Step 1	20	27 cm	540 cm
Step 2			
Step 3			
Step 40			

TABLE 12-5

6. Assume that the seed square of the *quadratic Koch fractal* has sides of length 1. Let M denote the number of sides, L the length of each side, and P the perimeter of the shape obtained at the indicated step of the construction. Complete the missing entries in Table 12-6.

	M	L	P
Start	4	1	4
Step 1	20	$\frac{1}{3}$	$\frac{20}{3}$
Step 2			
Step 3			
Step 4			

TABLE 12-6

7. Assume that the seed square of the *quadratic Koch fractal* has area $A = 81$. Let R denote the number of squares added at a particular step, S the area of each added square, T the total new area added, and Q the area of the shape obtained at a particular step of the construction. Complete the missing entries in Table 12-7.

	R	S	T	Q
Start	0	0	0	81
Step 1	4	9	36	117
Step 2	20	1	20	137
Step 3				
Step 4				

TABLE 12-7

8. Assume that the seed square of the *quadratic Koch fractal* has area $A = 243$. Let R denote the number of squares added at a particular step, S the area of each added square, T the total new area added, and Q the area of the shape obtained at a particular step of the construction. Complete the missing entries in Table 12-8.

	R	S	T	Q
Start	0	0	0	243
Step 1	4	27	108	351
Step 2	20	3	60	411
Step 3				
Step 4				

TABLE 12-8

Exercises 9 through 12 refer to a variation of the Koch snowflake called the ***Koch antisnowflake****. The Koch antisnowflake is much like the Koch snowflake, but it is based on a recursive rule that removes equilateral triangles. The recursive replacement rule for the Koch antisnowflake is as follows:*

Koch Antisnowflake

- **Start:** *Start with a solid seed equilateral triangle [Fig. 12-36(a)].*
- **Replacement rule:** *In each step replace any boundary line segment ——— with a ‾‾\/‾‾ (where the point is always facing toward the interior of the snowflake). [Figures 12-36(b) and (c) show the figures obtained at Steps 1 and 2, respectively.]*

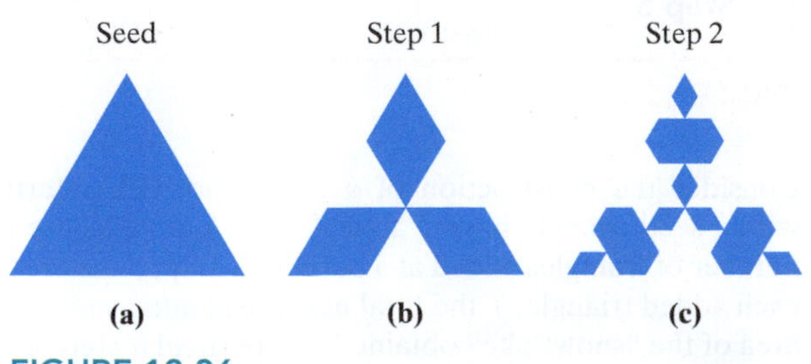

FIGURE 12-36

9. Assume that the seed triangle of the *Koch antisnowflake* has sides of length 81 cm. Let M denote the number of sides, L the length of each side, and P the perimeter of the shape obtained at the indicated step of the construction. Complete the missing entries in Table 12-9.

	M	L	P
Start	3	81 cm	243 cm
Step 1	12	27 cm	324 cm
Step 2			
Step 3			
Step 4			
Step 5			

TABLE 12-9

10. Assume that the seed triangle of the *Koch antisnowflake* has sides of length 18 cm. Let M denote the number of sides, L the length of each side, and P the perimeter of the shape obtained at the indicated step of the construction. Complete the missing entries in Table 12-10.

	M	L	P
Start	3	18 cm	54 cm
Step 1	12	6 cm	72 cm
Step 2			
Step 3			
Step 4			
Step 5			

TABLE 12-10

11. Assume that the seed triangle of the *Koch antisnowflake* has area $A = 81$. Let R denote the number of triangles subtracted at a particular step, S the area of each subtracted triangle, T the total area subtracted, and Q the area of the shape obtained at a particular step of the construction. Complete the missing entries in Table 12-11.

	R	S	T	Q
Start	0	0	0	81
Step 1	3	9	27	54
Step 2	12	1	12	42
Step 3				
Step 4				
Step 5				

TABLE 12-11

12. Assume that the seed triangle of the *Koch antisnowflake* has area $A = 729$. Let R denote the number of triangles subtracted at a particular step, S the area of each subtracted triangle, T the total area subtracted, and Q the area of the shape obtained at a particular step of the construction. Complete the missing entries in Table 12-12.

	R	S	T	Q
Start	0	0	0	729
Step 1	3	81	243	486
Step 2	12	9	108	378
Step 3				
Step 4				
Step 5				

TABLE 12-12

Exercises 13 through 16 refer to the construction of the ***quadratic Koch island****. The quadratic Koch island is defined by the following recursive replacement rule.*

Quadratic Koch Island

- **Start:** *Start with a seed square [Fig. 12-37(a)]. (Notice that here we are only dealing with the boundary of the square.)*
- **Replacement rule:** *In each step replace any horizontal boundary segment with the "sawtooth" version shown in Fig. 12-37(b) and any vertical line segment with the "sawtooth" version shown in Fig. 12-37(c).*

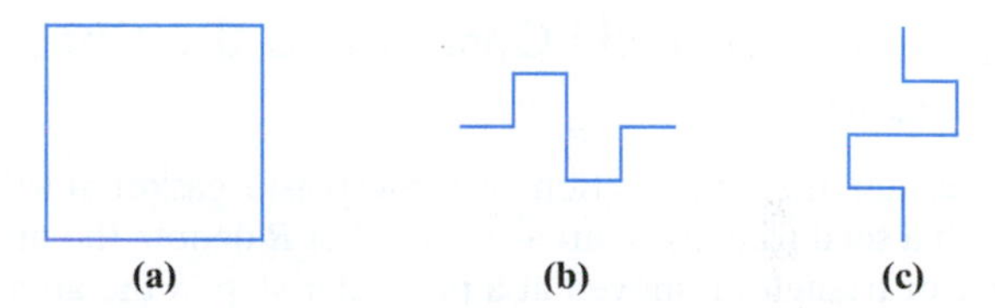

FIGURE 12-37

13. Assume that the seed square of the *quadratic Koch island* has sides of length 16.

    (a) Carefully draw the figures obtained in Steps 1 and 2 of the construction. (*Hint:* Use graph paper and make the seed square a 16 by 16 square.)

    (b) Find the perimeter of the figure obtained in Step 1 of the construction.

    (c) Find the perimeter of the figure obtained in Step 2 of the construction.

    (d) Explain why the quadratic Koch island has infinite perimeter.

14. Assume that the seed square of the *quadratic Koch island* has sides of length a.

(a) Carefully draw the figures obtained in Steps 1 and 2 of the construction. (*Hint:* Use graph paper and make the seed square a 16 by 16 square.)

(b) Find the perimeter of the figure obtained in Step 1 of the construction.

(c) Find the perimeter of the figure obtained in Step 2 of the construction.

(d) Explain why the quadratic Koch Island has infinite perimeter.

15. This exercise is a continuation of Exercise 13.

(a) Find the area of the figure obtained in Step 1 of the construction.

(b) Find the area of the figure obtained in Step 2 of the construction.

(c) Explain why the area of the quadratic Koch Island is the same as the area of the seed square.

16. This exercise is a continuation of Exercise 14.

(a) Find the area of the figure obtained in Step 1 of the construction.

(b) Find the area of the figure obtained in Step 2 of the construction.

(c) Explain why the area of the quadratic Koch Island is the same as the area of the seed square.

12.2 The Sierpinski Gasket and the Chaos Game

17. Consider the construction of a Sierpinski gasket starting with a seed triangle of area $A = 64$. Let R denote the number of triangles removed at a particular step, S the area of each removed triangle, T the total area removed, and Q the area of the "gasket" obtained at a particular step of the construction. Complete the missing entries in Table 12-13.

	R	S	T	Q
Start	0	0	0	64
Step 1	1	16	16	48
Step 2	3	4	12	36
Step 3				
Step 4				
Step 5				

TABLE 12-13

18. Consider the construction of a Sierpinski gasket starting with a seed triangle of area $A = 1$. Let R denote the number of triangles removed at a particular step, S the area of each removed triangle, T the total area removed, and Q the area of the "gasket" obtained at a particular step of the construction. Complete the missing entries in Table 12-14.

	R	S	T	Q
Start	0	0	0	1
Step 1	1	$\frac{1}{4}$	$\frac{1}{4}$	$\frac{3}{4}$
Step 2	3	$\frac{1}{16}$	$\frac{3}{16}$	$\frac{9}{16}$
Step 3				
Step 4				
Step 5				

TABLE 12-14

19. Assume that the seed triangle of the *Sierpinski gasket* has perimeter of length $P = 8$ cm. Let U denote the number of solid triangles at a particular step, V the perimeter of each solid triangle, and W the length of the boundary of the "gasket" obtained at a particular step of the construction. Complete the missing entries in Table 12-15.

	U	V	W
Start	1	8 cm	8 cm
Step 1	3	4 cm	12 cm
Step 2			
Step 3			
Step 4			
Step 5			

TABLE 12-15

20. Assume that the seed triangle of the *Sierpinski gasket* has perimeter $P = 20$. Let U denote the number of solid triangles at a particular step, V the perimeter of each solid triangle, and W the length of the boundary of the "gasket" obtained at a particular step of the construction. Complete the missing entries in Table 12-16.

	U	V	W
Start	1	20	20
Step 1	3	10	30
Step 2			
Step 3			
Step 4			
Step 5			

TABLE 12-16

Exercises 21 through 24 refer to the ***Sierpinski ternary gasket****, a variation of the Sierpinski gasket defined by the following recursive replacement rule.*

Sierpinski Ternary Gasket

- **Start:** *Start with a solid seed equilateral triangle [Fig. 12-38(a)].*
- **Replacement rule:** *In each step replace any solid triangle with a . [Figures 12-38(b) and (c) show Steps 1 and 2, respectively.]*

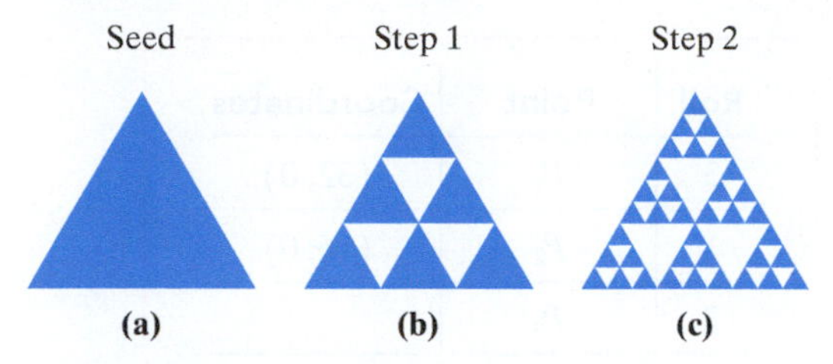

FIGURE 12-38

21. Assume that the seed triangle of the *Sierpinski ternary gasket* has area $A = 1$. Let R denote the number of triangles removed at a particular step, S the area of each removed triangle, T the total area removed, and Q the area of the "ternary gasket" obtained at a particular step of the construction. Complete the missing entries in Table 12-17.

	R	S	T	Q
Start	0	0	0	1
Step 1	3	$\frac{1}{9}$	$\frac{1}{3}$	$\frac{2}{3}$
Step 2				
Step 3				
Step 4				
Step *N*				

TABLE 12-17

22. Assume that the seed triangle of the *Sierpinski ternary gasket* has area $A = 81$. Let R denote the number of triangles removed at a particular step, S the area of each removed triangle, T the total area removed, and Q the area of the "gasket" obtained at a particular step of the construction. Complete the missing entries in Table 12-18.

	R	S	T	Q
Start	0	0	0	81
Step 1	3	9	27	54
Step 2				
Step 3				
Step 4				
Step *N*				

TABLE 12-18

23. Assume that the seed triangle of the *Sierpinski ternary gasket* has perimeter of length $P = 9$ cm. Let U denote the number of shaded triangles at a particular step, V the perimeter of each shaded triangle, and W the length of the boundary of the "ternary gasket" obtained at a particular step of the construction. Complete the missing entries in Table 12-19.

	U	V	W
Start	1	9 cm	9 cm
Step 1	6	3 cm	18 cm
Step 2			
Step 3			
Step 4			
Step *N*			

TABLE 12-19

24. Assume that the seed triangle of the *Sierpinski ternary gasket* has perimeter P. Let U denote the number of shaded triangles at a particular step, V the perimeter of each shaded triangle, and W the length of the boundary of the "gasket" obtained at a particular step of the construction. Complete the missing entries in Table 12-20.

	U	V	W
Start	1	P	P
Step 1	6	$\frac{P}{3}$	$2P$
Step 2			
Step 3			
Step 4			
Step *N*			

TABLE 12-20

Exercises 25 and 26 refer to a variation of the Sierpinski gasket called the ***box fractal****. The box fractal is defined by the following recursive rule:*

- **Start.** *Start with a solid seed square [Fig. 12-39(a)].*
- **Step 1.** *Subdivide the seed square into nine equal subsquares, and remove the center subsquare along each of the sides [Fig. 12-39(b)].*
- **Step 2.** *Subdivide each of the remaining solid squares into nine subsquares, and remove the center subsquare along each side [Fig. 12-39(c)]. Call this process (subdividing a solid square into nine subsquares and removing the central subsquares along the four sides) procedure BF.*
- **Steps 3, 4, etc.** *Apply procedure BF to each solid square of the "carpet" obtained in the previous step.*

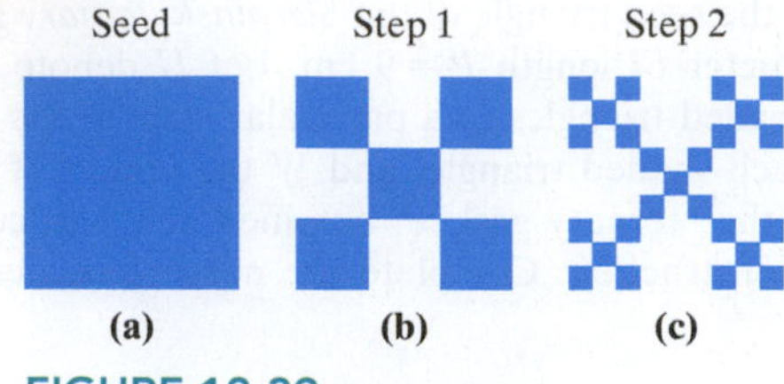

FIGURE 12-39

25. Assume that that the seed square for the *box fractal* has area $A = 1$.

(a) Find the area of the figure obtained in Step 1 of the construction.

(b) Find the area of the figure obtained in Step 2 of the construction.

(c) Find the area of the figure obtained in Step N of the construction.

26. Assume that the seed square for the *box fractal* has sides of length 1.

(a) Find the perimeter of the figure obtained in Step 1 of the construction.

(b) Find the perimeter of the figure obtained in Step 2 of the construction.

(c) Find the perimeter of the figure obtained in Step N of the construction.

Exercises 27 through 30 refer to the chaos game as described in Section 12.2. You should use graph paper for these exercises. Start with an isosceles right triangle ABC with AB = AC = 32, as shown in Fig. 12-40. Choose vertex A for a roll of 1 or 2, vertex B for a roll of 3 or 4, and vertex C for a roll of 5 or 6.

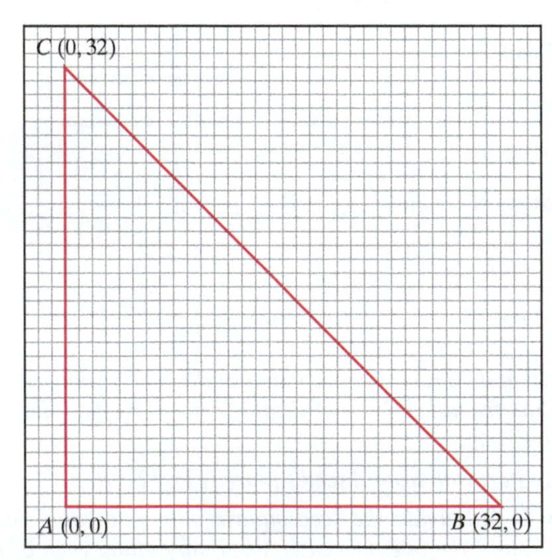

FIGURE 12-40

27. Suppose that the die is rolled six times and that the outcomes are 3, 1, 6, 4, 5, and 5. Carefully draw the points P_1 through P_6 corresponding to these outcomes. (*Note:* Each of the points P_1 through P_6 falls on a grid point of the graph. You should be able to identify the location of each point without using a ruler.)

28. Suppose that the die is rolled six times and that the outcomes are 2, 6, 1, 4, 3, and 6. Carefully draw the points P_1 through P_6 corresponding to these outcomes. (*Note:* Each of the points P_1 through P_6 falls on a grid point of the graph. You should be able to identify the location of each point without using a ruler.)

29. Using a rectangular coordinate system with A at (0, 0), B at (32, 0), and C at (0, 32), complete Table 12-21.

Roll	Point	Coordinates
3	P_1	(32, 0)
1	P_2	(16, 0)
2	P_3	
3	P_4	
5	P_5	
5	P_6	

TABLE 12-21

30. Using a rectangular coordinate system with A at (0, 0), B at (32, 0), and C at (0, 32), complete Table 12-22.

Roll	Point	Coordinates
2	P_1	(0, 0)
6	P_2	(0, 16)
5	P_3	
1	P_4	
3	P_5	
6	P_6	

TABLE 12-22

Exercises 31 through 34 refer to a variation of the chaos game. In this game you start with a square ABCD with sides of length 27 as shown in Fig. 12-41 and a fair die that you will roll many times. When you roll a 1, choose vertex A; when you roll a 2, choose vertex B; when you roll a 3, choose vertex C; and when you roll a 4 choose vertex D. (When you roll a 5 or a 6, disregard the roll and roll again.) A sequence of rolls will generate a sequence of points $P_1, P_2, P_3, \ldots$ inside or on the boundary of the square according to the following rules.

- **Start.** *Roll the die. Mark the chosen vertex and call it P_1.*
- **Step 1.** *Roll the die again. From P_1 move two-thirds of the way toward the new chosen vertex. Mark this point and call it P_2.*
- **Steps 2, 3, etc.** *Each time you roll the die, mark the point two-thirds of the way between the previous point and the chosen vertex.*

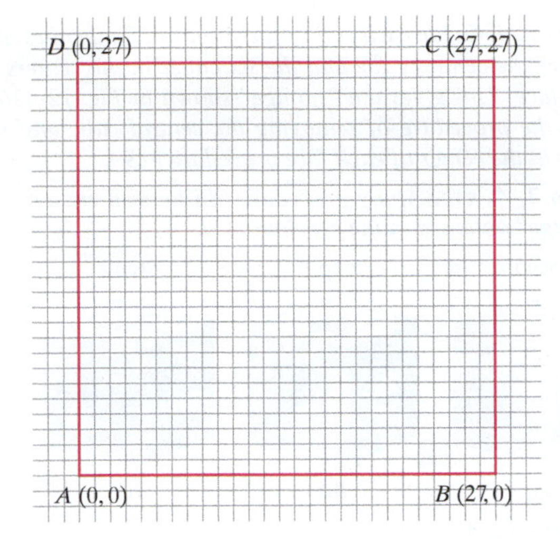

FIGURE 12-41

31. Using graph paper, find the points P_1, P_2, P_3, and P_4 corresponding to

(a) the sequence of rolls 4, 2, 1, 2.

(b) the sequence of rolls 3, 2, 1, 2.

(c) the sequence of rolls 3, 3, 1, 1.

32. Using graph paper, find the points P_1, P_2, P_3, and P_4 corresponding to

(a) the sequence of rolls 2, 2, 4, 4.

(b) the sequence of rolls 2, 3, 4, 1.

(c) the sequence of rolls 1, 3, 4, 1.

33. Using a rectangular coordinate system with A at (0, 0), B at (27, 0), C at (27, 27), and D at (0, 27), find the sequence of rolls that would produce the given sequence of marked points.

(a) P_1: (0, 27), P_2: (18, 9), P_3: (6, 3), P_4: (20, 1)

(b) P_1: (27, 27), P_2: (9, 9), P_3: (3, 3), P_4: (19, 19)

(c) P_1: (0, 0), P_2: (18, 18), P_3: (6, 24), P_4: (20, 8)

34. Using a rectangular coordinate system with A at (0, 0), B at (27, 0), C at (27, 27), and D at (0, 27), find the sequence of rolls that would produce the given sequence of marked points.

(a) P_1: (27, 0), P_2: (27, 18), P_3: (9, 24), P_4: (3, 8)

(b) P_1: (0, 27), P_2: (18, 9), P_3: (24, 3), P_4: (8, 19)

(c) P_1: (27, 27), P_2: (9, 9), P_3: (21, 3), P_4: (7, 19)

12.3 The Twisted Sierpinski Gasket

No exercises for this section.

12.4 The Mandelbrot Set

Exercises 35 through 40 are a review of complex number arithmetic. Recall that (1) to add two complex numbers you simply add the real parts and the imaginary parts: e.g., $(2+3i)+(5+2i)=7+5i$*; (2) to multiply two complex numbers you multiply them as if they were polynomials and use the fact that* $i^2=-1$*: e.g.,* $(2+3i)(5+2i)=10+4i+15i+6i^2=4+19i$*. Finally, if you know how to multiply two complex numbers then you also know how to square them, since* $(a+bi)^2=(a+bi)(a+bi)$*.*

35. Simplify each expression.

(a) $(-i)^2+(-i)$ **(b)** $(-1-i)^2+(-i)$

(c) $i^2+(-i)$

36. Simplify each expression.

(a) $(1+i)^2+(1+i)$ **(b)** $(1+3i)^2+(1+i)$

(c) $(-7+7i)^2+(1+i)$

37. Simplify each expression. (Give your answers rounded to three significant digits.)

(a) $(-0.25+0.25i)^2+(-0.25+0.25i)$

(b) $(-0.25-0.25i)^2+(-0.25-0.25i)$

38. Simplify each expression. (Give your answers rounded to three significant digits.)

(a) $(-0.25+0.125i)^2+(-0.25+0.125i)$

(b) $(-0.2+0.8i)^2+(-0.2+0.8i)$

39. (a) Plot the points corresponding to the complex numbers $(1+i)$, $i(1+i)$, $i^2(1+i)$, and $i^3(1+i)$.

(b) Plot the points corresponding to the complex numbers $(3-2i)$, $i(3-2i)$, $i^2(3-2i)$, and $i^3(3-2i)$.

(c) What geometric effect does multiplication by i have on a complex number?

40. (a) Plot the points corresponding to the complex numbers $(1+i)$, $-i(1+i)$, $(-i)^2(1+i)$, and $(-i)^3(1+i)$.

(b) Plot the points corresponding to the complex numbers $(0.8+1.2i)$, $-i(0.8+1.2i)$, $(-i)^2(0.8+1.2i)$, and $(-i)^3(0.8+1.2i)$.

(c) What geometric effect does multiplication by $-i$ have on a complex number?

41. Consider the Mandelbrot sequence with seed $s=-2$.

(a) Find s_1, s_2, s_3, and s_4.

(b) Find s_{100}.

(c) Is this Mandelbrot sequence *escaping*, *periodic*, or *attracted*? Explain.

42. Consider the Mandelbrot sequence with seed $s=2$.

(a) Find s_1, s_2, s_3, and s_4.

(b) Is this Mandelbrot sequence *escaping*, *periodic*, or *attracted*? Explain.

43. Consider the Mandelbrot sequence with seed $s = -0.5$.

(a) Using a calculator find s_1 through s_5, rounded to four decimal places.

(b) Suppose you are given $s_N = -0.366$. Using a calculator find s_{N+1}, rounded to four decimal places.

(c) Is this Mandelbrot sequence *escaping*, *periodic*, or *attracted*? Explain.

44. Consider the Mandelbrot sequence with seed $s = -0.25$.

(a) Using a calculator find s_1 through s_{10}, rounded to six decimal places.

(b) Suppose you are given $s_N = -0.207107$. Using a calculator find s_{N+1}, rounded to six decimal places.

(c) Is this Mandelbrot sequence *escaping*, *periodic*, or *attracted*? Explain.

45. Consider the Mandelbrot sequence with seed $s = -i$.

(a) Find s_1 through s_5. (*Hint*: Try Exercise 35 first.)

(b) Is this Mandelbrot sequence *escaping*, *periodic*, or *attracted*? Explain.

46. Consider the Mandelbrot sequence with seed $s = 1 + i$. Find s_1, s_2, and s_3. (*Hint*: Try Exercise 36 first.)

JOGGING

47. Let A denote the area of the seed triangle of the *Sierpinski gasket*.

(a) Find the area of the gasket at step N of the construction expressed in terms of A and N. (*Hint*: Try Exercises 17 and 18 first.)

(b) Explain why the area of the Sierpinski gasket is infinitesimally small (i.e., smaller than any positive quantity).

48. Let P denote the perimeter of the seed triangle of the *Sierpinski gasket*.

(a) Find the perimeter of the gasket at step N of the construction expressed in terms of P and N. (*Hint*: Try Exercises 19 and 20 first.)

(b) Explain why the Sierpinski gasket has an infinitely long perimeter.

Exercises 49 and 50 refer to the Menger sponge, a three-dimensional cousin of the Sierpinski gasket. The ***Menger sponge*** *is defined by the following recursive construction rule.*

Menger Sponge

- **Start.** *Start with a solid seed cube [Fig. 12-42(a)].*
- **Step 1.** *Subdivide the seed cube into 27 equal subcubes and remove the central cube and the six cubes in the centers of each face. This leaves a "sponge" consisting of 20 solid subcubes, as shown in Fig. 12-42(b).*
- **Step 2.** *Subdivide each solid subcube into 27 subcubes and remove the central cube and the six cubes in the centers of each face. This gives the "sponge" shown in Fig. 12-42(c). (Call the procedure of removing the central cube and the cubes in the center of each face procedure MS.)*
- **Steps 3, 4, etc.** *Apply procedure MS to each cube of the "sponge" obtained in the previous step.*

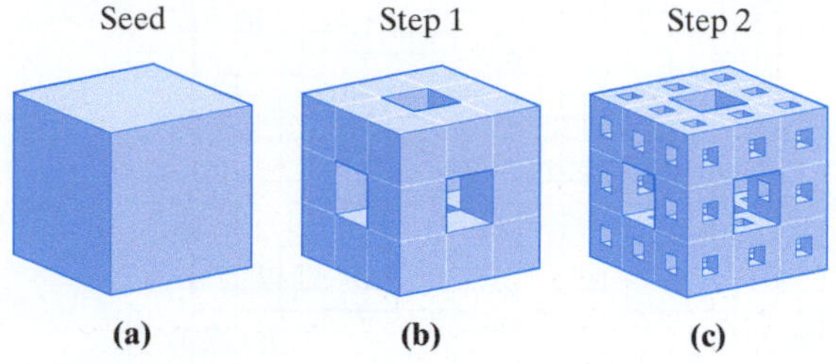

FIGURE 12-42

49. Assume that the seed cube of the Menger sponge has volume 1.

(a) Let C denote the total number of cubes removed at a particular step of the construction, U the volume of each removed cube, and V the volume of the sponge at that particular step of the construction. Complete the entries in the following table.

	C	U	V
Start	0	0	1
Step 1	7	$\frac{1}{27}$	$\frac{20}{27}$
Step 2			
Step 3			
Step 4			
Step *N*			

(b) Explain why the Menger sponge has infinitesimally small volume.

50. Let H denote total number of cubic holes in the "sponge" obtained at a particular step of the construction of the Menger sponge.

(a) Complete the entries in the following table.

	H
Start	0
Step 1	7
Step 2	$7 + 20 \times 7$
Step 3	
Step 4	
Step 5	

(b) Find a formula that gives the value of H for the "sponge" obtained at Step N of the construction. (*Hint*: You will need to use the geometric sum formula from Chapter 9.)

51. Consider the Mandelbrot sequence with seed $s = -0.75$. Show that this Mandelbrot sequence is attracted to the value -0.5. (*Hint*: Consider the quadratic equation $x^2 - 0.75 = x$, and consider why solving this equation helps.)

52. Consider the Mandelbrot sequence with seed $s = 0.25$. Is this Mandelbrot sequence *escaping*, *periodic*, or *attracted*? If attracted, to what number? (*Hint*: Consider the quadratic equation $x^2 + 0.25 = x$, and consider why solving this equation helps.)

53. Consider the Mandelbrot sequence with seed $s = -1.25$. Is this Mandelbrot sequence *escaping*, *periodic*, or *attracted*? If attracted, to what number?

54. Consider the Mandelbrot sequence with seed $s = \sqrt{2}$. Is this Mandelbrot sequence *escaping*, *periodic*, or *attracted*? If attracted, to what number?

RUNNING

55. Suppose that we play the chaos game using triangle ABC and that M_1, M_2, and M_3 are the midpoints of the three sides of the triangle. Explain why it is impossible at any time during the game to land inside triangle $M_1M_2M_3$.

56. Consider the following variation of the chaos game. The game is played just like with the ordinary chaos game but with the following change in rules: If you roll a 1, 2, or 3, move halfway toward vertex A; if you roll a 4, move halfway toward vertex B; and if you roll a 5 or 6, move halfway toward vertex C. What familiar geometric fractal is approximated by repeated rolls in this game? Explain.

57. (a) Show that the complex number $s = -0.25 + 0.25i$ is in the Mandelbrot set.

    (b) Show that the complex number $s = -0.25 - 0.25i$ is in the Mandelbrot set. [*Hint*: Your work for (a) can help you here.]

58. Show that the Mandelbrot set has a reflection symmetry. (*Hint*: Compare the Mandelbrot sequences with seeds $a + bi$ and $a - bi$.)

Exercises 59 and 60 refer to the concept of ***fractal dimension****. The fractal dimension of a geometric fractal consisting of N self-similar copies of itself each reduced by a scaling factor of S is* $D = \frac{\log N}{\log S}$.

59. Compute the fractal dimension of the Koch curve.

60. Compute the fractal dimension of the Sierpinski gasket.

APPLET BYTES MyMathLab®

These Applet Bytes are exercises built around the applet* ***Geometric Fractals*** *(available in MyMathLab in the Multimedia Library or Tools for Success.) Exercises 61 through 64 deal with the* ***Sierpinski carpet****, a geometric fractal that is a square version of the Sierpinski gasket. The applet allows you to see, step-by-step, how the Sierpinski carpet is generated.*

61. To familiarize yourself with the Sierpinski carpet, open the applet and click on the "Sierpinski Carpet" tab on the upper right of the window. After the new window opens with a blue square, click on the "Next Step" button to see Step 1 of the construction. Continue clicking on "Next Step" to see Steps 2 through 6. After carefully looking at the first few steps of the construction,

    (a) give a *step-by-step description* of the procedure for constructing the Sierpinski carpet. (*Hint:* See p.363 for the analogous description of the Sierpinski gasket.)

    (b) give a definition of the Sierpinski carpet using a *recursive replacement rule*. (*Hint:* See p.363 for the analogous replacement rule for the Sierpinski gasket.)

62. Suppose that the area of the seed square (Step 0) of the Sierpinski carpet is A.

    (a) Find the area (in terms of A) of the figures at Steps 1, 2, and 3.

    (b) Find the area of the figure at Step 6.

    (c) Give a general formula (in terms of A and N) for the area of the figure at Step N of the construction.

63. Think of the construction of the Sierpinski carpet as a process where you start with a solid blue square, punch a square "hole" in it in Step 1, and continue punching smaller square "holes" at each step of the construction. Using this interpretation,

    (a) find the number of square "holes" at Step 3.

    (b) find the number of square "holes" at Step 5.

    (c) give a general formula (in terms of N) for the number of square "holes" at Step N. [*Hint:* Here you will need to use the *geometric sum formula* (see p.275).]

64. Suppose that the seed square (Step 0) of the Sierpinski carpet has sides of length 1.

    (a) Find the length of the boundary of the figures at Steps 1, 2, and 3.

    (b) Find the length of the boundary of the figure at Step 5.

    (c) Find the length of the boundary of the figure at Step 6.

---

* MyMathLab code required.

13

Stock price chart broken down using Fibonacci time zones.

Fibonacci Numbers and the Golden Ratio

Tales of Rabbits and Gnomons

Fib retracements, arcs and fans, time zones, Elliott waves. Surfer's talk? No, Wall Street talk. All of the these terms describe some of the technical tools used by "chartists"—professional traders and hedge-funders who use the past up-and-down patterns in the chart of a stock to predict how the stock will perform in the future. Chartists live and die by their predictions: An accurate prediction is golden; a bad prediction can spell ruin. So it is quite remarkable that some of the most technical tools used by financial traders to predict market prices originated, more than 800 years ago, with, of all things, the behavior of a bunch of prolific bunny rabbits.

Rewind the clock to the year 1202. That's the year that a young Italian mathematician named Leonardo Pisano, known to the world as Fibonacci, published a book called *Liber Abaci* (literally translated as "The Book of Computation"). In *Liber Abaci*, Fibonacci introduced to the Europe of the Middle Ages the Hindu numerals and the Arabic algorithms for doing arithmetic and basic algebra—practically all of the school mathematics of today.

Because of its scope and influence, *Liber Abaci* was one of the most important books in the history of mathematics, but ironically, Fibonacci's modern fame (and the reason Wall Street traders talk about *Fibonacci time zones* and *Fibonacci analysis*) is a single example in the book involving a fairly simple rabbit–breeding problem:

> *A certain man put a pair of rabbits in a place surrounded in all sides by a wall. How many pairs of rabbits can be produced from that pair in a year if it is supposed that every month each pair begets a new pair, which from the second month on becomes productive?* (From Liber Abaci by Leonardo Pisano.)

The numbers that give the general solution to this problem are now called the *Fibonacci numbers*, and, as they say, the rest is history.

This chapter starts by revisiting Fibonacci's rabbit-breeding problem (first discussed in Chapter 9, Example 9.5), followed by an introduction to the *Fibonacci numbers* and some of their mathematical properties. In Section 13.2 we introduce and discuss the *golden ratio*, an important irrational number closely connected to the Fibonacci numbers and playing a significant role in art, architecture, music, and mathematics (of course!). In Section 13.3 we introduce *gnomons*—no, they are not little men living in trees in the forest—and look at various examples of geometric gnomons. In Section 13.4 we combine the different concepts introduced in the earlier sections—Fibonacci numbers, the golden ratio, gnomons, and gnomonic growth to illustrate the mechanisms by which nature generates some of its beautiful spirals.

13.1 Fibonacci Numbers

We start this section with an abbreviated discussion of the Fibonacci rabbit problem. A much more detailed discussion of the problem was given in Example 9.5, page 263.

EXAMPLE 13.1 FIBONACCI'S RABBITS REVISITED

The key elements of Fibonacci's rabbit problem are as follows:

- **Start.** The population count starts with one pair of baby rabbits ($P_0 = 1$). [*Note*: The rabbit count is by pairs, and each pair is assumed to be a male and a female.]
- **Month 1.** One month later the original pair is mature and able to produce offspring, but there is still one pair of rabbits: $P_1 = 1$.
- **Month 2.** The original pair produces a baby pair. There are now two pairs (one baby pair plus the parent pair): $P_2 = 1 + 1 = 2$.
- **Month 3.** The original pair produces another baby pair. There are now three pairs [the two pairs from the previous month (now both mature) plus the new baby pair]: $P_3 = 2 + 1 = 3$.
- **Month 4.** The two mature pairs in the previous month both have offspring. There are now five pairs [the three pairs from the previous month plus two new baby pairs]: $P_4 = 3 + 2 = 5$.

- **Month 5.** The three mature pairs in the previous month all have offspring. There are now eight pairs [the five pairs from the previous month plus three new baby pairs]: $P_4 = 5 + 3 = 8$.

  As long as the rabbits continue doing their thing and don't die, the pattern will continue: Each month the population will consist of the population in the previous month (mature pairs) plus the population in the previous previous month (baby pairs). It's a lot easier to express the idea in mathematical notation:
- **Month N.** $P_N = P_{N-1} + P_{N-2}$.

The month-by-month sequence for the growth of the rabbit population is given by 1, 1, 2, 3, 5, 8, 13, 21, ...

If we forget about rabbits and think of the sequence 1, 1, 2, 3, 5, 8, 13, 21, . . . as just an infinite sequence of numbers, we have what almost everyone calls *the Fibonacci sequence*.

FIBONACCI SEQUENCE (INFINITE LIST FORM)

1, 1, 2, 3, 5, 8, 13, 21, 34, 55, 89, 144, ...

The numbers in the Fibonacci sequence are called the **Fibonacci numbers**. (The conventional notation is to use F_N to describe the Nth Fibonacci number and to start the count at F_1, so we write $F_1 = 1, F_2 = 1, F_3 = 2, F_4 = 3$, etc.)

FIBONACCI NUMBERS (RECURSIVE FORMULA)

$$F_N = F_{N-1} + F_{N-2};\ F_1 = 1 \text{ and } F_2 = 1.$$

The recursive formula makes it very easy to find any Fibonacci number, as long as you know all the Fibonacci numbers that come before it. If you don't, you may have a problem.

EXAMPLE 13.2 FIBONACCI NUMBERS GET BIG FAST

Suppose you were given the following choice: You can have \$100 billion or a sum equivalent to F_{100} pennies. Which one would you choose? Surely, this is a no brainer—how could you pass on the \$100 billion? (By the way, that much money would make you considerably richer than Bill Gates.) But before you make a rash decision, let's see if we can figure out the dollar value of the second option. To do so, we need to compute the 100th Fibonacci number F_{100}.

How could one find the value of F_{100}? With a little patience (and a calculator) we could use the recursive formula for the Fibonacci numbers as a "crank" that we repeatedly turn to ratchet our way up the Fibonacci sequence: From the seeds F_1 and F_2 we compute F_3, then use F_3 and F_2 to compute F_4, and so on. If all goes well, after many turns of the crank (we will skip the details) you will eventually get to

$$F_{97} = 83{,}621{,}143{,}489{,}848{,}422{,}977$$

and

$$F_{98} = 135{,}301{,}852{,}344{,}706{,}746{,}049$$

One more turn of the crank gives

$$F_{99} = F_{98} + F_{97} = 218{,}922{,}995{,}834{,}555{,}169{,}026$$

and the last turn gives

$$F_{100} = F_{99} + F_{98} = 354{,}224{,}848{,}179{,}261{,}915{,}075$$

Thus, the F_{100} cents can be rounded nicely to \$3,542,248,481,792,619,150. How much money is that? If you take \$100 billion for yourself and then divide what's left evenly among every man, woman, and child on Earth (about 7.5 billion people), each person would get more than *\$450 million*!

The most obvious lesson to be drawn from Example 13.2 is that Fibonacci numbers grow very large very quickly. A more subtle lesson (less obvious because we cheated in Example 13.2 and skipped most of the work) is that computing Fibonacci numbers using the recursive formula takes an enormous amount of effort (each turn of the crank involves just one addition, but as we noted, the numbers being added get very large very quickly).

Is there a more convenient way to compute Fibonacci numbers—without the need to repeatedly turn the crank in the recursive formula? Yes and no. In 1736 Leonhard Euler (the same Euler behind the namesake theorems in Chapter 5) discovered a formula for the Fibonacci numbers that does not rely on previous Fibonacci numbers. The formula was lost and rediscovered 100 years later by French mathematician and astronomer Jacques Binet, who somehow ended up getting all the credit, as the formula is now known as *Binet's formula*. Now come the bad news.

BINET'S FORMULA (ORIGINAL VERSION)

$$F_N = \left[\left(\frac{1+\sqrt{5}}{2}\right)^N - \left(\frac{1-\sqrt{5}}{2}\right)^N\right] \Big/ \sqrt{5}.$$

Admittedly, Binet's original formula is quite complicated and intimidating, and even with a good calculator you might have trouble finding an exact value when N is large, but there is a simplified version of the formula that makes the calculations a bit easier. In this simplified version we essentially disregard the second half of the numerator (it is a very small number) and make up for it by rounding to the nearest integer.

BINET'S FORMULA (SIMPLIFIED VERSION)

$$F_N = \left[\!\left[\left(\frac{1+\sqrt{5}}{2}\right)^N \Big/ \sqrt{5}\right]\!\right],$$ where $[\![\ ]\!]$ means "rounded to the nearest integer."

Binet's simplified formula is an explicit formula (we don't have to know the previous Fibonacci numbers to use it), but it only makes sense to use it to compute very large Fibonacci numbers (for smaller Fibonacci numbers you are much better off using the recursive formula). For example, if you need to find F_{100} you might consider using Binet's simplified formula. To do so you will need a good calculator that can handle fairly large numbers.

EXAMPLE 13.3 COMPUTING F_{100} WITH BINET'S SIMPLIFIED FORMULA

Binet's simplified formula for F_{100} is $F_{100} = \left[\!\left[\left(\frac{1+\sqrt{5}}{2}\right)^{100} \Big/ \sqrt{5}\right]\!\right]$.

The key step is to compute the number inside the double square brackets. The last step is to round the number to the nearest integer, which is trivial. For this example, *web2.0calc* is used for the calculation. Figure 13-1(a) shows a screen shot of the input prior to the calculation; Fig. 13-1(b) shows the answer given by *web2.0calc*. The last step is rounding the answer to the nearest integer, but in this case the calculator shows no decimal part to the answer, so no rounding is needed. We are done:

$$F_{100} = 354{,}224{,}848{,}179{,}261{,}915{,}075.$$

Fibonacci Numbers in Nature

One of the major attractions of the Fibonacci numbers is how often they show up in natural organisms, particularly flowers and plants that grow as spirals. The petal counts of most varieties of daisies are Fibonacci numbers—most often 3, 5, 8, 13, 21, 34, or 55 (but giant daisies with 89 petals also exist).

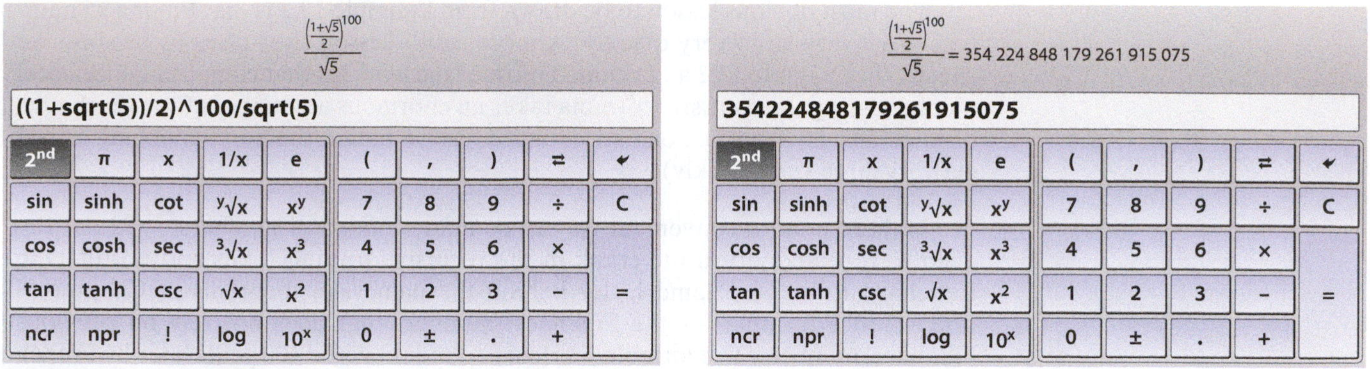

$$\frac{\left(\frac{1+\sqrt{5}}{2}\right)^{100}}{\sqrt{5}}$$

((1+sqrt(5))/2)^100/sqrt(5)

(a)

$$\frac{\left(\frac{1+\sqrt{5}}{2}\right)^{100}}{\sqrt{5}} = 354\ 224\ 848\ 179\ 261\ 915\ 075$$

354224848179261915075

(b)

FIGURE 13-1 Computing F_{100} with Binet's simplified formula.

Figure 13-2 shows three varieties of daisies with 13, 21, and 34 petals, respectively. The bracts of a typical pinecone are arranged in 5, 8, and 13 spiraling rows depending on the direction you count [Figs. 13-3(a) and (b)]; and the seeds on a sunflower head are arranged in 21 and 34 spiraling rows [Figs. 13-3(c) and (d)]. Why Fibonacci numbers? It is a bit of a mystery, but definitely related to the spiraling nature of the growth. We will come back to spiral growth and the connection with Fibonacci numbers in the last section of this chapter.

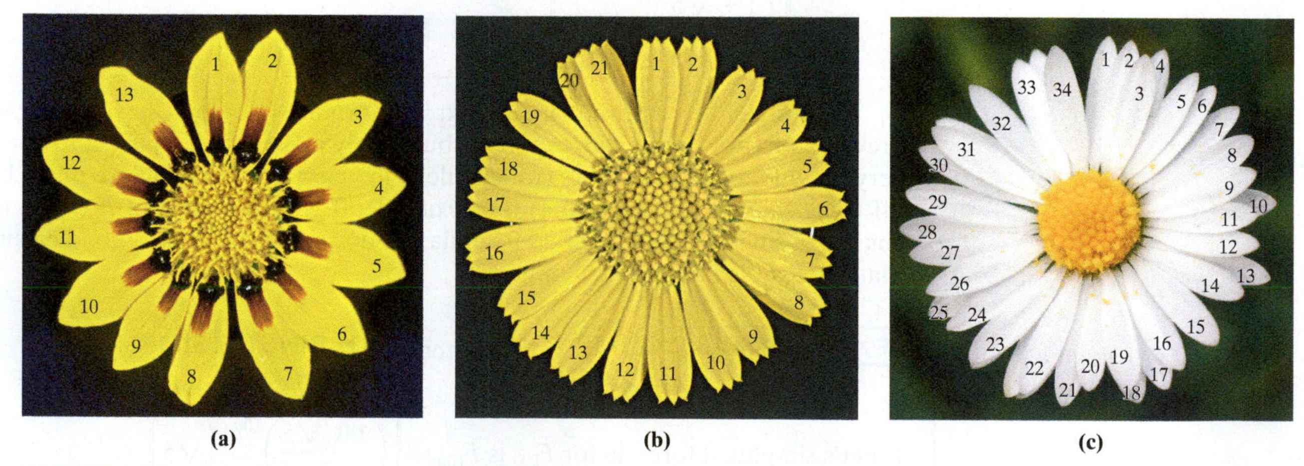

(a) (b) (c)

FIGURE 13-2 (a) Yellow daisy (13 petals). (b) English daisy (21 petals). (c) Oxeye daisy (34 petals).

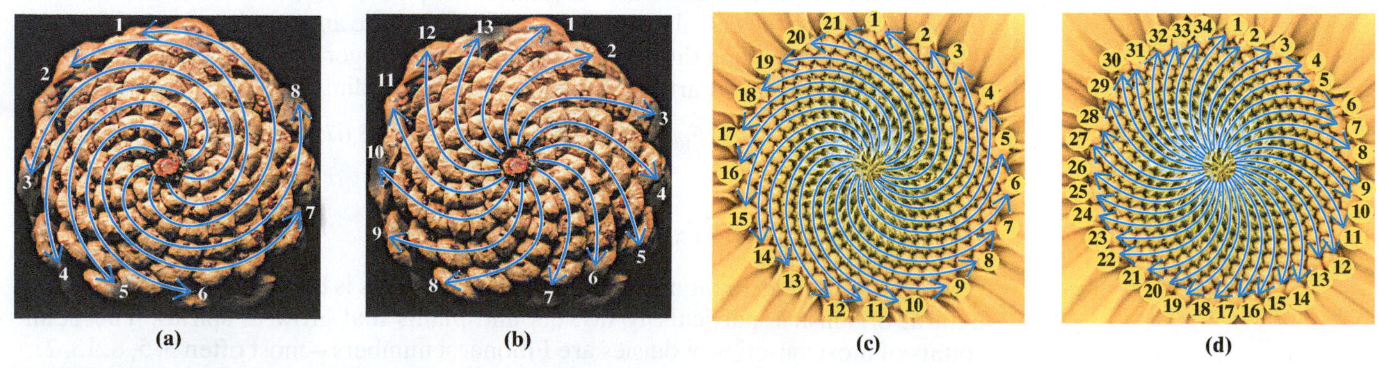

(a) (b) (c) (d)

FIGURE 13-3 (a) and (b) Arrangement of bracts in a pinecone. (c) and (d) Arrangement of seeds in a sunflower head.

13.2 The Golden Ratio

In this section we will introduce a remarkable number known as the *golden ratio* (sometimes also called the *golden mean* or the *golden section*). We will use the Greek letter ϕ ("phi"; pronounced "fi" with a long "i") to denote this special irrational number.

Irrational numbers are numbers that have an infinite, nonrepeating decimal expansion. This is not an actual definition (a proper mathematical definition is quite complicated and beyond the scope of this book), but it will serve our purposes for the moment. While π is undoubtedly the best known of all irrational numbers, the easiest place to find a mother lode of irrational numbers is to look at the square roots of positive integers: When a positive integer is *not a perfect square its square root is an irrational number*. So for the record, $\sqrt{2}$, $\sqrt{3}$, $\sqrt{5}$, $\sqrt{6}$, and $\sqrt{7}$ are all examples of irrational numbers.

Irrational numbers don't mix well with rational numbers. Say you take $\sqrt{7}$, add 1 to it, and then divide the result by 2. What do you have? You have another irrational number, but there is no way to write it as a single numerical entity—it is simply "the number" $\frac{1+\sqrt{7}}{2}$. There is one exception to this situation—the number $\frac{1+\sqrt{5}}{2}$. This number is so special and important that it has its own symbol and name. This number is, in fact, the golden ratio ϕ.

- **The Golden Ratio.** The golden ratio is the irrational number $\phi = \frac{1+\sqrt{5}}{2}$. A decimal approximation for ϕ accurate to 10 decimal places is $\phi \approx 1.6180339887$. (For the purposes of memorization, $\phi \approx 1.618$ is a convenient shortcut.)

What is it that makes the golden ratio such a special number? There are several different explanations, but in one way or another they all fall back to the following fact: $\phi^2 = \phi + 1$. For ease of reference we call this the *golden property*.

- **The Golden Property.** $\phi^2 = \phi + 1$. Restated in plain English, this property says that to square the golden ratio all you have to do is add one to it.

Our next example shows that the golden ratio is the only *positive* number satisfying the golden property.

EXAMPLE 13.4 SOLVING $x^2 = x + 1$

To find all numbers with the property that squaring the number is the same as adding one to it we set up the quadratic equation $x^2 = x + 1$. Solving this equation involves the use of the quadratic formula—standard fare in high school algebra. (For a review of the quadratic formula, see Exercises 25 through 28.)

To solve the equation we first rewrite it in the form $x^2 - x - 1 = 0$. The two solutions are $\frac{1 \pm \sqrt{5}}{2}$. The positive solution is $\phi = \frac{1+\sqrt{5}}{2} \approx 1.6180339887$. The second (negative) solution is $\frac{1-\sqrt{5}}{2} \approx -0.6180339887$. Both approximations are accurate to 10 decimal places.

Notice that the two solutions appear to have identical decimal parts. This is not a coincidence, since one is positive, the other is negative, and their sum equals 1. It follows that the negative solution can be written as $1 - \phi$. ■

Our next example illustrates why the golden property is relevant to our discussion.

EXAMPLE 13.5 THE DIVINE PROPORTION

Imagine that the line segment in Fig. 13-4(a) represents some undefined unit—a building, a work of art, a musical composition, whatever. This unit is to be split into two unequal sections in a nice, aesthetically pleasing and balanced proportion.

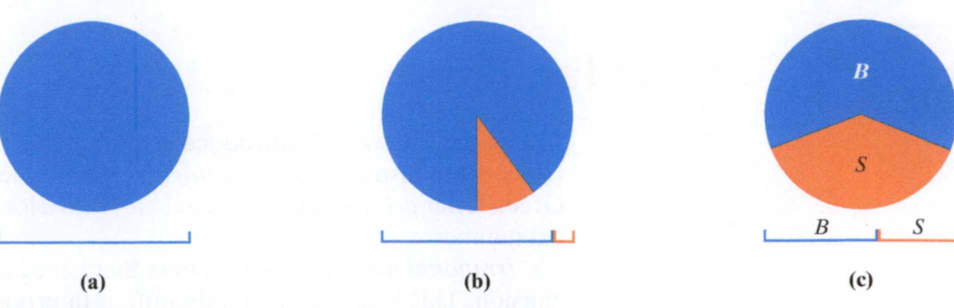

FIGURE 13-4 Searching for the golden split.

Figure 13-4(b) shows a split that most of us would consider pretty unbalanced—the larger piece is too large and out of proportion in relation to the shorter piece.

What kind of split would make for an ideal proportion? The ancient Greeks—masters of both geometry and aesthetics—came up with a very clever answer: *Make the split in such a way that the ratio of the bigger piece to the smaller piece is equal to the ratio of the whole unit to the bigger piece*. They called this proportion the **divine proportion**.

If we let B and S stand for the sizes of the bigger and smaller pieces respectively [Fig. 13-4(c)], the divine proportion is satisfied when $\frac{B}{S} = \frac{B+S}{B}$, or equivalently, $\frac{B}{S} = 1 + \frac{S}{B}$. If we now let x denote the ratio $\frac{B}{S}$, then $\frac{B}{S} = 1 + \frac{S}{B}$ becomes the equation $x = 1 + \frac{1}{x}$, or equivalently, $x^2 = x + 1$. Since $x = \frac{B}{S}$ has to be positive (B and S are both positive) and the only positive number satisfying $x^2 = x + 1$ is the golden ratio ϕ, we can conclude that $\frac{B}{S} = \phi$ (i.e., the divine proportion is satisfied only when the ratio of the bigger piece to the shorter piece equals the golden ratio ϕ).

Throughout history, many famous painters, sculptors, architects, and designers are said to have looked at the golden ratio $\phi = 1.618\ldots$ as the perfect ratio of big to small and used it in their works. How much of it is true and how much of it is hype is a matter of some debate, but there is no shortage of man-made structures and everyday objects with proportions that are close to golden (that is, the aspect ratio between the longer side and the shorter side is close to 1.618). Figure 13-5 shows some random examples: (a) a standard credit card measures 8.5 by 5.3 cm (aspect ratio of approximately 1.604); (b) the screen of a standard 13-inch laptop measures 11 5/16 by 7 1/16 in (aspect ratio of approximately 1.602); (c) the CN tower in Toronto (the tallest building in North America) has a height of 553 m with an observation deck located at a height of 342 m (a ratio of approximately 1.617). While none of these ratios is exactly the golden ratio, in the real world the difference between a ratio of 1.60 and 1.618 . . . is invisible to the human eye—so who cares? We can think of these ratios as *imperfectly golden*.

(a)

(b)

(c)

FIGURE 13-5 Imperfectly golden proportions.

Fibonacci Numbers and the Golden Ratio

Other than the fact that the Fibonacci numbers and the golden ratio share equal billing in the title of this chapter, there is no particular reason to guess that they are connected in any way. But they are, and their relationship is very tight. In this section we will briefly discuss just a few of the many ways that the Fibonacci numbers and the golden ratio come together.

1. **Binet's formula.** Take another look at the original version of Binet's formula. The numerator has two numbers raised to the Nth power: $\frac{1+\sqrt{5}}{2}$ and $\frac{1-\sqrt{5}}{2}$. We now know that the first of these numbers is ϕ and the second is $1-\phi$ (see Example 13.4). It follows that we can rewrite Binet's original formula in terms of ϕ: $F_N = [\phi^N - (1-\phi)^N]/\sqrt{5}$. Binet's simplified formula takes an even nicer form: $F_N = [\![\phi^N/\sqrt{5}]\!]$.

2. **Golden power formula: $\phi^N = F_N\phi + F_{N-1}$.** This formula is a generalization of the golden property. It expresses any power of ϕ in terms of ϕ and Fibonacci numbers. To see how this formula comes about, let's compute ϕ^2 through ϕ^5.

   - $\phi^2 = \phi + 1$. [The golden property].
   - $\phi^3 = 2\phi + 1$. [Multiply both sides of $\phi^2 = \phi + 1$ by ϕ and replace ϕ^2 by $\phi + 1$. This gives $\phi^3 = \phi^2 + \phi = (\phi + 1) + \phi = 2\phi + 1$.]
   - $\phi^4 = 3\phi + 2$. [Multiply both sides of $\phi^3 = 2\phi + 1$ by ϕ and replace ϕ^2 by $\phi + 1$. This gives $\phi^4 = 2\phi^2 + \phi = 2(\phi + 1) + \phi = 3\phi + 2$.]
   - $\phi^5 = 5\phi + 3$. [Multiply both sides of $\phi^4 = 3\phi + 2$ by ϕ and replace ϕ^2 by $\phi + 1$. This gives $\phi^5 = 3\phi^2 + 2\phi = 3(\phi + 1) + 2\phi = 5\phi + 3$.]

   If you look at the pattern that is emerging, you will notice that for each power of ϕ the coefficients on the right-hand side are consecutive Fibonacci numbers: $\phi^2 = F_2\phi + F_1$, $\phi^3 = F_3\phi + F_2$, $\phi^4 = F_4\phi + F_3$, and $\phi^5 = F_5\phi + F_4$. The general version of this observation gives the golden power formula $\phi^N = F_N\phi + F_{N-1}$.

3. **Ratio of consecutive Fibonacci numbers.** Let's look at the sequence of numbers obtained by dividing a Fibonacci number by the preceding Fibonacci number (in other words, the sequence defined by the fractions $\frac{F_{N+1}}{F_N}$). Writing this sequence in fractional form doesn't give us anything that looks very interesting:

   $$\frac{1}{1}, \frac{2}{1}, \frac{3}{2}, \frac{5}{3}, \frac{8}{5}, \frac{13}{8}, \frac{21}{13}, \frac{34}{21}, \frac{55}{34}, \frac{89}{55}, \frac{144}{89}, \frac{233}{144}, \frac{377}{233}, \frac{610}{377}, \frac{987}{610}, \frac{1597}{987}, \ldots$$

   If we write the same numbers as decimals (rounded to six decimal places when needed), things look a lot more interesting:

   1, 2, 1.5, 1.666667, 1.6, 1.625, 1.615385, 1.619048, 1.617647, 1.618182, 1.617978, 1.618056, 1.618026, 1.618037, 1.618033, 1.618034, ...

   After a little while, the numbers in this sequence start to look like they are being attracted toward some number, and yes, that number is ϕ. (In fact, rounded to six decimal places $\phi = 1.618034$, matching exactly the last number in our second list.) The fact that ratios of successive Fibonacci numbers get closer and closer to the golden ratio can be described symbolically by $(\frac{F_{N+1}}{F_N}) \to \phi$.

13.3 Gnomons

The most common usage of the word *gnomon* is to describe the pin of a sundial—the part that casts the shadow that shows the time of day. The original Greek meaning of the word *gnomon* is "one who knows," so it's not surprising that the word should find its way into the vocabulary of mathematics.

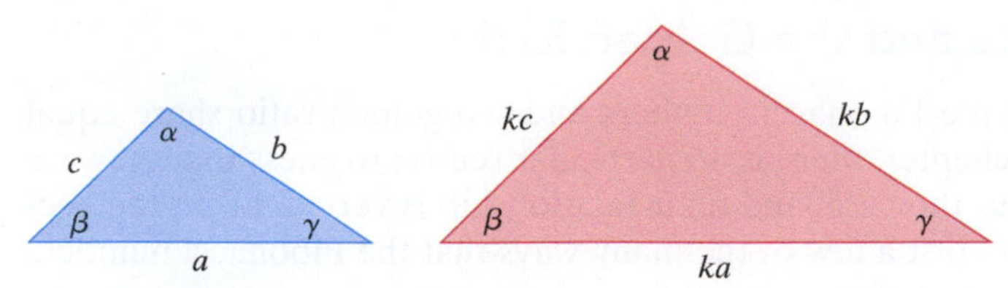

FIGURE 13-6 Similar triangles ($k = 1.6$).

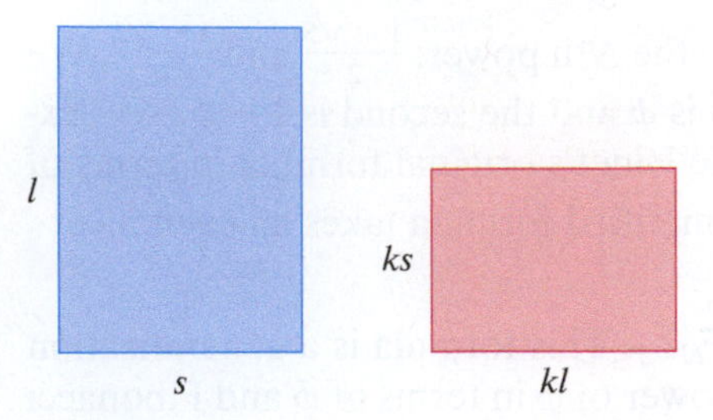

FIGURE 13-7 Similar rectangles ($k = 0.75$).

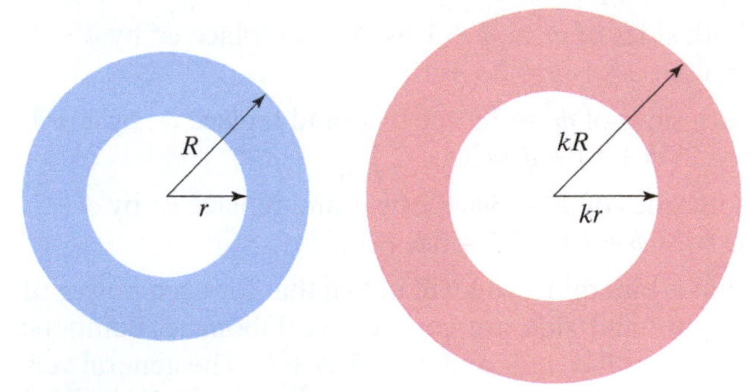

FIGURE 13-8 Similar rings ($k = 4/3$).

In this section we will discuss a different meaning for the word *gnomon*. Before we do so, we will do a brief review of a few facts from high school geometry.

Similarity

We know from geometry that two objects are said to be **similar** if one is a scaled version of the other. The following important facts about similarity of basic two-dimensional figures will come in handy later in the chapter:

- **Triangles:** Two triangles are similar if and only if the measures of their respective angles are the same. Alternatively, two triangles are similar if and only if corresponding sides are proportional. In other words, if triangle 1 has sides of length a, b, and c, then triangle 2 is similar to triangle 1 if and only if its sides have length ka, kb, and kc for some positive constant k called the *scaling factor* (Fig. 13-6). When $k > 1$, triangle 2 is larger than triangle 1; when $0 < k < 1$, triangle 2 is smaller than triangle 1.
- **Squares:** Two squares are always similar.
- **Rectangles:** Two rectangles are similar if their corresponding sides are proportional (Fig. 13-7).
- **Circles and disks:** Two circles are always similar. Any circular disk (a circle plus all its interior) is similar to any other circular disk.
- **Circular rings:** Two circular rings are similar if and only if their inner and outer radii are proportional (Fig. 13-8).

Gnomons

We will now return to the main topic of this section—gnomons. In geometry, a **gnomon** G to a figure A is a connected figure that, when suitably *attached* to A, produces a new figure similar to A. By "attached," we mean that the two figures are coupled into one figure without any overlap. Informally, we will describe it this way: G is a gnomon to A if G & A *is similar to* A (Fig. 13-9). Here the symbol "&" should be taken to mean "attached in some suitable way."

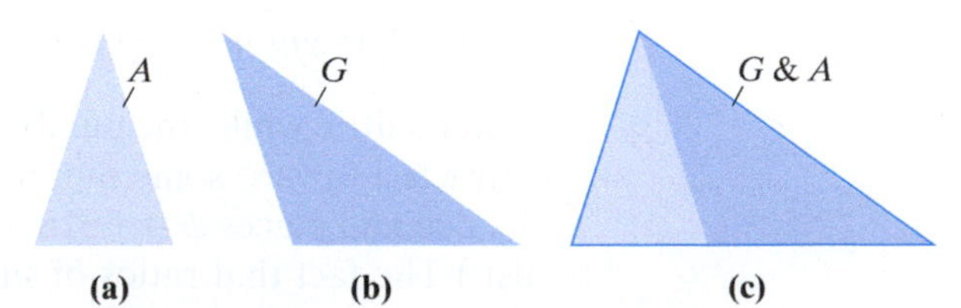

FIGURE 13-9 (a) The original object A. (b) The gnomon G. (c) G & A is similar to A.

EXAMPLE 13.6 GNOMONS TO SQUARES

Consider the square S in Fig. 13-10(a). The L-shaped figure G in Fig. 13-10(b) is a gnomon to the square—when G is attached to S as shown in Fig. 13-10(c), we get the square S'.

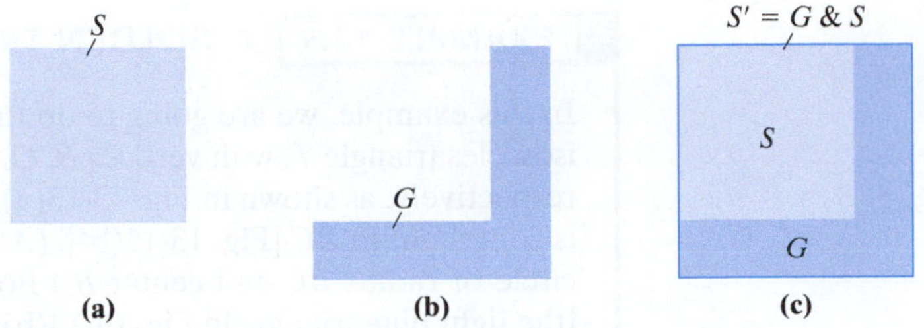

S G $S' = G \& S$ S G

(a) (b) (c)

FIGURE 13-10 (a) A square S. (b) The gnomon G. (c) G & S form a larger square.

Note that the wording is *not* reversible. The square S *is not* a gnomon to the L-shaped figure G, since there is no way to attach the two to form an L-shaped figure similar to G.

EXAMPLE 13.7 GNOMONS TO CIRCULAR DISKS

Consider the circular disk C with radius r in Fig. 13-11(a). The O-ring G in Fig. 13-11(b) with inner radius r is a gnomon to C. Clearly, G & C form the circular disk C' shown in Fig. 13-11(c). Since all circular disks are similar, C' is similar to C.

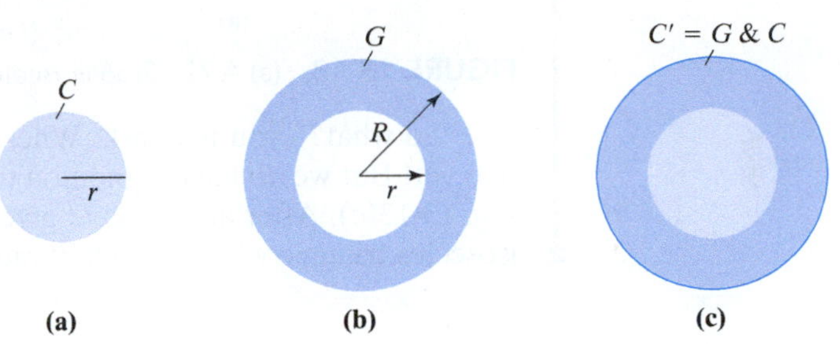

FIGURE 13-11 (a) A circular disk C. (b) The gnomon G. (c) G & C form a larger circular disk.

EXAMPLE 13.8 GNOMONS TO RECTANGLES

Consider a rectangle R of height h and base b as shown in Fig. 13-12(a). The L-shaped figure G shown in Fig. 13-12(b) can clearly be attached to R to form the larger rectangle R' shown in Fig. 13-12(c). This does not, in and of itself, guarantee that G is a gnomon to R. The rectangle R' [with height $(h+x)$ and base $(b+y)$] is similar to R if and only if their corresponding sides are proportional, which requires that $b/h = (b+y)/(h+x)$. With a little algebraic manipulation, this can be simplified to $b/h = y/x$.

There is a simple geometric way to determine if the L-shaped G is a gnomon to R—just extend the diagonal of R in G & R. If the extended diagonal passes through the outside corner of G, then G is a gnomon [Fig. 13-12(c)]; if it doesn't, then it isn't [Fig. 13-12(d)].

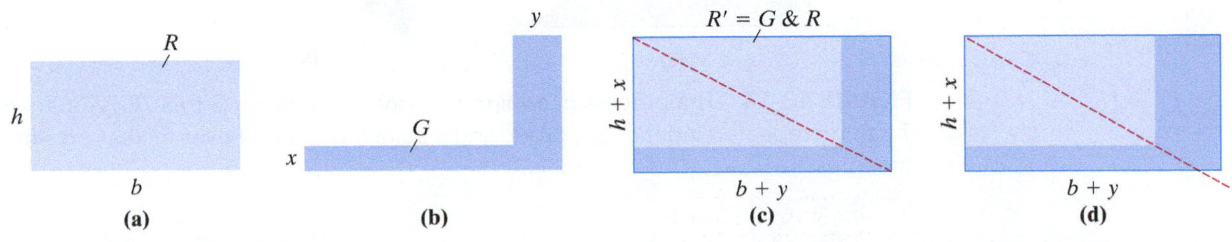

FIGURE 13-12 (a) A rectangle R. (b) A candidate for gnomon G. (c) G & R is similar to R. (d) G & R is not similar to R.

EXAMPLE 13.9 A GOLDEN TRIANGLE

In this example, we are going to do things a little bit backwards. Let's start with an isosceles triangle *T*, with vertices *B*, *C*, and *D* whose angles measure 72°, 72°, and 36°, respectively, as shown in Fig. 13-13(a). On side *CD* we mark the point *A* so that *BA* is congruent to *BC* [Fig. 13-13(b)]. (*A* is the point of intersection of side *CD* and the circle of radius *BC* and center *B*.) For convenience, we will call the triangle *ABC* [the light blue triangle in Fig. 13-13(b)] *T*′. Since *T*′ is an isosceles triangle, the measure of angle *BAC* equals the measure of angle *BCA* (72°), so it follows that angle *ABC* measures 36°. This implies that triangle *T*′ has angles equal to those of triangle *T*, and thus they are similar triangles.

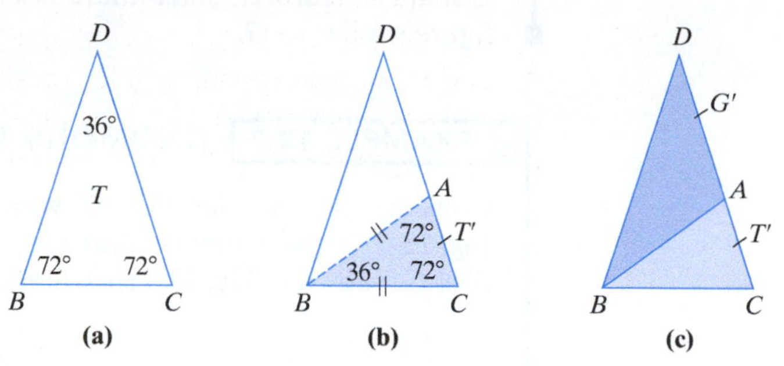

FIGURE 13-13 (a) A 72-72-36 isosceles triangle *T*. (b) *T*′ is similar to *T*. (c) *G*′ & *T*′ = *T*.

"So what?" you may ask. Where is the gnomon to triangle *T*? We don't have one yet! But we *do* have a gnomon to triangle *T*′—it is triangle *BAD*, labeled *G*′ in Fig. 13-13(c). After all, *G*′ & *T*′ give *T*—a triangle similar to *T*′. Note that *G*′ is an isosceles triangle with angles that measure 36°, 36°, and 108°.

We now know how to find a gnomon not only to triangle *T*′ but also to any 72-72-36 triangle, including the original triangle *T*: Attach a 36-36-108 triangle to one of the longer sides [Fig. 13-14(a)]. If we repeat this process indefinitely, we get a spiraling series of ever-increasing 72-72-36 triangles [Fig. 13-14(b)]. It's not too far-fetched to use a family analogy: Triangles *T* and *G* are the "parents," with *T* having the "dominant genes"; the "offspring" of their union looks just like *T* (but bigger). The offspring then has offspring of its own (looking exactly like its grandparent *T*), and so on ad infinitum.

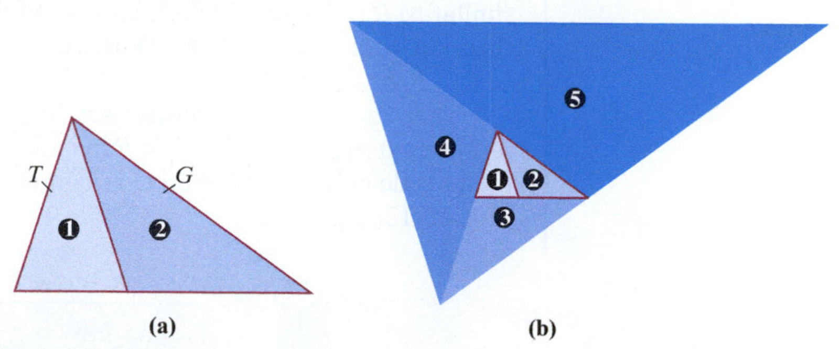

FIGURE 13-14 The process of adding a 36-36-108 gnomon *G* to a 72-72-36 triangle *T* can be repeated indefinitely, producing a spiraling chain of ever-increasing similar triangles.

Example 13.9 is of special interest to us for two reasons. First, this is the first time we have an example in which the figure and its gnomon are of the same type (isosceles triangles). Second, the isosceles triangles in this story (72-72-36 and 36-36-108) have a property that makes them unique: In both cases, the ratio of their sides (longer side over shorter side) is the golden ratio (see Exercise 68). These are the only two isosceles triangles with this property, and for this reason they are called **golden triangles**.

EXAMPLE 13.10 SQUARE GNOMONS TO RECTANGLES

We saw in Example 13.8 that *any* rectangle can have an L-shaped gnomon. Much more interesting is the case when a rectangle has a square gnomon. Not every rectangle can have a square gnomon, and the ones that do are quite special.

Consider a rectangle R with sides of length B and S [Fig. 13-15(a)], and suppose that the square G with sides of length B shown in Fig. 13-15(b) is a gnomon to R. If so, then the rectangle R' shown in Fig. 13-15(c) must be similar to R, which implies that their corresponding sides must be proportional: $\frac{B}{S} = \frac{B+S}{B}$. If this proportion looks familiar, it's because it is the *divine proportion* we first discussed in Example 13.5 and whose only possible solution is $B/S = \phi$.

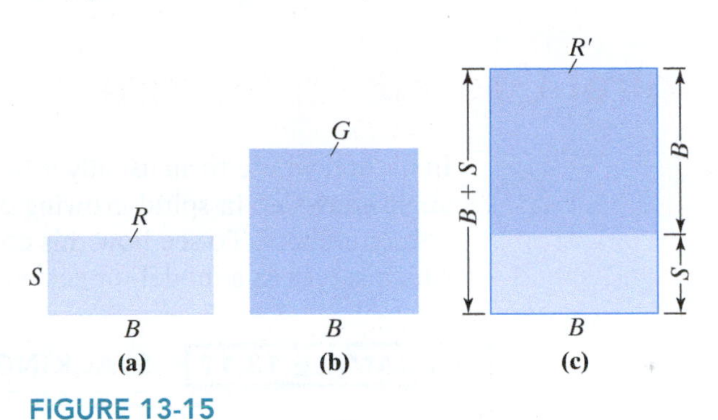

FIGURE 13-15

Example 13.10 tells us that the only way that a rectangle can have a square gnomon is if its sides are in a divine proportion (i.e., $B/S = \phi$ where B and S are the lengths of the bigger and shorter sides, respectively.

Golden and Fibonacci Rectangles

A rectangle whose sides are in the proportion of the golden ratio is called a **golden rectangle**. In other words, a golden rectangle is a rectangle with sides B and S satisfying $B/S = \phi$. A close relative to a golden rectangle is a **Fibonacci rectangle**—a rectangle whose sides are consecutive Fibonacci numbers.

EXAMPLE 13.11 GOLDEN AND ALMOST GOLDEN RECTANGLES

Figure 13-16 shows an assortment of rectangles (please note that the rectangles are not drawn to the same scale). Some are golden, some are close.

- The rectangle in Fig. 13-16(a) has $B = 1$ and $S = 1/\phi$. Since $B/S = 1/(1/\phi) = \phi$, this is a golden rectangle.
- The rectangle in Fig. 13-16(b) has $B = \phi + 1$ and $S = \phi$. Here $B/S = (\phi + 1)/\phi$. Since $\phi + 1 = \phi^2$, this is another golden rectangle.
- The rectangle in Fig. 13-16(c) has $B = 8$ and $S = 5$. This is a Fibonacci rectangle, since 5 and 8 are consecutive Fibonacci numbers. The ratio of the sides is $B/S = 8/5 = 1.6$, so this is not a golden rectangle. On the other hand, the ratio 1.6 is reasonably close to $\phi = 1.618\ldots$, so we will think of this rectangle as "imperfectly golden."
- The rectangle in Fig. 13-16(d) with $B = 89$ and $S = 55$ is another Fibonacci rectangle. Since $89/55 = 1.61818\ldots$, this rectangle is as good as golden—the ratio of the sides is the same as the golden ratio up to three decimal places.

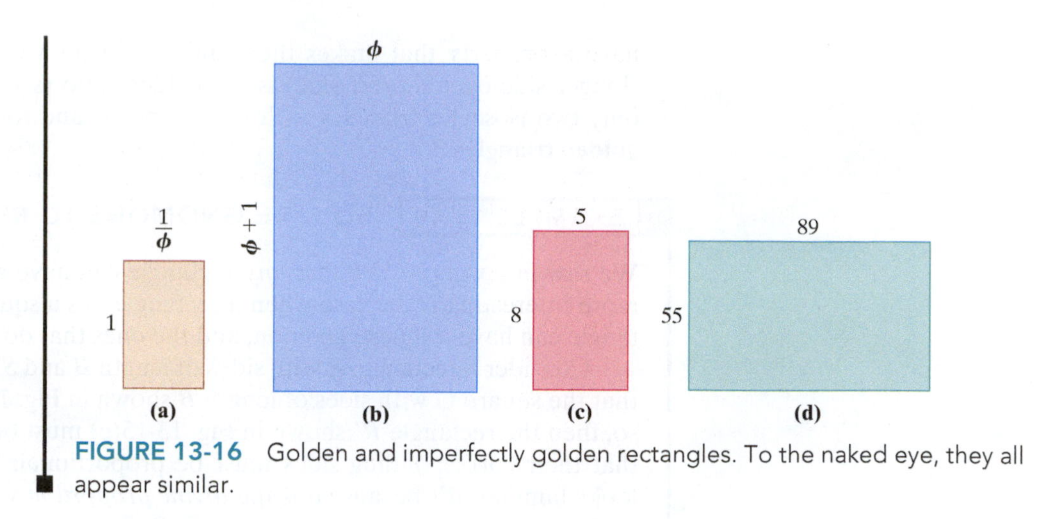

FIGURE 13-16 Golden and imperfectly golden rectangles. To the naked eye, they all appear similar.

13.4 Spiral Growth in Nature

In nature, where form usually follows function, the perfect balance of a golden rectangle shows up in spiral-growing organisms, often in the form of consecutive Fibonacci numbers. To see how this connection works, consider the following example, which serves as a model for certain natural growth processes.

EXAMPLE 13.12 STACKING SQUARES ON FIBONACCI RECTANGLES

Start with a 1 by 1 square [Fig. 13-17(a)] and attach to it another 1 by 1 square to form the 1 by 2 Fibonacci rectangle shown in Fig. 13-17(b). We will call this the "second-generation" rectangle. Next, add a 2 by 2 square. This gives the 3 by 2 Fibonacci rectangle shown in Fig. 13-17(c)—the "third generation" rectangle. Next, add a 3 by 3 square as shown Fig. 13-17(d), giving a 3 by 5 Fibonacci rectangle—the "fourth generation". Next, add a 5 by 5 square as shown in Fig. 13-17(e) giving an 8 by 5 Fibonacci rectangle. You get the picture—we can keep doing this as long as we want.

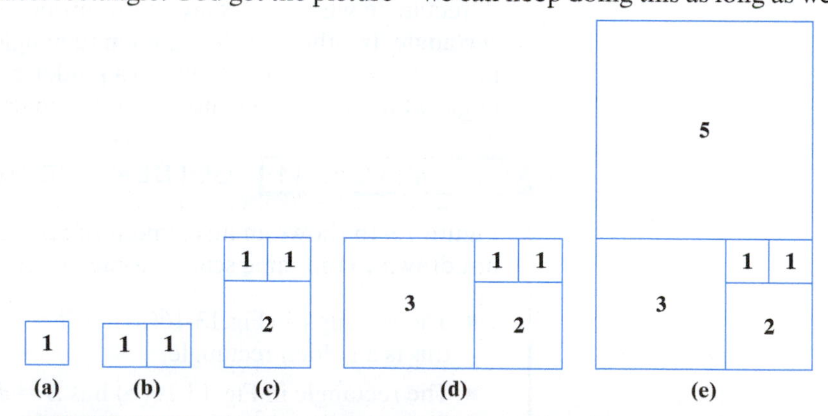

FIGURE 13-17 Fibonacci rectangles beget Fibonacci rectangles.

We might imagine these growing Fibonacci rectangles describing the growth of a living organism. In each generation, the organism grows by adding a square (a very simple, basic shape). The interesting feature of this growth is that as the Fibonacci rectangles grow larger, they become very close to golden rectangles, and as such, they become essentially similar to one another. This kind of growth— getting bigger while maintaining the same overall shape and proportion—is characteristic of the way many natural organisms grow.

The next example is a simple variation of Example 13.12.

EXAMPLE 13.13 THE GROWTH OF A "CHAMBERED" FIBONACCI RECTANGLE

Let's revisit the growth process of the previous example, except now let's create within each of the squares being added an interior "chamber" in the form of a quarter-circle. We need to be a little careful about how we attach the chambered square in each successive generation, but other than that, we can repeat the sequence of steps in Example 13.12 to get the sequence of shapes shown in Fig. 13-18. These figures depict the consecutive generations in the evolution of the *chambered Fibonacci rectangle*. The outer spiral formed by the circular arcs is often called a **Fibonacci spiral**, shown in Fig. 13-19.

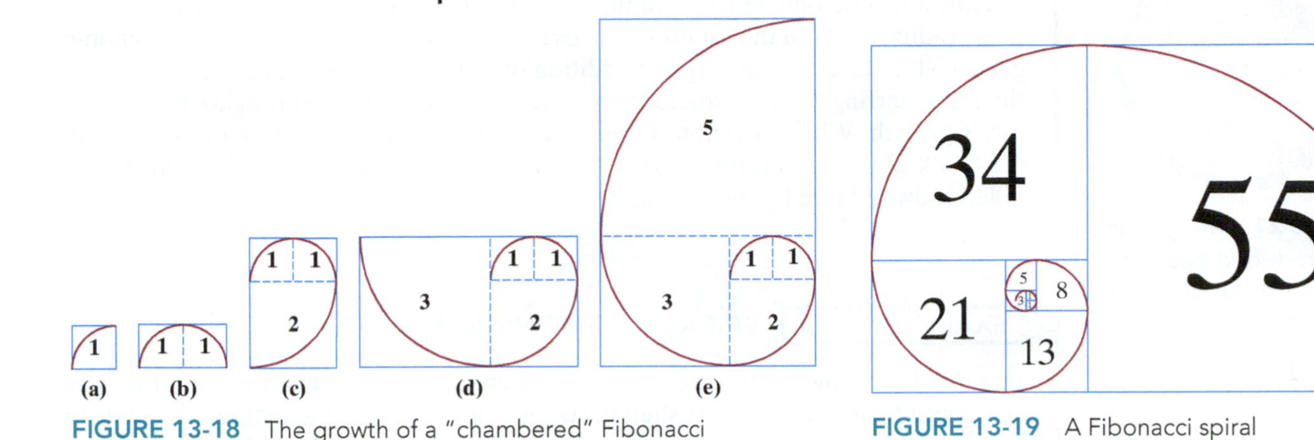

FIGURE 13-18 The growth of a "chambered" Fibonacci rectangle.

FIGURE 13-19 A Fibonacci spiral after 10 "generations."

Gnomonic Growth

Natural organisms grow in essentially two different ways. Humans, most animals, and many plants grow following what can informally be described as an *all-around growth* rule. In this type of growth, all living parts of the organism grow simultaneously—but not necessarily at the same rate. One characteristic of this type of growth is that there is no obvious way to distinguish between the newer and the older parts of the organism. In fact, the distinction between new and old parts does not make much sense. The historical record (so to speak) of the organism's growth is lost.

Contrast this with the kind of growth exemplified by the shell of a chambered nautilus, a ram's horn, or the trunk of a redwood tree (Fig. 13-20). These organisms grow following a *one-sided* or *asymmetric growth* rule, meaning that the organism has a new part added to it in such a way that the old organism together with the added part form the new organism. At any stage of the growth process, we can see not only the present form of the organism but also the organism's entire past. All the previous stages of growth are the building blocks that make up the present structure.

FIGURE 13-20 (a) Chambered nautilus (cross section). (b) Bighorn ram. (c) Redwood tree (cross section).

The other important aspect of natural growth is the principle of *self-similarity*: Organisms like to maintain their overall shape and proportions as they grow. This is where gnomons come into the picture. For the organism to retain its overall structure as it grows, the new growth must be a *gnomon* of the entire organism. We will call this kind of growth process **gnomonic growth**.

We have already seen abstract mathematical examples of gnomonic growth (Examples 13.12 and 13.13). Here is a pair of more realistic examples.

EXAMPLE 13.14 CIRCULAR GNOMONIC GROWTH

We know from Example 13.7 that the gnomon to a circular disk is an O-ring with an inner radius equal to the radius of the circle. We can, thus, have circular gnomonic growth (Fig. 13-21) by the regular addition of O-rings. O-rings added one layer at a time to a starting circular structure preserve the circular shape throughout the structure's growth. When carried to three dimensions, this is a good model for the way the trunk of a redwood tree grows. And this is why we can "read" the history of a felled redwood tree by studying its rings.

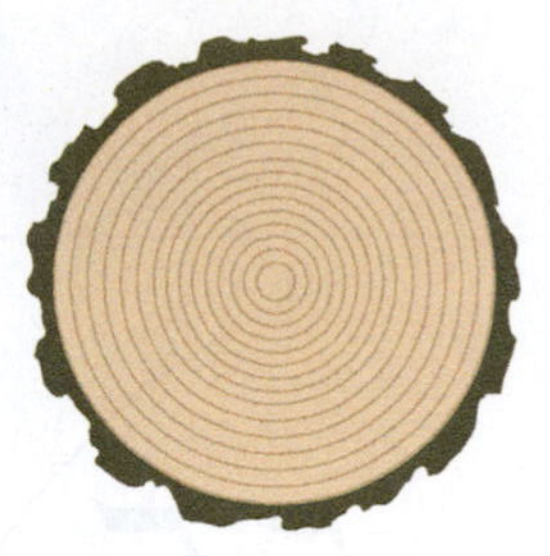

FIGURE 13-21 The growth rings in a redwood tree — an example of circular gnomonic growth.

EXAMPLE 13.15 SPIRAL GNOMONIC GROWTH

Figure 13-22 shows a diagram of a cross section of the chambered nautilus. The chambered nautilus builds its shell in stages, each time adding another chamber to the already existing shell. At every stage of its growth, the shape of the chambered nautilus shell remains the same—the beautiful and distinctive spiral shown in Fig. 13-20(a). This is a classic example of gnomonic growth—each new chamber added to the shell is a gnomon of the entire shell. The gnomonic growth of the shell proceeds, in essence, as follows: Starting with its initial shell (a tiny spiral similar in all respects to the adult spiral shape), the animal builds a chamber (by producing a special secretion around its body that calcifies and hardens). The resulting, slightly enlarged spiral shell is similar to the original one. The process then repeats itself over many stages, each one a season in the growth of the animal. Each new chamber adds a gnomon to the shell, so the shell grows and yet remains similar to itself. This process is a real-life variation of the mathematical spiral-building process discussed in Example 13.13. The curve generated by the outer edge of a nautilus shell's cross section is called a *logarithmic spiral.*

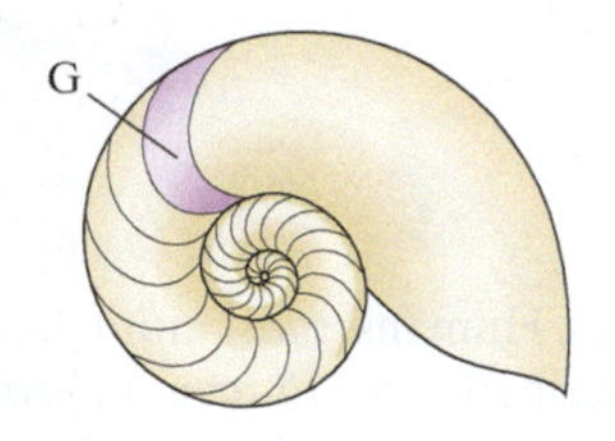

FIGURE 13-22 Gnomonic growth of a chambered nautilus.

More complex examples of gnomonic growth occur in sunflowers, daisies, pineapples, pinecones, and so on. Here, the rules that govern growth are somewhat more involved, but Fibonacci numbers and the golden ratio once again play a prominent role.

Conclusion

Some of the most beautiful shapes in nature arise from a basic principle of design: *form follows function.* The beauty of natural shapes is a result of their inherent elegance and efficiency, and imitating nature's designs has helped humans design and build beautiful and efficient structures of their own.

In this chapter, we examined a special type of growth—gnomonic growth—where an organism grows by the addition of gnomons, thereby preserving its basic shape even as it grows. Many beautiful spiral-shaped organisms, from seashells to flowers, exhibit this type of growth.

To us, understanding the basic principles behind spiral growth was relevant because it introduced us to some wonderful mathematical concepts that have been known and studied in their own right for centuries—*Fibonacci numbers, the golden ratio, the divine proportion*, and *gnomons*.

To humans, these abstract mathematical concepts have been, by and large, intellectual curiosities. To nature—the consummate artist and builder—they are the building tools for some of its most beautiful creations: flowers, plants, and seashells.

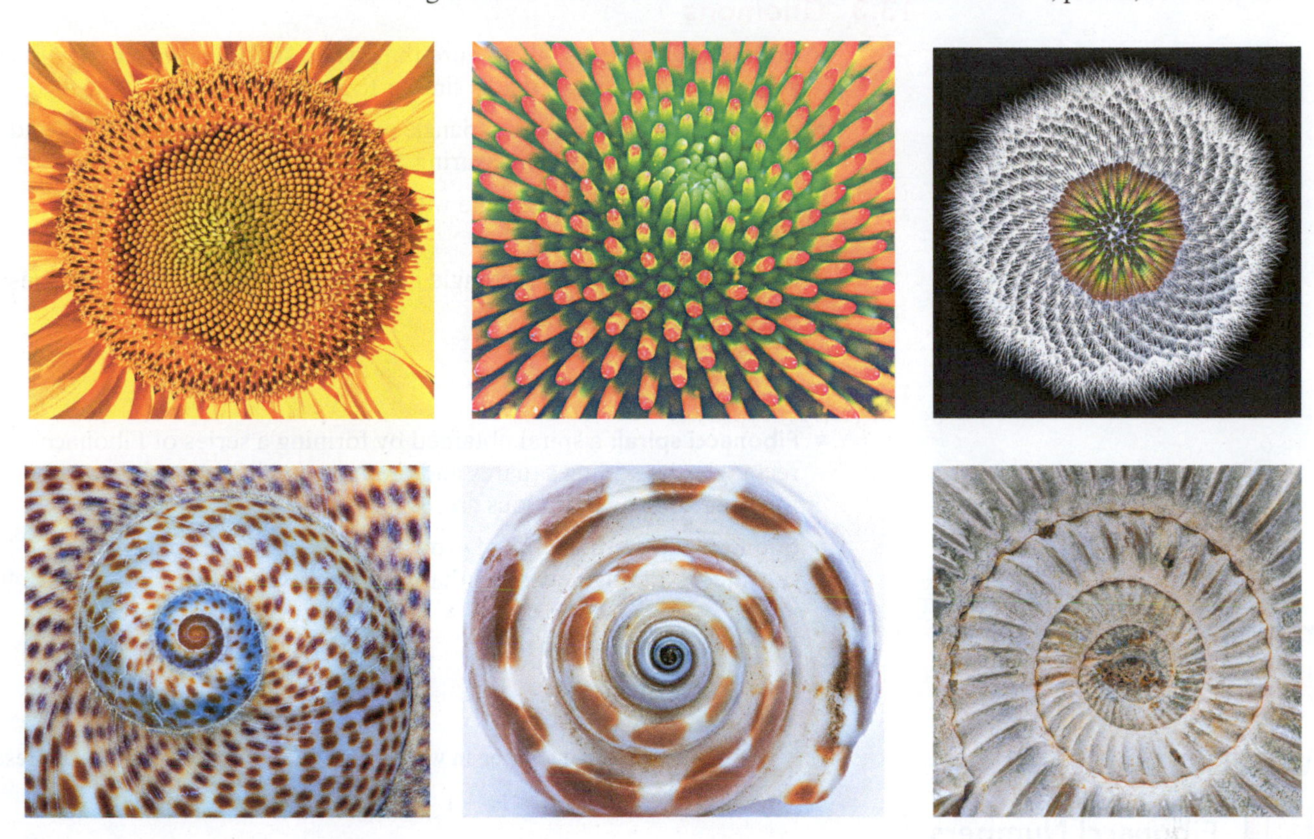

KEY CONCEPTS

13.1 Fibonacci Numbers

- **Fibonacci sequence:** the sequence 1, 1, 2, 3, 5, 8, 13, 21, 34, ... (each term of the sequence is the sum of the two preceding terms), **388**
- **Fibonacci numbers:** the terms of the Fibonacci sequence. The Nth Fibonacci number is denoted by F_N, **388**
- **Binet's formula:** $F_N = \left(\frac{1}{\sqrt{5}}\right)\left[\left(\frac{1+\sqrt{5}}{2}\right)^N - \left(\frac{1-\sqrt{5}}{2}\right)^N\right]$.

  A simplified version of the formula is given by

  $F_N = \left[\!\left[\left(\frac{1}{\sqrt{5}}\right)\left(\frac{1+\sqrt{5}}{2}\right)^N\right]\!\right]$ (where $[\![\;]\!]$ denotes round to the nearest integer), **389**

13.2 The Golden Ratio

- **golden ratio ϕ:** the irrational number $\frac{1+\sqrt{5}}{2}$, **391**
- **golden property:** $\phi^2 = \phi + 1$, **391**

- **divine proportion:** $\frac{B}{S} = \frac{B+S}{B}$, where B and S are, respectively the bigger and smaller parts of a whole that has been split into two unequal parts, **392**
- **golden power formula:** $\phi^N = F_N\phi + F_{N-1}$, **393**
- **ratio of consecutive Fibonacci numbers:** As Fibonacci numbers get bigger and bigger, the ratios F_{N+1}/F_N get closer and closer to ϕ, **393**

13.3 Gnomons

- **gnomon (to a figure *A*):** a figure that, when suitably combined with A, produces a new figure that is similar to A, **394**
- **golden triangle:** an isosceles triangle with angles measuring 72°, 72°, and 36°, or alternatively, angles measuring 36°, 36°, and 108°, **397**
- **golden rectangle:** a rectangle with sides of length B and S where $B/S = \phi$, **397**
- **Fibonacci rectangle:** a rectangle having sides whose lengths are consecutive Fibonacci numbers, **397**

13.4 Spiral Growth in Nature

- **Fibonacci spiral:** a spiral obtained by forming a series of Fibonacci rectangles by adding squares and then connecting opposite corners of the squares in a continuous arc, **399**
- **gnomonic growth:** a type of growth where an organism grows by repeatedly adding new parts that are gnomons to the old organism, **400**

EXERCISES

WALKING

13.1 Fibonacci Numbers

1. Compute the value of each of the following.

(a) F_{15} (b) $F_{15} - 2$

(c) F_{15-2} (d) $\frac{F_{15}}{5}$

(e) $F_{15/5}$

2. Compute the value of each of the following.

(a) F_{16} (b) $F_{16} + 1$

(c) F_{16+1} (d) $\frac{F_{16}}{4}$

(e) $F_{16/4}$

3. Compute the value of each of the following.

(a) $F_1 + F_2 + F_3 + F_4 + F_5$

(b) $F_{1+2+3+4+5}$

(c) $F_3 \times F_4$

(d) $F_{3\times4}$

4. Compute the value of each of the following.

(a) $F_1 + F_3 + F_5 + F_7$ (b) $F_{1+3+5+7}$

(c) F_{10}/F_5 (d) F_{10/F_5}

5. Describe in words what each of the expressions represents.

(a) $3F_N + 1$

(b) $3F_{N+1}$

(c) $F_{3N} + 1$

(d) F_{3N+1}

6. Describe in words what each of the expressions represents.

(a) $F_{2N} - 3$ (b) F_{2N-3}

(c) $2F_N - 3$ (d) $2F_{N-3}$

7. Given that $F_{36} = 14{,}930{,}352$ and $F_{37} = 24{,}157{,}817$,

(a) find F_{38}. (b) find F_{39}.

8. Given that $F_{32} = 2{,}178{,}309$ and $F_{33} = 3{,}524{,}578$,

(a) find F_{34}. (b) find F_{35}.

9. Given that $F_{36} = 14{,}930{,}352$ and $F_{37} = 24{,}157{,}817$,

(a) find F_{35}. (b) find F_{34}.

10. Given that $F_{32} = 2{,}178{,}309$ and $F_{33} = 3{,}524{,}578$,

(a) find F_{31}. (b) find F_{30}.

11. Using a good calculator (an online calculator if necessary) and Binet's simplified formula, compute F_{20}.

12. Using a good calculator (an online calculator if necessary) and Binet's simplified formula, compute F_{25}.

13. Consider the following sequence of equations involving Fibonacci numbers.

$$1 + 2 = 3$$
$$1 + 2 + 5 = 8$$
$$1 + 2 + 5 + 13 = 21$$
$$1 + 2 + 5 + 13 + 34 = 55$$
$$\vdots$$

(a) Write down a reasonable choice for the fifth equation in this sequence.

(b) Find the subscript that will make the following equation true.

$$F_1 + F_3 + F_5 + \cdots + F_{21} = F_?$$

(c) Find the subscript that will make the following equation true (assume N is odd).

$$F_1 + F_3 + F_5 + \cdots + F_N = F_?$$

14. Consider the following sequence of equations involving Fibonacci numbers.

$$2(2) - 3 = 1$$
$$2(3) - 5 = 1$$
$$2(5) - 8 = 2$$
$$2(8) - 13 = 3$$
$$\vdots$$

(a) Write down a reasonable choice for the fifth equation in this sequence.

(b) Find the subscript that will make the following equation true.

$$2(F_?) - F_{15} = F_{12}$$

(c) Find the subscript that will make the following equation true.

$$2(F_{N+2}) - F_{N+3} = F_?$$

15. Fact: *If we make a list of any four consecutive Fibonacci numbers, the first one times the fourth one is always equal to the third one squared minus the second one squared.*

(a) Verify this fact for the list F_8, F_9, F_{10}, F_{11}.

(b) Using the list F_N, F_{N+1}, F_{N+2}, F_{N+3}, write this fact as a mathematical formula.

16. Fact: *If we make a list of any 10 consecutive Fibonacci numbers, the sum of all these numbers divided by 11 is always equal to the seventh number on the list.*

(a) Verify this fact for the list $F_1, F_2, \ldots, F_{10}$.

(b) Using the list $F_N, F_{N+1}, \ldots, F_{N+9}$, write this fact as a mathematical formula.

17. Express each of the following as a single Fibonacci number.

(a) $F_{N+1} + F_{N+2} =$

(b) $F_N - F_{N-2} =$

(c) $F_N + F_{N+1} + F_{N+3} + F_{N+5} =$

18. Express each of the following as a single Fibonacci number.

(a) $F_{N-2} + F_{N-3} =$

(b) $F_{N+2} - F_N =$

(c) $F_{N-3} + F_{N-2} + F_N + F_{N+2} =$

19. Express each of the following as a ratio of two Fibonacci numbers.

(a) $1 + \dfrac{F_N}{F_{N-1}} =$

(b) $\dfrac{F_{N-1}}{F_N} - 1 =$

20. Express each of the following as a ratio of two Fibonacci numbers.

(a) $1 + \dfrac{F_{N-1}}{F_N} =$

(b) $1 - \dfrac{F_N}{F_{N-2}} =$

Exercises 21 through 24 refer to "Fibonacci-like" sequences. Fibonacci-like sequences are based on the same recursive rule as the Fibonacci sequence (from the third term on each term is the sum of the two preceding terms), but they are different in how they get started.

21. Consider the Fibonacci-like sequence 5, 5, 10, 15, 25, 40, . . . , and let A_N denote the Nth term of the sequence.

(a) Find A_{10}.

(b) Given that $F_{25} = 75{,}025$, find A_{25}.

(c) Express A_N in terms of F_N.

22. Consider the Fibonacci-like sequence 2, 4, 6, 10, 16, 26, . . . , and let B_N denote the Nth term of the sequence.

(a) Find B_9.

(b) Given that $F_{20} = 6765$, find B_{19}.

(c) Express B_N in terms of F_{N+1}.

23. Consider the Fibonacci-like sequence 1, 3, 4, 7, 11, 18, 29, 47, . . . , and let L_N denote the Nth term of the sequence. (*Note:* This sequence is called the *Lucas sequence*, and the terms of the sequence are called the *Lucas numbers*.)

(a) Find L_{12}.

(b) The Lucas numbers are related to the Fibonacci numbers by the formula $L_N = 2F_{N+1} - F_N$. Verify that this formula is true for $N = 1, 2, 3$, and 4.

(c) Given that $F_{20} = 6765$ and $F_{21} = 10{,}946$, find L_{20}.

24. Consider the Fibonacci-like sequence 1, 4, 5, 9, 14, 23, 37, . . . , and let T_N denote the Nth term of the sequence.

(a) Find T_{12}.

(b) The numbers in this sequence are related to the Fibonacci numbers by the formula $T_N = 3F_{N+1} - 2F_N$. Verify that this formula is true for $N = 1, 2, 3$, and 4.

(c) Given that $F_{20} = 6765$ and $F_{21} = 10{,}946$, find T_{20}.

13.2 The Golden Ratio

Exercises 25 through 29 involve solving quadratic equations using the quadratic formula. Here is an instant refresher on the ***quadratic formula*** *(for a more in-depth review, any high school algebra book should do):*

- *To use the quadratic formula the quadratic equation must be in the* ***standard form*** $ax^2 + bx + c = 0$*. If the equation is not in standard form, you need to get it into that form.*
- *The solutions of the quadratic equation* $ax^2 + bx + c = 0$ *are given by* $x = (-b \pm \sqrt{b^2 - 4ac})/2a$*. The formula gives two different solutions unless* $b^2 - 4ac = 0$*.*

25. Consider the quadratic equation $x^2 = x + 1$.

- **(a)** Use the quadratic formula to find the two solutions of the equation. (Remember that the equation has to be changed to standard form first.) Give the value of each solution rounded to five decimal places.
- **(b)** Find the sum of the two solutions in (a).
- **(c)** Explain why the decimal part has to be exactly the same in both solutions.

26. Consider the quadratic equation $x^2 = 3x + 1$.

- **(a)** Use the quadratic formula to find the two solutions of the equation. (Remember that the equation has to be changed to standard form first.) Give the value of each solution rounded to five decimal places.
- **(b)** Find the sum of the two solutions in (a).
- **(c)** Explain why the decimal part has to be exactly the same in both solutions.

27. Consider the quadratic equation $3x^2 = 8x + 5$.

- **(a)** Use the quadratic formula to find the two solutions of the equation. Give the value of each solution rounded to five decimal places.
- **(b)** Find the sum of the two solutions found in (a).

28. Consider the quadratic equation $8x^2 = 5x + 2$.

- **(a)** Use the quadratic formula to find the two solutions of the equation. Give the value of each solution rounded to five decimal places.
- **(b)** Find the sum of the two solutions found in (a).

29. Consider the quadratic equation $55x^2 = 34x + 21$.

- **(a)** Without using the quadratic formula, show that $x = 1$ is one of the two solutions of the equation.
- **(b)** Without using the quadratic formula, find the second solution of the equation. (*Hint*: The sum of the two solutions of $ax^2 + bx + c = 0$ is given by $-b/a$.)

30. Consider the quadratic equation $89x^2 = 55x + 34$.

- **(a)** Without using the quadratic formula, show that $x = 1$ is one of the two solutions of the equation.
- **(b)** Without using the quadratic formula, find the second solution of the equation. (*Hint*: The sum of the two solutions of $ax^2 + bx + c = 0$ is given by $-b/a$.)

31. Consider the quadratic equation $21x^2 = 34x + 55$.

- **(a)** Without using the quadratic formula, show that $x = -1$ is one of the two solutions of the equation.
- **(b)** Without using the quadratic formula, find the second solution of the equation. (*Hint*: The sum of the two solutions of $ax^2 + bx + c = 0$ is given by $-b/a$.)

32. Consider the quadratic equation $34x^2 = 55x + 89$.

- **(a)** Without using the quadratic formula, show that $x = -1$ is one of the two solutions of the equation.
- **(b)** Without using the quadratic formula, find the second solution of the equation. (*Hint*: The sum of the two solutions of $ax^2 + bx + c = 0$ is given by $-b/a$.)

33. Consider the quadratic equation $(F_N)x^2 = (F_{N-1})x + F_{N-2}$, where F_{N-2}, F_{N-1}, and F_N are consecutive Fibonacci numbers.

- **(a)** Show that $x = 1$ is one of the two solutions of the equation. [*Hint*: Try Exercises 29(a) or 30(a) first.]
- **(b)** Find the second solution of the equation expressed in terms of Fibonacci numbers. [*Hint*: Try Exercises 29(b) or 30(b) first.]

34. Consider the quadratic equation $(F_{N-2})x^2 = (F_{N-1})x + F_N$, where F_{N-2}, F_{N-1}, and F_N are consecutive Fibonacci numbers.

- **(a)** Show that $x = -1$ is one of the two solutions of the equation. [*Hint*: Try Exercises 31(a) or 32(a) first.]
- **(b)** Find the second solution of the equation expressed in terms of Fibonacci numbers. [*Hint*: Try Exercises 31(b) or 32(b) first.]

35. The *reciprocal* of $\phi = \frac{1+\sqrt{5}}{2}$ is the irrational number $\frac{1}{\phi} = \frac{2}{1+\sqrt{5}}$.

- **(a)** Using a calculator, compute $\frac{1}{\phi}$ to 10 decimal places.
- **(b)** Explain why $\frac{1}{\phi}$ has exactly the same decimal part as ϕ. (*Hint*: Show that $\frac{1}{\phi} = \phi - 1$.)

36. The square of the golden ratio is the irrational number $\phi^2 = \left(\frac{1+\sqrt{5}}{2}\right)^2 = \frac{3+\sqrt{5}}{2}$.

- **(a)** Using a calculator, compute ϕ^2 to 10 decimal places.
- **(b)** Explain why ϕ^2 has exactly the same decimal part as ϕ.

37. Given that $F_{499} \approx 8.617 \times 10^{103}$,

- **(a)** find an approximate value for F_{500} in scientific notation. (*Hint*: $F_N \approx \phi F_{N-1}$.)
- **(b)** find an approximate value for F_{498} in scientific notation.

38. Given that $F_{1002} \approx 1.138 \times 10^{209}$,

- **(a)** find an approximate value for F_{1003} in scientific notation. (*Hint*: $F_N \approx \phi F_{N-1}$.)
- **(b)** find an approximate value for F_{1001} in scientific notation.

39. The *Fibonacci sequence of order 2* is the sequence of numbers 1, 2, 5, 12, 29, 70, ... Each term in this sequence (from the third term on) equals two times the term before it plus the term two places before it; in other words, $A_N = 2A_{N-1} + A_{N-2}$ $(N \geq 3)$.

(a) Compute A_7.

(b) Use your calculator to compute to five decimal places the ratio A_7/A_6.

(c) Use your calculator to compute to five decimal places the ratio A_{11}/A_{10}.

(d) Guess the value (to five decimal places) of the ratio A_N/A_{N-1} when $N > 11$.

40. The *Fibonacci sequence of order 3* is the sequence of numbers 1, 3, 10, 33, 109, ... Each term in this sequence (from the third term on) equals three times the term before it plus the term two places before it; in other words, $A_N = 3A_{N-1} + A_{N-2}$ $(N \geq 3)$.

(a) Compute A_6.

(b) Use your calculator to compute to five decimal places the ratio A_6/A_5.

(c) Guess the value (to five decimal places) of the ratio A_N/A_{N-1} when $N > 6$.

13.3 Gnomons

41. R and R' are similar rectangles. Suppose that the width of R is a and the width of R' is $3a$.

(a) If the perimeter of R is 41.5 in., what is the perimeter of R'?

(b) If the area of R is 105 sq. in., what is the area of R'?

42. O and O' are similar O-rings. The inner radius of O is 5 ft, and the inner radius of $O'$is 15 ft.

(a) If the circumference of the outer circle of O is 14π ft, what is the circumference of the outer circle of O'?

(b) Suppose that it takes 1.5 gallons of paint to paint the O-ring O. If the paint is used at the same rate, how much paint is needed to paint the O-ring O'?

43. Triangles T and T' shown in Fig. 13-23 are similar triangles. (Note that the triangles are not drawn to scale.)

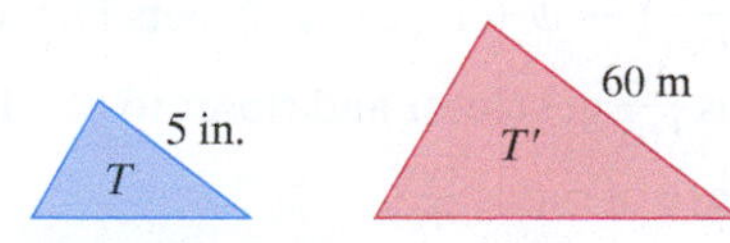

FIGURE 13-23

(a) If the perimeter of T is 13 in., what is the perimeter of T' (in meters)?

(b) If the area of T is 20 sq. in., what is the area of T' (in square meters)?

44. Polygons P and P' shown in Fig. 13-24 are similar polygons.

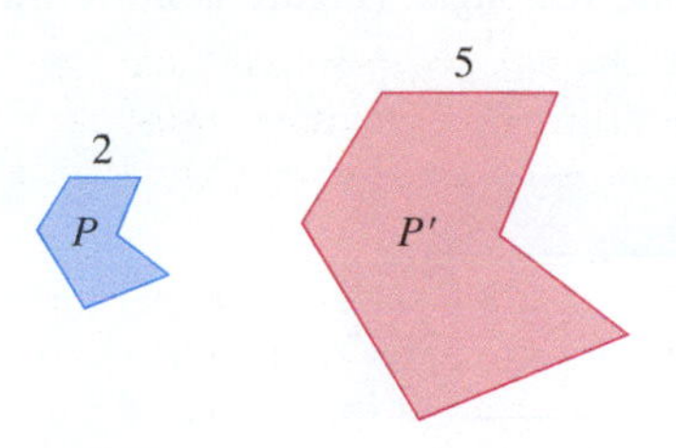

FIGURE 13-24

(a) If the perimeter of P is 10, what is the perimeter of P'?

(b) If the area of P is 30, what is the area of P'?

45. Find the value of x so that the shaded rectangle in Fig. 13-25 is a gnomon to the white 2 by 10 rectangle. (Figure is not drawn to scale.) [*Hint:* The aspect ratio of the entire rectangle is $(x+2)/10$. Match it with the aspect ratio of the white rectangle.]

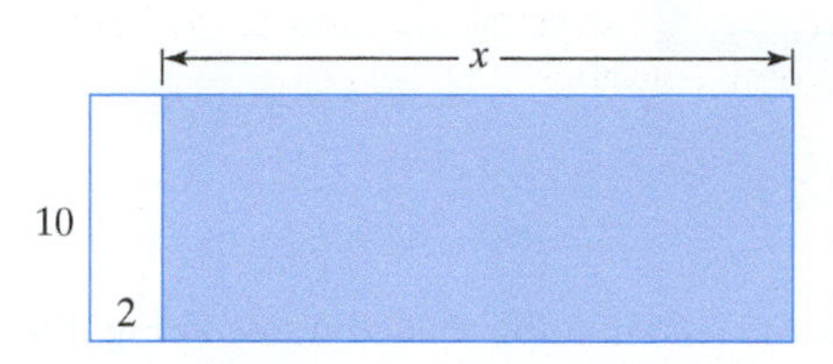

FIGURE 13-25

46. Find the value of x so that the shaded figure in Fig. 13-26 is a gnomon to the white rectangle. (Figure is not drawn to scale.)

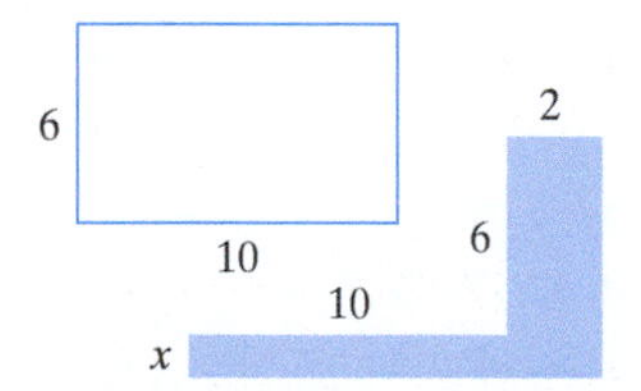

FIGURE 13-26

47. Find the value of x so that the shaded figure in Fig. 13-27 is a gnomon to the white rectangle. (Figure is not drawn to scale.)

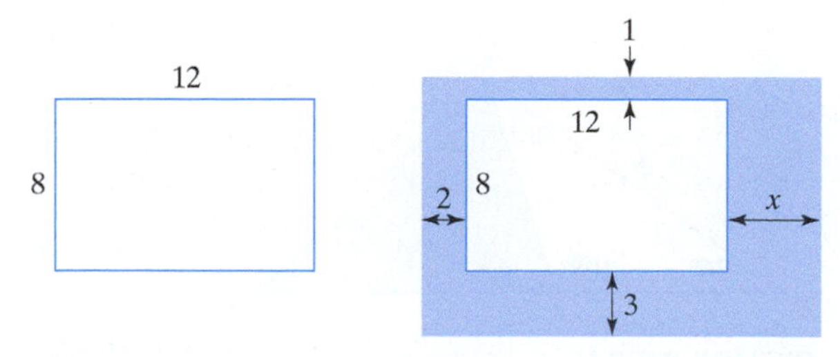

FIGURE 13-27

48. Find the value of x so that the shaded figure in Fig. 13-28 is a gnomon to the white rectangle. (Figure is not drawn to scale.)

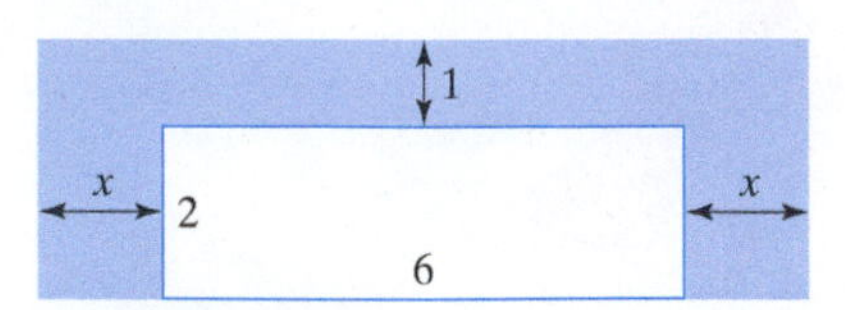

FIGURE 13-28

49. Find the value of x so that the shaded frame in Fig. 13-29 is a gnomon to the white x by 5 rectangle. (Figure is not drawn to scale.)

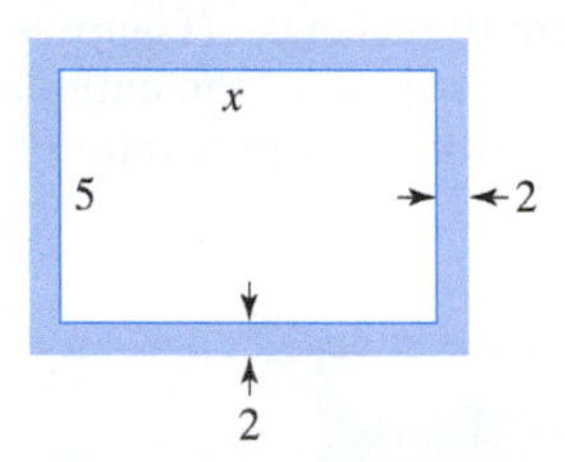

FIGURE 13-29

50. Find the value of x so that the shaded frame in Fig. 13-30 is a gnomon to the white x by 8 rectangle. (Figure is not drawn to scale.)

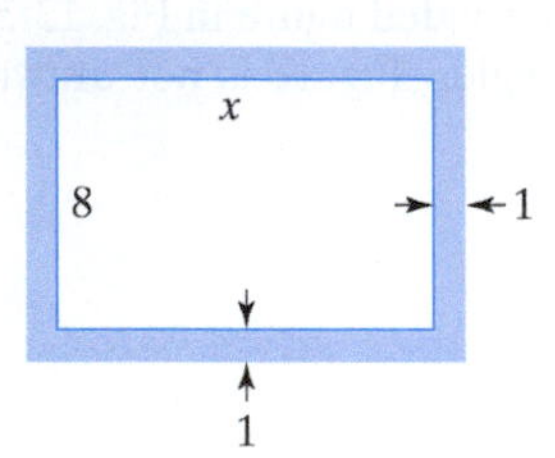

FIGURE 13-30

51. In Fig. 13-31 triangle BCA is a 36-36-108 triangle with sides of length ϕ and 1. Suppose that triangle ACD is a gnomon to triangle BCA.

(a) Find the measure of the angles of triangle ACD.

(b) Find the length of the three sides of triangle ACD.

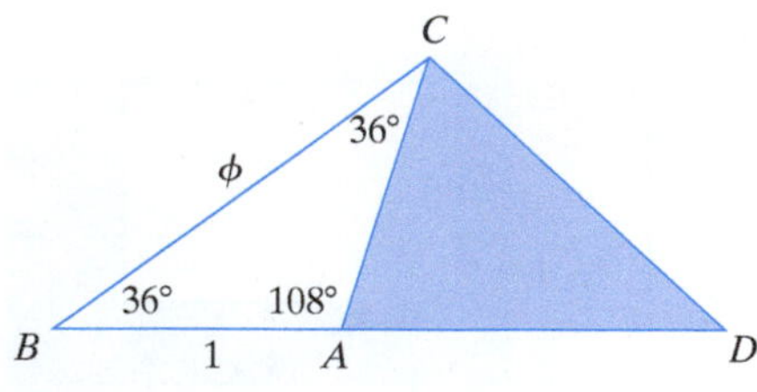

FIGURE 13-31

52. Find the values of x and y so that in Fig. 13-32 the shaded triangle is a gnomon to the white triangle ABC.

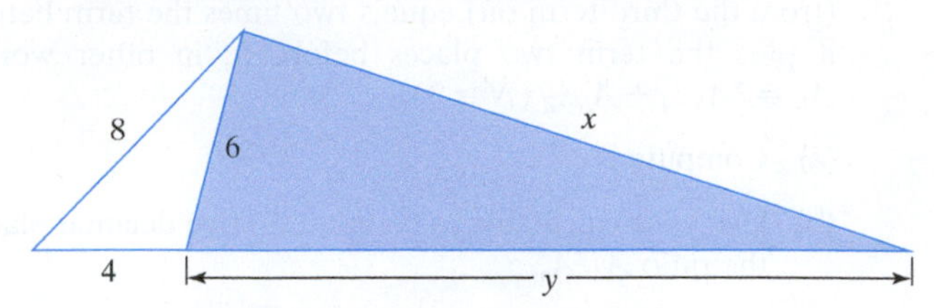

FIGURE 13-32

53. Find the values of x and y so that in Fig. 13-33 the shaded figure is a gnomon to the white triangle.

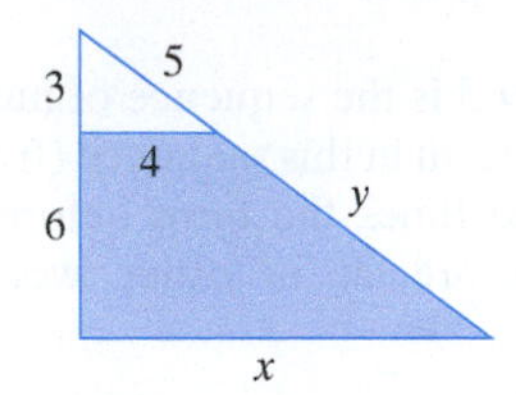

FIGURE 13-33

54. Find the values of x and y so that in Fig. 13-34 the shaded triangle is a gnomon to the white triangle.

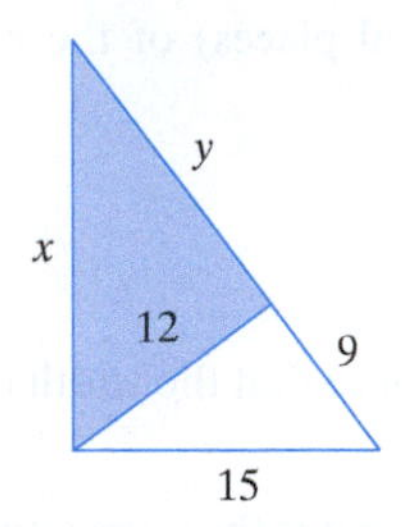

FIGURE 13-34

13.4 Spiral Growth in Nature

No exercises for this section.

JOGGING

55. Consider the sequence of ratios $\frac{F_N}{F_{N+1}}$.

(a) Using a calculator compute the first 14 terms of this sequence in decimal form (rounded to six decimal places when needed).

(b) Explain why $\left(\frac{F_N}{F_{N+1}}\right) \to \phi - 1$ (i.e., as N gets larger and larger, the ratios $\frac{F_N}{F_{N+1}}$ get closer and closer to $\phi - 1$).

56. Consider the sequence of ratios $\frac{F_{N+2}}{F_N}$.

(a) Using a calculator compute the first 15 terms of this sequence in decimal form (rounded to six decimal places when needed).

(b) Explain why $\left(\frac{F_{N+2}}{F_N}\right) \to \phi + 1$ (i.e., as N gets larger and larger, the ratios $\frac{F_{N+2}}{F_N}$ get closer and closer to $\phi + 1$).

57. Consider the sequence T given by the following recursive definition: $T_{N+1} = 1 + \frac{1}{T_N}$, and $T_1 = 1$.

(a) Find the first six terms of the sequence, and leave the terms in fractional form.

(b) Explain why $T_N \to \phi$ (i.e., as N gets larger and larger, T_N gets closer and closer to ϕ).

58. Consider the sequence U given by the following recursive definition: $U_{N+1} = \frac{1}{1 + U_N}$, and $U_1 = 1$. As N gets larger and larger, the terms of this sequence get closer and closer to some number. Give the number expressed in terms of the golden ratio ϕ. (*Hint*: Try Exercises 55 and 57 first.)

59. **Lucas numbers.** The *Lucas sequence* is the Fibonacci-like sequence 1, 3, 4, 7, 11, 18, 29, 47, . . . (first introduced in Exercise 23). The numbers in the Lucas sequence are called the *Lucas numbers*, and we will use L_N to denote the Nth Lucas number. The Lucas numbers satisfy the recursive rule $L_N = L_{N-1} + L_{N-2}$ (just like the Fibonacci numbers), but start with the initial values $L_1 = 1$, $L_2 = 3$.

(a) Show that the Lucas numbers are related to the Fibonacci numbers by the formula $L_N = 2F_{N+1} - F_N$. [*Hint*: Let $K_N = 2F_{N+1} - F_N$, and show that the numbers K_N satisfy exactly the same definition as the Lucas numbers (same initial values and same recursive rule).]

(b) Show that $\left(\frac{L_{N+1}}{L_N}\right) \to \phi$. [Hint: Use (a) combined with the fact that $\left(\frac{F_{N+1}}{F_N}\right) \to \phi$.]

60. (a) Explain what happens to the values of $\left(\frac{1-\sqrt{5}}{2}\right)^N$ as N gets larger. (*Hint*: Get a calculator and experiment with $N = 6, 7, 8, \ldots$ until you get the picture.)

(b) Explain why $F_N \to \frac{\phi^N}{\sqrt{5}}$. [Hint: Use (a) and the original Binet's formula.]

(c) Using (b), explain why $\left(\frac{F_{N+1}}{F_N}\right) \to \phi$.

61. Explain why the only even Fibonacci numbers are those having a subscript that is a multiple of 3.

62. Show that $F_{N+1}^2 - F_N^2 = (F_{N-1})(F_{N+2})$.

63. Rectangle A is 10 by 20. Rectangle B is a gnomon to rectangle A. What are the dimensions of rectangle B?

64. Explain why the shaded figure in Fig. 13-35 cannot have a square gnomon.

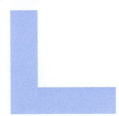

FIGURE 13-35

65. Suppose you have a picture of dimensions x by y and you put a frame of a fixed width w around the picture (Fig. 13-36). Is it possible for the frame to be a gnomon to the original picture? If so, under what circumstances? Explain your answer.

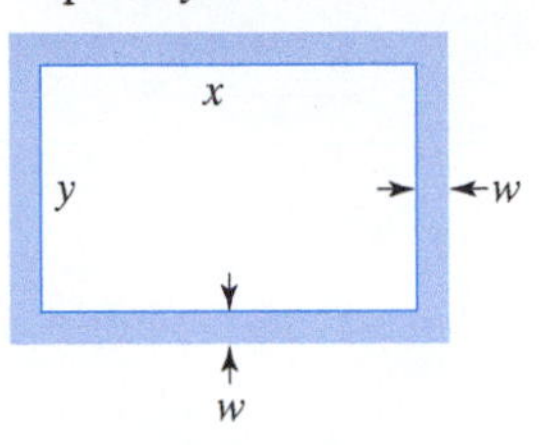

FIGURE 13-36

66. Find the values of x and y so that in Fig. 13-37 the shaded triangle is a gnomon to the white triangle. (Figure is not drawn to scale.)

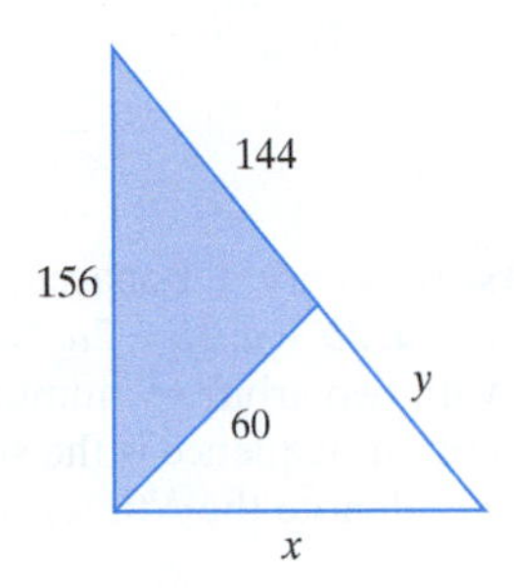

FIGURE 13-37

67. Let $ABCD$ be an arbitrary rectangle as shown in Fig. 13-38. Let AE be perpendicular to the diagonal BD and EF perpendicular to AB as shown. Show that the rectangle $BCEF$ is a gnomon to the rectangle $ADEF$.

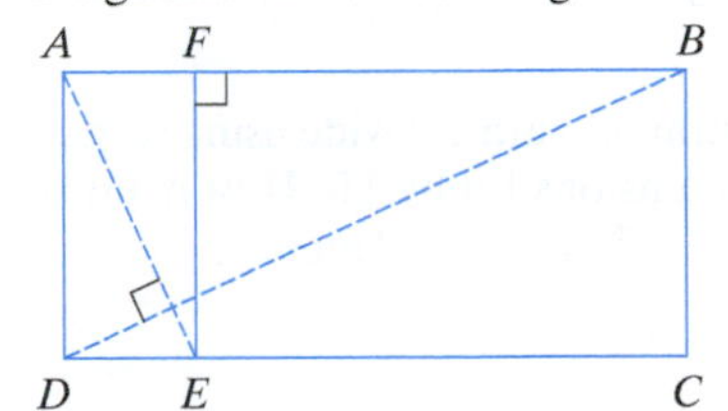

FIGURE 13-38

68. In Fig. 13-39 triangle BCD is a 72-72-36 triangle with base of length 1 and longer side of length x. (Using this choice of values, the ratio of the longer side to the shorter side is $x/1 = x$.)

(a) Show that $x = \phi$. (*Hint*: Triangle ACB is similar to triangle BCD.)

(b) What are the interior angles of triangle DAB?

(c) Show that in the isosceles triangle DAB, the ratio of the longer to the shorter side is also ϕ.

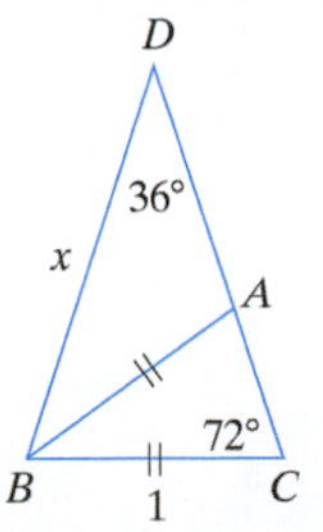

FIGURE 13-39

69. Show that each of the diagonals of the regular pentagon shown in Fig. 13-40 has length ϕ.

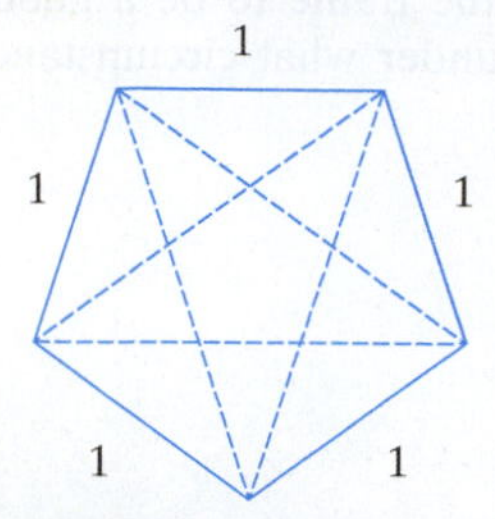

FIGURE 13-40

70. **(a)** A regular decagon (10 sides) is inscribed in a circle of radius 1. Find the perimeter in terms of ϕ.

(b) Repeat (a) with radius r. Find the perimeter in terms of ϕ and r.

RUNNING

71. **Generic Fibonacci-like numbers.** A generic Fibonacci-like sequence has the form $a, b, b+a, 2b+a, 3b+2a, 5b+3a, \ldots$ (i.e., the sequence starts with two arbitrary numbers a and b and after that each term of the sequence is the sum of the two previous terms). Let G_N denote the Nth term of this sequence.

(a) Show that generic Fibonacci-like numbers are related to the Fibonacci numbers by the formula $G_N = bF_{N-1} + aF_{N-2}$. [*Hint*: Try Exercise 59(a) first.]

(b) Show that $(G_{N+1}/G_N) \to \phi$. [*Hint*: Try Exercise 59(b) first.]

72. You are designing a straight path 2 ft wide using rectangular paving stones with dimensions 1 ft by 2 ft. How many different designs are possible for a path of length

(a) 4 ft?

(b) 8 ft?

(c) N ft? (*Hint*: Give the answer in terms of Fibonacci numbers.)

73. Show that $F_1 + F_2 + F_3 + \cdots + F_N = F_{N+2} - 1$.

74. Show that $F_1 + F_3 + F_5 + \cdots + F_N = F_{N+1}$. (Note that on the left side of the equation we are adding the Fibonacci numbers with odd subscripts up to N.)

75. Show that every positive integer greater than 2 can be written as the sum of distinct Fibonacci numbers.

76. In Fig. 13-41, $ABCD$ is a square and the three triangles I, II, and III have equal areas. Show that $x/y = \phi$.

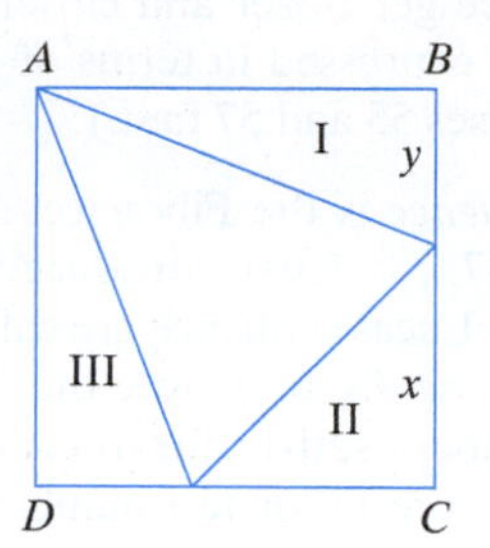

FIGURE 13-41

77. During the time of the Greeks the star pentagram shown in Fig. 13-42 was a symbol of the Brotherhood of Pythagoras. Consider the three segments of lengths x, y, and z shown in the figure.

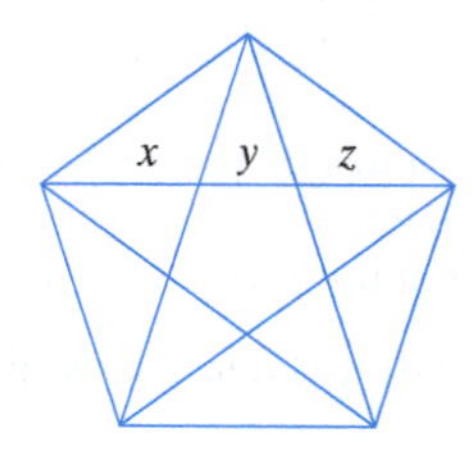

FIGURE 13-42

(a) Show that $x/y = \phi$, $(x+y)/z = \phi$, and $(x+y+z)/(x+y) = \phi$.

(b) Show that if $y = 1$, then $x = \phi$, $(x+y) = \phi^2$, and $x+y+z = \phi^3$.

Type of match	Payoff	Probability
5 + 1	$J	$1/({}_{69}C_5 \times 26)$
5 + 0	$1,000,000	$25/({}_{69}C_5 \times 26)$
4 + 1	$50,000	$({}_5C_4 \times {}_{64}C_1)/({}_{69}C_5 \times 26)$
4 + 0	$100	$({}_5C_4 \times {}_{64}C_1 \times 25)/({}_{69}C_5 \times 26)$
3 + 1	$100	$({}_5C_3 \times {}_{64}C_2)/({}_{69}C_5 \times 26)$
2 + 1	$7	$({}_5C_2 \times {}_{64}C_3)/({}_{69}C_5 \times 26)$
3 + 0	$7	$({}_5C_3 \times {}_{64}C_2 \times 25)/({}_{69}C_5 \times 26)$
1 + 1	$4	$({}_5C_1 \times {}_{64}C_4)/({}_{69}C_5 \times 26)$
0 + 1	$4	${}_{64}C_5/({}_{69}C_5 \times 26)$

PART 5 Statistics

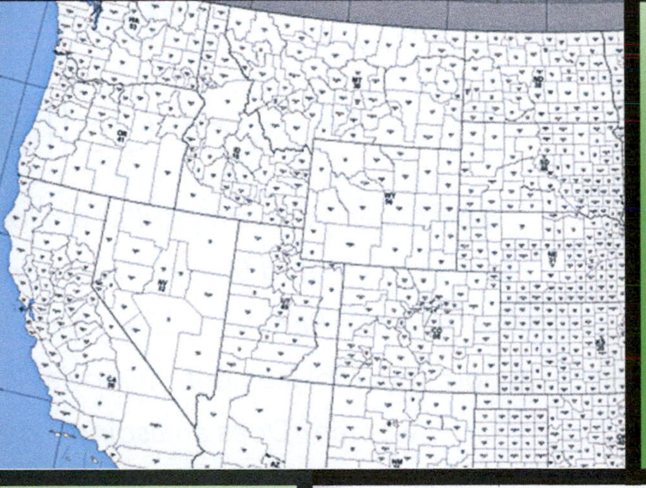

Heads	Frequency	Heads	Frequency	Heads	Frequency
33	3	45	467	57	301
34	4	46	589	58	233
35	9	47	661	59	155
36	14	48	719	60	108
37	30	49	774	61	71
38	49	50	802	62	45
39	67	51	777	63	26
40	114	52	736	64	16
41	159	53	672	65	9
42	224	54	579	66	4
43	305	55	487	67	3
44	401	56	387		

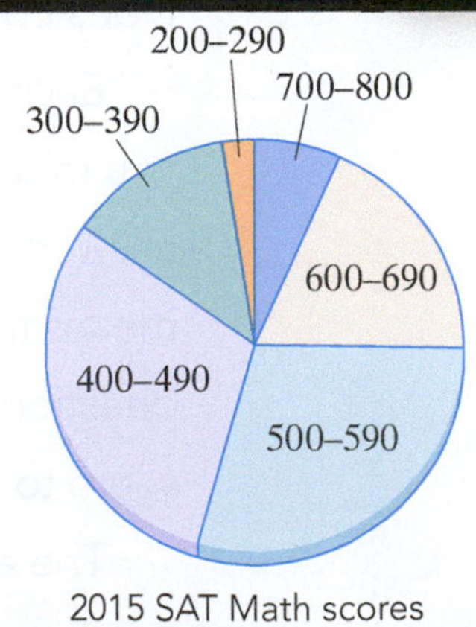

2015 SAT Math scores

Does drinking coffee help you live longer? Yes. No. Maybe. For details, see Example 14.12.

Censuses, Surveys, Polls, and Studies

The Joys of Collecting Data

How many mountain gorillas are left in the wild? How many people live in the United States? What percentage of the voters plan to vote for the Republican Party candidate in the next election? Is hormone replacement therapy a risk factor for breast cancer in older women? Does an aspirin a day help prevent heart disease? Does coffee make you live longer?

Each one of these questions has an answer, and if you dig around you might get it. And if you dig a little harder you might find other answers, and most likely, the answers will contradict each other. Which means that often, the answers you get are wrong, or at the very least, questionable. But why? Why can't we get definitive answers to what seem to be clearly answerable questions?

The answer is in the data.

Right or wrong, there are some questions in life that cannot be answered without data. Data are the clay with which we build the statistical bricks that inform much of our modern world. The purpose of collecting data is to make hypotheses, draw conclusions, and find explanations for what is happening around us. Our conclusions and explanations are only as good as the data we collect, and collecting good reliable data is a lot harder than it seems.

> "Data! Data! Data!" he cried impatiently. "I can't make bricks without clay."
>
> – *Sherlock Holmes*

An interesting data collection story lurks behind every piece of statistical information. How was the data collected? By whom? Why? When? For how long? Who funded the effort? In this chapter we will explore a few basic ideas and issues in data collection. Most of the relevant issues and ideas are introduced through real-life examples and case studies. The chapter has lots of stories and no mathematical formulas (well, maybe two little ones).

One of the most basic purposes of data collection is *enumeration*: "How many members of a given population are there?" (for example, "How many people are currently residing in the United States?") or "How many members of a given population share a particular feature?" (for example, "How many of the people currently residing in the United States are college graduates?"). Answers to these types of questions can be answered by a census, or estimated by using samples. We discuss this topic in Section 14.1. In Section 14.2 we discuss *measurement* questions. These are questions for which some characteristic or behavior must be measured, including opinions and intentions. Typically, these questions are answered through surveys and public opinion polls. The most difficult data collection questions are those involving cause and effect: Does *X* cause *Y*? These are the types of questions that affect us most—diet, medications, exercise, behavior, are all stepchildren of cause and effect. We discuss data collection methods for cause and effect questions in Section 14.3.

14.1 Enumeration

> enu•mer•ate: to ascertain the number of . . . .
>
> – *Merriam-Webster Dictionary**

The most basic enumeration questions are questions of the form *How many . . . are there?* To have a definitive, unequivocal answer to a question of this type, it is essential that the set of "things" being counted—called the *population*—be clearly defined.

- **Population.** This is the term we use to describe the set of "individuals" being *enumerated* (i.e., counted). The individuals may be people, animals, plants, or inanimate objects. A population is well-defined when given an individual we can clearly determine if the individual belongs to the population (and, therefore, should be counted) or not.

Typically, the members of the population are indivisible things, so the answer to an enumeration question of the type *How many . . . are there?* is a positive integer or zero (for enumeration purposes we don't count parts of a person or an animal, and even inanimate objects are counted in terms of whole units). The standard convention is to use the letter *N* to denote the size of a population and to refer to a population count as the *N-value* of the population.

- ***N*-value.** A positive integer (or zero) that is alleged to be a count of the number of individuals in a given population.

> “cen•sus: a usually complete enumeration of a population . . . . .”
> – Merriam-Webster Dictionary*

There are essentially two ways to find the *N*-value of a given population: do a *complete head count* (this is just a commonly used metaphor—we can do “head counts” even for objects that don’t have heads), or do only a *partial head count* and use this partial information to estimate the *N*-value of the population. The first approach is called a **census**; the second approach is called (in very broad terms) a **survey**. We will illustrate both of these strategies next.

Censuses for Enumeration

EXAMPLE 14.1 HUNTING FOR BUFFALO (NICKELS)

A buffalo nickel is a historic, collectible coin whose value is definitely more than the five-cent face value of the coin—depending on its condition and the year it was minted, a buffalo nickel can be worth anywhere from a few dollars to several thousand dollars. If you have some in your possession, it makes sense to set them aside and not comingle them with ordinary nickels.

Imagine that you have a very large coin jar where you have thrown your spare change over the years, and you know for a fact that there are some buffalo nickels in that jar (you remember throwing some in the jar when you were a kid). Imagine that you have some spare time and you need some extra cash, so you decide to hunt down the buffalo nickels, enumerate them, and eventually sell them to a coin dealer.

What is the population for this example? As is often the case, here the population is defined by the purpose of the enumeration: Since you are only interested in the buffalo nickels, they constitute the population of interest, and the number of buffalo nickels in the jar is the *N*-value you are after. (Had you been interested in how much money is in the jar, the population of interest would be the entire set of coins in the jar, with a separate tally for nickels, dimes, quarters, etc.)

To find your *N*-value, your only reasonable option in this situation is to conduct a *census* of the buffalo nickels in your coin jar. This may be a little time-consuming, but not hard—you are counting inanimate objects, and unlike humans and animals (more on that soon) inanimate objects are quite cooperative when it comes to a head count. To conduct your census you dump all the coins out of the jar and go through them one by one. Each time you find a buffalo nickel you set it aside and add one to your tally. When you have gone through all the coins in the jar you have completed the census. Ka-ching! ■

Censuses become considerably more complicated when you are enumerating animal populations. Unlike buffalo nickels or other inanimate species, most animal species live in large and tough habitats, move around, hide from humans, and some species—tigers, for example—like to snack on the people trying to enumerate them. Moreover, a full census for an entire animal species requires a lot of planning and tends to be quite expensive. Currently, efforts to accurately enumerate the members of a species by means of a census are very rare and reserved for species that are critically endangered. One such species is the mountain gorilla.

EXAMPLE 14.2 HOW MANY MOUNTAIN GORILLAS ARE LEFT IN THE WILD?

Mountain gorillas (*Gorilla beringei beringei*) in the wild can only be found in central Africa, in a relatively small natural habitat consisting of four contiguous national parks (one in Rwanda, two in Uganda and one in the Democratic Republic of Congo).

*By permission. From *Merriam-Webster’s Collegiate® Dictionary, 11th Edition* ©2016 by Merriam-Webster, Inc. (*www.Merriam-Webster.com*).

Due to loss of habitat and poaching, the mountain gorilla population had been in decline for decades and became (and still is) critically endangered— in 1985 there were less than 300 mountain gorillas left living in the wild. Since then, thanks to a world-wide conservation effort (nudged by the success of the movie *Gorillas in the Mist*), poaching has been greatly reduced and the mountain gorilla population is slowly coming back. A census conducted in 2011 had a total count of 880 mountain gorillas living in the wild. A new census (the most detailed and comprehensive so far) is being conducted at the time of the writing of this chapter (June, 2016); the final results are due to come out in 2017.

To understand how much effort goes into an animal population census, consider the details of the 2011 mountain gorilla census.

- The census was conducted by seven different teams of researchers. Each team was responsible for conducting a full sweep of a specific region within the gorilla's natural habitat.
- Because mountain gorillas move around in small groups, the strategy was for a team to follow behind a particular group for three days and enumerate the individuals in the group.
- To enumerate the individuals in a group, the research team would use their nests as a *proxy*. This means that, rather than counting the gorillas directly (that would require following the group too closely), the team would count the number of nests left behind (each gorilla builds his or her own new nest to sleep in each night).
- To make sure that the counts were completely accurate, samples of fecal matter (i.e., gorilla poop) were collected from each nest for later analysis. The poop sample serves as a *second proxy*—it provides the DNA markers that uniquely identify each individual. To insure accuracy, samples were collected from nests on three consecutive nights.
- It took about nine months after the field work was completed to analyze the collected samples, organize the data, and release the final count: $N = 880$.

One would think that accurately enumerating human populations should be a lot easier than enumerating animals in the wild, but this is not always the case. What is the resident population of the United States? We have a rough idea but not an exact count, in spite of the fact that we spend billions of dollars trying.

EXAMPLE 14.3 HOW MANY PEOPLE LIVE IN THE UNITED STATES?

This is an example of an enumeration question whose answer changes by the minute, and yet there is a current, up-to-date answer that you can easily find: Go to *www.census.gov/popclock*. This Web site is the virtual version of the United States population clock maintained by the U.S. Census Bureau. (Check it out now and see what the current number is!) To be clear about what that number represents: It is the official N-value for the *resident population* of the United States, and it enumerates everyone—citizens, permanent residents, illegal aliens, visitors, and tourists—physically present in the United States at that moment in time.

How is the Census Bureau able to keep track of the N-value of a population that essentially changes by the minute? There are two parts to the answer: (1) establishing a baseline, and (2) using that baseline as the initial population together with a mathematical model that projects how that population changes over time. We discussed part (2) in detail in Chapter 9, Example 9.8 (*A Short-Term Model of the U.S. Population*), so in this example we will focus on describing how the baseline is set. This is where *the Census* comes in.

Every 10 years, the United States conducts a census of its population, officially called the United States Population Census but most commonly referred to (at least in the United States) as *the Census*. While most countries conduct periodic censuses of their populations, the United States is the only country where the population census is required by the country's constitution.

> "... this was an outstanding census. When this fact is added to prior positive evaluations, the American public can be proud of the *2010 Census* their participation made possible."
>
> – *Robert Graves, Director of the Census Bureau*

Accurately enumerating a population as large and diverse as that of the United States is practically impossible, but we try it anyway (it is required by law). The amount of effort and money spent on this enumeration is staggering. The 2010 U.S. Census involved hiring 635,000 enumerators and cost around $13 billion—roughly $42 per head counted. (We will describe in more detail the inner workings of the U.S. Census in Case Study 1 at the end of this section.)

The final official tally for the 2010 Census: On April 1, 2010 (Census Day), the resident population of the United States was $N = 308{,}745{,}538$. While nobody expected this number to be right on, post-enumeration surveys (using a more sophisticated methodology than just counting heads) showed that the official count was off by roughly 36,000 heads too many. This may seem like a lot of heads to overcount after spending that much money, but it represents a relative error of just 0.012%. By traditional U.S. Census standards the 2010 Census was a big success. ■

Surveys for Enumeration

In this chapter we use the word *survey* in its broadest meaning: Any strategy that uses a *sample* from a *population* to draw conclusions about the entire population is a survey (even when there are no questionnaires to fill or phone calls to answer). When the nurse draws a sample of blood from your arm and the sample is used to measure your white cell count, in the broadest sense of the word she is in fact conducting a survey. The working assumption is that the blood drawn from your arm is a good representative of the rest of your blood.

One way to think of a survey is as the flip side of a census—a census involves *every* individual in the population, a survey only involves *some* individuals. Before we look at some examples of how surveys are used to estimate population counts, we introduce a few useful terms.

- **Survey.** A data collection strategy that uses a sample to draw inferences about a population.
- **Sample.** A subset of the population chosen to be the providers of information in a survey.
- **Sampling.** The act of selecting a sample.
- **Statistic.** A numerical estimate of some measurable characteristic of a population obtained from a sample.
- **Parameter.** A true measurement of some characteristic of a population. In general, a *parameter* is an unknown quantity, and a *statistic* is an educated guess as to what that unknown quantity might be.

Surveys are used to measure characteristics of a population or to predict the future actions of a group of individuals (we will cover these uses in the next section), but surveys can also be used to estimate the N-value of a population. We will discuss this particular use of surveys in the next couple of examples.

EXAMPLE 14.4 DEFECTIVE LIGHTBULBS: ONE-SAMPLE ESTIMATION

Have you ever bought an electronic product, taken it home, and found out it is defective? When the product is a big-ticket item like a computer or a TV, buying a lemon is especially frustrating; but, even for minor items like a lightbulb, it can be aggravating (you climb on a step ladder, remove the old lightbulb, screw in the new lightbulb, flip the switch, no light, @%#!).

To minimize the problem of factory defectives, manufacturers do quality control tests on their products before shipping. For big-ticket items, every single item is factory tested for quality control, but for cheaper items like lightbulbs this is not cost-effective, so the quality control testing is done using a sample.

Imagine you are quality control manager for a lightbulb manufacturer. The lightbulbs come out of the assembly line in batches of 100,000. The manufacturer's specs are that if the batch has more than 2500 defective bulbs (2.5%) the batch cannot be shipped. In this situation the population of interest consists of the defective lightbulbs in the batch, but testing the entire batch of 100,000 to enumerate the defectives is too time-consuming and expensive. Instead, you have a sample of 800 selected and tested. Out of the 800 lightbulbs tested, 17 turn out to be defective.

We can now estimate the total number of defective lightbulbs in the batch by assuming that *the proportion of defectives in the entire batch is approximately the same as the proportion of defectives in the sample.* In other words, $\frac{17}{800} \approx \frac{N}{100{,}000}$. Solving for N gives an estimate (i.e., a statistic) of $N \approx 2125$ defectives in the batch, well under the allowed quota of 2500. We will never know the exact number of defective lightbulbs in the batch (the parameter) unless we test the entire batch, but we feel confident that this shipment is good to go. (We will revisit the issue of confidence in our statistic in Chapter 17, but that's another story.) ■

The underlying assumption of the sampling method used in Example 14.4 [sometimes called *one-sample* (or *single-sample*) *estimation*] is that the percentage of defectives in the sample is roughly the same as the percentage of defectives in the entire batch. To insure that this assumption is valid, the sample has to be carefully chosen and be sufficiently large. We will discuss what "carefully chosen" and "sufficiently large" mean in greater detail in Section 14.2.

The general description of **one-sample estimation** is as follows: Suppose we have a general population of known size P (in our last example, $P = 100{,}000$ lightbulbs) and we want to find the N-value of a subpopulation having some specified characteristic (for example, being a defective lightbulb). We can estimate this N-value by carefully choosing a sample of size n (in our last example $n = 800$) and counting the number of individuals in that sample having that particular characteristic. Call that number k (in our last example, $k = 17$). Then, if we assume that the percentage of individuals with the desired characteristic is roughly the same in both the sample and the general population, we get $\frac{k}{n} \approx \frac{N}{P}$. Solving for N gives $N \approx \left(\frac{k}{n}\right)P$.

We can only estimate the N-value of a subpopulation using a sample if we know P, the size of the general population. In many real-life applications we don't. There is, however, an extremely useful sampling method that allows us to estimate the N-value of any population called *two-sample estimation*, and more commonly as the *capture-recapture* method. The most common application of this method is in the study of fish and wildlife populations (thus the name *capture-recapture*), and it involves taking two consecutive samples of the population (thus the name *two-sample estimation*). We will illustrate the method with an example first and then give a general description.

EXAMPLE 14.5 COUNTING FISH POPULATIONS: CAPTURE-RECAPTURE

Imagine you are a wildlife biologist studying the ecosystem in a small lake. A major player in this lake is the northern pike—a voracious carnivore that eats the hatchlings of other fish such as trout and salmon. To maintain a healthy ecosystem, the population of pike has to be controlled, and this requires taking regular measurements of the N-value of the pike population in the lake. Here's how you do this.

- **Step 1 (the capture).** You capture the first sample of pike, tag them (gently—you want to avoid harming them in any way), and release them back into the lake. Let's say in this case the first sample consists of $n_1 = 200$ pike. Once you release the tagged pike back into the lake, you can assume that the percentage of tagged pike in the lake is given by the ratio $\frac{200}{N}$ (remember, N is your unknown).
- **Step 2 (the recapture).** After waiting awhile (you want the released fish to disperse naturally throughout the lake) you capture a second sample of pike and count the number of pike in the second sample that have tags. Let's say that the second sample consisted of $n_2 = 150$ pike, of which $k = 21$ had tags. This means that the percentage of tagged pike in the sample is given by the ratio $\frac{21}{150}$ (we won't worry about the computations yet).

The working assumption now is that *the percentage of tagged pike in the sample is roughly the same as the percentage of tagged pike in the lake.* In other words, $\frac{21}{150} \approx \frac{200}{N}$. Solving for N gives $N \approx \frac{200 \times 150}{21} = 1428.57$. Since the number of pike in the lake has to be a whole number we round the estimate to $N = 1429$.

The underlying assumption of the capture-recapture method is that both samples (the capture sample and the recapture sample) are good representatives of the entire population, and for this to happen, several requirements have to be met:

(1) The chances of being captured are the same for all members of the population.

(2) The chances of being recaptured are the same for both tagged and untagged individuals.

(3) The general population remains unchanged between the capture and the recapture (i.e., no births, deaths, or escapes).

(4) The tags do not come off.

When these requirements are met (or mostly met), the capture-recapture method gives very good estimates of a population size.

CAPTURE-RECAPTURE METHOD

- **Step 1 (the capture).** Choose a sample of the population consisting of n_1 individuals. "Tag" the individuals in the sample. Release the tagged individuals back into the general population.

  ["Tag" is a metaphor for giving the individuals an identifying mark—it could be anything from an ink dot on a dorsal fin to an asterisk on a spreadsheet. The two most important things about the tag are that (a) it should not harm the individual and (b) it should not come off in the time between the capture and the recapture.]
- **Step 2 (the recapture).** Choose a second sample of the population consisting of n_2 individuals and count the number of tagged individuals in the second sample. Let k denote the number of tagged individuals in the second sample.
- **Step 3.** Set up the proportion $\frac{k}{n_2} \approx \frac{n_1}{N}$, and solve for N. When the value of N is not a whole number, round it to the nearest whole number.

The capture-recapture method has many other applications beyond wildlife ecology. It is used in epidemiology to estimate the number of individuals infected with a particular disease, and it is used in public health to estimate birth rates and death rates in a particular population.

We conclude this section with Case Study 1, a general discussion of the United States Census, the role it plays in our lives, and the statistical issues that are involved.

CASE STUDY 1 THE UNITED STATES CENSUS

Article 1, Section 2, of the U.S. Constitution mandates that a national census be conducted every 10 years. The original purpose of the national census was to "count heads" for a twofold purpose: *taxes* and *political representation* (i.e., the apportionment of seats in the House of Representatives to the states based on their populations—a topic we discussed at length in Chapter 4). Like many other parts of the Constitution, Article 1, Section 2, was a compromise of many conflicting interests: The census count was to exclude "Indians not taxed" and to count slaves as "three-fifths of a free Person."

> "[An] Enumeration shall be made within three Years after the first Meeting of the Congress of the United States, and within every subsequent Term of ten Years, in such Manner as they [Congress] shall by Law direct."
>
> *– Article 1, Section 2, U.S. Constitution*

Today, the scope of the U.S. Census has been greatly expanded by the Fourteenth Amendment and by the courts to count the full "resident population" of the United States on Census Day (April 1 of every year that ends with 0). The modern census does a lot more than give the national population count: It provides a complete breakdown of the population by state, county, city, etc.; it collects demographic information about the population (gender, age, ethnicity, marital status, etc.); and it collects economic information (income, employment, housing, etc.).

The data collected by the U.S. Census is considered a vital part of the nation's economic and political infrastructure—it's impossible to imagine the United States functioning without the census data. Among other things, the census data is used to

- apportion the seats in the House of Representatives,
- redraw legislative districts within each state,
- allocate federal tax dollars to states, counties, cities, and municipalities,
- develop plans for the future by federal, state, and local governments,
- collect vital government data such as the Consumer Price Index and the Current Population Survey, and
- develop strategic plans for production and services by business and industry.

All the data collected by the Census is obtained through some form of sampling (they are all statistics, rather than parameters) except for one—the national and individual state populations are required by law to be collected by means of a full census. (So ruled the Supreme Court in 1999 in *Department of Commerce et al. v. United States House of Representatives et al.*)

Table 14-1 shows the population count from the last three national censuses, the approximate cost of each, the cost per head counted, and the estimated error (undercount/overcount) in each case (+ indicates an overcount, – an undercount). All it takes is a look at Table 14-1 to see that the U.S. Census is an incredibly expensive data collection effort that (except for the 2010 Census) does not produce very good data and misses a lot of people. (Ironically, to obtain a more accurate count and estimate the error in the official count, the Census Bureau uses a *post-enumeration* survey based on the capture-recapture method—cheaper, more efficient, and more accurate). So, why do we continue spending billions on a flawed data collection system? Good question

Year	N	Cost	Cost per person	Error
1990	248,709,873	~$2.5 billion	~$10	−4 million
2000	281,421,906	~$4.5 billion	~$16	−1.4 million
2010	308,745,538	~$13 billion	~$42	+36,000

TABLE 14-1 ■ Cost and performance of U.S. Census (last three censuses)

14.2 Measurement

Finding the *N*-value of a population is, in a sense, the simplest data collection problem there is (not to say that it is easy, as we have learned by now). In most situations we don't need to perform any measurements or ask any questions—we just count (either the sample or the entire population). Things become quite a bit trickier when the information we want requires more than just counting heads: What is the incoming freshman class GPA at the State University? What is the median home price in California? What is the level of customer satisfaction with a company's customer service? What percentage of voters will vote Republican in the next election?, etc. The data needed to answer these types of questions require either some sort of *measurement* (GPA's, home prices) or some form of question/answer interaction (*Overall, how satisfied or dissatisfied are you with our company's customer service? If the election were held today, would you vote for the Republican Party's candidate?* etc.). We will use the term **measurement problem** to describe both situations (the point being that asking questions to assess people's opinions or intentions is also a form of measurement).

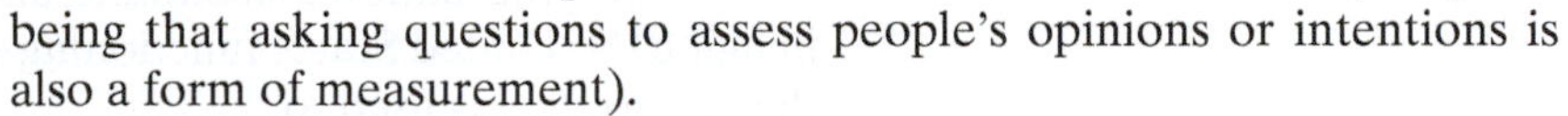

The typical way to answer a measurement problem is to conduct a survey (called a **poll** when the measurements require asking questions and recording answers). The three basic steps in any survey or poll are: (1) sampling (i.e., choosing the sample—or samples if more than one sample is required), (2) "measuring" the individuals in the sample, and (3) drawing inferences about the population from the measurements in the sample. Each of these three steps sounds simple enough, but there is a lot of devil in the details. For the rest of this chapter we will focus on step 1. We will discuss steps 2 and 3 in Chapters 15 and 17.

Sampling

The basic philosophy behind sampling is simple and well understood—if you choose a "good" sample you can get reasonably reliable data about the population by measuring just the individuals in the sample (you can never draw perfectly accurate data from a sample). Conversely, if your sample is not good (statisticians call it *biased*), then any conclusions you draw from the sample are unreliable.

A good sample is one that is "representative" of the population and is large enough to cover the variability in the population. The preceding sentence is a bit vague to say the least, and we will try to clarify its meaning in the remainder of this section, but let's start with the issue of sample size.

When the population is highly homogeneous, then a small sample is good enough to represent the entire population. Take, for example, blood samples for lab testing: All it takes is a small blood sample drawn from an arm to get reliable data about all kinds of measures—white and red cell counts, cholesterol levels, sugar levels, etc. A small blood sample is a good enough sample because a person's blood

> "Whether you poll the United States or New York State or Baton Rouge . . .you need . . . the same number of interviews or sample [size]. It's no mystery really—if a cook has two pots of soup on the stove, one far larger than the other, and thoroughly stirs them both, he doesn't have to take more spoonfuls from one than the other to sample the taste accurately."
>
> *– George Gallup*

is essentially the same throughout the body. At the other end of the spectrum, when a population is very heterogeneous, a large sample is necessary (but not sufficient) if the sample is going to represent all the variability in the population. What's important to keep in mind is that it's not the size of the population that dictates the size of a good sample but rather the variability in the population. That's why a sample of 1500 people can be good enough to represent the population of a small city, a large city, or the whole country.

It is a customary to use n to denote the size of a sample (to distinguish it from N, the size of the population). The ratio n/N is called the **sampling proportion**, and is typically expressed as a percentage.

EXAMPLE 14.6 DEFECTIVE LIGHTBULBS: SAMPLING PROPORTION

In Example 14.4 we used a sample of $n = 800$ lightbulbs taken from a batch of $N = 100{,}000$ to determine the number of defective lightbulbs in the batch. In this example we are lucky—we know both n and N and we can compute the sampling proportion easily: $n/N = 0.008 = 0.8\%$.

While a sampling proportion of 0.8% may seem very small (and in many applications it would be), it is more than adequate in this situation. Lightbulbs are fairly homogeneous objects (no moving parts and just a very few elements that could go bad), so a sample of $n = 800$ is large enough for this population. A more relevant question is, How was the sample chosen?

Choosing a "good" sample is the most important and complex part of the data collection process. But how do we know if a sample is a "good" sample or not? The key lies in the idea of "equal opportunity"—we want every member of the population to have an equal chance of being included in the sample. When some members of the population are less likely to be selected for the sample than others (even if it is unintentional) we have a **biased sample** (or **sampling bias**), and a biased sample is an unreliable sample. To restate the point, in a survey, reliable measurements of a population are only possible when using an *unbiased* sample.

Depending on the nature of the population, choosing an unbiased sample can be difficult, sometimes impossible. There are many reasons why a sample can be biased, and we will illustrate some of them next.

Selection Bias

When the selection of a survey sample has a systematic tendency (whether intentional or not) to consistently favor some elements or groups within the population over others, we say that the survey suffers from **selection bias**.

One significant example of selection bias occurs when the *sampling frame* for a survey is different from the *target population*. As the name indicates, the **target population** is the population to which the conclusions of the survey apply—in other words the population the survey is talking about. The **sampling frame** is the population from which the sample is drawn. Ideally, the two should be the same but sometimes they are not, and in that case there will be individuals in the population that have zero chance of being selected for the sample (Fig. 14-1).

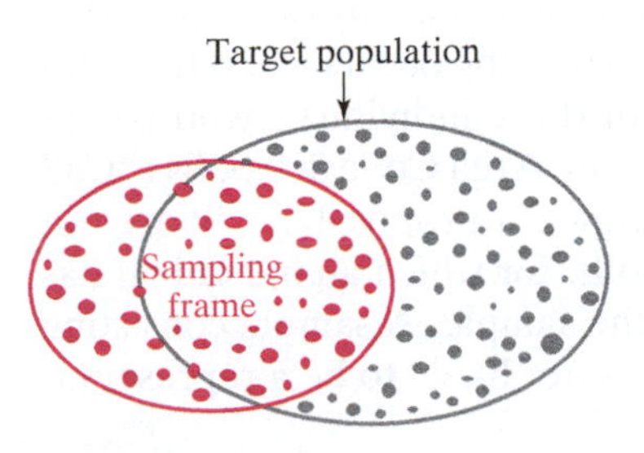

FIGURE 14-1 The sample is selected from the red part of the population. Gray individuals have zero chance of being in the sample.

The distinction between the target population and the sampling frame presents a very significant problem for polls that try to predict the results of an election. The target population for such a poll consists of the people that are going to vote in the election, but how do you identify that group? The conventional approach is to use registered voters as the sampling frame, but using registered voters does not always work out very well, as shown in the next example.

EXAMPLE 14.7 PRE-ELECTION POLLS: REGISTERED V. LIKELY VOTERS

A CNN/USA Today/Gallup poll conducted right before the November 2, 2004, national election asked the following question: "If the election for Congress were being held today, which party's candidate would you vote for in your congressional district, the Democratic Party's candidate or the Republican Party's candidate?"

When the question was asked of 1866 *registered* voters nationwide, the results of the poll were 49% for the Democratic Party candidate, 47% for the Republican Party candidate, 4% undecided.

When exactly the same question was asked of 1573 *likely* voters nationwide, the results of the poll were 50% for the Republican Party candidate, 47% for the Democratic Party candidate, 3% undecided.

Clearly, one of the two polls had to be wrong, because in the first poll the Democrats beat out the Republicans, whereas in the second poll it was the other way around. The only significant difference between the two polls was the choice of the sampling frame—in the first poll the sampling frame consisted of *all registered voters*, and in the second poll the sampling frame consisted of *all likely voters*. Although neither one faithfully represents the target population of *actual voters*, using likely voters instead of registered voters for the sampling frame gives much more reliable data. (The second poll predicted very closely the average results of the 2004 congressional races across the nation.)

So, why don't all pre-election polls use likely voters as a sampling frame instead of registered voters? The answer is economics. Registered voters are relatively easy to identify—every county registrar can produce an accurate list of registered voters. Not every registered voter votes, though, and it is much harder to identify those who are "likely" to vote. (What is the definition of *likely* anyway?) Typically, one has to look at demographic factors (age, ethnicity, etc.) as well as past voting behavior to figure out who is likely to vote and who isn't. Doing that takes a lot more effort, time, and money.

Convenience Sampling

There is always a cost (effort, time, money) associated with collecting data, and it is a truism that this cost is proportional to the quality of the data collected—the better the data, the more effort required to collect it. It follows that the temptation to take shortcuts when collecting data is always there and that data collected "on the cheap" should always be scrutinized carefully. One commonly used shortcut in sampling is known as **convenience sampling**. In convenience sampling the selection of which individuals are in the sample is dictated by what is easiest or cheapest for the people collecting the data.

A classic example of convenience sampling is when interviewers set up at a fixed location such as a mall or outside a supermarket and ask passersby to be part of a public opinion poll. A different type of convenience sampling occurs when the sample is based on self-selection—the sample consists of those individuals who volunteer to be in it. *Self-selection bias* is the reason why many Area Code 800 polls ("Call 1-800-YOU-NUTS to express your opinion on the new tax proposal . . . ") are not to be trusted. Even worse are the Area Code 900 polls, for which an individual has to actually pay (sometimes as much as $2) to be in the sample. A sample consisting entirely of individuals who paid to be in the sample is not likely to be a representative sample of general public opinion.

Convenience sampling is not always bad—at times there is no other choice or the alternatives are so expensive that they have to be ruled out. We should keep in mind, however, that data collected through convenience sampling are naturally tainted and should always be scrutinized (that's why we always want to get to the details of *how* the data were collected).

Our next case study is a famous case in the history of pre-election polls, and it illustrates the many things that can go wrong when a biased sample is selected—even when it is a very large sample.

CASE STUDY 2 THE 1936 *LITERARY DIGEST* POLL

The U.S. presidential election of 1936 pitted Alfred Landon, the Republican governor of Kansas, against the incumbent Democratic President, Franklin D. Roosevelt. At the time of the election, the nation had not yet emerged from the Great Depression, and economic issues such as unemployment and government spending were the dominant themes of the campaign.

The *Literary Digest*, one of the most respected magazines of the time, conducted a poll a couple of weeks before the election. The magazine had used polls to accurately predict the results of every presidential election since 1916, and their 1936 poll was the largest and most ambitious poll ever. The *sampling frame* for the *Literary Digest* poll consisted of an enormous list of names that included (1) every person listed in a telephone directory anywhere in the United States, (2) every person on a magazine subscription list, and (3) every person listed on the roster of a club or professional association. From this sampling frame a list of about 10 million names was created, and every name on this list was mailed a mock ballot and asked to mark it and return it to the magazine.

One cannot help but be impressed by the sheer scope and ambition of the 1936 *Literary Digest* poll, as well as the magazine's unbounded confidence in its accuracy. In its issue of August 22, 1936, the *Literary Digest* crowed:

> *Once again, [we are] asking more than ten million voters—one out of four, representing every county in the United States—to settle November's election in October.*
>
> *Next week, the first answers from these ten million will begin the incoming tide of marked ballots, to be triple-checked, verified, five-times cross-classified and totaled. When the last figure has been totted and checked, if past experience is a criterion, the country will know to within a fraction of 1 percent the actual popular vote of forty million [voters].*

Based on the poll results, the *Literary Digest* predicted a landslide victory for Landon with 57% of the vote, against Roosevelt's 43%. Amazingly, the election turned out to be a landslide victory for Roosevelt with 62% of the vote, against 38% for Landon. The difference between the poll's prediction and the actual election results was a whopping 19%, the largest error ever in a major public opinion poll. The results damaged the credibility of the magazine so much so that soon after the election its sales dried up and it went out of business—the victim of a major statistical blunder.

For the same election, a young pollster named George Gallup was able to predict accurately a victory for Roosevelt using a sample of "only" 50,000 people. In fact, Gallup *also* publicly predicted, to within 1%, the incorrect results that the *Literary Digest* would get using a sample of just 3000 people taken from the same sampling frame the magazine was using. What went wrong with the *Literary Digest* poll and why was Gallup able to do so much better?

The first thing seriously wrong with the *Literary Digest* poll was the sampling frame, consisting of names taken from telephone directories, lists of magazine subscribers, rosters of club members, and so on. Telephones in 1936 were something of a luxury, and magazine subscriptions and club memberships even more so, at a time when 9 million people were unemployed. When it came to economic status the *Literary Digest* sample was far from being a representative cross section of the voters. This was a critical problem, because voters often vote on economic issues, and given the economic conditions of the time, this was especially true in 1936.

The second serious problem with the *Literary Digest* poll was the issue of *nonresponse bias*. In a typical poll it is understood that not every individual is willing to respond to the request to participate (and in a democracy we cannot force them to do so). Those individuals who do not respond to the poll are called *nonrespondents*, and those who do are called *respondents*. The percentage of respondents out of the total sample is called the **response rate**. For the *Literary Digest* poll, out of a sample of 10 million people who were mailed a mock ballot only about 2.4 million mailed a ballot back, resulting in a 24% response rate. When the response rate to a poll is low, the poll is said to suffer from **nonresponse bias**. (Exactly at what point the response rate is to be considered low depends on the circumstances and nature of the poll, but a response rate of 24% is generally considered very low.)

Nonresponse bias can be viewed as a special type of selection bias—it excludes from the sample reluctant and uninterested people. Since reluctant and uninterested people can represent a significant slice of the population, we don't want them excluded from the sample. But getting reluctant, uninterested, and apathetic slugs to participate in a survey is a conundrum—in a free country we cannot force people to participate, and bribing them with money or chocolate chip cookies is not always a practical solution.

One of the significant problems with the *Literary Digest* poll was that the poll was conducted by mail. This approach is the most likely to magnify nonresponse bias, because people often consider a mailed questionnaire just another form of junk mail. Of course, given the size of their sample, the *Literary Digest* hardly had a choice. This illustrates another important point: Bigger is not better, and a big sample can be more of a liability than an asset.

Quota Sampling

Quota sampling is a systematic effort to force the sample to be representative of a given population through the use of quotas—the sample should have so many women, so many men, so many blacks, so many whites, so many people living in urban areas, so many people living in rural areas, and so on. The proportions in each category in the sample should be the same as those in the population. If we can assume that every important characteristic of the population is taken into account when the quotas are set up, it is reasonable to expect that the sample will be representative of the population and produce reliable data.

Our next historical example illustrates some of the difficulties with the assumptions behind quota sampling.

EXAMPLE 14.8 THE 1948 PRESIDENTIAL ELECTION: QUOTA SAMPLING

George Gallup had introduced quota sampling as early as 1935 and had used it successfully to predict the winner of the 1936, 1940, and 1944 presidential elections. Quota sampling thus acquired the reputation of being a "scientifically reliable" sampling method, and by the 1948 presidential election all three major national polls—the Gallup poll, the Roper poll, and the Crossley poll—used quota sampling to make their pre-election predictions.

For the 1948 election between Thomas Dewey and Harry Truman, Gallup conducted a poll with a sample of approximately 3250 people. Each individual in the sample was interviewed in person by a professional interviewer to minimize nonresponse bias, and each interviewer was given a very detailed set of quotas to meet—for example, 7 white males under 40 living in a rural area, 5 black males over 40 living in a rural area, 6 white females under 40 living in an urban area, and so on. By the time all the interviewers met their quotas, the entire sample was expected to

"Ain't the way I heard it," Truman gloats while holding an early edition of the *Chicago Daily Tribune* in which the headline erroneously claimed a Dewey victory based on the predictions of all the polls.

accurately represent the entire population in every respect: gender, race, age, and so on.

Based on his sample, Gallup predicted that Dewey, the Republican candidate, would win the election with 49.5% of the vote to Truman's 44.5% (with third-party candidates Strom Thurmond and Henry Wallace accounting for the remaining 6%). The Roper and Crossley polls also predicted an easy victory for Dewey. (In fact, after an early September poll showed Truman trailing Dewey by 13 percentage points, Roper announced that he would discontinue polling since the outcome was already so obvious.) The actual results of the election turned out to be almost the exact reverse of Gallup's prediction: Truman got 49.9% and Dewey 44.5% of the national vote.

Truman's victory was a great surprise to the nation as a whole. So convinced was the *Chicago Daily Tribune* of Dewey's victory that it went to press on its early edition for November 4, 1948, with the headline "Dewey defeats Truman." The picture of Truman holding aloft a copy of the *Tribune* and his famous retort "Ain't the way I heard it" have become part of our national folklore.

To pollsters and statisticians, the erroneous predictions of the 1948 election had two lessons: (1) *Poll until election day*, and (2) *quota sampling is intrinsically flawed*.

What's wrong with quota sampling? After all, the basic idea behind it appears to be a good one: Force the sample to be a representative cross section of the population by having each important characteristic of the population proportionally represented in the sample. Since income is an important factor in determining how people vote, the sample should have all income groups represented in the same proportion as the population at large. The same should be true for gender, race, age, and so on. Right away, we can see a potential problem: Where do we stop? No matter how careful we might be, we might miss some criterion that would affect the way people vote, and the sample could be deficient in this regard.

An even more serious flaw in quota sampling is that, other than meeting the quotas, the interviewers are free to choose whom they interview. This opens the door to selection bias. Looking back over the history of quota sampling, we can see a clear tendency to overestimate the Republican vote. In 1936, using quota sampling, Gallup predicted that the Republican candidate would get 44% of the vote, but the actual number was 38%. In 1940, the prediction was 48%, and the actual vote was 45%; in 1944, the prediction was 48%, and the actual vote was 46%. Gallup was able to predict the winner correctly in each of these elections, mostly because the spread between the candidates was large enough to cover the error. In 1948, Gallup (and all the other pollsters) simply ran out of luck.

The failure of quota sampling as a method for getting representative samples has a simple moral: *Even with the most carefully laid plans, human intervention in choosing the sample can result in selection bias.*

Random Sampling

The best alternative to human selection is to let the *laws of chance* determine the selection of a sample. Sampling methods that use randomness as part of their design are known as **random sampling** methods, and any sample obtained through random sampling is called a **random sample** (or a *probability sample*).

The idea behind random sampling is that the decision as to which individuals should or should not be in the sample is best left to chance because the laws of chance are better than human design in coming up with a representative sample. At first, this idea seems somewhat counterintuitive. How can a process based on random selection guarantee an unbiased sample? Isn't it possible to get by sheer bad luck a sample that is very biased (say, for example, a sample consisting of males only)? In theory, such an outcome is possible, but in practice, when the sample is large enough, the odds of it happening are so low that we can pretty much rule it out.

Most present-day methods of quality control in industry, corporate audits in business, and public opinion polling are based on random sampling. The reliability of data collected by random sampling methods is supported by both practical experience and mathematical theory. (We will discuss some of the details of this theory in Chapter 17.)

The most basic form of random sampling is called **simple random sampling**. It is based on the same principle as a lottery: Any set of numbers of a given size has an equal chance of being chosen as any other set of numbers of the same size. Thus, if a lottery ticket consists of six winning numbers, a fair lottery is one in which any combination of six numbers has the same chance of winning as any other combination of six numbers. In sampling, this means that any group of members of the population should have the same chance of being the sample as any other group of the same size.

EXAMPLE 14.9 DEFECTIVE LIGHTBULBS: SIMPLE RANDOM SAMPLING

We introduced the idea of sampling as a tool for quality control in Example 14.4, and in Example 14.6 we followed up with a discussion of the sample size. The last part of the story is the method for choosing the sample, and for most quality-control testing situations, simple random sampling is the method of choice.

Suppose we want to choose a sample of $n = 800$ lightbulbs out of a batch of $N = 100{,}000$ lightbulbs coming out of an assembly line. The first step is to use a computer program to randomly draw 800 different numbers between 1 and 100,000. This can be done in a matter of seconds. Say that the 800 numbers chosen by the computer are 74, 159, 311, etc. Then, as the lightbulbs move out of the production line the 74th, 159th, 311th, etc. bulbs are selected and tested. With robotic arms, the whole process can be implemented seamlessly and without any human intervention.

In theory, simple random sampling is easy to implement. We put the name of each individual in the population in "a hat," mix the names well, and then draw as many names as we need for our sample. Of course "a hat" is just a metaphor. If our population is 100 million voters and we want to choose a simple random sample of 1500, we will not be putting all 100 million names in a real hat and then drawing 1500 names one by one. These days, the "hat" is a computer database containing a list of members of the population. A computer program then randomly selects the names. This is a fine idea for easily accessible populations such as lightbulbs coming off of an assembly line, but a hopeless one when it comes to national surveys and public opinion polls.

Implementing simple random sampling in national public opinion polls raises problems of expediency and cost. Interviewing hundreds of individuals chosen by simple random sampling means chasing people all over the country, a task that requires an inordinate amount of time and money. For most public opinion polls—especially those done on a regular basis—the time and money needed to do this are simply not available.

The alternative to simple random sampling used nowadays for national surveys and public opinion polls is a sampling method known as **stratified sampling**. The basic idea of stratified sampling is to break the sampling frame into categories, called **strata**, and then (unlike quota sampling) *randomly* choose a sample from these strata. The chosen strata are then further divided into categories, called substrata, and a random sample is taken from these substrata. The selected substrata are further subdivided, a random sample is taken from them, and so on. The process goes on for a predetermined number of steps (usually four or five).

Our next example illustrates how stratified sampling is used to conduct *national public opinion polls*. Basic variations of the same idea can be used at the state, city, or local level. The specific details, of course, will be different.

EXAMPLE 14.10 PUBLIC OPINION POLLS: STRATIFIED SAMPLING

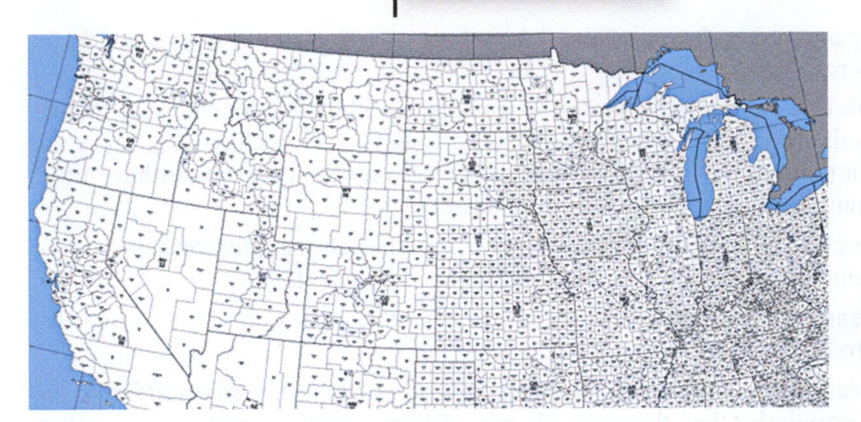

In national public opinion polls the *strata* and *substrata* are defined by a combination of geographic and demographic criteria. For example, the nation is first divided into "size of community" *strata* (big cities, medium cities, small cities, villages, rural areas, etc.). The strata are then subdivided by geographical region (New England, Middle Atlantic, East Central, etc.). This is the first layer of substrata. Within each geographical region and within each size of community stratum some communities (called *sampling locations*) are selected by simple random sampling. The selected sampling locations are the only places where interviews will be conducted. Next, each of the selected sampling locations is further subdivided into geographical units called *wards*. This is the second layer of substrata. Within each sampling location some of the wards are selected using simple random sampling. The selected wards are then divided into smaller units, called *precincts* (third layer), and within each ward some of its precincts are selected by simple random sampling. At the last stage, *households* (fourth layer) are selected from within each precinct by simple random sampling. The interviewers are then given specific instructions as to which households in their assigned area they must conduct interviews in and the order that they must follow.

The efficiency of stratified sampling compared with simple random sampling in terms of cost and time is clear. The members of the sample are clustered in well-defined and easily manageable areas, significantly reducing the cost of conducting interviews as well as the response time needed to collect the data. For a large, heterogeneous nation like the United States, stratified sampling has generally proved to be a reliable way to collect national data. ■

We conclude this section with a list of basic concepts concerning surveys and sampling.

- **Survey.** A data collection strategy based on using information from a sample to draw conclusions about a general population.
- **Sample.** A subset of the population chosen to be the providers of information in a survey.
- **Sampling.** The act of choosing a sample.
- **Sampling proportion.** The proportion of the population represented by the sample.
- **Statistic.** A numerical measurement of some characteristic of a population obtained using a sample. A statistic is always an estimate of the real measurement.
- **Parameter.** An exact (accurate) measurement of some characteristic of a population. A parameter is the real measurement we would like to have.
- **Sampling error.** The difference between a parameter and the estimate for that parameter (i.e., the statistic) obtained from a sample.

- **Sampling variability.** The natural variability found between different samples of the same population, even when the samples are chosen using the same methodology.
- **Sampling bias.** This occurs when some members of the population are less likely to be selected for the sample than others (even if it is unintentional).
- **Sampling frame.** The pool of individuals from which the sample is drawn (not necessarily the same as the target population).
- **Response rate.** The percentage of respondents in a poll out of the total sample size.
- **Nonresponse bias.** A type of bias that is the result of too many nonrespondents (i.e., low response rates).
- **Convenience sampling.** A sampling strategy based on the convenience factor: Individuals that can be reached conveniently have a high chance of being included in the sample; individuals whose access is more inconvenient have little or no chance of being included in the sample.
- **Self-selection bias.** A type of bias that occurs when the sample consists of individuals that volunteer to be in the sample (i.e., self-select).
- **Quota sampling.** A sampling method that uses quotas as a way to force the sample to be representative of the population.
- **Simple random sampling.** A sampling method in which any group of individuals in a population has the same chance of being in the sample as any other group of equal size.
- **Stratified sampling.** A sampling method that uses several layers of strata and substrata and chooses the sample by a process of random selection within each layer.
- **Systematic sampling.** A random sampling method that can be used only when the individuals in the population can be listed in numerical order (1, 2, 3, etc.). The first member of the sample is chosen at random (that's the starting point) and the rest are chosen at fixed, regular intervals from the starting point. For example, in a population of $N = 100$ individuals numbered 1 through 100, the sample might consist of individuals numbered 7 (the starting point), 17, 27, 37, . . . , 97.

14.3 Cause and Effect

Good news, bad news. When it comes to the connections between our lifestyle habits (diet, exercise, smoking, pill popping, etc.) and our health and well-being (longevity, weight control, chances of disease, etc.), the news is full of both good and bad: "coffee is good for you—you will live longer if you drink a few cups a day"; "Hormone replacement therapy (HRT) is bad for women after menopause—it increases their chances of getting breast cancer."

How much should we trust these types of pronouncements? And what should we do when the information is conflicting ("HRT is good for you; Oops! On second thought, HRT is really bad for you; No, actually HRT is kind of good for you after all!"). Our next example illustrates how serious this issue can be.

EXAMPLE 14.11 HORMONE REPLACEMENT THERAPY

Hormone replacement therapy (HRT) is basically a form of estrogen replacement and is an accepted and widely used therapy to treat the symptoms of menopause in women. No controversy there. The question is whether women after menopause should continue HRT, and if so, what are the benefits of doing so? Here is a chronology of the answers:

- *Nurses' Health Study*, 1985: Good news! Women who took estrogen had one-third as many heart attacks as women who didn't. Hormone replacement therapy (estrogen plus progestin) becomes a highly recommended treatment for preventing heart disease (as well as osteoporosis) in postmenopausal women. By 2001, 15 million prescriptions for HRT are filled annually and about 5 million are for postmenopausal women.
- *Women's Health Initiative Study*, 2002: Bad news! Hormone replacement therapy significantly increases the risk of heart disease and breast cancer, and is a risk factor for stroke. Its only health benefit is for preventing osteoporosis and possibly colorectal cancer. Hormone replacement therapy is no longer recommended as a general therapy for older women (it is still recommended as an effective therapy for women during menopause).
- *Women's Health Initiative Follow-up*, 2007: Mixed news! Hormone replacement therapy offers protection against heart disease for women who start taking it during menopause but increases the risk of heart disease for women who start taking it after menopause.
- *Women's Health Initiative Follow-up II*, 2009: Bad news! Hormone replacement therapy significantly increases the risk of breast cancer in menopausal and postmenopausal women.

> "There is very strong evidence that estrogen plus progestin causes breast cancer. You start women on hormones and within five years their risk of breast cancer is clearly elevated. You stop the hormones and within one year their risk is essentially back to normal. It's reasonably convincing cause and effect data."
>
> – *Marcia Stefanik, Women's Health Initiative Follow-up II*

The relation between HRT and heart disease or breast cancer in older women is one of thousands of important cause-and-effect questions that still remain unanswered, and women are left to struggle weighing the good and the bad. Why?

A typical cause-and-effect statement (at least the kind that makes the news) takes the form *treatment X* causes *result Y*. The treatment *X* might be a food, a drug, a therapy, or a lifestyle habit such as exercise, and the result *Y* a disease, a health benefit, or a behavioral change. These types of questions are incredibly difficult to answer definitively because there is a huge difference between *correlation* and *causation*, and yet it is very difficult to separate one from the other. A **correlation** (or **association**) between two events *X* and *Y* occurs when there is an observed mutual relationship between the two events. A **causation** (or **causal relationship**) between *X* and *Y* occurs when one event is the cause of the other one.

An observed correlation between two events *X* and *Y* [Fig. 14-2(a)] can occur for many reasons: Maybe *X* is the cause and *Y* is the effect [Fig. 14-2(b)], or Y is the cause and *X* is the effect [Fig. 14-2(c)], or both *X* and *Y* are effects of a common cause C [Fig. 14-2(d)], or *X* and *Y* are the effects of two different correlated causes [Fig. 14-2(e)]. It follows that trying to establish a cause-and-effect relationship between two events just because a correlation is observed is tricky to say the least.

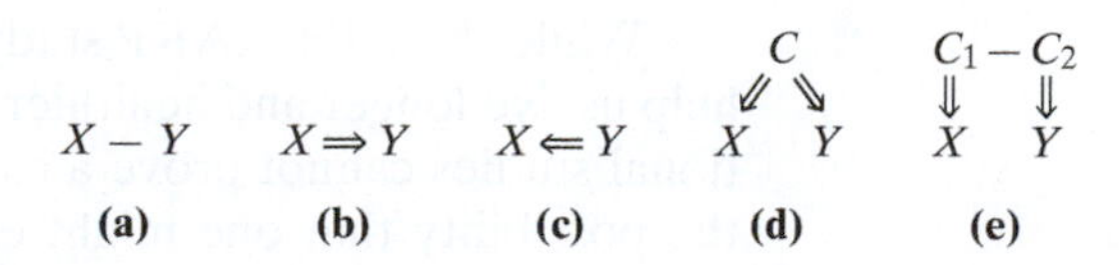

FIGURE 14-2 (a) *X* and *Y* are correlated. (b) *X* is the cause of *Y*. (c) *Y* is the cause of *X*. (d) *X* and *Y* have a common cause. (e) *X* and *Y* have separate causes.

In general, finding a correlation between a treatment and an observed result is easy. The difficult part is establishing that the correlation is indeed due to some cause-and-effect relation. As statisticians are fond of reminding us, "correlation is not causation."

In the rest of this section we will illustrate the complexities of establishing cause and effect from observed correlations.

EXAMPLE 14.12 DOES DRINKING COFFEE HELP YOU LIVE LONGER?

In mid May 2012 a big story made the headlines: a major study showing that drinking coffee helps you live longer. For those of us who often wonder—as we sip another double espresso—what our coffee addiction does to us, this was a great bit of news. Can it really be true?

The coffee finding came from the *NIH-AARP Diet and Health Study*—a very large **observational study** sponsored by the National Institutes of Health (NIH). In an observational study individuals are tracked for an extensive period of time and records are kept of their lifestyle factors (diet, physical activity, smoking, alcohol consumption, etc.) as well as disease rates and general health. Whenever a correlation between a lifestyle factor X and an observed change Y is found, a hypothesis is formulated: *Could X be a cause of Y?* Let's check out the 'coffee causes longevity hypothesis' in a little more detail.

The NIH-AARP study involved 400,000 subjects living in the United States aged 50 to 71. Initially, the subjects were given an extensive survey concerning their health, nutrition, and lifestyle habits and then were tracked for 12 years for health status, life span, and causes of death if they had indeed died.

One of the lifestyle factors considered was coffee consumption, and the initial correlation found by the researchers was that the non-coffee drinkers lived longer than coffee drinkers. (During the 12-year period of the study, 19% of the men and 15% of the women coffee drinkers died, but among non-coffee drinkers the death rates dropped to 13% men and 10% women.) These correlations would point towards the hypothesis that coffee is bad for you, but it turned out that coffee drinking is also positively correlated with smoking: Coffee drinkers are more likely to be smokers than non-coffee drinkers. This raises the possibility that it is really the smoking that is bad for you and not the coffee. Similar correlations were found between coffee drinking and a host of other lifestyle factors: Coffee drinkers consume more alcohol, consume more red meat, have lower levels of physical activity, and consume fewer fruits and vegetables. Any one of these (or a combination of several) could be the cause of the observed effect.

However, once the researchers *controlled* for smoking (comparing longevity among non-smoking coffee drinkers with longevity among non-smoking non-coffee drinkers) the tables were turned: Coffee consumption was correlated with living longer and with a reduced incidence of cancer and heart disease. The same positive correlation showed up when the researchers controlled for other *confounding variables*—alcohol consumption, body mass index (BMI), age, ethnicity, marital status, physical activity, consumption of red meat, and (for postmenopausal women) use or nonuse of HRT. The general conclusion of the study was that men who drank two or more cups of a day had a 10% better chance of living through the study than those who didn't, and for women the advantage was 13%. ■

While the NIH-AARP study raises the tantalizing possibility that coffee might help us live longer and healthier lives, the evidence is far from conclusive. Observational studies cannot prove a cause-and-effect relationship—they can only suggest the possibility that one might exist. The NIH-AARP study had many flaws—the lifestyle factors were self-reported and people are known to fudge the truth (if you smoked three packs a day, would your really put that down in a questionnaire?), the lifestyle data was collected only at the beginning of the study (what about those who changed their lifestyle in the middle of the study?), and there were many other possible *confounding variables* (high cholesterol, high blood pressure, type of health insurance, etc.) that were not considered. No doubt there will be follow-up studies to clarify things a bit. In the meantime, if you are a coffee drinker, enjoy your cup of java. If you are not, don't start now.

One of the problems with cause-and-effect findings is that they often get a lot of coverage in the press—the more surprising and unexpected the finding the more coverage it gets. Some findings can result in a major shift in public health policy, and this is a serious issue when the finding turns out to be false. This is illustrated in our next example.

EXAMPLE 14.13 THE ALAR SCARE

Alar is a chemical used by apple growers to regulate the rate at which apples ripen. Until 1989, practically all apples sold in grocery stores were sprayed with Alar. But in 1989 Alar became bad news, denounced in newspapers and on TV as a potent cancer-causing agent and a primary cause of cancer in children. As a result of these reports, people stopped buying apples, schools all over the country removed apple juice from their lunch menus, and the Washington State apple industry lost an estimated $375 million.

The case against Alar was based on a single 1973 study in which laboratory mice were exposed to the active chemicals in Alar. The dosage used in the study was eight times greater than the maximum tolerated dosage—a concentration at which even harmless substances can produce tissue damage. In fact, a child would have to eat about 200,000 apples a day to be exposed to an equivalent dosage of the chemical. Subsequent studies conducted by the National Cancer Institute and the Environmental Protection Agency failed to show any correlation between Alar and cancer in children.

While it is generally accepted now that Alar does not cause cancer, because of potential legal liability, it is no longer used. The Alar scare turned out to be a false alarm based on a poor understanding of the statistical evidence. Unfortunately, it left in its wake a long list of casualties, among them the apple industry, the product's manufacturer, the media, and the public's confidence in the system.

Clinical Studies

A second approach to establishing a cause-and-effect relationship between a treatment and a result is a **clinical study** (or **clinical trial**). Clinical studies are used to demonstrate that a treatment X (usually a drug, a vaccine, or a therapy) is *effective*. (For simplicity we will call X an *effective treatment* if X helps cure a disease or improves the condition of a patient.) The only way to be sure that X is an effective treatment is to isolate X from other possible causes (**confounding variables**) that could explain the same effect.

The classic way to isolate a treatment X from all other possible confounding variables is to use a **controlled study**. In a controlled study the subjects are divided into groups—some groups (called the *treatment groups*) are the ones that get the treatment; the other groups (called the *control groups*) don't get the treatment. The control groups are there for comparison purposes only—they give the experimenters a baseline to see if the treatment groups do better or not. If the treatment groups show better results than the control groups, then there is good reason to suspect that the treatment might be an effective treatment.

To eliminate the many potential confounding variables that can bias its results, a well-designed controlled study should have control and treatment groups that are similar in every characteristic other than the fact that one group is being treated and the other one is not. (It would be a very bad idea, for example, to have a treatment group that is all female and a control group that is all male.) The most reliable way to get equally representative treatment and control groups is to use a *randomized controlled study*. In a **randomized controlled study**, the subjects are assigned to the treatment group or the control group randomly.

When the randomization part of a randomized controlled study is properly done, treatment and control groups can be assumed to be statistically similar. But there is still one major difference between the two groups that can significantly affect the validity of the study—a critical confounding variable known as the *placebo effect*. The **placebo effect** follows from the generally accepted principle that *just the idea that one is getting a treatment can produce positive results*. Thus, when subjects in a study are getting a pill or a vaccine or some other kind of treatment, how can the researchers separate positive results that are consequences of the treatment itself from those that might be caused by the placebo effect? When possible, the standard way to handle this problem is to give the control group a *placebo*. A **placebo** is a *make-believe* form of treatment—a harmless pill, an injection of saline solution, or any other fake type of treatment intended to look like the real treatment. A controlled study in which the subjects in the control group are given a placebo is called a **controlled placebo study**.

By giving all subjects a seemingly equal treatment (the treatment group gets the real treatment and the control group gets a placebo that looks like the real treatment), we do not eliminate the placebo effect but rather control it—whatever its effect might be, all subjects (be they in the control group or the treatment group) are experiencing the placebo effect to an equal extent. It goes without saying that the use of placebos is pointless if the subject knows he or she is getting a placebo. Thus, a second key element of a good controlled placebo study is that all subjects be kept in the dark as to whether they are being treated with a real treatment or a placebo. A study in which neither the members of the treatment group nor the members of the control group know to which of the two groups they belong is called a **blind study**.

Blindness is a key requirement of a controlled placebo study but not the only one. To keep the interpretation of the results (which can often be ambiguous) totally objective, it is important that the scientists conducting the study and collecting the data also be in the dark when it comes to who got the treatment and who got the placebo. A controlled placebo study in which neither the subjects nor the scientists conducting the experiment know which subjects are in the treatment group and which are in the control group is called a **double-blind study**.

Our next case study illustrates one of the most famous and important double-blind studies in the annals of clinical research.

CASE STUDY 3 THE 1954 SALK POLIO VACCINE FIELD TRIALS

Polio (infantile paralysis) has been practically eradicated in the Western world. In the first half of the twentieth century, however, it was a major public health problem. Over one-half million cases of polio were reported between 1930 and 1950, and the actual number may have been considerably higher.

Because polio attacks mostly children and because its effects can be so serious (paralysis or death), eradication of the disease became a top public health priority in the United States. By the late 1940s, it was known that polio is a virus and, as such, can best be treated by a vaccine that is itself made up of a virus. The vaccine virus can be a closely related virus that does not have the same harmful effects, or it can be the actual virus that produces the disease but that has been killed by a special treatment. The former is known as a *live-virus vaccine*, the latter as a *killed-virus vaccine*. In response to either vaccine, the body is known to produce *antibodies* that remain in the system and give the individual immunity against an attack by the real virus.

Both the live-virus and the killed-virus approaches have their advantages and disadvantages. The live-virus approach produces a stronger reaction and better immunity, but at the same time, it is also more likely to cause a harmful reaction and, in some cases, even to produce the very disease it is supposed to prevent. The killed-virus approach is safer in terms of the likelihood of producing a harmful reaction, but it is also less effective in providing the desired level of immunity.

These facts are important because they help us understand the extraordinary amount of caution that went into the design of the study that tested the effectiveness of the polio vaccine. By 1953, several potential vaccines had been developed, one of the more promising of which was a killed-virus vaccine developed by Jonas Salk at the University of Pittsburgh. The killed-virus approach was chosen because there was a great potential risk in testing a live-virus vaccine in a large-scale study. (A large-scale study was needed to collect enough information on polio, which, in the 1950s, had a rate of incidence among children of about 1 in 2000.)

The testing of any new vaccine or drug creates many ethical dilemmas that have to be taken into account in the design of the study. With a killed-virus vaccine the risk of harmful consequences produced by the vaccine itself is small, so one possible approach would have been to distribute the vaccine widely among the population and then follow up on whether there was a decline in the national incidence of polio in subsequent years. This approach, which was not possible at the time because supplies were limited, is called the *vital statistics* approach and is the simplest way to test a vaccine. This is essentially the way the smallpox vaccine was determined to be effective. The problem with such an approach for polio is that polio is an epidemic type of disease, which means that there is a great variation in the incidence of the disease from one year to the next. In 1952, there were close to 60,000 reported cases of polio in the United States, but in 1953, the number of reported cases had dropped to almost half that (about 35,000). Since no vaccine or treatment was used, the cause of the drop was the natural variability typical of epidemic diseases. But if an ineffective polio vaccine had been tested in 1952 without a control group, the observed effect of a large drop in the incidence of polio in 1953 could have been incorrectly interpreted as statistical evidence that the vaccine worked.

The final decision on how best to test the effectiveness of the Salk vaccine was left to an advisory committee of doctors, public officials, and statisticians convened by the National Foundation for Infantile Paralysis and the Public Health Service. It was a highly controversial decision, but at the end, a large-scale, randomized, double-blind, controlled placebo study was chosen. Approximately 750,000 children were randomly selected to participate in the study. Of these, about 340,000 declined to participate, and another 8500 dropped out in the middle of the experiment. The remaining children were randomly divided into two groups—a treatment group and a control group—with approximately 200,000 children in each group. Neither the families of the children nor the researchers collecting the data knew if a particular child was getting the actual vaccine or a shot of harmless solution. The latter was critical because polio is not an easy disease to diagnose—it comes in many different forms and degrees. Sometimes it can be a borderline call, and if the doctor collecting the data had prior knowledge of whether the subject had received the real vaccine or the placebo, the diagnosis could have been subjectively tipped one way or the other.

A summary of the results of the Salk vaccine field trials is shown in Table 14-2. These data were taken as conclusive evidence that the Salk vaccine was an effective treatment for polio, and on the basis of this study, a massive inoculation campaign was put into effect. Today, all children are routinely inoculated against polio, and the disease has essentially been eradicated in the United States.

	Number of children	Number of reported cases of polio	Number of paralytic cases of polio	Number of fatal cases of polio
Treatment group	200,745	82	33	0
Control group	201,229	162	115	4
Declined to participate in the study	338,778	182*	121*	0*
Dropped out in the middle	8,484	2*	1*	0*
Total	749,236	428	270	4

*These figures are not a reliable indicator of the actual number of cases—they are only self-reported cases.
[Data from Thomas Francis, Jr., et al., "An Evaluation of the 1954 Poliomyelitis Vaccine Trials—Summary Report," *American Journal of Public Health*, 45 (1955), 25.]

TABLE 14-2 ■ Results of the Salk vaccine field trials

Conclusion

In this chapter we have discussed different methods for collecting data. In principle, the most accurate method is a *census*, a method that relies on collecting data from each member of the population. In most cases, because of considerations of cost and time (as well as the fact that sometimes members of the population don't want to be counted), a census is an unrealistic strategy. When data are collected from only a subset of the population (called a *sample*), the data collection method is called a *survey*. The most important rule in designing good surveys is to eliminate or minimize *sample bias*. Today, almost all strategies for collecting data are based on surveys in which the laws of chance are used to determine how the sample is selected, and these methods for collecting data are called *random* (or *probability*) *sampling* methods. Random sampling is the best way known to minimize or eliminate sample bias. Two of the most common random sampling methods are *simple random sampling* and *stratified sampling*. In some special situations, other more complicated types of random sampling can be used.

Sometimes identifying the sample is not enough. In cases in which cause-and-effect questions are involved, the data may come to the surface only after an extensive study has been carried out. The critical issue in establishing that there is a true cause-and-effect relation between two events is to distinguish between *correlations* and *causations*; the fact that two variables or behaviors are *correlated* does not mean that there is necessarily a *cause-and-effect* relation between them. The two most commonly used strategies for distinguishing true cause-and-effect relations from observed correlations are *observational studies* and *controlled double-blind placebo studies*. Both of these strategies are nowadays used (and sometimes abused) to settle issues affecting every aspect of our lives. We can thank this area of statistics for many breakthroughs in social science, medicine, and public health, as well as for the constant and dire warnings about our health, our diet, and practically anything that's delicious or fun.

KEY CONCEPTS

14.1 Enumeration

- **population:** The set of individuals (humans, animals or inanimate objects) being enumerated, **411**
- ***N*-value:** The count giving the number of individuals in a given population, **411**
- **census:** a complete enumeration of a population, **412**
- **survey:** a data collection strategy that uses a sample to draw inferences about a population, **412, 414**
- **sample:** a subset of the population chosen to be the providers of information in a survey, **414**
- **sampling:** the act of selecting a sample, **414**
- **statistic:** a numerical estimate of some measurable characteristic of a population obtained from a sample, **414**
- **parameter:** a true measurement of some characteristic of a population, **414**
- **one-sample estimation:** a method for estimating the size of a subpopulation using a sample, **415**
- **two-sample estimation (capture-recapture):** a method for estimating the size of a population using two samples, **416**

14.2 Measurement

- **poll:** a survey in which the data collection involves asking questions and recording answers, **418**
- **sampling proportion:** the percentage or proportion of the population represented by the sample, **419**
- **biased sample (sampling bias):** a sample in which not every member of the population had an equal chance of being included, **419**
- **selection bias:** a bias that occurs when some members of the population have no chance of being included in the sample, **419**
- **target population:** the population to which the conclusions of the survey apply, **419**
- **sampling frame:** the population from which the sample is drawn, **419**
- **convenience sampling:** a sampling strategy based on selecting the most convenient individuals to be in the sample, **420**
- **self-selection bias:** a type of bias that results when the sample consists of individuals who volunteer to be in the sample, **420**
- **quota sampling:** a sampling method that uses quotas as a way to force the sample to be representative of the population, **422**
- **response rate:** the percentage of respondents in a poll out of the total sample size, **422**
- **nonresponse bias:** a type of bias that results from having low response rates, **422**
- **simple random sampling:** a sampling method where any group of individuals in a population has the same chance of being in the sample as any other group of equal size, **424**
- **stratified sampling:** a sampling method that uses several layers of strata and substrata and chooses the sample by a process of random selection within each layer, **424**
- **sampling error:** the difference between a parameter and a statistic obtained from a sample, **425**
- **sampling variability:** the natural variability in the statistics obtained by different samples of the same population, even when the samples are chosen using the same methodology, **426**
- **systematic sampling:** A random sampling method that can be used only when the individuals in the population can be listed in numerical order (1, 2, 3, etc.). The first member of the sample is chosen at random (that's the starting point) and the rest are chosen at fixed, regular intervals from the starting point, **426**

14.3 Cause and Effect

- **correlation (association):** between two events *X* and *Y*; occurs when there is an observed mutual relationship between the two events, **427**
- **causation (causal, or cause-and-effect, relationship):** between two events *X* and *Y*; occurs when one event is the cause of the other one, **427**
- **observational study:** a study where individuals are tracked for an extensive period of time to look for correlations between lifestyle factors and disease or wellness factors, **428**
- **confounding variable:** an alternative possible cause for an observed effect; not the hypothetical cause, **429**

- **clinical study (clinical trial):** a study intended to demonstrate that a specific treatment is effective in treating some disease or symptom, **429**
- **controlled study:** a study in which the subjects are divided into treatment groups and control groups, **429**
- **randomized controlled study:** the assignment of subjects to treatment and control groups is done at random, **429**
- **placebo:** a fake treatment; intended to make the subject believe he or she is being treated, **430**
- **controlled placebo study:** a study in which the subjects in the control groups are given a placebo, **430**
- **blind study:** a study in which the subjects don't know if they are getting the treatment or the placebo, **430**
- **double-blind study:** a study in which neither the subjects nor the experimenters know who is getting the real treatment and who is getting the placebo, **430**

EXERCISES

WALKING

14.1 Enumeration

1. As part of a sixth-grade class project the teacher brings to class a large jar containing 200 gumballs of two different colors: red and green. Andy is asked to draw a sample of his own choosing and estimate the number of red gumballs in the jar. Andy draws a sample of 25 gumballs, of which 8 are red and 17 are green. Use Andy's sample to estimate the number of red gumballs in the jar.

2. As part of a sixth-grade class project the teacher brings to class a large jar containing 200 gumballs of two different colors: red and green. Brianna is asked to draw a sample of her own choosing and estimate the number of red gumballs in the jar. Brianna draws a sample of 40 gumballs, of which 14 are red and 26 are green. Use Brianna's sample to estimate the number of red gumballs in the jar.

3. Madison County has a population of 34,522 people. The county hospital is interested in estimating the number of people in the county with blood-type A−. To do this they test blood samples from 253 patients. Out of this group, 17 have blood-type A−. Use this sample to estimate the number of people in Madison County with blood-type A−.

4. Madison County has a population of 34,522 people. The county hospital is interested in estimating the number of people in the county with blood-type AB+. To do this they test blood samples from 527 patients. Out of this group, 22 have blood-type AB+. Use this sample to estimate the number of people in Madison County with blood-type AB+.

5. A big concert was held at the Bowl. Men and women had to go through separate lines to get into the concert (the women had to have their purses checked). Once everyone was inside, total attendance at the concert had to be recorded. The turnstile counters on the female entrance showed a total count of 1542 females, but the turnstile counters on the male entrance were broken and there was no exact record of how many males attended. A sample taken from the 200 seats in Section A showed 121 females and 79 males in that section. Using the numbers from Section A, estimate the total attendance at the concert. (*Hint:* The proportion of females at the concert should be roughly the same as the proportion of females in Section A.)

6. A large jar contains an unknown number of red gumballs and 150 green gumballs. As part of a seventh-grade class project the teacher asks Carlos to estimate the total number of gumballs in the jar using a sample. Carlos draws a sample of 50 gumballs, of which 19 are red and 31 are green. Use Carlos' sample to estimate the number of gumballs in the jar.

7. You want to estimate how many fish there are in a small pond. Let's suppose that you first capture $n_1 = 500$ fish, tag them, and throw them back into the pond. After a couple of days you go back to the pond and capture $n_2 = 120$ fish, of which $k = 30$ are tagged. Estimate the number of fish in the pond.

8. To estimate the population in a rookery, 4965 fur seal pups were captured and tagged in early August. In late August, 900 fur seal pups were captured. Of these, 218 had been tagged. Based on these figures, estimate the population of fur seal pups in the rookery. [*Source:* Chapman and Johnson, "Estimation of Fur Seal Pup Populations by Randomized Sampling," *Transactions of the American Fisheries Society*, 97 (July 1968), 264–270.]

9. To count whale populations, the "capture" is done by means of a photograph, and the "tagging" is done by identifying each captured whale through their unique individual pigmentation and markings. To estimate the population of gray whales in a region of the Pacific between Northern California and Southeast Alaska, 121 gray whales were

"captured" and "tagged" in 2007. In 2008, 172 whales were "recaptured." Of these, 76 had been "tagged" in the 2007 survey. Based on these figures, estimate the population of gray whales in the region. [*Source:* Calambokidis, J., J.L. Laake and A. Klimek, "Abundance and population structure of seasonal gray whales in the Pacific Northwest, 1998–2008." Paper IWC/62/BRG32 submitted to the *International Whaling Commission Scientific Committee*, 2010.]

10. The critically endangered Maui's dolphin is currently restricted to a relatively small stretch of coastline along the west coast of New Zealand's North Island. The dolphins are "captured" by just collecting samples of DNA and "tagged" by identifying their DNA fingerprint. A 2010–2011 capture-recapture study "captured" and "tagged" 26 Maui's dolphins in 2010. In 2011, 27 Maui's dolphins were "recaptured" and through their DNA, 12 were identified as having been "tagged" in 2010. Based on these figures, estimate the population of Maui's dolphins in 2011. [*Source:* Oremus, M., et al, "Distribution, group characteristics and movements of the critically endangered Maui's Dolphin (*Cephalorhynchus hectori maui*)." *Endangered Species Research*, preprint.]

Exercises 11 and 12 refer to Chapman's correction. Chapman's correction is a small tweak on the formula used in the capture-recapture method. Using the same three input variables n_1 *(size of the first sample),* n_2 *(size of the second sample), and k (number of tagged individuals in the second sample), Chapman's correction is given by the formula* $\frac{k+1}{n_2+1} \approx \frac{n_1+1}{N+1}$. *Solving for N gives Chapman's correction estimate for the size of the population.*

11. Use Chapman's correction to estimate the population of gray whales described in Exercise 9. Compare the two answers.

12. Use Chapman's correction to estimate the population of Maui's dolphins described in Exercise 10. Compare the two answers.

13. Starting in 2004, a study to determine the number of lake sturgeon on Rainy River and Lake of the Woods on the United States–Canada border was conducted by the Canadian Ministry of Natural Resources, the Minnesota Department of Natural Resources, and the Rainy River First Nations. Using the capture-recapture method, the size of the population of lake sturgeon on Rainy River and Lake of the Woods was estimated at $N = 160{,}286$. In the capture phase of the study, 1700 lake sturgeon were caught, tagged, and released. Of these tagged sturgeon, seven were recaptured during the recapture phase of the study. Based on these figures, estimate the number of sturgeon caught in the recapture phase of the study. [*Source:* Dan Gauthier, "Lake of the Woods Sturgeon Population Recovering," *Daily Miner and News* (Kenora, Ont.), June 11, 2005, p. 31.]

14. A 2004 study conducted at Utah Lake using the capture-recapture method estimated the carp population in the lake to be about 1.1 million. Over a period of 15 days, workers captured, tagged, and released 24,000 carp. In the recapture phase of the study 10,300 carp were recaptured. Estimate how many of the carp that were recaptured had tags. [*Source:* Brett Prettyman, "With Carp Cooking Utah Lake, It's Time to Eat," *Salt Lake Tribune*, July 15, 2004, p. D3.]

14.2 Measurement

15. Name the sampling method that best describes each situation. Choose your answer from the following (A) simple random sampling, (B) convenience sampling, (C) quota sampling, (D) stratified sampling, (E) census.

(a) George wants to know how the rest of the class did on the last quiz. He peeks at the scores of a few students sitting right next to him. Based on what he sees, he concludes that nobody did very well.

(b) Eureka High School has 400 freshmen, 300 sophomores, 300 juniors, and 200 seniors. The student newspaper conducts a poll asking students if the football coach should be fired. The student newspaper selects 20 freshmen, 15 sophomores, 15 juniors, and 10 seniors for the poll.

(c) For the last football game of the season, the coach chooses the three captains by putting the names of all the players in a hat and drawing three names. (Maybe that's why they are trying to fire him!)

(d) For the last football game of the season, the coach chooses the three captains by putting the names of all the *seniors* in a hat and drawing three names.

16. An audit is performed on last year's 15,000 student-aid packages given out by the financial aid office at Tasmania State University. Roughly half of the student-aid packages were less than $1000 (Category 1), about one-fourth were between $1000 and $5000 (Category 2), and another quarter were over $5000 (Category 3). For each audit described below, name the sampling method that best describes it. Choose your answer from the following: (A) simple random sampling, (B) convenience sampling, (C) quota sampling, (D) stratified sampling, (E) census.

(a) The auditor reviews all 15,000 student-aid packages.

(b) The auditor selects 200 student-aid packages in Category 1, 100 student-aid packages in Category 2, and 100 student-aid packages in Category 3.

(c) The auditor reviews the first 500 student-aid packages that he comes across.

(d) The auditor first separates the student- aid packages by school (Agriculture, Arts and Humanities, Engineering, Nursing, Social Science, Science, and Mathematics). Three of these schools are selected at random and further subdivided by major. Ten majors are randomly selected within each selected school, and then 20 students are randomly selected from each of the selected majors.

Exercises 17 through 20 refer to the following story: The city of Cleansburg has 8325 registered voters. There is an election for mayor of Cleansburg, and there are three candidates for the position: Smith, Jones, and Brown. The day before the election a telephone poll of 680 randomly chosen registered voters produced the following results: 306 people surveyed indicated that they would vote for Smith, 272 indicated that they would vote for Jones and 102 indicated that they would vote for Brown.

17. (a) Describe the population for this survey.

(b) Describe the sample for this survey.

(c) Name the sampling method used for this survey.

18. (a) Give the sampling proportion for this survey.

(b) Give the sample statistic estimating the percentage of the vote going to Smith.

19. Given that in the actual election Smith received 42% of the vote, Jones 43% of the vote, and Brown 15% of the vote, find the sampling errors in the survey expressed as percentages.

20. Do you think that the sampling error in this example was due primarily to sampling bias or to chance? Explain your answer.

Exercises 21 through 24 refer to the following story: The 1250 students at Eureka High School are having an election for Homecoming King. The candidates are Tomlinson (captain of the football team), Garcia (class president), and Marsalis (member of the marching band). At the football game a week before the election, a pre-election poll was taken of students as they entered the stadium gates. Of the students who attended the game, 203 planned to vote for Tomlinson, 42 planned to vote for Garcia, and 105 planned to vote for Marsalis.

21. (a) Describe the sample for this survey.

(b) Give the sampling proportion for this survey.

22. Name the sampling method used for this survey.

23. (a) Compare and contrast the population and the sampling frame for this survey.

(b) Is the sampling error a result of sampling variability or of sample bias? Explain

24. (a) Give the sample statistics estimating the percentage of the vote going to each candidate.

(b) A week after this survey, Garcia was elected Homecoming King with 51% of the vote, Marsalis got 30% of the vote, and Tomlinson came in last with 19% of the vote. Find the sampling errors in the survey expressed as percentages.

Exercises 25 through 28 refer to the following story: The Cleansburg Planning Department is trying to determine what percent of the people in the city want to spend public funds to revitalize the downtown mall. To do so, the department decides to conduct a survey. Five professional interviewers are hired. Each interviewer is asked to pick a street corner of his or her choice within the city limits, and every day between 4:00 P.M. and 6:00 P.M. the interviewers are supposed to ask each passerby if he or she wishes to respond to a survey sponsored by Cleansburg City Hall. If the response is yes, the follow-up question is asked: Are you in favor of spending public funds to revitalize the downtown mall? The interviewers are asked to return to the same street corner as many days as are necessary until each has conducted a total of 100 interviews. The results of the survey are shown in Table 14-3.

Interviewer	Yes[a]	No[b]	Nonrespondents[c]
A	35	65	321
B	21	79	208
C	58	42	103
D	78	22	87
E[d]	12	63	594

[a]In favor of spending public funds to revitalize the downtown mall.
[b]Opposed to spending public funds to revitalize the downtown mall.
[c]Declined to be interviewed or had no opinion.
[d]Got frustrated and quit.

TABLE 14-3

25. (a) Describe as specifically as you can the target population for this survey.

(b) Compare and contrast the target population and the sampling frame for this survey.

26. (a) What is the size of the sample?

(b) Calculate the response rate in this survey. Was this survey subject to nonresponse bias?

27. (a) Can you explain the big difference in the data from interviewer to interviewer?

(b) One of the interviewers conducted the interviews at a street corner downtown. Which interviewer? Explain.

(c) Do you think the survey was subject to selection bias? Explain.

(d) Was the sampling method used in this survey the same as quota sampling? Explain.

28. Do you think this was a good survey? If you were a consultant to the Cleansburg Planning Department, could you suggest some improvements? Be specific.

Exercises 29 and 30 refer to the following story: An orange grower wishes to compute the average yield from his orchard. The orchard contains three varieties of trees: 50% of his trees are of variety A, 25% of variety B, and 25% of variety C.

29. (a) Suppose that the grower samples randomly from 300 trees of variety A, 150 trees of variety B, and 150 trees of variety C. What type of sampling is being used?

(b) Suppose that the grower selects for his sample a 10 by 30 rectangular block of 300 trees of variety A, a 10 by 15 rectangular block of 150 trees of variety B, and a 10 by 15 rectangular block of 150 trees of variety C. What type of sampling is being used?

30. (a) Suppose that in his survey, the grower found that each tree of variety A averages 100 oranges, each tree of variety B averages 50 oranges, and each tree of variety C averages 70 oranges. Estimate the average yield per tree of his orchard.

(b) Is the yield you found in (a) a parameter or a statistic? Explain.

31. You are a fruit wholesaler. You have just received 250 crates of pineapples: 75 crates came from supplier A, 75 crates from supplier B, and 100 crates from supplier C. You wish to determine if the pineapples are good enough to ship to your best customers by inspecting a sample of $n = 20$ crates. Describe how you might implement each of the following sampling methods.

- **(a)** Simple random sampling
- **(b)** Convenience sampling
- **(c)** Stratified sampling
- **(d)** Quota sampling

32. The Dean of Students at Tasmania State University wants to determine how many undergraduates at TSU are familiar with a new financial aid program offered by the university. There are 15,000 undergraduates at TSU, so it is too expensive to conduct a census. Instead, the dean decides to conduct a survey using a sample of 150 undergraduates. Describe how the dean might implement each of the following sampling methods. (See also Exercises 57 through 60.)

- **(a)** Simple random sampling
- **(b)** Convenience sampling
- **(c)** Stratified sampling
- **(d)** Quota sampling

14.3 Cause and Effect

Exercises 33 through 36 refer to the following story: The manufacturer of a new vitamin (vitamin X) decides to sponsor a study to determine the vitamin's effectiveness in curing the common cold. Five hundred college students having a cold were recruited from colleges in the San Diego area and were paid to participate as subjects in this study. The subjects were each given two tablets of vitamin X a day. Based on information provided by the subjects themselves, 457 of the 500 subjects were cured of their colds within 3 days. (The average number of days a cold lasts is 4.87 days.) As a result of this study, the manufacturer launched an advertising campaign based on the claim that "vitamin X is more than 90% effective in curing the common cold."

33. **(a)** Describe as specifically as you can the target population for the study.

(b) Describe the sampling frame for the study.

(c) Describe the sample used for the study.

34. **(a)** Was the study a controlled study? Explain.

(b) List four possible causes other than the effectiveness of vitamin X itself that could have confounded the results of the study.

35. List four different problems with the study that indicate poor design.

36. Make some suggestions for improving the study.

Exercises 37 through 40 refer to a clinical study conducted at the Houston Veterans Administration Medical Center on the effectiveness of knee surgery to cure degenerative arthritis (osteoarthritis) of the knee. Of the 324 individuals who met the inclusion criteria for the study, 144 declined to participate. The researchers randomly divided the remaining 180 subjects into three groups: One group received a type of arthroscopic knee surgery called debridement; a second group received a type of arthroscopic knee surgery called lavage; and a third group received skin incisions to make it look like they had had arthroscopic knee surgery, but no actual surgery was performed. The patients in the study did not know which group they were in and in particular did not know if they were receiving the real surgery or simulated surgery. All the patients who participated in the study were evaluated for two years after the procedure. In the two-year follow-up, all three groups said that they had slightly less pain and better knee movement, but the "fake" surgery group often reported the best results. [Source: *New England Journal of Medicine,* 347, no. 2 (July 11, 2002): 81–88.]

37. **(a)** Describe as specifically as you can the target population for this study.

(b) Describe the sample.

38. **(a)** Was the sample chosen by random sampling? Explain.

(b) Was this study a controlled placebo experiment? Explain.

(c) Describe the treatment group(s) in this study.

39. **(a)** Could this study be considered a randomized controlled study? Explain.

(b) Was this study blind, double blind, or neither?

40. As a result of this study, the Department of Veterans Affairs issued an advisory to its doctors recommending that they stop using arthroscopic knee surgery for patients suffering from osteoarthritis. Do you agree or disagree with the advisory? Explain your answer.

Exercises 41 through 44 refer to a clinical trial named APPROVe designed to determine whether Vioxx, a medication used for arthritis and acute pain, was effective in preventing the recurrence of colorectal polyps in patients with a history of colorectal adenomas. APPROVe was conducted between 2002 and 2003 and involved 2586 participants, all of whom had a history of colorectal adenomas. The participants were randomly divided into two groups: 1287 were given 25 milligrams of Vioxx daily for the duration of the clinical trial (originally intended to last three years), and 1299 patients were given a placebo. Neither the participants nor the doctors involved in the clinical trial knew who was in which group. During the trial, 72 of the participants had cardiovascular events (mostly heart attacks or strokes). Later it was found that 46 of these people were from the group taking the Vioxx and only 26 were from the group taking the placebo. Based on these results, the clinical trial was stopped in 2003 and Vioxx was taken off the market in 2004.

41. Describe as specifically as you can the target population for APPROVe.

42. Describe the sample for APPROVe.

43. **(a)** Describe the control and treatment groups in APPROVe.

(b) APPROVe can be described as a *double-blind, randomized controlled placebo study*. Explain why each of these terms applies.

44. What conclusions would you draw from APPROVe?

Exercises 45 through 48 refer to a study on the effectiveness of an HPV (human papilloma virus) vaccine conducted between October 1998 and November 1999. HPV is the most common sexually transmitted infection—more than 20 million Americans are infected with HPV—but most HPV infections are benign, and in most cases infected individuals are not even aware they are infected. (On the other hand, some HPV infections can lead to cervical cancer in women.) The researchers recruited 2392 women from 16 different centers across the United States to participate in the study through advertisements on college campuses and in the surrounding communities. To be eligible to participate in the study, the subjects had to meet the following criteria: (1) be a female between 16 and 23 years of age, (2) not be pregnant, (3) have no prior abnormal Pap smears, and (4) report to have had sexual relations with no more than five men. At each center, half of the participants were randomly selected to receive the HPV vaccine, and the other half received a placebo injection. After 17.4 months, the incidence of HPV infection was 3.8 per 100 woman-years at risk in the placebo group and 0 per 100 woman-years at risk in the vaccine group. In addition, all nine cases of HPV-related cervical precancerous growths occurred among the placebo recipients. [Source: *New England Journal of Medicine*, 347, no. 21 (November 21, 2002): 1645–1651.]

45. **(a)** Describe as specifically as you can the target population for the study.

(b) Describe the sampling frame for the study.

46. **(a)** Describe the sample for the study.

(b) Was the sample chosen using random sampling? Explain.

47. **(a)** Describe the treatment group in the study.

(b) Could this study be considered a double-blind, randomized controlled placebo study? Explain.

48. Carefully state what a legitimate conclusion from this study might be.

Exercises 49 through 52 refer to a landmark study conducted in 1896 in Denmark by Dr. Johannes Fibiger, who went on to receive the Nobel Prize in Medicine in 1926. The purpose of the study was to determine the effectiveness of a new serum for treating diphtheria, a common and often deadly respiratory disease in those days. Fibiger conducted his study over a one-year period (May 1896–April 1897) in one particular Copenhagen hospital. New diphtheria patients admitted to the hospital received different treatments based on the day of admission. In one set of days (call them "even" days for convenience), the patients were treated with the new serum daily and received the standard treatment. Patients admitted on alternate days (the "odd" days) received just the standard treatment. Over the one-year period of the study, eight of the 239 patients admitted on the "even" days and treated with the serum died, whereas 30 of the 245 patients admitted on the "odd" days died.

49. **(a)** Describe as specifically as you can the target population for Fibiger's study.

(b) Describe the sampling frame for the study.

50. **(a)** Describe the sample for Fibiger's study.

(b) Is selection bias a possible problem in this study? Explain.

51. **(a)** Describe the control and treatment groups in Fibiger's study.

(b) What conclusions would you draw from Fibiger's study? Explain.

52. In a different study on the effectiveness of the diphtheria serum conducted prior to Fibiger's study, patients in one Copenhagen hospital were chosen to be in the treatment group and were given the new serum, whereas patients in a different Copenhagen hospital were chosen to be in the control group and were given the standard treatment. Fibiger did not believe that the results of this earlier study could be trusted. What are some possible confounding variables that may have affected the results of this earlier study?

Exercises 53 through 56 refer to a study conducted between 2008 and 2010 on the effectiveness of saw palmetto fruit extracts at treating lower urinary tract symptoms in men with prostate enlargement. (Saw palmetto is a widely used over-the-counter supplement for treating urinary tract symptoms.) In the study, 369 men aged 45 years or older were randomly divided into a group taking a daily placebo and a group taking saw palmetto. Participants were nonpaid volunteers recruited at 11 North American sites. All had moderately impaired urinary flow. Because the saw palmetto extract has a mild odor, the doses were administered using gelcaps to eliminate the odor. In an analysis of the 306 men who completed the 72-week trial, both groups had similar small improvements in mean symptom scores, but saw palmetto conferred no benefit over placebo on symptom scores or on any secondary outcomes. [Source: *Journal of the American Medical Association*, 306(12), 2011, 1344–1351.]

53. **(a)** Describe as specifically as you can the target population for the study.

(b) Compare and contrast the sampling frame and target population for the study.

54. **(a)** Describe the sample for the study.

(b) Was the sample chosen using random sampling? Explain.

55. **(a)** Describe the treatment group in the study.

(b) Explain why the experimenters took the trouble to cover the mild odor of saw palmetto to the point of packaging the doses in the form of gelcaps.

(c) Was this study a blind, randomized, controlled placebo study? Explain.

56. If you were a 55-year-old male with an enlarged prostate taking saw palmetto daily, how might you react to this study?

JOGGING

Exercises 57 through 60 refer to the following story (see also Exercise 32): The Dean of Students at Tasmania State University wants to determine how many undergraduates at TSU are familiar with a new financial aid program offered by the university. There are 15,000 undergraduates at TSU, so it is too expensive to conduct a census. The following sampling method, known as ***systematic sampling,*** *is used to choose a representative sample of undergraduates to poll. Start with the registrar's alphabetical listing containing the names of all undergraduates. Randomly pick a number between 1 and 100, and count that far down the list. Take that name and every 100th name after it. For example, if the random number chosen is 73, then pick the 73rd, 173rd, 273rd, and so forth, names on the list.*

57. (a) Compare and contrast the sampling frame and the target population for this survey.

(b) Give the exact N-value of the population.

58. (a) Find the sampling proportion.

(b) Suppose that the survey had a response rate of 90%. Find the size n of the sample.

59. (a) Explain why the method used for choosing the sample is not simple random sampling.

(b) If 100% of those responding claimed that they were not familiar with the new financial aid program offered by the university, is this result more likely due to sampling variability or to sample bias? Explain.

60. (a) Suppose that the survey had a response rate of 90% and that 108 students responded that they were not familiar with the new financial aid program. Give a statistic for the total number of students at the university who were not familiar with the new financial aid program.

(b) Do you think the results of this survey will be reliable? Explain.

61. Imagine you have a very large coin jar full of nickels, dimes, and quarters. You would like to know how much money you have in the jar, but you don't want to go through the trouble of counting all the coins. You decide to estimate how many nickels, dimes, and quarters are in the jar using the capture-recapture method. After shaking the jar well, you draw a first sample of 150 coins and get 36 quarters, 45 nickels, and 69 dimes. Using a permanent ink marker you tag each of the 150 coins with a black dot and put the coins back in the jar, shake the jar really well to let the tagged coins mix well with the rest, and draw a second sample of 100 coins. The second sample has 28 quarters, 29 nickels, and 43 dimes. Of these, 4 quarters, 5 nickels, and 8 dimes have black dots. Estimate how much money is in the jar. (*Hint*: You will need a separate calculation for estimating the quarters, nickels, and dimes in the jar.)

62. One implicit assumption when using the capture-recapture method to estimate the size of a population is that the capture process is truly random, with all individuals having the same likelihood of being captured. Sometimes that is not true, and some populations have a large number of individuals that are "trap-happy" individuals (more prone to capture than others, more likely to take the bait, less cagey, slower, dumber, etc.). If that were the case, would the capture-recapture method be likely to *underestimate* or *overestimate* the size of the population? Explain your answer.

63. One implicit assumption when using the capture-recapture method to estimate the size of a population is that when individuals are tagged in the capture stage, these individuals are not affected in any harmful way by the tags. Sometimes, though, tagged individuals become affected, with the tags often making them more likely prey to predators (imagine, for example, tagging fish with bright yellow tags that make them stand out or tagging a bird on a wing in such a way that it affects its ability to fly). If that were the case, would the capture-recapture method be likely to *underestimate* or *overestimate* the size of the population? Explain your answer.

64. Informal surveys. In everyday life we are constantly involved in activities that can be described as *informal surveys*, often without even realizing it. Here are some examples.

(i) Al gets up in the morning and wants to know what kind of day it is going to be, so he peeks out the window. He doesn't see any dark clouds, so he figures it's not going to rain.

(ii) Betty takes a sip from a cup of coffee and burns her lips. She concludes that the coffee is too hot and decides to add a tad of cold water to it.

(iii) Carla got her first Math 101 exam back with a C grade on it. The students sitting on each side of her also received C grades. She concludes that the entire Math 101 class received a C on the first exam.

For each of the preceding examples,

(a) describe the population.

(b) discuss whether the sample is random or not.

(c) discuss the validity of the conclusions drawn. (There is no right or wrong answer to this question, but you should be able to make a reasonable case for your position.)

65. Read the examples of informal surveys given in Exercise 64. Give three new examples of your own. Make them as different as possible from the ones given in Exercise 64 [changing coffee to soup in (ii) is not a new example].

66. Leading-question bias. The way the questions in many surveys are phrased can itself be a source of bias. When a question is worded in such a way as to predispose the respondent to provide a particular response, the results of the survey are tainted by a special type of bias called *leading-question bias*. The following is an extreme hypothetical situation intended to drive the point home.

In an effort to find out how the American taxpayer feels about a tax increase, the Institute for Tax Reform conducts a "scientific" one-question poll.

Are you in favor of paying higher taxes to bail the federal government out of its disastrous economic policies and its mismanagement of the federal budget? Yes____. No____.

Ninety-five percent of the respondents answered no.

(a) Explain why the results of this survey might be invalid.

(b) Rephrase the question in a neutral way. Pay particular attention to highly charged words.

(c) Make up your own (more subtle) example of leading-question bias. Analyze the critical words that are the cause of bias.

67. Question order bias. In July 1999, a Gallup poll of 1061 people asked the following two questions:

- *As you may know, former Major League Baseball player Pete Rose is ineligible for baseball's Hall of Fame because of charges that he gambled on baseball games. Do you think he should or should not be eligible for admission to the Hall of Fame?*
- *As you may know, former Major League Baseball player Shoeless Joe Jackson is ineligible for baseball's Hall of Fame because of charges that he took money from gamblers in exchange for fixing the 1919 World Series. Do you think he should or should not be eligible for admission to the Hall of Fame?*

The order in which the questions were asked was random: Approximately half of the people polled were asked about Rose first and Jackson second; the other half were asked about Jackson first and Rose second. When the order of the questions was Rose first and Jackson second, 64% of the respondents said that Rose should be eligible for admission to the Hall of Fame and 33% said that Jackson should be eligible for admission to the Hall of Fame. When the order of the questions was Jackson first and Rose second, 52% said that Rose should be eligible for admission to the Hall of Fame and 45% said that Jackson should be eligible for admission to the Hall of Fame. Explain why you think each player's support for eligibility was less (by 12% in each case) when the player was second in the order of the questions.

68. Today, most consumer marketing surveys are conducted by telephone. In selecting a sample of households that are representative of all the households in a given geographical area, the two basic techniques used are (1) randomly selecting telephone numbers to call from the local telephone directory or directories and (2) using a computer to randomly generate seven-digit numbers to try that are compatible with the local phone numbers.

(a) Briefly discuss the advantages and disadvantages of each technique. In your opinion, which of the two will produce the more reliable data? Explain.

(b) Suppose that you are trying to market burglar alarms in New York City. Which of the two techniques for selecting the sample would you use? Explain your reasons.

69. The following two surveys were conducted in January 1991 to assess how the American public viewed media coverage of the Persian Gulf war. Survey 1 was an Area Code 900 telephone poll survey conducted by *ABC News*. Viewers were asked to call a certain 900 number if they believed that the media were doing a good job of covering the war and a different 900 number if they believed that the media were not doing a good job in covering the war. Each call cost 50 cents. Of the 60,000 respondents, 83% believed that the media were not doing a good job. Survey 2 was a telephone poll of 1500 randomly selected households across the United States conducted by the *Times-Mirror* survey organization. In this poll, 80% of the respondents indicated that they approved of the press coverage of the war.

(a) Briefly discuss survey 1, indicating any possible types of bias.

(b) Briefly discuss survey 2, indicating any possible types of bias.

(c) Can you explain the discrepancy between the results of the two surveys?

(d) In your opinion, which of the two surveys gives the more reliable data?

70. An article in the *Providence Journal* about automobile accident fatalities includes the following observation: "Forty-two percent of all fatalities occurred on Friday, Saturday, and Sunday, apparently because of increased drinking on the weekends."

(a) Give a possible argument as to why the conclusion drawn may not be justified by the data.

(b) Give a different possible argument as to why the conclusion drawn may be justified by the data after all.

71. (a) For the capture-recapture method to give a reasonable estimate of N, what assumptions about the two samples must be true?

(b) Give reasons why the assumptions in (a) may not hold true in many situations.

72. Consider the following hypothetical survey designed to find out what percentage of people cheat on their income taxes.

Fifteen hundred taxpayers are randomly selected from the Internal Revenue Service (IRS) rolls. These individuals are then interviewed in person by representatives of the IRS and read the following statement.

"This survey is for information purposes only. Your answer will be held in strict confidence. Have you ever cheated on your income taxes? Yes____. No____."

Twelve percent of the respondents answered yes.

(a) Explain why the above figure might be unreliable.

(b) Can you think of ways in which a survey of this type might be designed so that more reliable information could be obtained? In particular, discuss who should be sponsoring the survey and how the interviews should be carried out.

RUNNING

73. One of the problems with the capture-recapture method is that in some animal populations there are individuals that are trap-happy (easy to trap) and others that are more cagey and hard to trap. Too many trap-happy individuals can skew the data (see Exercise 62). A **removal method** is a method for estimating the N-value of a population that takes into account the existence of trap-happy individuals by trapping them and removing them. In the first "capture," individuals from the general population are trapped, counted, and removed from the habitat so that they can't be trapped again. In the "recapture," individuals from the remaining population (those that had not been trapped before) are trapped and counted. The number of individuals trapped in the capture can be denoted by pN, where p denotes the fraction of the population trapped and N is the size of the population. The number of individuals left after the removal is $(1-p)N$. If we assume that the number of individuals trapped in each capture represents the same fraction of the population, then the number of individuals trapped in the recapture should be $p(1-p)N$. From the two equations (pN = number of individuals trapped in the capture; $p(1-p)N$ = number of individuals trapped in the recapture) we can solve for N and get an estimate of the population.

Suppose 250 individuals are trapped in the capture stage and removed from the population, and 150 individuals are trapped in the recapture stage. Estimate the size of the population.

74. Darroch's method. is a method for estimating the size of a population using multiple (more than two) captures. For example, suppose that there are four captures of sizes n_1, n_2, n_3, and n_4, respectively, and let M be the total number of *distinct* individuals caught in the four captures (i.e., an individual that is captured in more than one capture is counted only once). Darroch's method gives the estimate for N as the unique solution of the equation $(1-\frac{M}{N}) = (1-\frac{n_1}{N})(1-\frac{n_2}{N})(1-\frac{n_3}{N})(1-\frac{n_4}{N})$.

(a) Suppose that we are estimating the size of a population of fish in a pond using four separate captures. The sizes of the captures are $n_1 = 30, n_2 = 15, n_3 = 22$, and $n_4 = 45$. The number of distinct fish caught is $M = 75$. Estimate the size of the population using Darroch's formula.

(b) Show that with just two captures Darroch's method gives the same answer as the capture-recapture method.

15

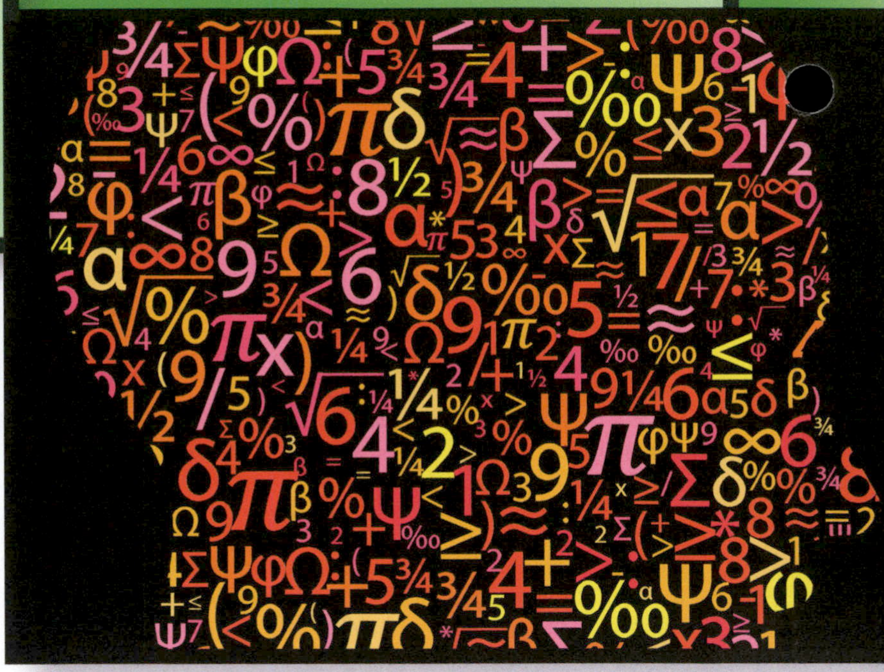

SAT test scores—enough data to blow your mind . . . unless you know how to organize it, package it, and summarize it. (For details, see Examples 15.6, 15.13, and 15.16.)

Graphs, Charts, and Numbers

The Data Show and Tell

In 2015 a total of 1,698,521 college-bound students took the SAT. Put them all in one place and you would have the fifth largest city in the United States—larger than Boston and San Francisco put together. Each of the 1,698,521 tests taken generated several numbers: individual sub-scores in three separate subject areas (Mathematics, Critical Thinking, and Writing), and percentiles for each subject area plus a composite percentile for the entire test. Put it all together and you have more than 10 million numbers to play with—a vast sea of mysterious but potentially useful data.

When it comes to rapid mental processing, we humans do well with images and words, but not so well with numbers. Psychologists have found that a typical human mind can juggle the relationships between six, at most seven, numbers at one time. Anything beyond that requires a more deliberate and purposeful effort. Certainly, by the time you get to a dozen or more numbers, some organization and management of the *data* will be required. And what about managing and organizing a *data set* of more than 10 million numbers? No worries.

There are essentially two major strategies for describing a large data set. One of them is to take advantage of the human mind's great talent for visualization and describe the data using pictures (graphs and charts). In Section 15.1 we cover some of the standard graphical tools for describing data—*bar charts, pictograms, pie charts, and histograms*—and discuss a few basic rules for when and how to use each. The second strategy for describing a large data set is to use *numerical summaries*. While a few numbers can't tell the full story behind 10 million numbers, it's amazing how much mileage one can get from a few well-chosen numerical summaries. In Section 15.2 we cover the classic numerical summaries for a data set—*averages* (*means*), *medians*, *quartiles*, *percentiles*, and *five-number summaries*. All of these numerical summaries help us identify where the data set sits in relation to the possible values that the data could take and are called *measures of location*. A second group of useful numbers consists of those that provide information about the *spread* of the data set. In Section 15.3 we briefly discuss the three most commonly used *measures of spread*—the range, the interquartile range, and the standard deviation.

> "I've come loaded with statistics, for I've noticed that a man can't prove anything without statistics. No man can."
>
> – Mark Twain

15.1 Graphs and Charts

The old adage "a picture is worth 1000 words" is even more valid when applied to numbers instead of words. A single, well-chosen graphical display can say a lot about the patterns that lie hidden within a bunch of numbers, especially when there are a lot of them numbers. OK, we are going to need a slightly more formal terminology than "them numbers," so we start with a couple of basic definitions.

- **Data set.** A *data set* is a collection of numbers that represent all the values obtained when measuring some characteristic of a population (such as a test score, a stock price, a sales figure, etc.). [*Note*: Technically speaking a data set is not a set because repeated numbers are included—in a real set, repeats are not included.]

- **Data point.** Each of the individual numbers in a data set is called a *data point*. In all the data sets we will consider in this chapter, the data points will be rational numbers (either whole numbers, decimals, or percentages). We will let N denote the total number of data points in the data set, and we will refer to N as the *size* of the data set.

Creating a graphical representation of a data set is somewhat like taking a photograph, except that instead of taking snapshots of people or nature we are creating a picture of the data. And, just like photography is a skill that can be elevated into an art form, creating the most appropriate and efficient graph or chart for a data set is both a skill and an art. In this section we are only going to cover the most basic types of graphs and charts, and discuss when to use one or the other.

The choice of what is the most appropriate type of graph or chart to use in graphically displaying a data set depends on many criteria, and many of these criteria are not quantifiable: Who is your audience? Do you have a message? Are you trying to sell a product or idea? Are you looking for particular patterns? To keep things simple, in this section we will focus on two basic but critical questions: (1) How large is the data set? and (2) Does the data represent a *discrete* or a *continuous* variable?

- **Discrete variable.** A discrete variable is one in which the values of the variable can only change by minimum increments. In other words, two different values of the variable cannot be arbitrarily close to each other—there is a minimum gap required.

A classic example of a discrete variable is a test score. If the test is multiple-choice then it is obvious—the minimum gap between two scores is one point (assuming each question is worth a point). Even when the scoring is more refined, such as in an essay, there is a minimum gap between scores—half a point, a quarter of a point, a tenth of a point. There is always a limit as to how far the person doing the scoring can go in finessing the differences between essays.

- **Continuous variables.** Continuous variables are variables that can take infinitely many values, and those values can differ by arbitrarily small amounts.

A good example of a continuous variable is the measured distance between two points. In theory, two different distance measurements could differ by an inch, a tenth of an inch, or one-millionth of an inch—there is no minimum gap that must separate the two. (In practice, of course, there is the problem of how refined can our measurement be—if the difference between the two distances is one-millionth of an inch we probably won't be able to measure the difference.)

The case of money (especially when we are dealing with large sums) is particularly interesting because money is really a discrete variable (the minimum gap between two sums of money is one cent), but it is usually described as if it were a continuous variable. In other words, when two large sums of money differ by a penny we will think of that difference as being "infinitely small."

Bar Graphs, Pictograms, and Line Graphs

We will start our exploration of data sets and their graphical representations with a fictitious example. Except for the details, this example represents a situation familiar to every college student.

EXAMPLE 15.1 STAT 101 MIDTERM SCORES

As usual, the day after the midterm exam in his Stat 101 class, Dr. Blackbeard has posted the scores online (Table 15-1). The data set consists of $N = 75$ data points (the number of students who took the test). Each data point (listed under the "Score" columns) is an integer between 0 and 25 (Dr. Blackbeard gives no partial credit). Note that the numbers listed under the "ID" columns are not data points—they are the last four digits of the student IDs used as substitutes for their names to protect the students' rights of privacy.

Like students everywhere, the students in the Stat 101 class have one question foremost on their mind when they look at Table 15-1: How did I do? Each student can answer this question directly from the table. It's the next question that is statistically much more interesting. How did the class as a whole do?

The first step in organizing the data set in Table 15-1 is to create a **frequency table** such as Table 15-2. In this table, the number below each score gives the *frequency* of the score—that is, the number of students getting that particular score. We can readily see from Table 15-2 that there was one student with a score of 1, one with a score of 6, two with a score of 7, six with a score of 8, and so on. Notice that the scores with a frequency of zero are not listed in the table.

ID	Score	ID	Score	ID	Score	ID	Score	ID	Score
1257	12	2651	10	4355	8	6336	11	8007	13
1297	16	2658	11	4396	7	6510	13	8041	9
1348	11	2794	9	4445	11	6622	11	8129	11
1379	24	2795	13	4787	11	6754	8	8366	13
1450	9	2833	10	4855	14	6798	9	8493	8
1506	10	2905	10	4944	6	6873	9	8522	8
1731	14	3269	13	5298	11	6931	12	8664	10
1753	8	3284	15	5434	13	7041	13	8767	7
1818	12	3310	11	5604	10	7196	13	9128	10
2030	12	3596	9	5644	9	7292	12	9380	9
2058	11	3906	14	5689	11	7362	10	9424	10
2462	10	4042	10	5736	10	7503	10	9541	8
2489	11	4124	12	5852	9	7616	14	9928	15
2542	10	4204	12	5877	9	7629	14	9953	11
2619	1	4224	10	5906	12	7961	12	9973	10

TABLE 15-1 ■ Stat 101 midterm scores ($N = 75$)

Score	1	6	7	8	9	10	11	12	13	14	15	16	24
Frequency	1	1	2	6	10	16	13	9	8	5	2	1	1

TABLE 15-2 ■ Frequency table for the Stat 101 midterm scores

Table 15-2 shows that the $N = 75$ scores fall into $M = 13$ different **actual values** (1, 6, 7, . . . , 15, 16, 24). If we think of the actual values as "bins" where we can place the data, we get a graph such as the one shown in Fig. 15-1(a). A slightly simpler version of the same idea is shown in Fig. 15-1(b).

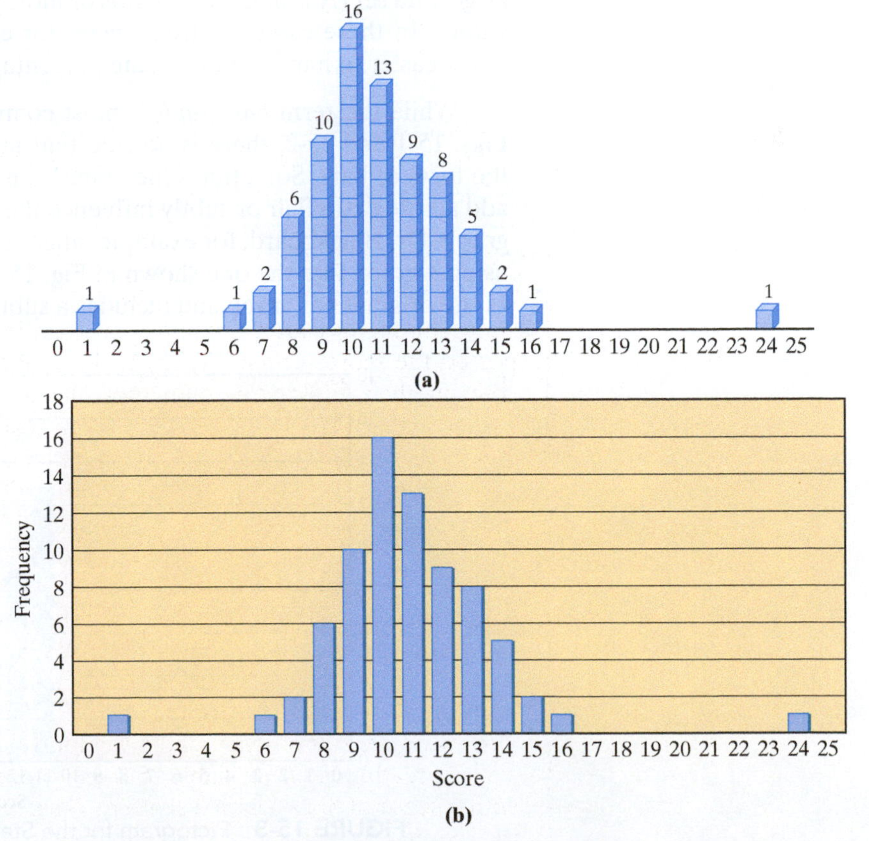

FIGURE 15-1 Bar graphs for the Stat 101 data set.

Figures 15-1(a) and (b) are both examples of a type of graph called a **bar graph**. Figure 15-1(a) is the fancier, 3D version; Fig. 15-1(b) is a more plain 2D version, but they both illustrate the data set equally well. When looking at either one, one of the first things we would notice are the two *outliers* (i.e., data points that do not fit-in with the rest of the data). In the Stat 101 data set there are two obvious outliers—the score of 24 (head and shoulders above the rest of the class) and the score of 1 (lagging way behind the pack).

Sometimes it is useful to draw a bar graph using *relative frequencies*—that is, making the heights of the bars represent percentages of the population. Figure 15-2 shows a *relative frequency bar graph* for the Stat 101 data set. Notice that we indicated on the graph that we are dealing with percentages rather than absolute frequencies and that the size of the data set is $N = 75$. This allows anyone who wishes to do so to compute the actual frequencies. For example, Fig. 15-2 indicates that 12% of the 75 students scored a 12 on the exam, so the actual frequency is given by $75 \times 0.12 = 9$ students. The change from actual frequencies to percentages (or vice versa) does not change the shape of the graph—it is basically a change of scale.

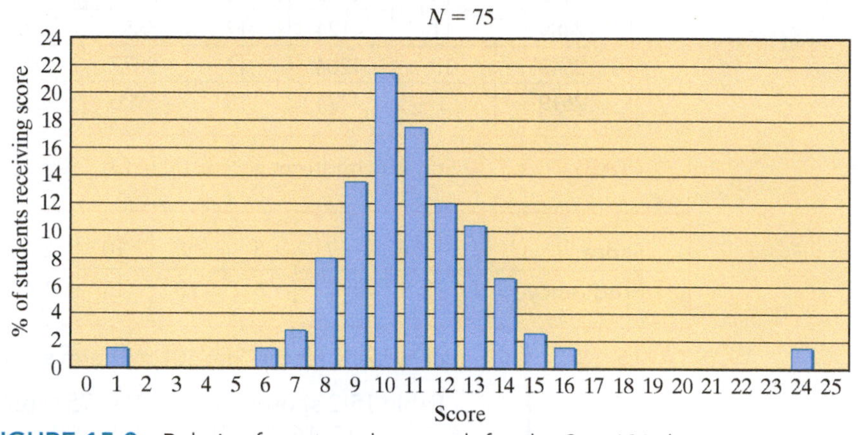

FIGURE 15-2 Relative frequency bar graph for the Stat 101 data set.

Relative frequency bar graphs are especially convenient when we have a very large data set (N is in the thousands or more) but a relatively small number of actual values. In these cases the frequencies for each bar tend to be very large and it is much easier to handle things using percentages.

While the term *bar graph* is most commonly used for graphs like the ones in Figs. 15-1 and 15-2, there is no rule that mandates that the columns have to take the form of bars. Sometimes the "bins" can be filled with symbolic images that can add a little extra flair or subtly influence the content of the information given by the graph. Dr. Blackbeard, for example, might have chosen to display the midterm data using a graph like the one shown in Fig. 15-3, which conveys the same information as the original bar graph and includes a subtle individual message to each student.

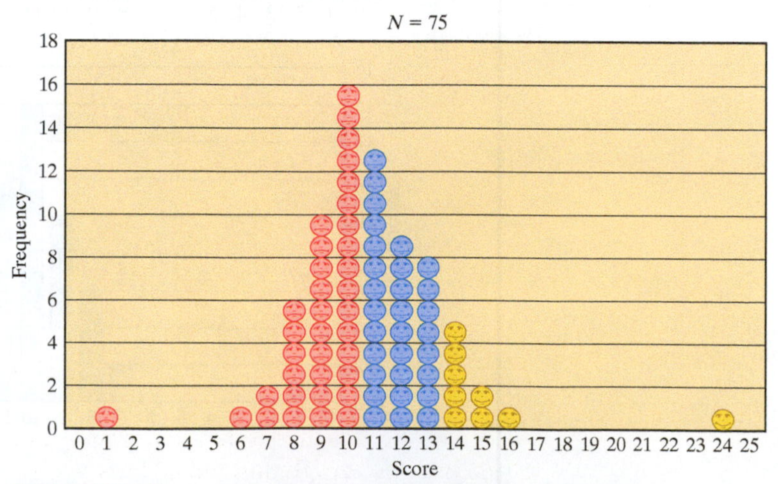

FIGURE 15-3 Pictogram for the Stat 101 data set.

Frequency charts that use icons or symbolic images instead of bars to display the frequencies are commonly referred to as **pictograms**. The point of a pictogram is that a graph is often used not only to inform but also to impress and persuade, and, in such cases, a well-chosen icon or image can be a more effective tool than just a bar.

EXAMPLE 15.2 SELLING THE XYZ CORPORATION

Figure 15-4(a) is a pictogram showing the growth in yearly sales of the XYZ Corporation between 2011 and 2016. It's a good picture to show at a shareholders meeting, but the picture is actually quite misleading. Figure 15-4(b) shows a pictogram for exactly the same data with a much more accurate picture of how well the XYZ Corporation had been doing.

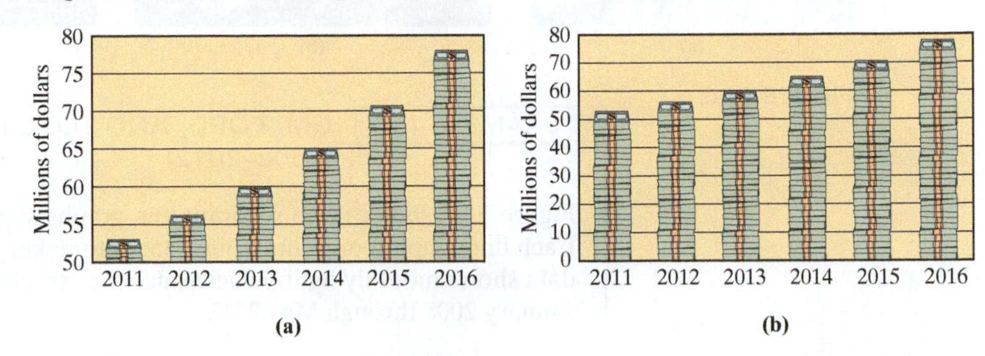

FIGURE 15-4 XYZ Corp. annual sales. Two pictograms for the same data set.

The difference between the two pictograms can be attributed to a couple of standard tricks: (1) stretching the scale of the vertical axis and (2) "cheating" on the choice of starting value on the vertical axis. As an educated consumer, you should always be on the lookout for these tricks. In graphical descriptions of data, a fine line separates objectivity from propaganda.

For convenience let's call the number of bars in a bar graph M. In a typical bar graph M is a number in single or double digits, and rarely will you see a bar graph with $M > 100$. Even if you push the columns together and make them really skinny you can only fit so many in a reasonably sized window. Figure 15-5 shows a bar graph with $M = 80$, and that is already pushing the upper limit of what a reasonably sized bar graph can accommodate.

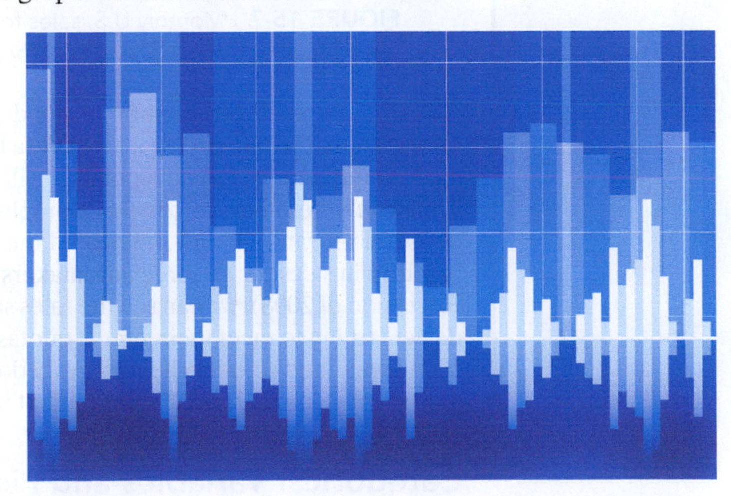

FIGURE 15-5

An alternative to a bar graph that eliminates the need for bars is a line graph. In a **line graph**, instead of bars we use points (small circles or small squares) with adjacent points connected by lines [Fig. 15-6(a)]. Line graphs are particularly useful when the number of values plotted is large [Fig. 15-6(b)], or when the graph represents a *time-series* (i.e., the horizontal axis represents a time variable) such as when tracking the price of a company's stock in a stock market [Fig. 15-6(c)].

(a) (b) (c)

FIGURE 15-6

EXAMPLE 15.3 GM, FORD, AND TOYOTA MONTHLY SALES (2008–2012)

Figure 15-7 shows three separate line graphs superimposed on a single time line. Each line graph represents a different automaker (GM, Ford, and Toyota), and the data shows monthly light-vehicle sales (i.e., trucks are not included) for the period January 2008 through May 2012.

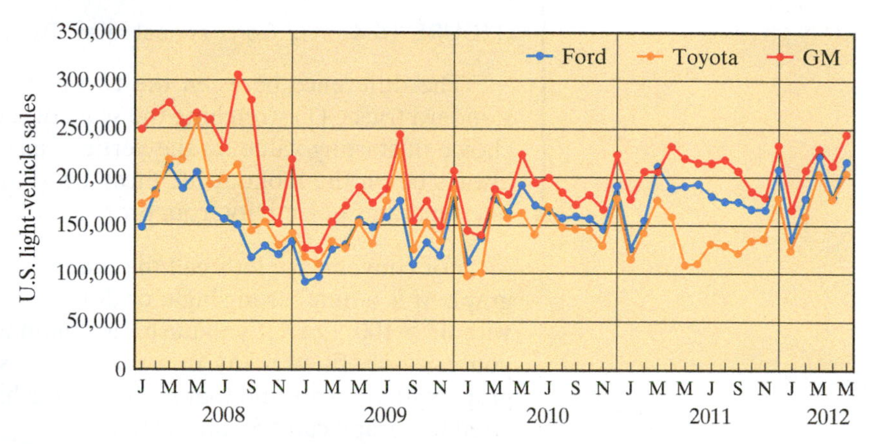

FIGURE 15-7 Monthly U.S. sales for GM, Ford, and Toyota light vehicles (Jan. 2008–May 2012). Data from *The Wall Street Journal*.

This graph illustrates how a good graphical display can pack a lot of useful information. First, there are a lot of data being displayed—53 different data points for each manufacturer, 159 data points in all, and yet, the graph does not appear crowded. This would be much harder to accomplish using bar graphs. Secondly, there is a lot of useful information lurking behind the visible patterns in the data—we can see the big drop in sales for all three automakers at the end of 2008 (the aftermath of the Big Recession of 2008), we can see the sales spiking every December (Christmas bonuses, dealers trying to meet their annual quotas), we can see the 2012 sales climbing back to 2008 levels, and so on. When properly done, a statistical graph can be a great exploratory tool—all you have to do is look at it carefully and ask the right questions.

Categorical Variables and Pie Charts

A variable need not always represent a measurable quantity—variables can also describe non-numerical characteristics of a population such as gender, ethnicity, nationality, emotions, feelings, actions, etc. Variables of this type are called **categorical** (or **qualitative**) variables.

In some ways, categorical variables must be treated differently from numerical variables—you can't add, rescale, or average categories—but when it comes to graphical displays of data, categorical variables can be treated much like discrete numerical variables.

EXAMPLE 15.4 HOW DO COLLEGE STUDENTS SPEND THEIR DAY?

Activity	Hours
Sleep	8.7
Leisure/Sports	4.1
Work	2.4
Education	3.3
Eat/Drink	1.0
Grooming	0.8
Travel	1.4
Other	2.3
Total	**24**

TABLE 15-3 ■ Time use of a typical college student

Table 15-3 shows a breakdown of how a typical full-time university or college student spends a typical nonholiday weekday. Time use is broken into eight major categories, including the catch-all category "Other." (*Source:* Bureau of Labor Statistics, *American Time Use Survey*. Data include individuals, ages 15 to 49, who were enrolled full time at a university or college. Data include nonholiday weekdays and are averages for 2010–2014.)

Figure 15-8(a) is a bar graph for the data in Table 15-3. The only difference between this bar graph and the others we saw earlier is the absence of a numerical horizontal axis. Since the data is categorical, there is no number line or time line for the placement of the columns. In fact, other than making sure that the bottoms of the columns are aligned, you can position the columns any way you want.

Figure 15-8(b) is a *pie chart* representing the data in Table 15-3. In a **pie chart**, the categories take the form of wedges of pie (or slices of pizza, if you prefer), with the size of each wedge (measured by the central angle of the wedge) proportional to the frequency of that category. To calculate the central angle for a category (in degrees) we first convert the frequency of that category into a relative frequency (i.e., a percentage) and then use the fact that $1\% = 3.6°$ (which follows from $100\% = 360°$).

Here is the computation of the central angle corresponding to the category "Sleep" (8.7 hours): First, $8.7/24 = 0.3625 = 36.25\%$; then, $36.25 \times 3.6° = 130.5°$. [These two steps can be nicely combined into a single calculation: $(8.7/24) \times 360° = 130.5°$. This works because converting the 0.3625 to 36.25% and multiplying by 3.6 is the same as leaving the 0.3625 alone and multiplying by 360.]

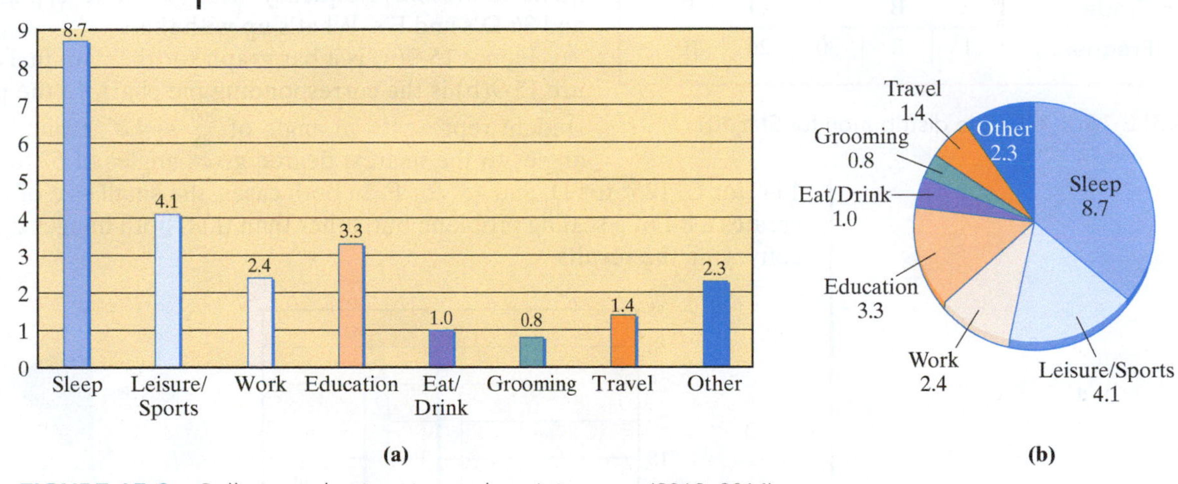

FIGURE 15-8 College and university students' time use (2010–2014).

When the computation of the central angle for a category gives you decimal parts of a degree, as it did for the Sleep category, you can round the angle to the nearest degree. In general, that is more than good enough. Table 15-4 shows the calculations for the central angles of all eight categories in Table 15-3.

Activity	Hours	Central angle
Sleep	8.7	$(8.7/24) \times 360° \approx 130°$
Leisure/Sports	4.1	$(4.1/24) \times 360° \approx 61°$
Work	2.4	$(2.4/24) \times 360° = 36°$
Education	3.3	$(3.3/24) \times 360° \approx 50°$
Eat/Drink	1.0	$(1.0/24) \times 360° = 15°$
Grooming	0.8	$(0.8/24) \times 360° = 12°$
Travel	1.4	$(1.4/24) \times 360° = 21°$
Other	2.3	$(2.3/24) \times 360° \approx 35°$
Total	**24**	**360°**

(*Note*: Decimal parts of .5 are rounded both up and down to make the total come out to 360.)

TABLE 15-4 ■ Central angle on pie chart for Table 15-3

Sometimes a variable starts out as a numerical variable, but then, as a matter of convenience, it is converted into a categorical variable. You are certainly familiar with one of the classic examples of this—your numerical scores (in an exam, or in a course) converted into categories (grades A, B, C, D, or F) according to some artificial scale [e.g., A = 90–100, B = 80–89, C = 70–79, D = 60–69, F = 0–59 or some variation thereof. (Have you ever wondered why it's not A = 80–100, B = 60–79, C = 40–59, D = 20–39, F = 0–19?)].

EXAMPLE 15.5 STAT 101 MIDTERM GRADES

We are going to take a second look at the midterm scores in Prof. Blackbeard's Stat 101 section (Example 15.1). The midterm score is a numerical variable that can take integer values between 0 and 25. Prof. Blackbeard's grading scale is A = 18–25, B = 14–17, C = 11–13, D = 9–10, and F = 0–8. (This is a somewhat unusual way to scale for grades, but Prof. Blackbeard is a pretty unusual guy and has his own ideas about how to do things.)

Grade	A	B	C	D	F
Frequency	1	8	30	26	10

TABLE 15-5 ■ Grade distribution for Stat 101 midterm

Table 15-5 is the categorical (grade) version of the original numerical (score) frequency table (Table 15-2, page 445). One A and 36 D's and F's. What's up with that?

Figure 15-9(a) is a bar graph for the data in Table 15-5. Figure 15-9(b) is the corresponding pie chart. In the pie chart, each student represents an angle of $\frac{360°}{75} = 4.8°$. Rounding the central angles to the nearest degree gives angles of 5° for A, 38° for B, 144° for C, 125° for D, and 48° for F. In both cases, the small size of the A category creates a bit of a scaling problem; but, other than that, both images do a good job of conveying the results.

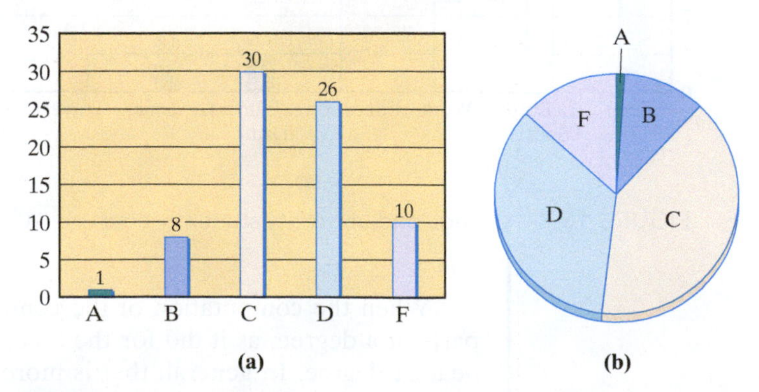

FIGURE 15-9 Grade distribution for Stat 101 midterm.

Converting numerical data into categorical data is particularly useful in situations involving large populations and large numbers of possible values for the measurements of the population. Our next example illustrates an important application of this idea.

EXAMPLE 15.6 SAT MATH SCORES (2015)

The college dreams and aspirations of millions of high school seniors often ride on their SAT scores. The SAT is divided into three subject areas (Math, Critical Reading, and Writing), with separate scores in each subject area that range from a minimum of 200 to a maximum of 800 points and going up in increments of 10 points. There is a total of 61 possible such scores (200, 210, 220, . . . , 790, 800) for each subject area—that's a lot of possibilities. In this example, we will look at the 2015 Math scores only (see Exercises 15 and 16 for the Critical Reading and Writing scores).

A total of $N = 1{,}698{,}521$ college-bound seniors took the SAT in 2015. A full breakdown of the number of students by individual score would require a frequency table with 61 entries, or a bar graph with 61 columns, or a pie chart with 61 wedges. For most purposes, that's a case of too much information! As an alternative, the results are broken into six score ranges called **class intervals**: 200–290, 300–390, . . . , 600–690, and 700–800. (Think of these class intervals as analogous to letter grades except that there are no letters.) The first five class intervals contain 10 possible scores; the last class interval contains 11 possible scores.

The second column of Table 15-6 shows the aggregate results of the 2015 SAT Mathematics section. (*Source*: Data from 2015 College Bound Seniors: Total Group Profile Report. Copyright © 2015. The College Board.) With numbers as large as these, it is somewhat easier to work with percentages of the population in each class interval (column 3). Figure 15-10(a) is a relative frequency bar graph for the data in column 3 of Table 15-6; Fig. 15-10(b) shows the corresponding pie chart.

Score range	Number of test-takers	Percentage of test-takers
700–800	121,057	7.1%
600–690	299,537	17.6%
500–590	488,277	28.7%
400–490	501,418	29.6%
300–390	239,599	14.1%
200–290	48,633	2.9%
Total	**$N = 1{,}698{,}521$**	**100%**

TABLE 15-6 ■ 2015 SAT Mathematics test scores.

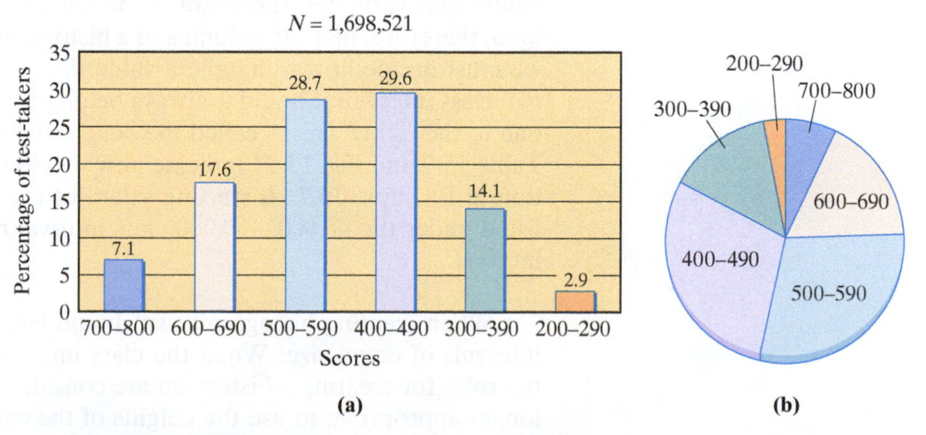

FIGURE 15-10 2015 SAT Mathematics test scores.

Continuous Variables and Histograms

When a numerical variable is continuous, its possible values can vary by infinitesimally small increments. As a consequence, there are no gaps between the class intervals, and our old way of doing things (using separate columns for each actual value or category) will no longer work. In this case we use a variation of a bar graph called a **histogram**. We illustrate the concept of a histogram in the next example.

EXAMPLE 15.7 STARTING SALARIES OF TSU GRADUATES

Suppose we want to use a graph to display the distribution of starting salaries of the 2016 graduating class at Tasmania State University ($N = 3258$).

The starting salaries range from a low of $40,350 to a high of $74,800. Based on this range and the amount of detail we want to show, we must decide on the length of the class intervals. A reasonable choice would be to use class intervals defined in increments of $5000. Table 15-7 is a frequency table for the salaries based on these class intervals. We chose a starting value of $40,000 for convenience. The third column in the table shows the data as a percentage of the population (rounded to the nearest percentage point).

The histogram showing the relative frequency of each class interval is shown in Fig. 15-11.

Salary	Frequency	Percentage
$40{,}000^{+}$–45,000	228	7%
$45{,}000^{+}$–50,000	456	14%
$50{,}000^{+}$–55,000	1043	32%
$55{,}000^{+}$–60,000	912	28%
$60{,}000^{+}$–65,000	391	12%
$65{,}000^{+}$–70,000	163	5%
$70{,}000^{+}$–75,000	65	2%
Total	**3258**	**100%**

TABLE 15-7 ■ Starting salaries of 2016 TSU graduates

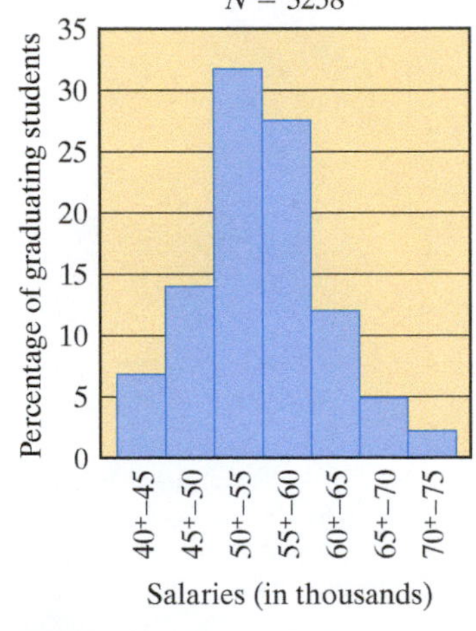

FIGURE 15-11 Histogram for data in Table 15-7.

As we can see, a histogram is very similar to a bar graph. Several important distinctions must be made, however. To begin with, because a histogram is used for continuous variables, there can be no gaps between the class intervals, and it follows, therefore, that the columns of a histogram must touch each other. In addition, we must decide how to handle a value that falls exactly on the boundary between two class intervals. Should it always belong to the class interval to the left or to the one to the right? This is called the *endpoint convention*. The "plus" superscripts in Table 15-7 and Fig. 15-11 indicate how we chose to deal with the endpoint convention in Example 15.7. A starting salary of exactly $50,000, for example, would be listed under the $45{,}000^{+}$–50,000 class interval rather than the $50{,}000^{+}$–55,000 class interval.

When creating histograms, we should try, as much as possible, to define class intervals of equal size. When the class intervals are significantly different in size, the rules for creating a histogram are considerably more complicated, since it is no longer appropriate to use the heights of the columns to indicate the frequencies of the class intervals.

Means, Medians, and Percentiles

As useful as bar graphs, pie charts, and histograms can be in describing data, in some circumstances (for example, in everyday conversation) a graph is not readily available and we have to resort to other means of description. One of the most convenient and commonly used devices is to use numerical summaries of the data. In this section we will discuss a few of the more frequently used numerical summaries for a data set.

The Average

The most commonly used number for summarizing a data set is the *average*, also called the *mean*. (There is no universal agreement as to which of these names is a better choice—in some settings *mean* is a better choice than *average*, in other settings it's the other way around. In this chapter we will use whichever seems the better choice at the moment.)

- **Average (mean).** The average A of a set of N numbers is found by adding the numbers and dividing the total by N. In other words, the average of the data set $d_1, d_2, d_3, \ldots, d_N$, is $A = (d_1 + d_2 + \cdots + d_N)/N$.

EXAMPLE 15.8 AVERAGE SCORE IN THE STAT 101 MIDTERM

In this example we will find the average test score in the Stat 101 exam first introduced in Example 15.1. To find this average we need to add all the test scores and divide by 75. The addition of the 75 test scores can be simplified considerably if we use a frequency table. (Table 15-8 is the same as Table 15-2, shown again for the reader's convenience.)

Score	1	6	7	8	9	10	11	12	13	14	15	16	24
Frequency	1	1	2	6	10	16	13	9	8	5	2	1	1

TABLE 15-8 ■ Frequency table for the Stat 101 data set

From the frequency table we can find the sum S of all the test scores as follows: Multiply each test score by its corresponding frequency, and then add these products. Thus, the sum of all the test scores is

$$S = (1 \times 1) + (6 \times 1) + (7 \times 2) + (8 \times 6) + \cdots + (16 \times 1) + (24 \times 1) = 814$$

If we divide this sum by $N = 75$, we get the average test score A (rounded to two decimal places): $A = 814/75 \approx 10.85$ points. ■

Value	Frequency
d_1	f_1
d_2	f_2
$\vdots$	$\vdots$
d_k	f_k

TABLE 15-9

In general, to find the average A of a data set given by a frequency table such as Table 15-9 we do the following:

- **Step 1.** $S = d_1 f_1 + d_2 f_2 + \cdots + d_k f_k$.
- **Step 2.** $N = f_1 + f_2 + \cdots + f_k$.
- **Step 3.** $A = S/N$.

When dealing with data sets that have outliers, averages can be quite misleading. As our next example illustrates, even a single outlier can have a big effect on the average.

EXAMPLE 15.9 STARTING SALARIES OF PHILOSOPHY MAJORS

Imagine that you just read in the paper the following remarkable tidbit: *The average starting salary of 2016 Tasmania State University graduates with a philosophy major was \$76,400 a year!* This is quite an impressive number, but before we all rush out to change majors, let's point out that one of those philosophy majors happened to be basketball star "Hoops" Tallman, who is now doing his thing in the NBA for a starting salary of \$3.5 million a year.

If we were to take this one outlier out of the population of 75 philosophy majors, we would have a more realistic picture of what philosophy majors are making. Here is how we can do it:

- The total of all 75 salaries is 75 times the average salary: $75 \times \$76{,}400 = \$5{,}730{,}000$
- The total of the other 74 salaries (excluding Hoops's cool 3.5 mill): $\$5{,}730{,}000 - \$3{,}500{,}000 = \$2{,}230{,}000$
- The average of the remaining 74 salaries: $\$2{,}230{,}000/74 \approx \$30{,}135$

Percentiles

While a single numerical summary—such as the average—can be useful, it is rarely sufficient to give a meaningful description of a data set. A better picture of the data set can be presented by using a well-organized cadre of numerical summaries. The most common way to do this is by means of *percentiles*.

- **Percentiles.** The *pth percentile* of a data set is a number X_p such that $p\%$ of the data is smaller or equal to X_p and $(100 - p)\%$ of the data is bigger or equal to X_p.

Many college students are familiar with percentiles, if for no other reason than the way they pop up in SAT reports. In all SAT reports, a given score—say a score of 590 in the Mathematics section—is identified with a percentile—say the 75th percentile. This can be interpreted to mean that 75% of those taking the test scored 590 *or less*, or, looking up instead of down, that 25% of those taking the test scored 590 *or more*.

There are several different ways to compute percentiles that will satisfy the definition, and different statistics books describe different methods. We will illustrate one such method below.

The first step in finding the pth *percentile* of a data set of N numbers is to *sort the numbers from smallest to largest*. Let's denote the sorted data values by $d_1, d_2, d_3, \ldots, d_N$, where d_1 represents the smallest number in the data set, d_2 the second smallest number, and so on. Sometimes we will also need to talk about the average of two consecutive numbers in the sorted list, so we will use more unusual subscripts such as $d_{3.5}$ to represent the average of the data values d_3 and d_4, $d_{7.5}$ to represent the average of the data values d_7 and d_8, and so on.

The next, and most important, step is to identify which d represents the pth percentile of the data set. To do this, we compute the pth *percent of N*, which we will call the **locator** and denote it by the letter L. [In other words, $L = (\frac{p}{100})N$.] If L happens to be a whole number, then the pth *percentile* will be $d_{L.5}$ (the average of d_L and d_{L+1}). If L is not a whole number, then the pth *percentile* will be d_{L^+}, where L^+ represents the value of L *rounded up*.

The procedure for finding the pth *percentile* of a data set is summarized as follows:

FINDING THE *p*TH PERCENTILE OF A DATA SET

- **Step 0.** Sort the data set from smallest to largest. Let $d_1, d_2, d_3, \ldots, d_N$ represent the sorted data.
- **Step 1.** Find the locator $L = \left(\frac{p}{100}\right)N$ (that is, L is $p\%$ of the size of the data set).
- **Step 2.** Depending on whether L is a whole number or not, the pth *percentile* is given by
  - $d_{L.5}$ if L is a whole number.
  - d_{L^+} if L is not a whole number (L^+ is L rounded up).

The following example illustrates the procedure for finding percentiles of a data set.

EXAMPLE 15.10 SCHOLARSHIPS BY PERCENTILE

To reward good academic performance from its athletes, Tasmania State University has a program in which athletes with GPAs in the 80th or higher percentile of their team's GPAs get a \$5000 scholarship and athletes with GPAs in the 55th or higher percentile of their team's GPAs who did not get the \$5000 scholarship get a \$2000 scholarship.

The women's volleyball team has $N = 15$ players on the roster. A list of their GPAs is as follows:

3.42, 3.91, 3.33, 3.65, 3.57, 3.45, 4.0, 3.71, 3.35, 3.82, 3.67, 3.88, 3.76, 3.41, 3.62

The sorted list of GPAs (from smallest to largest) is:

3.33, 3.35, 3.41, 3.42, 3.45, 3.57, 3.62, 3.65, 3.67, 3.71, 3.76, 3.82, 3.88, 3.91, 4.0

- **\$5000 scholarships:** The locator for the 80th percentile is $(0.8) \times 15 = 12$. Here the locator is a whole number, so the 80th percentile is given by $d_{12.5} = 3.85$ (the average between $d_{12} = 3.82$ and $d_{13} = 3.88$). Thus, three students (the ones with GPAs of 3.88, 3.91, and 4.0) get \$5000 scholarships.
- **\$2000 scholarships:** The locator for the 55th percentile is $(0.55) \times 15 = 8.25$. This locator is not a whole number, so we round it up to 9, and the 55th percentile is given by $d_9 = 3.67$. Thus, the students with GPAs of 3.67, 3.71, 3.76, and 3.82 (all students with GPAs of 3.67 or higher except the ones that already received \$5000 scholarships) get \$2000 scholarships.

The Median and the Quartiles

The 50th percentile of a data set is known as the **median** and denoted by M. The median splits a data set into two halves—half of the data is smaller or equal to the median and half of the data is bigger or equal to the median.

We can find the median by simply applying the definition of percentile with $p = 50$, but the bottom line comes down to this: (1) when N is *odd*, the median is the data point in position $(N + 1)/2$ of the sorted data set; (2) when N is *even*, the median is the average of the data points in positions $N/2$ and $N/2 + 1$ of the sorted data set. [All of the preceding follows from the fact that the locator for the median is $L = \frac{N}{2}$. When N is even, $\frac{N}{2}$ is a whole number; when N is odd, $\frac{N}{2}$ is not a whole number.]

FINDING THE MEDIAN OF A DATA SET

- Sort the data set from smallest to largest. Let $d_1, d_2, d_3, \ldots, d_N$ represent the sorted data.
- If N is odd, the median is $d_{(N+1)/2}$.
- If N is even, the median is the average of $d_{N/2}$ and $d_{(N/2)+1}$.

After the median, the next most commonly used set of percentiles are the **quartiles**. The *first quartile* (denoted by Q_1) is the 25th percentile, and the *third quartile* (denoted by Q_3) is the 75th percentile of the data set.

EXAMPLE 15.11 HOME PRICES IN GREEN HILLS

During the last year, 11 homes sold in the Green Hills subdivision. The selling prices, in chronological order, were $267,000, $252,000, $228,000, $234,000, $292,000, $263,000, $221,000, $245,000, $270,000, $238,000, and $255,000. We are going to find the *median* and the *quartiles* of the $N = 11$ home prices.

Sorting the home prices from smallest to largest (and dropping the 000's) gives the sorted list

221, 228, 234, 238, 245, 252, 255, 263, 267, 270, 292

The locator for the median is $(0.5) \times 11 = 5.5$, the locator for the first quartile is $(0.25) \times 11 = 2.75$, and the locator for the third quartile is $(0.75) \times 11 = 8.25$. Since these locators are not whole numbers, they must be rounded up: 5.5 to 6, 2.75 to 3, and 8.25 to 9. Thus, the median home price is given by $d_6 = 252$ (i.e., $M = \$252{,}000$), the first quartile is given by $d_3 = 234$ (i.e., $Q_1 = \$234{,}000$), and the third quartile is given by $d_9 = 267$ (i.e., $Q_3 = \$267{,}000$).

Oops! Just this morning a home sold in Green Hills for $264,000. We need to recalculate the median and quartiles for what are now $N = 12$ home prices.

We can use the sorted data set that we already had—all we have to do is insert the new home price (264) in the right spot (remember, we drop the 000's!). This gives

221, 228, 234, 238, 245, 252, 255, 263, **264**, 267, 270, 292

Now $N = 12$ and in this case the median is the average of $d_6 = 252$ and $d_7 = 255$. It follows that the median home price is $M = \$253{,}500$. The locator for the first quartile is $0.25 \times 12 = 3$. Since the locator is a whole number, the first quartile is the average of $d_3 = 234$ and $d_4 = 238$ (i.e., $Q_1 = \$236{,}000$). Similarly, the third quartile is $Q_3 = 265{,}500$ (the average of $d_9 = 264$ and $d_{10} = 267$).

EXAMPLE 15.12 MEDIAN AND QUARTILES FOR THE STAT 101 MIDTERM

We will now find the median and quartile scores for the Stat 101 data set (shown again in Table 15-10).

Score	1	6	7	8	9	10	11	12	13	14	15	16	24
Frequency	1	1	2	6	10	16	13	9	8	5	2	1	1

TABLE 15-10 ■ Frequency table for the Stat 101 data set

Having the frequency table available eliminates the need for sorting the scores—the frequency table has, in fact, done this for us. Here $N = 75$ (odd), so the median [with locator $L = (75 + 1)/2 = 38$] is the thirty-eighth score (counting from the left) in the frequency table. To find the thirty-eighth number in Table 15-10, we tally frequencies as we move from left to right: $1+1=2$; $1+1+2=4$; $1+1+2+6=10$; $1+1+2+6+10=20$; $1+1+2+6+10+16=36$. At this point, we know that the 36th test score on the list is a 10 (the last of the 10's) and the next 13 scores are all 11's. We can conclude that the 38th test score is 11. Thus, $M=11$.

The locator for the first quartile is $L=(0.25)\times 75=18.75$. Thus, $Q_1=d_{19}$. To find the nineteenth score in the frequency table, we tally frequencies from left to right: $1+1=2; 1+1+2=4; 1+1+2+6=10; 1+1+2+6+10=20$. At this point we realize that $d_{10}=8$ (the last of the 8's) and that d_{11} through d_{20} all equal 9. Hence, the first quartile of the Stat 101 midterm scores is $Q_1=d_{19}=9$.

Since the first and third quartiles are at an equal "distance" from the two ends of the sorted data set, a quick way to locate the third quartile now is to look for the nineteenth score in the frequency table when we count frequencies *from right to left*. We leave it to the reader to verify that the third quartile of the Stat 101 data set is $Q_3=12$.

EXAMPLE 15.13 MEDIAN AND QUARTILES FOR 2015 SAT MATH SCORES

In this example we continue the discussion of the 2015 SAT Math scores introduced in Example 15.6. Recall that the number of college-bound students taking the test was $N=1{,}698{,}521$. As reported by the College Board, the median score in the test was $M=510$, the first quartile score was $Q_1=430$, and the third quartile score was $Q_3=590$. What can we make of this information?

Let's start with the median $M=510$. The data set consists of $N=1{,}698{,}521$ scores, so the locator for the median is $L=(N+1)/2=1{,}698{,}522/2=849{,}261$. This means that there were *at least* 849,261 students (probably many more) with SAT Math scores of 510 *or less* and, likewise, *at least* 849,261 students with SAT Math scores of 510 *or more*.

Why did we use "at least" twice in the preceding paragraph? Imagine the group of all 510 scores lined up from left to right as part of the ordered data set, with the median 849,261st score somewhere in that group. What are the chances that it just happens to be the very last one (or very first one) in the group? Extremely small, given that there are thousands of 510 scores in the data set. So if it is not the last one, then there are more 510 scores to its right, and if it is not the first one, then there are more 510 scores to its left. Using the words "at least" covers us in these cases.

In a similar vein, we can conclude that there were at least 424,631 scores of $Q_1=430$ or less [the locator for the first quartile is $(0.25)\times 1{,}698{,}521=424{,}630.25$ and at least 1,273,891 scores of $Q_3=590$ or less [the locator for the third quartile is $(0.75)\times 1{,}698{,}521=1{,}273{,}890.75$].

A note of warning: Medians, quartiles, and general percentiles are often computed using statistical calculators or statistical software packages, which is all well and fine since the whole process can be a bit tedious. The problem is that there is no universally agreed upon procedure for computing percentiles, so different types of calculators and different statistical packages may give different answers from each other and from those given in this book for quartiles and other percentiles (everyone agrees on the median). *Keep this in mind when doing the exercises*—the answer given by your calculator may be slightly different from the one you would get from the procedure we use in the book.

The Five-Number Summary

A common way to summarize a large data set is by means of its *five-number summary*. The **five-number summary** is given by (1) the smallest value in the data set (called the *Min*), (2) the *first quartile* Q_1, (3) the *median* M, (4) the *third quartile* Q_3, and (5) the largest value in the data set (called the *Max*). These five numbers together often tell us a great deal about the data.

EXAMPLE 15.14 FIVE-NUMBER SUMMARY FOR THE STAT 101 MIDTERM

For the Stat 101 data set, the five-number summary is $Min = 1$, $Q_1 = 9$, $M = 11$, $Q_3 = 12$, $Max = 24$ (see Example 15.12). What useful information can we get out of this?

Right away we can see that the $N = 75$ test scores were not evenly spread out over the range of possible scores. For example, given that $M = 11$ and $Q_3 = 12$, we can conclude that 25% or more of the class (that means at least 19 students) scored either 11 or 12 on the test. At the same time, given that $Q_3 = 12$ and $Max = 24$, we can conclude that at most 18 students had scores in the 13–24 point range. Using similar arguments, we can conclude that at least 19 students had scores between $Q_1 = 9$ and $M = 11$ points and at most 18 students scored in the 1–8 point range.

The "big picture" we get from the five-number summary of the Stat 101 test scores is that there was a lot of bunching up in a narrow band of scores (at least half of the students in the class scored in the range 9–12 points), and the rest of the class was all over the place. In general, this type of "lumpy" distribution of test scores is indicative of a test with an uneven level of difficulty—a bunch of easy questions and a bunch of really hard questions with little in between. (Having seen the data, we know that the *Min* and *Max* scores were both outliers and that if we disregard these two outliers, the test results don't look quite so bad. Of course, there is no way to pick this up from just the five-number summary.)

Box Plots

Invented by the American statistician John W. Tukey (1915–2000), a *box plot* (also known as a *box-and-whisker* plot) is a picture of the five-number summary of a data set. The **box plot** consists of a rectangular box that sits above a number line representing the data values and extends from the first quartile Q_1 to the third quartile Q_3 on that number line. A vertical line crosses the box, indicating the position of the median M. On both sides of the box are "whiskers" extending to the smallest value, *Min*, and largest value, *Max*, of the data. Figure 15-12 shows a generic box plot for a data set.

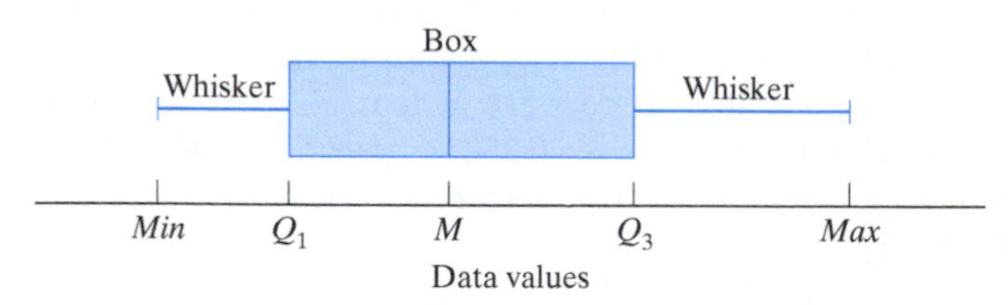

FIGURE 15-12

Figure 15-13(a) shows a box plot for the Stat 101 data set (see Example 15.12). The long whiskers in this box plot are largely due to the outliers 1 and 24. Figure 15-13(b) shows a variation of the same box plot, but with the two outliers, marked with two crosses, segregated from the rest of the data. (When there are outliers, it is useful to segregate them from the rest of the data set—we think of outliers as "anomalies" within the data set.)

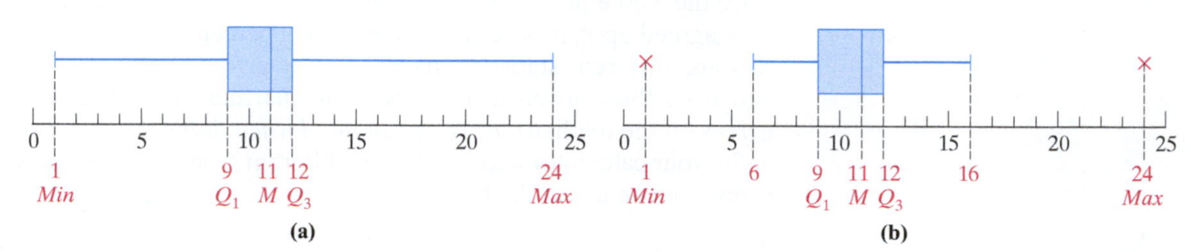

FIGURE 15-13 (a) Box plot for the Stat 101 data set. (b) Same box plot with the outliers separated from the rest of the data.

Box plots are particularly useful when comparing similar data for two or more populations. This is illustrated in the next example.

EXAMPLE 15.15 COMPARING AGRICULTURE AND ENGINEERING SALARIES

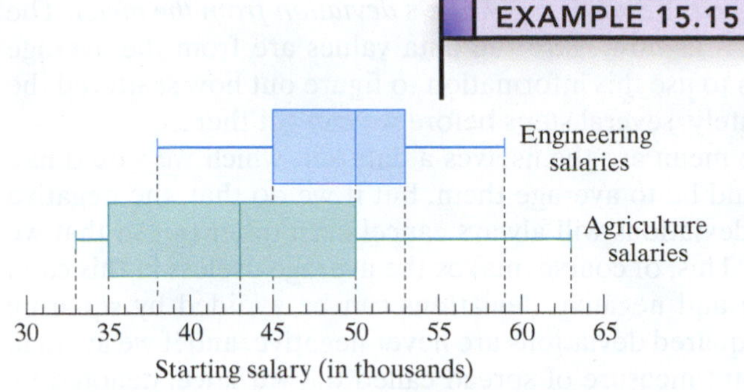

FIGURE 15-14 Comparison of starting salaries of first-year graduates in agriculture and engineering.

Figure 15-14 shows box plots for the starting salaries of two different populations: first-year agriculture and engineering graduates of Tasmania State University.

Superimposing the two box plots on the same scale allows us to make some useful comparisons. It is clear, for instance, that engineering graduates are doing better overall than agriculture graduates, even though at the very top levels agriculture graduates are better paid. Another interesting point is that the median salary of agriculture graduates ($43,000) is less than the first quartile of the salaries of engineering graduates ($45,000). The very short whisker on the left side of the agriculture box plot tells us that the bottom 25% of agriculture salaries are concentrated in a very narrow salary range ($33,000–$35,000). We can also see that agriculture salaries are much more spread out than engineering salaries, even though most of the spread occurs at the higher end of the salary scale.

15.3 Ranges and Standard Deviations

There are several different ways to describe the spread of a data set; in this section we will describe the three most commonly used ones.

- **The range.** The range R of a data set is given by the difference between the highest and lowest values of the data $(R = Max - Min)$.

The range is most useful when there are no outliers in the data. In the presence of outliers the range tells a distorted story. For example, the range of the test scores in the Stat 101 exam (Example 15.1) is $24 - 1 = 23$ points, an indication of a big spread within the scores (i.e., a very heterogeneous group of students). True enough, but if we discount the two outliers, the remaining 73 test scores would have a much smaller range of $16 - 6 = 10$ points.

To eliminate the possible distortion caused by outliers, a common practice when measuring the spread of a data set is to use the *interquartile range*, denoted by the acronym *IQR*.

- **The interquartile range.** The interquartile range IQR is the difference between the third quartile and the first quartile $(IQR = Q_3 - Q_1)$. The IQR tells us how spread out the middle 50% of the data values are. For many types of real-world data, the IQR is a useful measure of spread.

EXAMPLE 15.16 RANGE AND IQR FOR 2015 SAT MATH SCORES

The five-number summary for the 2015 SAT Math scores (see Example 15.13) was $Min = 200$ (yes, there were a few jokers who missed every question!), $Q_1 = 430$, $M = 510$, $Q_3 = 590$, $Max = 800$ (there are still a few geniuses around!). It follows that the 2015 SAT Math scores had a range of 600 points $(800 - 200 = 600)$ and an IQR of 160 points $(IQR = 590 - 430 = 160)$.

The Standard Deviation

The most important and most commonly used measure of spread for a data set is the *standard deviation*. The key concept for understanding the standard deviation is the concept of *deviation from the mean*. If A is the mean (average) of the data set and

x is an arbitrary data value, the difference $x - A$ is x's *deviation from the mean*. The deviations from the mean tell us how "far" the data values are from the average value of the data. The idea is to use this information to figure out how scattered the data is. There are, unfortunately, several steps before we can get there.

The deviations from the mean are themselves a data set, which we would like to summarize. One way would be to average them, but if we do that, the negative deviations and the positive deviations will always cancel each other out so that we end up with an average of 0. This, of course, makes the average useless in this case. The cancellation of positive and negative deviations can be avoided by squaring each of the deviations. The squared deviations are never negative, and if we average them out, we get an important measure of spread called the **variance**, denoted by V. Finally, we take the square root of the variance and get the **standard deviation**, denoted by the Greek letter σ (and sometimes by the acronym SD).

The following is an outline of the definition of the standard deviation of a data set.

THE STANDARD DEVIATION OF A DATA SET

- Let A denote the mean of the data set. For each number x in the data set, compute its *deviation from the mean* $(x - A)$ and *square* each of these numbers. These numbers are called the *squared deviations*.
- Find the average of the squared deviations. This number is called the *variance* V.
- The *standard deviation* is the square root of the variance $(\sigma = \sqrt{V})$.

Standard deviations of large data sets are not fun to calculate by hand, and they are rarely found that way. The standard procedure for calculating standard deviations is to use a computer or a good scientific or business calculator, often preprogrammed to do all the steps automatically. Be that as it may, it is still important to understand what's behind the computation of a standard deviation, even when the actual grunt work is going to be performed by a machine. (As a matter of fact, the applet ***Numerical Summaries of Data***, available in *MyMathLab* in the *Multimedia Library* or in *Tools for Success*, does all the grunt work involved in computing standard deviations as well as some of the other numerical summaries discussed in this section.)

EXAMPLE 15.17 CALCULATION OF A SD

Over the course of the semester, Angela turned in all of her homework assignments. Her grades in the 10 assignments (sorted from lowest to highest) were 85, 86, 87, 88, 89, 91, 92, 93, 94, and 95. Our goal in this example is to calculate the standard deviation of this data set the old-fashioned way (i.e., doing our own grunt work).

x	$(x - 90)$	$(x - 90)^2$
85	−5	25
86	−4	16
87	−3	9
88	−2	4
89	−1	1
91	1	1
92	2	4
93	3	9
94	4	16
95	5	25

TABLE 15-11

The first step is to find the mean A of the data set. It's not hard to see that $A = 90$. We are lucky—this is a nice round number! The second step is to calculate the *deviations from the mean* and then the *squared deviations*. The details are shown in the second and third columns of Table 15-11. When we average the squared deviations, we get $(25 + 16 + 9 + 4 + 1 + 1 + 4 + 9 + 16 + 25)/10 = 11$. This means that the variance is $V = 11$ and, thus, the standard deviation (rounded to one decimal place) is $\sigma = \sqrt{11} \approx 3.3$ points. ■

Standard deviations are measured in the same units as the original data, so in Example 15.17 the standard deviation of Angela's homework scores was roughly 3.3 points. What should we make of this fact? It is clear from just a casual look at Angela's homework scores that she was pretty consistent in her homework, never straying too much above or below her average score of 90 points. The standard deviation is, in effect, a way to measure this degree of consistency (or lack thereof). A small standard deviation tells us that the data are consistent and the spread of the data is small, as is the case with Angela's homework scores.

The ultimate in consistency within a data set is when all the data values are the same (like Angela's friend Chloe, who got a 20 in every homework assignment). When this happens the standard deviation is 0. On the other hand, when there is a lot of inconsistency within the data set, we are going to get a large standard deviation. This is illustrated by Angela's other friend, Tiki, whose homework scores were 5, 15, 25, 35, 45, 55, 65, 75, 85, and 95. We would expect the standard deviation of this data set to be quite large—in fact, it is almost 29 points.

The standard deviation is arguably the most important and frequently used measure of data spread. Yet it is not a particularly intuitive concept. Here are a few basic guidelines that recap our preceding discussion:

- The standard deviation of a data set is measured in the same units as the original data. For example, if the data are points on a test, then the standard deviation is also given in points. Conversely, if the standard deviation is given in dollars, then we can conclude that the original data must have been money—home prices, salaries, or something like that. For sure, the data couldn't have been test scores on an exam.
- It is pointless to compare standard deviations of data sets that are given in different units. Even for data sets that are given in the same units—say, for example, test scores—the underlying scale should be the same. We should not try to compare standard deviations for SAT scores measured on a scale of 200–800 points with standard deviations of a set of homework assignments measured on a scale of 0–100 points.
- For data sets that are based on the same underlying scale, a comparison of standard deviations can tell us something about how much the data are scattered. If the standard deviation is small, we can conclude that the data points are all bunched together—there is little scatter. As the standard deviation increases, we can conclude that the data points are beginning to scatter and spread out. The more spread out they are, the larger the standard deviation becomes. A standard deviation of 0 means that all data values are the same.

As a measure of spread, the standard deviation is particularly useful for analyzing real-life data. We will come to appreciate its importance in this context in Chapter 17.

Conclusion

Graphical summaries of data can be produced by bar graphs, pictograms, pie charts, histograms, and so on. (There are many other types of graphical descriptions that we did not discuss in the chapter.) The kind of graph that is the most appropriate for a situation depends on many factors, and creating a good "picture" of a data set is as much an art as a science.

"Statistical reasoning will one day be as necessary for efficient citizenship as the ability to read and write."

– H. G. Wells

Numerical summaries of data, when properly used, help us understand the overall pattern of a data set without getting bogged down in the details. They fall into two categories: (1) measures of location, such as the *average*, the *median*, and the *quartiles*, and (2) measures of spread, such as the *range*, the *interquartile range*, and the *standard deviation*. Sometimes we even combine numerical summaries and graphical displays, as in the case of the *box plot*. We touched upon all of these in this chapter, but the subject is a big one, and by necessity we only scratched the surface.

In this day and age, we are all consumers of data, and at one time or another, we are likely to be providers of data as well. Thus, understanding the basics of how data are organized and summarized has become an essential requirement for personal success and good citizenship.

KEY CONCEPTS

15.1 Graphs and Charts

- **data set:** a collection of numbers that represent all the values obtained when measuring some characteristic of a population, **443**
- **data point:** an individual number in a data set, **443**
- **discrete variable:** a variable that can only take on a discrete set of values—there is a minimum gap between two possible values of the variable, **444**
- **continuous variable:** a variable that can take on infinitely many values and those values can differ by arbitrarily small amounts, **444**
- **frequency table:** a table showing the frequency of each actual value in a data set, **444**
- **actual values:** the distinct values taken on by the variable in a data set, **445**
- **bar graph:** a graph with bars (columns) representing each of the actual values in the data set. The height of each column represents the frequency of that value in the data set, **446**
- **pictogram:** a variation of a bar graph that uses icons or symbolic images to represent the frequencies of the actual values in a data set, **447**
- **line graph:** a graph where the values of the data are given by points with adjacent points connected by a line, **447**
- **categorical (qualitative) variable:** a variable that takes on non-numerical values, **448**
- **pie chart:** a chart consisting of a circle broken up into wedges. Each wedge represents a category, with the size of the wedge proportional to the relative frequency of the corresponding category, **449**
- **class interval:** a category consisting of a range of numerical values, **451**
- **histogram:** a variation of a bar graph used to describe frequencies of class intervals in the case of a continuous variable, **452**

15.2 Means, Medians, and Percentiles

- **average (mean):** given a data set consisting of the numbers $d_1, d_2, \ldots, d_N$, their average is the number $A = (d_1 + d_2 + \cdots + d_N)/N$, **453**
- **percentile:** given a data set, the pth percentile of the data set is a number X_p such that $p\%$ of the numbers in the data set are smaller or equal to X_p and $(100 - p)\%$ of the numbers in the data set are bigger or equal to X_p, **454**
- **locator:** the locator for the pth percentile of a data set consisting of N numbers is the number $L = (\frac{p}{100})N$, **454**
- **median:** the median M of a data set is the 50th percentile of the data set—half the numbers in the data set are smaller or equal to M, half are bigger or equal to M, **455**

- **quartiles:** the first quartile Q_1 is the 25th percentile of a data set; the third quartile Q_3 is the 75th percentile, **456**
- **five-number summary:** a summary of the data set consisting of the *Min* (smallest value in the data set), the first quartile Q_1, the median M, the third quartile Q_3, and the *Max* (largest value in the data set), **457**
- **box plot (box and whisker plot):** **458**

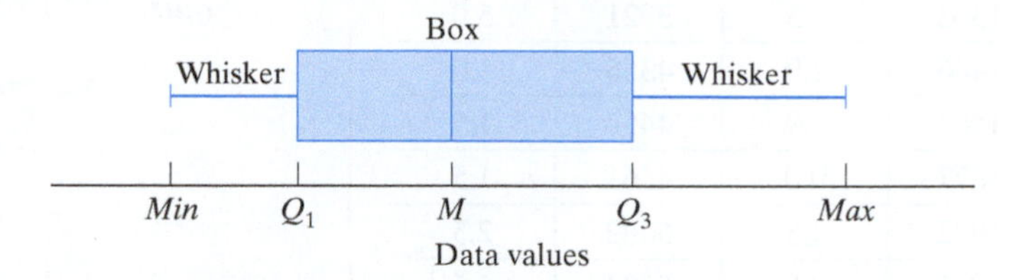

15.3 Ranges and Standard Deviations

- **range:** the range R of a data set is the difference between the largest and the smallest values in the data set $(R = Max - Min)$, **459**
- **interquartile range:** the interquartile range IQR of a data set is the difference between the third and first quartiles of the data set $(IQR = Q_3 - Q_1)$, **459**
- **variance:** the variance V of a data set is the average of the squared differences between the data points and the average of the data set [i.e., for each data point d, compute $(d - A)^2$; V is the average of these numbers], **460**
- **standard deviation:** the standard deviation σ of a data set is the square root of the variance of the data set, $(\sigma = \sqrt{V})$, **460**

EXERCISES

WALKING

15.1 Graphs and Charts

Exercises 1 through 4 refer to the data set shown in Table 15-12. The table shows the scores on a Chem 103 test consisting of 10 questions worth 10 points each.

Student ID	Score	Student ID	Score
1362	50	4315	70
1486	70	4719	70
1721	80	4951	60
1932	60	5321	60
2489	70	5872	100
2766	10	6433	50
2877	80	6921	50
2964	60	8317	70
3217	70	8854	100
3588	80	8964	80
3780	80	9158	60
3921	60	9347	60
4107	40		

TABLE 15-12 ■ Chem 103 test scores

1. **(a)** Make a frequency table for the Chem 103 test scores.
   (b) Draw a bar graph for the data in Table 15-12.
2. Draw a line graph for the data in Table 15-12.
3. Suppose that the grading scale for the test is A: 80–100; B: 70–79; C: 60–69; D: 50–59; and F: 0–49.
   (a) Make a frequency table for the distribution of the test grades.
   (b) Draw a relative frequency bar graph for the test grades.
4. Suppose that the grading scale for the test is A: 80–100; B: 70–79; C: 60–69; D: 50–59; and F: 0–49.
   (a) What percentage of the students who took the test got a grade of D?
   (b) In a pie chart showing the distribution of the test grades, what is the size of the central angle (in degrees) of the "wedge" representing the grade of D?
   (c) Draw a pie chart showing the distribution of the test grades. Give the central angles for each wedge in the pie chart (round your answer to the nearest degree).

Exercises 5 through 10 refer to Table 15-13, which gives the home-to-school distance d (rounded to the nearest half-mile) for each of the 27 kindergarten students at Cleansburg Elementary School.

Student ID	d	Student ID	d
1362	1.5	3921	5.0
1486	2.0	4355	1.0
1587	1.0	4454	1.5
1877	0.0	4561	1.5
1932	1.5	5482	2.5
1946	0.0	5533	1.5
2103	2.5	5717	8.5
2877	1.0	6307	1.5
2964	0.5	6573	0.5
3491	0.0	8436	3.0
3588	0.5	8592	0.0
3711	1.5	8964	2.0
3780	2.0	9205	0.5
		9658	6.0

TABLE 15-13 ■ Home-to-school distance

5. **(a)** Make a frequency table for the distances in Table 15-13.

(b) Draw a line graph for the data in Table 15-13.

6. Draw a bar graph for the data in Table 15-13.

7. Draw a bar graph for the home-to-school distances for the kindergarteners at Cleansburg Elementary School using the following class intervals:

Very close: Less than 1 mile

Close: 1 mile up to and including 1.5 miles

Nearby: 2 miles up to and including 2.5 miles

Not too far: 3 miles up to and including 4.5 miles

Far: 5 miles or more

8. Draw a bar graph for the home-to-school distances for the kindergarteners at Cleansburg Elementary School using the following class intervals:

Zone A: 1.5 miles or less

Zone B: more than 1.5 miles up to and including 2.5 miles

Zone C: more than 2.5 miles up to and including 3.5 miles

Zone D: more than 3.5 miles

9. Using the class intervals given in Exercise 7, draw a pie chart for the home-to-school distances for the kindergarteners at Cleansburg Elementary School. Give the central angles for each wedge of the pie chart. Round your answer to the nearest degree.

10. Using the class intervals given in Exercise 8, draw a pie chart for the home-to-school distances for the kindergarteners at Cleansburg Elementary School. Give the central angles for each wedge of the pie chart. Round your answer to the nearest degree.

Exercises 11 and 12 refer to the bar graph shown in Fig. 15-15 describing the scores of a group of students on a 10-point math quiz.

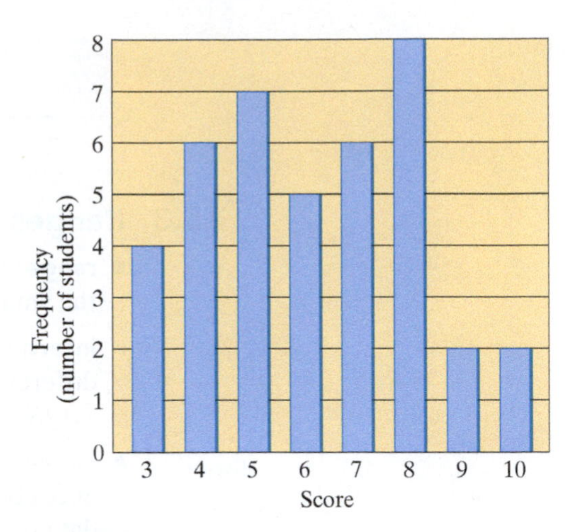

FIGURE 15-15

11. **(a)** How many students took the math quiz?

(b) What percentage of the students scored 2 points?

(c) If a grade of 6 or more was needed to pass the quiz, what percentage of the students passed? (Round your answer to the nearest percent.)

12. Draw a relative frequency bar graph showing the results of the quiz.

Exercises 13 and 14 refer to the pie chart in Fig. 15-16.

Cause of Death in U.S. Among 18- to 22-Year-Olds (2005)
$N = 19{,}548$

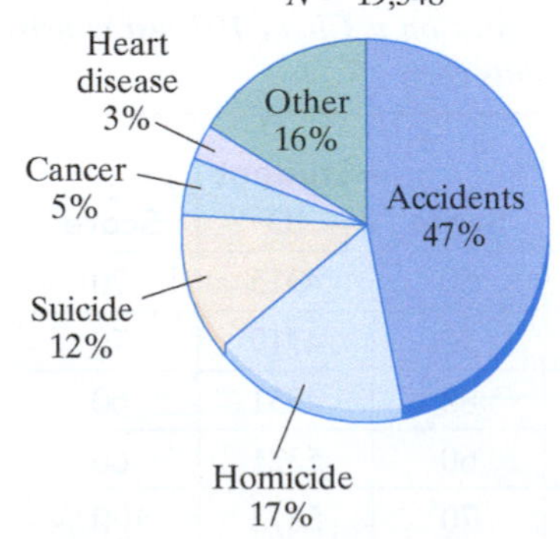

Source: Centers for Disease Control and Prevention, *www.cdc.gov*.

FIGURE 15-16

13. **(a)** Is cause of death a quantitative or a qualitative variable?

(b) Use the data provided in the pie chart to estimate the number of 18- to 22-year-olds who died in the United States in 2005 due to an accident.

14. Use the data provided in the pie chart to estimate the number of 18- to 22-year-olds who died in the United States in 2005 for each category shown in the pie chart.

15. Table 15-14 shows the class interval frequencies for the 2015 Critical Reading scores on the SAT. Draw a relative frequency bar graph for the data in Table 15-14. (Round the relative frequencies to the nearest tenth of a percent.)

Score range	Number of test-takers
700–800	75,659
600–690	257,184
500–590	495,917
400–490	540,157
300–390	264,155
200–290	65,449
Total	$N = 1{,}698{,}521$

Source: Data from 2015 College Bound Seniors: Total Group Profile Report. Copyright © 2015. The College Board.

TABLE 15-14 ■ 2015 SAT Critical Reading Scores

16. Table 15-15 shows the class interval frequencies for the 2015 Writing scores on the SAT. Draw a pie chart for the data in Table 15-15. Indicate the degree of the central angle for each wedge of the pie chart (rounded to the nearest degree).

Score range	Number of test-takers
700–800	70,216
600–690	229,224
500–590	445,181
400–490	575,463
300–390	311,883
200–290	66,554
Total	$N = 1{,}698{,}521$

Source: Data from 2015 College Bound Seniors: Total Group Profile Report. Copyright © 2015. The College Board.

TABLE 15-15 ■ 2015 SAT Writing scores

17. Table 15-16 shows the percentage of U.S. working married couples in which the wife's income is higher than the husband's (1999–2009).

(a) Draw a pictogram for the data in Table 15-16. Assume you are trying to convince your audience that things are looking great for women in the workplace and that women's salaries are catching up to men's very quickly.

(b) Draw a different pictogram for the data in Table 15-16, where you are trying to convince your audience that women's salaries are catching up with men's very slowly.

Year	1999	2000	2001	2002	2003	2004
Percent	28.9	29.9	30.7	31.9	32.4	32.6
Year	2005	2006	2007	2008	2009	
Percent	33.0	33.4	33.5	34.5	37.7	

Source: Bureau of Labor Statistics, *www.bls.gov*.

TABLE 15-16

18. Table 15-17 shows the percentage of U.S. workers who are members of unions (2000–2011).

(a) Draw a pictogram for the data in Table 15-17. Assume you are trying to convince your audience that unions are holding their own and that the percentage of union members in the workforce is steady.

(b) Draw a different pictogram for the data in Table 15-17, where you are trying to convince your audience that there is a steep decline in union membership in the U.S. workforce.

Year	2000	2001	2002	2003	2004	2005
Percent	13.4	13.3	13.3	12.9	12.5	12.5
Year	2006	2007	2008	2009	2010	2011
Percent	12.0	12.1	12.4	12.3	11.9	11.8

Source: Bureau of Labor Statistics, *www.bls.gov*.

TABLE 15-17 ■ Percentage of unionized U.S. workers

Exercises 19 and 20 refer to Table 15-18, which shows the birth weights (in ounces) of the 625 babies born in Cleansburg hospitals in 2016.

More than	Less than or equal to	Number of babies	More than	Less than or equal to	Number of babies
48	60	15	108	120	184
60	72	24	120	132	142
72	84	41	132	144	26
84	96	67	144	156	5
96	108	119	156	168	2

TABLE 15-18 ■ Cleansburg birth weights (in ounces)

19. (a) Give the length of each class interval (in ounces).

(b) Suppose that a baby weighs exactly 5 pounds 4 ounces. To what class interval does she belong? Describe the endpoint convention.

(c) Draw the histogram describing the 2016 birth weights in Cleansburg using the class intervals given in Table 15-18.

20. **(a)** Write a new frequency table for the birth weights in Cleansburg using class intervals of length equal to 24 ounces. Use the same endpoint convention as the one used in Table 15-18.

(b) Draw the histogram corresponding to the frequency table found in (a).

Exercises 21 and 22 refer to the two histograms shown in Fig. 15-17 summarizing the 2016 payrolls of the 30 teams in Major League Baseball. The two histograms are based on the same data set but use slightly different class intervals. (You can assume that no team had a payroll that was exactly equal to a whole number of millions of dollars.)

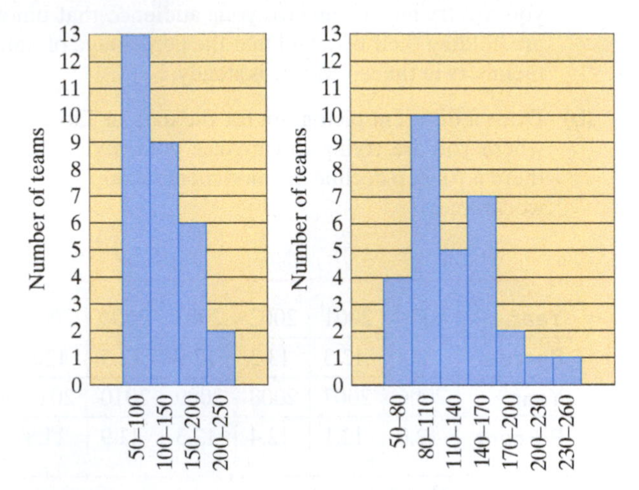

FIGURE 15-17 2016 MLB team payrolls (in millions). $N = 30$. (Data from Major League Baseball, *MLB.com*.)

21. **(a)** How many teams had a 2016 payroll of more than \$100 million?

(b) How many teams had a 2016 payroll of less than \$110 million?

(c) How many teams had a 2016 payroll between \$100 and \$110 million?

22. **(a)** How many teams had a 2016 payroll of less than \$150 million?

(b) How many teams had a 2016 payroll of more than \$140 million?

(c) How many teams had a 2016 payroll between \$140 and \$150 million?

15.2 Means, Medians, and Percentiles

23. Consider the data set $\{3, -5, 7, 4, 8, 2, 8, -3, -6\}$.

(a) Find the average A of the data set.

(b) Find the median M of the data set.

(c) Consider the data set $\{3, -5, 7, 4, 8, 2, 8, -3, -6, 2\}$ obtained by adding one more data point to the original data set. Find the average and median of this data set.

24. Consider the data set $\{-4, 6, 8, -5.2, 10.4, 10, 12.6, -13\}$

(a) Find the average A of the data set.

(b) Find the median M of the data set.

(c) Consider the data set $\{-4, 6, 8, -5.2, 10.4, 10, 12.6\}$ having one less data point than the original set. Find the average and the median of this data set.

25. Find the average A and the median M of each data set.

(a) $\{0, 1, 2, 3, 4, 5, 6, 7, 8, 9\}$

(b) $\{1, 2, 3, 4, 5, 6, 7, 8, 9\}$

(c) $\{1, 2, 3, 4, 5, 6, 7, 8, 9, 10\}$

26. Find the average A and the median M of each data set.

(a) $\{1, 2, 1, 2, 1, 2, 1, 2, 1, 2\}$

(b) $\{1, 2, 3, 4, 1, 2, 3, 4, 1, 2, 3, 4, 1, 2, 3, 4\}$

(c) $\{1, 2, 3, 4, 5, 5, 4, 3, 2, 1\}$

27. Find the average A and the median M of each data set.

(a) $\{5, 10, 15, 20, 25, 60\}$

(b) $\{105, 110, 115, 120, 125, 160\}$

28. Find the average A and the median M of each data set.

(a) $\{5, 10, 15, 20, 25, 30, 35, 40, 45, 50\}$

(b) $\{55, 60, 65, 70, 75, 80, 85, 90, 95, 100\}$

29. Table 15-19 shows the results of a 5-point musical aptitude test given to a group of first-grade students.

(a) Find the average aptitude score.

(b) Find the median aptitude score.

Aptitude score	0	1	2	3	4	5
Frequency	24	16	20	12	5	3

TABLE 15-19

30. Table 15-20 shows the ages of the firefighters in the Cleansburg Fire Department.

Age	25	27	28	29	30
Frequency	2	7	6	9	15
Age	31	32	33	37	39
Frequency	12	9	9	6	4

TABLE 15-20

(a) Find the average age of the Cleansburg firefighters rounded to two decimal places.

(b) Find the median age of the Cleansburg firefighters.

31. Table 15-21 shows the relative frequencies of the scores of a group of students on a philosophy quiz.

Score	4	5	6	7	8
Relative frequency	7%	11%	19%	24%	39%

TABLE 15-21

(a) Find the average quiz score.

(b) Find the median quiz score.

32. Table 15-22 shows the relative frequencies of the scores of a group of students on a 10-point math quiz.

Score	3	4	5	6	7	8	9
Relative frequency	8%	12%	16%	20%	18%	14%	12%

TABLE 15-22

(a) Find the average quiz score rounded to two decimal places.

(b) Find the median quiz score.

33. Consider the data set $\{-5, 7, 4, 8, 2, 8, -3, -6\}$.

(a) Find the first quartile Q_1 of the data set.

(b) Find the third quartile Q_3 of the data set.

(c) Consider the data set $\{-5, 7, 4, 8, 2, 8, -3, -6, 2\}$ obtained by adding one more data point to the original data set. Find the first and third quartiles of this data set.

34. Consider the data set $\{-4, 6, 8, -5.2, 10.4, 10, 12.6, -13\}$.

(a) Find the first quartile Q_1 of the data set.

(b) Find the third quartile Q_3 of the data set.

(c) Consider the data set $\{-4, 6, 8, -5.2, 10.4, 10, 12.6\}$ obtained by deleting one data point from the original data set. Find the first and third quartiles of this data set.

35. For each data set, find the 75th and the 90th percentiles.

(a) $\{1, 2, 3, 4, \ldots, 98, 99, 100\}$

(b) $\{0, 1, 2, 3, 4, \ldots, 98, 99, 100\}$

(c) $\{1, 2, 3, 4, \ldots, 98, 99\}$

(d) $\{1, 2, 3, 4, \ldots, 98\}$

36. For each data set, find the 10th and the 25th percentiles.

(a) $\{1, 2, 3, \ldots, 49, 50, 50, 49, \ldots, 3, 2, 1\}$

(b) $\{1, 2, 3, \ldots, 49, 50, 49, \ldots, 3, 2, 1\}$

(c) $\{1, 2, 3, \ldots, 49, 49, \ldots, 3, 2, 1\}$

37. This exercise refers to the age distribution in the Cleansburg Fire Department shown in Table 15-20 (Exercise 30).

(a) Find the first quartile of the data set.

(b) Find the third quartile of the data set.

(c) Find the 90th percentile of the data set.

38. This exercise refers to the math quiz scores shown in Table 15-22 (Exercise 32).

(a) Find the first quartile of the data set.

(b) Find the third quartile of the data set.

(c) Find the 70th percentile of the data set.

Exercises 39 and 40 refer to SAT test scores for 2014. A total of $N = 1{,}672{,}395$ college-bound students took the SAT in 2014. Assume that the test scores are sorted from lowest to highest and that the sorted data set is $\{d_1, d_2, \ldots, d_{1,672,395}\}$.

39. (a) Determine the position of the median M.

(b) Determine the position of the first quartile Q_1.

(c) Determine the position of the 80th percentile.

40. (a) Determine the position of the third quartile Q_3.

(b) Determine the position of the 60th percentile.

41. Consider the data set $\{-5, 7, 4, 8, 2, 8, -3, -6\}$.

(a) Find the five-number summary of the data set. (*Hint*: see Exercise 33).

(b) Draw a box plot for the data set.

42. Consider the data set $\{-4, 6, 8, -5.2, 10.4, 10, 12.6, -13\}$.

(a) Find the five-number summary of the data set. (*Hint*: see Exercise 34).

(b) Draw a box plot for the data set.

43. This exercise refers to the distribution of ages in the Cleansburg Fire Department shown in Table 15-20 (see Exercises 30 and 37).

(a) Find the five-number summary of the data set.

(b) Draw a box plot for the data set.

44. This exercise refers to the distribution of math quiz scores shown in Table 15-22 (see Exercises 32 and 38).

(a) Find the five-number summary of the data set.

(b) Draw a box plot for the data set.

Exercises 45 and 46 refer to the two box plots in Fig. 15-18 showing the starting salaries of Tasmania State University first-year graduates in agriculture and engineering. (These are the two box plots discussed in Example 15.15.)

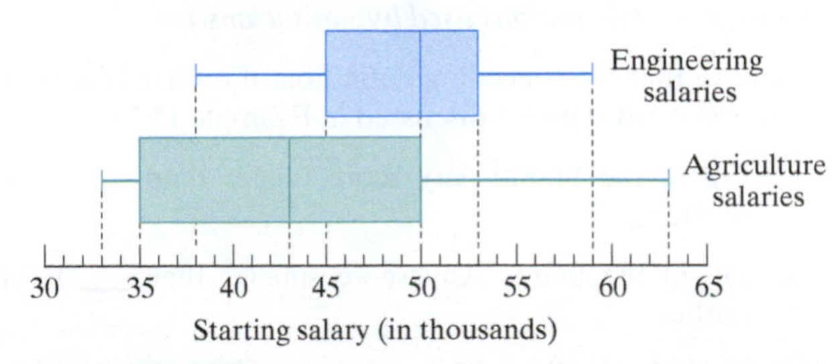

FIGURE 15-18

45. (a) What is the median salary for agriculture majors?

(b) What is the median salary for engineering majors?

(c) Explain how we can tell that the median salary for engineering majors is the same as the third quartile salary for agriculture majors.

46. (a) Fill in the blank: Of the 612 engineering graduates, at most ____ had a starting salary greater than $45,000.

(b) Fill in the blank: If there were 240 agriculture graduates with starting salaries of $35,000 or less, the total number of agriculture graduates is approximately ____.

15.3 Ranges and Standard Deviations

47. For the data set $\{-5, 7, 4, 8, 2, 8, -3, -6\}$, find

(a) the range.

(b) the interquartile range (see Exercise 33).

48. For the data set $\{-4, 6, 8, -5.2, 10.4, 10, 12.6, -13\}$, find

(a) the range.

(b) the interquartile range (see Exercise 34).

49. A realty company has sold $N = 341$ homes in the last year. The five-number summary for the sale prices is $Min = \$97{,}000$, $Q_1 = \$115{,}000$, $M = \$143{,}000$, $Q_3 = \$156{,}000$, and $Max = \$249{,}000$.

(a) Find the interquartile range of the home sale prices.

(b) How many homes sold for a price between $115,000 and $156,000 (inclusive)? (*Note*: If you don't believe that you have enough information to give an exact answer, you should give the answer in the form of "at least ___" or "at most ___.")

50. This exercise refers to the starting salaries of Tasmania State University first-year graduates in agriculture and engineering discussed in Exercises 45 and 46.

(a) Estimate the range for the starting salaries of agriculture majors.

(b) Estimate the interquartile range for the starting salaries of engineering majors.

For Exercises 51 through 54, you should use the following definition of an outlier: An ***outlier*** *is any data value that is above the third quartile by more than 1.5 times the IQR [Outlier $> Q_3 + 1.5(IQR)$] or below the first quartile by more than 1.5 times the IQR [Outlier $< Q_1 - 1.5(IQR)$]. (Note: There is no one universally agreed upon definition of an outlier; this is but one of several definitions used by statisticians.)*

51. Suppose that the preceding definition of outlier is applied to the Stat 101 data set discussed in Example 15.14.

(a) Fill in the blank: Any score bigger than ___ is an outlier.

(b) Fill in the blank: Any score smaller than ___ is an outlier.

(c) Find the outliers (if there are any) in the Stat 101 data set.

52. Using the preceding definition, find the outliers (if there are any) in the City of Cleansburg Fire Department data set discussed in Exercises 30 and 37. (*Hint*: Do Exercise 37 first.)

53. The distribution of the heights (in inches) of 18-year-old U.S. males has first quartile $Q_1 = 67$ in. and third quartile $Q_3 = 71$ in. Using the preceding definition, determine which heights correspond to outliers.

54. The distribution of the heights (in inches) of 18-year-old U.S. females has first quartile $Q_1 = 62.5$ in. and third quartile $Q_3 = 66$ in. Using the preceding definition, determine which heights correspond to outliers.

The purpose of Exercises 55 and 56 is to practice computing standard deviations the old fashioned way (by hand). Granted, computing standard deviations this way is not the way it is generally done in practice; a good calculator (or a computer package) will do it much faster and more accurately. The point is that computing a few standard deviations the old-fashioned way should help you understand the concept a little better. If you use a calculator or a computer to answer these exercises, you are defeating their purpose.

55. Find the standard deviation of each of the following data sets.

(a) $\{5, 5, 5, 5\}$

(b) $\{0, 5, 5, 10\}$

(c) $\{0, 10, 10, 20\}$

(d) $\{1, 2, 3, 4, 5\}$

56. Find the standard deviation of each of the following data sets.

(a) $\{3, 3, 3, 3\}$

(b) $\{0, 6, 6, 8\}$

(c) $\{-6, 0, 0, 18\}$

(d) $\{6, 7, 8, 9, 10\}$

JOGGING

Exercises 57 and 58 refer to the mode of a data set. The ***mode*** *of a data set is the data point that occurs with the highest frequency. When there are several data points (or categories) tied for the most frequent, each of them is a mode, but if all data points have the same frequency, rather than say that every data point is a mode, it is customary to say that there is no mode.*

57. (a) Find the mode of the data set given by Table 15-20 (Exercise 30).

(b) Find the mode of the data set given by Fig. 15-15 (Exercises 11 and 12).

58. (a) Find the mode category for the data set described by the pie chart in Fig. 15-19(a).

(b) Find the mode category for the data set shown in Fig. 15-19(b). If there is no mode, your answer should indicate so.

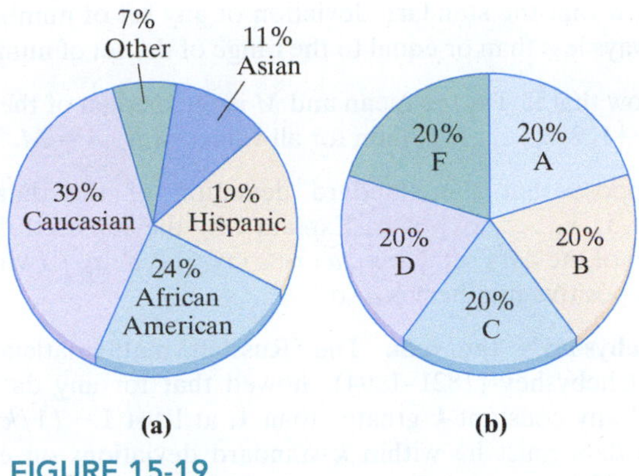

FIGURE 15-19

59. Mike's average on the first five exams in Econ 1A is 88. What must he earn on the next exam to raise his overall average to 90?

60. Explain each of the following statements regarding the median score in one of the SAT sections:

(a) If the number of test-takers N is odd, then the median score must end in 0.

(b) If the number of test-takers N is even, then the median score can end in 0 or 5, but the chances that it will end in 5 are very low.

61. In 2006, the median SAT score was the average of $d_{732,872}$ and $d_{732,873}$, where $\{d_1, d_2, \ldots, d_N\}$ denotes the data set of all SAT scores ordered from lowest to highest. Determine the number of students N who took the SAT in 2006.

62. In 2004, the third quartile of the SAT scores was $d_{1,064,256}$, where $\{d_1, d_2, \ldots, d_N\}$ denotes the data set of all SAT scores ordered from lowest to highest. Determine the number of students N who took the SAT in 2004.

63. **(a)** Give an example of 10 numbers with an average less than the median.

(b) Give an example of 10 numbers with a median less than the average.

(c) Give an example of 10 numbers with an average less than the first quartile.

(d) Give an example of 10 numbers with an average more than the third quartile.

64. Suppose that the average of 10 numbers is 7.5 and that the smallest of them is $Min = 3$.

(a) What is the smallest possible value of Max?

(b) What is the largest possible value of Max?

65. Figure 15-20 shows two different histograms summarizing the 2008 payrolls of the 30 teams in Major League Baseball. Using the information shown in the figure, it can be determined that the median payroll in Major League Baseball in 2008 falls somewhere between \$70 million and \$80 million. Explain how.

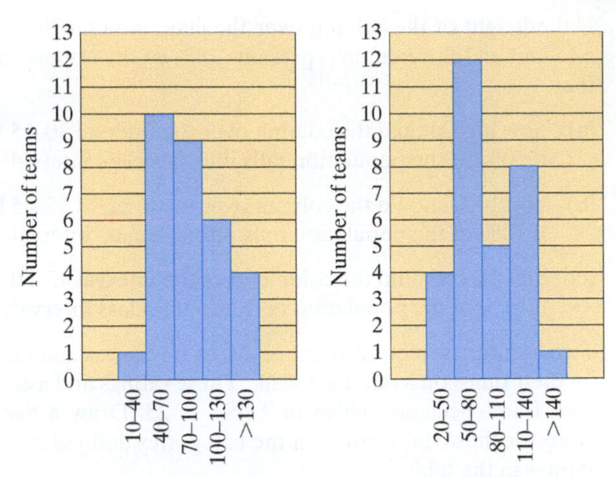

FIGURE 15-20 2008 MLB team payrolls (in millions). $N = 30$. (Data from Major League Baseball, *MLB.com*.)

66. What happens to the five-number summary of the Stat 101 data set (see Example 15.14) if

(a) two points are added to each score?

(b) 10% is added to each score?

67. Let A denote the average and M the median of the data set $\{x_1, x_2, x_3, \ldots, x_N\}$.

Let c be any constant.

(a) Find the average of the data set $\{x_1 + c, x_2 + c, x_3 + c, \ldots, x_N + c\}$ expressed in terms of A and c.

(b) Find the median of the data set $\{x_1 + c, x_2 + c, x_3 + c, \ldots, x_N + c\}$ expressed in terms of M and c.

68. Explain why the data sets $\{x_1, x_2, x_3, \ldots, x_N\}$ and $\{x_1 + c, x_2 + c, x_3 + c, \ldots, x_N + c\}$ have

(a) the same range.

(b) the same standard deviation.

Exercises 69 and 70 refer to histograms with unequal class intervals. When sketching such histograms, the columns must be drawn so that the frequencies or percentages are proportional to the area of the column. Figure 15-21 illustrates this. If the column over class interval 1 represents 10% of the population, then the column over class interval 2, also representing 10% of the population, must be one-third as high, because the class interval is three times as large (Fig. 15-21).

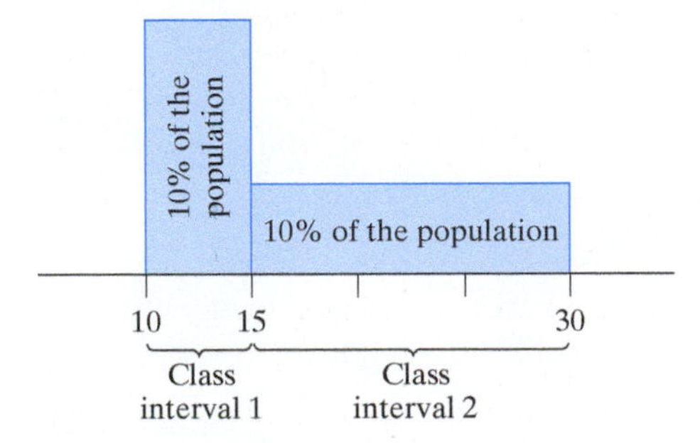

FIGURE 15-21

69. If the height of the column over the class interval 20–30 is one unit and the column represents 25% of the population, then

(a) how high should the column over the interval 30–35 be if 50% of the population falls into this class interval?

(b) how high should the column over the interval 35–45 be if 10% of the population falls into this class interval?

(c) how high should the column over the interval 45–60 be if 15% of the population falls into this class interval?

70. Two hundred senior citizens are tested for fitness and rated on their times on a one-mile walk. These ratings and associated frequencies are given in Table 15-23. Draw a histogram for these data based on the categories defined by the ratings in the table.

Time	Rating	Frequency
6^+ to 10 minutes	Fast	10
10^+ to 16 minutes	Fit	90
16^+ to 24 minutes	Average	80
24^+ to 40 minutes	Slow	20

TABLE 15-23

RUNNING

71. A data set is called **constant** if every value in the data set is the same. Explain why any data set with standard deviation 0 must be a constant data set.

72. Show that the standard deviation of any set of numbers is always less than or equal to the range of the set of numbers.

73. Show that if A is the mean and M is the median of the data set $\{1, 2, 3, \ldots, N\}$, then for all values of N, $A = M$.

74. Suppose that the standard deviation of the data set $\{x_1, x_2, x_3, \ldots, x_N\}$ is σ. Explain why the standard deviation of the data set $\{a \cdot x_1, a \cdot x_2, a \cdot x_3, \ldots, a \cdot x_N\}$ (where a is a positive number) is $a \cdot \sigma$.

75. Chebyshev's theorem. The Russian mathematician P. L. Chebyshev (1821–1894) showed that for any data set and any constant k greater than 1, at least $1 - (1/k^2)$ of the data must lie within k standard deviations on either side of the mean A. For example, when $k = 2$, this says that $1 - \frac{1}{4} = \frac{3}{4}$ (i.e., 75%) of the data must lie within two standard deviations of A (i.e., somewhere between $A - 2\sigma$ and $A + 2\sigma$).

(a) Using Chebyshev's theorem, what percentage of a data set must lie within three standard deviations of the mean?

(b) How many standard deviations on each side of the mean must we take to be assured of including 99% of the data?

(c) Suppose that the average of a data set is A. Explain why there is no number k of standard deviations for which we can be certain that 100% of the data lies within k standard deviations on either side of the mean A.

January 2016: Powerball Lottery jackpot hits $1.5 billion. (For more on the Powerball Lottery, see Examples 16.16 and 16.35.)

Probabilities, Odds, and Expectations

Measuring Uncertainty and Risk

In the first days of January, 2016, the Powerball Lottery jackpot crossed into billion-dollar territory, reaching $1.58 billion on January 12—the largest jackpot ever recorded in the United States—and lottery fever swept the nation once again. Buy a $2 lottery ticket and make *one and a half billion dollars*? Who could argue with that deal? Waiting in line for hours to buy lottery tickets? A small price to pay.

The lottery frenzy created by the January, 2016, Powerball jackpot (by the way, the final jackpot of $1.58 billion was split among three different winners) is an extreme example of the combination that makes big prize lotteries so appealing all over the world: low risk–high reward. Spend $2 for a lottery ticket, make millions (or even billions) if you win. But there is a key missing element in this equation: What are the chances (probabilities, odds) that you *will* win? Without taking into account this third variable, the low risk–high reward lottery formula is just fool's gold.

By the time we are finished with this chapter we will be able to understand how risk, rewards, and probabilities are combined in a precise mathematical way that allows us to answer major existential questions like, "Should I really spend my money buying lottery tickets?" To get there, we will have to cover a lot of ground quickly—by necessity, this chapter covers a lot of important ideas in just a few pages. The chapter starts with an introduction to *sample spaces* and *events*—the basic building blocks for computing probabilities (Section 16.1). Section 16.2 is a quick introduction (for many readers a review) of the *multiplication rule*, *permutations*, and *combinations*—these are key mathematical tools for probability calculations. In Section 16.3 we finally get to *probabilities* and *odds* (they are two different formulations of the same idea) with plenty of examples that illustrate how to calculate both. In Section 16.4 we introduce the concept of *expectation* (or *expected value*), the key concept in understanding the relationship between risk and reward. In Section 16.5 we use several real-world examples (including lotteries) to illustrate how to use expectation as a tool for measuring risk.

> "His sacred majesty, Chance, decides everything."
>
> – *Voltaire*

Disclaimer: Throughout the chapter we will use many examples of games of chance—craps, poker, horse races, lotteries, etc. Games of chance often serve as the ideal setting in which to analyze and deal with interesting probability questions. This should not be taken in any way, shape, or form as an encouragement to gamble or an endorsement of gambling—on the contrary, several of the examples in the chapter are there to illustrate mathematically why gambling, in general, is not such a good idea.

Sample Spaces and Events

In broad terms, probability is the *quantification of uncertainty*. To understand what that means, we need to start by formalizing the notion of uncertainty.

We will use the term **random experiment** to describe an activity or a process whose outcome cannot be predicted ahead of time. Typical examples of random experiments are tossing a coin, rolling a pair of dice, drawing cards out of a deck of cards, predicting the result of a football game, and forecasting the path of a hurricane.

Associated with every random experiment is the *set* of all of its possible outcomes, called the **sample space** of the experiment. For the sake of simplicity, we will concentrate on random experiments for which there is only a finite set of outcomes, although sample spaces with infinitely many outcomes are both possible and important.

We illustrate the concept of a sample space by means of several examples. Since the sample space of any experiment is a set of individual outcomes, we will use set notation to describe it. We will consistently use the letter S to denote a sample space and the letter N to denote the *size* of the sample space S (i.e., the number of outcomes in S).

EXAMPLE 16.1 TOSSING A COIN

One simple random experiment is to *toss a quarter* and *observe whether it lands heads or tails*. The sample space can be described by $S = \{H, T\}$, where H stands for *Heads* and T for *Tails*. Here $N = 2$.

A couple of comments about coins are in order here. First, the fact that the coin in Example 16.1 is a quarter is essentially irrelevant. Practically all coins have an obvious "heads" side (and, thus, a "tails" side), and even when they don't—as in a "buffalo nickel"—we can agree ahead of time which side is which. Second, there are fake coins out there where both sides are "heads." Tossing such a coin does not fit our definition of a random experiment, so from now on, we will assume that all coins used in our experiments have two different sides, which we will call H and T.

EXAMPLE 16.2 TOSSING A COIN TWICE: EPISODE 1

Suppose we toss a coin *twice* and *record* the outcome of each toss (H or T) in the order it happens. The sample space now is $S = \{HH, HT, TH, TT\}$, where HT means that the first toss came up H and the second toss came up T, which is a different outcome from TH (first toss T and second toss H). In this sample space $N = 4$.

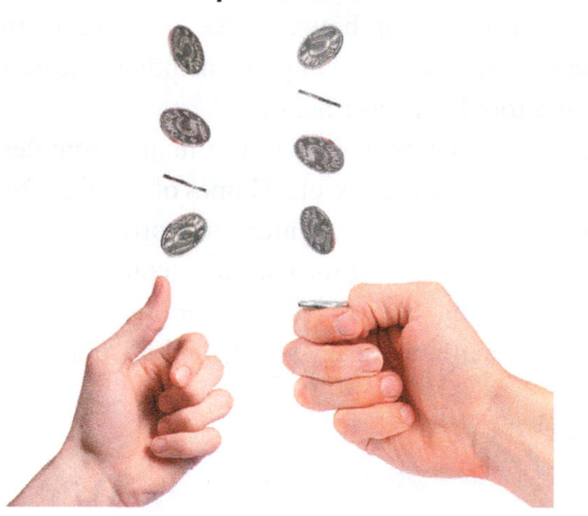

Suppose now we *toss two coins* at the same time (tricky but definitely possible). This random experiment appears different from the one where we toss one coin twice, but the sample space is still $S = \{HH, HT, TH, TT\}$. (Here we must agree what the order of the symbols is—for example, the first symbol describes the coin tossed by the left hand and the second the coin tossed by the right hand.)

Since they have the same sample space, we will consider the two random experiments just described as equivalent, so whether you toss two coins at once or one coin twice, from our point of view there is no difference between the two.

Now let's consider a different random experiment. We are still tossing a coin twice, but we only care now about the *number of heads* that come up. Here there are only three possible outcomes (no heads, one head, or both heads), and symbolically we might describe this sample space as $S = \{0, 1, 2\}$.

The important point made in Example 16.2 is that a random experiment is defined by two things: what the *act* is (such as tossing coins) and what the *observation* that follows the act is.

EXAMPLE 16.3 SHOOTING FREE THROWS: EPISODE 1

Here is a familiar scenario: Your favorite basketball team is down by 1, clock running out, and one of the players in your team is fouled and goes to the line to shoot free throws, with the game riding on the outcome. It's not a good time to think of sample spaces, but let's do it anyway.

Clearly the shooting of free throws is a random experiment, but what is the sample space? As in Example 16.2, the answer depends on a few subtleties.

In one scenario (the *penalty situation*) the player is going to shoot two free throws no matter what. In this case one could argue that what really matters is how many free throws he makes (make both and win the game, miss one and tie and

go to overtime, miss both and lose the game). When we look at it this way the sample space is $S = \{0, 1, 2\}$.

A somewhat more stressful scenario is when the player is shooting a *one-and-one*. This means that the player gets to shoot the second free throw only if he makes the first one. In this case there are also three possible outcomes, but the circumstances are different because the sequence of events is relevant (miss the first free throw and lose the game, make the first free throw but miss the second one and tie the game, make both and win the game). We can describe this sample space as $S = \{f, sf, ss\}$, where we use f to indicate *failure* (missed the free throw) and s to indicate *success*.

We will now discuss a couple of examples of random experiments involving dice. A die is a cube, usually made of plastic, whose six faces are marked with dots (from 1 to 6) called "pips." Random experiments using dice have a long-standing tradition in our culture and are a part of both gambling and recreational games such as Monopoly or Yahtzee.

EXAMPLE 16.4 ROLLING A PAIR OF DICE: EPISODE 1

The most general scenario when rolling a pair of dice is to consider each die separately (in this regard, it helps to think of them as being of different colors—say one is white and the other one is red). The sample space consists of the 36 outcomes shown in Fig. 16-1.

Throughout this chapter we will see several examples of rolling a pair of dice, so for convenience we will use the notation $(1, 1), (1, 2), \ldots, (6, 6)$ to denote the possible outcomes when you roll a pair of dice. The first number represents one die (say white) and the second number the other die (say red).

FIGURE 16-1

An alternative scenario when rolling a pair of dice is to only consider the *total* of the two numbers rolled. In this situation we don't really care how a particular total comes about. We can "roll a seven" in various paired combinations—$(1, 6)$, $(2, 5)$, $(3, 4)$, $(4, 3)$, $(5, 2)$ and $(6, 1)$—but no matter how the individual dice come up, the only thing that matters is the total rolled.

The possible outcomes in this scenario range from "rolling a two" to "rolling a twelve," and the sample space can be described by $S = \{2, 3, 4, 5, 6, 7, 8, 9, 10, 11, 12\}$.

EXAMPLE 16.5 A CAR AND TWO GOATS: EPISODE 1

Imagine you are a contestant in a television game show. The host of the show shows you three doors (we'll call them Door #1, Door #2 and Door #3). You get to pick one of the three doors and keep the prize hidden behind it—a fancy new sports car behind one of the three doors and a goat behind each of the other two. (If this story sounds familiar to you, then you are not too young to remember Monty Hall and the long-running game show *Let's Make a Deal.*)

This game gives us a slightly more exotic example of a random experiment than the more mundane tossing of a pair of dice. The randomness comes from the fact that the sports car is randomly placed behind one of the three doors and you must guess which one (we are assuming here that you want to win the sports car and you are not too interested in taking home a goat).

As you blindly confront the decision on which door to choose you think of sample spaces. What is the proper sample space that describes your predicament? If we use the notation (x, y) to denote the fact that you choose Door #x and the car is behind Door #y, we can describe the sample space as consisting of the following nine outcomes:

$$S = \{(1,1), (1,2), (1,3), (2,1), (2,2), (2,3), (3,1), (3,2), (3,3)\}.$$

Of the above, $(1,1)$, $(2,2)$ and $(3,3)$ mean you are going home driving a fancy new sports car, the other six mean you get a goat. (In the *Let's Make a Deal* show the undesirable prizes were called Zonks, and participants could trade the Zonk for a more practical consolation prize like a toaster or a television set.)

Imagine now that you make your decision and choose one of the three doors—say you choose Door #1. The host, who knows where the car is, opens Door #3 to show there is a goat behind it and offers you the option of switching. Should you stay with your original choice or switch to Door #2? This is an interesting question, and we will return to it later in the chapter.

EXAMPLE 16.6 HORSE RACES AND TRIFECTAS: EPISODE 1

In the language of horse racing, a *trifecta* is a bet on predicting the first-, second-, and third-place finishers in a horse race in the right order. A trifecta ticket has boxes for the first-, second-, and third-place finishers in which you can enter the numbers of the three horses you predict will finish first, second, and third, respectively.

Imagine you bought a trifecta ticket in a horse race with 10 horses numbered 1 through 10. You marked horse number 4 to finish first, number 6 to finish second, and number 5 to finish third. We will use the notation (4, 6, 5) to denote your trifecta ticket. Note that if horse number 6 finishes first, horse number 4 second, and horse number 5 third you are out of luck—close but no cigar.

The random experiment in this example is the running of the race, and the observation of interest is which horses finished first, second, and third, respectively. The sample space consists of all possible outcomes of the form (x, y, z) where x, y, and z, are three different numbers between 1 and 10. There are 720 outcomes in this sample space (we will see why this is the case in Example 16.12), so writing them all down, one by one, is not such a good idea. There are a couple of options, the simplest one is using the "..." notation, as in $S = \{(1,2,3), (1,3,2), (2,1,3), (2,3,1), (3,1,2), \ldots, (10,9,8)\}$. The "..." is a way of saying "and so on," somewhat along the same lines as the "..." notation used in Chapter 9 for describing sequences.

Example 16.6 illustrates the point that sample spaces can be quite large and have a lot of outcomes and that we are often justified in not writing each and every one of these outcomes down. This is where the "..." comes in handy. The key thing is to understand what the sample space looks like and most importantly, to find N, the size of the sample space. If we can do it without having to list all the outcomes, then so much the better. We will discuss how this is done in the next section.

Events

An **event** is a *subset* of the sample space, and any *subset* of the sample space can be considered an event. By definition, events are sets, so we will use set notation to describe them and the basic rules of set theory to work with them.

In Section 2.4 we introduced the fact that a subset with N elements has 2^N subsets—it follows that in a sample space with N outcomes there are 2^N different events.

EXAMPLE 16.7 TOSSING A COIN TWICE: EPISODE TWO

If you toss a coin twice and record the result of each toss, the sample space is $S = \{HH, HT, TH, TT\}$—we discussed this in Example 16.2. This sample space has 4 outcomes and $2^4 = 16$ different subsets, each representing a different event, as shown in Table 16-1.

1. $\{\ \}$	5. $\{TT\}$	9. $\{HT, TH\}$	13. $\{HH, HT, TT\}$
2. $\{HH\}$	6. $\{HH, HT\}$	10. $\{HT, TT\}$	14. $\{HH, TH, TT\}$
3. $\{HT\}$	7. $\{HH, TH\}$	11. $\{TH, TT\}$	15. $\{HT, TH, TT\}$
4. $\{TH\}$	8. $\{HH, TT\}$	12. $\{HH, HT, TH\}$	16. $\{HH, HT, TH, TT\}$

TABLE 16-1 ■ The 16 possible events when tossing a coin twice.

Several of the events listed in Table 16-1 deserve special attention.

- The empty set $\{\ \}$ is considered an event, essentially an event with no outcomes. At first glance, the idea of an event with no outcomes might seem a little weird, but we will soon see that it is very convenient to have it in our toolkit. The event with no outcomes is called the **impossible event**.
- The entire sample space S is an event, the event where every possible outcome is included. When viewed as an event, the entire sample space is called the **certain event**.
- The events $\{HH\}$, $\{HT\}$, $\{TH\}$, and $\{TT\}$ consist of just one outcome. An event consisting of a single outcome is called a **simple event**. Simple events are just the individual outcomes in a sample space reinterpreted in the language of events.

A convenient way to think of an event is as a "package" in which outcomes with some common characteristic are bundled together, but the "package" idea has to be interpreted very broadly: The package can have nothing in it, everything in it, or just a single item in it.

EXAMPLE 16.8 TOSSING A COIN THREE TIMES

If you toss a coin three consecutive times and record the result of each toss, the sample space is $S = \{HHH, HHT, HTH, HTT, THH, THT, TTH, TTT\}$, $N = 8$, and the number of possible events is $2^8 = 256$—too many to list. Table 16-2 shows a few of these events, described both in plain English and in the language of sets.

Event (plain English description)	Event (set description)
All tosses come up heads	$\{HHH\}$
All tosses come up the same	$\{HHH, TTT\}$
Toss one head and two tails	$\{HTT, THT, TTH\}$
First toss comes up heads	$\{HTT, HTH, HHT, HHH\}$
The tosses don't come up all the same	$\{HHT, HTH, HTT, THH, THT, TTH\}$
Toss more heads than tails	$\{HHT, HTH, THH, HHH\}$
Toss an equal number of heads and tails	$\{\ \}$ (the *impossible* event)
Toss three or fewer heads	S (the certain event)

TABLE 16-2 ■ Some of the 256 possible events when tossing a coin three times.

16.2 The Multiplication Rule, Permutations, and Combinations

In this section we will introduce some of the basic mathematical tools that are used to determine the number of outcomes in a sample space (or an event) without having to list them all and then tallying them one by one. We will use the word **counting** to describe the process of determining the number of elements in a set, and our counting methods will be considerably more sophisticated than the kind of counting we learned in kindergarten.

You are probably already familiar with the first rule of counting—*the multiplication rule.*

- **Multiplication Rule.** If there are m different ways to do X and n different ways to do Y, then X and Y together (and in that order) can be done in $m \times n$ different ways.

Informally, the multiplication rule simply says that *when something is done in multiple stages, the number of ways it can be done is found by multiplying the number of ways each of the individual stages can be done.* The multiplication rule is a simple but powerful idea that we can best understand by looking at some examples.

EXAMPLE 16.9 ICE CREAM: EPISODE 1

There is nothing sweeter than a great ice cream parlor, so we will use this theme in a few of our examples. Your local ice cream parlor offers 2 different choices of cones and 3 different flavors of ice cream (this is a boutique ice cream parlor!). Using the multiplication rule we can conclude that there are $2 \times 3 = 6$ different choices for a *single* (cone and one scoop of ice cream) order. The choices are illustrated in Figure 16-2.

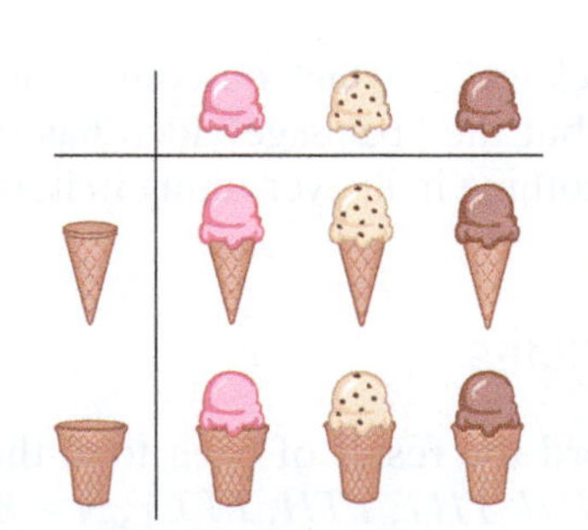

FIGURE 16-2 The scoop on the multiplication rule.

Figure 16-2 also clarifies the logic behind the multiplication rule: If we had m different types of cones and n different flavors of ice cream, then the corresponding arrangement would show the cone/flavor combinations laid out in m rows and n columns, illustrating the $m \times n$ possibilities.

EXAMPLE 16.10 ICE CREAM: EPISODE 2

A bigger ice cream parlor around the corner offers 5 different choices of cones, 31 different flavors of ice cream, and 8 different choices of topping. The question is, If you are going to choose a cone and a single scoop of ice cream but then add a topping for good measure, how many orders are possible? The number of choices is easy to find using the multiplication rule twice: First, there are $5 \times 31 = 155$ different cone/flavor combinations, and each of these can be combined with one of the 8 toppings into a grand total of $155 \times 8 = 1240$ different cone/flavor/topping combinations. A slightly different way to think about the question is to think of making your decision in three stages: (1) choose the type of cone (5 choices), (2) choose the flavor of your scoop (31 choices), and (3) choose your topping (8 choices). The combined decision can be made in $5 \times 31 \times 8 = 1240$ different ways.

EXAMPLE 16.11 THE MAKING OF A WARDROBE

Dolores is a young saleswoman planning her next business trip. She is thinking about packing three different pairs of shoes, four skirts, six blouses, and two jackets. If all the items are color coordinated, how many different *outfits* will she be able to create by combining these items?

To answer this question, we must first define what we mean by an "outfit." Let's assume that an outfit consists of one pair of shoes, one skirt, one blouse, and one jacket. Then to make an outfit Dolores must choose a pair of shoes (three choices), a skirt (four choices), a blouse (six choices), and a jacket (two choices). By the multiplication rule the total number of possible outfits is $3 \times 4 \times 6 \times 2 = 144$. (Think about it—Dolores can be on the road for over four months and never have to wear the same outfit twice! And it all fits in a small suitcase.)

Let's consider now a similar wardrobe question but with a few twists. Once again, Dolores is packing for a business trip. This time, she packs three pairs of shoes, four skirts, three pairs of slacks, six blouses, three turtlenecks, and two jackets. As before, we can assume that she coordinates the colors so that everything goes with everything else. This time, we will define an outfit as consisting of a pair of shoes, a choice of "lower wear" (either a skirt *or* a pair of slacks), and a choice of "upper wear" (it could be a blouse *or* a turtleneck *or both*), and, finally, she may or may not choose to wear a jacket. How many different such outfits are possible?

Our strategy will be to think of an outfit as being put together in stages and to draw a box for each of the stages. We then separately count the number of choices at each stage and enter that number in the corresponding box. (Some of these calculations can themselves be mini-counting problems.) The last step is to multiply the numbers in each box. The details are illustrated in Fig. 16-3. The final count for the number of different outfits is $N = 3 \times 7 \times 27 \times 3 = 1701$.

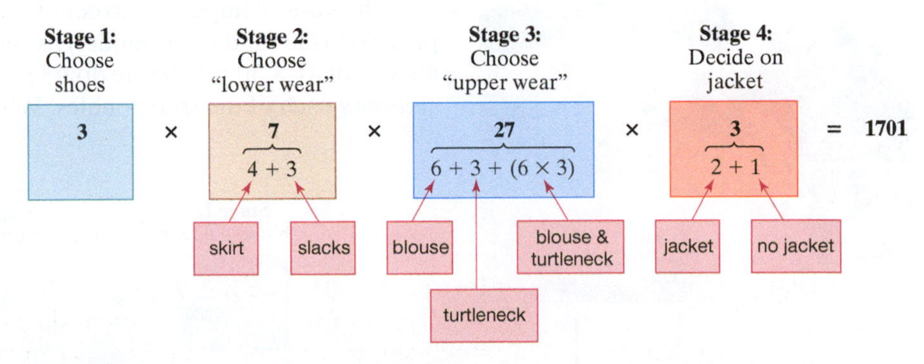

FIGURE 16-3

The method of drawing boxes representing the successive stages in a process, and putting the number of choices for each stage inside the box is a convenient strategy that often helps clarify one's thinking. Silly as it may seem, we strongly recommend it. For ease of reference, we will call it the *box model* for counting.

EXAMPLE 16.12 HORSE RACES AND TRIFECTAS: EPISODE 2

This example is a follow-up to Example 16.6. A trifecta ticket consists of three imaginary boxes labeled First place, Second place, and Third place in which you can enter the numbers of the three horses you are betting will come in first, second, and third in the race. If there are 10 horses entered in the race, there are $10 \times 9 \times 8 = 720$ different outcomes in the sample space. Figure 16-4 illustrates the reasoning behind this count.

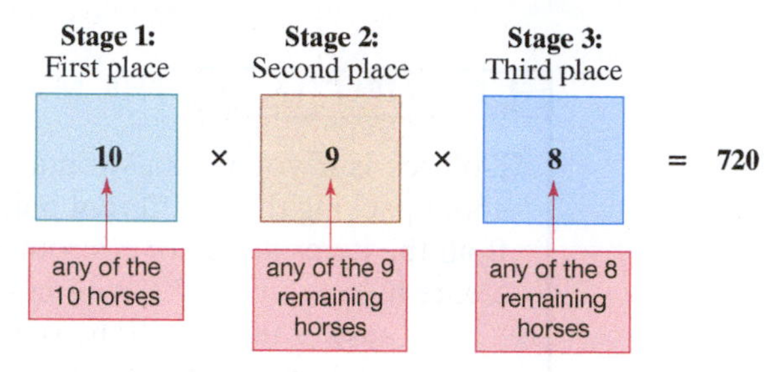

FIGURE 16-4

Permutations and Combinations

Many counting problems boil down to counting the number of ways in which we can select groups of objects from a larger group of objects. Often these problems require somewhat more sophisticated counting methods than the plain vanilla multiplication rule. In this section we will discuss the dual concepts of *permutation*—a group of objects in which the ordering of the objects within the group *makes a difference*—and *combination*—a group of objects in which the ordering of the objects is *irrelevant*. A note of warning about the terminology: In Chapter 2 we defined the term *permutation* as a reordering or shuffling of a set of objects. We now stretch the meaning of the word.

EXAMPLE 16.13 ICE CREAM: EPISODE 3

The ice cream parlor around the corner offers 31 different flavors of ice cream. One of their specials is the "true double"—two scoops of ice cream of *different* flavors served in a bowl and with chocolate sprinkles on top. Say you want to order a true double—how many different choices do you have?

The natural impulse is to count the number of choices using the multiplication rule (and a box model) as shown in Fig. 16-5. This would give an answer of $31 \times 30 = 930$ true doubles. Unfortunately, this answer is *double counting* each of the true doubles. Why?

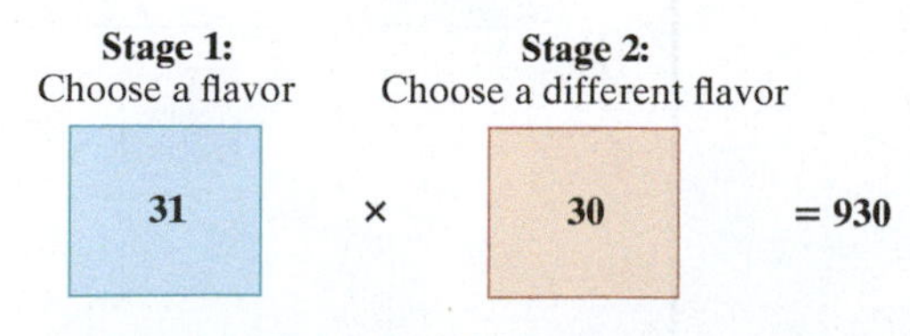

FIGURE 16-5

When we use the multiplication rule, *there is a well-defined order to things*, and a scoop of vanilla followed by a scoop of chocolate is counted separately from a scoop of chocolate followed by a scoop of vanilla. But by all reasonable standards, a bowl with a scoop of vanilla plus a scoop of chocolate is the same regardless of which scoop goes in the bowl first. (This is not necessarily true when the ice cream is served in a cone. Fussy people can be very picky about the order of the flavors when the scoops are stacked up.)

The good news is that now that we understand why the count of 930 is wrong we can fix it. All we have to do to get the correct answer is to divide the original count by 2. It follows that the number of possible true doubles is $(31 \times 30)/2 = 465$.

Example 16.13 is an important one. It warns us that we have to be careful about how we use the multiplication rule and box models in counting problems where the order in which we choose the objects (ice cream flavors) does not affect the answer. Let's take this idea to the next level.

EXAMPLE 16.14 ICE CREAM: EPISODE 4

Some days you have a real craving for ice cream, and on such days you like to go to the ice cream parlor around the corner and order a "true triple." A "true triple" consists of three scoops of ice cream each of a different flavor. How many different "true triples" are there?

Starting with the multiplication rule, we have 31 choices for the "first" flavor, 30 choices for the "second" flavor, and 29 choices for the "third" flavor, for an apparent grand total of $31 \times 30 \times 29 = 26{,}970$. But once again this answer counts each true triple more than once; in fact, it does so *six times*! (More on that shortly.) If we accept this, the correct answer must be $26{,}970/6 = 4495$.

Why is it that in the total of 26,970 we are counting each true triple *six* times? The answer comes down to this: Any three flavors (call them X, Y, and Z) can be listed in 6 different ways (XYZ, XZY, YXZ, YZX, ZXY, and ZYX). The multiplication rule counts each of these separately, but when you think in terms of scoops of ice cream in a bowl, they are the same true triple regardless of the order in which the three flavors are listed.

The bottom line is that with 31 different flavors we can create 4495 different true triples. We can better understand where this number comes from by looking at it in its raw, uncalculated form $[(31 \times 30 \times 29)/(3 \times 2 \times 1) = 4495]$. The numerator $(31 \times 30 \times 29)$ comes from counting *ordered* triples using the multiplication rule; the denominator $(3 \times 2 \times 1)$ comes from counting the number of ways in which three things (in this case, the three flavors in a triple) can be rearranged. The denominator $3 \times 2 \times 1$ is already familiar to us—it is the *factorial* of 3. (We discussed the factorial in Chapters 2 and 6, so we won't dwell on it here.)

Our next example will deal with the game of poker. Despite the great deal of exposure poker gets on television these days, there are a lot of misconceptions about the mathematics behind poker hands. For readers not familiar with the game, poker is a betting game played with a standard deck of cards [a standard deck of cards has 52 cards divided into 4 *suits* (clubs, diamonds, hearts, and spades) and with 13 *values* in each suit $(2, 3, \ldots, 10, \text{J}, \text{Q}, \text{K}, \text{A})$]. There are many variations of poker depending on how many cards are dealt and whether all the cards are *down* cards (only the player receiving the card can see it) or some of the cards are *up* cards (dealt face up so that they can be seen by all the players).

EXAMPLE 16.15 FIVE-CARD POKER HANDS: EPISODE 1

In this example we will compare two types of games: five-card *stud poker* and five-card *draw poker*. In both of these games a player ends up with five cards, but there is an important difference when analyzing the mathematics behind the games: In five-card *draw* [Fig. 16-6(a)] the order in which the cards are dealt is irrelevant; in five-card *stud* [Figs. 16-6(b) and (c)] the order in which the cards come up is extremely relevant. The reason for this is that in five-card draw all cards are dealt down, but in five-card stud only the first card is dealt down—the remaining four cards are dealt up, one at a time. This means that players can assess the relative strengths of the other players' hands as the game progresses and play their hands accordingly.

(a)

(b)

(c)

FIGURE 16-6 (a) *Draw* poker: All cards are down cards. The order in which the cards are dealt is irrelevant. (b) *Stud* poker: Some cards are up. (c) *Stud* poker: Same up cards as in (b) but dealt in a different order. It makes a difference.

Counting the number of five-card *stud* poker hands is a direct application of the multiplication rule: 52 possibilities for the first card, 51 for the second card, 50 for the third card, 49 for the fourth card, and 48 for the fifth card, for an awesome total of $52 \times 51 \times 50 \times 49 \times 48 = 311{,}875{,}200$ possible hands.

Counting the number of five-card *draw* poker hands requires a little more finesse. Here a player gets five down cards and the hand is the same regardless of the order in which the cards are dealt. There are $5! = 120$ different ways in which the same set of five cards can be ordered, so that one draw hand corresponds to 120 different stud hands. Thus, the stud hands count is exactly 120 times bigger than the draw hands count. Great! All we have to do then is divide the 311,875,200 (number of stud hands) by 120 and get our answer: There are 2,598,960 possible five-card draw hands. [As before, it's more telling to look at this answer in the uncalculated form $(52 \times 51 \times 50 \times 49 \times 48)/5! = 2{,}598{,}960$.] ■

We are now ready to generalize the ideas developed in Examples 16.14 and 16.15. Suppose that we have a set of n distinct objects and we want to choose r different objects from this set. The number of ways that this can be done depends on whether the choices are *ordered* or *unordered*. Ordered choices are the generalization of stud poker hands—when the same objects are chosen in a different order you get a different outcome. Unordered choices are the generalization of draw poker hands—the order in which the objects are chosen does not affect the outcome. To distinguish between these two scenarios, we use the terms **permutation** to describe an ordered selection and **combination** to describe an unordered selection.

- **Permutation.** An *ordered* selection of r objects chosen from a set of n objects (r can be any integer between 0 and n). In the special case where $r = n$ all the objects are selected and the permutation is just a rearrangement or shuffing of the objects.
- **Combination.** An *unordered* selection of r objects chosen from a set of n objects (r can be any integer between 0 and n).

We will use the standard notation ${}_nP_r$ to denote the *number of permutations* of r out of n objects and ${}_nC_r$ to denote the number of *combinations* of r out of n objects. (Some calculators use variations of this notation, such as $P_{n,r}$ and $C_{n,r}$, respectively.)

The numbers ${}_nP_r$ and ${}_nC_r$ are typically found using a scientific calculator, but it is always a good idea to know the basic formulae behind the computations:

NUMBER OF PERMUTATIONS (OF *r* OUT OF *n* OBJECTS)

$${}_nP_r = n(n-1)(n-2)\ldots(n-r+1), \text{ or, equivalently, } {}_nP_r = \frac{n!}{(n-r)!}$$

NUMBER OF COMBINATIONS (OF *r* OUT OF *n* OBJECTS)

$${}_nC_r = {}_nP_r/r!, \text{ or, equivalently, } {}_nC_r = \frac{n!}{r!(n-r)!}$$

EXAMPLE 16.16 THE POWERBALL LOTTERY: EPISODE 1

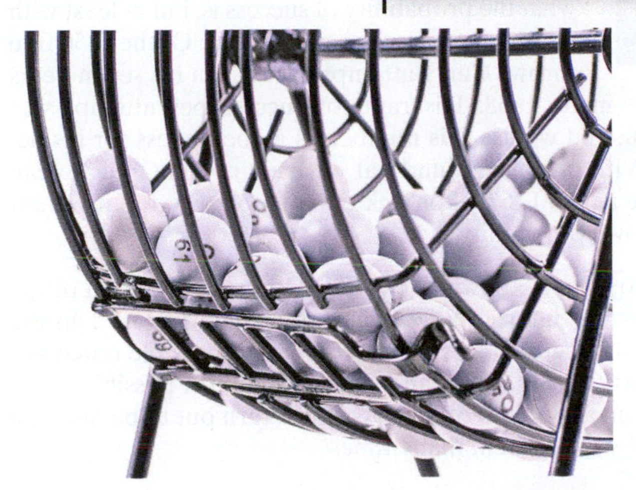

A typical lottery drawing consists of taking a few balls out of a larger set of numbered balls inside a spinning container called a *hopper*. The range of numbers on the balls and the number of balls drawn from the hopper can vary with various kinds of lotteries. If the numbers drawn from the hopper match the numbers on your lottery ticket, you win the jackpot. You can also win smaller prizes if some (but not all) of your numbers match some of the numbers drawn.

For the most recent version of the Powerball lottery, the drawing comes in two parts. First, five balls are drawn from a hopper containing white balls numbered 1 through 69. Then, a separate "powerball" is drawn from a different hopper containing red balls numbered 1 through 26. In this example we will accurately count all possible Powerball lottery drawings, even though this is a very large number.

We start counting the number of drawings of five balls from the first hopper, containing balls numbered 1 through 69. Since the order in which the numbered balls come out of the hopper is irrelevant (the only thing that matters is whether you have a match), we are dealing with *combinations* of five numbers taken from the set of numbers 1 through 69. The number of such combinations is ${}_{69}C_5 = 11{,}238{,}513$.

On top of that, we have the 26 possibilities for the red "powerball" (only one ball is drawn out of the second hopper containing the 26 red balls). Using the multiplication rule, we can conclude that the number of possible Powerball drawings is $({}_{69}C_5) \times 26 = 292{,}201{,}338$.

16.3 Probabilities and Odds

If we toss a coin in the air, *what is the probability that it will land heads up?* This one is not a very profound question, and almost everybody agrees on the answer, although not necessarily for the same reason. The standard answer given is 1 out of 2, or 1/2. But why is the answer 1/2, and what does such an answer mean?

One common explanation given for the answer of 1/2 is that when we toss a coin, there are two possible outcomes (H and T), and since H represents one of the two possibilities, the probability of the outcome H must be 1 out of 2 or 1/2. This logic, while correct in the case of an honest coin, has a lot of holes in it.

Consider how the same argument would sound in a different scenario.

EXAMPLE 16.17 SHOOTING FREE THROWS: EPISODE 2

Close your eyes and imagine a basketball player in the act of shooting a free throw. Just as in a coin toss, there are two possible outcomes to the free-throw: make or miss; success or failure. Unlike the case of a coin toss, however, we just can't conclude that since the sample space has two outcomes, the probability of success is 1 out of 2, or 1/2.

In shooting free throws, just as in many other human activities, the probability of success very much depends on the abilities of the player—it makes a difference whether the shooter is Stephen Curry of the Golden State Warriors—the number one free-throw shooter in the NBA—or some dude shooting free throws on the playground. In either case, we can't know for sure what the probability of success is, but at least with Curry we can make a good guess: Of the 1850 free throws Curry attempted so far in his seven years in the NBA (2009–2016) he made 1668. His ratio of success per attempt was $1668/1850 \approx 0.9016 = 90.16\%$, and we use this number as a good guess for his actual probability of success. With our playground pal we have no clue, but we can safely guess that it is nowhere near 90.16%. For lack of a number, we will just punt and call it p (short for "unknown probability").

Example 16.17 leads us to what is known as the *empirical* interpretation of the concept of probability. Under this interpretation when tossing an honest coin the probability of *heads* is 1/2 not because *heads* is one out of two possible outcomes but because, if we were to toss the coin over and over—hundreds, possibly thousands, of times—in the long run about half of the tosses will turn out to be heads, a fact that has been confirmed by experiment many times.

The argument as to exactly how to interpret the statement "the probability of X is such and such" goes back to the late 1600s, and it wasn't until the 1930s that a formal theory for dealing with probabilities was developed by the Russian mathematician A. N. Kolmogorov (1903–1987). This theory has made probability one of the most useful and important concepts in modern mathematics. In this section we will introduce some of the basic ideas behind the mathematical theory of probability.

Probability Assignments

Let's return to free-throw shooting, as it is a useful metaphor for many probability questions.

EXAMPLE 16.18 SHOOTING FREE THROWS: EPISODE 3

Think of a generic basketball player shooting a free throw. We know nothing about his or her abilities—for all we know, the player could be Stephen Curry, or Joe Schmoe.

How can we describe the probability that this unknown free-throw shooter will make that free throw? It seems that there is no way to answer this question, since we know nothing about the ability of the shooter. We could argue that the probability could be just about any number between 0 and 1. No problem—we make our unknown probability a variable, say p.

Event	Probability
$\{ \}$	0
$\{s\}$	p
$\{f\}$	$1-p$
$\{s, f\}$	1

TABLE 16-3

What can we say about the probability that our shooter misses the free throw? A lot. Since there are only two possible outcomes in the sample space $S = \{s, f\}$, the probability of success (s) and the probability of failure (f) must complement each other—in other words, must add up to 1. This means that the probability of missing the free throw must be $1-p$.

Table 16-3 shows the probability of each possible event in the sample space $S = \{s, f\}$. Basically, Table 16-3 gives a generic model for free-throw shooting. It works when the free-throw shooter is Stephen Curry or Joe Schmoe. Each one of the choices results in a different assignment of numbers to the events in the sample space.

Example 16.18 illustrates the concept of a *probability assignment.* A **probability assignment** is a function that assigns to each event E a number between 0 and 1, which represents the probability of the event E and which we denote by $\Pr(E)$. A probability assignment always assigns probability 0 to the *impossible event* $[\Pr(\{ \}) = 0]$ and probability 1 to the whole sample space $[\Pr(S) = 1]$.

With finite sample spaces a probability assignment is defined by assigning probabilities to just the simple events (that is, to the individual outcomes) in the sample space. Once we do this, we can find the probability of any event by simply *adding the probabilities of the individual outcomes that make up that event.* There are only two requirements for a valid probability assignment:

- All probabilities are numbers between 0 and 1
- The sum of the probabilities of the simple events equals 1.

The next example illustrates how a probability assignment works.

EXAMPLE 16.19 HANDICAPPING A TENNIS TOURNAMENT: EPISODE 1

There are six players playing in a tennis tournament: A (Russian, female), B (Croatian, male), C (Australian, male), D (Swiss, male), E (American, female), and F (American, female).

To handicap the winner of the tournament we need a probability assignment on the sample space $S = \{A, B, C, D, E, F\}$. With sporting events the probability assignment is subjective (it reflects an opinion), but professional odds-makers are usually very good at getting close to the right probabilities. For example, imagine that a professional odds-maker comes up with the following probability assignment: $\Pr(A) = 0.08$, $\Pr(B) = 0.16$, $\Pr(C) = 0.20$, $\Pr(D) = 0.25$, $\Pr(E) = 0.16$. [We are missing $\Pr(F)$ from the list, but since the probabilities of the simple events must add up to 1, we can do the arithmetic: $\Pr(F) = 0.15$.]

Once we have the probabilities of the simple events, the probabilities of all other events follow by addition. For example, the probability that an American will win the tournament is given by $\Pr(E) + \Pr(F) = 0.16 + 0.15 = 0.31$. Likewise, the probability that a male will win the tournament is given by $\Pr(B) + \Pr(C) + \Pr(D) = 0.16 + 0.20 + 0.25 = 0.61$. The probability that an American male will win the tournament is $\Pr(\{ \}) = 0$, since this one is an impossible event—there are no American males in the tournament!

The probability assignment discussed in Example 16.19 reflects the opinion of one specific observer. A different odds-maker might have a slightly different perspective and come up with a different probability assignment. This underscores the fact that sometimes there is no one single "correct" probability assignment on a sample space.

Once a specific probability assignment is made on a sample space, the combination of the sample space and the probability assignment is called a **probability space**. The following summarizes the key elements of a probability space.

- **Sample space:** $S = \{o_1, o_2, \ldots, o_N\}$.
- **Probability assignment:** $\Pr(o_1), \Pr(o_2), \ldots, \Pr(o_N)$.
  [Each of these is a number between 0 and 1 satisfying $\Pr(o_1) + \Pr(o_2) + \cdots + \Pr(o_N) = 1$.]
- **Events:** These are all the subsets of S, including $\{\ \}$ and S itself. The probability of an event is given by the sum of the probabilities of the individual outcomes that make up the event. [In particular, $\Pr(\{\ \}) = 0$ and $\Pr(S) = 1$.]

Equiprobable Spaces

One of the most common uses of randomness in the real world is as a mechanism to guarantee fairness. That's why it is universally accepted that tossing a coin, rolling a die, or drawing cards is a fair way of choosing among equally deserving choices. This is true as long as the coin, die, or deck of cards is "honest."

What does *honesty* mean when applied to coins, dice, or decks of cards? It essentially means that all individual outcomes in the sample space are equally probable. Thus, an *honest* coin is one in which H and T have the same probability of coming up, and an *honest* die is one in which each of the numbers 1 through 6 is equally likely to be rolled. A probability space in which each simple event has an equal probability is called an **equiprobable space**. (Informally, you can think of an equiprobable space as an "equal opportunity" probability space.)

In equiprobable spaces, calculating probabilities of events becomes simply a matter of counting. First, we need to find N, the size of the sample space. Each individual outcome in the sample space will have probability equal to $1/N$ (they all have the same probability, and the sum of the probabilities must equal 1). To find the probability of the event E we then find k, the number of outcomes in E. Since each of these outcomes has probability $1/N$, the probability of E is then k/N.

- **Probability of an event:** If k denotes the size of an event E and N denotes the size of the sample space S, then in an equiprobable space $\Pr(E) = k/N$.

EXAMPLE 16.20 ROLLING A PAIR OF DICE: EPISODE 2

This example is a follow-up of Example 16.4. Imagine that you are playing a game that involves rolling a pair of *honest* dice, and the only thing that matters is the total of the two numbers rolled. As we saw in Example 16.4, the sample space in this situation is $S = \{2, 3, 4, 5, 6, 7, 8, 9, 10, 11, 12\}$, where the outcomes are the possible totals that could be rolled. This sample space has $N = 11$ possible outcomes, but the outcomes are not equally likely, so it would be wrong to assign to each outcome the probability $1/11$. In fact, most people with some experience rolling a pair of dice know that the likelihood of rolling a 7 is much higher than that of rolling a 12.

To determine the probability of each total, we consider the roll of each die separately and look at the sample space consisting of the 36 possible ways in which one can roll a pair of dice (one white and one red) as discussed in Example 16.4 (see Fig. 16-1, page 475). Because the dice are honest, each of these 36 possible outcomes is equally likely to occur, so the probability of each is $1/36$.

Table 16-4 shows the probability of rolling a 2, 3, 4, . . . , 12. In each case the numerator represents the number of ways that particular total can be rolled. For example, the event "roll a 7" consists of six distinct possible rolls:
{⚀⚅, ⚁⚄, ⚂⚃, ⚃⚂, ⚄⚁, ⚅⚀}. Thus, the probability of "rolling a 7" is $6/36 = 1/6$. The other probabilities are computed in a similar manner.

Event	Probability
"Roll a 2": {⚀⚀}	1/36
"Roll a 3": {⚀⚁, ⚁⚀}	2/36
"Roll a 4": {⚀⚂, ⚁⚁, ⚂⚀}	3/36
"Roll a 5": {⚀⚃, ⚁⚂, ⚂⚁, ⚃⚀}	4/36
"Roll a 6": {⚀⚄, ⚁⚃, ⚂⚂, ⚃⚁, ⚄⚀}	5/36
"Roll a 7": {⚀⚅, ⚁⚄, ⚂⚃, ⚃⚂, ⚄⚁, ⚅⚀}	6/36
"Roll an 8": {⚁⚅, ⚂⚄, ⚃⚃, ⚄⚂, ⚅⚁}	5/36
"Roll a 9": {⚂⚅, ⚃⚄, ⚄⚃, ⚅⚂}	4/36
"Roll a 10": {⚃⚅, ⚄⚄, ⚅⚃}	3/36
"Roll an 11": {⚄⚅, ⚅⚄}	2/36
"Roll a 12": {⚅⚅}	1/36

TABLE 16-4

EXAMPLE 16.21 ROLLING A PAIR OF DICE: EPISODE 3

Once again, we are rolling a pair of *honest* dice. We are now going to find the probability of the event E: "at least one of the dice comes up an ace." (In dice jargon, a ⚀ is called an ace.)

This is a slightly more sophisticated counting question than the ones in Example 16.20. We will show three different ways to answer the question.

- **Tallying.** We can just write down all the individual outcomes in the event E and tally their number. This approach gives

  E = {⚀⚀, ⚀⚁, ⚀⚂, ⚀⚃, ⚀⚄, ⚀⚅, ⚁⚀, ⚂⚀, ⚃⚀, ⚄⚀, ⚅⚀}

  and $\Pr(E) = 11/36$. There is not much to it, but in a larger sample space it will take a lot of time and effort to list every individual outcome in the event.

- **Complementary event.** Behind this approach is the germ of a really important idea. Imagine that you are playing a game, and you win if at least one of the two dice comes up an ace (that's event E). Otherwise you lose (call that event F). The two events E and F are called **complementary events**, and E is called the **complement** of F (and vice versa). The key idea is that *the probabilities of complementary events add up to 1*. Thus, when you want $\Pr(E)$ but it's actually easier to find $\Pr(F)$, then you first do that. Once you have $\Pr(F)$, you are in business: $\Pr(E) = 1 - \Pr(F)$.

  Here we can find $\Pr(F)$ by a direct application of the multiplication principle. There are 5 possibilities for the first die (it can't be an ace but it can be anything else) and likewise there are five possibilities for the second die. This means that there are 25 different outcomes in the event F, and thus $\Pr(F) = 25/36$. It follows that $\Pr(E) = 1 - (25/36) = 11/36$.

- **Independent events.** We will let F_1 denote the event "the white die does *not* come up an ace" and F_2 denote the event "the red die does *not* come up an ace." Clearly, $\Pr(F_1) = 5/6$ and $\Pr(F_2) = 5/6$. Now comes a critical observation: The probability that both events F_1 and F_2 happen can be found by *multiplying* their respective probabilities. This means that $\Pr(F) = (5/6) \times (5/6) = 25/36$. We can now find $\Pr(E)$ exactly as before: $\Pr(E) = 1 - \Pr(F) = 11/36$.

Of the three approaches used in Example 16.21, the last approach appears to be the most convoluted, but, in fact, it is the one with the most promise. It is based on the concept of *independence* of events. Two events are said to be **independent events** if the occurrence of one event does not affect the probability of the occurrence of the other. When events E and F are independent, the probability that both occur is the product of their respective probabilities; in other words, $\Pr(E \text{ and } F) = \Pr(E) \cdot \Pr(F)$. This is called the **multiplication principle for independent events**.

The multiplication principle for independent events in an important and useful rule, but be forewarned—it works *only* with independent events! For events that are *not* independent, multiplying their respective probabilities gives us a bunch of nonsense.

Our next example illustrates the real power of the multiplication principle for independent events. As we just mentioned, this principle can only be applied when the events in question are independent, and in many circumstances this is not the case. Fortunately, in the examples we are going to consider the independence of the events in question is intuitively obvious. For example, if we roll an honest die several times, what happens on each roll has no impact whatsoever on what is going to happen on subsequent rolls (i.e., the rolls are independent events).

EXAMPLE 16.22 ROLLING MORE DICE

Imagine a game in which you roll an honest die four times. If at least one of your rolls comes up an ace, you are a winner. Let E denote the event "you win" and F denote the event "you lose." We will find $\Pr(E)$ by first finding $\Pr(F)$, using the same ideas we developed in Example 16.21.

We will let F_1, F_2, F_3, and F_4 denote the events "first roll is not an ace," "second roll is not an ace," "third roll is not an ace," and "fourth roll is not an ace," respectively. Then

$$\Pr(F_1) = 5/6, \quad \Pr(F_2) = 5/6, \quad \Pr(F_3) = 5/6, \quad \Pr(F_4) = 5/6$$

Now we use the multiplication principle for independent events:

$$\Pr(F) = (5/6) \times (5/6) \times (5/6) \times (5/6) = (5/6)^4 \approx 0.482$$

Finally, we find $\Pr(E)$:

$$\Pr(E) = 1 - \Pr(F) \approx 0.518$$

EXAMPLE 16.23 FIVE-CARD POKER HANDS: EPISODE 2

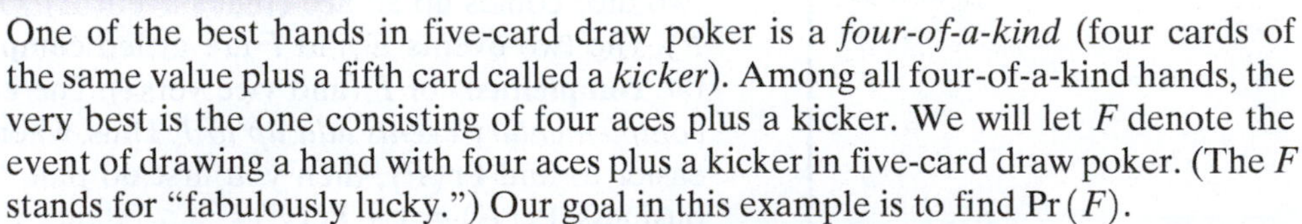

One of the best hands in five-card draw poker is a *four-of-a-kind* (four cards of the same value plus a fifth card called a *kicker*). Among all four-of-a-kind hands, the very best is the one consisting of four aces plus a kicker. We will let F denote the event of drawing a hand with four aces plus a kicker in five-card draw poker. (The F stands for "fabulously lucky.") Our goal in this example is to find $\Pr(F)$.

The size of our sample space is $N = {}_{52}C_5 = 2{,}598{,}960$ (see Example 16.15). Of these roughly 2.6 million possible hands, there are only 48 hands in F: Four of the five cards are the aces; the kicker can be any one of the other 48 cards in the deck. Thus, $\Pr(F) = 48/2{,}598{,}960 \approx 0.0000185$, roughly 1 in 50,000.

EXAMPLE 16.24 A CAR AND TWO GOATS: EPISODE 2

This example is a follow-up of Example 16.5. You are a contestant in a television game show and you have to choose one of three doors to win either a fancy sports car (behind one of the doors) or a goat (behind each of the other two). For starters, the sample space for the game consists of nine outcomes: $S = \{(1,1), (1,2), (1,3), (2,1), (2,2), (2,3), (3,1), (3,2), (3,3)\}$, where (x, y) denotes that you chose Door #*x*, and the car is behind Door #*y*. Assuming that the car is randomly placed behind one of the three doors and that you make a random choice of door, each of the nine outcomes has an equal probability of $1/9$. The set of outcomes $E = \{(1,1), (2,2), (3,3)\}$ represents the event $E =$ "you win the car"; the remaining set of outcomes $F = \{(1,2), (1,3), (2,1), (2,3), (3,1), (3,2)\}$ represents the complementary event $F =$ "you win a goat." This formalizes what is everyone's intuition: Your probability of winning a car is $\Pr(E) = 1/3$; your probability of winning a goat is $\Pr(F) = 2/3$.

So let's say that you choose Door #1. At this point nothing has changed, so your probability of winning a car is still 1/3. Now comes a new wrinkle: before showing what is behind Door #1 the host—who knows where the car is—opens Door #3 to show there is a goat behind it, and offers you the option of switching. Should you *stick* with your original choice, or *switch* to Door #2? Most people's intuition is that it makes no difference—stick or switch, your probability of winning the car is the same. This is a notorious example where people's intuitions turn out to be wrong.

If you stick with Door #1 then you are essentially disregarding the information provided by the host (that the car is not behind Door #3). Your probability of winning the car is still $1/3$, and your probability of winning a goat is still $2/3$.

Now let's find the probabilities if you switch to Door #2. The key here is to consider the reasons why the host opened Door #3. Obviously the host would never open the door showing the car, so if the car happens to be behind Door #2, the host is forced to open Door #3. In this case, the host opens Door #3 with probability 1 (2 out of 2 times). On the other hand, if the car is behind Door #1 the host could have chosen to open either Door #3 or Door #2. Assuming that he chooses each door with probability $1/2$, when the car is behind Door #1 he opens Door #3 half of the time (1 out of 2 times). This means that once the host opened Door #3, 2 out of every 3 times he does it because he has no choice (as the car is behind Door #2), and 1 out of 3 times he does it because he randomly chooses it (and the car is behind Door #1). It follows that when you switch, the probability of winning the car is 2/3; the probability of winning a goat is $1/3$. You should definitely switch!

Odds

Dealing with probabilities as numbers that are always between 0 and 1 is the mathematician's way of having a consistent terminology. To the everyday user, consistency is not that much of a concern, and we know that people often express the likelihood of an event in terms of *odds*. Probabilities and odds are not identical concepts but are closely related.

- **Odds.** Let E be an arbitrary event. If F denotes the number of ways that event E can occur (the *favorable* outcomes or *hits*) and U denotes the number of ways that event E does not occur (the *unfavorable* outcomes, or *misses*), then the **odds of** (also called the **odds in favor of**) the event E are given by the ratio F to U, and the **odds against** the event E are given by the ratio U to F. When F and U have a common factor, it is customary (and desirable) to simplify the numbers before giving the odds. One rarely hears odds of 15 to 5 — they are much better expressed as odds of 3 to 1.

EXAMPLE 16.25 SHOOTING FREE THROWS: EPISODE 4

In Example 16.17 we discussed the act of shooting a free throw, and we used Stephen Curry—a truly great free-throw shooter—to illustrate how we can estimate probabilities of random events. We can apply the same ideas to estimate odds.

Between 2009 (the year Curry joined the NBA) and the end of the 2015–2016 NBA season (when this example was written), Curry made 1668 free throws and missed 182. We can use these numbers to estimate the *odds in favor* of Curry making a random free throw as 1668 (favorable outcomes) to 182 (unfavorable outcomes). These are fairly large numbers and are not commonly used to express odds. To simplify matters we divide 1668 by 182 ($1668/182 \approx 9.16$) and, with very little loss of accuracy, restate the odds as 9 to 1— numbers that are much easier to digest.

EXAMPLE 16.26 ODDS OF ROLLING A "NATURAL"

Suppose that you are playing a game in which you roll a pair of dice, presumably honest. In this game, when you roll a "natural" (i.e., roll a 7 or an 11) you automatically win.

If we let E denote the event "roll a natural," we can check that out of 36 possible outcomes 8 are favorable (6 ways to "roll a 7" and two ways to "roll an 11"—see Table 16-4) and the other 28 are unfavorable. It follows that the odds of rolling a "natural" are 2 to 7 (originally 8 to 28 and then simplified to 2 to 7).

We can convert odds into probabilities, and probabilities into odds using the following two rules:

- If the odds of the event E are F to U, then $\Pr(E) = F/(F+U)$.
- If $\Pr(E) = A/B$, then the odds of E are A to $B - A$. (When the probability is given in decimal form, the best thing to do is to first convert the decimal form into fractional form.)

EXAMPLE 16.27 HANDICAPPING A TENNIS TOURNAMENT: EPISODE 2

This example is a follow-up to Example 16.19. Recall that the probability assignment for the tennis tournament was as follows: $\Pr(A) = 0.08$, $\Pr(B) = 0.16$, $\Pr(C) = 0.20$, $\Pr(D) = 0.25$, $\Pr(E) = 0.16$, and $\Pr(F) = 0.15$. We will now express each of these probabilities as odds. (Notice that to do so we first convert the decimals into fractions in reduced form.)

- $\Pr(A) = 0.08 = 8/100 = 2/25$. Thus, the odds of A winning the tournament are 2 to 23.
- $\Pr(B) = 0.16 = 16/100 = 4/25$. The odds of B winning the tournament are 4 to 21.
- $\Pr(C) = 0.20 = 20/100 = 1/5$. The odds of C winning the tournament are 1 to 4.
- $\Pr(D) = 0.25 = 25/100 = 1/4$. The odds of D winning the tournament are 1 to 3.
- $\Pr(E) = \Pr(B) = 4/25$. The odds of E winning the tournament are the same as B's (4 to 21).
- $\Pr(F) = 0.15 = 15/100 = 3/20$. The odds of F winning the tournament are 3 to 17.

A final word of caution: There is a difference between odds as discussed in this section and the *payoff odds* posted by casinos or bookmakers in sports gambling situations. Suppose we read in the newspaper, for example, that the Las Vegas bookmakers have established that "the odds that the Chicago Cubs will win the World Series are 5 to 2." What this means is that if you want to bet in favor of the Cubs, for every \$2 that you bet, you can win \$5 if the Cubs win the World Series. This ratio may be taken as some indication of the actual odds in favor of the Cubs winning, but several other factors affect payoff odds, and the connection between payoff odds and actual odds is tenuous at best.

16.4 Expectations

In this section we will use probabilities to analyze mathematically the measurement of risk. We start with an introduction to the important concept of a weighted average.

Weighted Averages

EXAMPLE 16.28 COMPUTING CLASS SCORES

Imagine that you are a student in Prof. Goodie's Stat 100 class. The grading for the class is based on two midterms, homework, and a final exam. The breakdown for the scoring is given in the first two rows of Table 16-5. Your scores are given in the last row. You needed to average 90% or above to get an A in the course. Did you?

	Midterm 1	Midterm 2	Homework	Final exam
Weight	20%	20%	25%	35%
Possible points	100	100	200	200
Your scores	82	88	182	190

TABLE 16-5

One of your friends says you should add all your scores and divide by the total number of possible points to compute your final average. This gives $(82 + 88 + 182 + 190)/600 = 0.9033\cdots \approx 90.33\%$.

If your friend is right you would end up with an A. A second friend says that you are going to miss an A by just one percentage point, and isn't that too bad! Your second friend's calculation is as follows: You got 82% on the first midterm, 88% on the second midterm, 91% (182/200) on the homework, and 95% (190/200) on the final exam. The average of these four percentages is $(82\% + 88\% + 95\% + 91\%)/4 = 89\%$.

Fortunately, you know more math than either one of your friends. The correct computation, you explain to them patiently, requires that we take into account that (1) the scores are based on different scales (100 points, 200 points) and (2) the weights of the scores (20%, 20%, 25%, 35%) are not all the same. To take care of (1) we express the numerical scores as percentages (82%, 88%, 91%, 95%), and to take care of (2) we multiply these percentages by their respective weights $(20\% = 0.20, 25\% = 0.25, 35\% = 0.35)$. We then add all of these numbers. Thus, your correct average is

$$(0.2 \times 82\%) + (0.2 \times 88\%) + (0.25 \times 91\%) + (0.35 \times 95\%) = 90\%$$

How sweet it is—you are getting that A after all!

EXAMPLE 16.29 AVERAGE GPAs AT TASMANIA STATE UNIVERSITY

Table 16-6 shows GPAs at Tasmania State University broken down by class. Our task is to compute the overall GPA of *all undergraduates* at TSU.

Class	Freshman	Sophomore	Junior	Senior
Average GPA	2.75	3.08	2.94	3.15

TABLE 16-6 ■ GPAs at Tasmania State University by class

The information given in Table 16-6 is not enough to compute the overall GPA because the number of students in each class is not the same. After a little digging we find out that the 15,000 undergraduates at Tasmania State are divided by class as follows: 4800 freshmen, 4200 sophomores, 3300 juniors, and 2700 seniors.

To compute the overall school GPA we need to "weigh" the average GPA for each class with the relative size of that class. The relative size of the freshman class is $4800/15{,}000 = 0.32$, the relative size of the sophomore class is $4200/15{,}000 = 0.28$, and so on (0.22 for the junior class and 0.18 for the senior class). When all is said and done, the overall school GPA is

$$(0.32 \times 2.75) + (0.28 \times 3.08) + (0.22 \times 2.94) + (0.18 \times 3.15) = 2.9562$$

Examples 16.28 and 16.29 illustrate how to average numbers that have different relative *weights* using the notion of a *weighted average*. In Example 16.28 the scores (82%, 88%, 91%, and 95%) were multiplied by their corresponding "weights" (0.20, 0.20, 0.25, and 0.35) and then added to compute the *weighted average*. In Example 16.29 the class GPAs (2.75, 3.08, 2.94, and 3.15) were multiplied by the corresponding class "weight" (0.32, 0.28, 0.22, 0.18) to compute the *weighted average*. Note that in both examples the *weights* add up to 1, which is not surprising since they represent percentages that must add up to 100%.

WEIGHTED AVERAGE

Let X be a variable that takes the values $x_1, x_2, \ldots, x_N$, and let $w_1, w_2, \ldots, w_N$ denote the respective weights for these values, with $w_1 + w_2 + \cdots + w_N = 1$. The **weighted average** for X is given by

$$(w_1 \cdot x_1) + (w_2 \cdot x_2) + \cdots + (w_N \cdot x_N)$$

The idea of a weighted average is particularly useful when the weights represent probabilities.

EXAMPLE 16.30 GUESSING ANSWERS IN THE SAT: EPISODE 1

In the multiple-choice sections of the SAT each question has five possible answers (A, B, C, D, and E). A correct answer is worth 1 point, and, to discourage indiscriminate guessing, an incorrect answer carries a penalty of $1/4$ point (i.e., it counts as $-1/4$ point).

Imagine you are taking the SAT and are facing a multiple-choice question for which you have no clue as to which of the five choices might be the right answer—they all look equally plausible. You can either play it safe and leave it blank, or you can gamble and take a wild guess. In the latter case, the upside of getting 1 point must be measured against the downside of getting a penalty of $1/4$ point. What should you do?

Table 16-7 summarizes the possible options and their respective payoffs (for the purposes of illustration assume that the correct answer is B, but of course, you don't know that when you are taking the test).

Option	Leave blank	A	B	C	D	E
Payoff	0	−0.25	1	−0.25	−0.25	−0.25

TABLE 16-7

If you are randomly guessing the answer to this multiple-choice question you are unwittingly conducting a *random experiment* with sample space $S = \{A, B, C, D, E\}$. Since each of the five choices is equally likely to be the correct answer (remember, you are clueless on this one), you assign equal probabilities of $p = 1/5 = 0.2$ to each outcome. This probability assignment, combined with the information in Table 16-7, gives us Table 16-8.

Outcome	Correct answer (B)	Incorrect answer (A, C, D, or E)
Point payoff	1	−0.25
Probability	0.2	0.8

TABLE 16-8

Using Table 16-8 we can compute something analogous to a weighted average for the point payoffs with the *probabilities as weights*. We will call this the *expected payoff*. Here the expected payoff E is

$$E = (0.2 \times 1) + (0.8 \times -0.25) = 0.2 - 0.2 = 0 \text{ points}$$

The fact that the expected payoff is 0 points implies that this guessing game is a "fair" game—in the long term (if you were to make these kinds of guesses many times) the penalties that you would accrue for wrong guesses are neutralized by the benefits that you would get for your lucky guesses. This knowledge will not impact what happens with an individual question [the possible outcomes are still a +1 (lucky) or a −0.25 (wrong guess)], but it gives you some strategically useful information: On the average, totally random guessing on the multiple-choice section of the SAT neither helps nor hurts! ■

EXAMPLE 16.31 GUESSING ANSWERS IN THE SAT: EPISODE 2

Example 16.30 illustrated what happens when the multiple-choice question has you completely stumped—each of the five possible choices (A, B, C, D, or E) looks equally likely to be the right answer. To put it bluntly, you are clueless! At a slightly better place on the ignorance scale is a question for which you can definitely rule out one or two of the possible answers. Under these circumstances we must do a different calculation for the risk/benefits of guessing.

Let's consider first a multiple-choice question for which you can safely rule out one of the five choices. For the purposes of illustration let's assume that the correct answer is B and that you can definitely rule out choice E. Among the four possible choices (A, B, C, or D) you have no idea which is most likely to be the correct answer, so you are going to randomly guess. In this scenario we assign the same probability (0.25) to each of the four choices, and the guessing game is described in Table 16-9.

Outcome	Correct answer (B)	Incorrect answer (A, C, or D)
Point payoff	1	−0.25
Probability	0.25	0.75

TABLE 16-9

Once again, the *expected payoff E* can be computed as a weighted average:

$$E = (0.25 \times 1) + (0.75 \times -0.25) = 0.25 - 0.1875 = 0.0625$$

An expected payoff of 0.0625 points is very small—it takes 16 guesses of this type to generate an expected payoff equivalent to 1 correct answer (1 point). At the same time, the fact that it is a positive expected payoff means that the benefit justifies (barely) the risk.

A much better situation occurs when you can rule out two of the five possible choices (say, D and E). Now the random experiment of guessing the answer is described in Table 16-10. [Here it's a little more convenient to express everything in terms of fractions.]

Outcome	Correct answer (B)	Incorrect answer (A or C)
Point payoff	1	$-1/4$
Probability	$1/3$	$2/3$

TABLE 16-10

In this situation the expected payoff E is given by

$$E = \left(\tfrac{1}{3} \times 1\right) + \left(\tfrac{2}{3} \times -\tfrac{1}{4}\right) = \tfrac{1}{3} - \tfrac{1}{6} = \tfrac{1}{6}$$

Examples 16.30 and 16.31 illustrate the mathematical reasoning behind a commonly used piece of advice given to SAT-takers: *Guess the answer if you can rule out some of the options; otherwise, don't bother.*

The basic idea illustrated in Examples 16.30 and 16.31 is that of an *expectation* (or *expected value*).

EXPECTATION

Suppose X is a variable that takes on numerical values $x_1, x_2, x_3, \ldots, x_N$, with probabilities $p_1, p_2, p_3, \ldots, p_N$, respectively (with $p_1 + p_2 + \cdots + p_N = 1$). The *expectation* (or *expected value*) of X is given by

$$E = (p_1 \cdot x_1) + (p_2 \cdot x_2) + (p_3 \cdot x_3) + \cdots + (p_N \cdot x_N)$$

16.5 Measuring Risk

> "A lot of people approach risk as if it's the enemy, when it's really fortune's accomplice."
>
> – Sting

In many real-life situations, we face decisions that can have many different potential consequences—some good, some bad, some neutral. These kinds of decisions are often quite hard to make because there are so many intangibles, but sometimes the decision comes down to a simple question: Is the reward worth the risk? If we can quantify the risks and the rewards, then we can compute the *expectation* (expected value) of the outcome to help us make the right decision. The classic illustration of this type of situation is provided by "games" in which there is money riding on the outcome. This includes not only typical gambling situations (dice games, card games, lotteries, etc.) but also buying life insurance or playing the stock market.

Playing the stock market is a legalized form of gambling in which people make investment decisions (buy? sell? when? what?) instead of rolling the dice or drawing cards. Some people have a knack for making good investment decisions and in the long run can do very well—others, just the opposite. A lot has to do with what drives the decision-making process—gut feelings, rumors, and can't-miss tips from your cousin Vinny at one end of the spectrum; reliable information combined with sound mathematical principles at the other end. Which approach would you rather trust your money to? (If your answer is the former, then you might as well skip the next example.)

EXAMPLE 16.32 TO BUY OR NOT TO BUY?

Fibber Pharmaceutical is a new start-up in the pharmaceutical business. Its stock is currently selling for $10. Fibber's future hinges on a new experimental vaccine it believes has great promise for the treatment of the Zika virus. Before the vaccine can be approved by the Food and Drug Administration (FDA) for use with the general public it must pass a series of clinical trials known as Phase I, Phase II, and Phase III trials. If the vaccine passes all three trials and is approved by the FDA, shares of Fibber are predicted to jump tenfold to $100 a share. At the other end of the spectrum, the vaccine may turn out to be a complete flop and fail Phase I trials. In this case, Fibber's shares will be worthless. In between these two extremes are two other possibilities: The vaccine will pass Phase I trials but fail Phase II trials or will pass Phase I and Phase II trials but fail Phase III trials. In the former case shares of Fibber are expected to drop to $5 a share; in the latter case shares of Fibber are expected to go up to $15 a share.

Table 16-11 summarizes the four possible outcomes of the clinical trials. The last row of Table 16-11 gives the probability of each outcome based on previous experience with similar types of vaccines.

Outcome of trials	Fail Phase I	Fail Phase II	Fail Phase III	FDA approval
Estimated share price	$0	$5	$15	$100
Probability	0.25	0.45	0.20	0.10

TABLE 16-11

Combining the second and third rows of Table 16-11, we can compute the expected value E of a future share of Fibber Pharmaceutical:

$$E = (0.25 \times \$0) + (0.45 \times \$5) + (0.20 \times \$15) + (0.10 \times \$100) = \$15.25$$

To better understand the meaning of the $15.25 expected value of a $10 share of Fibber Pharmaceutical, imagine playing the following game: For a cost of $10 you get to roll a die. This is no ordinary die—this die has only four faces (labeled I, II, III, and IV), and the probabilities of each face coming up are different (0.25, 0.45, 0.20, and 0.10, respectively). If you roll a I, your payoff is $0 (your money is gone); if you roll a II, your payoff is $5 (you are still losing $5); if you roll a III, your payoff is $15 (you made $5); and if you roll a IV, your payoff is $100 (jackpot!). The beauty of this random experiment is that it can be played over and over, hundreds or even thousands of times. If you do this, sometimes you'll roll a I and poof—your money is gone, a few times you'll roll a IV and make out like a bandit, other times you'll roll a III or a II and make or lose a little money. The key observation is that if you play this game long enough, the average payoff is going to be $15.25 per roll of the die, giving you an average profit (gain) of $5.25 per roll. In purely mathematical terms, this game is a game well worth playing (but since this book does not condone gambling, this is just a theoretical observation).

Is Fibber Pharmaceutical then a good or a bad bet? At first glance, an expected value of $15.25 per $10 invested looks like a pretty good bet, but we have to balance this with the several years that it might take to collect on the investment (unlike the die game that pays off right away). For example, assuming seven years to complete the clinical trials, an investment in Fibber shares has a comparable expected payoff as a seven-year bond with a fixed annual yield of about 6.25%. ■

EXAMPLE 16.33 RAFFLES AND FUND-RAISERS

A common event at many fund-raisers is to conduct a raffle—another form of legalized gambling. At this particular fund-raiser, the raffle tickets are going for $2. In this raffle, the organizers will draw one grand-prize winner worth $500, four second-prize winners worth $50 each, and fifteen third-prize winners worth $20 each. Sounds like a pretty good deal for a $2 investment, but is it? The answer, of course, depends on how many tickets are sold in this raffle.

Suppose that 2000 raffle tickets are sold, and let's assume a raffle ticket can only win one of the prizes. Table 16-12 shows the payoffs and the respective probabilities for each of the four possible "prizes" (including the "no prize" scenario).

	Grand prize	Second prize	Third prize	No prize
Payoff	$500	$50	$20	$0
Probability	1/2000	4/2000	15/2000	1980/2000

TABLE 16-12

From Table 16-12 we can compute the expected value E of a $2 raffle ticket:

$$E = \left(\frac{1}{2000} \times \$500\right) + \left(\frac{4}{2000} \times \$50\right) + \left(\frac{15}{2000} \times \$20\right) + \left(\frac{1980}{2000} \times \$0\right) = \$0.50$$

If we subtract the cost of a raffle ticket ($2) from the expected value of that ticket ($0.50) we get an *expected gain* of $-\$1.50$. The negative expected gain is an indication that this game favors the "house"—in this case the people running the raffle. We would expect this—it is, after all, a fund-raiser. But the computation gives us a precise measure of the extent to which the game favors the house: On the average we should expect to lose $1.50 for every $2 raffle ticket purchased. If we purchased, say, 100 raffle tickets, we would in all likelihood have a few winning tickets and plenty of losing tickets, but who cares about the details—at the end we would expect a net loss of about $150. ■

The moral of Example 16.33 is that you should buy raffle tickets at fund-raisers because you want to support a good cause—as an investment, raffle tickets are usually very bad investments. What about gambling for the sake of gambling? We will discuss this idea in the next two examples.

EXAMPLE 16.34 CHUCK-A-LUCK

Chuck-a-luck is an old game, played mostly in carnivals and county fairs. We will discuss it here because it involves some interesting probability calculations. To play chuck-a-luck you place a bet, say $1, on one of the numbers 1 through 6. Say that you bet on the number 4. You then roll three dice (presumably honest). If you roll three 4's, you win $3.00; if you roll just two 4's, you win $2; if you roll just one 4, you win $1 (and, in addition to your winnings you get your original $1 back). If you roll no 4's, you lose your $1. Sounds like a pretty good game, doesn't it?

To compute the expected payoff for chuck-a-luck we will first need to compute some probabilities. When we roll three dice, the sample space consists of $6 \times 6 \times 6 = 216$ outcomes. The different events we will consider are shown in the first row of Table 16-13 (the * indicates any number other than a 4). The second row of Table 16-13 shows the size (number of outcomes) of that event, and the third row shows the respective probabilities.

Event	(4, 4, 4)	(4, 4, *)	(4, *, 4)	(*, 4, 4)	(4, *, *)	(*, 4, *)	(*, *, 4)	(*, *, *)
Size	1	5	5	5	5×5	5×5	5×5	$5 \times 5 \times 5$
Probability	1/216	5/216	5/216	5/216	25/216	25/216	25/216	125/216

TABLE 16-13

Combining columns in Table 16-13, we can deduce the probability of rolling three 4's, two 4's, one 4, or no 4's when we roll three dice (row 3 of Table 16-14). The payoffs in row 2 of Table 16-14 are based on a $1 bet.

Roll	Three 4's	Two 4's	One 4	No 4's
Payoff	$4	$3	$2	$0
Probability	1/216	15/216	75/216	125/216

TABLE 16-14

The *expected value* of a $1 bet on chuck-a-luck is given by

$$E = \left(\tfrac{1}{216} \times \$4\right) + \left(\tfrac{15}{216} \times \$3\right) + \left(\tfrac{75}{216} \times \$2\right) \approx \$0.92,$$

and the *expected gain* is $\$0.92 - \$1 = -\$0.08$.

Essentially, this means that in the long run, for every $1 bet on chuck-a-luck, the player will lose on the average about 8 cents, or 8%. This, of course, represents the "house" profit and in gambling is commonly referred to as the *house margin* or *house advantage*.

A game is considered a **fair game** (or an **even game**) when no player has a built-in advantage over another player in the game. In the case of a game between a player and the "house" (such as raffles, lotteries, and casino games) a fair game is one where the *expected gain* (expected value of a bet minus the amount of the bet) is $0.

We will now dive into the fascinating topic of big prize lotteries. National and state lotteries are a big business, and millions of people all over the world buy lottery tickets—mostly for the wrong reasons. In our next example we will discuss the Powerball lottery, a multi-state lottery played in 44 states, the District of Columbia, Puerto Rico, and the U.S. Virgin Islands. We will use the Powerball lottery for the purposes of illustration, but the same ideas can be used to analyze other big prize lotteries. When we buy a $2 lottery ticket, our heads are filled with fantasies of how we will spend all that money when we win, but the real question we should be thinking about is much less glamorous: What is the expected value of that $2 lottery ticket?

EXAMPLE 16.35 THE POWERBALL LOTTERY: EPISODE 2

We introduced a few basic details of the Powerball lottery in Example 16.16. Powerball drawings are held twice a week (Wednesdays and Saturdays), and in each drawing five white balls are randomly drawn from a hopper containing 69 white balls numbered 1 through 69, and a single "powerball" is drawn from a second hopper containing 26 red balls numbered 1 through 26. (This means that the lucky numbers must all be different, but the mega number could be the same as one of the lucky numbers.) A perfect match (i.e., correctly guessing all five of the lucky numbers and the mega number—described in lottery parlance as "5 + 1") wins the jackpot. You can also win lesser prizes by matching some but not all of the numbers. (We will use lottery notation to describe the possible types of matches: "$x + 0$" means x hits on the "white" numbers plus a miss on the "powerball" (red) number; "$x + 1$" means x hits on the "white" numbers plus a hit on the "powerball" number.)

Table 16-15 shows all the types of matches that have positive payoffs (column 1), the payoffs (column 2), the formal probabilities given in mathematical notation (column 3), and an approximation of the odds corresponding to those probabilities (column 4). Notice that the payoffs for all prizes except the jackpot are fixed amounts. The payoff for the jackpot is a variable sum J that depends on the size of the betting pool—how many people bought tickets for that drawing—as well as on whether the jackpot was rolled over from previous drawings. In any case, J is always in the tens of millions and often in the hundreds of millions. In January, 2016, it hit 1.58 billion.

Type of match	Payoff	Probability	Odds
5 + 1	$J	$1/({}_{69}C_5 \times 26)$	1 in 292,201,000
5 + 0	$1,000,000	$25/({}_{69}C_5 \times 26)$	1 in 11,688,000
4 + 1	$50,000	$({}_5C_4 \times {}_{64}C_1)/({}_{69}C_5 \times 26)$	1 in 913,100
4 + 0	$100	$({}_5C_4 \times {}_{64}C_1 \times 25)/({}_{69}C_5 \times 26)$	1 in 36,500
3 + 1	$100	$({}_5C_3 \times {}_{64}C_2)/({}_{69}C_5 \times 26)$	1 in 14,500
3 + 0	$7	$({}_5C_3 \times {}_{64}C_2 \times 25)/({}_{69}C_5 \times 26)$	1 in 580
2 + 1	$7	$({}_5C_2 \times {}_{64}C_3)/({}_{69}C_5 \times 26)$	1 in 700
1 + 1	$4	$({}_5C_1 \times {}_{64}C_4)/({}_{69}C_5 \times 26)$	1 in 91
0 + 1	$4	${}_{64}C_5/({}_{69}C_5 \times 26)$	1 in 37

TABLE 16-15 ■ Powerball prizes, payoffs, probabilities, and approximate odds. The size of the sample space is ${}_{69}C_5 \times 26 = 292{,}201{,}338$. [Note: There are no payoffs for the "2 + 0" and "1 + 0" matches.]

All the probabilities in the third column of Table 16-15 are based on the same counting principles, so we will illustrate just one of the entries under the Probability column and leave the rest to the reader. As discussed in Example 16.16, there are ${}_{69}C_5 \times 26 = 292{,}201{,}338$ possible Powerball drawings. This is the denominator for all the probabilities in column 3. Now, let's count the number of outcomes in the event "3 + 0" (exactly 3 out of 5 hits on the winning "white" numbers plus a miss on the red powerball number. For this to happen, a person must have chosen 3 out of the 5 "good" (i.e., winning) white numbers, 2 out of the 64 "bad" white numbers, plus 1 out of the 25 "bad" red powerball numbers. There are ${}_5C_3$ ways to do the first, ${}_{64}C_2$ ways to do the second, and 25 ways to do the last, giving a total of ${}_5C_3 \times {}_{64}C_2 \times 25$ ways for all of these things to happen together. (We left all probabilities in column 3 in uncalculated form to show where the numbers come from, but with a good calculator you can always convert these to actual numbers.The probability in the "3 + 0" row turns out to be $504{,}000/292{,}201{,}338 \approx 1/580$.)

For any value of the jackpot J we can use Table 16-15 to calculate the expected payoff of a Powerball ticket (multiply each payoff times its corresponding probability and add). Table 16-16 shows several possible jackpots, their expected payoffs, and the expected values for a $2 ticket (rounded to the nearest penny).

One useful lesson that can be drawn from the data in Table 16-16 is that to get a positive expected value from a Powerball lottery ticket, the jackpot must be at least close to $500 million. So, as rational individuals, should we go out and buy Powerball tickets anytime the jackpot is $500 million or higher? Not so fast. Here are a few additional wrinkles that have to be considered. First of all, you only get the advertised jackpot prize if you take the money in the form of an annuity paid annually over 30 years. If you want to get all the money up front, (called a "lump-sum" payment), you have to settle for less—roughly 35% less. Second, you have to pay federal income taxes on your lottery winnings, to the tune of 35% to 40% (and in some states, also state income taxes). Between these two bites (a 35% bite if you want the lump-sum payment followed by a 35%–40% bite for taxes), you are really going to end up with less than half of the advertised jackpot.

Jackpot	Expected payoff	Expected value
$100 million	$0.66	−$1.34
$200 million	$1.01	−$0.99
$300 million	$1.35	−$0.65
$400 million	$1.69	−$0.31
$500 million	$2.03	$0.03
$1 billion	$3.74	$1.74
$1.5 billion	$5.45	$3.45

TABLE 16-16 ■ Expected payoffs and values of a $2 Powerball ticket

To estimate the true expected value of your Powerball ticket, you need to think in terms of the actual take-home payoff rather than the advertised "official" payoff for the jackpot. For example, if you choose to take the lump-sum payment, a jackpot prize of $1 billion becomes, after the lottery and the government take their bite, a take-home payoff of less than $500 million. But look at Table 16-16 once again: A payoff of $500 million has an expected value of basically zero!

What if the jackpot is more than $1 billion? Surely, at this point buying Powerball tickets makes good mathematical sense. Not so fast. At this point you have another problem—the possibility that you may have to split the jackpot with other winners. As jackpots get larger, the probability of more than one winning ticket increases—for jackpots of $1 billion or greater, the probability is quite high that there will be two, sometimes three winners dividing the money. For example, for the January 13, 2016, record-busting $1.58 billion jackpot there were three winning tickets—and each ended up with a tad under $328 million before taxes (the lump-sum prize of $983.5 million divided three ways). By the time they paid their federal and state income taxes, these poor folks ended up with about $197 million each—a lot of money, yes, but a far cry from the $1.58 billion that started it all.

The main lesson to be drawn from Example 16.35 is that there is no mathematical sweet spot for buying a lottery ticket: You need official jackpots of over $1 billion to have a positive expected gain (that's taking into account taxes and the fact that the cash value of a jackpot is much less than the official advertised value), but at that point there is a good chance that if you win you will have to split the prize with others.

Unlike a lottery, buying insurance is a smart and rational form of "gambling." If you think about it carefully, it is also quite twisted. When you buy insurance you are making a bet that something bad is going to happen to you. The "bad" could be that you have to go to the hospital (that's why you buy health insurance), or you total your car (auto insurance to the rescue), or your computer goes on the blink (extended warranty—hello!). The twisted part is that you win your bet when that bad thing happens to you. In that case the "house" (in this case the insurance company) has to pay off. On the other hand, if nothing bad happens to you (you don't have to go to the hospital, you don't total your car, your computer doesn't go on the blink) the insurance company wins the bet and gets to pocket your premium. Buying insurance is the kind of bet you always hope to lose. Other than that, buying insurance is just like any other bet. It is not surprising then that the concept of expected value is used by insurance companies to set the price of premiums (the process is called *expectation based pricing*). Our last example is an oversimplification of how the process works, but it illustrates the key ideas behind the setting of life insurance premiums.

EXAMPLE 16.36 SETTING LIFE INSURANCE PREMIUMS

A life insurance company is offering a $100,000 one-year term life insurance policy to Janice, a 55-year-old nonsmoking female in moderately good health. What should be a reasonable premium for this policy?

For starters, we will let $P denote the *break-even*, or *fair*, premium that the life insurance company should charge Janice if it were not in it to make a profit. This can be done by setting the expected value to be 0. Using mortality tables, the life insurance company can determine that the probability that someone in Janice's demographic group will die within the next year is 1 in 500, or 0.002. The second row of Table 16-17 gives the payoffs to the life insurance company for each of the two possible outcomes, and the third row gives their respective probabilities.

Outcome	Janice dies	Janice doesn't die
Payoff	$\$(P - 100{,}000)$	$\$P$
Probability	0.002	0.998

TABLE 16-17

Setting the expected payoff equal to 0 and solving for P gives $P = \$200$. This is the annual premium the insurance company should charge to break even, but of course, insurance companies are in business to make a profit. A standard gross profit margin in the insurance industry is 20%, which in this case would tack on $40 to the premium. We can conclude that a reasonable annual premium for Janice's life insurance policy should be about $240. ■

Conclusion

While the average citizen thinks of chance and risk as vague, informal concepts that are useful primarily when discussing the weather or playing the lottery, scientists and mathematicians think of the theory of probability as a formal framework within which the laws that govern chance events can be understood. The basic elements of this theory are *sample spaces* (which represent a precise mathematical description of all the possible outcomes of a *random experiment*), *events* (collections of these outcomes), and *probability assignments* (which assign a numerical value to each of the events in the sample space).

Of the many ways in which probabilities can be assigned, a particularly important case is the one in which all simple outcomes have the same probability (*equiprobable spaces*). When this happens, the critical steps in calculating probabilities revolve around two basic (but not necessarily easy) questions: (1) What is the size of the sample space, and (2) what is the size of the event in question? To answer these kinds of questions, knowing the basic principles of "counting" is critical.

Probabilities are an essential element in the calculus of risk. Often the calculus of risk is informal and fuzzy ("I have a gut feeling my number is coming up," "It's a good time to invest in real estate," etc.), but in many situations both the risk and the associated payoffs can be quantified in precise mathematical terms. In these situations the relationship between the risks we take and the payoffs we expect can be measured using an important mathematical tool called the *expected value*. The notion of expected value gives us the ability to make rational decisions—what risks are worth taking and what risks are just plain foolish.

When we stop to think how much of our lives is ruled by fate and chance, the importance of probability theory in almost every walk of life is hardly surprising. Understanding the basic mathematical principles behind this theory can help us better judge when taking a chance is a smart move and when it is not. In the long run, this will make us not only better card players but also better and more successful citizens.

KEY CONCEPTS

16.1 Sample Spaces and Events

- **random experiment:** an activity or process whose outcome cannot be predicted ahead of time, **473**
- **sample space:** the set of all possible outcomes of a random experiment, **473**
- **event:** any subset of the sample space, **477**
- **impossible event:** the empty set { } (i.e., an event with no outcomes), **477**
- **certain event:** the entire sample space S (i.e., the event consisting of all outcomes), **477**
- **simple event:** any event consisting of a single outcome, **477**

16.2 The Multiplication Rule, Permutations, and Combinations

- **multiplication rule:** if there are m different ways to do X and n different ways to do Y, then X and Y together (and in that order) can be done in $m \times n$ different ways, **478**
- **permutation:** an *ordered* selection of r objects chosen from a set of n objects (r can be any integer between 0 and n), **482**
- **combination:** an *unordered* selection of r objects chosen from a set of n objects (r can be any integer between 0 and n), **482**
- **$_nP_r$:** the *number* of permutations of r objects chosen from a set of n objects, **483**
- **$_nC_r$:** the *number* of combinations of r objects chosen from a set of n objects, **483**

16.3 Probabilities and Odds

- **probability assignment:** a function that assigns to each event E a number $\Pr(E)$ between 0 and 1 that represents the probability of the event. (The *impossible* event $\{\ \}$ is always assigned probability 0 and the *certain* event S is always assigned probability 1), **485**
- **probability space:** a sample space together with a probability assignment on it, **486**
- **equiprobable space:** a probability space where each simple event has an equal probability, **486**
- **complementary events:** two events that are negations (complements) of each other (i.e., if one of the events is E, the other event is "not E"), **487**
- **independent events:** two events such that the occurrence of one does not affect the probability of occurrence of the other one, **488**
- **multiplication principle for independent events:** when E and F are independent events $\Pr(E \text{ and } F) = \Pr(E) \cdot \Pr(F)$, **488**

16.4 Expectations

- **weighted average:** $w_1 \cdot x_1 + w_2 \cdot x_2 + \cdots + w_N \cdot x_N$, where the x's are distinct numerical values of some variable and the w's are their respective *weights* (with the sum of the weights equal to 1), **492**
- **expectation (expected value):** $p_1 \cdot x_1 + p_2 \cdot x_2 + \cdots + p_N \cdot x_N$, where the x's are the numerical values of some variable and the p's are their respective probabilities, **494**

16.5 Measuring Risk

- **fair (even) game:** a game where no player (including "the house" when there is one) has a built-in advantage over the other players, **497**

EXERCISES

WALKING

16.1 Sample Spaces and Events

1. Using set notation, write out the sample space for each of the following random experiments.

    (a) A coin is tossed three times in a row. The observation is how the coin lands (H or T) on each toss.

    (b) A basketball player shoots three consecutive free throws. The observation is the result of each free throw (s for success, f for failure).

    (c) A coin is tossed three times in a row. The observation is the number of times the coin comes up tails.

    (d) A basketball player shoots three consecutive free throws. The observation is the number of successes.

2. Using set notation, write out the sample space for each of the following random experiments.

   (a) A coin is tossed four times in a row. The observation is how the coin lands (H or T) on each toss.

   (b) A student randomly guesses the answers to a four-question true-or-false quiz. The observation is the student's answer (T or F) for each question.

   (c) A coin is tossed four times in a row. The observation is the percentage of tosses that are heads.

   (d) A student randomly guesses the answers to a four-question true-or-false quiz. The observation is the percentage of correct answers in the test.

3. Using set notation, write out the sample space for each of the following random experiments:

   (a) Roll three dice. The observation is the total of the three numbers rolled.

   (b) Toss a coin five times. The observation is the difference (# of heads – # of tails) in the five tosses.

4. Using set notation, write out the sample space for each of the following random experiments:

   (a) Three runners (A, B, and C) are running in a race (assume that there are no ties). The observation is the order in which the three runners cross the finish line.

   (b) Four runners (A, B, C, and D) are running in a qualifying race (assume that there are no ties). The top two finishers qualify for the finals. The observation is the pair of runners that qualify for the finals.

5. The board of directors of Fibber Corporation has five members (A, B, C, D, and E). Using set notation write out the sample space for each of the following random experiments:

   (a) A chairman and a treasurer are elected.

   (b) Three directors are selected to form a search committee to hire a new CEO.

6. You reach into a large jar containing jelly beans of four different flavors [Juicy Pear (J), Kiwi (K), Licorice (L), and Mango (M)] and grab two jelly beans at random. The observation is the flavors of the two jelly beans. Using set notation, write out the sample space for this experiment.

In Exercises 7 through 10, the sample spaces are too big to write down in full. In these exercises, you should describe the sample space either by describing a generic outcome or by listing some outcomes and then using the . . . notation. In the latter case, you should write down enough outcomes to make the description reasonably clear.

7. A coin is tossed 10 times in a row. The observation is how the coin lands (H or T) on each toss. Describe the sample space.

8. A student randomly guesses the answers to a 10-question true-or-false quiz. The observation is the student's answer (T or F) for each question. Describe the sample space.

9. A die is rolled four times in a row. The observation is the number that comes up on each roll. Describe the sample space.

10. A student randomly guesses the answers to a multiple-choice quiz consisting of 10 questions. The observation is the student's answer (A, B, C, D, or E) for each question. Describe the sample space.

11. A coin is tossed three times in a row. The observation is how the coin lands (heads or tails) on each toss [see Exercise 1(a)]. Write out the event described by each of the following statements as a set.

   (a) E_1: "the coin comes up heads exactly twice."

   (b) E_2: "all three tosses come up the same."

   (c) E_3: "exactly half of the tosses come up heads."

   (d) E_4: "the first two tosses come up tails."

12. A student randomly guesses the answers to a four-question true-or-false quiz. The observation is the student's answer (T or F) for each question [see Exercise 2(b)]. Write out the event described by each of the following statements as a set.

   (a) E_1: "the student answers T to two out of the four questions."

   (b) E_2: "the student answers T to *at least* two out of the four questions."

   (c) E_3: "the student answers T to *at most* two out of the four questions."

   (d) E_4: "the student answers T to the first two questions."

13. A pair of dice is rolled. The observation is the number that comes up on each die (see Example 16.4). Write out the event described by each of the following statements as a set.

   (a) E_1: "*pairs* are rolled." (A "pair" is a roll in which both dice come up the same number.)

   (b) E_2: "*craps* are rolled." ("Craps" is a roll in which the sum of the two numbers rolled is 2, 3, or 12.)

   (c) E_3: "a *natural* is rolled." (A "natural" is a roll in which the sum of the two numbers rolled is 7 or 11.)

14. A card is drawn out of a standard deck of 52 cards. Each card can be described by giving its "value" (A, 2, 3, 4, . . . , 10, J, Q, K) and its "suit" (H for hearts, C for clubs, D for diamonds, and S for spades). For example, $2D$ denotes the two of diamonds and JH denotes the jack of hearts. Write out the event described by each of the following statements as a set.

   (a) E_1: "draw a queen."

   (b) E_2: "draw a heart."

   (c) E_3: "draw a *face* card." (A "face" card is a jack, queen, or king.)

15. A coin is tossed 10 times in a row. The observation is how the coin lands (H or T) on each toss (see Exercise 7). Write out the event described by each of the following statements as a set.

   (a) E_1: "none of the tosses comes up tails."

   (b) E_2: "exactly one of the 10 tosses comes up tails."

   (c) E_3: "nine or more of the tosses come up tails."

16. Five candidates (A, B, C, D, and E) have a chance to be selected to be on *American Idol*. Any subset of them (including none of them or all of them) can be selected. The observation is which subset of individuals is selected. Write out the event described by each of the following statements as a set.

 (a) E_1: "two candidates get selected."

 (b) E_2: "three candidates get selected."

 (c) E_3: "three candidates get selected, and A is not one of them."

16.2 The Multiplication Rule, Permutations, and Combinations

17. A California license plate starts with a digit other than 0 followed by three capital letters followed by three more digits (0 through 9).

 (a) How many different California license plates are possible?

 (b) How many different California license plates start with a 5 and end with a 9?

 (c) How many different California license plates have no repeated symbols (all the digits are different and all the letters are different)?

18. A computer password consists of four letters (A through Z) followed by a single digit (0 through 9). Assume that the passwords are not case sensitive (i.e., that an uppercase letter is the same as a lowercase letter).

 (a) How many different passwords are possible?

 (b) How many different passwords end in 1?

 (c) How many different passwords do not start with Z?

 (d) How many different passwords have no Z's in them?

19. Jack packs two pairs of shoes, one pair of boots, three pairs of jeans, four pairs of dress pants, and three dress shirts for a trip.

 (a) How many different outfits can Jack make with these items?

 (b) If Jack were also to bring along two jackets so that he could wear either a dress shirt or a dress shirt plus a jacket, how many outfits could Jack make?

20. A French restaurant offers a menu consisting of three different appetizers, two different soups, four different salads, nine different main courses, and five different desserts.

 (a) A fixed-price lunch meal consists of a choice of appetizer, salad, and main course. How many different fixed-price lunch meals are possible?

 (b) A fixed-price dinner meal consists of a choice of appetizer, a choice of soup or salad, a main course, and a dessert. How many different fixed-price dinner meals are possible?

 (c) A dinner special consists of a choice of soup, salad, or both, plus a main course. How many dinner specials are possible?

21. A set of reference books consists of eight volumes numbered 1 through 8.

 (a) In how many ways can the eight books be arranged on a shelf?

 (b) In how many ways can the eight books be arranged on a shelf so that at least one book is out of order?

22. Nine people (four men and five women) line up at a checkout stand in a grocery store.

 (a) In how many ways can they line up?

 (b) In how many ways can they line up if the first person in line must be a woman?

23. Nine people (four men and five women) line up at a checkout stand in a grocery store.

 (a) In how many ways can they line up if all five women must be at the front of the line?

 (b) In how many ways can they line up if they must alternate woman, man, woman, man, and so on?

24. A set of reference books consists of eight volumes numbered 1 through 8.

 (a) In how many ways can the eight books be arranged so that Volume 8 is in the correct position on the shelf (i.e., the last one from left to right)?

 (b) In how many ways can the eight books be arranged so that Volumes 1 and 2 are in their correct positions on the shelf?

25. Determine the number of outcomes N in each sample space.

 (a) A coin is tossed 10 times in a row. The result of each toss (H or T) is observed.

 (b) A die is rolled four times in a row. The number that comes up on each roll is observed.

 (c) A die is rolled four times in a row. The sum of the numbers rolled is observed.

26. Determine the number of outcomes N in each sample space.

 (a) A student randomly answers a 10-question true-false quiz. The student's answer (T or F) to each question is observed.

 (b) A student randomly answers a 10-question multiple-choice quiz. The student's answer (A, B, C, D, or E) to each question is observed.

In Exercises 27 through 30, you are asked to give your answer using the notation ${}_nP_r$ or ${}_nC_r$. You do not have to give an actual numerical answer, so you won't need a calculator.

27. The board of directors of the XYZ Corporation has 15 members.

 (a) How many different slates of four officers (a President, a Vice President, a Treasurer, and a Secretary) can be chosen?

 (b) A four-person committee needs to be selected to conduct a search for a new CEO. In how many ways can the search committee be selected?

28. There are 10 athletes entered in an Olympic event.

(a) In how many ways can one pick the winners of the gold, silver, and bronze medals?

(b) In how many ways can one pick the seven athletes who will not earn any medals?

29. The Brute Force Bandits is a punk rock band planning their next concert tour. The band has a total of 30 new songs in their repertoire.

(a) How many different set lists of 18 songs can the band select to play on the tour? (*Hint*: Assume that the order in which the songs are listed is irrelevant.)

(b) How many different ways are there for the band to record a CD consisting of 18 songs chosen from the 30 new songs? (*Hint*: In recording a CD, the order in which the songs appear on the CD is relevant.)

30. There are 347 NCAA Division I college basketball teams.

(a) How many different top-25 rankings are possible? [Assume that every team has a chance to be a top-25 team.]

(b) How many ways are there to choose 64 teams (unseeded) to make it to the NCAA tournament? [Assume every combination of 64 teams is possible.]

31. A major league baseball team roster consists of 40 players of which 25 are considered active.

(a) How many ways are there for a manager to select 25 active players from a major league roster?

(b) How many ways can a manager select a nine-player batting lineup from the active roster for opening day?

32. Bob has 20 different dress shirts in his wardrobe.

(a) In how many ways can Bob select seven shirts to pack for a business trip?

(b) In how many ways can Bob select 5 of the 7 dress shirts he packed for the business trip—one for the Monday meeting, one for the Tuesday dinner, one for the Wednesday party, one for the Thursday conference, and one for the Friday date?

16.3 Probabilities and Odds

33. Consider the sample space $S = \{o_1, o_2, o_3, o_4, o_5\}$. Suppose that $\Pr(o_1) = 0.22$ and $\Pr(o_2) = 0.24$.

(a) Find the probability assignment for the probability space when o_3, o_4, and o_5 all have the same probability.

(b) Find the probability assignment for the probability space when $\Pr(o_5) = 0.1$ and o_3 has the same probability as o_4 and o_5 combined.

34. Consider the sample space $S = \{o_1, o_2, o_3, o_4\}$. Suppose that $\Pr(o_1) + \Pr(o_2) = \Pr(o_3) + \Pr(o_4)$ and that $\Pr(o_1) = 0.15$.

(a) Find the probability assignment for the probability space when o_2 and o_3 have the same probability.

(b) Find the probability assignment for the probability space when $\Pr(o_3) = 0.22$.

35. Four candidates are running for mayor of Happyville. According to the polls candidate A has a "one in five" probability of winning [i.e., $\Pr(A) = 1/5$]. Of the other three candidates, all we know is that candidate C is twice as likely to win as candidate B and that candidate D is three times as likely to win as candidate B. Find the probability assignment for this probability space.

36. Seven horses (A, B, C, D, E, F, and G) are running in the Boonsville Sweepstakes. According to the oddsmakers, A has a "one in four" probability of winning [i.e., $\Pr(A) = 1/4$], B has a "three in ten" probability of winning, and C has a "one in twenty" probability of winning. The remaining four horses all have the same probability of winning. Find the probability assignment for the probability space.

37. An honest coin is tossed three times in a row. Find the probability of each of the following events. (*Hint*: Do Exercise 11 first.)

(a) E_1: "the coin comes up heads exactly twice."

(b) E_2: "all three tosses come up the same."

(c) E_3: "exactly half of the tosses come up heads."

(d) E_4: "the first two tosses come up tails."

38. A student randomly guesses the answers to a four-question true-or-false $(T - F)$ quiz. Find the probability of each of the following events. (*Hint:* Do Exercise 12 first.)

(a) E_1: "the student answers F on two of the four questions."

(b) E_2: "the student answers F on *at least* two of the four questions."

(c) E_3: "the student answers F on *at most* two of the four questions."

(d) E_4: "the student answers F to the first two questions."

39. A pair of honest dice is rolled. Find the probability of each of the following events. (*Hint:* Do Exercise 13 first.)

(a) E_1: "*pairs* are rolled." (A "pair" is a roll in which both dice come up the same number.)

(b) E_2: "*craps* are rolled." ("Craps" is a roll in which the sum of the two numbers rolled is 2, 3, or 12.)

(c) E_3: "a *natural* is rolled." (A "natural" is a roll in which the sum of the two numbers rolled is 7 or 11.)

40. A card is drawn at random out of a well-shuffled deck of 52 cards. Find the probability of each of the following events. (*Hint*: Do Exercise 14 first.)

(a) E_1: "draw a queen."

(b) E_2: "draw a heart."

(c) E_3: "draw a *face* card." (A "face" card is a jack, queen, or king.)

41. An honest coin is tossed 10 times in a row. Find the probability of each of the following events. (*Hint*: Do Exercise 15 first.)

(a) E_1: "none of the tosses comes up tails."

(b) E_2: "exactly one of the 10 tosses comes up tails."

(c) E_3: "nine or more of the tosses come up tails."

42. Five candidates (A, B, C, D, and E) have a chance to be selected to be on *American Idol*. Any subset of them (including none of them or all of them) can be selected, and assume that the selection process is completely random (the subsets of candidates are all equally likely). Find the probability of each of the following events. (*Hint*: Do Exercise 16 first.)

(a) E_1: "two candidates get selected."

(b) E_2: "three candidates get selected."

(c) E_3: "three candidates get selected, and A is not one of them."

43. A student takes a 10-question true-or-false quiz and randomly guesses the answer to each question. Suppose that a correct answer is worth 1 point and an incorrect answer is worth -0.5 points. Find the probability that the student

(a) gets 10 points.

(b) gets -5 points.

(c) gets 8.5 points.

(d) gets 8 or more points.

(e) gets 5 points.

(f) gets 7 or more points.

44. Suppose that the probability of giving birth to a boy and the probability of giving birth to a girl are both 0.5. Find the probability that in a family of four children,

(a) all four children are girls.

(b) there are two girls and two boys.

(c) the youngest child is a girl.

(d) the oldest child is a boy.

45. The Tasmania State University glee club has 15 members. A *quartet* of four members must be chosen to sing at the university president's reception. Assume that the quartet is chosen randomly by drawing the names out of a hat. Find the probability that

(a) Alice (one of the members of the glee club) is chosen to be in the quartet.

(b) Alice is not chosen to be in the quartet.

(c) the four members chosen for the quartet are Alice, Bert, Cathy, and Dale.

46. Ten professional basketball teams are participating in a draft lottery. (A draft lottery is a lottery to determine the order in which teams get to draft players.) Ten balls, each containing the name of one team (call them $A, B, C, D, E, F, G, H, I$, and J for short), are placed in an urn and thoroughly mixed. Four balls are drawn, one at a time, from the urn. The four teams chosen get to draft first and in the order they are chosen. The remaining six teams have to draft in reverse order of season records. Find the probability that

(a) A is the first team chosen.

(b) A is one of the four teams chosen.

(c) A is not one of the four teams chosen.

47. An honest coin is tossed 10 times in a row. The result of each toss (H or T) is observed. Find the probability of the event E = "a T comes up at least once." (*Hint:* Find the probability of the complementary event.)

48. Imagine a game in which you roll an honest die three times. Find the probability of the event E = "at least one of the rolls of the dice comes up a 6." (*Hint:* See Example 16.22.)

49. Find the odds of each of the following events.

(a) an event E with $\Pr(E) = 4/7$

(b) an event E with $\Pr(E) = 0.6$

50. Find the odds of each of the following events.

(a) an event E with $\Pr(E) = 3/11$

(b) an event E with $\Pr(E) = 0.375$

51. In each case, find the probability of an event E having the given odds.

(a) The odds in favor of E are 3 to 5.

(b) The odds against E are 8 to 15.

52. In each case, find the probability of an event E having the given odds.

(a) The odds in favor of E are 4 to 3.

(b) The odds against E are 12 to 5.

(c) The odds in favor of E are the same as the odds against E.

16.4 Expectations

53. The scoring for a Psych 101 final grade is shown in Table 16-18. The last row of the table shows Paul's individual scores. Find Paul's score in the course, expressed as a percent.

	Test 1	Test 2	Test 3	Quizzes	Paper	Final
Weight	15%	15%	15%	10%	25%	20%
Points possible	100	100	100	120	100	180
Paul's score	77	83	91	90	87	144

TABLE 16-18

54. Table 16-19 shows the aggregate scores of a golf player over an entire tournament.

Score	2	3	4	5	6
Percentage of holes	1.4%	36.1%	50%	11.1%	1.4%

TABLE 16-19

What was the player's average score per hole?

55. At Thomas Jefferson High School, the student body is divided by age as follows: 7% of the students are 14, 22% of the students are 15, 24% of the students are 16, 23% of the students are 17, 19% of the students are 18, and the rest of the students are 19. Find the average age of the students at Thomas Jefferson High School.

56. In 2005 the Middletown Zoo averaged 4000 visitors on sunny days, 3000 visitors on cloudy days, 1500 visitors on rainy days, and only 100 visitors on snowy days. The percentage of days of each type in 2005 is shown in Table 16-20. Find the average daily attendance at the Middletown Zoo for 2005.

Weather condition	Sunny	Cloudy	Rainy	Snowy
Percentage of days	47%	27%	19%	7%

TABLE 16-20

57. A box contains twenty $1 bills, ten $5 bills, five $10 bills, four $20 bills, and one $100 bill. You blindly reach into the box and draw a bill at random. What is the expected value of your draw?

58. A basketball player shoots two consecutive free throws. Each free-throw is worth 1 point and has probability of success $p = 3/4$. Let X denote the number of points scored. Find the expected value of X.

59. A fair coin is tossed three times. Find the expected number of heads that come up.

60. A pair of honest dice is rolled once. Find the expected value of the sum of the two numbers rolled. (*Hint*: See Example 16.20).

16.5 Measuring Risk

61. Suppose that you roll a pair of honest dice. If you roll a total of 7, you win $18; if you roll a total of 11, you win $54; if you roll any other total, you lose $9. Find the expected payoff for this game.

62. On an American roulette wheel, there are 18 red numbers, 18 black numbers, plus 2 green numbers (0 and 00). If you bet $*N* on red, you win $*N* if a red number comes up (i.e., you get $2*N* back—your original bet plus your $*N* profit); if a black or green number comes up, you lose your $*N* bet. Find the expected payoff of a $1 bet on red.

63. On an American roulette wheel, there are 38 numbers: 00, 0, 1, 2, . . . , 36. If you bet $*N* on any one number—say, for example, on 10—you win $36*N* if 10 comes up (i.e., you get $37*N* back—your original bet plus your $36*N* profit); if any other number comes up, you lose your $*N* bet. Find the expected payoff of a $1 bet on 10 (or any other number).

64. Suppose that you roll a single die. If an odd number (1, 3, or 5) comes up, you win the amount of your roll ($1, $3, or $5, respectively). If an even number (2, 4, or 6) comes up, you have to pay the house the amount of your roll ($2, $4, or $6, respectively).

(a) Find the expected payoff for this game.

(b) Is this a fair game? Explain.

65. Joe is buying a new plasma TV at Circuit Town. The salesman offers Joe a three-year extended warranty for $80. The salesman tells Joe that 24% of these plasma TVs require repairs within the first three years, and the average cost of a repair is $400. Should Joe buy the extended warranty? Explain your reasoning.

66. Jackie is buying a new MP3 player from Better Buy. The store offers her a two-year extended warranty for $19. Jackie read in a consumer magazine that for this model MP3, 5% require repairs within the first two years at an average cost of $50. Should Jackie buy the extended warranty? Explain your reasoning.

67. The service history of the Prego SUV is as follows: 50% will need no repairs during the first year, 35% will have repair costs of $500 during the first year, 12% will have repair costs of $1500 during the first year, and the remaining SUVs (the real lemons) will have repair costs of $4000 during their first year. Determine the price that the insurance company should charge for a one-year extended warranty on a Prego SUV if it wants to make an average profit of $50 per policy.

68. An insurance company plans to sell a $250,000 one-year term life insurance policy to a 60-year-old male. Of 2.5 million men having similar risk factors, the company estimates that 7500 of them will die in the next year. What is the premium that the insurance company should charge if it would like to make a profit of $50 on each policy?

JOGGING

69. The ski club at Tasmania State University has 35 members (15 females and 20 males). A committee of three members—a President, a Vice President, and a Treasurer—must be chosen.

(a) How many different three-member committees can be chosen?

(b) How many different three-member committees can be chosen in which the committee members are all females?

(c) How many different three-member committees can be chosen in which the committee members are all the same gender?

(d) How many different three-member committees can be chosen in which the committee members are *not* all the same gender?

70. The ski club at Tasmania State University has 35 members (15 females and 20 males). A committee of four members of equal standing must be chosen.

(a) How many different four-member committees can be chosen?

(b) How many different four-member committees can be chosen consisting of two males and two females?

71. Andy and Roger are playing in a tennis match. (A tennis match is a best-of-five contest: The first player to win three games wins the match, and there are no ties.) We can describe the outcome of the tennis match by a string of letters (A or R) that indicate the winner of each game. For example, the string $RARR$ represents an outcome where Roger wins games 1, 3, and 4, at which point the match is over (game 5 is not played).

(a) Describe the event "Roger wins the match in game 5."

(b) Describe the event "Roger wins the match."

(c) Describe the event "the match goes five games."

72. Two teams (call them X and Y) play against each other in the World Series. (The World Series is a best-of-seven series: The first team to win four games wins the series, and games cannot end in a tie.) We can describe an outcome for the World Series by writing a string of letters that indicate (in order) the winner of each game. For example, the string $XYXXYX$ represents the outcome: X wins game 1, Y wins game 2, X wins game 3, and so on.

(a) Describe the event "X wins in five games."

(b) Describe the event "the series is over in game 5."

(c) Describe the event "the series is over in game 6."

(d) Find the size of the sample space.

73. An urn contains seven red balls and three blue balls.

(a) If three balls are selected all at once, what is the probability that two are blue and one is red?

(b) If three balls are selected by pulling out a ball, noting its color, and putting it back in the urn before the next selection, what is the probability that only the first and third balls drawn are blue?

(c) If three balls are selected one at a time without putting them back in the urn, what is the probability that only the first and third balls drawn are blue?

74. If we toss an honest coin 10 times, what is the probability of

(a) getting 5 heads and 5 tails?

(b) getting 3 heads and 7 tails?

75. If an honest coin is tossed N times, what is the probability of getting the same number of heads as tails? (*Hints:* 1. Try Exercise 74(a) first. 2. Consider two cases: N even and N odd.)

76. A draw poker hand consists of 5 cards taken from a deck of 52 cards where the order of the cards is irrelevant (see Examples 16.15 and 16.23). Assuming you are playing with an honest deck, find the probability of each of the following hands.

(a) A "four of a kind" (four cards of the same value and a "kicker" of any other value)

(b) A "flush" (all five cards of the same suit)

(c) An "ace-high straight" (10, J, Q, K, A of any suit but not all the same suit)

(d) A "full house" (three cards of equal value and two other cards of equal value)

77. There are 500 tickets sold in a raffle. If you have three of these tickets and five prizes are to be given, what is the probability that you will win at least one prize? (Give your answer in symbolic notation.)

78. This exercise refers to the game of chuck-a-luck discussed in Example 16.34. Explain why, when you roll three dice,

(a) the probability of rolling two 4's plus another number (not a 4) is $15/216$.

(b) the probability of rolling one 4 plus two other numbers (not 4's) is $75/216$.

(c) the probability of rolling no 4's is $125/216$.

79. Yahtzee. Yahtzee is a dice game in which five standard dice are rolled at one time.

(a) What is the probability of scoring "Yahtzee" with one roll of the dice? (You score Yahtzee when all five dice match.)

(b) What is the probability of a *four of a kind* with one roll of the dice? (A *four of a kind* is rolled when four of the five dice match.)

(c) What is the probability of rolling a *large straight* in one roll of the dice? (A *large straight* consists of five numbers in succession on the dice.)

80. In head-to-head, 7-card stud poker you make your hand by selecting your 5 best cards from the 2 in your hand and 3 from the 5 common cards showing on the table (the "board"). Suppose you are holding A ♠, 2 ♠, in your hand and Q ♠, K ♥, 3 ♠, 7 ♦, are showing on the board (this is called a "flush draw")—if the last card ("river" card) is a spade you will have an ace high flush and a guaranteed win. Assume that your opponent has a decent hand and if you don't get the spade on the river card you will lose the hand.

(a) Suppose there is \$100 in the pot and your opponent moves "all-in" with a \$50 bet. Should you call the bet or fold? Explain.

(b) Suppose there is \$100 in the pot and your opponent moves "all-in" with a \$20 bet. Should you call the bet or fold? Explain.

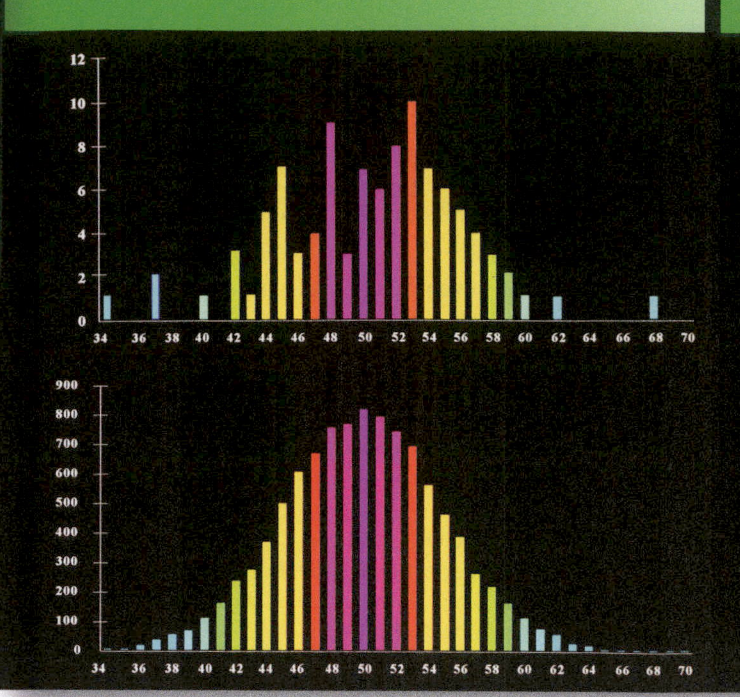

Top: Ten thousand coin tosses (100 reps of 100 tosses). Adapted from Kerrich, J., *An Experimental Introduction to the Theory of Probability.* **Bottom:** One million coin tosses (10,000 reps of 100 tosses). Computer simulation. (For more details, see Example 17.4.)

17

The Mathematics of Normality

The Call of the Bell

On April 9, 1940, the German army invaded Denmark and Norway. For John Kerrich—a mathematician who happened to be in Denmark at the time but was supposed to return to England a few days later—the timing could not have been worse. Like many other British subjects, he was arrested and interned in a prisoner-of-war camp where he remained for the rest of the war. With plenty of time on his hands, Kerrich embarked on a series of probability experiments, which he summarized in a short book he published at the end of the war. In the best known of these experiments, Kerrich tossed a coin 10,000 times and kept a meticulous record of the outcome of each toss.

By the time he was done, Kerrich had tossed 5067 heads and 4933 tails. In and of itself, this is not the kind of fact that will turn too many heads. Then Kerrich separated the results of his coin-tossing experiment into groups of 100 tosses (1–100, 101–200, . . . , 9901–10,000) and tallied the number of heads in each group. He recorded tallies of as many as 68 heads, as few as 34 heads, and plenty of others in between. (The top bar graph on the opposite page summarizes Kerrich's data—the height of each bar representing the number of times that particular number of heads came up). What is not clear from the bar graph is whether Kerrich was on to something, and if he was, what was it?

Kerrich was indeed on to something, and the only reason it's hard to tell what it was is that he stopped tossing a little too soon. Had he tossed the coin one million times or so, he would have ended up with a gorgeous bell-shaped bar graph like the bottom graph on the opposite page. How can we be so sure? Tossing a coin is unpredictable—wouldn't tossing a coin a million times be even more so? Surprisingly, no. We will learn the reasons why we are guaranteed to end up with a *bell-shaped distribution* of head tallies later in this chapter.

Normal (i.e., bell-shaped) distributions of data play a special role in the world of statistics. In Section 17.1 we ease into the topic of normal distributions by discussing several real-life examples of *approximately normal* (i.e., roughly bell-shaped) data sets. In Section 17.2 we discuss the mathematical properties of *normal curves* and the normal distributions they describe. In Section 17.3 we apply the mathematical properties we introduced in Section 17.2 to analyze and interpret approximately normal real-world data sets. In Section 17.4 we discuss the connection between *random events* and normal distributions.

17.1 Approximately Normal Data Sets

We start this section by looking at four different real-world data sets.

EXAMPLE 17.1 HEIGHT DISTRIBUTION OF NBA PLAYERS (2016 PLAYOFFS)

Table 17-1 is a frequency table showing the heights (in inches) of the 215 National Basketball Association players in the 2016 playoff team rosters (data from *www.nba.com*). The table shows that there were just two players under 6 feet tall (71 in. or less), and four players over 7 feet tall (85 in. or more). Not surprisingly, the bulk of the players were between 6 and 7 feet tall. Within this group we can see (reading the table from left to right) that the frequencies steadily go up, reach a maximum at 81 in., and then rapidly go down. (There is one exception to this pattern at 80 in., where there is an unexpected drop in frequency—we'll think of it as an anomaly.)

Height (inches)	69	70	71	72	73	74	75	76	77	78	79	80	81	82	83	84	85	86	87
Number of players	1	0	1	4	6	9	18	18	19	20	22	17	26	19	18	13	2	1	1

TABLE 17-1 ■ Height distribution of $N = 215$ NBA players

Figure 17-1(a) is a bar graph for the heights in Table 17-1. The bar graph illustrates the general pattern of the data—flat early on, a steady increase in the frequencies between 72 in. and 81 in. (with the exception of the drop at 80 in.), a rapid decrease in the frequencies between 81 in. and 85 in., and flat again at the end. Very roughly speaking, the bar graph is shaped like a bell, but a bit off center, skewed to the left and with a quirky dent at 80 in. An idealized version of the NBA player's heights would look like the bar graph in Fig. 17-1(b)—smoother and more bell shaped, but still a bit off center and skewed to the left.

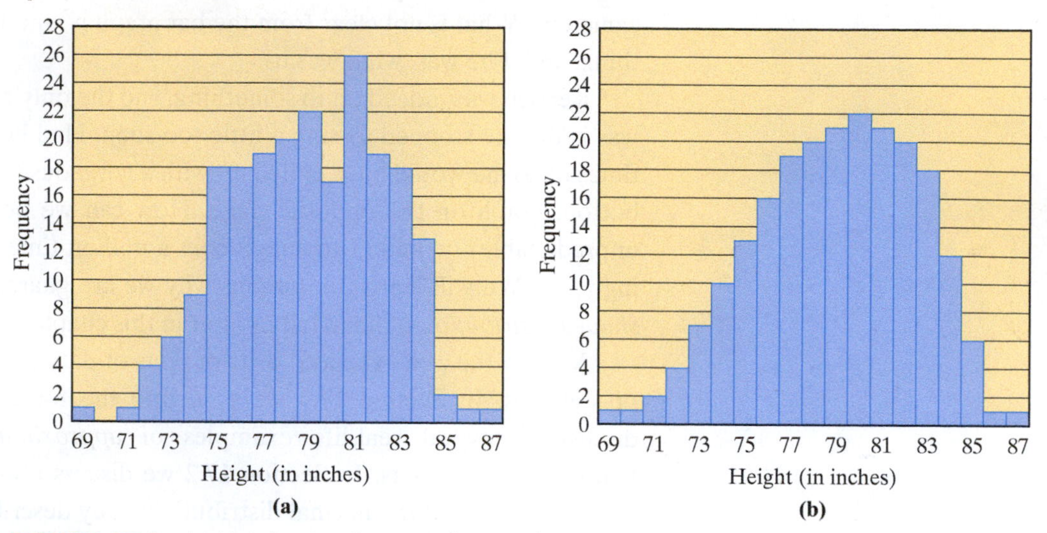

FIGURE 17-1 (a) Height distribution of NBA players (2016 playoff team rosters) $N = 215$; (b) an idealized height distribution.

EXAMPLE 17.2 METACRITIC MOVIE RATINGS

Metacritic (*www.metacritic.com*) is a Web site where one can find, arguably, the most objective ratings of movies, music, video games, and television shows. Metacritic combines all the available reviews of a movie, album, video game, etc. into a single "metascore" between 1 and 100 using a proprietary *weighted averaging method* (i.e., a review from the *New York Times* carries more weight than a review from the *Village Voice*). In this example, we will take a look at the $N = 1384$ metascores for all "Action" movies listed in the Metacritic database as of July 2016.

For convenience, metascores were grouped into 20 classes (1–5, 6–10, . . . , 96–100). Table 17-2 shows the number of movies in each class, and Fig. 17-2(a) shows a bar graph of the data in Table 17-2. This bar graph is closer to a bell shape than the NBA heights bar graph in Fig. 17-1(a), but it is far from perfect. The peak of the bell is "off-center" at the 36–40 class, and the distribution is *skewed to the*

Metascore	1–5	6–10	11–15	16–20	21–25	26–30	31–35	36–40	41–45	46–50
Frequency	0	5	16	35	57	71	112	150	149	143
Metascore	51–55	56–60	61–65	66–70	71–75	76–80	81–85	86–90	91–95	96–100
Frequency	125	128	112	91	82	48	32	16	6	6

TABLE 17-2 ■ Metascores of $N = 1384$ Action movies on the *Metacritic* database (July 2016)

right, meaning that the bulk of the scores fall to the right of the peak. Figure 17-2(b) shows an idealized version of the metascore data, with the peak of the bell in the middle and the data more or less equally distributed to the left and right of center.

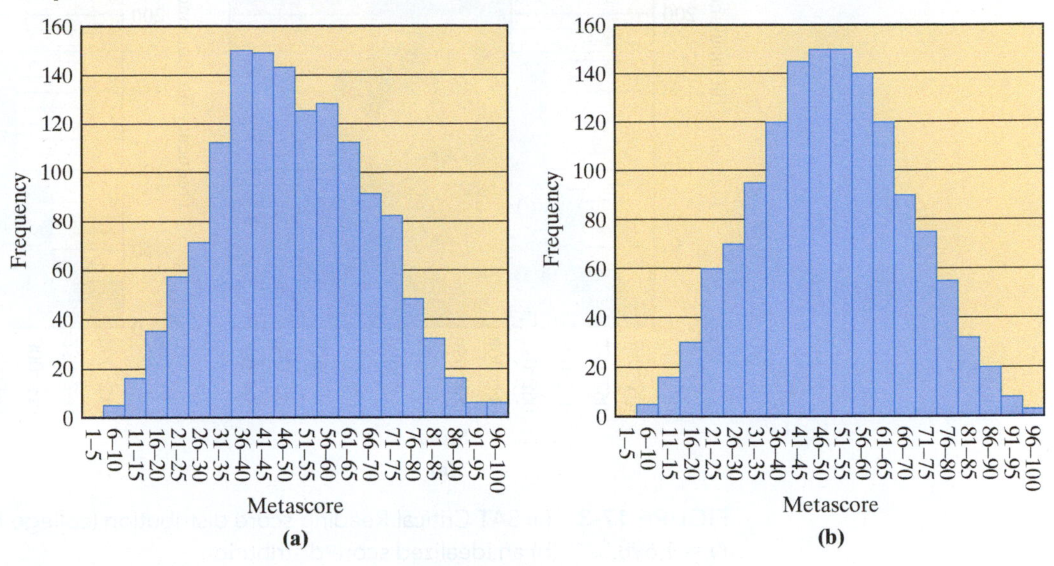

FIGURE 17-2 (a) Distribution of "Action" movie metascores in Metacritic database (July 2016) $N = 1384$; (b) an idealized distribution of metascores.

EXAMPLE 17.3 **SAT CRITICAL READING SCORES (2015)**

Table 17-3 shows the Critical Reading scores on the SAT for college-bound seniors in 2015 (Data from *SAT Data Tables*, 2015, The College Board.). For convenience, the scores (ranging between 200 and 800 points and going up in increments of 10) are grouped into 12 score ranges. The frequency column gives the number of students scoring in each range. The total number of test-takers was $N = 1{,}698{,}521$.

Score range	Frequency	Score range	Frequency
200–240	29,306	500–540	263,052
250–290	36,143	550–590	232,865
300–340	89,114	600–640	163,644
350–390	175,041	650–690	93,540
400–440	256,858	700–740	45,264
450–490	283,299	750–800	30,395

TABLE 17-3 ■ SAT Critical Reading scores for college-bound seniors in 2015 ($N = 1{,}698{,}521$)

Figure 17-3(a) shows a bar graph for the score frequencies in Table 17-3. The bar graph looks a lot like a bell, slightly skewed to the right. The bar graph in Fig. 17-3(b) shows an idealized, bell-shaped distribution of the SAT scores. The two bar graphs are certainly not the same, but they are quite similar. Informally, we can interpret this to mean that the real SAT score data is very well behaved.

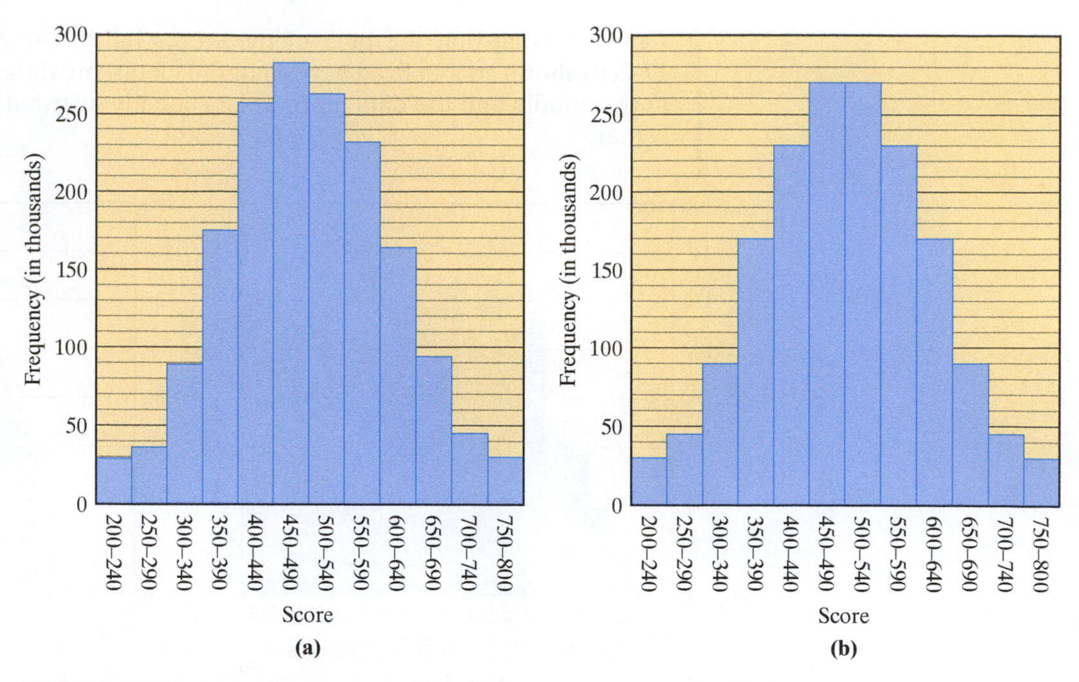

FIGURE 17-3 (a) SAT Critical Reading score distribution (college-bound seniors in 2015), $N = 1{,}698{,}521$; (b) an idealized score distribution.

EXAMPLE 17.4 COIN-TOSSING (COMPUTER SIMULATION)

We started this chapter with the story of the mathematician John Kerrich and his coin-tossing experiment. While a prisoner of war in Denmark during World War II, Kerrich tossed a coin 10,000 times and kept careful records of the outcome of each toss. When Kerrich broke down the 10,000 tosses into groups of 100 tosses and tallied the number of heads in each group, he got the results shown in Table 17-4. Figure 17-4 is the bar graph showing the results of Kerrich's coin-tossing experiment. It would be quite a stretch to say that this bar graph is bell shaped.

Heads	Frequency	Heads	Frequency
34	1	51	6
37	2	52	8
40	1	53	10
42	3	54	7
43	1	55	6
44	5	56	5
45	7	57	4
46	3	58	3
47	4	59	2
48	9	60	1
49	3	62	1
50	7	68	1

TABLE 17-4 ■ Kerrich's coin-tossing experiment (100 tosses repeated 100 times)

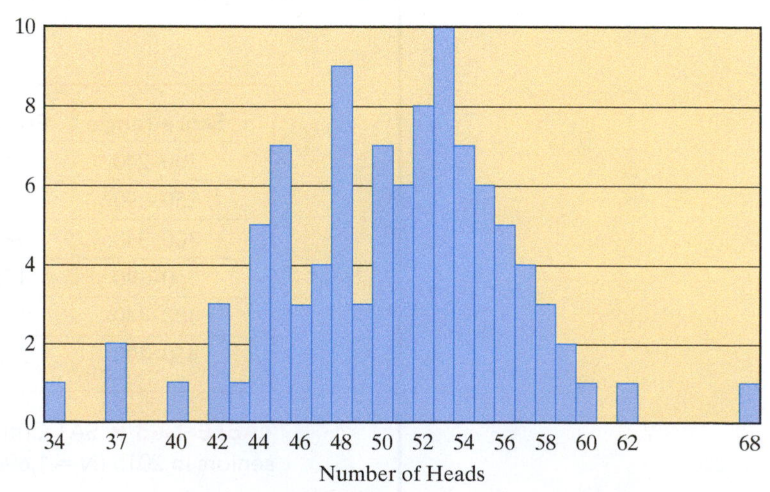

FIGURE 17-4 Number of heads in 100 tosses (100 repetitions). (*Source*: Kerrich, J., *An Experimental Introduction to the Theory of Probability*, 1946.)

Imagine now a more ambitious coin-tossing experiment: one million coin tosses broken down into 10,000 sets of 100 tosses. Not something you'd want to do the old-fashioned way, but with a computer and the right software it's an experiment that can be done in a matter of seconds. The results of such an experiment are given in Table 17-5 and summarized by the bar graph in Fig. 17-5. The bar graph has an almost perfect bell shape—about as close to perfection as a real-world data set can get.

Heads	Frequency	Heads	Frequency	Heads	Frequency
33	3	45	467	57	301
34	4	46	589	58	233
35	9	47	661	59	155
36	14	48	719	60	108
37	30	49	774	61	71
38	49	50	802	62	45
39	67	51	777	63	26
40	114	52	736	64	16
41	159	53	672	65	9
42	224	54	579	66	4
43	305	55	487	67	3
44	401	56	387		

TABLE 17-5 ■ Computer-generated coin-tossing experiment (100 tosses repeated 10,000 times)

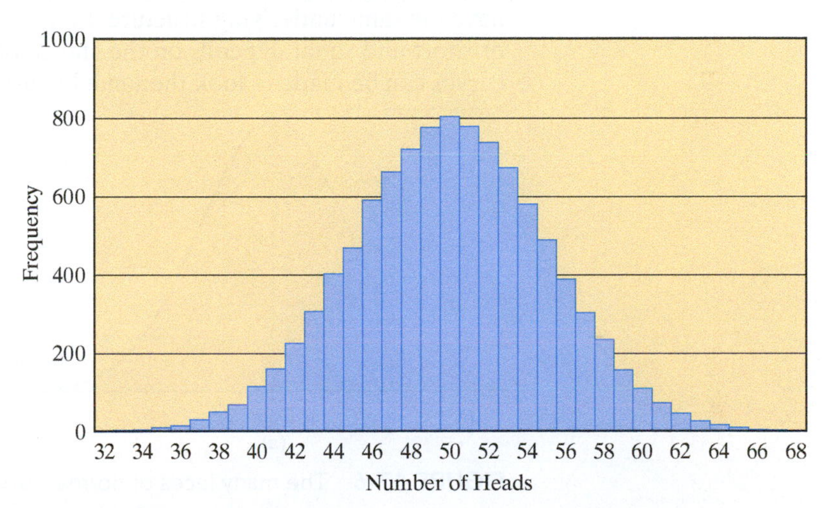

FIGURE 17-5 Number of heads in 100 tosses (10,000 repetitions). (*Source*: Computer simulation.)

In the preceding examples we saw four very different data sets, each representing a different source of data: In Example 17.1 the data measured a human *physical characteristic* (height), in Example 17.2 the data measured a form of human *judgment* (metascore), in Example 17.3 the data measured a form of human *performance* (test score), and in Example 17.4 the data measured *random variability*. In spite of these differences, the data sets shared one common characteristic—their bar graphs (with the horizontal axis representing the data and the vertical axis representing the frequencies) were all approximately bell shaped. In some cases (e.g., Example 17.1), the approximation is pretty rough, and in other cases (e.g., Example 17.4), the approximation is almost perfect. We will describe such data sets as being *approximately normal* (hanging a fair amount of leeway on the word "approximately").

- **Approximately normal distribution.** A distribution of data having a bar graph that is *approximately* bell shaped.
- **Normal distribution.** A distribution of data having a bar graph that is perfectly bell shaped.

When a large real-life data set has an approximately normal distribution, we can get a lot of useful information about the data by modeling the data set as if it were an idealized normal distribution (i.e., we pretend that the data is bell shaped even when it is not quite so). Normal distributions share a lot of nice properties, and the idea is to take advantage of these properties, even when the distribution is only approximately normal. In the next two sections, we will discuss how this is done.

17.2 Normal Curves and Normal Distributions

German bank note honoring Carl Friedrich Gauss (1777–1855), with a normal curve showing to the left of his image.

A **normal curve** is a perfect bell-shaped curve. It is the continuous version of a perfectly bell-shaped bar graph, and we can use the properties of normal curves to analyze the properties of normal (and, therefore, approximately normal) distributions. The study of normal curves can be traced back to the great German mathematician Carl Friedrich Gauss, and for this reason, normal curves are sometimes known as *Gaussian curves*. Normal curves all share the same basic shape—that of a bell—but otherwise they can differ widely in their appearance. Figure 17-6(a) shows a typical normal curve (with the data values on the horizontal axis and their frequencies on the vertical axis); Fig. 17-6(b) shows three different normal curves superimposed on the same set of axes—one of them is short and squat, another one is tall and skinny, and the third one falls somewhere in between. Mathematically speaking, however, they all have the same underlying structure. In fact, whether a normal curve is skinny and tall or short and squat depends on the choice of units on the axes, and any two normal curves can be made to look the same by just fiddling with the scales of the axes.

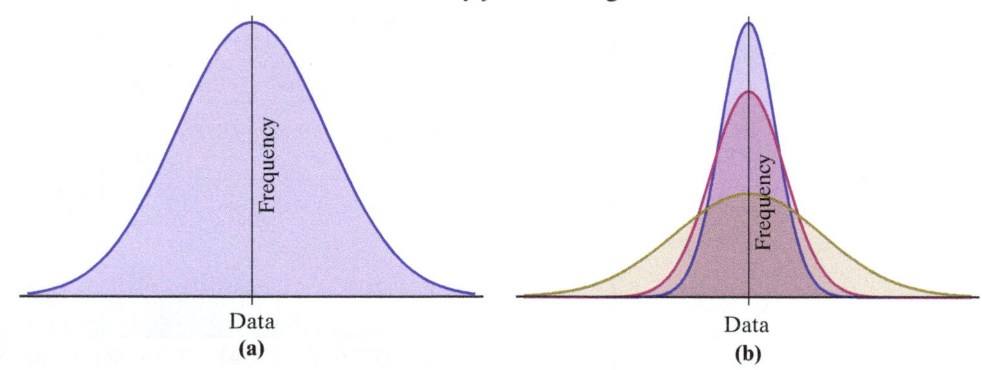

FIGURE 17-6 The many faces of normal curves.

What follows is a summary of some of the key mathematical properties of normal curves. These properties are going to help us greatly later on in the chapter.

1. **Symmetry.** Every normal curve has a vertical axis of symmetry, splitting the bell-shaped region outlined by the curve into two identical halves. This is the only line of symmetry of a normal curve, so we can refer to it without ambiguity as *the axis of symmetry*. This means that in a normal distribution the right half of the distribution is a mirror image of the left half.
2. **Median and mean.** We will call the point of intersection of the horizontal (data) axis and the line of symmetry of the normal curve the **center** of the distribution. The center corresponds to both the *median* M and the *mean* (average) μ of a normal distribution. It follows that in a normal distribution, the average μ and the median M are equal, and in an approximately normal distribution M and μ are close, but not necessarily equal.
   - In a normal distribution, $M = \mu$.
   - In an approximately normal distribution, $M \approx \mu$.
3. **Standard deviation.** The **standard deviation**—traditionally denoted by the Greek letter σ (sigma)—is an important measure of the spread of a data set, and

it is particularly useful when dealing with normal (or approximately normal) distributions, as we will see shortly. The easiest way to describe the standard deviation of a normal distribution is visually. If you were to bend a piece of wire into a bell-shaped normal curve, at the very top you would be bending the wire downward [Fig. 17-7(a)], but at the bottom you would be bending the wire upward [Fig. 17-7(b)]. As you move your hands down the wire, the curvature gradually changes, and there is one point on each side of the curve where the transition from being bent downward to being bent upward takes place. Such a point [P in Fig. 17-7(c)] is called a **point of inflection** of the curve. The standard deviation σ of a normal distribution is the *horizontal distance* between the axis of symmetry of the curve and one of the two points of inflection [P or P' in Fig. 17-7(d)].

- In a *normal distribution*, the standard deviation σ equals the horizontal distance between a point of inflection (either one) and the axis of symmetry of the curve.

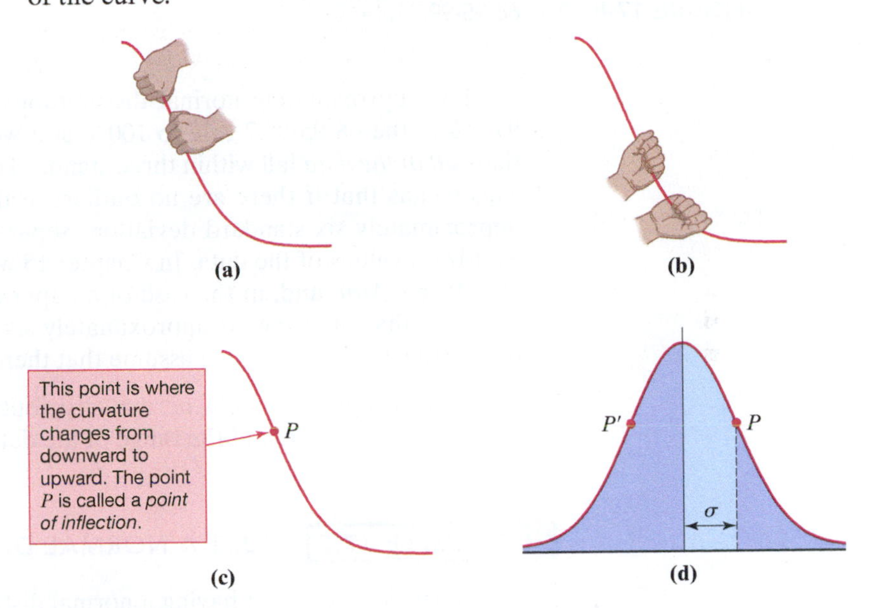
(a)
(b)
This point is where the curvature changes from downward to upward. The point P is called a point of inflection.
P
(c)
P'
P
σ
(d)

FIGURE 17-7

4. **Quartiles.** We learned in Chapter 15 how to find the quartiles of a data set. When the data set has a normal distribution, the first and third quartiles can be approximated using the mean μ and the standard deviation σ. The magic number to memorize is 0.675. Multiplying the standard deviation by 0.675 tells us how far to go to the right or left of the mean to locate the quartiles.

   - In a *normal distribution*, $Q_3 \approx \mu + (0.675)\sigma$, and $Q_1 \approx \mu - (0.675)\sigma$.

5. **The 68-95-99.7 Rule.** When we look at a typical bell-shaped distribution, we can see that most of the data are concentrated near the center. As we move away from the center the heights of the columns drop rather fast, and if we move far enough away from the center, there are essentially no data to be found. These observations can be formalized using the following three facts that together form what is usually known as the *68-95-99.7 rule*.

   - In a normal distribution, approximately 68% of all the data fall within one standard deviation above and below the mean $(\mu \pm \sigma)$. The remaining 32% are split equally between values smaller or equal to $\mu - \sigma\,(16\%)$ and values bigger or equal to $\mu + \sigma\,(16\%)$. Using P_k to denote the kth percentile, this means that $\mu - \sigma \approx P_{16}$, and $\mu + \sigma \approx P_{84}$ [Fig. 17-8(a)].

   - In a normal distribution, approximately 95% of all the data fall within two standard deviations above and below the mean $(\mu \pm 2\sigma)$. This implies that $\mu - 2\sigma \approx P_{2.5}$, and $\mu + 2\sigma \approx P_{97.5}$ [Fig. 17-8(b)].

- In a normal distribution, approximately 99.7% (i.e., practically 100%) of all the data fall within three standard deviations above and below the mean ($\mu \pm 3\sigma$). This implies that $\mu - 3\sigma \approx P_{0.15}$, and $\mu + 3\sigma \approx P_{99.85}$ [Fig. 17-8(c)].

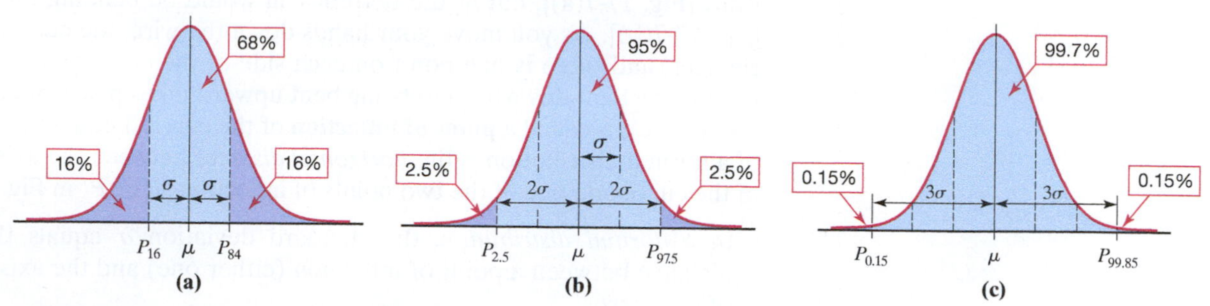

FIGURE 17-8 The 68-95-99.7 rule.

For approximately normal distributions, it is often convenient to round the 99.7% in the 68-95-99.7 rule to 100% and work under the assumption that essentially *all of the data* fall within three standard deviations above and below the mean. This means that if there are no outliers in the data, we can figure that there are approximately six standard deviations separating the smallest (*Min*) and the largest (*Max*) values of the data. In Chapter 15 we defined the range R of a data set as $R = Max - Min$, and, in the case of an approximately normal distribution, we can conclude that the range is approximately six standard deviations. Remember that this is true as long as we can assume that there are no outliers.

- In an approximately normal distribution with no outliers, $Max \approx \mu + 3\sigma$, $Min \approx \mu - 3\sigma$, and the range of the data is $R \approx 6\sigma$.

EXAMPLE 17.5 JUST A NORMAL DATA SET ($\mu = 495$, $\sigma = 116$)

Let's consider a data set having a normal distribution with mean $\mu = 495$ and standard deviation $\sigma = 116$. For now let's not worry about what this data set represents—we'll come back to it later. We are going to use the various properties of normal distributions we have just learned to analyze this data set in some detail.

- **Median.** In a normal distribution the median is equal to the mean, so we have $M = 495$. This means that half of the data set consists of values of 495 or less, and the other half of the data set consists of values of 495 or more.
- **Quartiles.** The first quartile is $Q_1 \approx \mu - (0.675)\sigma = 495 - 0.675 \times 116 \approx 416.7$; the third quartile is $Q_3 \approx \mu + (0.675)\sigma = 495 + 0.675 \times 116 \approx 573.3$. This means that the bottom 25% of the data have values of 416.7 or less, the next 25% of the data fall between 416.7 and 495 (inclusive), the third 25% of the data fall between 495 and 573.3 (inclusive), and the top 25% of the data have values of 573.3 or higher [Fig. 17-9(a)].
- **Middle 68%.** The first part of the 68-95-99.7 rule tells us that 68% of the data fall between $\mu - \sigma = 495 - 116 = 379$ and $\mu + \sigma = 495 + 116 = 611$. From this we can conclude that 16% of the data have values of 379 or less and 16% of the data have values of 611 or more (i.e., 84% have values of 611 or less) [Fig. 17-9(b)]. Using P_k to denote the kth percentile, we have $P_{16} = 379$ and $P_{84} = 611$.
- **Middle 95%.** The second part of the 68-95-99.7 rule tells us that 95% of the data values fall between $\mu - 2\sigma = 495 - 232 = 263$ and $\mu + 2\sigma = 495 + 232 = 727$. It follows that 2.5% of the data has a value of 263 or less ($P_{2.5} = 263$), and 97.5% of the data has a value of 727 or less ($P_{97.5} = 727$) [Fig. 17-9(c)].

- **Middle 99.7%.** The last part of the 68-95-99.7 rule tells us that 99.7% of the data values fall between $\mu - 3\sigma = 495 - 348 = 147$ and $\mu + 3\sigma = 495 + 348 = 843$. It follows that 0.15% of the data (a tiny fraction) has a value of 147 or less ($P_{0.15} = 147$), and 99.85% of the data (practically all of it) has a value of 843 or less ($P_{99.85} = 843$) [Fig. 17-9(d)]. For all practical purposes (and unless the data set is very large), we can assume that 100% of the data falls between 147 and 843.

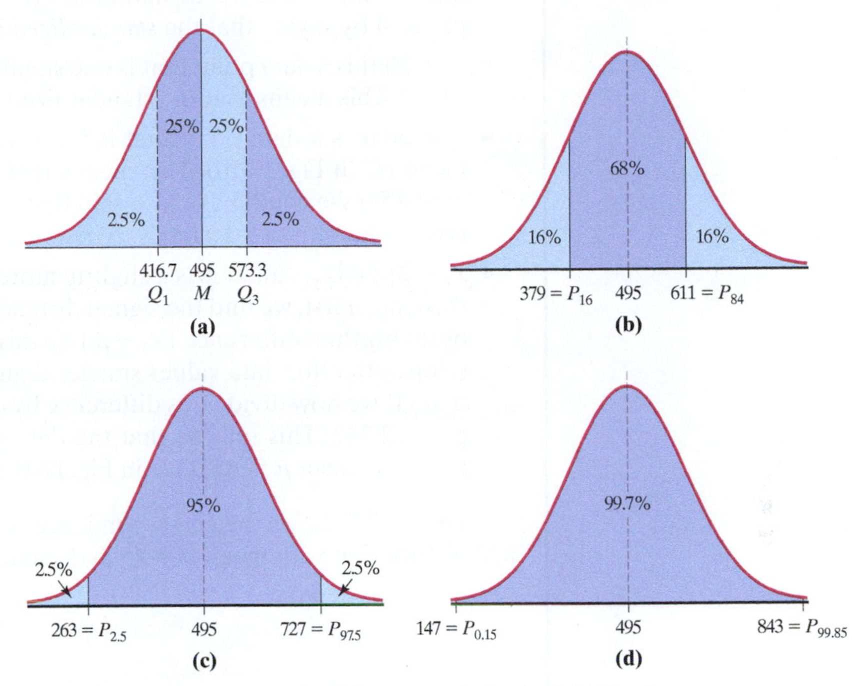

FIGURE 17-9 (a) Quartiles; (b), (c), and (d) 68-95-99.7 rule.

Table 17-6 summarizes the various benchmarks that we have found (expressed as percentiles) for the distribution of this normal data set.

Percentile	99.85	97.5	84	75	50	25	16	2.5	0.15
Value	843	727	611	573.3	495	416.7	379	263	147

TABLE 17-6

Standardizing Normal Data

We have seen that normal curves don't all look alike, but this is only a matter of perception. In fact, all normal distributions tell the same underlying story but use slightly different dialects to do it. One way to understand the story of any given normal distribution is to rephrase it in a simple common language—a language that uses the mean μ and the standard deviation σ as its only vocabulary. This process is called *standardizing* the data.

To standardize a data value x, *we measure how far x has strayed from the mean μ using the standard deviation σ as the unit of measurement.* A standardized data value is often referred to as a ***z*-value.**

The best way to illustrate the process of standardizing normal data is by means of a few examples.

EXAMPLE 17.6 STANDARDIZING NORMAL DATA

Let's consider a normally distributed data set with mean $\mu = 45$ ft and standard deviation $\sigma = 10$ ft. We will standardize several data values, starting with a couple of easy cases.

- $x_1 = 55$ ft is a data point located 10 ft *above* the mean $\mu = 45$ ft. Lucky for us, 10 ft happens to be exactly *one* standard deviation. The fact that $x_1 = 55$ ft is located one standard deviation *above* the mean (*A* in Fig. 17-10) can be rephrased by saying that the *standardized value* of $x_1 = 55$ is $z_1 = 1$.
- $x_2 = 35$ ft is a data point that is one standard deviation *below* the mean (*B* in Fig. 17-10). This means that the standardized value of $x_2 = 35$ is $z_2 = -1$.
- $x_3 = 50$ ft is a data point that is 5 ft (i.e., half a standard deviation) above the mean (*C* in Fig. 17-10). This means that the standardized value of $x_3 = 50$ ft is $z_3 = 0.5$. (We could have also said that the standardized value is $z_3 = 1/2$, but it is customary to use decimals to describe standardized values.)
- $x_4 = 21.58$ is . . . uh, this is a slightly more complicated case. How do we handle this one? First, we find the signed distance between the data value and the mean by taking their difference $(x_4 - \mu)$. In this case we get 21.58 ft $-$ 45 ft $= -23.42$ ft. (Notice that for data values smaller than the mean this difference will be negative.) If we now divide this difference by $\sigma = 10$ ft, we get the standardized value $z_4 = -2.342$. This tells us that the data point x_4 is -2.342 standard deviations from the mean $\mu = 45$ ft (*D* in Fig. 17-10).

Table 17-7 shows the correspondence between x-values and z-values for a normal distribution with mean $\mu = 45$ and standard deviation $\sigma = 10$.

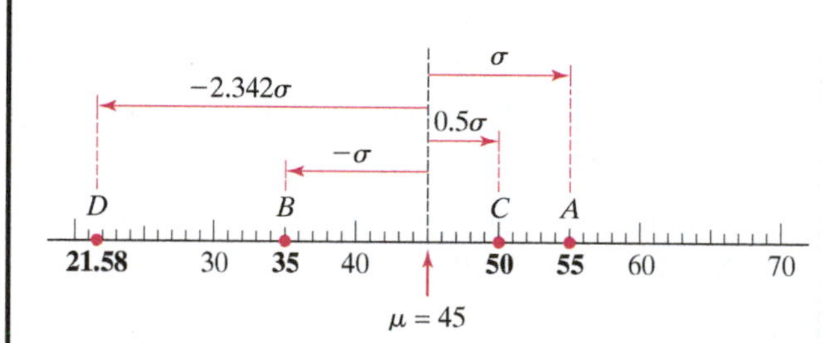

FIGURE 17-10

x	z
55	1
35	−1
50	0.5
21.58	−2.342
a	$(a - 45)/10$

TABLE 17-7

In Example 17.6 we were somewhat fortunate in that the standard deviation was $\sigma = 10$, an especially easy number to work with. It helped us get our feet wet. What do we do in more realistic situations, when the mean and standard deviation may not be such nice round numbers? Other than the fact that we may need a calculator to do the arithmetic, the basic idea we used in Example 17.6 remains the same.

STANDARDIZING RULE

In a normal distribution with mean μ and standard deviation σ, the standardized value of a data point x is $z = (x - \mu)/\sigma$.

EXAMPLE 17.7 FROM X TO Z

Consider a normally distributed data set with mean $\mu = 63.18$ lb and standard deviation $\sigma = 13.27$ lb. What is the z-value of $x = 91.54$ lb? With a calculator, this is not nearly as hard as it looks:

$$z = (x - \mu)/\sigma = (91.54 - 63.18)/13.27 = 28.36/13.27 \approx 2.14$$

One important point to note is that while the original data is given in pounds, there are no units given for the z-value. (The units for the z-value are standard deviations, and this is implicit in the very fact that it is a z-value.)

The process of standardizing data can also be reversed, and given a z-value we can go back and find the corresponding x-value. All we have to do is take the formula $z = (x - \mu)/\sigma$ and solve for x in terms of z. When we do this we get the equivalent formula $x = \mu + \sigma \cdot z$. Given μ, σ, and a value for z, this formula allows us to "unstandardize" z and find the original data value x.

EXAMPLE 17.8 **FROM Z TO X**

Consider a normal distribution with mean $\mu = 235.7$ inches and standard deviation $\sigma = 41.58$ inches. What is the data value x that corresponds to the standardized z-value $z = -3.45$?

We first compute the value of -3.45 standard deviations:

$$-3.45\sigma = -3.45 \times 41.58 = -143.451 \text{ inches.}$$

The negative sign indicates that the data point is to be located below the mean. Thus, $x = 235.7 - 143.451 = 92.249$ inches.

17.3 Modeling Approximately Normal Distributions

In this section we will use what we learned in Section 17.2 to analyze some real-world data sets, including a couple we introduced in Section 17.1.

EXAMPLE 17.9 **2015 SAT CRITICAL READING SCORES REVISITED**

In this example we are taking another look at the 2015 scores in the Critical Reading section of the SAT, first introduced in Example 17.3. We are now going to analyze the data set using only three simple pieces of information: (1) the total number of test-takers was $N = 1{,}698{,}521$, (2) the mean (average) score was $\mu = 495$ points, and the standard deviation was $\sigma = 116$ points. We will model the data set using a normal distribution (we saw in Example 17.3 that the data set is nicely behaved and looks pretty close to a bell-shaped distribution). There is one important consideration specific to this example that will have an impact on our calculations: SAT section scores go in increments of 10 points and range between 200 and 800 points.

- **Median.** The median should be close to the mean of 495 points. Since N is odd, the median must be one of the test scores (we learned this in Chapter 15), and this gives us two reasonable guesses for the value of the median score: $M = 490$, or $M = 500$.
- **First Quartile.** An estimate for the first quartile score is $Q_1 \approx \mu - (0.675)\sigma = 495 - 0.675 \times 116 = 416.7$. Here again, since the first quartile must be one of the test scores, we have two reasonable guesses for what it might be: $Q_1 = 410$, or $Q_1 = 420$ (we should not assume that the answer is to round to the nearest multiple of 10.)

- **Third Quartile.** An estimate for the third quartile score is $Q_3 \approx \mu + (0.675)\sigma = 495 + 0.675 \times 116 = 573.3$. Our two best guesses should be $Q_3 \approx 570$ or $Q_3 = 580$.

For the remaining calculations, the reader is referred to Example 17.5, where we discussed a normal distribution with mean $\mu = 495$ and standard deviation $\sigma = 116$ (and no, this was not a coincidence!).

- **16th and 84th percentiles.** The 16th percentile score is one standard deviation below the mean. Since $\mu - \sigma = 495 - 116 = 379$ the best guess for the 16th percentile score is either 380 or 370. Likewise, the 84th percentile score is one standard deviation above the mean. Since $\mu + \sigma = 495 + 116 = 611$, the best guess for the 84th percentile score is either 610 or 620.
- **2.5th and 97.5th percentiles.** A similar analysis shows that a good guess for the 2.5th percentile score is 260 or 270, $(\mu - 2\sigma = 263)$ and that a good guess for the 97.5th percentile is 720 or 730 $(\mu + 2\sigma = 727)$.
- **Min and Max.** Ordinarily, the 99.7 part of the 68-95-99.7 rule would give us an estimate for the smallest score $(Min = \mu - 3\sigma = 147)$ and the highest score $(Max = \mu + 3\sigma = 843)$, but these values are useless in this case because SAT section scores fall between 200 and 800 points anyway.

Table 17-8 shows a comparison between the educated guesses we came up with using a normal distribution model and the actual values given in the College Board's 2015 national report. By all accounts, our guesses turned out to be pretty accurate.

Percentile	Guess	Actual value
$M = p_{50}$	490 or 500	490
$Q_1 = p_{25}$	410 or 420	410
$Q_3 = p_{75}$	570 or 580	570
p_{16}	370 or 380	380
p_{84}	610 or 620	620
$p_{2.5}$	260 or 270	270
$p_{97.5}$	720 or 730	730

TABLE 17-8

EXAMPLE 17.10 HEIGHTS OF NBA PLAYOFF PLAYERS (2016) REVISITED

In Example 17.1 we introduced the distribution of heights of 215 NBA players on teams that played in the 2016 playoffs. The bar graph for the data set was very roughly bell shaped, but we are going to try to model it with a normal distribution anyway. We will use the mean height $\mu = 78.9$ in. and the standard deviation $\sigma = 3.373$ in. (computed using a statistical package) to estimate a few other values and see how they compare with the actual values.

- **Median.** Since $M \approx \mu = 78.9$ in. and the median should be a number in the data set ($N = 215$ is odd), a good guess for the median height is 79 in. That's exactly right: $M = 79$ in.
- **First Quartile.** The estimate for the first quartile height is $Q_1 \approx \mu - (0.675)\sigma = 78.9 - 0.675 \times 3.373 \approx 76.6$ in. Since Q_1 is also a number in the data set, we have two good guesses: 76 in. or 77 in. The actual value is $Q_1 = 76$ in.

- **Third Quartile.** The estimate for the third quartile height is $Q_3 \approx \mu + (0.675)\sigma = 78.9 + 0.675 \times 3.373 \approx 81.2$ in. Our two best guesses are 81 in. or 82 in. The actual value is $Q_3 = 82$ in.
- **16th percentile.** The estimate for the 16th percentile height is $\mu - \sigma = 78.9 - 3.373 \approx 75.5$ in. The actual value is 75 in.
- **84th percentile.** The estimate for the 84th percentile height is $\mu + \sigma = 78.9 + 3.373 \approx 82.3$ in. The actual value is 83 in.
- **2.5th percentile.** The estimate for the 2.5th percentile height is $p_{2.5} = \mu - 2\sigma = 78.9 - 2 \times 3.373 \approx 72.2$ in. The actual value is 72 in.
- **97.5th percentile.** The estimate for the 97.5th percentile height is $p_{97.5} = \mu + 2\sigma = 78.9 + 2 \times 3.373 \approx 85.6$ in. The actual value is 84 in.
- **Minimum.** A rough estimate for the minimum height is $Min = \mu - 3\sigma = 78.9 - 3 \times 3.373 \approx 68.8$ in. This is right on target, because the smallest height in the data set is 69 in.
- **Maximum.** A rough estimate for the maximum height is $Max = \mu + 3\sigma = 78.9 + 3 \times 3.373 \approx 89$ in. The tallest height in the data set is 87 in.

Example 17.10 shows that even in the case of a data set that was very roughly bell shaped, we can do pretty well estimating the distribution of the data using a normal distribution model. In fact, most of our estimates were within an inch of the correct height, and the two cases where the estimates were off by more than an inch (the 97.5th percentile and the maximum height) can be explained by the fact that the data set was skewed to the left.

Our next example shows an application of modeling with normal distributions to determine the percentile ranks in weight and length of babies and infants.

EXAMPLE 17.11 LENGTHS OF NEWBORN BABY BOYS

According to data published by the Centers for Disease Control and Prevention (CDC), the median length of newborn baby boys in the United States is $M = 49.98888$ cm, and the first quartile length is $Q_1 = 48.18937$ cm. For convenience, we round these to $M = 50$ cm and $Q_1 = 48.2$ cm. We will use these two values ($M = 50$; $Q_1 = 48.2$) to make some estimates and predictions about the length of newborn baby boys using the assumption that the distribution is approximately normal.

The first two numbers we are going to need are the mean and standard deviation of the distribution. The mean is easy, since we can assume that it is approximately the same as the median, so let's say $\mu \approx 50$ cm. To find the standard deviation, we will use the fact that $Q_1 \approx \mu - (0.675)\sigma$—in this case, $48.2 \approx 50 - (0.675)\sigma$. Solving for σ gives $\sigma \approx 1.8/0.675 \approx 2.67$ cm.

We will now use $\mu \approx 50$ cm and $\sigma \approx 2.67$ cm, to estimate other data values and percentiles:

- **Third Quartile.** $Q_3 \approx \mu + (0.675)\sigma \approx 50 + 0.675 \times 2.67 \approx 51.8$ cm. (The exact value of the third quartile listed in the CDC chart is 51.77126 cm—not too bad!)
- **97.5th percentile.** The 97.5th percentile is estimated to be two standard deviations above the mean. In this case $p_{97.5} \approx 50 + 2 \times 2.67 \approx 55.34$ cm. The best to which we can compare it in the CDC tables is the 97th percentile, listed as 54.919 cm. Since we expect the 97th percentile to be a little bit smaller than the 97.5th percentile, the two values are perfectly consistent.
- **52.7 cm.** A baby boy measures 52.7 cm at birth. Without using a chart, can we estimate the percentile rank for his length? In this case yes—we are lucky: 52.7 cm. is roughly one standard deviation above the mean (just a tad over). We can estimate this places him in the 84th percentile.

17.4 Normality in Random Events

In the opening pages of this chapter we briefly introduced John Kerrich's original coin-tossing experiment (10,000 coin tosses). In Example 17.4 we expanded the coin-tossing experiment to 1 million tosses (using a computer, of course). In this section we will return to coin-tossing experiments and a few more interesting variations of that theme (large-scale coin tossing, after all, has little importance in its own right, but it serves as an excellent metaphor for much more meaningful applications).

When John Kerrich tossed a coin 10,000 times, he broke up the tosses into sets of 100 and recorded the number of heads in each set (see Example 17.4). When you toss a coin 100 times the number of heads is a *variable* (you don't expect to get the same number of heads every time you do it) that to a large extent depends on the laws of *chance*. We call such a variable a **random variable**.

EXAMPLE 17.12 COIN TOSSING 100

Imagine that you are about to toss a fair coin 100 times. For ease of reference, we call this a *trial*. We let X denote the number of heads that could come up in such a trial. X is a random variable that can take any value between 0 and 100, but the possible values of X are not all equally likely: $X = 0$ and $X = 100$ are extremely unlikely (their probability is practically 0), $X = 50$ is much more likely (although the probability is less than most people would guess—about 8%). Since the coin is fair, heads and tails are equally likely, and this implies that $X = 49$ (49 heads, 51 tails) has the same probability as $X = 51$ (51 heads, 49 tails), and the same goes for $X = 48$ and $X = 52$, etc. This is all useful information, but it still doesn't tell us what is going to happen when we toss the coin 100 times.

One way to get a feel for the probabilities of the different values of X is to repeat the trial (100 coin tosses) many times and check the frequencies of the different values of X. I did this (using a computer to do the coin tossing). Figure 17-11 shows the evolution over time of the results (imagine it as time-lapse photography): 10 trials [Fig. 17-11(a)], 100 trials [Fig. 17-11(b)], 500 trials [Fig. 17-11(c)], 1000 trials [Fig. 17-11(d)], 5000 trials [Fig. 17-11(e)], and finally, 10,000 trials [Fig. 17-11(f)].

Figure 17-11 paints a pretty clear picture of what happens: As the number of trials increases, the distribution of the data becomes more and more bell shaped. At the end, we have data from 10,000 trials, and the bar graph gives an almost perfect normal distribution!

What would happen if someone else decided to repeat the experiment—toss a fair coin (be it by hand or by computer) 100 times, count the number of heads, and repeat this for a while? The first 10 trials are likely to produce results very different from ours, but as the number of trials increases, the results and our results will begin to look more and more alike. After 10,000 trials, the bar graph will be almost identical to the bar graph shown in Fig. 17-11(f). In a sense, this says that doing the experiments a second time is a total waste of time—in fact, it was even a waste the first time! The outline of the final distribution could have been predicted without ever tossing a coin!

As we saw in Section 17.3, knowing that the random variable X has an approximately normal distribution is quite useful. The clincher would be to find out the values of the mean μ and the standard deviation σ of this distribution. Looking at Fig. 17-11(f), we can pretty much see where the mean is—right at 50. (This is not surprising, since the axis of symmetry of the distribution has to pass through 50 as a simple consequence of the fact that the coin is fair.) The value of the standard deviation σ is less obvious. For now, let's accept the fact that $\sigma = 5$. (We will explain how we got this value shortly.)

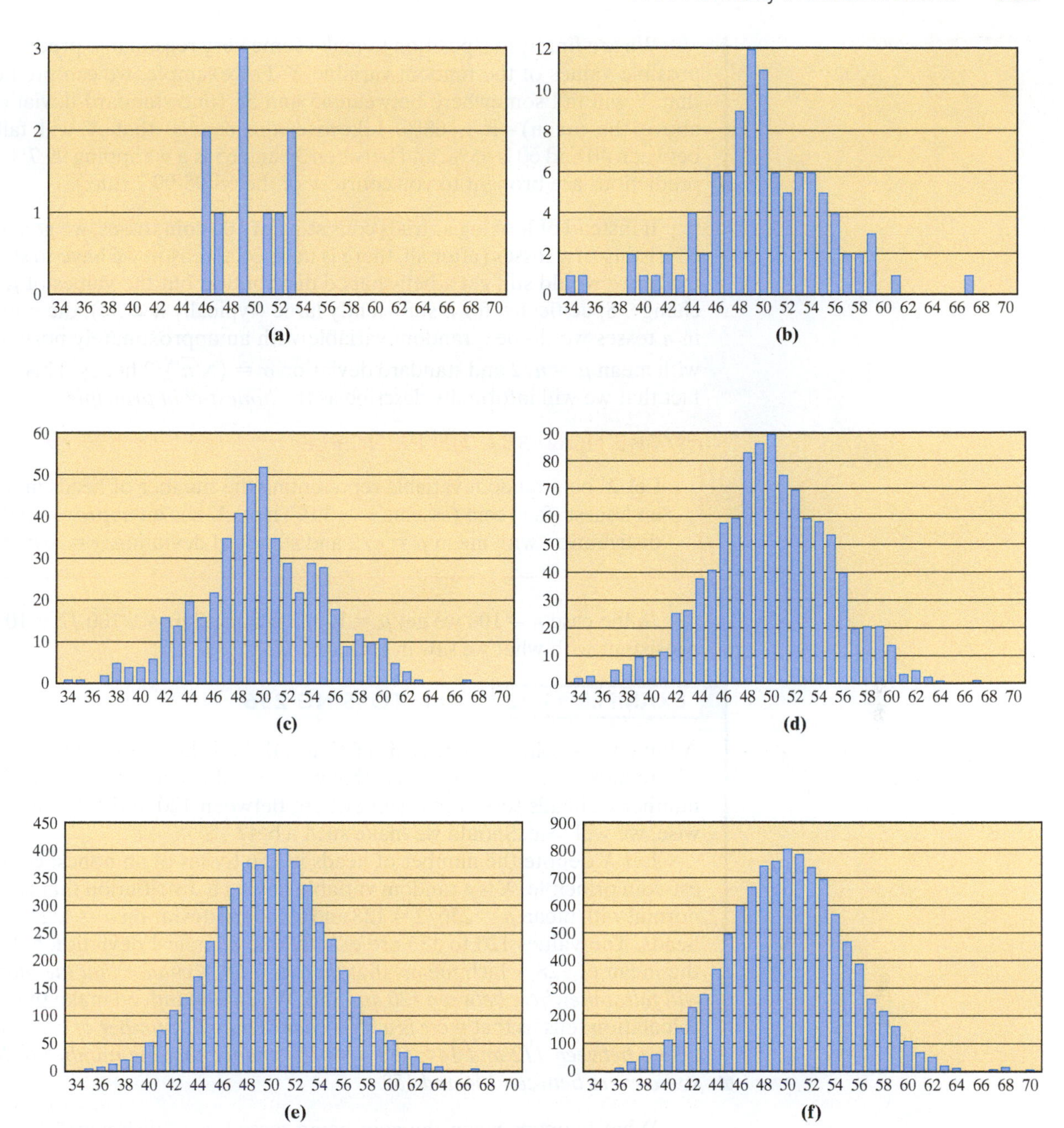

FIGURE 17-11 Distribution of X (number of heads in 100 coin tosses). (a) 10 trials, (b) 100 trials, (c) 500 trials, (d) 1000 trials, (e) 5000 trials, and (f) 10,000 trials.

Let's summarize what we now know. An honest coin is tossed 100 times, and X is the number of heads in the 100 tosses. When we repeat this a large number of times (say N), the random variable X will have an approximately normal distribution with mean $\mu = 50$ and standard deviation $\sigma = 5$, and the larger the value of N is, the better this approximation will be.

The real significance of these facts is that they are true not because we took the trouble to toss a coin a million times. Even if we did not toss a coin at all, all of these statements would still be true. *For a sufficiently large number of repetitions of a trial consisting of 100 tosses of a fair coin the number of heads X is a random variable that has an approximately normal distribution with mean $\mu = 50$ and standard deviation $\sigma = 5$ heads.*

We are finally in a position to make some very reasonable predictions about the possible values of the random variable X. For example, we can predict the chance that X will fall somewhere between 45 and 55 (one standard deviation below and above the mean)—it is 68%. Likewise, the chance that X will fall somewhere between 40 and 60 is 95%, and between 35 and 65 is a whopping 99.7%. (All of these predictions are brought to you courtesy of the 68-95-99.7 rule.)

If instead of looking at trials consisting of 100 coin tosses, we generalize to trials consisting of n tosses (after all, there is no special reason we have to stick to 100-toss trials) we would still get a bell-shaped distribution but the values of μ and σ would change. Specifically, for n sufficiently large (typically $n \geq 30$), the number of heads in n tosses would be a random variable with an approximately normal distribution with mean $\mu = n/2$ and standard deviation $\sigma = (\sqrt{n})/2$ heads. This is an important fact that we will informally describe as the *honest-coin principle*.

THE HONEST-COIN PRINCIPLE

Let X be a random variable representing the number of heads in n tosses of an honest (fair) coin (assume $n \geq 30$). Then, X has an *approximately normal* distribution with mean $\mu = n/2$ and standard deviation $\sigma = \sqrt{n}/2$.

In the case $n = 100$ we get $\mu = 100/2 = 50$ and $\sigma = \sqrt{100}/2 = 10/2 = 5$ heads, consistent with what we saw in Example 17.12.

EXAMPLE 17.13 COIN-TOSSING 256

A fair coin is going to be tossed 256 times. Before this is done, we have the opportunity to make some bets. Let's say that we can make a bet (with even odds) that if the number of heads tossed falls somewhere between 120 and 136, we will win; otherwise, we will lose. Should we make such a bet?

Let X denote the number of heads in 256 tosses of an honest coin. By the honest-coin principle, X is a random variable having a distribution that is approximately normal with mean $\mu = 256/2 = 128$ and standard deviation $\sigma = \sqrt{256}/2 = 16/2 = 8$ heads. The values 120 to 136 are exactly one standard deviation below and above the mean of 128, which means that *there is a 68% chance that the number of heads will fall somewhere between 120 and 136*. We should indeed make this bet! A similar calculation tells us that *there is a 95% chance that the number of heads will fall somewhere between 112 and 144*, and *a 99.7% chance that the number of heads will fall somewhere between 104 and 152*.

What happens when the coin being tossed is a "dishonest" (i.e., unbalanced) coin? Surprisingly, the distribution of the number of heads X in n tosses of a dishonest coin is still approximately normal, as long as the number n is not too small (a good rule of thumb is $n \geq 30$). All we need now is a *dishonest-coin principle* to tell us how to find the mean and the standard deviation.

THE DISHONEST-COIN PRINCIPLE

Let X be a random variable representing the number of heads in n tosses of a coin (assume $n \geq 30$), and let p denote the probability of heads on each toss of the coin. Then, X has an approximately normal distribution with mean $\mu = np$ and standard deviation $\sigma = \sqrt{np(1-p)}$.

EXAMPLE 17.14 COIN-TOSSING WITH A DISHONEST COIN

A coin is rigged so that it comes up heads only 20% of the time (i.e., $p = 0.20$), and let X represent the number of heads in 100 tosses of the coin. What can we say about X?

By the dishonest-coin principle, X has an approximately normal distribution with mean $\mu = 100 \times 0.20 = 20$ and standard deviation $\sigma = \sqrt{100 \times 0.20 \times 0.80} = 4$. Applying the 68-95-99.7 rule with $\mu = 20$ and $\sigma = 4$ gives the following facts:

1. There is about a 68% chance that X will fall somewhere between 16 and 24.
2. There is about a 95% chance that X will fall somewhere between 12 and 28.
3. It is almost guaranteed (about 99.7% chance) that X will fall somewhere between 8 and 32.

Note that in this example, heads and tails are no longer interchangeable concepts—heads is an outcome with probability $p = 0.2$, while tails is an outcome with much higher probability (0.8). We can, however, apply the principle equally well to describe the distribution of the number of tails in 100 coin tosses of the same dishonest coin: The distribution for the number of tails is approximately normal with mean $\mu = 100 \times 0.80 = 80$ and standard deviation $\sigma = \sqrt{100 \times 0.80 \times 0.20} = 4$. Note that the standard deviation is the same for heads and tails, but the mean is not.

The dishonest-coin principle can be applied to any coin, even one that is fair (i.e., with $p = 1/2$). In the case $p = 1/2$, the honest- and dishonest-coin principles say the same thing (Exercise 74).

The dishonest-coin principle is a special version of one of the most important laws in statistics, a law generally known as the *central limit theorem*. We will now briefly illustrate why the importance of the dishonest-coin principle goes beyond the tossing of coins.

EXAMPLE 17.15 SAMPLING FOR DEFECTIVE LIGHTBULBS

An assembly line produces 100,000 lightbulbs a day, 20% of which generally turn out to be defective. Suppose that we draw a random sample of $n = 100$ lightbulbs. Let X represent the *number of defective lightbulbs* in the sample. What can we say about X?

A moment's reflection will show that, in a sense, this example is Example 17.14 in disguise (think of selecting defective lightbulbs as analogous to tossing heads with a dishonest coin). We can use the dishonest-coin principle to infer that X has an approximately normal distribution with mean 20 lightbulbs and standard deviation equal to 4 lightbulbs. Using these facts, we can draw the following conclusions:

1. There is a 68% chance that the number of defective lightbulbs in the sample will fall somewhere between 16 and 24.
2. There is a 95% chance that the number of defective lightbulbs in the sample will fall somewhere between 12 and 28.
3. The number of defective lightbulbs in the sample is practically guaranteed (a 99.7% chance) to fall somewhere between 8 and 32.

Probably the most important point here is that each of the preceding facts can be rephrased in terms of sampling errors, a concept we first discussed in Chapter 14. For example, say we had 24 defective lightbulbs in the sample; in other words, 24% of the sample (24 out of 100) are defective lightbulbs. If we use this statistic to estimate the percentage of defective lightbulbs overall, then the sampling error would be 4% (because the estimate is 24% and the value of the parameter is 20%). By the same token, if we had 16 defective lightbulbs in the sample, the sampling error would be −4%. Coincidentally, the standard deviation is $\sigma = 4$ lightbulbs, or 4% of the sample. (We computed it in Example 17.14.) Thus, we can rephrase our previous assertions about sampling errors as follows:

1. When estimating the proportion of defective lightbulbs coming out of the assembly line by using a sample of 100 lightbulbs, there is a 68% chance that the sampling error will fall somewhere between −4% and 4%.
2. When estimating the proportion of defective lightbulbs coming out of the assembly line by using a sample of 100 lightbulbs, there is a 95% chance that the sampling error will fall somewhere between −8% and 8%.
3. When estimating the proportion of defective lightbulbs coming out of the assembly line by using a sample of 100 lightbulbs, there is a 99.7% chance that the sampling error will fall somewhere between −12% and 12%.

EXAMPLE 17.16 SAMPLING WITH LARGER SAMPLES

Suppose that we have the same assembly line as in Example 17.15, but this time we are going to choose a really big sample of $n = 1600$ lightbulbs. Before we even count the number of defective lightbulbs in the sample, let's see how much mileage we can get out of the dishonest-coin principle. The standard deviation for the distribution of defective lightbulbs in the sample is $\sigma = \sqrt{1600 \times 0.2 \times 0.8} = 16$. This is great, because $\sigma = 16$ just happens to be exactly 1% of the sample $(16/1600 = 1\%)$. So, $\sigma = 1\%$ of the sample. We can now describe the number of defective lightbulbs coming out of the assembly line in terms of the sampling error.

1. We can say with some confidence (a 68% chance) that the sampling error will fall somewhere between −1% and 1%.
2. We can say with a lot of confidence (a 95% chance) that the sampling error will fall somewhere between −2% and 2%.
3. We can say with tremendous confidence (a 99.7% chance) that the sampling error will fall somewhere between −3% and 3%.

The next and last example shows how the dishonest-coin principle can be used to estimate the margin of error in a public opinion poll, an issue of considerable importance in modern statistics.

EXAMPLE 17.17 THE 2012 PRESIDENTIAL ELECTION AND THE POLLS

In the days preceding the 2012 presidential election most reputable national polls were predicting a very tight race in terms of the popular vote. Some polls showed President Obama a percentage point or two ahead of Mitt Romney, some polls showed Romney a percentage point or two ahead of Obama, and some polls showed a dead heat. The final results—roughly 51% of the popular vote for Obama, 49% for Romney—fell well within the margin of error of most polls, confirming that overall the polls did a pretty good job. So, what exactly is the *margin of error* of a poll?

To answer this question let's look at a poll conducted by the Monmouth University Polling Institute the weekend of Nov. 1–Nov. 4, 2012, just days before the election. The poll consisted of a national random sample of $n = 1417$ *likely voters* who were asked: "if the election for President were today, would you vote for Mitt Romney the Republican, or Barack Obama the Democrat, or some other candidate?" The results were 48% for Obama, 48% for Romney, 2% other, 2% undecided (*Source*: *www.monmouth.edu/polling*).

Statistically, a poll in an election between *two* candidates is like a series of coin tosses of an imaginary coin. The number n of people sampled is the number of coin tosses, a vote for one candidate means the coin came up one way (say Romney is heads), a vote for the other candidate means the coin came up the other way (say Obama is tails). The probability p that our imaginary coin comes up heads is given by the relative percentage of people in the poll favoring the "heads" candidate. In the case of the Monmouth poll, when we disregard the few "undecided" and "other" votes, half of the "coin tosses" came up heads and half came up tails, so we make $p = \frac{1}{2}$—the coin happens to be an honest coin. (Keep in mind that this is unusual, and in general the coin is likely to be dishonest.)

We start our calculation using the *dishonest-coin principle* (in this example it turns out to be the honest-coin principle), with $n = 1417$ and $p = \frac{1}{2}$ to get $\sigma = \sqrt{1417 \times 1/2 \times 1/2} \approx 18.8$ votes. This gives the *standard deviation* of the sample distribution. For convenience, we express this standard deviation as a percentage of the sample size: $18.8/1417 \approx 0.0132675 \approx 1.33\%$. This means that one standard deviation is equal to roughly 1.33% of the sample.

The standard deviation of the sample distribution, expressed as a percentage of the sample, is called the **standard error.** The standard error is a yardstick that, in combination with the 68-95-99.7 rule gives us a measure of the predictive power of a poll: (1) roughly 68% of the time the results of the election are going to be within *one* standard error of those predicted by the poll $(\pm\sigma)$; (2) roughly 95% of the time the results of the election are going to be within *two* standard errors of those predicted by the poll $(\pm 2\sigma)$; and (3) almost always (99.7% of the time) the results of the election are going to be within *three* standard errors of those predicted by the poll $(\pm 3\sigma)$.

In the Monmouth poll, the poll statistic was 50% for Romney ("heads"), and 50% for Obama ("tails"), with a standard error of 1.33%. (Remember that we are disregarding the "undecided" and "other" votes.) This means that a **95% confidence interval** for the Romney popular vote (and in this case also for the Obama popular vote) would be 50% ± 2.66%, i.e., in the interval 47.34% to 52.66%. This is what is generally considered the safe margin of error for a poll, and most polls report margins of error in terms of a 95% confidence interval. If you want to be really confident of your prediction you can use a **99.7% confidence interval** $(\pm 3\sigma)$. For the Monmouth poll the 99.7% confidence interval would be 50% ± 3.99%.

Conclusion

From basketball players' heights to movie reviews to test scores and coin tosses, many real-life data sets follow the call of the bell. In this chapter we studied normal and approximately normal data sets, some of their mathematical properties, and how these properties can be used to analyze real-life, bell-shaped data sets, even when the data set is very roughly bell shaped. It was a brief introduction to what is undoubtedly one of the most widely used and sophisticated tools of modern mathematical statistics.

In this chapter we also got a brief glimpse of how statistical inferences can be drawn based on limited data. The ability to draw reliable statistical inferences gives us a way not only to analyze what has already taken place but also to make reasonably accurate large-scale predictions of what will happen in certain random situations. Casinos know, without any shadow of a doubt, that in the long run they will make a profit—the dishonest-coin principle is on their side! The same principle gives us the confidence to trust the results of surveys and public opinion polls (when the samples are unbiased), the quality of the products we buy, and even the statistical data our government uses to make many of its decisions. In all of these cases, bell-shaped distributions of data and the laws of probability come together to give us insight into what was, is, and most likely will be.

KEY CONCEPTS

17.1 Approximately Normal Data Sets

- **approximately normal distribution:** a distribution of data having a bar graph that is approximately bell shaped, **513**
- **normal distribution:** a distribution of data having a bell-shaped graph, **513**

17.2 Normal Curves and Normal Distributions

- **normal curve:** a perfect bell-shaped curve representing a normal distribution, **514**
- **center:** in a normal curve, the point of intersection of the data axis and the vertical axis of symmetry of the curve, **514**
- **point of inflection:** in a normal curve, the two places on the curve where the curvature changes, **515**
- **standard deviation:** in a normal distribution, the distance between a point of inflection and the axis of symmetry of the curve, **514**
- **68-95-99.7 rule:** in a normal distribution, 68% of the data fall within one standard deviation above and below the mean; 95% of the data fall within two standard deviations above and below the mean; and 99.7% of the data fall within three standard deviations above and below the mean, **515**
- **z-value:** for a given data value x in a normal distribution with mean μ and standard deviation σ, the value $z = (x - \mu)/\sigma$, **517**

17.3 Modeling Approximately Normal Distributions

No key concepts in this section.

17.4 Normality in Random Events

- **random variable:** a variable whose values depend on chance, **522**
- **the honest-coin principle:** when tossing an honest (fair) coin n times (assume $n \geq 30$), the number of heads tossed is a random variable with mean $\mu = n/2$ and standard deviation $\sigma = \sqrt{n}/2$, **524**
- **the dishonest-coin principle:** when tossing a dishonest coin (call the probability of heads p) n times (assume $n \geq 30$), the number of heads tossed is a random variable with mean $\mu = np$ and standard deviation $\sigma = \sqrt{np(1-p)}$, **524**
- **standard error:** the standard deviation of a sampling distribution expressed as a percentage of the size of the sample, **527**
- **95% confidence interval:** two standard errors below and above the statistic obtained from the sample, **527**
- **99.7% confidence interval:** three standard errors below and above the statistic obtained from the sample, **527**

EXERCISES

WALKING

17.1 Approximately Normal Data Sets

No exercises for this section.

17.2 Normal Curves and Normal Distributions

1. Consider the normal distribution represented by the normal curve in Fig. 17-12. Assume that P is a point of inflection of the curve.

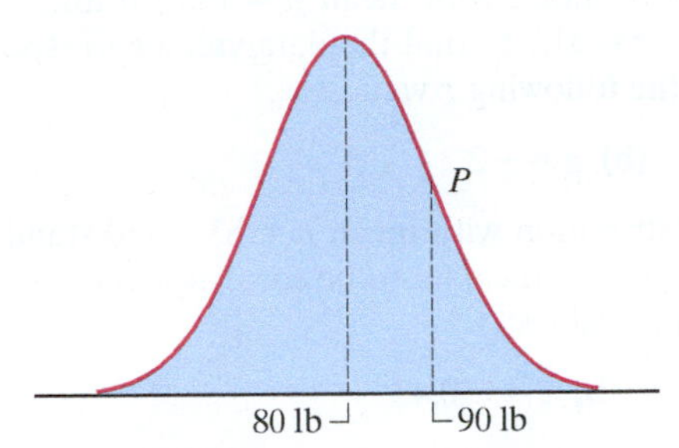

FIGURE 17-12

   (a) Find the mean μ of the distribution.

   (b) Find the median M of the distribution.

   (c) Find the standard deviation σ of the distribution.

2. Consider the normal distribution represented by the normal curve in Fig. 17-13. Assume that P is a point of inflection of the curve.

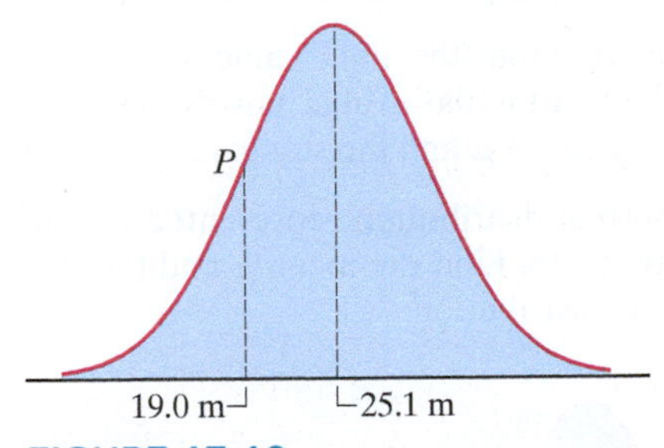

FIGURE 17-13

   (a) Find the mean μ of the distribution.

   (b) Find the median M of the distribution.

   (c) Find the standard deviation σ of the distribution.

3. Consider the normal distribution represented by the normal curve in Fig. 17-14. Assume that P and P' are the two points of inflection of the curve.

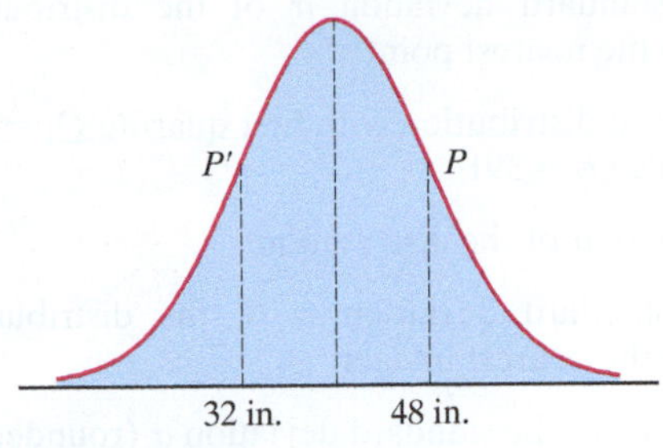

FIGURE 17-14

   (a) Find the median M of the distribution.

   (b) Find the standard deviation σ of the distribution.

   (c) Find the third quartile Q_3 of the distribution.

   (d) Find the first quartile Q_1 of the distribution.

4. Consider the normal distribution represented by the normal curve in Fig. 17-15. Assume that P and P' are the two points of inflection of the curve.

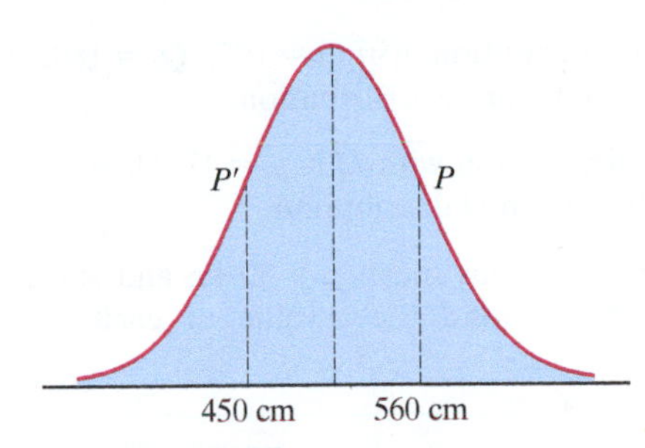

FIGURE 17-15

   (a) Find the median M of the distribution.

   (b) Find the standard deviation σ of the distribution.

   (c) Find the third quartile Q_3 of the distribution rounded to the nearest centimeter.

   (d) Find the first quartile Q_1 of the distribution rounded to the nearest centimeter.

5. Consider a normal distribution with mean $\mu = 81.2$ lb and standard deviation $\sigma = 12.4$ lb.

   (a) Find the third quartile Q_3 of the distribution rounded to the nearest tenth of a pound.

   (b) Find the first quartile Q_1 of the distribution rounded to the nearest tenth of a pound.

6. Consider a normal distribution with mean $\mu = 2354$ points and standard deviation $\sigma = 468$ points.

   (a) Find the third quartile Q_3 of the distribution rounded to the nearest point.

   (b) Find the first quartile Q_1 of the distribution rounded to the nearest point.

7. Consider a normal distribution with first quartile $Q_1 = 432.5$ points and third quartile $Q_3 = 567.5$ points.

   (a) Find the mean μ of the distribution.

   (b) Find the standard deviation σ of the distribution rounded to the nearest point.

8. Consider a normal distribution with first quartile $Q_1 = 229$ and third quartile $Q_3 = 391$.

   (a) Find the mean μ of the distribution.

   (b) Find the standard deviation σ of the distribution rounded to the nearest integer.

9. Estimate the value of the standard deviation σ (rounded to the nearest tenth of an inch) of a normal distribution with mean $\mu = 81.2$ in. and third quartile $Q_3 = 94.7$ in.

10. Estimate the value of the standard deviation σ (rounded to the nearest dollar) of a normal distribution with mean $\mu = \$18{,}565$ and first quartile $Q_1 = \$15{,}514$.

11. Explain why a distribution with median $M = 82$, mean $\mu = 71$, and standard deviation $\sigma = 11$ cannot be a normal distribution.

12. Explain why a distribution with median $M = 453$, mean $\mu = 453$, first quartile $Q_1 = 343$, and third quartile $Q_3 = 553$ cannot be a normal distribution.

13. Explain why a distribution with $\mu = 195$, $Q_1 = 180$, and $Q_3 = 220$ cannot be a normal distribution.

14. Explain why a distribution with $M = \mu = 47$, $Q_1 = 35$, and $\sigma = 10$ cannot be a normal distribution.

15. A normal distribution has mean $\mu = 30$ kg and standard deviation $\sigma = 15$ kg. Find the z-value of each of the following:

   (a) $x = 45$ kg

   (b) $x = 0$ kg

   (c) $x = 54$ kg

   (d) $x = 3$ kg

16. A normal distribution has mean $\mu = 110$ points and standard deviation $\sigma = 12$ points. Find the z-value of each of the following:

   (a) $x = 98$ points.

   (b) $x = 110$ points.

   (c) $x = 128$ points.

   (d) $x = 71$ points.

17. A normal distribution has mean $\mu = 310$ points and third quartile $Q_3 = 391$ points. Find the z-value of each of the following:

   (a) $x = 490$ points.

   (b) $x = 250$ points.

   (c) $x = 220$ points.

   (d) $x = 442$ points.

18. A normal distribution has mean $\mu = 49.5$ lb and first quartile $Q_1 = 44.1$ lb. Find the z-value of each of the following.

   (a) $x = 41.5$ lb

   (b) $x = 61.5$ lb

   (c) $x = 35.1$ lb

   (d) $x = 67.5$ lb

19. In a normal distribution with mean $\mu = 183.5$ ft and standard deviation $\sigma = 31.2$ ft, find the data value corresponding to each of the following z values:

   (a) $z = 0$ (b) $z = -1$

20. In a normal distribution with mean $\mu = 83.2$ and standard deviation $\sigma = 4.6$, find the data value corresponding to each of the following z values:

   (a) $z = 0$ (b) $z = 2$

21. In a normal distribution with mean $\mu = 183.5$ ft and standard deviation $\sigma = 31.2$ ft, find the data value corresponding to each of the following z values:

   (a) $z = 0.5$ (b) $z = -2.3$

22. In a normal distribution with mean $\mu = 83.2$ and standard deviation $\sigma = 4.6$, find the data value corresponding to each of the following z values:

   (a) $z = -1.5$ (b) $z = -0.43$

23. In a normal distribution with mean $\mu = 50$ lb, a weight of $x = 84$ lb. has a z-value of 2. Find the standard deviation σ.

24. In a normal distribution with mean $\mu = 30$, the data value $x = -6$ has a z-value of -3. Find the standard deviation σ.

25. In a normal distribution with standard deviation $\sigma = 15$, the data value $x = 50$ has a z-value of 3. Find the mean μ.

26. In a normal distribution with standard deviation $\sigma = 20$, the data value $x = 10$ has a z-value of $z = -2$. Find the mean μ.

27. In a normal distribution, the data value $x_1 = 20$ has a z-value (z_1) of -2 and the data value $x_2 = 100$ has a z-value (z_2) of 3. Find the mean μ and the standard deviation σ.

28. In a normal distribution, the data value $x_1 = -10$ has a z-value (z_1) of 0 and the data value $x_2 = 50$ has a z-value (z_2) of 2. Find the mean μ and the standard deviation σ.

29. Consider the normal distribution represented by the normal curve in Fig. 17-16. Find the mean μ and the standard deviation σ of the distribution.

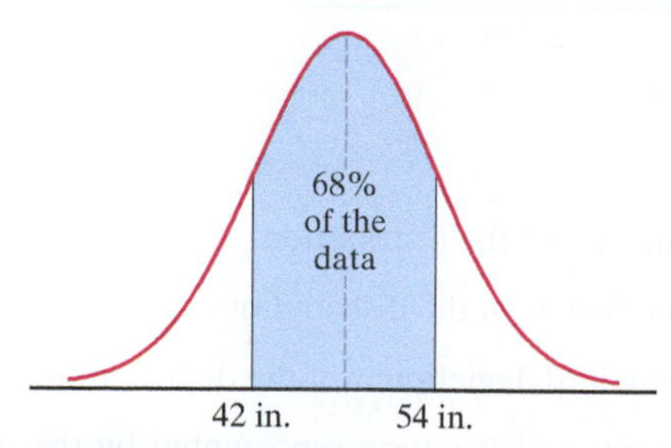

FIGURE 17-16

30. Consider the normal distribution represented by the normal curve in Fig. 17-17. Find the mean μ and the standard deviation σ of the distribution.

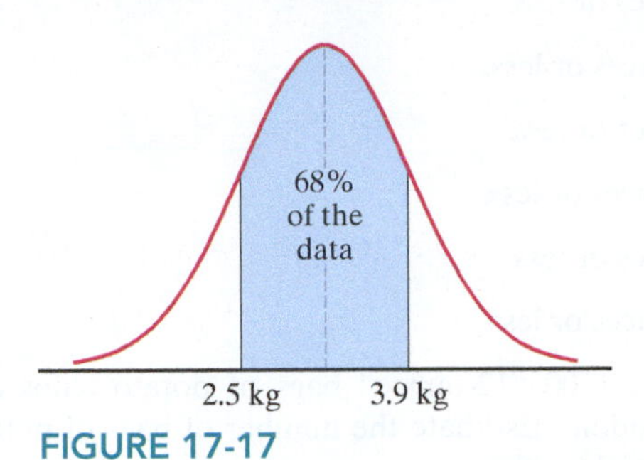

FIGURE 17-17

31. Consider the normal distribution represented by the normal curve in Fig. 17-18.

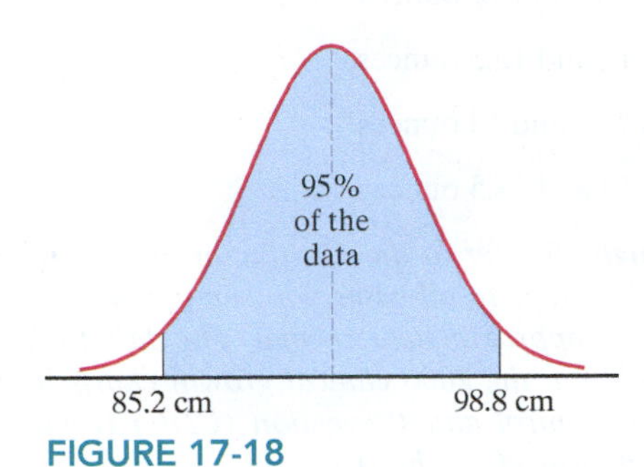

FIGURE 17-18

(a) Find the mean μ and the standard deviation σ of the distribution.

(b) Find the first quartile Q_1 and the third quartile Q_3 of the distribution rounded to the nearest tenth of a centimeter.

32. Consider the normal distribution defined by Fig. 17-19.

(a) Find the mean μ and the standard deviation σ of the distribution.

(b) Find the first quartile Q_1 and the third quartile Q_3 of the distribution rounded to the nearest tenth of an ounce.

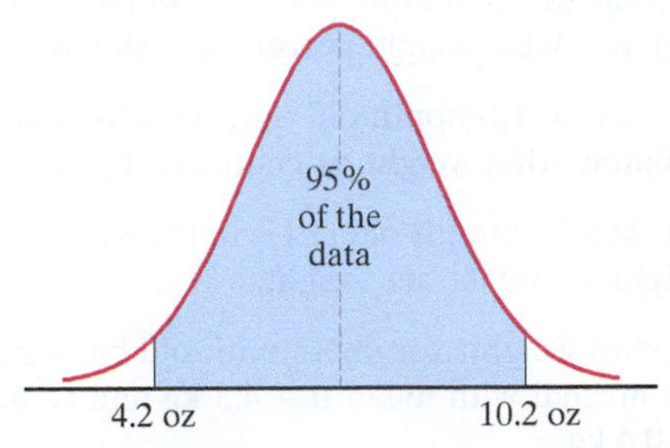

FIGURE 17-19

33. A normal distribution has mean $\mu = 71.5$ in., and the 16th percentile of the distribution is $P_{16} = 65.2$ in.

(a) Find the standard deviation σ.

(b) Find the 84th percentile P_{84}.

34. A normal distribution has standard deviation $\sigma = 12.3$ points and the 84th percentile of the distribution is $P_{84} = 66.7$ points.

(a) Find the median μ.

(b) Find the 16th percentile P_{16}.

35. A normal distribution has standard deviation $\sigma = 6.1$ cm, and the 97.5th percentile of the distribution is $P_{97.5} = 81.5$ cm. Find the mean μ.

36. A normal distribution has mean $\mu = 56.3$ cm, and the 2.5th percentile of the distribution is $P_{2.5} = 33.7$ cm. Find the standard deviation σ.

37. A normal distribution has mean $\mu = 12.6$ and standard deviation $\sigma = 4.0$. Approximately what percent of the data fall between 9.9 and 16.6?

38. A normal distribution has mean $\mu = 500$ and standard deviation $\sigma = 35$. Approximately what percent of the data fall between 465 and 605?

39. In a normal distribution, what percent of the data have z-values satisfying

(a) $z \leq 2$?

(b) $1 \leq z \leq 2$?

40. In a normal distribution, what percent of the data have z-values satisfying

(a) $z \geq -3$

(b) $-3 \leq z \leq 2$

17.3 Modeling Approximately Normal Distributions

Exercises 41 through 44 refer to the following: 2500 students take a college entrance exam. The scores on the exam have an approximately normal distribution with mean $\mu = $ 52 points and standard deviation $\sigma = 11$ points.

41. (a) Estimate the average score on the exam.

(b) Estimate the percentage of students scoring 52 points or more.

(c) Estimate the percentage of students scoring between 41 and 63 points.

(d) Estimate the percentage of students scoring 63 points or more.

42. (a) Estimate the percentage of students scoring between 30 and 74 points.

(b) Estimate the percentage of students scoring 30 points or less.

(c) Estimate the percentage of students scoring 19 points or less.

43. (a) Estimate the first-quartile score for the exam.

(b) Estimate the third-quartile score for the exam.

44. For each score, estimate the percentile corresponding to the score.

(a) 52 points

(b) 63 points

(c) 60 points

(d) 85 points

Exercises 45 through 48 refer to the following: As part of a research project, the blood pressures of 2000 patients in a hospital are recorded. The systolic blood pressures (given in millimeters) have an approximately normal distribution with mean $\mu = 125$ and standard deviation $\sigma = 13$.

45. (a) Estimate the number of patients whose blood pressure was between 99 and 151 millimeters.

(b) Estimate the number of patients whose blood pressure was between 112 and 151 millimeters.

46. (a) Estimate the number of patients whose blood pressure was 99 millimeters or higher.

(b) Estimate the number of patients whose blood pressure was between 99 and 138 millimeters.

47. For each of the following blood pressures, estimate the corresponding percentile.

(a) 112 millimeters

(b) 138 millimeters

(c) 164 millimeters

48. (a) Assuming that there were no outliers, estimate the value of the lowest (*Min*) and the highest (*Max*) blood pressures.

(b) Assuming that there were no outliers, give an estimate of the five-number summary (*Min*, Q_1, μ, Q_3, *Max*) of the distribution of blood pressures.

Exercises 49 through 52 refer to the following: Packaged foods sold at supermarkets are not always the weight indicated on the package. Variability always crops up in the manufacturing and packaging process. Suppose that the exact weight of a "12-ounce" bag of potato chips is a random variable that has an approximately normal distribution with mean $\mu = 12$ ounces and standard deviation $\sigma = 0.5$ ounce.

49. If a "12-ounce" bag of potato chips is chosen at random, what are the chances that

(a) it weighs somewhere between 11 and 13 ounces?

(b) it weighs somewhere between 12 and 13 ounces?

(c) it weighs more than 11 ounces?

50. If a "12-ounce" bag of potato chips is chosen at random, what are the chances that

(a) it weighs somewhere between 11.5 and 12.5 ounces?

(b) it weighs somewhere between 12 and 12.5 ounces?

(c) it weighs more than 12.5 ounces?

51. Suppose that 500 "12-ounce" bags of potato chips are chosen at random. Estimate the number of bags with weight

(a) 11 ounces or less.

(b) 11.5 ounces or less.

(c) 12 ounces or less.

(d) 12.5 ounces or less.

(e) 13 ounces or less.

(f) 13.5 ounces or less.

52. Suppose that 1500 "12-ounce" bags of potato chips are chosen at random. Estimate the number of bags of potato chips with weight

(a) between 11 and 11.5 ounces.

(b) between 11.5 and 12 ounces.

(c) between 12 and 12.5 ounces.

(d) between 12.5 and 13 ounces.

(e) between 13 and 13.5 ounces.

Exercises 53 through 56 refer to the distribution of weights for infants by age and gender. In all cases you can assume that the weight distribution is approximately normal. The data for these exercises are taken from the 2000 clinical growth charts by the Centers of Disease Control and Prevention (CDC) (www.cdc.gov/growthcharts/clinical_charts.htm).

53. The distribution of weights for six-month-old baby boys has mean $\mu = 8.16$ kg and standard deviation $\sigma = 0.95$ kg.

(a) Suppose that a six-month-old boy weighs 7.21 kg. Approximately what weight percentile is he in?

(b) Suppose that a six-month-old boy weighs 10 kg. Approximately what weight percentile is he in?

(c) Suppose that a six-month-old boy is in the 75th percentile in weight. Estimate his weight to the nearest tenth of a kilogram.

54. The distribution of weights for 12-month-old baby girls has mean $\mu = 9.7$ kg and standard deviation $\sigma = 1.0$ kg.

(a) Suppose that a 12-month-old girl weighs 7.7 kg. Approximately what weight percentile is she in?

(b) Suppose that a 12-month-old girl weighs 10.4 kg. Approximately what weight percentile is she in?

(c) Suppose that a 12-month-old girl is in the 84th percentile in weight. Estimate her weight.

55. The distribution of weights for one-month-old baby girls is approximately normal with mean $\mu = 4.5$ kg and standard deviation $\sigma = 0.4$ kg.

(a) Suppose that a one-month-old girl weighs 5.3 kg. Approximately what weight percentile is she in?

(b) Suppose that a one-month-old girl weighs 5.7 kg. Approximately what weight percentile is she in?

(c) Suppose that a one-month-old girl is in the 25th percentile in weight. Estimate her weight.

56. The distribution of weights for 12-month-old baby boys has mean $\mu = 10.5$ kg and standard deviation $\sigma = 1.2$ kg.

(a) Suppose that a 12-month-old boy weighs 11.3 kg. Approximately what weight percentile is he in?

(b) Suppose that a 12-month-old boy weighs 8.1 kg. Approximately what weight percentile is he in?

(c) Suppose that a 12-month-old boy is in the 84th percentile in weight. Estimate his weight.

17.4 Normality in Random Events

57. An honest coin is tossed $n = 3600$ times. Let the random variable Y denote the number of tails tossed.

(a) Find the mean and the standard deviation of the distribution of the random variable Y.

(b) Estimate the chances that Y will fall somewhere between 1770 and 1830.

(c) Estimate the chances that Y will fall somewhere between 1800 and 1830.

(d) Estimate the chances that Y will fall somewhere between 1830 and 1860.

58. An honest coin is tossed $n = 6400$ times. Let the random variable X denote the number of heads tossed.

(a) Find the mean and the standard deviation of the distribution of the random variable X.

(b) Estimate the chances that X will fall somewhere between 3120 and 3280.

(c) Estimate the chances that X will fall somewhere between 3080 and 3200.

(d) Estimate the chances that X will fall somewhere between 3240 and 3280.

59. Suppose that a random sample of $n = 7056$ adults is to be chosen for a survey. Assume that the gender of each adult in the sample is equally likely to be male as it is female. Estimate the probability that

(a) the number of females in the sample is between 3486 and 3570.

(b) the number of females in the sample is less than 3486.

(c) the percentage of females in the sample is below 50.6%. (*Hint*: Find 50.6% of 7056 first.)

60. An honest die is rolled. If the roll comes out even (2, 4, or 6), you will win \$1; if the roll comes out odd (1, 3, or 5), you will lose \$1. Suppose that in one evening you play this game $n = 2500$ times in a row.

(a) Estimate the probability that by the end of the evening you will not have lost any money.

(b) Estimate the probability that the number of "even rolls" (roll a 2, 4, or 6) will fall between 1250 and 1300.

(c) Estimate the probability that you will win \$100 or more.

61. A dishonest coin with probability of heads $p = 0.4$ is tossed $n = 600$ times. Let the random variable X represent the number of times the coin comes up heads.

(a) Find the mean and standard deviation for the distribution of X.

(b) Find the first and third quartiles for the distribution of X.

(c) Find the probability that the number of heads will fall somewhere between 216 and 264.

62. A dishonest coin with probability of heads $p = 0.75$ is tossed $n = 1200$ times. Let the random variable X represent the number of times the coin comes up heads.

(a) Find the mean and standard deviation for the distribution of X.

(b) Find the first and third quartiles for the distribution of X.

(c) Find the probability that the number of heads will fall somewhere between 900 and 945.

63. Suppose that an honest die is rolled $n = 180$ times. Let the random variable X represent the number of times the number 6 is rolled.

(a) Find the mean and standard deviation for the distribution of X. (*Hint:* Use the dishonest-coin principle with $p = 1/6$ to find μ and σ.)

(b) Find the probability that a 6 will be rolled more than 40 times.

(c) Find the probability that a 6 will be rolled somewhere between 30 and 35 times.

64. Suppose that 1 out of every 10 plasma televisions shipped has a defective speaker. Out of a shipment of $n = 400$ plasma televisions, find the probability that there are

(a) at most 40 with defective speakers. (*Hint:* Use the dishonest-coin principle with $p = 1/10 = 0.1$ to find μ and σ.)

(b) more than 52 with defective speakers.

JOGGING

65. Find the z-value of the median of a normal distribution.

66. Find the z-value of

(a) the first quartile of a normal distribution.

(b) the third quartile of a normal distribution.

Percentiles. *The* ***pth percentile*** *of a sorted data set is a number x_p such that p% of the data fall at or below x_p and $(100 - p)$% of the data fall at or above x_p. (For details, see Chapter 15, Section 15.2.) For normally distributed data sets, there are detailed statistical tables that give the location of the pth percentile for every possible p between 1 and 99. The following table is an abbreviated version giving the approximate location of some of the more frequently used percentiles in a normal distribution with mean μ and standard deviation σ. For approximately normal distributions, Table 17-9 can be used to estimate these percentiles.*

Percentile	Approximate location	Percentile	Approximate location
99th	$\mu + 2.33\sigma$	1st	$\mu - 2.33\sigma$
95th	$\mu + 1.65\sigma$	5th	$\mu - 1.65\sigma$
90th	$\mu + 1.28\sigma$	10th	$\mu - 1.28\sigma$
80th	$\mu + 0.84\sigma$	20th	$\mu - 0.84\sigma$
75th	$\mu + 0.675\sigma$	25th	$\mu - 0.675\sigma$
70th	$\mu + 0.52\sigma$	30th	$\mu - 0.52\sigma$
60th	$\mu + 0.25\sigma$	40th	$\mu - 0.25\sigma$
50th	μ		

TABLE 17-9

In Exercises 67 through 73, you should use the table to make your estimates.

67. The distribution of weights for six-month-old baby boys has mean $\mu = 8.16$ kg and standard deviation $\sigma = 0.95$ kg (see Exercise 53).

(a) Suppose that a six-month-old baby boy weighs in the 95th percentile of his age group. Estimate his weight in kilograms rounded to two decimal places.

(b) Suppose that a six-month-old baby boy weighs in the 20th percentile of his age group. Estimate his weight in kilograms rounded to two decimal places.

68. Several thousand students took a college entrance exam. The scores on the exam have an approximately normal distribution with mean $\mu = 55$ points and standard deviation $\sigma = 12$ points.

(a) For a student who scored in the 99th percentile, estimate the student's score on the exam.

(b) For a student who scored in the 30th percentile, estimate the student's score on the exam.

69. Consider again the distribution of weights of six-month-old baby boys discussed in Exercise 67.

(a) Jimmy is a six-month-old baby who weighs 8.4 kg. Estimate the percentile corresponding to Jimmy's weight.

(b) David is a six-month-old baby who weighs $7\frac{2}{3}$ kg. Estimate the percentile corresponding to David's weight.

70. Consider again the college entrance exam discussed in Exercise 68.

(a) Mary scored 83 points on the exam. Estimate the percentile in which this score places her.

(b) Adam scored 45 points on the exam. Estimate the percentile in which this score places him.

(c) If there were 250 students that scored 35 points or less on the exam, estimate the total number of students who took the exam.

71. In 2015, a total of 1,698,521 college-bound seniors took the SAT exam. The distribution of scores in the Mathematics section of the SAT was approximately normal with mean $\mu = 511$ and standard deviation $\sigma = 120$. Data from SAT Data Tables, 2015, The College Board.

(a) Estimate the third quartile score on the exam. (*Note*: SAT scores come in multiples of 10.)

(b) Estimate the 84th percentile score on the exam. (*Hint*: Use the 68-95-99.7 rule.)

(c) Estimate the 70th percentile score on the exam. (*Hint*: Use Table 17-9.)

72. Consider a normal distribution with mean $\mu = 0$ and standard deviation $\sigma = 1$.

(a) Find the 90th percentile (rounded to two decimal places).

(b) Find the 10th percentile (rounded to two decimal places).

(c) Find the 80th percentile (rounded to two decimal places).

(d) Find the 20th percentile (rounded to two decimal places).

(e) Suppose that you are given that the 85th percentile is approximately 1.04. Find the 15th percentile.

73. The grade breakdown in Dr. Blackbeard's Stat 101 class is: 10% A's, 20% B's, 40% C's, 25% D's, and 5% F's. The numerical class scores had an approximately normal distribution with mean $\mu = 65.2$ and standard deviation $\sigma = 10$.

(a) What is the minimum numerical score needed to get an A?

(b) What is the minimum numerical score needed to get a B?

(c) What is the minimum numerical score needed to get a C?

(d) What is the minimum numerical score needed to get a D?

74. Explain why when the dishonest-coin principle is applied to an honest coin ($p = 1/2$), we get the honest-coin principle.

RUNNING

75. An honest coin is tossed n times. Let the random variable Y denote the number of tails tossed. Find the value of n so that there is a 16% chance that Y will be at least $(n/2) + 10$.

76. An honest coin is tossed n times. Let the random variable X denote the number of heads tossed. Find the value of n so that there is a 95% chance that X will be between $(n/2) - 15$ and $(n/2) + 15$.

77. A dishonest coin with probability of heads $p = 0.1$ is tossed n times. Let the random variable X denote the number of heads tossed. Find the value of n so that there is a 95% chance that X will be between $(n/10) - 30$ and $(n/10) + 30$.

78. An honest pair of dice is rolled n times. Let the random variable Y denote the number of times a total of 7 is rolled. Find the value of n so that there is a 95% chance that Y will be between $(n/6) - 20$ and $(n/6) + 20$.

79. In American roulette there are 18 red numbers, 18 black numbers, and 2 white numbers (0 and 00) as illustrated in Fig. 17-20. The probability of a red number coming up on a single play of roulette is $p = 18/38 \approx 0.47$. Suppose that we go on a binge and bet \$1 on red 10,000 times in a row. (A \$1 bet on red wins \$1 if a red number comes up, if a black or white number comes up, the bettor loses \$1.)

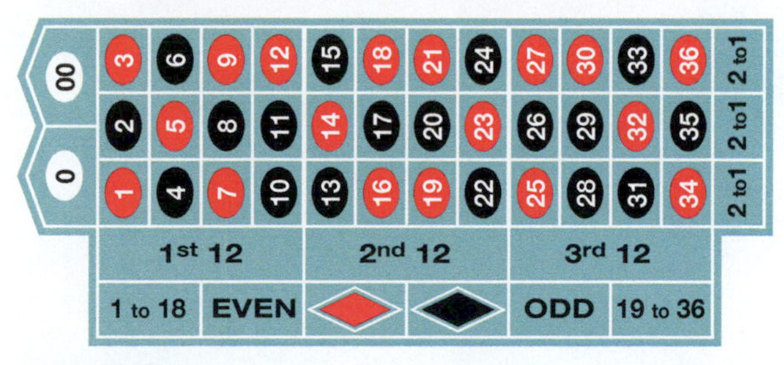

FIGURE 17-20

(a) Let Y represent the number of times we lose (i.e., the number of times that red does not come up). Use the dishonest-coin principle to describe the distribution of the random variable Y.

(b) Approximately what are the chances that we will lose 5300 times or more?

(c) Approximately what are the chances that we will lose somewhere between 5150 and 5450 times?

(d) Explain why the chances that we will break even or win in this situation are essentially zero.

80. After polling a random sample of 800 voters during the most recent gubernatorial race, the *Tasmania Gazette* reports the following:

> As the race for governor of Tasmania heads into its final days, our most recent poll shows Mrs. Butterworth ahead of the incumbent Mrs. Cubbison by 6 percentage points—53% to 47%. The results of the poll indicate with near certainty that if the election had been held at the time the poll was taken, Mrs. Butterworth would be the next governor of Tasmania.

(a) Estimate the standard error for this poll.

(b) Compute a 95% confidence interval for this poll.

(c) Compute a 99.7% confidence interval for this poll.

ANSWERS TO SELECTED EXERCISES

CHAPTER 1 The Mathematics of Elections

WALKING

1.1 Ballots and Preference Schedules

1.

Number of voters	5	3	5	3	2	3
1st	*A*	*A*	*C*	*D*	*D*	*B*
2nd	*B*	*D*	*E*	*C*	*C*	*E*
3rd	*C*	*B*	*D*	*B*	*B*	*A*
4th	*D*	*C*	*A*	*E*	*A*	*C*
5th	*E*	*E*	*B*	*A*	*E*	*D*

3. **(a)** 21 **(b)** 11 **(c)** *C*

5.

	37	36	24	13	5
1st	*B*	*A*	*B*	*E*	*C*
2nd	*E*	*B*	*A*	*B*	*E*
3rd	*A*	*D*	*D*	*C*	*A*
4th	*C*	*C*	*E*	*A*	*D*
5th	*D*	*E*	*C*	*D*	*B*

7.

	14	10	8	7	4
Ana	2	3	1	5	3
Bob	1	1	2	3	2
Cat	5	5	5	2	4
Dee	4	2	4	1	5
Eli	3	4	3	4	1

9.

Number of voters	255	480	765
1st	*L*	*C*	*M*
2nd	*M*	*M*	*L*
3rd	*C*	*L*	*C*

1.2 Plurality Method

11. **(a)** *C* **(b)** *C*, *B*, *A*, *D*

13. **(a)** *C* **(b)** *C*, *B*, *A*, *D*

15. **(a)** *D* **(b)** *D*, *C*, *B*, *A*, *E*

17. **(a)** *A* **(b)** *A*, *C*, *D*, *B*, *E*

19. **(a)** *A* **(b)** *A*, *C*, *D*, *B*, *E*

1.3 Borda Count

21. **(a)** *B* **(b)** *B*, *D*, *A*, *C*

23. **(a)** *C* **(b)** *C*, *D*, *B*, *A* **25.** *A*, *C*, *D*, *B*, *E*

27. Winner: Marcus Mariota (2534 points); 2nd place: Melvin Gordon (1250 points); 3rd place: Amari Cooper (1023 points).

29. *A*, *D*, *B*, *C* (*D* had 310 points).

1.4 Plurality with Elimination

31. **(a)** *A* **(b)** *A*, *C*, *B*, *D*

33. **(a)** *C* **(b)** *C*, *B*, *A*, *D*

35. *B*, *C*, *A*, *D*, *E*

37. **(a)** *D* **(b)** *D*, *C*, *B*, *A*, *E*

39. *A*

1.5 Pairwise Comparisons

41. **(a)** *D* **(b)** *D*, *B*, *A*, *C*

43. **(a)** *C* **(b)** *C*, *D*, *B*, *A*

45. *D*

47. *A*, *B*, *C*, *E*, *D* (*B*, *C*, and *E* have two points each, but in their head-to-head comparisons *B* beats *C* and *E*, and *C* beats *E*)

49. **(a)** 3 **(b)** *C*

1.6 Fairness Criteria

51. The winner under the Borda count method is *B*; the Condorcet candidate is *A*.

53. The winner under plurality is *R*; when *F* is eliminated *H* becomes the winner.

55. The winner under plurality-with-elimination is *A*; when *C* is eliminated *E* becomes the winner.

57. If *X* is the Condorcet candidate, then by definition *X* wins every pairwise comparison and is, therefore, the winner under the method of pairwise comparisons.

59. When a voter moves a candidate up in his or her ballot that candidate's Borda points increase. It follows that if *X* had the most Borda points and a voter changes his or her ballot to rank *X* higher, then *X* still has the most Borda points.

JOGGING

61. Suppose that *A* gets *a* first-place votes and *B* gets *b* first-place votes, where $a > b$. It is clear that candidate *A* wins the election under the plurality method, the plurality-with-elimination method, and the method of pairwise comparisons. Under the Borda count method, *A* gets $2a + b$ points while *B* gets $2b + a$ points. Since $a > b$, $2a + b > 2b + a$, and so again *A* wins the election.

63. The number of points under this variation is complementary to the number of points under the standard Borda count method: A first place is worth 1 point instead of *N*, a second place is worth 2 points instead of $N - 1, \ldots,$ a last place is worth *N* points instead of 1. It follows that having the fewest points here is equivalent to having the most points under the standard Borda count method, having the second fewest is equivalent to having the second most, and so on.

65. **(a)** 65

(b) 34 second-place votes and 31 third-place votes

(c) 31 second-place votes and 34 third-place votes

67. 5 points for each first-place vote, 3 points for each second-place vote, 1 point for each third-place vote

69. **(a)** C

(b) *A* is a Condorcet candidate but is eliminated in the first round.

Number of voters	10	6	6	3	3
1st	*B*	*A*	*A*	*D*	*C*
2nd	*C*	*B*	*C*	*A*	*A*
3rd	*D*	*D*	*B*	*C*	*B*
4th	*A*	*C*	*D*	*B*	*D*

(c) *B* wins under the Coombs method. However, if 8 voters move *B* from their third choice to their second choice, then *C* wins.

Number of voters	12	8	7	2	2
1st	*B*	*C*	*C*	*A*	*A*
2nd	*A*	*A*	*B*	*B*	*C*
3rd	*C*	*B*	*A*	*C*	*B*

CHAPTER 2 The Mathematics of Power

WALKING

2.1 Weighted Voting

1. **(a)** $[26: 15, 12, 10, 10, 3]$ **(b)** $[34: 15, 12, 10, 10, 3]$

3. **(a)** 11 **(b)** 20 **(c)** 15 **(d)** 16

5. **(a)** None **(b)** P_1 **(c)** P_1, P_2, P_3 **(d)** P_1, P_2, P_3, P_4

7. **(a)** 13 **(b)** 11

9. **(a)** $[49: 48, 24, 12, 12]$ **(b)** $[49: 36, 18, 9, 9]$
(c) $[49: 32, 16, 8, 8]$

2.2 Banzhaf Power

11. **(a)** 10 **(b)** 8, 9, 10 **(c)** 11, 12, 13, 14, 15

13. $\beta_1 = \frac{3}{8}; \beta_2 = \frac{3}{8}; \beta_3 = \frac{1}{8}; \beta_4 = \frac{1}{8}$

15. **(a)** P_1, P_2
(b) $\{P_1, P_2\}, \{P_1, P_3\}, \{P_1, P_2, P_3\}, \{P_1, P_2, P_4\}, \{P_1, P_3, P_4\}, \{P_2, P_3, P_4\}, \{P_1, P_2, P_3, P_4\}$
(c) $\beta_1 = \frac{5}{12}; \beta_2 = \frac{3}{12} = \frac{1}{4}; \beta_3 = \frac{3}{12} = \frac{1}{4}; \beta_4 = \frac{1}{12}$

17. **(a)** $\beta_1 = \frac{3}{5}; \beta_2 = \frac{1}{5}; \beta_3 = \frac{1}{5}$
(b) $\beta_1 = \frac{3}{5}; \beta_2 = \frac{1}{5}; \beta_3 = \frac{1}{5}$. The Banzhaf power distributions in (a) and (b) are the same.

19. **(a)** $\beta_1 = \frac{8}{24}; \beta_2 = \frac{6}{24}; \beta_3 = \frac{4}{24}; \beta_4 = \frac{4}{24}; \beta_5 = \frac{2}{24}$
(b) $\beta_1 = \frac{7}{19}; \beta_2 = \frac{5}{19}; \beta_3 = \frac{3}{19}; \beta_4 = \frac{3}{19}; \beta_5 = \frac{1}{19}$
(c) $\beta_1 = \frac{5}{15}; \beta_2 = \frac{5}{15}; \beta_3 = \frac{3}{15}; \beta_4 = \frac{1}{15}; \beta_5 = \frac{1}{15}$
(d) $\beta_1 = \frac{1}{5}; \beta_2 = \frac{1}{5}; \beta_3 = \frac{1}{5}; \beta_4 = \frac{1}{5}; \beta_5 = \frac{1}{5}$

21. **(a)** $\{\underline{P_1}, \underline{P_2}\}, \{\underline{P_1}, \underline{P_3}\}, \{\underline{P_1}, P_2, P_3\}$
(b) $\beta_1 = \frac{3}{5}; \beta_2 = \frac{1}{5}; \beta_3 = \frac{1}{5}$

23. **(a)** $\{\underline{P_1}, \underline{P_2}\}, \{\underline{P_1}, \underline{P_3}\}, \{\underline{P_2}, \underline{P_3}\}, \{P_1, P_2, P_3\}, \{\underline{P_1}, \underline{P_2}, P_4\}, \{\underline{P_1}, \underline{P_2}, P_5\}, \{\underline{P_1}, \underline{P_2}, P_6\}, \{\underline{P_1}, \underline{P_3}, P_4\}, \{\underline{P_1}, \underline{P_3}, P_5\}, \{\underline{P_1}, \underline{P_3}, P_6\}, \{\underline{P_2}, \underline{P_3}, P_4\}, \{\underline{P_2}, \underline{P_3}, P_5\}, \{\underline{P_2}, \underline{P_3}, P_6\}$
(b) $\{\underline{P_1}, \underline{P_2}, P_4\}, \{\underline{P_1}, \underline{P_3}, P_4\}, \{\underline{P_2}, \underline{P_3}, P_4\}, \{P_1, P_2, P_3, P_4\}, \{\underline{P_1}, \underline{P_2}, P_4, P_5\}, \{\underline{P_1}, \underline{P_2}, P_4, P_6\}, \{\underline{P_1}, \underline{P_3}, P_4, P_5\}, \{\underline{P_1}, \underline{P_3}, P_4, P_6\}, \{\underline{P_2}, \underline{P_3}, P_4, P_5\}, \{\underline{P_2}, \underline{P_3}, P_4, P_6\}, \{P_1, P_2, P_3, P_4, P_5\}, \{P_1, P_2, P_3, P_4, P_6\}, \{\underline{P_1}, \underline{P_2}, P_4, P_5, P_6\}, \{\underline{P_1}, \underline{P_3}, P_4, P_5, P_6\}, \{\underline{P_2}, \underline{P_3}, P_4, P_5, P_6\}, \{P_1, P_2, P_3, P_4, P_5, P_6\}$
(c) $\beta_4 = 0$
(d) $\beta_1 = \frac{1}{3}; \beta_2 = \frac{1}{3}; \beta_3 = \frac{1}{3}; \beta_4 = 0; \beta_5 = 0; \beta_6 = 0$

25. **(a)** $\{\underline{A}, \underline{B}\}, \{\underline{A}, \underline{C}\}, \{\underline{B}, \underline{C}\}, \{A, B, C\}, \{\underline{A}, \underline{B}, D\}, \{\underline{A}, \underline{C}, D\}$ $\{\underline{B}, \underline{C}, D\}, \{A, B, C, D\}$
(b) A, B, and C have Banzhaf power index of $\frac{4}{12}$ each; D is a dummy.

2.3 Shapley-Shubik Power

27. $\sigma_1 = \frac{10}{24}; \sigma_2 = \frac{10}{24}; \sigma_3 = \frac{2}{24}; \sigma_4 = \frac{2}{24}$

29. **(a)** $\langle P_1, \underline{P_2}, P_3\rangle, \langle P_1, \underline{P_3}, P_2\rangle, \langle P_2, \underline{P_1}, P_3\rangle, \langle P_2, P_3, \underline{P_1}\rangle, \langle P_3, \underline{P_1}, P_2\rangle, \langle P_3, P_2, \underline{P_1}\rangle$
(b) $\sigma_1 = \frac{4}{6}; \sigma_2 = \frac{1}{6}; \sigma_3 = \frac{1}{6}$

31. **(a)** $\sigma_1 = 1; \sigma_2 = 0; \sigma_3 = 0; \sigma_4 = 0$
(b) $\sigma_1 = \frac{4}{6}; \sigma_2 = \frac{1}{6}; \sigma_3 = \frac{1}{6}; \sigma_4 = 0$
(c) $\sigma_1 = \frac{1}{2}; \sigma_2 = \frac{1}{2}; \sigma_3 = 0; \sigma_4 = 0$
(d) $\sigma_1 = \frac{1}{3}; \sigma_2 = \frac{1}{3}; \sigma_3 = \frac{1}{3}; \sigma_4 = 0$

33. **(a)** $\sigma_1 = \frac{10}{24}; \sigma_2 = \frac{6}{24}; \sigma_3 = \frac{6}{24}; \sigma_4 = \frac{2}{24}$
(b) $\sigma_1 = \frac{10}{24}; \sigma_2 = \frac{6}{24}; \sigma_3 = \frac{6}{24}; \sigma_4 = \frac{2}{24}$
(c) $\sigma_1 = \frac{10}{24}; \sigma_2 = \frac{6}{24}; \sigma_3 = \frac{6}{24}; \sigma_4 = \frac{2}{24}$

35. **(a)** $\langle P_1, \underline{P_2}, P_3\rangle, \langle P_1, \underline{P_3}, P_2\rangle, \langle P_2, \underline{P_1}, P_3\rangle, \langle P_2, P_3, \underline{P_1}\rangle, \langle P_3, \underline{P_1}, P_2\rangle, \langle P_3, P_2, \underline{P_1}\rangle$
(b) $\sigma_1 = \frac{4}{6}; \sigma_2 = \frac{1}{6}; \sigma_3 = \frac{1}{6}$

37. $\sigma_1 = \frac{10}{24}; \sigma_2 = \frac{6}{24}; \sigma_3 = \frac{6}{24}; \sigma_4 = \frac{2}{24}$

2.4 Subsets and Permutations

39. **(a)** $2^{10} = 1024$ **(b)** 1023 **(c)** 10 **(d)** 1013

41. **(a)** 1023 **(b)** 1013

43. **(a)** 63 **(b)** 31 **(c)** 31 **(d)** 15 **(e)** 16

45. **(a)** 6,227,020,800 **(b)** 6,402,373,705,728,000
(c) 15,511,210,043,330,985,984,000,000 **(d)** 491,857 years

47. **(a)** 1,037,836,800 **(b)** 286 **(c)** 715 **(d)** 1287

49. **(a)** 362,880 **(b)** 11 **(c)** 110 **(d)** 504 **(e)** 10,100

51. **(a)** $7! = 5040$ **(b)** $6! = 720$ **(c)** $6! = 720$ **(d)** 4320

53. **(a)** 4320 **(b)** $\frac{4320}{5040} = \frac{6}{7}$
(c) $\sigma_1 = \frac{6}{7}; \sigma_2 = \sigma_3 = \sigma_4 = \sigma_5 = \sigma_6 = \sigma_7 = \frac{1}{42}$

JOGGING

55. **(a)** Suppose that a winning coalition that contains P is not a winning coalition without P. Then P would be a critical player in that coalition, contradicting the fact that P is a dummy.
(b) P is a dummy $\Leftrightarrow$ P is never critical $\Leftrightarrow$ the numerator of its Banzhaf power index is 0 $\Leftrightarrow$ its Banzhaf power index is 0.
(c) Suppose that P is not a dummy. Then P is critical in some winning coalition. Let S denote the other players in that winning coalition. The sequential coalition with the players in S first (in any order) followed by P and then followed by the remaining players has P as its pivotal player. Thus, P's Shapley-Shubik power index is not zero. Conversely, if P's Shapley-Shubik power index is not zero, then P is pivotal in some sequential coalition. A coalition consisting of P together with the players preceding P in that sequential coalition is a winning coalition, and P is a critical player in it. Thus, P is not a dummy.

57. **(a)** $7 \le q \le 13$
(b) For $q = 7$ or $q = 8$, P_1 is a dictator because $\{P_1\}$ is a winning coalition.
(c) For $q = 9$, only P_1 has veto power because P_2 and P_3 together have only five votes.
(d) For $10 \le q \le 12$, both P_1 and P_2 have veto power because no motion can pass without both of their votes. For $q = 13$, all three players have veto power.
(e) For $q = 7$ or $q = 8$, both P_2 and P_3 are dummies because P_1 is a dictator. For $10 \le q \le 12$, P_3 is a dummy because all winning coalitions contain $\{P_1, P_2\}$, which is itself a winning coalition.

59. **(a)** The winning coalitions are $\{P_1, P_2\}$ and $\{P_1, P_2, P_3\}$.
(b) The winning coalitions are $\{P_1, P_2\}, \{P_1, P_2, P_3\}, \{P_1, P_2, P_4\}, \{P_1, P_3, P_4\}$, and $\{P_1, P_2, P_3, P_4\}$.
(c) The winning coalitions for both weighted voting systems are those consisting of any three of the five players, any four of the five players, and the grand coalition.

(d) If a player is critical in a winning coalition, then that coalition is no longer winning if the player is removed. An equivalent system (with the same winning coalitions) will find that player critical in the same coalitions.

(e) If a player, P, is pivotal in a sequential coalition, then the players to P's left in the sequential coalition do not form a winning coalition, but including P does make it a winning coalition. An equivalent system (with the same winning coalitions) will find the player P pivotal in that same sequential coalition.

61. **(a)** $\frac{V}{2} < q \leq V - w_1$ (where V denotes the sum of all the weights)
(b) $V - w_N < q \leq V$
(c) $V - w_i < q \leq V - w_{i+1}$

63. You should buy from P_3.

65. **(a)** You should buy from P_2.
(b) You should buy two votes from P_2.
(c) Buying a single vote from P_2 raises your power from $\frac{1}{25} = 4\%$ to $\frac{3}{25} = 12\%$. Buying a second vote from P_2 raises your power to $\frac{2}{13} \approx 15.4\%$.

67. **(a)** Both have $\beta_1 = \frac{2}{5}$, $\beta_2 = \frac{1}{5}$, $\beta_3 = \frac{1}{5}$, and $\beta_4 = \frac{1}{5}$.
(b) In the weighted voting system $[q: w_1, w_2, \ldots, w_N]$, if P_k is critical in a coalition, then the sum of the weights of all the players in that coalition (including P_k) is at least q but the sum of the weights of all the players in the coalition except P_k is less than q. Consequently, if the weights of all the players in that coalition are multiplied by $c > 0$ ($c = 0$ would make no sense), then the sum of the weights of all the players in the coalition (including P_k) is at least cq but the sum of the weights of all the players in the coalition except P_k is less than cq. Therefore, P_k is critical in the same coalitions in the weighted voting system $[cq: cw_1, cw_2, \ldots, cw_N]$.

69. **(a)** $\beta_1 = \frac{2}{5}$, $\beta_2 = \beta_3 = \beta_4 = \frac{1}{5}$
(b) $\sigma_1 = \frac{1}{2}$, $\sigma_2 = \sigma_3 = \sigma_4 = \frac{1}{6}$

CHAPTER 3 The Mathematics of Sharing

WALKING

3.1 Fair-Division Games

1. **(a)** s_2 and s_3 **(b)** s_2 and s_3 **(c)** s_1, s_2, and s_3
(d) 1. Henry gets s_2; Tom gets s_3; Fred gets s_1.
2. Henry gets s_3; Tom gets s_2; Fred gets s_1.
(e) Henry gets s_2; Tom gets s_3; Fred gets s_1.

3. **(a)** s_2, s_3 **(b)** s_1, s_2, s_4 **(c)** s_2, s_3 **(d)** s_1, s_2, s_3, s_4
(e) 1. Angie gets s_2; Bev gets s_1; Ceci gets s_3; Dina gets s_4.
2. Angie gets s_2; Bev gets s_4; Ceci gets s_3; Dina gets s_1.
3. Angie gets s_3; Bev gets s_1; Ceci gets s_2; Dina gets s_4.
4. Angie gets s_3; Bev gets s_4; Ceci gets s_2; Dina gets s_1.

5. **(a)** s_2, s_3, s_4 **(b)** s_3, s_4 **(c)** s_1, s_2 **(d)** s_1, s_4
(e) 1. Allen gets s_2; Brady gets s_3; Cody gets s_1; Diane gets s_4.
2. Allen gets s_3; Brady gets s_4; Cody gets s_2; Diane gets s_1.
3. Allen gets s_4; Brady gets s_3; Cody gets s_2; Diane gets s_1.

7. **(a)** s_1, s_4 **(b)** s_1, s_2 **(c)** s_1, s_3 **(d)** s_4
(e) Adams: s_1; Benson: s_2; Cagle: s_3; Duncan: s_4

9. **(a)** \$16 **(b)** \$6 **(c)** \$3.20

11. **(a)** \$4.80 **(b)** \$14.40 **(c)** \$0.96

13. **(a)** s_1: \$2.50; s_2: \$3.75; s_3: \$5.00; s_4: \$6.25; s_5: \$7.50; s_6: \$5.00
(b) s_3, s_4, s_5, s_6

3.2 The Divider-Chooser Method

15. **(a)** s_1: $[0, 8]$; s_2: $[8, 12]$ **(b)** s_1; \$8.00

17. **(a)** s_1: $[0, 10]$; s_2: $[10, 28]$ **(b)** s_2; \$6.50

19. **(a)** yes; s_2 (75%) **(b)** no (David values s_2 at 60%, s_1 at 40%.)
(c) yes; s_2 (75%)

3.3 The Lone-Divider Method

21. **(a)** C_1: s_2, C_2: s_3, D: s_1 or C_1: s_3, C_2: s_2, D: s_1
(b) C_1: s_2, C_2: s_1, D: s_3 or C_1: s_2, C_2: s_3, D: s_1 or C_1: s_3, C_2: s_1, D: s_2

23. **(a)** C_1: s_2, C_2: s_1, C_3: s_3, D: s_4 (First, C_1 must receive s_2. Then, C_3 must receive s_3. So, C_2 receives s_1 and D receives s_4.)
(b) C_1: s_2, C_2: s_3, C_3: s_1, D: s_4 or C_1: s_3, C_2: s_1, C_3: s_2, D: s_4
(c) C_1: s_2, C_2: s_1, C_3: s_4, D: s_3 or C_1: s_2, C_2: s_3, C_3: s_1, D: s_4 or C_1: s_2, C_2: s_3, C_3: s_4, D: s_1

25. **(a)** Tim **(b)** Mark gets s_2; Tim gets s_1; Maia gets s_3; Kelly gets s_4

27. **(a)** fair
(b) not fair ($\{s_2, s_3, s_4\}$ may be worth less than 75% to C_2 and C_3).
(c) fair
(d) not fair ($\{s_1, s_4\}$ may be worth less than 50% to C_2 and C_3).

29. **(a)** C_1: s_2, C_2: s_4, C_3: s_5, C_4: s_3, D: s_1; C_1: s_4, C_2: s_2, C_3: s_5, C_4: s_3, D: s_1 (If C_1 is to receive s_2, then C_2 must receive s_4 (and conversely). So, C_4 must receive s_3. It follows that C_3 must receive s_5. This leaves D with s_1.)
(b) C_1: s_2, C_2: s_4, C_3: s_5, C_4: s_3, D: s_1 (If C_1 is to receive s_2, then C_2 must receive s_4. So, C_4 must receive s_3. It follows that C_3 must receive s_5. This leaves D with s_1.)

31. **(a)** Gong (he is the only player who could value each piece equally).
(b) Egan: $\{s_3, s_4\}$; Fine: $\{s_1, s_3, s_4\}$; Hart: $\{s_3\}$
(c) Egan: s_4; Fine: s_1; Gong: s_2; Hart: s_3

33. **(a)** s_1: $[0, 4]$; s_2: $[4, 8]$; s_3: $[8, 12]$ **(b)** s_1, s_2 **(c)** s_2, s_3
(d) Three possible fair divisions are
1. Jackie: s_2; Karla; s_1; Lori: s_3
2. Jackie: s_1; Karla; s_2; Lori: s_3
3. Jackie: s_3; Karla: s_1; Lori: s_2

3.4 The Lone-Chooser Method

35. **(a)** 40° strawberry; 40° strawberry; 90° vanilla–10° strawberry
(b) 75° vanilla; 15° vanilla–40° strawberry; 50° strawberry
(c) Angela: 80° strawberry; Boris: 15° vanilla–90° strawberry; Carlos: 165° vanilla–10° strawberry
(d) Angela: \$12.00; Boris: \$10.00; Carlos: \$22.67

37. **(a)** s_2; 90° vanilla; 90° vanilla; 60° strawberry
(b) 40° strawberry; 40° strawberry; 40° strawberry
(c) Angela: 80° strawberry; Boris: 90° vanilla–60° strawberry; Carlos: 90° vanilla–40° strawberry
(d) Angela: \$12.00; Boris: \$12.00; Carlos: \$14.67

39. **(a)** s_1 (60° chocolate; 30° chocolate–30° strawberry; 60° strawberry)
(b) 30° orange; 30° orange; 30° orange–90° vanilla
(c) Arthur: 60° orange; Brian: 90° strawberry–30° chocolate; Carl: 90° vanilla–30° orange–60° chocolate
(d) Arthur: $33\frac{1}{3}\%$; Brian: $66\frac{2}{3}\%$; Carl: $83\frac{1}{3}\%$

41. **(a)** s_1: $[0, 3]$; s_2: $[3, 12]$
(b) Jackie picks s_2 (J_1: $[3, 6]$; J_2: $[6, 9]$; J_3: $[9, 12]$)
(c) K_1: $[0, 1]$; K_2: $[1, 2]$; K_3: $[2, 3]$
(d) Jackie: $[3, 9]$; Karla: $[0, 2]$; Lori: $[2, 3]$ and $[9, 12]$
Jackie: 50%; Karla: $33\frac{1}{3}\%$; Lori: $38\frac{8}{9}\%$

3.5 The Method of Sealed Bids

43. **(a)** Ana: \$300; Belle: \$300; Chloe: \$400
(b) In the first settlement, Ana gets the desk and receives \$120 in cash, Belle gets the dresser, and Chloe gets the vanity and the tapestry and pays \$360.
(c) \$240
(d) In the final settlement, Ana gets the desk and receives \$200 in cash, Belle gets the dresser and receives \$80, and Chloe gets the vanity and the tapestry and pays \$280.

45. **(a)** A: \$600; B: \$582; C: \$618; D: \$606; E : \$600
(b) In the first settlement, A gets items 4 and 5 and pays \$765, B gets \$582, C gets items 1 and 3 and pays \$287, D gets \$606, and E gets items 2 and 6 and pays \$266.
(c) \$130
(d) In the final settlement, A gets items 4 and 5 and pays \$739, B gets \$608, C gets items 1 and 3 and pays \$261, D gets \$632, and E gets items 2 and 6 and pays \$240.

47. Anne gets \$75,000 and Chia gets \$80,000.

49. Ali gets to do Chore 4; Briana gets to do Chore 2; Caren gets to do Chores 1 and 3. Ali has to pay \$15 to Caren; Briana is even (neither pays nor gets money).

3.6 The Method of Markers

51. **(a)** 10, 11, 12, 13 **(b)** 1, 2, 3 **(c)** 5, 6, 7 **(d)** 4, 8, 9

53. **(a)** 1, 2 **(b)** 10, 11, 12 **(c)** 4, 5, 6, 7 **(d)** 3, 8, 9

55. **(a)** A: 19, 20; B: 15, 16, 17; C: 1, 2, 3; D: 11, 12, 13; E: 5, 6, 7, 8
(b) 4, 9, 10, 14, 18

57. **(a)** **W S S G S W W B G G G S G S G S B B**

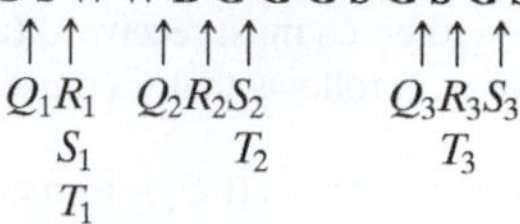

(b) Quintin gets a W, two S's, and a G; Ramon gets two W's and a B; Tim gets three G's and two S's; and Stephone gets an S and two B's. An S and two G's are left over.

59. **(a)**

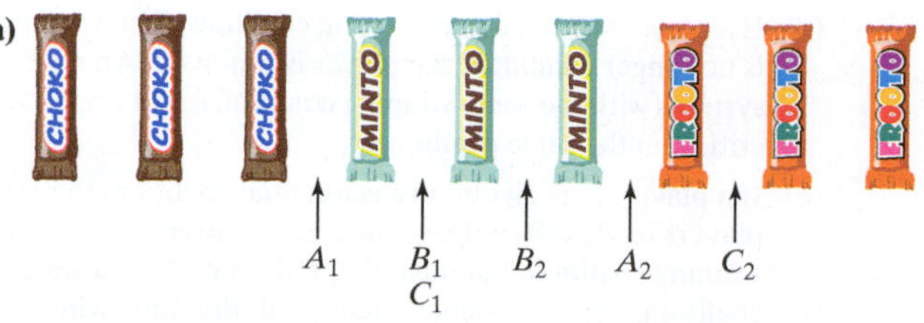

(b) Ana gets three Choko bars, Belle gets one Minto bar, and Chloe gets two Frooto bars. Two Minto bars and one Frooto bar are left over.
(c) Belle would select one of the Minto bars, Chloe would then select the Frooto bar, and then Ana would be left with the last Minto bar.

JOGGING

61. **(a)** C gets to choose one of the three pieces. One of the three must be worth at least $\frac{1}{3}$ of S.
(b) If C chooses either s_{21} or s_{22}, D_1 gets to choose s_1 (a fair share to him). If C chooses s_1, D_1 gets to choose between s_{21} and s_{22}. Together they are worth $\frac{2}{3}$ of S, so one of the two must be worth at least $\frac{1}{3}$ of S.
(c) Suppose that in D_2's opinion, the first cut by D_1 splits S into a 40%–60% split. In this case the value of the three pieces to D_2 is $s_1 = 40\%$, $s_{21} = 30\%$, and $s_{22} = 30\%$. If C or D_1 choose s_1, D_2 will not get a fair share.

63. B must pay $\dfrac{x+y}{4}$ dollars to A.

65. B receives the strawberry part. C will receive $\frac{5}{6}$ of the chocolate part. A receives the vanilla part and $\frac{1}{6}$ of the chocolate part.

67. **(a)** two, three, or four choosers bidding on the same item; three or four choosers bidding on the same two items; four choosers bidding on the same three items
(b) 10
(c) $\dfrac{(N-1)(N-2)}{2}$

CHAPTER 4 The Mathematics of Apportionment

WALKING

4.1 Apportionment Problems and Apportionment Methods

1. **(a)** 50,000
(b) Apure: 66.2; Barinas: 53.4; Carabobo: 26.6; Dolores: 13.8

3. **(a)** 1040
(b) the daily average ridership per bus
(c) A: 43.558; B: 29.875; C: 19.702; D: 13.615; E: 9.865; F: 8.385

5. **(a)** 137
(b) 200,000
(c) A: 8,240,000; B: 6,380,000; C: 4,960,000; D: 4,520,000; E: 3,300,000

7. 35.41

9. **(a)** 0.5%
(b) A: 22.74; B: 16.14; C: 77.24; D: 29.96; E: 20.84; F: 33.08

4.2 Hamilton's Method

11. A: 66; B: 53; C: 27; D: 14

13. A: 43; B: 30; C: 20; D: 14; E: 10; F: 8

15. A: 41; B: 32; C: 25; D: 23; E: 16

17. A: 23; B: 16; C: 77; D: 30; E: 21; F: 33

19. **(a)** Dunes: 3; Smithville: 6; Johnstown: 15
(b) Dunes: 2; Smithville: 7; Johnstown: 16
(c) Dunes's apportionment went from 3 social workers to 2 as the total number of social workers increased from 24 to 25.

4.3 Jefferson's Method

21. **(a)** $SD = 200{,}000$; standard quotas are 22.5 (Arcadia), 24.5 (Belarmine), 19.5 (Crowley), 33.5 (Dandia).
(b) 98 **(c)** 99 **(d)** 102 **(e)** 100 **(f)** 100
(g) We know that $d = 195{,}800$ and $d = 196{,}000$ work from parts (e) and (f) respectively. Any other divisor in between (for example $d = 195{,}900$) will also work.

23. A: 67; B: 54; C: 26; D: 13

25. A: 44; B: 30; C: 20; D: 13; E: 10; F: 8

27. A: 41; B: 32; C: 25; D: 23; E: 16

29. A: 22; B: 16; C: 78; D: 30; E: 21; F: 33

4.4 Adams's and Webster's Methods

31. A: 41; B: 32; C: 25; D: 22; E: 17

33. A: 23; B: 16; C: 77; D: 30; E: 21; F: 33

35. A: 66; B: 53; C: 27; D: 14

37. A: 41; B: 32; C: 25; D: 23; E: 16

39. A: 23; B: 16; C: 77; D: 30; E: 21; F: 33

4.5 The Huntington-Hill Method

41. **(a)** 2 **(b)** 1 **(c)** 1 **(d)** 2 **(e)** 2

43. 2. (Under the Huntington-Hill method the cutoff for rounding quotas between 1 and 2 is $\sqrt{2} \approx 1.41421$.)

45. **(a)** 40 **(b)** A: 4; B: 10; C: 1; D: 13; E: 12

47. **(a)** 40 **(b)** A: 3; B: 11; C: 2; D: 12; E: 12

49. **(a)** A: 34; B: 41; C: 22; D: 59; E: 15; F: 29
(b) A: 34; B: 41; C: 22; D: 59; E: 15; F: 29
(c) The apportionments are the same under both methods.

4.6 The Quota Rule and Apportionment Paradoxes

51. **(c)** Hamilton's method satisfies the quota rule. The only possible apportionment to X is 35 or 36.

53. **(e)** Only upper-quota violations are possible under Jefferson's method.

55. **(e)** Under Webster's method both lower- and upper-quota violations are possible.

57. Upper-quota violations are possible under Jefferson's method.

59. The Alabama paradox. (The apportionment to Dunes went from 3 social workers to 2 for no other reason than the fact that the county hired an additional social worker.)

61. **(a)** Bob: 0; Peter: 3; Ron: 8
(b) Bob: 1; Peter: 2; Ron: 8
(c) The population paradox. (For studying 3.70% more. Bob gets a piece of candy. However, Peter, who studies 4.94% more, has to give up a piece.)

63. **(a)** $SD = 618{,}000$
(b) Aila: 8; Balin: 25; Cona: 17
(c) Aila: 9; Balin: 24; Cona: 17; Dent: 15
(d) The new-states paradox. [The addition of Dent (and the addition of the exact number of seats that Dent is entitled to) still impacts the other apportionments: Balin has to give up one seat to Aila.]

65. **(a)** A: 88; B: 4; C: 5; D: 3 **(b)** A: 88; B: 4; C: 5; D:3
(c) Both Webster's method and the Huntington-Hill method violate the quota rule. (An upper-quota violation occurs with A's apportionment.)

JOGGING

67. **(a)** an upper-quota violation by state X
(b) a lower-quota violation by state X
(c) The number of seats that state X has received is within 0.5 of its standard quota.
(d) The number of seats that state X has received satisfies the quota rule but is not the result of conventional rounding of the standard quota.

69. If the modified divisor is d, then state A has modified quota $p_A/d \geq 2.5$. Also, state B has modified quota $p_B/d < 0.5$. So, $p_A \geq 2.5d > 5p_B$. That is, more than five-sixths of the population lives in state A.

71. **(a)** In Jefferson's method the modified quotas are larger than the standard quotas, so rounding down will give each state at least the integer part of the standard quota for that state.
(b) In Adams's method the modified quotas are smaller than the standard quota, so rounding up will give each state at most one more than the integer part of the standard quota for that state.
(c) If there are only two states, an upper-quota violation for one state results in a lower-quota violation for the other state. Neither Jefferson's nor Adams's method can have both upper- and lower-quota violations. So, when there are only two states, neither method can violate the quota rule.

73. **(a)** A: 33; B: 138; C: 4; D: 41; E: 14; F: 20
(b) State D is apportioned 41 seats under Lowndes's method (all the methods discussed in the chapter apportion 42 seats to state D). States C and E do better under Lowndes's method than under Hamilton's method.

CHAPTER 5 The Mathematics of Getting Around

WALKING

5.2 An Introduction to Graphs

1. **(a)** $\{A, B, C, X, Y, Z\}$
(b) $AX, AY, AZ, BB, BC, BX, CX, XY$
(c) $\deg(A) = 3, \deg(B) = 4, \deg(C) = 2, \deg(X) = 4,$ $\deg(Y) = 2, \deg(Z) = 1$
(d)

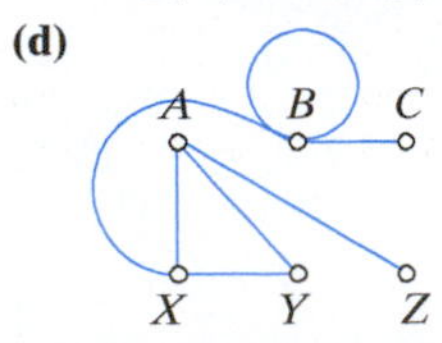

3. **(a)** $\{A, B, C, D, X, Y, Z\}$
(b) $AX, AX, AY, BX, BY, DZ, XY$
(c) $\deg(A) = 3, \deg(B) = 2, \deg(C) = 0, \deg(D) = 1,$ $\deg(X) = 4, \deg(Y) = 3, \deg(Z) = 1$ **(d)** 3

5.

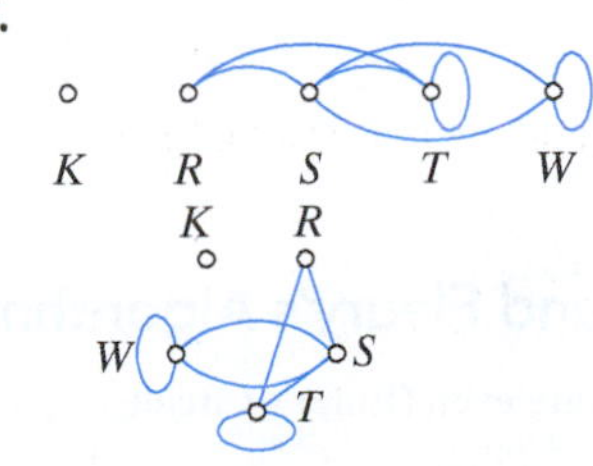

7. **(a)** A, B, D, E **(b)** AD, BC, DD, DE **(c)** 5 **(d)** 12

9. **(a)** **(b)** **(c)**

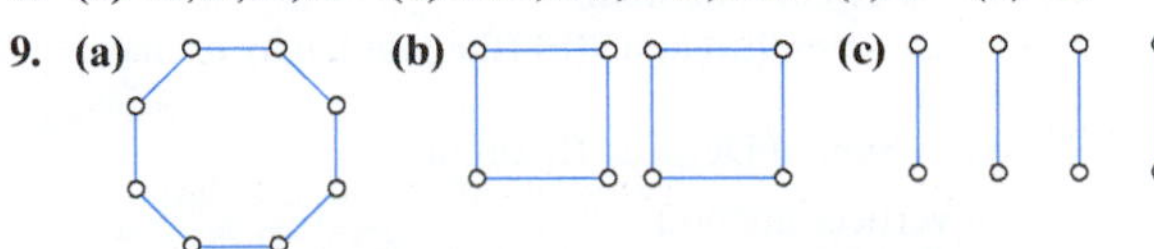

11. **(a)** C, B, A, H, F **(b)** C, B, D, A, H, F **(c)** C, B, A, H, F
(d) C, D, B, A, H, G, G, F **(e)** 4 **(f)** 3 **(g)** 12

13. **(a)** G, G **(b)** There are none.
(c) A, B, D, A; B, C, D, B; F, G, H, F
(d) A, B, C, D, A; F, G, G, H, F

15. **(a)** AH, EF **(b)** There are none. **(c)** AB, BC, BE, CD

17. **(a)** CI and HJ **(b)** 3 **(c)** The shortest path is C, I, H, J, of length 3. **(d)** The longest path is I, G, F, I, H, J of length 5. (I, F, G, I, H, J is another answer.)

19.

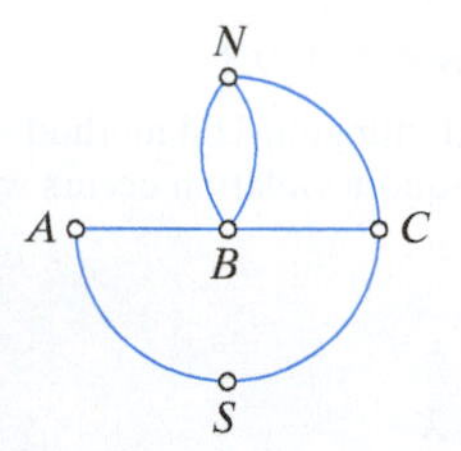

21.

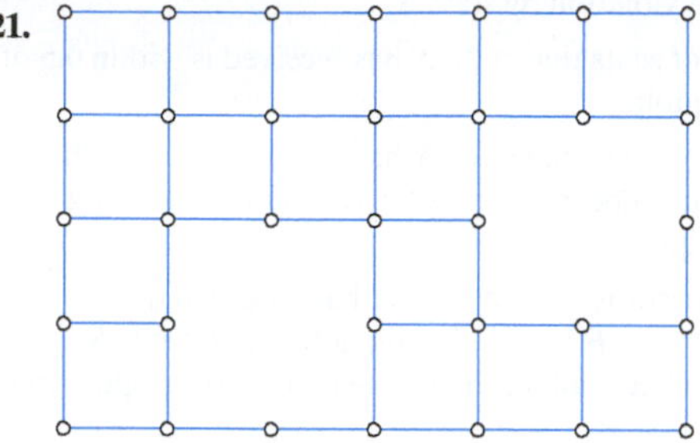

23

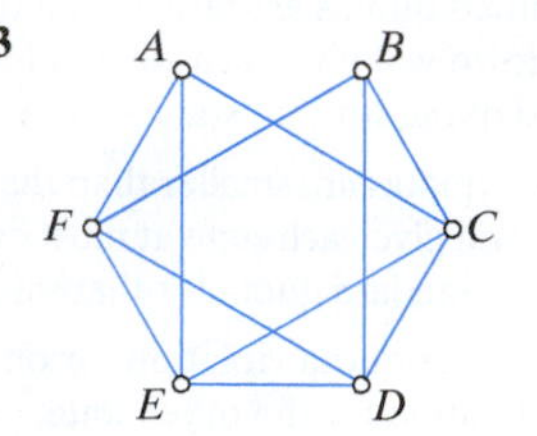

25

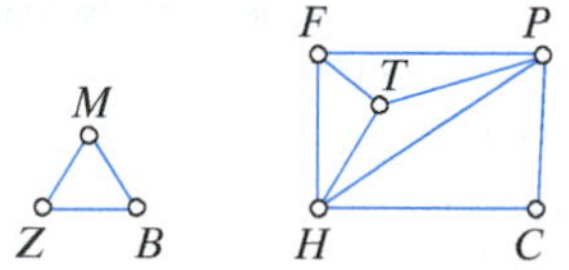

27

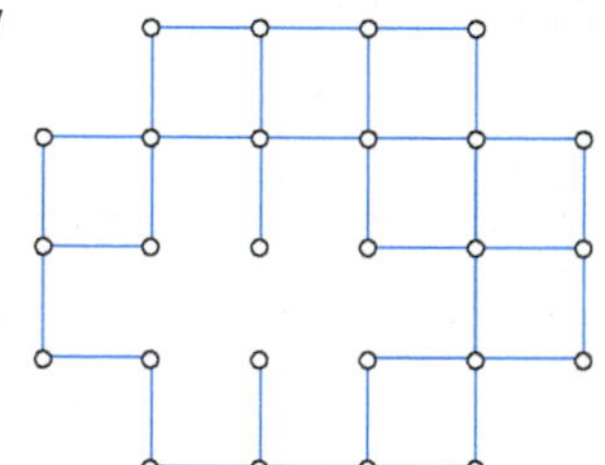

5.3 Euler's Theorems and Fleury's Algorithm

29 **(a)** (A); connected, all vertices are even (Euler's Circuit Theorem).
(b) (C); four vertices are odd (Euler's Path Theorem).
(c) (A); Euler's Circuit Theorem.
(d) (D); see Exercises 9(a) and 9(b) (the graph may or may not be connected).
(e) (F); Euler's Sum of Degrees Theorem

31 **(a)** (C); four vertices are odd.
(b) (A); connected, all vertices are even.
(c) (F); Euler's Sum of Degrees Theorem.
(d) (C); Euler's Path Theorem.

33 **(a)** (B); exactly two vertices are odd.
(b) (A); all vertices are even.
(c) (C); the vertex of degree 0 makes the graph disconnected.

35

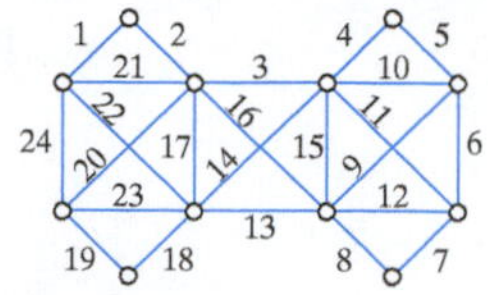

Other answers are possible.

37

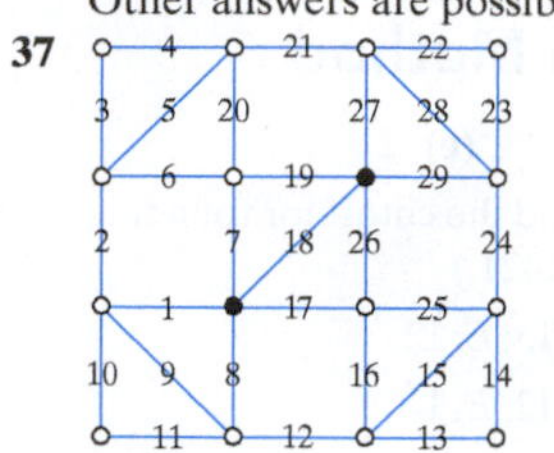

Other answers are possible.

39.

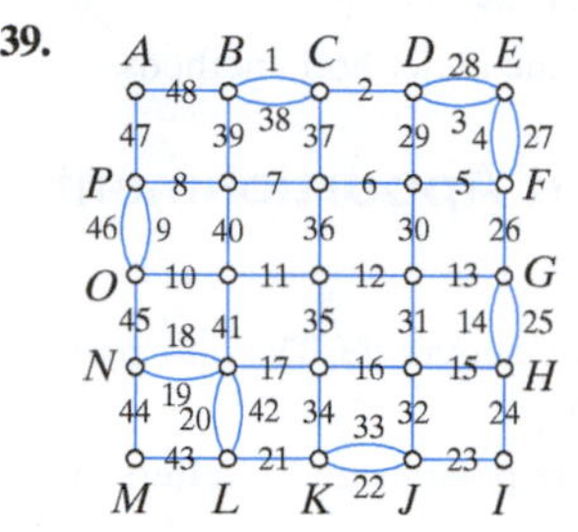

Many other answers are possible.

41. **(a)** PD or PB **(b)** BF

5.4 Eulerizing and Semi-Eulerizing Graphs

43.

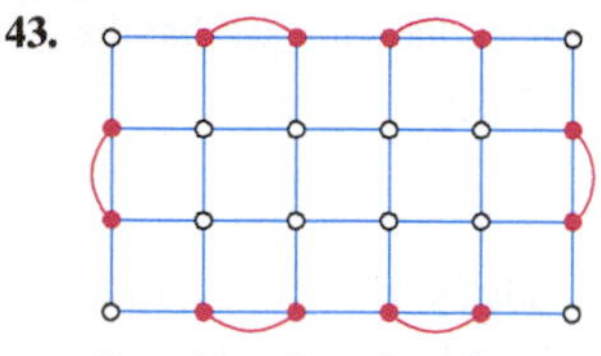

45.

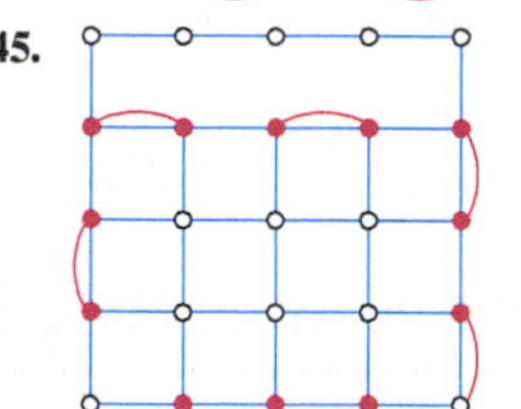

47.

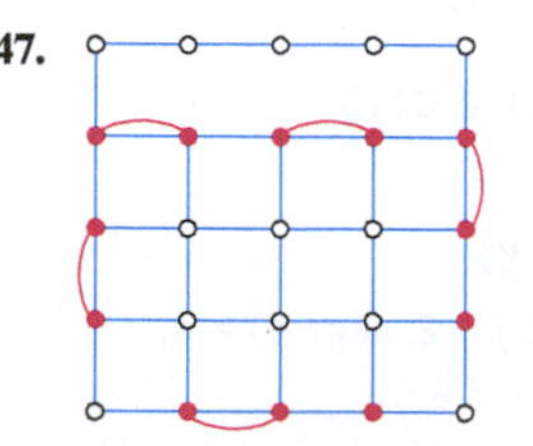

Other answers are possible.

49.

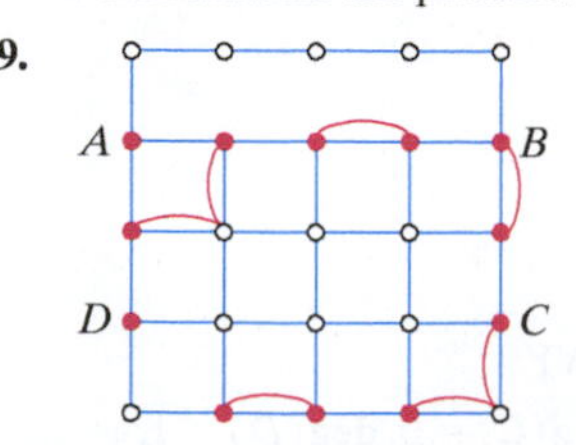

Other answers are possible.

51.

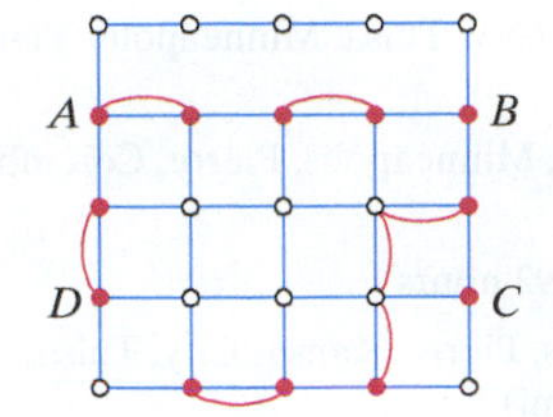

Other answers are possible.

53.

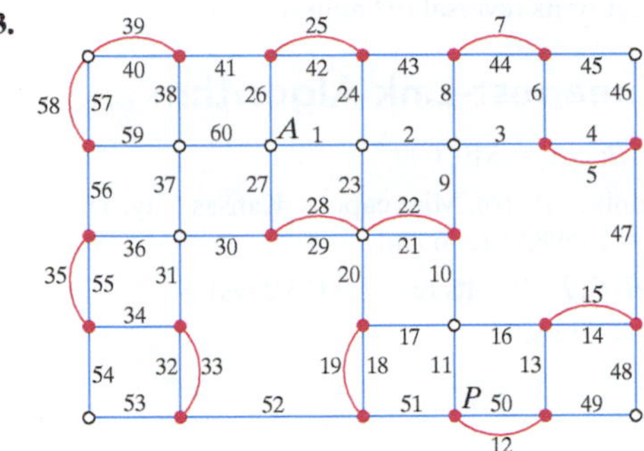

Many other answers are possible.

55.

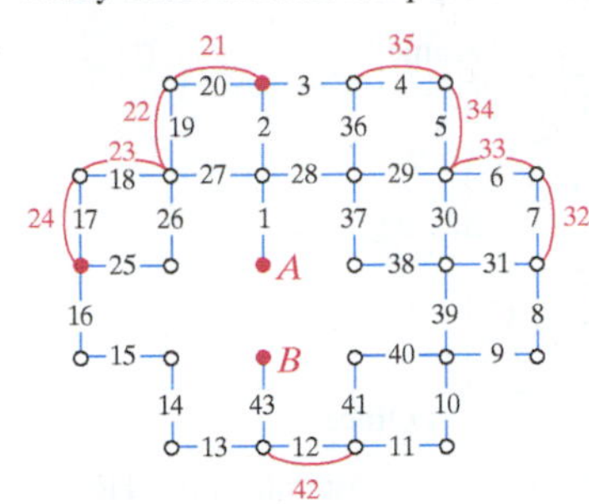

Many other answers are possible.

JOGGING

57. The minimum number of times one must lift the pencil is 5. There are 12 odd "vertices" in the figure; two can be used as the starting and ending vertices; the remaining 10 can be paired so that each pair forces one lifting of the pencil.

59. None. If a vertex had degree 1, then the edge incident to that vertex would be a bridge.

61. Each component of the graph is a separate connected graph. If the two odd vertices were in two different components, then each of those two components would have just one odd vertex, in violation of Euler's Sum of Degrees Theorem.

63. **(a)** *NK, C, B, NK, B, A, SK, C*
(b) *NK, C, B, NK, B, A, SK, C, SK*
(c) *NK, C, B, NK, B, A, SK, C, B*
(d) *NK, C, B, NK, B, A, SK, C, B, A*

65. **(a)** There are many ways to implement this algorithm. The following figures illustrate one possibility.

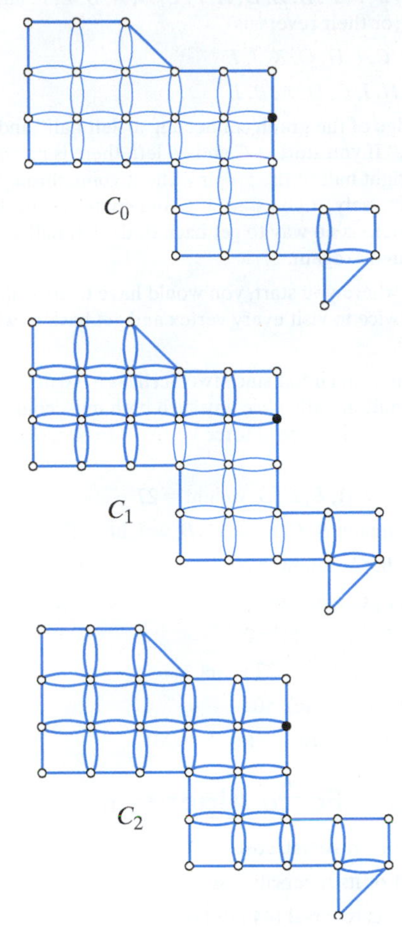

(b) **Step 1:** Find a path C_0 between the two vertices *v* and *w* of odd degree.

Step 2: Find a circuit C_0^* that kisses C_0 at any vertex that is on C_0. The path and the circuit can be combined into a larger path C_1 between the two odd vertices. If there are no kissing circuits to the larger path, we are done.

Step 3: Repeat Step 2 until an Euler path is found.

67. *D, L, R, L, C, A, R, B, R, B, D, C, L, D.* Others answers are possible.

69. Yes; one of the many possible journeys is given by crossing the bridges in the following order: *a, b, c, d, e, f, g, h, i, l, m, n, o, p, k.* (*Note*: Euler did not ask that the journey start and end at the same place.)

CHAPTER 6 The Mathematics of Touring

WALKING

6.2 Hamilton Paths and Circuits

1. **(a)** **1.** *A, B, D, C, E, F, G, A*;
2. *A, D, C, E, B, G, F, A*;
3. *A, D, B, E, C, F, G, A*
(b) *A, G, F, E, C, D, B*
(c) *D, A, G, B, C, E, F*

3. *A, B, C, D, E, F, G, A*; *A, B, E, D, C, F, G, A* and their mirror images

5. **(a)** *A, F, B, C, G, D, E* **(b)** *A, F, B, C, G, D, E, A*
(c) *E, D, G, C, B, F, A* **(d)** *E, D, G, C, B, F, A, E*

7. **(a)** 8 **(b)** *A, H, C, B, F, D, G, E, A*
(c) *A, H, C, B, F, D, G, E* and *A, E, G, D, F, B, C, H*

9. **(a)** *B, A, D, E, C*; *A, D, E, C, B*; *D, E, C, B, A*; *E, C, B, A, D*; *C, B, A, D, E*
(b) *A, B, E, D, C*; *A, D, E, B, C*; *B, A, E, C, D*; *B, C, E, A, D*; *C, B, E, D, A*; *C, D, E, B, A*; *D, C, E, A, B*; *D, A, E, C, B*

11. **(a)** A, B, C, D, E, F, A and $A, D, C, B\ E, F, A$ (or their reversals)

(b) C, B, A, D, E, F; C, B, E, D, A, F; C, D, A, B, E, F; and C, D, E, B, A, F (or their reversals)

13. **(a)** $A, B, E, D, C, I, H, G, K, J, F$

(b) $K, J, F, G, H, I, C, D, A, B, E$

(c) CI is a bridge of the graph connecting a "left half" and a "right half." If you start at C and go left, there is no way to get to the right half of the graph without going through C again. Conversely, if you start at C and cross over to the right half first, there is no way to get back to the left half without going through C again.

(d) No matter where you start, you would have to cross the bridge CI twice to visit every vertex and get back to where you started.

15. There is no Hamilton circuit since two vertices have degree 1. There is no Hamilton path since any such path must contain edges AB, BE, and BC, which would force vertex B to be visited more than once.

17. **(a)** 6 **(b)** B, D, A, E, C, B; weight = 27

(c) The mirror image B, C, E, A, D, B; weight = 27

19. **(a)** A, D, F, E, B, C; weight = 29

(b) A, B, E, D, F, C; weight = 30

(c) A, D, F, E, B, C; weight = 29

21. **(a)** ≈77 years **(b)** ≈1622 years

23. **(a)** 190 **(b)** 210 **(c)** 50

25. **(a)** $N = 6$ **(b)** $N = 10$ **(c)** $N = 201$

6.3 The Brute-Force Algorithm

27. A, C, B, D, A or its reversal; cost = 102

29. A, D, C, B, E, A or its reversal; cost = 92 hours

31. B, A, C, D, B or its reversal (44 hours)

6.4 The Nearest-Neighbor and Repetitive Nearest-Neighbor Algorithms

33. **(a)** B, C, A, E, D, B; cost = \$722

(b) C, A, E, D, B, C; cost = \$722

(c) D, B, C, A, E, D; cost = \$722

(d E, C, A, D, B, E; cost = \$741

35. **(a)** A, D, E, C, B, A; cost = \$11,656

(b) A, D, B, C, E, A; cost = \$9,760

37. **(a)** A, E, B, C, D, A (92 hours)

(b) A, D, B, C, E, A (93 hours)

39. **(a)** Atlanta, Columbus, Kansas City, Tulsa, Minneapolis, Pierre, Atlanta; cost = \$2915.25

(b) Atlanta, Kansas City, Tulsa, Minneapolis, Pierre, Columbus, Atlanta; cost = \$2804.25

41. A, E, B, C, D, A or its reversal (92 hours)

43. Atlanta, Columbus, Minneapolis, Pierre, Kansas City, Tulsa, Atlanta; cost = \$2439.00 (3252 mi)

45. 12.5%

47. A, D, C, B, E, A or its reversal (92 hours)

6.5 The Cheapest-Link Algorithm

49. B, E, C, A, D, B; cost = \$10, 000

51. Atlanta, Columbus, Pierre, Minneapolis, Kansas City, Tulsa, Atlanta; cost = \$2598.75 (3465 mi)

53. **(a)** A, E, D, B, C, F, A or its reversal (75 days)

(b) $\varepsilon = 5/70 \approx 7.14\%$

JOGGING

55. **(a)** $7! = 5040$ **(b)** 5040 **(c)** $6! = 720$ **57.** \$1500

59. **(a)**

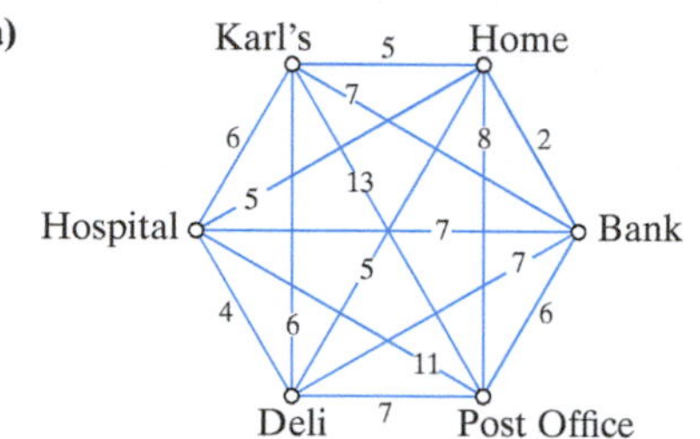

(b) Home, Bank, Post Office, Deli, Hospital, Karl's, Home

61. The graph describing the friendships among the guests does not have a Hamilton circuit. Thus, it is impossible to seat everyone around the table with friends on both sides.

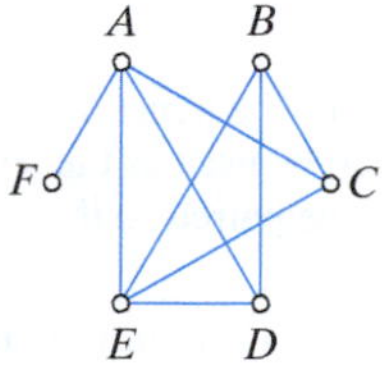

63. Suppose that the cheapest edge in a graph is the edge joining vertices X and Y. Using the nearest-neighbor algorithm, we will eventually visit one of these vertices—suppose that the first one of these vertices we visit is X. Then, since edge XY is the cheapest edge in the graph and since we have not yet visited vertex Y, the nearest-neighbor algorithm will take us to Y.

CHAPTER 7 The Mathematics of Networks

WALKING

7.1 Networks and Trees

1. No, the graph is disconnected. A, D, and G are connected to each other but not to the other four.

3. **(a)** 1 **(b)** 3 **(c)** 4

5. **(a)** A and G (or A and I). **(b)** A and H.

(c) There aren't any. A and H are the two vertices "farthest apart" in the network.

(d) 6.

7. **(a)** N and J; O and J; O and K.

(b) There aren't any. The largest possible separation is between N (or O) and the cluster of vertices H, I, J, K, L, and M at the bottom right. To get to any of those vertices, one must go through H. That takes four links. From H it takes only one link to get to I, L, and M and two links to get to J and K.

(c) 6

9. **(a)** 3 **(b)** 5 **(c)** 8 **(d)** 8

11. (B) The network violates the $N - 1$ *edges* property of trees.

13. (A) The network satisfies the $N-1$ *edges* property of trees.

15. (B) A tree must have redundancy $R=0$.

17. (C) We don't know if other pairs of vertices are connected by single or multiple paths.

19. (B) This network has 5 vertices and 10 edges (the sum of the degrees is 20). It violates the $N-1$ *edges* property of trees.

21. (A) This network has 5 vertices and 4 edges (the sum of the degrees is 8). It satisfies the $N-1$ *edges* property of trees.

23. (B) This network has an Euler circuit (see Section 5.3). It can't be a tree.

7.2 Spanning Trees, MSTs, and MaxSTs

25. **(a)** A B E C D (There are many other possible answers.)

(b) 6 **(c)** 1

27. **(a)** The network is its own spanning tree.

(b) 0 **(c)** 6

29. **(a)**

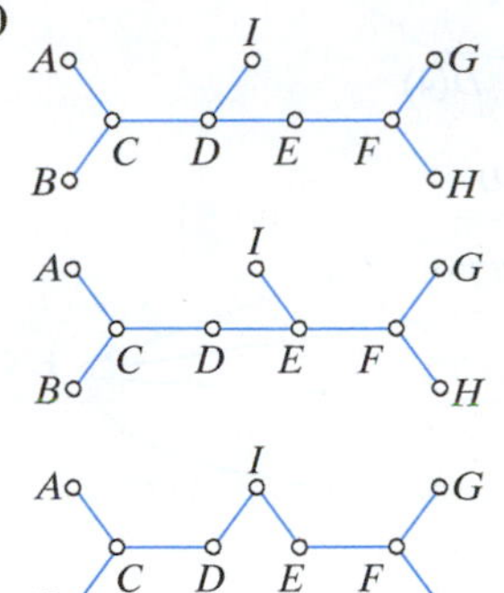

(b)

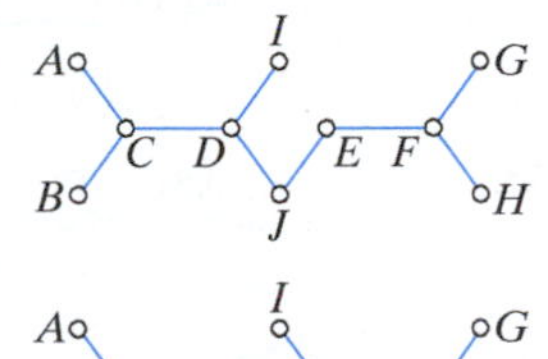

(c) 6

31. **(a)** 12 **(b)** 54

33. **(a)** 14 $(6+2\times4)$

(b)

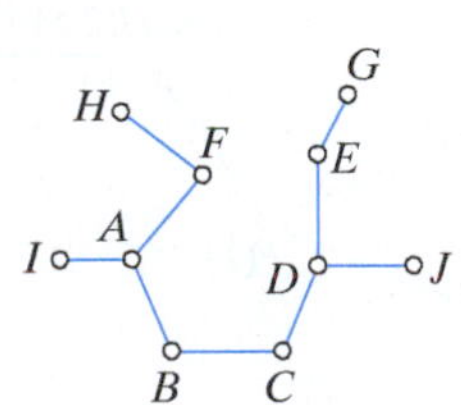

(c)

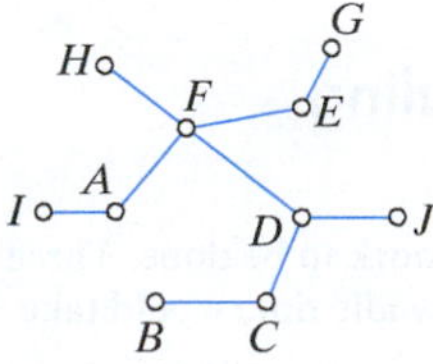

Several other answers are possible.

7.3 Kruskal's Algorithm

35. $380,000; the MST will have 19 edges at a cost of $20,000 each.

37. (Weight = 855)

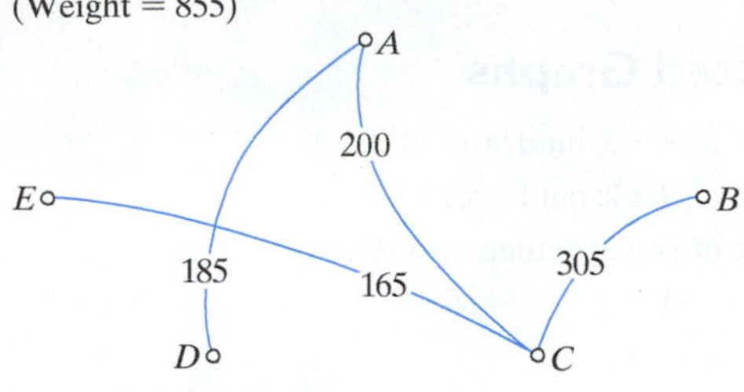

39. (Weight = 9.3)

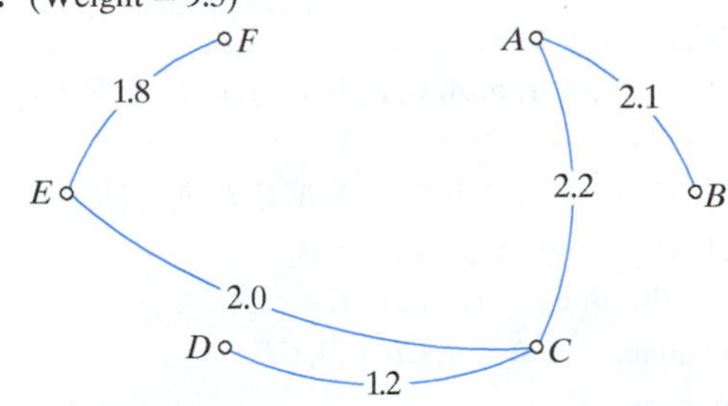

41. (Weight = 1520)

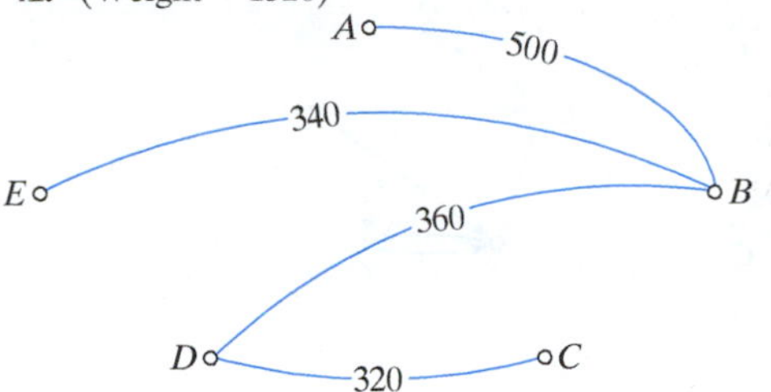

43. (Weight = 14.2)

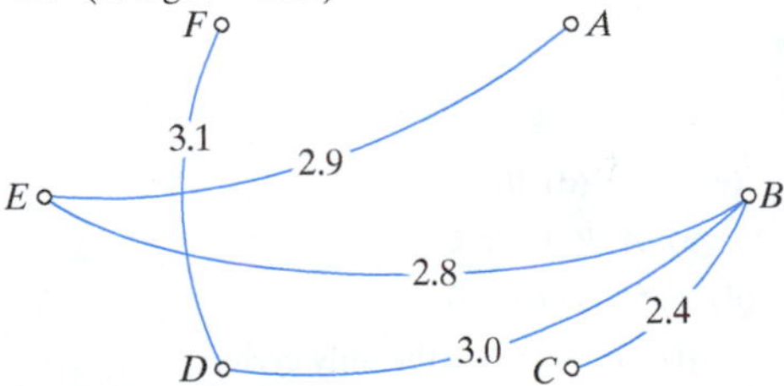

JOGGING

45. Kansas City–Tulsa, Pierre–Minneapolis, Minneapolis–Kansas City, Atlanta–Columbus, Columbus–Kansas City

47. **(a)** $k=0, 1, 2, 5$ are all possible.

(b) $0\le k\le 120$ and $k=123$ are all possible.

49. At each step of Kruskal's algorithm we choose (from the available edges that do not close circuits) the edge of least weight. When there is only one choice at each step, there is only one MST possible. The same argument applies to a MaxST.

51. Step 1 of Kruskal's algorithm will always choose XY. There is no other possibility.

53. Since the graph has no circuits, each component has to be a tree. Thus, in each component the number of edges is one less than the number of vertices. It follows that $M=N-K$.

CHAPTER 8 The Mathematics of Scheduling

WALKING

8.2 Directed Graphs

1. **(a)** indegree of $A = 3$; outdegree of $A = 2$

(b) indegree of $B = 2$; outdegree of $B = 2$

(c) indegree of $D = 3$; outdegree of $D = 0$

(d) 10 **(e)** 10

3. **(a)** $A, B, D; A, C, D$.

(b) $A, C, B, D; A, C, E, D$.

(c) A, B, A, C, D (or A, C, A, B, D).

(d) A, B, A, C, B, D (or A, B, A, C, E, D; or A, C, B, A, B, D).

(e) There are none.

5. **(a)** $A, B, A; A, C, A$. **(b)** $A, C, B, A; A, C, E, A$.

(c) A, C, A, B, A. **(d)** A, C, E, A, B, A.

7. **(a)** C, B, E **(b)** B, C **(c)** B, C, E

(d) There are none. **(e)** CA, CB, CD, CE

(f) There are none.

9. **(a)**

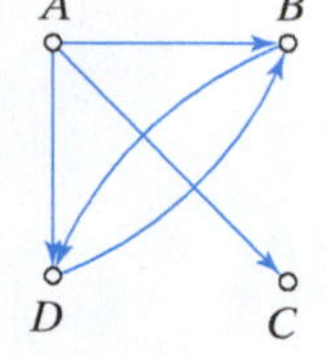

(b)

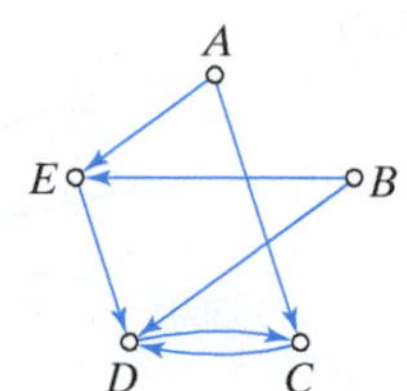

(c)

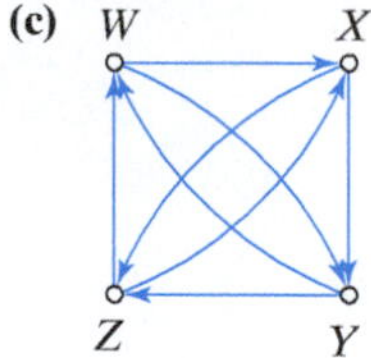

11. **(a)** 2 **(b)** 1 **(c)** 1 **(d)** 0

13. **(a)** A, B, D, E, F **(b)** A, B, D, E, C, F

(c) B, D, E, B **(d)** outdeg $(F) = 0$

(e) indeg$(A) = 0$ **(f)** B, D, E, B is the only cycle.

15.

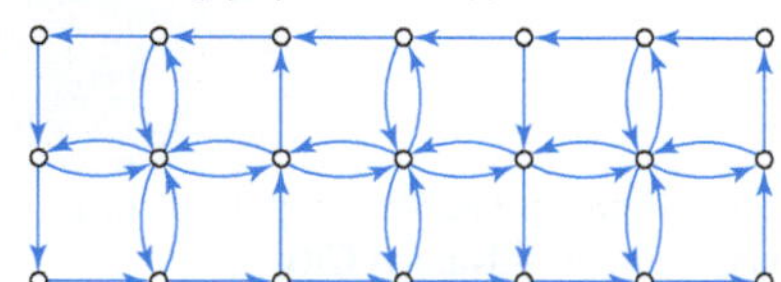

17. **(a)** B; that is the only person that everyone respects.

(b) A; that is the only person that no one respects.

(c) The individual corresponding to the vertex having the largest indegree.

19. **(a)** and **(b)** The band's Web site and the TicketMonster Web site are likely to have large indegree and zero or very small outdegree (the local Smallville Web sites are likely to have hyperlinks to both but are not likely to have hyperlinks coming from either). This makes X and Y the two likely choices. It is much more likely that the rock band's Web site will have a hyperlink to the ticket-selling Web site than the other way around. Given that there is a hyperlink from Y to X but not the other way, the most likely choice is Y for the rock band and X for the TicketMonster Web site.

(c) The radio station is likely to be a major source of information on anything related to the concert (the band, where to buy tickets, where to stay if you are from out of town, etc.), so it should be a vertex with high outdegree. V is the most likely choice.

(d) Z is a vertex linked to only the band's Web site and to the ticket-selling Web site. None of the other Web sites are paying any attention to Z. This is most likely Joe Fan's blog.

(e) The two sister hotels would clearly have links to each other's Web sites, and since they are offering a special package for the concert, they are likely to have a link to the ticket-selling Web site. The logical choices are U and W.

21.

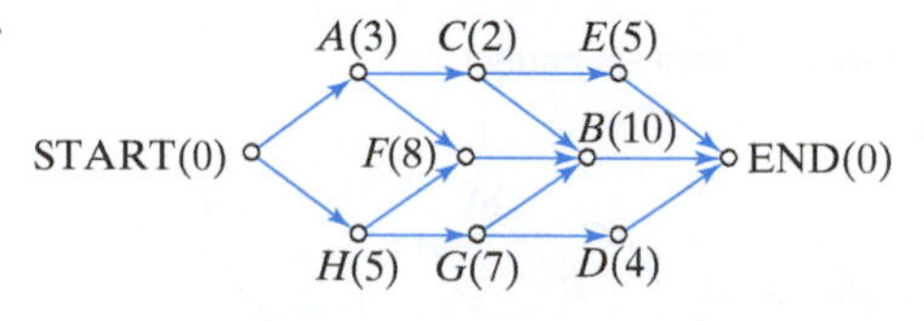

23.

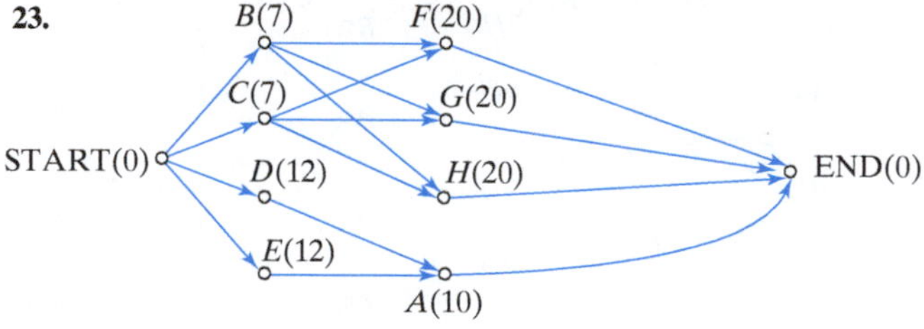

25.

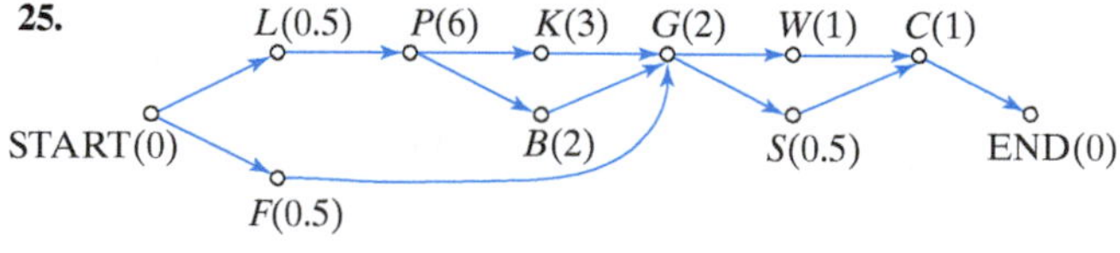

27.

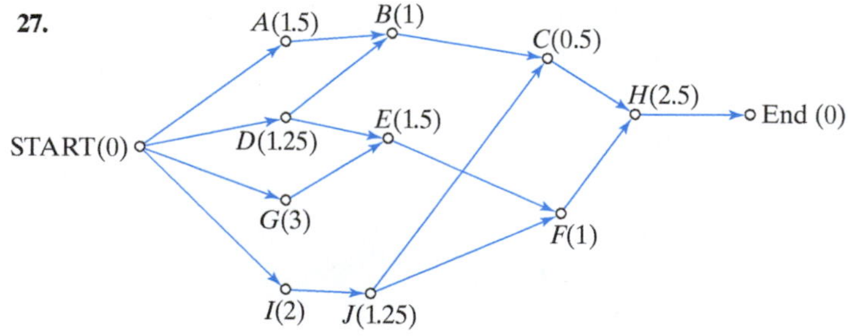

8.3 Priority-List Scheduling

29. **(a)** 18 hours

(b) There is a total of 75 hours of work to be done. Three processors working without any idle time would take $\frac{75}{3} = 25$ hours to complete the project.

31. There is a total of 75 hours of work to be done. Dividing the work equally between the six processors would require each processor to do $\frac{75}{6} = 12.5$ hours of work. Since there are no $\frac{1}{2}$-hour jobs, the completion time could not be less than 13 hours.

33. $AD(8)$, $AW(6)$, $AF(5)$, $IF(5)$, $AP(7)$, $IW(7)$, $ID(5)$, $IP(4)$, $PL(4)$, $PU(3)$, $HU(4)$, $IC(1)$, $PD(3)$, $EU(2)$, $FW(6)$

35. **(a)** No. *B* must be completed before *C* can be started.

(b) No. *B* must be completed before *D* can be started.

(c) No. *G* is a ready task and it is ahead of *H* in the priority list.

37

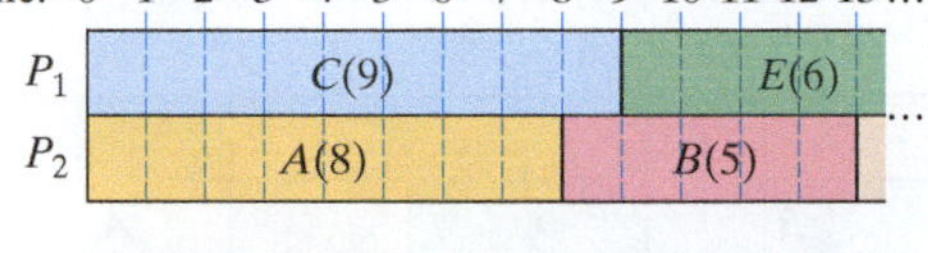

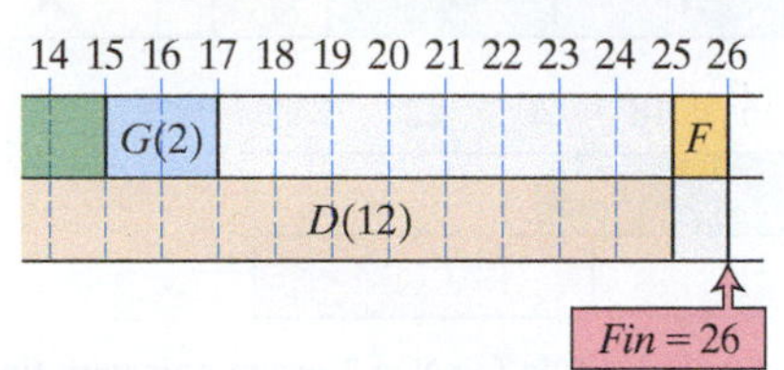

39.

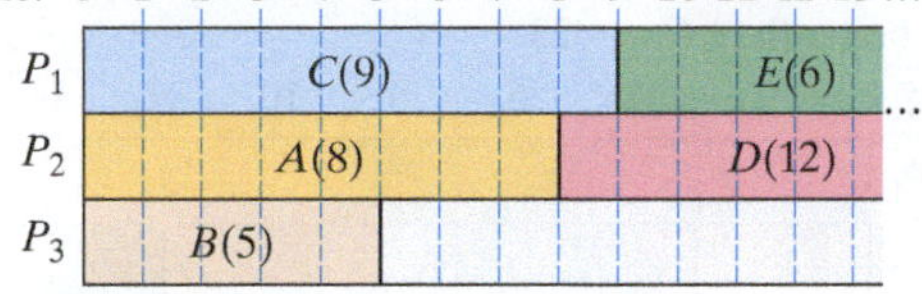

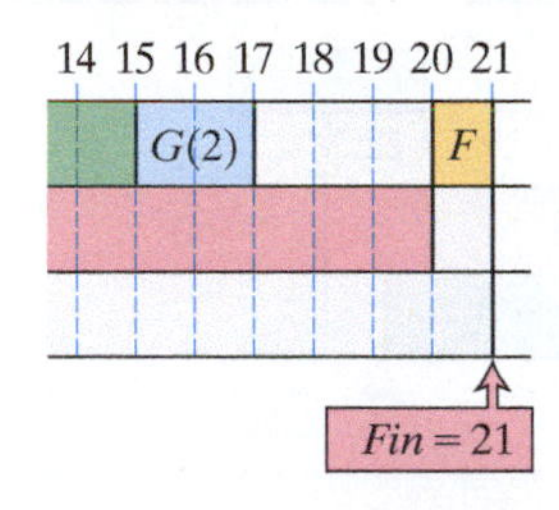

41. Both priority lists have all tasks in the top two paths of the digraph listed before any task in the bottom path.

8.4 The Decreasing-Time Algorithm

43.

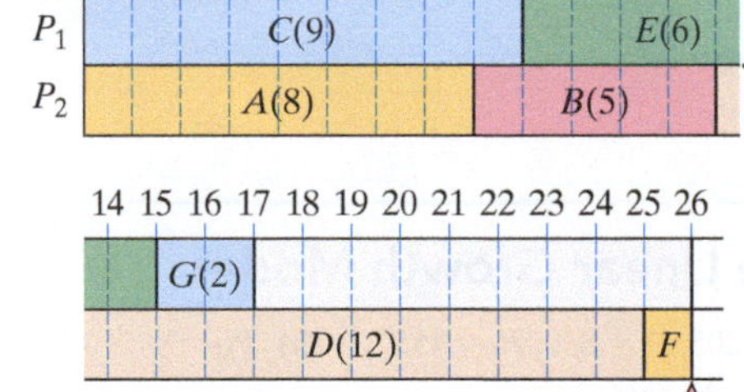

45. Time: 0 2 4 6 8 10 12 14 16 18 20 22 24 26 28 30 32...

P_1: D(15), H(5), J(5)

P_2: C(10), G(4)

P_3: B(4), F(3), A, E(7), I(15)

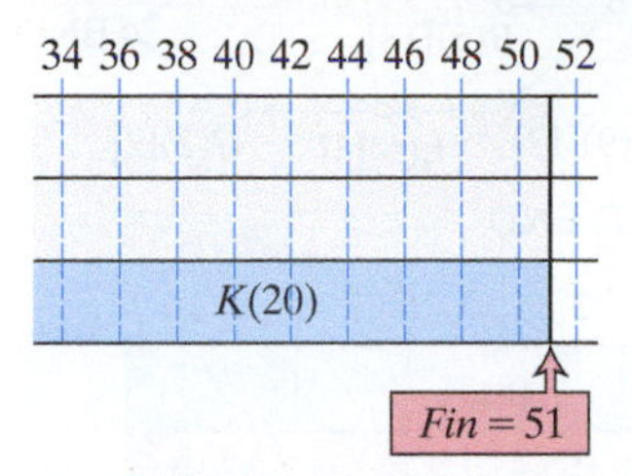

47. **(a)**

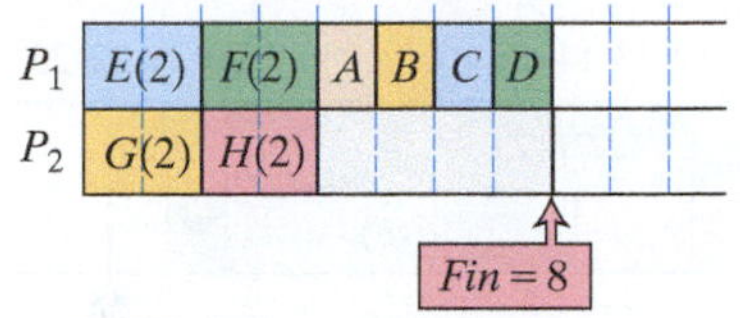

(b) Time: 0 1 2 3 4 5 6 7 8

P_1: A, B, C, D, F(2)

P_2: E(2), G(2), H(2)

Opt = 6

(c) $\frac{2}{6} = 33\frac{1}{3}\%$

49. **(a)** Time: 0 1 2 3 4 5 6 7 8 9 10 11...

P_1: E(5), F(5)

P_2: G(5), H(5)

P_3: I(5), J(5)

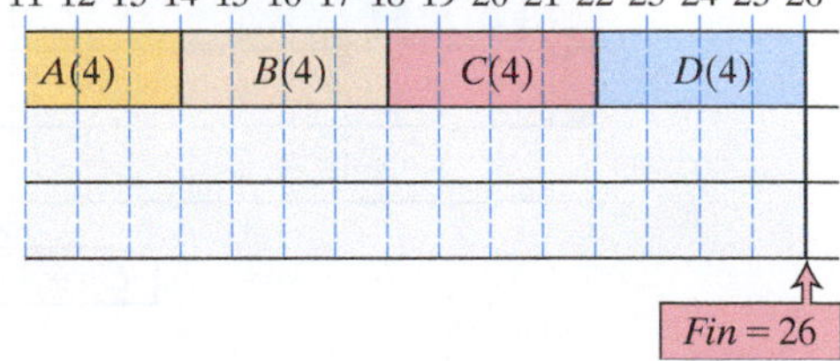

(b)

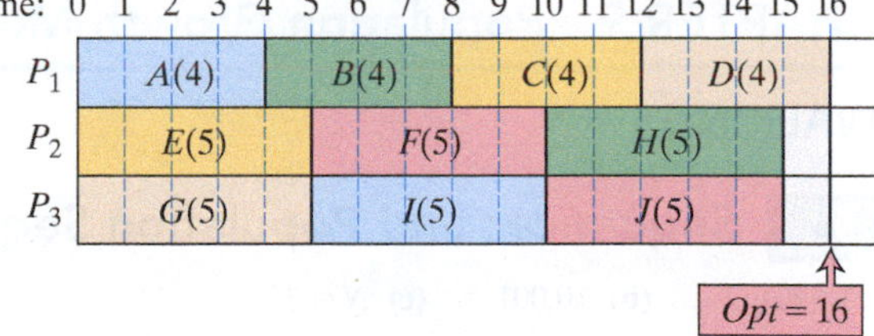

(c) $\frac{10}{16} = 62.5\%$

8.5 Critical Paths and the Critical-Path Algorithm

51. (a)

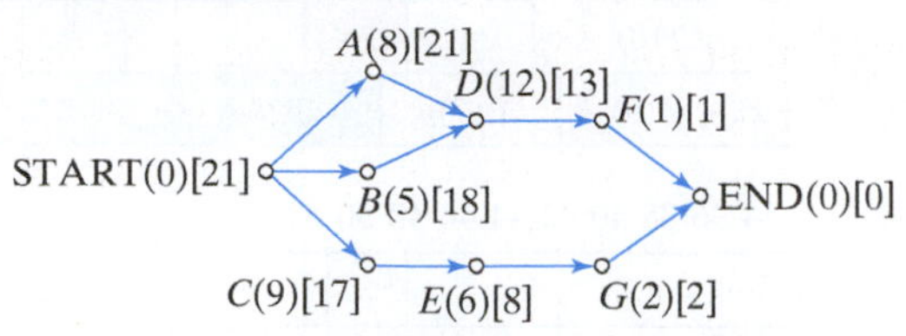

(b) START, A, D, F, END

(c) Time: 0 1 2 3 4 5 6 7 8 9 10 11…

P_1 A(8) …

P_2 B(5) C(9)

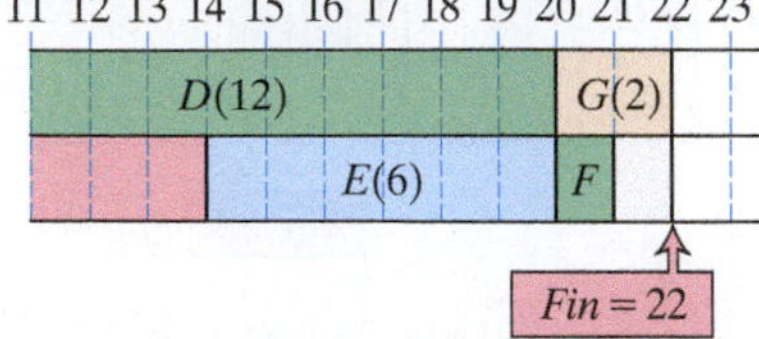

(d) There are a total of 43 work units, so the shortest time the project can be completed by two workers is $\frac{43}{2} = 21.5$ time units. Since there are no tasks taking less than one time unit, the shortest time in which the project can actually be completed is 22 hours.

53. (a) START, B, E, I, K, END

(b)

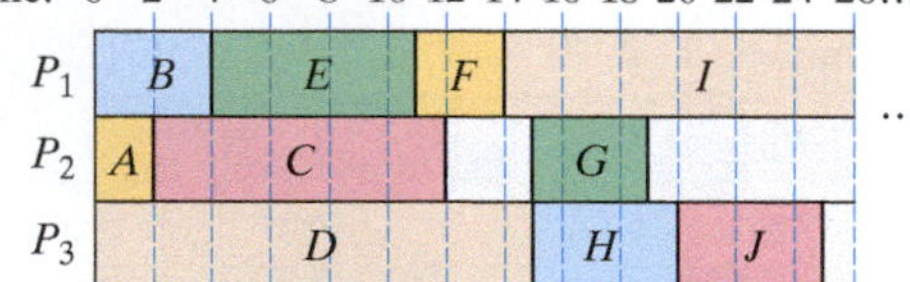

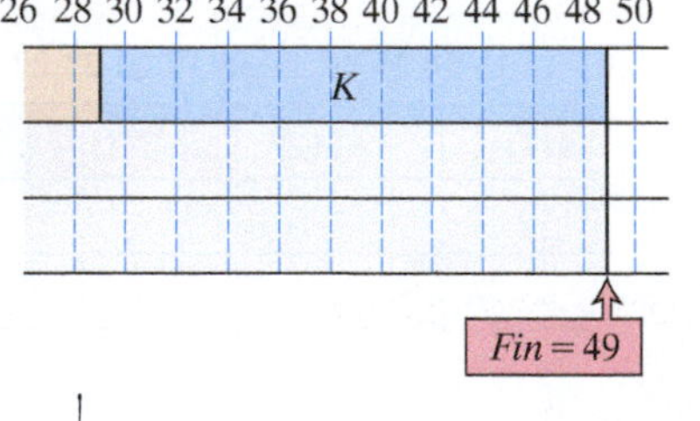

55.

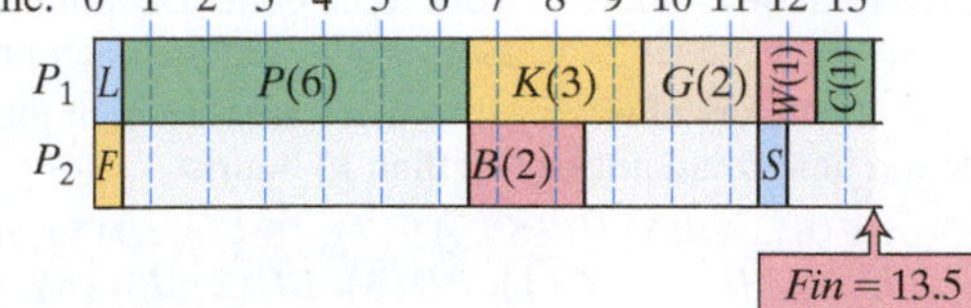

57. (a) The critical path is START, F, H, G, D, E, END; critical time = 20.

(b) The critical-time priority list is F, B, H, G, C, A, D, E, J.

(c) A timeline for the project is shown below. $Fin = 25$.

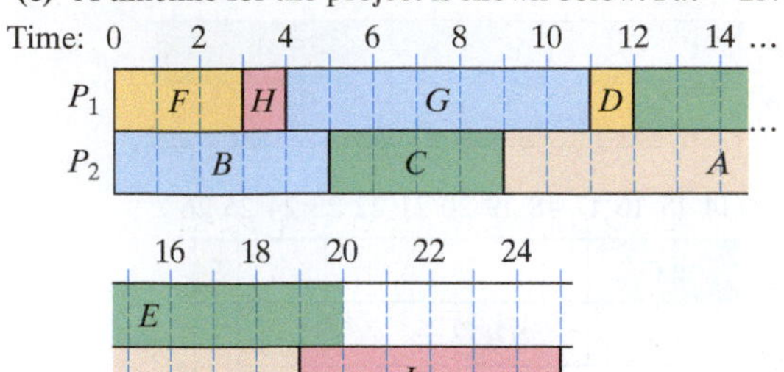

(d) An optimal schedule for $N = 2$ processors with finishing time $Opt = 23$ is shown below. With two processors you can't do any better than $Opt = 23$.

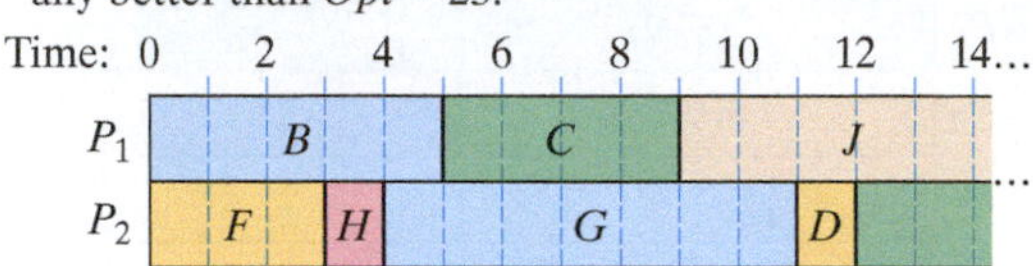

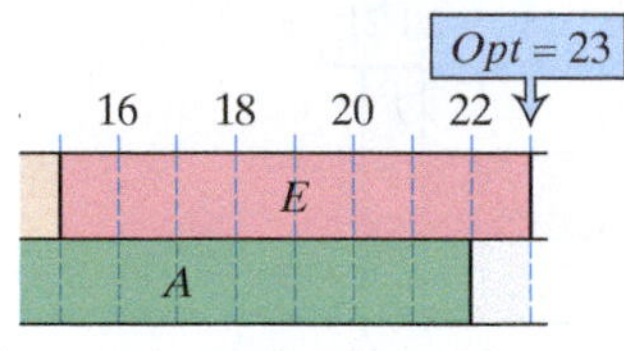

(e) $\varepsilon = \frac{2}{23} = 8.7\%$

JOGGING

59. Each arc of the graph contributes 1 to the indegree sum and 1 to the outdegree sum.

61. $A, B, C, E, G, H, D, F, I$

CHAPTER 9 Population Growth Models

WALKING

9.1 Sequences and Population Sequences

1. (a) 2 **(b)** 10,001 **(c)** $N = 3$

3. (a) 1 **(b)** 3 **(c)** $N = 5$

5. (a) 1 **(b)** −1 **(c)** $N = 1, 3, 5, 7, \ldots$

7. (a) $A_3 = 3, A_4 = 7, A_5 = 17, A_6 = 41$ **(b)** $A_8 = 239$

9. (a) $A_3 = -3, A_4 = -1, A_5 = 5, A_6 = 7$ **(b)** $A_8 = -17$

11. (a) 36, 49 **(b)** $A_N = N^2$ **(c)** $P_N = (N+1)^2$

13. (a) 28, 36 **(b)** $A_N = N(N-1)/2$ **(c)** $P_N = N(N+1)/2$

15. (a) $\frac{24}{9}, \frac{28}{10}$ **(b)** $A_N = \frac{4N}{N+3}$

17. (a) $\frac{1}{720}$ **(b)** $\frac{1}{12!} = \frac{1}{479{,}001{,}600}$

9.2 The Linear Growth Model

19. (a) $P_1 = 205, P_2 = 330, P_3 = 455$ **(b)** $P_N = 80 + 125N$

(c) $P_{100} = 12{,}580$

21. (a) $P_1 = 553, P_2 = 528, P_3 = 503$ **(b)** $P_N = 578 - 25N$

(c) $P_{23} = 3$

23. (a) $P_N = 8 + 3N$ **(b)** $P_{50} = 158$

25. (a) 5.9%

(b) Sometime in July, 2015 (the unemployment rate for June, 2015, would be 0.1%).

27. (a) 82.49 years

(b) the life expectancy for someone born in 2020 would be 90.17 years

29. 24,950 **31. (a)** 100th term **(b)** 16,050

33. **(a)** 5,625 **(b)** 10,000

35. **(a)** 213 **(b)** $137 + 2N$ **(c)** \$7124 **(d)** \$2652

9.3 The Exponential Growth Model

37. **(a)** 13.75 **(b)** $11 \times (1.25)^9 \approx 82$ **(c)** $P_N = 11 \times (1.25)^N$

39. **(a)** $P_1 = 20, P_2 = 80, P_3 = 320$ **(b)** $P_N = 5 \times 4^N$ **(c)** 9

41. **(a)** $P_N = (1.5) \times P_{N-1}$ with $P_0 = 200$ **(b)** $P_N = 200 \times (1.5)^N$
(c) approximately 11,500

43. ~8.94% **45** ~–11.97%

47. **(a)** $R = 3$ **(b)** $[(3)^{21} - 1] \times \frac{2}{2} = 10{,}460{,}353{,}202$

49. **(a)** $R = 0.5$ **(b)** $[(0.5)^{12} - 1] \times \frac{4}{-0.5} \approx 7.998$

51. **(a)** $2^{16} - 1 = 65{,}535$ **(b)** $2^N - 1$

9.4 The Logistic Growth Model

53. **(a)** $p_1 = 0.1680$ **(b)** $p_2 \approx 0.1118$ **(c)** 7.945%

55. **(a)** $p_1 = 0.1680, p_2 \approx 0.1118, p_3 \approx 0.0795,$
$p_4 \approx 0.0585, p_5 \approx 0.0441, p_6 \approx 0.0337,$
$p_7 \approx 0.0261, p_8 \approx 0.0203, p_9 \approx 0.0159, p_{10} \approx 0.0125$
(b) extinction

57. **(a)** $p_1 = 0.4320, p_2 \approx 0.4417, p_3 \approx 0.4439,$
$p_4 \approx 0.4443, p_5 \approx 0.4444, p_6 \approx 0.4444,$
$p_7 \approx 0.4444, p_8 \approx 0.4444, p_9 \approx 0.4444, p_{10} \approx 0.4444$
(b) It stabilizes at 44.44% of the habitat's carrying capacity.

59. **(a)** $p_1 = 0.3570, p_2 \approx 0.6427, p_3 \approx 0.6429,$
$p_4 \approx 0.6428, p_5 \approx 0.6429, p_6 \approx 0.6428,$
$p_7 \approx 0.6429, p_8 \approx 0.6428, p_9 \approx 0.6429, p_{10} \approx 0.6428$
(b) It stabilizes at $\frac{9}{14} \approx 64.29\%$ of the habitat's carrying capacity.

61. **(a)** $p_1 = 0.5200, p_2 = 0.8112, p_3 \approx 0.4978, p_4 \approx 0.8125,$
$p_5 \approx 0.4952, p_6 \approx 0.8124, p_7 \approx 0.4953, p_8 \approx 0.8124,$
$p_9 \approx 0.4953, p_{10} \approx 0.8124$
(b) The population settles into a two-period cycle alternating between a high-population period at 81.24% and a low-population period at 49.53% of the habitat's carrying capacity.

JOGGING

63. **(a)** exponential **(b)** linear **(c)** logistic **(d)** exponential
(e) logistic **(f)** linear **(g)** linear, exponential, and logistic

65. The first N terms of the arithmetic sequence are $P_0, P_0 + d, P_0 + 2d, \ldots, P_0 + (N - 1)d$. Their sum is

$$\frac{(P_0 + [P_0 + (N-1)d]) \times N}{2} = \frac{N}{2}[2P_0 + (N-1)d].$$

67. $1 + 3 + \cdots + (2N - 1) = \frac{2N \cdot N}{2} = N^2$

69. $8, 4, 2, 1, \frac{1}{2}, \frac{1}{4}, \ldots$ (Many other answers are possible.)

CHAPTER 10 Financial Mathematics

WALKING

10.1 Percentages

1. **(a)** 0.0625 **(b)** 0.0375 **(c)** 0.007

3. **(a)** 0.0082 **(b)** 0.0005

5. Lab 2 (85%), Lab 1 ($81\frac{1}{3}\%$), Lab 3 ($78\frac{2}{3}\%$)

7. 215 **9.** 9.5% **11.** 39.15%

13. 10.88% **15.** an increase of 4.4%

10.2 Simple Interest

17. \$1024.80 **19.** \$4000 **21.** 6.75% **23.** 8.33%

25. $r = 4.432 = 443.2\%$.

10.3 Compound Interest

27. **(a)** $\$3250(1.09)^4 = \$4587.64$
(b) $\$3250(1.09)^5 = \$5000.53$.

29. $\$25{,}000(1.035)^{20} = \$49{,}744.72$.

31. **(a)** 0.25% (or 0.0025)
(b) $\$1580(1.0025)^{36} = \$1728.60$.

33. **(a)** 0.01% (or 0.0001)
(b) $\$1580(1.0001)^{3 \times 365} = \$1762.83$.

35. $\$1580e^{(0.03) \times 3} = \$1728.80$.

37. $\$1580e^{(0.0365) \times \frac{500}{365}} = \$1661$. **39.** \$800

41. **(a)** 6% **(b)** 6.09% **(c)** ≈6.1678% **(d)** ≈6.1837%

43. 12 years

10.4 Retirement Savings

45. **(a)** \$380,566.52 **(b)** \$190,283.26 **(c)** \$570,849.78

47. **(a)** \$270,251.07 **(b)** \$532,419.17 **(c)** \$608,116.71

49. **(a)** \$416,680.55 **(b)** \$208,340.28 **(c)** \$625,020.83

51. **(a)** \$293,459.20 **(b)** \$587,922.62 **(c)** \$697,143.48

53. \$212.98

55. \$293,459.20 + \$49,634.67 + \$12,399.90 = \$355,493.77

10.5 Consumer Debt

57. $\$14{,}500 \times 0.005 \times \frac{1.005^{36}}{(1.005^{36} - 1)} = \$441.12$.

59. $\$400 \times \frac{(1.003^{48} - 1)}{(0.003 \times 1.003^{48})} = \$17{,}857$.

61. **(a)** \$933.72 **(b)** \$176,139.20 **63.** \$163,104

JOGGING

65. **(a)** \$1372.17 **(b)** \$21.99 **(c)** \$1039.92

67. 100%

69. **(a)** $\$\left(1 - \frac{x}{100}\right)\left[\left(1 - \frac{y}{100}\right)P\right]$
(b) $\$\left(1 - \frac{y}{100}\right)\left[\left(1 - \frac{x}{100}\right)P\right]$
(c) Multiplication is commutative.

71. Bank B offers the better deal. Monthly payments for the loan with Bank A are \$789.81; monthly payments for the loan with Bank B are \$780.17. [To calculate the monthly payments for Bank A use the amortization formula with $P = \$90{,}000$, $p = \frac{0.1}{12}$, $T = 360$. To calculate the monthly payments for Bank B use the amortization formula with $P = \$92{,}783.51$ (that's what you need to borrow so that when you subtract the 3% loan fee you end up with the \$90,000 you need), $p = \frac{0.095}{12}$, $T = 360$.]

73. **(a)** Monthly payments on a 30-year loan for $P = \$100{,}000$ with an APR of 6% are \$599.55 (call it \$600). Monthly payments on a 21-year loan for the same principal and APR are \$698.86 (call it \$700). Total interest paid on the 30-year loan (total payments − principal borrowed) = \$600 × 360 − \$100,000 = \$116,000. Total interest paid on the 21-year loan = \$700 × 252 − \$100,000 = \$76,400.

(b) Monthly payments on a 15-year loan for $P = \$100{,}000$ with an APR of 6% are \$843.86. Total interest paid on the 15-year loan = \$843.86 × 180 − \$100,000 = \$51,894.80 (call it \$51,895). Total interest saved in going from a 30- to a 15-year loan = \$116,000 − \$51,895 = 64,105.

(c) Monthly payments on a t-year loan for $P = \$100{,}000$ with an APR of 6% are given by $M = \$100{,}000 \times 0.005 \times \frac{1.005^{12t}}{1.005^{12t} - 1} = \$500 \times \frac{1.005^{12t}}{1.005^{12t} - 1}$

75. We'll use the annual contribution version of the retirement savings formula. The argument is exactly the same for the monthly contribution version. For all the retirement savings accounts, the APR is r and the number of years is Y.

(a) Let $V = P\left[\frac{(1+r)^Y - 1}{r}\right]$ be the value of the retirement savings account with annual contribution P. Then the value of a retirement savings account with annual contribution cP is $(cP)\left[\frac{(1+r)^Y - 1}{r}\right] = c\left(P\left[\frac{(1+r)^Y - 1}{r}\right]\right) = cV$.

(b) Let $V = P\left[\frac{(1+r)^Y - 1}{r}\right]$ be the value of the retirement savings account with annual contribution P, and let $W = Q\left[\frac{(1+r)^Y - 1}{r}\right]$ be the value of the retirement savings account with annual contribution Q. Then the value of a retirement savings account with annual contribution $(P + Q)$ is $(P+Q)\left[\frac{(1+r)^Y - 1}{r}\right] = P\left[\frac{(1+r)^Y - 1}{r}\right] + Q\left[\frac{(1+r)^Y - 1}{r}\right] = V + W$.

CHAPTER 11 The Mathematics of Symmetry

WALKING

11.2 Reflections

1. **(a)** C **(b)** F **(c)** E **(d)** B

3. **(a)**

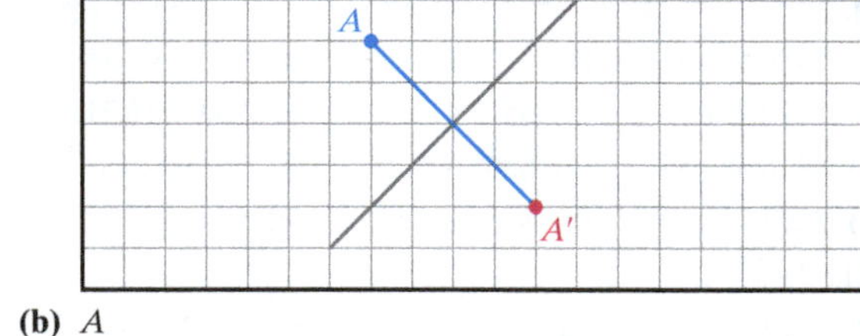

(b) A

5. **(a)** **(b)**

7. **(a)** **(b)** **(c)** **(d)**

Fixed points on $PQRS$

9. **(a)** **(b)**

11.

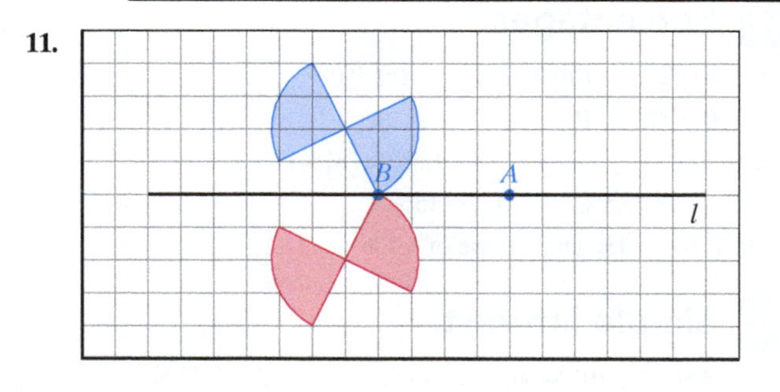

11.3 Rotations

13. **(a)** I **(b)** G **(c)** A **(d)** F **(e)** E

15. **(a)** 10° **(b)** 350° **(c)** 100° **(d)** 80°

17. **(a)** **(b)**

19. **(a)**
(b)

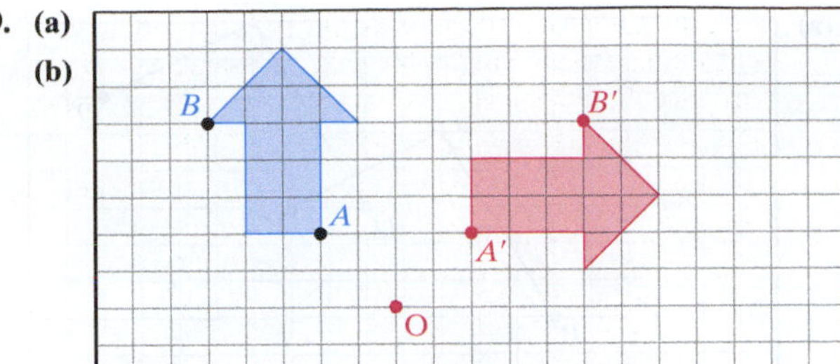

21.

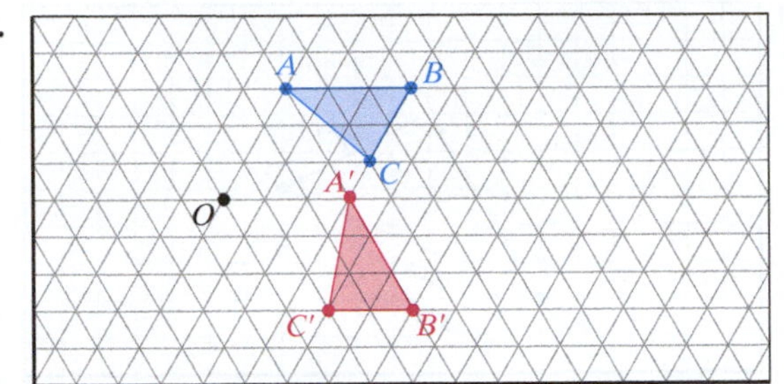

11.4 Translations

23. **(a)** *C* **(b)** *C* **(c)** *A* **(d)** *D*

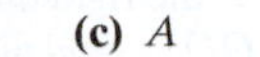

25. **(a)**
(b)
(c)

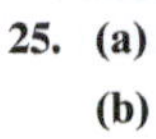

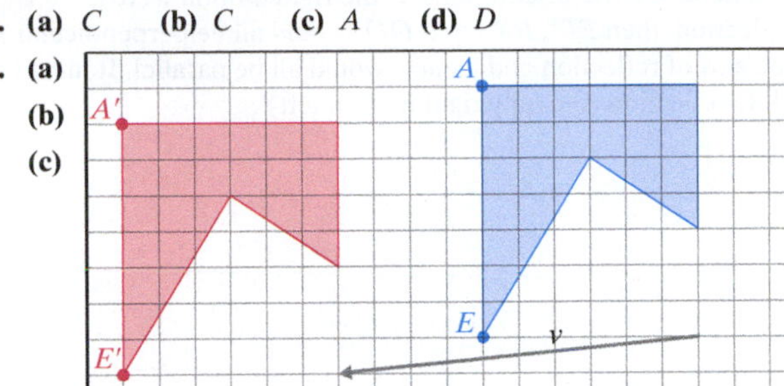

27.

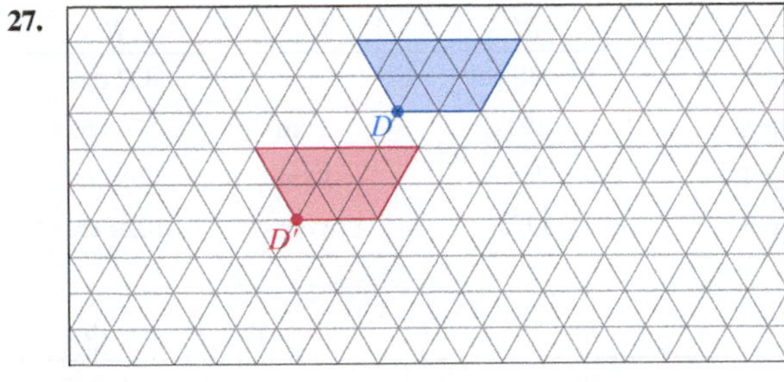

11.5 Glide Reflections

29.

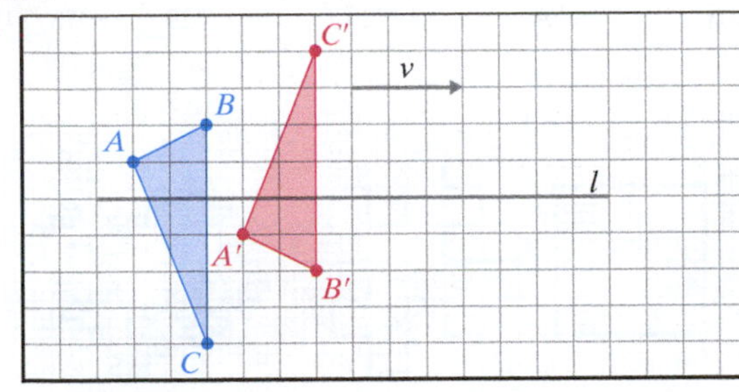

31.

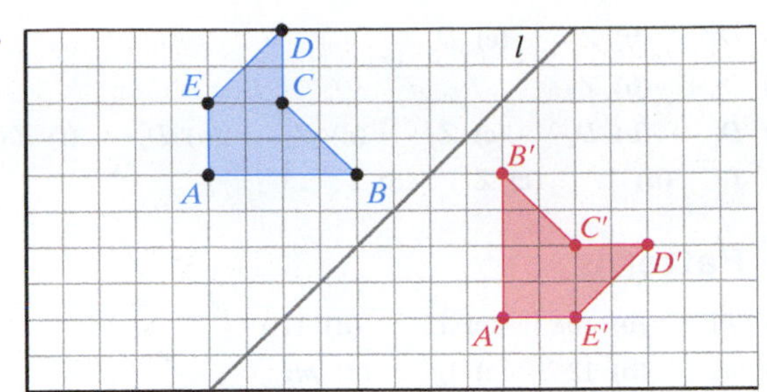

33. **(a)**
(b)
(c)

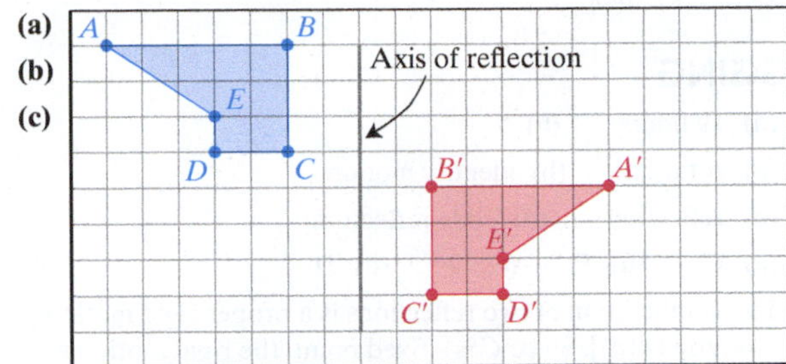

35.

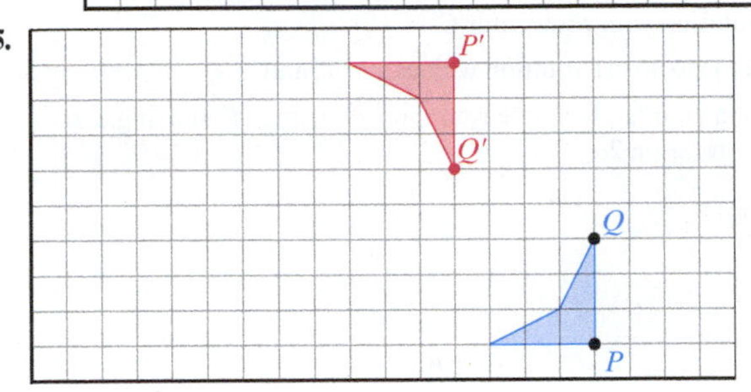

37.

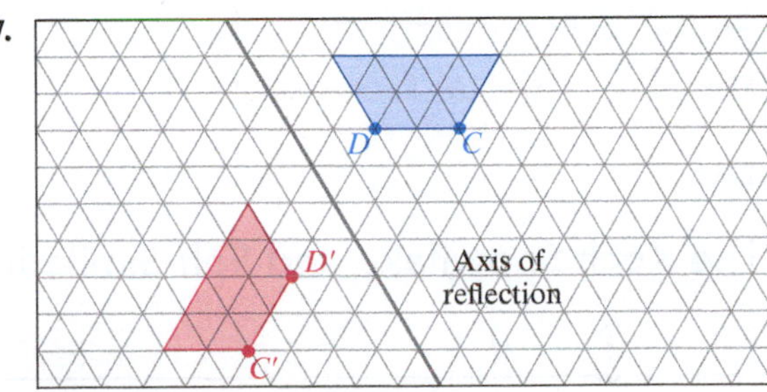

11.6 Symmetries and Symmetry Types

39. **(a)** reflection with axis going through the midpoints of AB and DC; reflection with axis going through the midpoints of AD and BC; identity and half-turn with rotocenter the center of the rectangle

(b) no reflections; identity and half-turn with rotocenter the center of the parallelogram

(c) reflection with axis going through the midpoints of AB and DC; identity

41. **(a)** reflections (three of them) with axis going through pairs of opposite vertices; reflections (three of them) with axis going through the midpoints of opposite sides of the hexagon; rotations of 60°, 120°, 180°, 240°, and 300° with rotocenter the center of the hexagon, plus the identity

(b) no reflections; rotations of 72°, 144°, 216°, 288° with rotocenter the center of the star, plus the identity

43. **(a)** D_2 **(b)** Z_2 **(c)** D_1

45. **(a)** D_6 **(b)** Z_5

47. **(a)** D_1 **(b)** D_1 **(c)** Z_1 **(d)** Z_2 **(e)** D_2 **(f)** Z_2

49. **(a)** J **(b)** T **(c)** Z **(d)** I

11.7 Patterns

51. **(a)** $m1$ **(b)** $1m$ **(c)** 12 **(d)** 11

53. **(a)** $m1$ **(b)** 12 **(c)** $1g$ **(d)** mg

55. 12 **57.** mm

JOGGING

59. **(a)** 19 hours **(b)** 5

61. **(a)** rotation **(b)** identity motion
(c) reflection **(d)** glide reflection

63. **(a)** C **(b)** P **(c)** D **(d)** D

65. The combination of two reflections is a proper rigid motion [see Exercise 64(d)]. Since C is a fixed point, the rigid motion must be a rotation with rotocenter C.

67. **(a)** a clockwise rotation with center C and angle of rotation 2α
(b) a counterclockwise rotation with center C and angle of rotation 2α

69. **(a)**

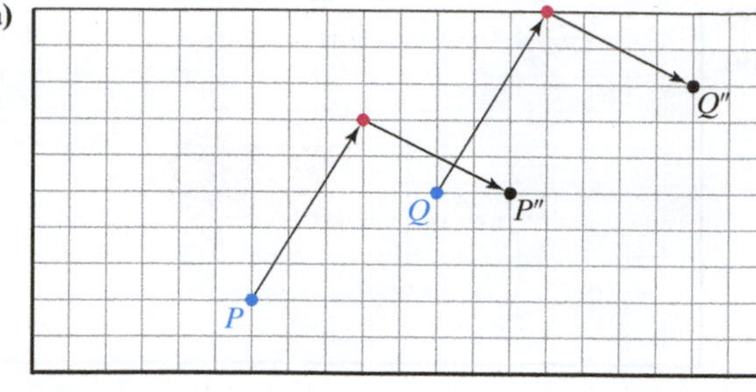

(b)

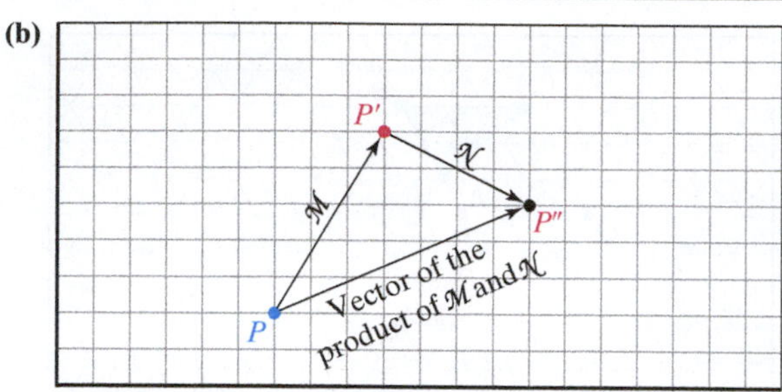

71. Rotations and translations are proper rigid motions. The given motion is an improper rigid motion (it reverses the clockwise-counterclockwise orientation). If the rigid motion were a reflection, then PP', RR', and QQ' would all be perpendicular to the axis of reflection and, hence, would all be parallel. It must be a glide reflection (the only rigid motion left).

CHAPTER 12 Fractal Geometry

WALKING

12.1 The Koch Snowflake and Self-Similarity

1.

	M	L	P
Step 2	48	9 cm	432 cm
Step 3	192	3 cm	576 cm
Step 4	768	1 cm	768 cm
Step 5	3072	1/3 cm	1024 cm

3.

	R	S	T	Q
Step 3	48	1/9	16/3	376/3
Step 4	192	1/81	64/27	3448/27
Step 5	768	1/729	256/243	31288/243

5.

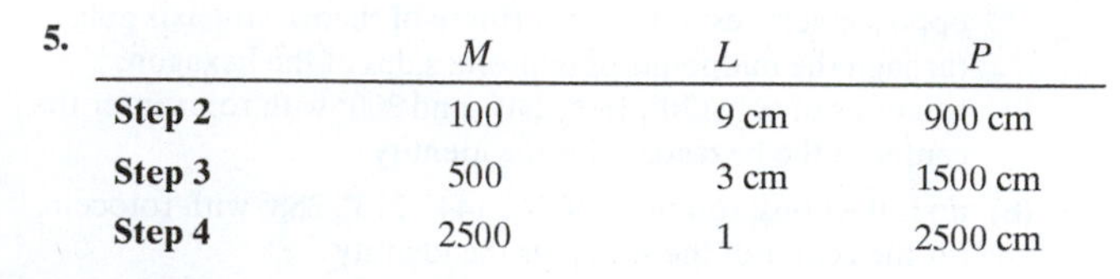

	M	L	P
Step 2	100	9 cm	900 cm
Step 3	500	3 cm	1500 cm
Step 4	2500	1	2500 cm

7.

	R	S	T	Q
Step 3	100	1/9	100/9	1333/9
Step 4	500	1/81	500/81	12,497/81

9.

	M	L	P
Step 2	48	9 cm	432 cm
Step 3	192	3 cm	576 cm
Step 4	768	1 cm	768 cm
Step 5	3072	1/3 cm	1024 cm

11.

	R	S	T	Q
Step 3	48	1/9	48/9	110/3
Step 4	192	1/81	192/81	926/27
Step 5	768	1/729	768/729	8078/243

13. **(a)**

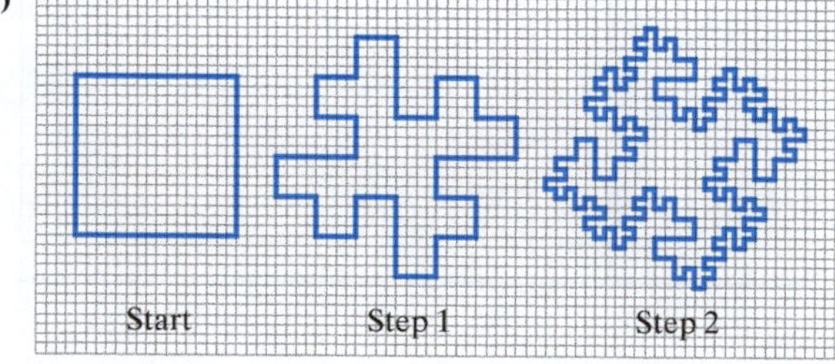

(b) 128 **(c)** 256
(d) The perimeter doubles at each step of the construction.

15. **(a)** 256 **(b)** 256
(c) At each step, the area added is the same as the area subtracted.

12.2 The Sierpinski Gasket and the Chaos Game

17.

	R	S	T	Q
Step 3	9	1	9	27
Step 4	27	$1/4$	$27/4$	$81/4$
Step 5	81	$1/16$	$81/16$	$243/16$

19.

	U	V	W
Step 2	9	2 cm	18 cm
Step 3	27	1 cm	27 cm
Step 4	81	$1/2$ cm	$81/2$ cm
Step 5	243	$1/4$ cm	$243/4$ cm

21.

	R	S	T	Q
Step 2	18	$1/81$	$2/9$	$4/9$
Step 3	108	$1/729$	$4/27$	$8/27$
Step 4	648	$1/6561$	$8/81$	$16/81$
Step N	$3 \times 6^{N-1}$	$1/9^N$	$3 \times 6^{N-1}/9^N$	$(2/3)^N$

23.

	U	V	W
Step 2	36	1 cm	36 cm
Step 3	216	$1/3$ cm	72 cm
Step 4	1296	$1/9$ cm	144 cm
Step N	6^N	$1/3^{N-2}$ cm	$9 \cdot 2^N$ cm

25. **(a)** $5/9$ **(b)** $(5/9)^2$ **(c)** $(5/9)^N$

27.

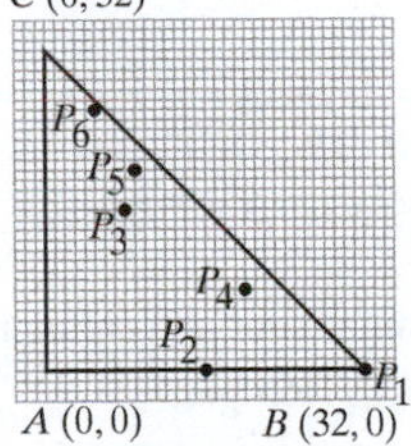

29.

Roll	Point	Coordinates
3	P_1	$(32, 0)$
1	P_2	$(16, 0)$
2	P_3	$(8, 0)$
3	P_4	$(20, 0)$
5	P_5	$(10, 16)$
5	P_6	$(5, 24)$

31. **(a)** **(b)**

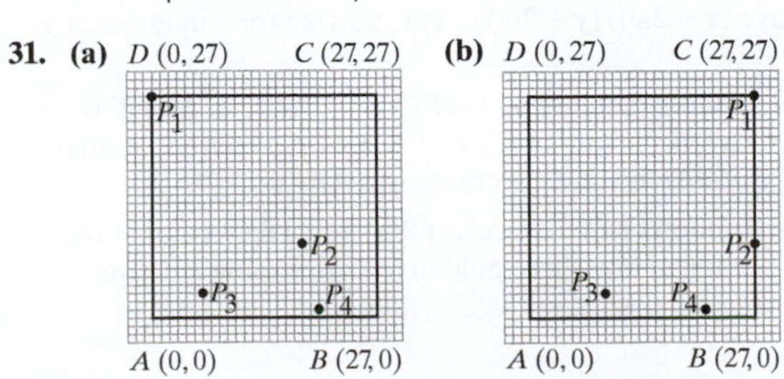

(c)

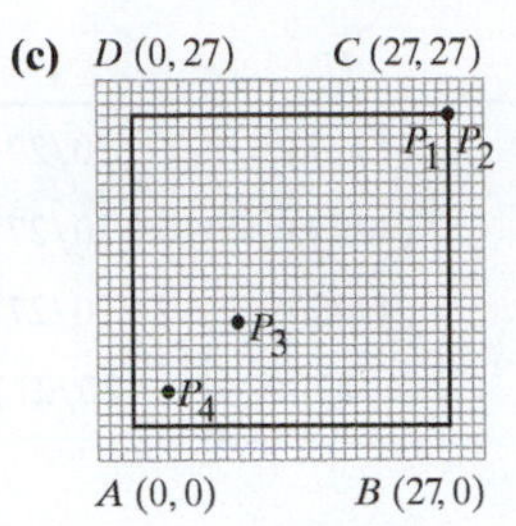

33. **(a)** 4, 2, 1, 2 **(b)** 3, 1, 1, 3 **(c)** 1, 3, 4, 2

12.4 The Mandelbrot Set

35. **(a)** $-1-i$ **(b)** i **(c)** $-1-i$

37. **(a)** $-0.25+0.125i$ **(b)** $-0.25-0.125i$

39. **(a)**

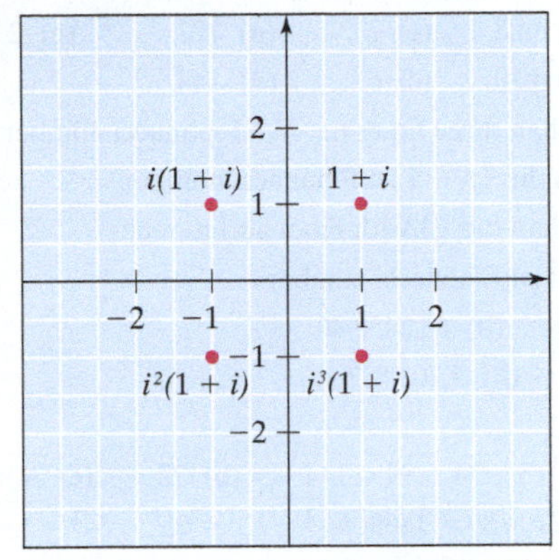

(b)

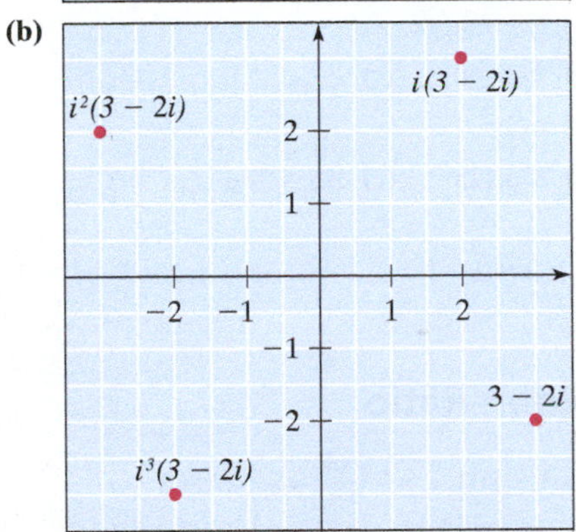

(c) It is a 90-degree counterclockwise rotation.

41. **(a)** 2, 2, 2, 2 **(b)** 2

(c) Attracted to 2. Each number in the sequence is 2.

43. **(a)** $s_1 = -0.25$; $s_2 = -0.4375$; $s_3 \approx -0.3086$; $s_4 \approx -0.4048$; $s_5 \approx -0.3362$

(b) -0.3660

(c) Attracted to -0.3660. From (a) and (b), the sequence gets close to -0.3660 and then $s_N = s_{N+1}$ (approximately).

45. **(a)** $s_1 = -1-i$; $s_2 = i$; $s_3 = -1-i$; $s_4 = i$; $s_5 = -1-i$

(b) Periodic; the odd terms are $-1-i$ and the even terms are i.

JOGGING

47. **(a)** $(3/4)^N A$

(b) The area of the Sierpinski gasket is smaller than the area of the gasket formed at any step of construction. As N gets larger and larger, $(3/4)^N$ gets smaller and smaller (i.e., approaches 0), and so does $(3/4)^N A$.

49. **(a)**

	C	U	V
Step 2	7×20	$(1/27)^2$	$(20/27)^2$
Step 3	7×20^2	$(1/27)^3$	$(20/27)^3$
Step 4	7×20^3	$(1/27)^4$	$(20/27)^4$
Step N	$7 \times 20^{N-1}$	$(1/27)^N$	$(20/27)^N$

(b) Since $20/27 < 1$, as N gets larger and larger, $(20/27)^N$ approaches 0.

51. -0.5 is a solution to the quadratic equation $x^2 - 0.75 = x$. This means that if $x_n = -0.5$, then $x_{n+1} = x + 0.75 = x_n = -0.5$. It follows that the Mandelbrot sequence is attracted to -0.5.

53. periodic

CHAPTER 13 Fibonacci Numbers and the Golden Ratio

WALKING

13.1 Fibonacci Numbers

1. **(a)** 610 **(b)** 608 **(c)** 233 **(d)** 122 **(e)** 2

3. **(a)** 12 **(b)** 610 **(c)** 6 **(d)** 144

5. **(a)** one more than three times the Nth Fibonacci number
(b) three times the $(N+1)$st Fibonacci number
(c) one more than the $(3N)$th Fibonacci number
(d) the $(3N+1)$st Fibonacci number

7. **(a)** 39,088,169 **(b)** 63,245,986

9. **(a)** 9,227,465 **(b)** 5,702,887

11. 6765

13. **(a)** $1 + 2 + 5 + 13 + 34 + 89 = 144$ **(b)** 22 **(c)** $N + 1$

15. **(a)** $F_8 \times F_{11} = 21 \times 89 = 1869$; $F_{10}{}^2 - F_9{}^2 = 55^2 - 34^2 = 1869$
(b) $F_N F_{N+3} = F_{N+2}^2 - F_{N+1}^2$

17. **(a)** F_{N+3} **(b)** F_{N-1} **(c)** F_{N+6}

19. **(a)** $\frac{F_{N+1}}{F_{N-1}}$ **(b)** $-\frac{F_{N-2}}{F_N}$

21. **(a)** 275 **(b)** 375,125 **(c)** $A_N = 5F_N$

23. **(a)** 322
(b) $2 \times 1 - 1 = 1; 2 \times 2 - 1 = 3; 2 \times 3 - 2 = 4; 2 \times 5 - 3 = 7$
(c) $L_{20} = 15{,}127$

13.2 The Golden Ratio

25. **(a)** $x = \frac{1+\sqrt{5}}{2} \approx 1.61803$, $x = \frac{1-\sqrt{5}}{2} \approx -0.61803$
(b) 1
(c) One solution is positive, the second solution is negative, and their sum equals 1. This implies that when you add the two solutions, their decimal parts must cancel each other out.

27. **(a)** $x = \frac{8+\sqrt{124}}{6} \approx 3.18925$, $x = \frac{8-\sqrt{124}}{6} \approx -0.52259$
(b) $\frac{8}{3}$

29. **(a)** Substituting $x = 1$ into the equation gives $55 = 34 + 21$.
(b) The sum of the two solutions is $\frac{34}{55}$. It follows that the second solution is $x = \frac{34}{55} - 1 = -\frac{21}{55}$.

31. **(a)** Substituting $x = -1$ into the equation gives $21 = -34 + 55$.
(b) The sum of the two solutions is $\frac{34}{21}$. It follows that the second solution is $x = \frac{34}{21} + 1 = \frac{55}{21}$.

33. **(a)** Substituting $x = 1$ into the equation gives $F_N = F_{N-1} + F_{N-2}$, true for Fibonacci numbers.
(b) The sum of the two solutions is $\frac{F_{N-1}}{F_N}$. It follows that the second solution is $x = \frac{F_{N-1}}{F_N} - 1 = -\frac{F_{N-2}}{F_N}$.

35. **(a)** $\frac{1}{\phi} \approx 0.6180339887$
(b) Start with the golden property: $\phi^2 = \phi + 1$. Divide both sides by ϕ and get $\phi = 1 + \frac{1}{\phi}$. This gives $\frac{1}{\phi} = \phi - 1$. It follows that ϕ and $\frac{1}{\phi}$ must have exactly the same decimal part.

37. **(a)** 1.394×10^{104} **(b)** 5.326×10^{103}

39. **(a)** 169 **(b)** 2.41429 **(c)** 2.41421 **(d)** 2.41421

13.3 Gnomons

41. **(a)** 124.5 in. **(b)** 945 in.^{2}

43. **(a)** 156 m **(b)** 2880 m^2

45. $x = 48$ **47.** $x = 4$ **49.** $x = 5$

51. **(a)** $m(\angle ACD) = 72°; m(\angle CAD) = 72°; m(\angle CDA) = 36°$
(b) $AC = 1; CD = \phi; AD = \phi$

53. $x = 12, y = 10$

JOGGING

55. **(a)** 1, 0.5, 0.666667, 0.6, 0.625, 0.615385, 0.619048, 0.617647, 0.618182, 0.61798, 0.618056, 0.618026, 0.618037, 0.618033, . . .
(b) We know that $\left(\frac{F_{N+1}}{F_N}\right) \to \phi$. It follows that $\left(\frac{F_N}{F_{N+1}}\right) \to \frac{1}{\phi}$. But $\frac{1}{\phi} = \phi - 1$. [This last assertion follows from taking $\phi^2 = \phi + 1$ and dividing both sides by ϕ. See Exercise 35(b).]

57. **(a)** $1, 1 + \frac{1}{1} = 2, 1 + \frac{1}{2} = \frac{3}{2}, 1 + \frac{1}{3/2} = \frac{5}{3}, 1 + \frac{1}{5/3} = \frac{8}{5}, 1 + \frac{1}{8/5} = \frac{13}{8}$
(b) $T_N = \frac{F_{N+1}}{F_N}$

59. **(a)** $K_1 = 2F_2 - F_1 = 2 \times 1 - 1 = 1$, $K_2 = 2F_3 - F_2 = 2 \times 2 - 1 = 3$, and $K_{N-1} + K_{N-2} = (2F_N - F_{N-1}) + (2F_{N-1} - F_{N-2}) = 2(F_N + F_{N-1}) - (F_{N-1} + F_{N-2}) = 2F_{N+1} - F_N = K_{N+1}$. It follows that the $K_N = L_N$.
(b) $\frac{L_{N+1}}{L_N} = \frac{2F_{N+2} - F_{N+1}}{2F_{N+1} - F_N}$. If you divide the numerator and denominator of the right-hand expression by F_{N+1} you get $\frac{2(F_{N+2}/F_{N+1}) - 1}{2 - (F_N/F_{N+1})}$. As N gets larger and larger, $(F_{N+2}/F_{N+1}) \to \phi$ and $(F_N/F_{N+1}) \to \frac{1}{\phi}$, and it follows that $\frac{2(F_{N+2}/F_{N+1}) - 1}{2 - (F_N/F_{N+1})} \to \frac{2\phi - 1}{2 - 1/\phi} = \phi$.

61. The sum of two odd numbers is always even, and the sum of an odd number and an even number is always odd. Since the seeds in the Fibonacci sequence are both odd, every third number is even and the others are all odd.

63. 20 by 30

65. Yes, but only when $x = y$ (i.e if the picture is a square). The reason is that for the frame to be a gnomon to the picture, we must have $(x + 2w)/(y + 2w) = x/y$. Solving for x in terms of y gives $x = y$.

67. From elementary geometry, $\angle AEF \cong \angle DBA$, so ΔAEF is similar to ΔDBA. Thus, $AF/FE = DA/AB$, which shows that rectangle $ADEF$ is similar to rectangle $ABCD$.

69. The result follows from Exercise 68(a) and a dissection of the regular pentagon into three golden triangles using any two nonintersecting diagonals.

CHAPTER 14 Censuses, Surveys, Polls, and Studies

WALKING

14.1 Enumeration

1. 64 **3.** 2320 **5.** 2549

7. 2000 **9.** 274

11. 273 (one less than under the standard capture-recapture formula).

13. 660

14.2 Measurement

15. **(a)** B **(b)** D **(c)** A **(d)** C

17. **(a)** the registered voters in Cleansburg

(b) the 680 registered voters polled by telephone

(c) simple random sampling

19. Smith: 3%; Jones: 3%; Brown: 0%

21. **(a)** The sample consisted of the 350 students attending the Eureka High School football game the week before the election.

(b) $\frac{350}{1250} = 28\%$

23. **(a)** The population consists of all 1250 students at Eureka High School, whereas the sampling frame consists only of those 350 students who attended the football game the week prior to the election.

(b) Mainly sampling bias. The sampling frame is not representative of the population.

25. **(a)** the citizens of Cleansburg

(b) The sampling frame is limited to that part of the target population that passes by a city street corner between 4:00 P.M. and 6:00 P.M.

27. **(a)** The choice of street corner could make a great deal of difference in the responses collected.

(b) Interviewer *D*. We are assuming that people who live or work downtown are much more likely to answer yes than people in other parts of town.

(c) Yes, for two main reasons: (1) People out on the street between 4 P.M. and 6 P.M. are not representative of the population at large. For example, office and white-collar workers are much more likely to be in the sample than homemakers and schoolteachers. (2) The five street corners were chosen by the interviewers, and the passersby are unlikely to represent a cross section of the city.

(d) No. No attempt was made to use quotas to get a representative cross section of the population.

29. **(a)** Stratified sampling.

(b) Quota sampling.

31. **(a)** Label the crates with numbers 1 through 250. Select 20 of these numbers at random (put the 250 numbers in a hat and draw out 20). Sample those 20 crates.

(b) Select the top 20 crates in the shipment (those easiest to access).

(c) Randomly sample 6 crates from supplier A, 6 crates from supplier B, and 8 crates from supplier C.

(d) Choose 6 crates from supplier A, 6 crates from supplier B, and 8 crates from supplier C.

14.3 Cause and Effect

33. **(a)** The target population consisted of anyone who could have a cold and would consider buying vitamin *X* (i.e., pretty much all adults).

(b) The sampling frame consisted of college students in the San Diego area having a cold at the time.

(c) The sample consisted of the 500 students that took vitamin *X*.

35. 1. Using college students—they are not a representative cross section of the population.

2. Using subjects only from the San Diego area.

3. Offering money as an incentive to participate.

4. Allowing self-reporting (the subjects themselves determine when their colds are over).

37. **(a)** All potential knee surgery patients.

(b) The 180 potential knee surgery patients at the Houston VA Medical Center who volunteered to be in the study

39. **(a)** Yes. The 180 patients in the study were assigned to a treatment group at random.

(b) Blind

41. The target population consisted of all people having a history of colorectal adenomas.

43. **(a)** The treatment group consisted of the 1287 patients who were given 25 milligrams of Vioxx daily. The control group consisted of the 1299 patients who were given a placebo.

(b) This clinical study was a controlled placebo study because there was a control group that did not receive the treatment, but instead received a placebo. It was a randomized controlled study because the 2586 participants were randomly divided into the treatment and control groups. The study was double blind because neither the participants nor the doctors involved in the study knew who was in each group.

45. **(a)** The target population consisted of women (particularly young women).

(b) The sampling frame consists of those women between 16 and 23 years of age who are not at high risk for HPV infection (i.e., those women having no prior abnormal Pap smears and at most five previous male sexual partners). Pregnant women are also excluded from the sampling frame due to the risks involved.

47. **(a)** The treatment group consisted of the women who received the HPV vaccine.

(b) This was a controlled placebo study because there was a control group that did not receive the treatment, but instead received a placebo injection. It was a randomized controlled study because the 2392 participants were randomly divided into the treatment and control groups. The study was likely double blind because neither the participants nor the medical personnel giving the injections knew who was in each group.

49. **(a)** The target population consisted of all people suffering from diphtheria.

(b) The sampling frame consisted of the individuals admitted to that particular Copenhagen hospital between May 1896 and April 1897 and having serious diptheria symptoms.

51. **(a)** The control group consisted of those patients admitted on the "odd" days who received just the standard treatment for diphtheria at the time. The treatment group consisted of those patients admitted on the "even" days who received both the new serum and the standard treatment.

(b) When the new serum was combined with the standard treatment, it proved effective in treating diphtheria.

53. (a) The target population consisted of men with prostate enlargement; particularly older men.

(b) The sampling frame consists of men aged 45 years or older having moderately impaired urinary flow. It appears to be a representative subset of the target population.

55. (a) The treatment group consisted of the group of volunteers that received saw palmetto daily.

(b) The saw palmetto odor would be a dead give away as to who was getting the treatment and who was getting a placebo. To make the study truly blind, the odor had to be covered up.

(c) Yes. There was a control group that received a placebo instead of the treatment, and the treatment group was randomized. We can infer the study was blind from the fact that an effort was made to hide the odor of the saw palmetto. From the description of the study there is no way to tell if the study was also double blind.

JOGGING

57. (a) The target population and the sampling frame both consist of all TSU undergraduates.

(b) 15,000

59. (a) In simple random sampling, any two members of the population have as much chance of both being in the sample as any other two, but in this sample, two people with the same last name—say Len Euler and Linda Euler—have no chance of being in the sample together.

(b) Sampling variability. The students sampled appear to be a representative cross section of all TSU undergraduates.

61. About $113.15 (252 quarters, 261 nickels, 371 dimes)

63. Capture-recapture would overestimate the true population. If the fraction of those tagged in the recapture appears lower than it is in reality, then the fraction of those tagged in the initial capture will also be computed as lower than the truth. This makes the total population appear larger than it really is.

65. Answers will vary.

67. Consideration of one exception made them less likely to consider a second exception.

69. (a) An Area Code 900 telephone poll represents an extreme case of selection bias. People who respond to these polls usually represent the extreme viewpoints (strongly for or strongly against), leaving out much of the middle of the road point of view. Economics also plays some role in the selection bias. (While 50 cents is not a lot of money anymore, poor people are much more likely to think twice before spending the money to express their opinion.)

(b) This survey was based on fairly standard modern-day polling techniques (random sample telephone interviews, etc.) but it had one subtle flaw. How reliable can a survey about the conduct of the newsmedia be when the survey itself is conducted by a newsmedia organization? ("The fox guarding the chicken-coop" syndrome.)

(c) Both surveys seem to have produced unreliable data—survey 1 overestimating the public's dissapproval of the role played by the newsmedia and survey 2 overestimating the public's support for the press coverage of the war.

(d) Any reasoned out answer should be acceptable. (Since Area Code 900 telephone polls are particularly unreliable, survey 2 gets our vote.)

71. (a) Both samples should be a representative cross section of the same population. In particular, it is essential that the first sample, after being released, be allowed to disperse evenly throughout the population, and that the population should not change between the time of the capture and the time of the recapture.

(b) It is possible (especially when dealing with elusive types of animals) that the very fact that the animals in the first sample allowed themselves to be captured makes such a sample biased (they could represent a slower, less cunning group). This type of bias is compounded with the animals that get captured the second time around. A second problem is the effect that the first capture can have on the captured animals. Sometimes the animal may be hurt (physically or emotionally), making it more (or less) likely to be captured the second time around. A third source of bias is the possibility that some of the tags will come off.

CHAPTER 15 Graphs, Charts, and Numbers

WALKING

15.1 Graphs and Charts

1. (a)

Score	10	40	50	60	70	80	100
Frequency	1	1	3	7	6	5	2

(b)

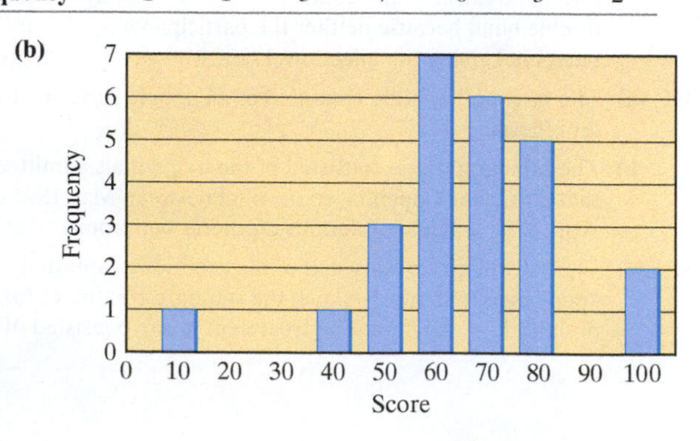

3. (a)

Grade	A	B	C	D	F
Frequency	7	6	7	3	2

(b)

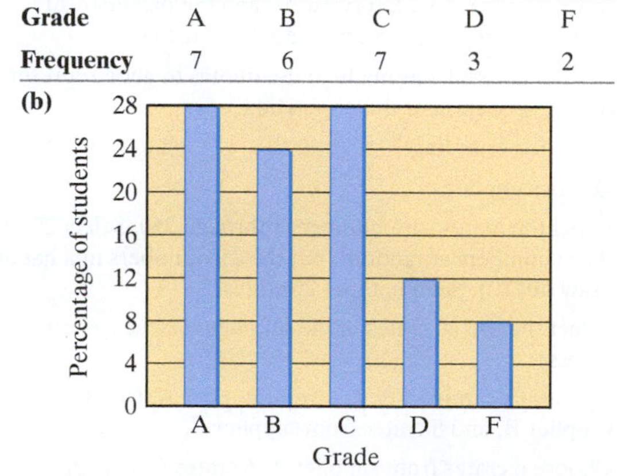

5. (a)

Distance	0.0	0.5	1.0	1.5	2.0	2.5	3.0	5.0	6.0	8.5
Frequency	4	4	3	7	3	2	1	1	1	1

(b)

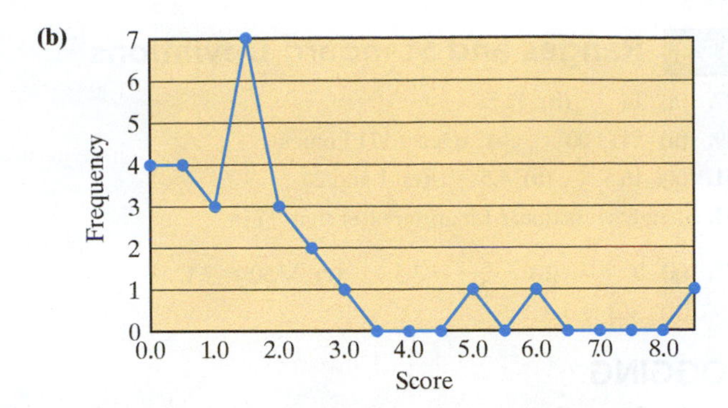

7.

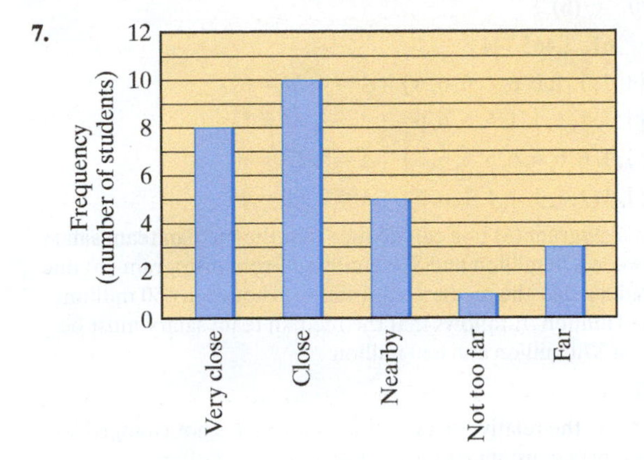

9.

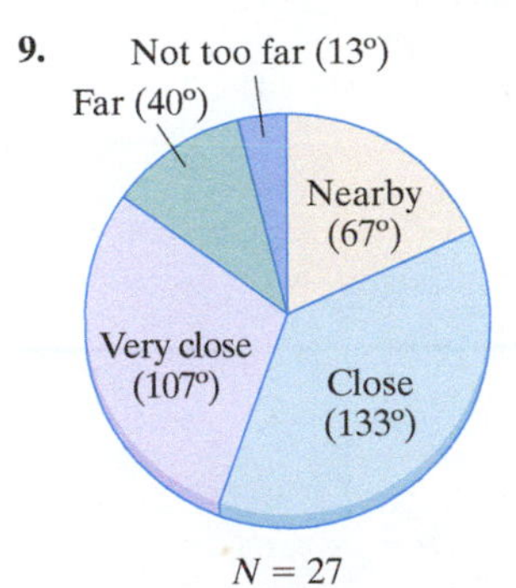

$N = 27$

11. (a) 40 **(b)** 0% **(c)** 57.5%

13. (a) qualitative **(b)** $0.47 \times 19{,}548 \approx 9188$

15.

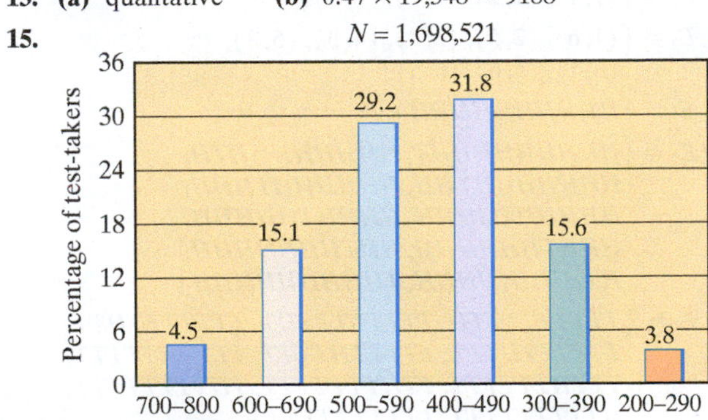

17. (a)

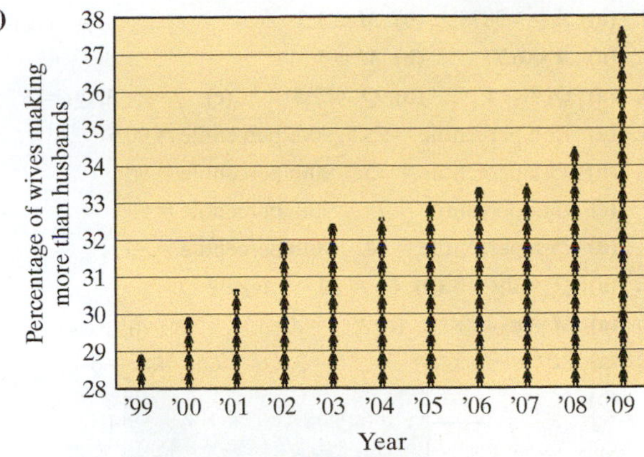

(b)

Percentage of wives making more than husbands

Year

19. (a) 12 ounces

(b) The third class interval: "more than 72 ounces and less than or equal to 84 ounces." A value that falls exactly on the boundary between two class intervals belongs to the class interval to the left.

(c)

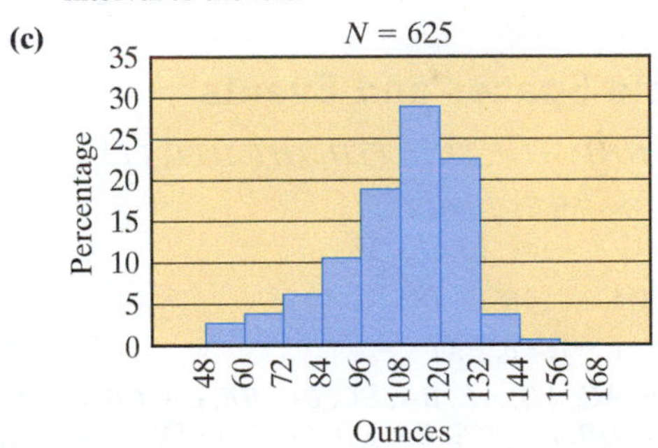

21. (a) 17 **(b)** 14 **(c)** 1

15.2 Means, Medians, and Percentiles

23. (a) $A = 2$ **(b)** $M = 3$ **(c)** $A = 2, M = 2.5$

25. (a) $A = 4.5, M = 4.5$ **(b)** $A = 5, M = 5$

(c) $A = 5.5, M = 5.5$

27. (a) $A = 22.5, M = 17.5$ **(b)** $A = 122.5, M = 117.5$

29. **(a)** $A = 1.5875$ **(b)** $M = 1.5$

31. **(a)** $A = 6.77$ **(b)** $M = 7$

33. **(a)** $Q_1 = -4$ **(b)** $Q_3 = 7.5$ **(c)** $Q_1 = -3, Q_3 = 7$

35. **(a)** 75th percentile = 75.5, 90th percentile = 90.5

(b) 75th percentile = 75, 90th percentile = 90

(c) 75th percentile = 75, 90th percentile = 90

(d) 75th percentile = 74, 90th percentile = 89

37. **(a)** $Q_1 = 29$ **(b)** $Q_3 = 32$ **(c)** 37

39. **(a)** $M = d_{836,198}$ **(b)** $Q_1 = d_{418,099}$ **(c)** $d_{1,337,916.5}$

41. **(a)** $Min = -6, Q_1 = -4, M = 3, Q_3 = 7.5, Max = 8$

(b)

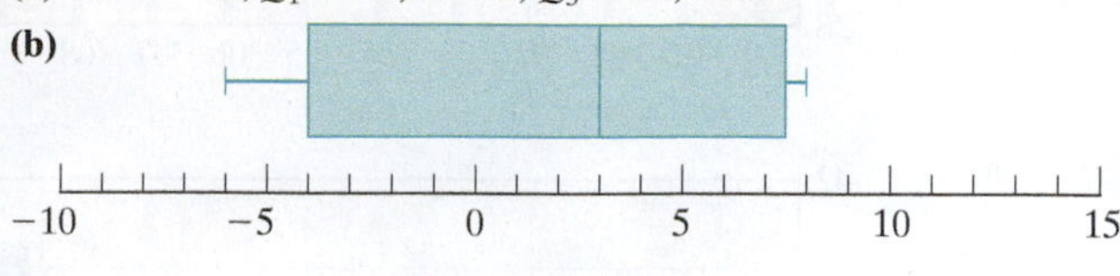

43. **(a)** $Min = 25, Q_1 = 29, M = 31, Q_3 = 32, Max = 39$

(b)

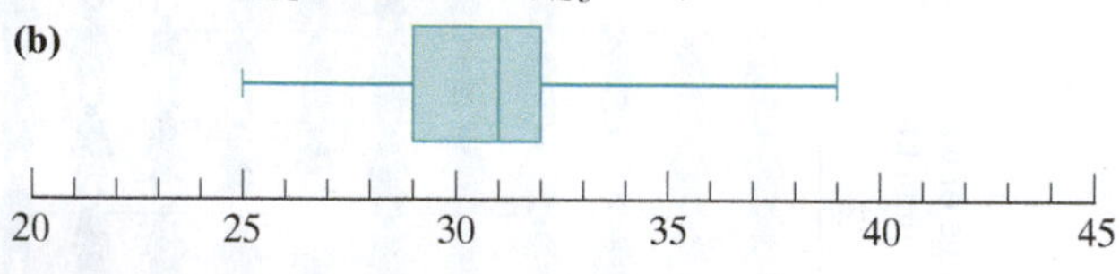

45. **(a)** \$43,000 **(b)** \$50,000

(c) The vertical line indicating the median salary in the engineering box plot is aligned with the right end of the box in the agriculture box plot.

15.3 Ranges and Standard Deviations

47. **(a)** 14 **(b)** 11.5

49. **(a)** \$41,000 **(b)** at least 171 homes

51. **(a)** 16.5 **(b)** 4.5 **(c)** 1 and 24

53. heights less than 61 in. or greater than 77 in.

55. **(a)** 0 **(b)** $\frac{5\sqrt{2}}{2} \approx 3.5$ **(c)** $\sqrt{50} \approx 7.1$

(d) $\sqrt{2} \approx 1.4$

JOGGING

57. **(a)** 30 **(b)** 8

59. 100 **61.** 1,465,744

63. **(a)** $\{1, 1, 1, 1, 6, 6, 6, 6, 6, 6\}$ $(A = 4, M = 6)$

(b) $\{1, 1, 1, 1, 1, 1, 6, 6, 6, 6\}$ $(A = 3, M = 1)$

(c) $\{1, 1, 6, 6, 6, 6, 6, 6, 6, 6\}$ $(A = 5, Q_1 = 6)$

(d) $\{1, 1, 1, 1, 1, 1, 1, 1, 6, 6\}$ $(A = 2, Q_3 = 1)$

65. From histogram (a) one can deduce that the median team salary is between \$70 million and \$100 million. From histogram (b) one can deduce that the median team salary is between \$50 million and \$80 million. It follows that the median team salary must be between \$70 million and \$80 million.

67. **(a)** $A + c$

(b) $M + c$; the relative sizes of the numbers are not changed by adding a constant c to every number.

69. **(a)** 4 **(b)** 0.4 **(c)** 0.4

CHAPTER 16 Probabilities, Odds, and Expectations

WALKING

16.1 Sample Spaces and Events

1. **(a)** $\{HHH, HHT, HTH, THH, TTH, THT, HTT, TTT\}$

(b) $\{sss, ssf, sfs, fss, ffs, fsf, sff, fff\}$

(c) $\{0, 1, 2, 3\}$ **(d)** $\{0, 1, 2, 3\}$

3. **(a)** $S = \{3, 4, 5, \ldots, 16, 17, 18\}$

(b) $S = \{5, 3, 1, -1, -3, -5\}$

5. **(a)** $S = \{AB, AC, AD, AE, BA, BC, BD, BE, CA, CB, CD, CE, DA, DB, DC, DE, EA, EB, EC, ED\}$ [XY denotes that X was elected chairman and Y was elected treasurer.]

(b) $S = \{ABC, ABD, ABE, ACD, ACE, ADE, BCD, BCE, BDE, CDE\}$

7. Answers will vary. A typical outcome is a string of 10 letters each of which can be either an H or a T. An answer like $\{HHHHHHHHHH, \ldots, TTTTTTTTTT\}$ is not sufficiently descriptive. An answer like $\{\ldots HTTHHHTHTH, \ldots, TTHTHHTTHT, \ldots, HHHTHTTHHT, \ldots\}$ is better. An answer like $\{(X_1X_2X_3X_4X_5X_6X_7X_8X_9X_{10})$: each X_i is either H or $T\}$ is best.

9. Answers will vary. An outcome is an ordered sequence of four numbers, each of which is an integer between 1 and 6 inclusive. The best answer would be something like $\{(n_1, n_2, n_3, n_4)$: each n_i is 1, 2, 3, 4, 5, or 6$\}$. An answer such as $\{(1, 1, 1, 1), \ldots, (1, 1, 1, 6), \ldots, (1, 2, 3, 4), \ldots, (3, 2, 6, 2), \ldots, (6, 6, 6, 6)\}$ showing a few typical outcomes is possible but not as good. An answer like $\{(1, 1, 1, 1), \ldots, (2, 2, 2, 2), \ldots, (6, 6, 6, 6)\}$ is not descriptive enough.

11. **(a)** $E_1 = \{HHT, HTH, THH\}$

(b) $E_2 = \{HHH, TTT\}$

(c) $E_3 = \{\ \}$

(d) $E_4 = \{TTH, TTT\}$

13. **(a)** $E_1 = \{(1, 1), (2, 2), (3, 3), (4, 4), (5, 5), (6, 6)\}$

(b) $E_2 = \{(1, 1), (1, 2), (2, 1), (6, 6)\}$

(c) $E_3 = \{(1, 6), (2, 5), (3, 4), (4, 3), (5, 2), (6, 1), (5, 6), (6, 5)\}$

15. **(a)** $E_1 = \{HHHHHHHHHH\}$

(b) $E_2 = \{HHHHHHHHHT, HHHHHHHHTH, HHHHHHHTHH, HHHHHHTHHH, HHHHHTHHHH, HHHHTHHHHH, HHHTHHHHHH, HHTHHHHHHH, HTHHHHHHHH, THHHHHHHHH\}$

(c) $E_3 = \{TTTTTTTTTH, TTTTTTTTHT, TTTTTTTHTT, TTTTTTHTTT, TTTTTHTTTT, TTTTHTTTTT, TTTHTTTTTT, TTHTTTTTTT, THTTTTTTTT, HTTTTTTTTT, TTTTTTTTTT\}$

16.2 The Multiplication Rule, Permutations, and Combinations

17. **(a)** 158,184,000 **(b)** 1,757,600 **(c)** 70,761,600

19. **(a)** $(2 + 1) \times (3 + 4) \times 3 = 63$

(b) $(2 + 1) \times (3 + 4) \times (3 + 3 \times 2) = 189$

21. **(a)** 40,320 **(b)** 40,319

23. **(a)** 2880 **(b)** 2880

25. **(a)** 1024 **(b)** 1296 **(c)** 21

27. **(a)** ${}_{15}P_4$ **(b)** ${}_{15}C_4$

29. **(a)** ${}_{30}C_{18}$ **(b)** ${}_{30}P_{18}$

31. **(a)** ${}_{40}C_{25} = 40{,}225{,}345{,}056$ **(b)** ${}_{25}P_9 = 741{,}354{,}768{,}000$

16.3 Probabilities and Odds

33. **(a)** $\Pr(o_1) = 0.22$, $\Pr(o_2) = 0.24$, $\Pr(o_3) = 0.18$, $\Pr(o_4) = 0.18$, $\Pr(o_5) = 0.18$
(b) $\Pr(o_1) = 0.22$, $\Pr(o_2) = 0.24$, $\Pr(o_3) = 0.27$, $\Pr(o_4) = 0.17$, $\Pr(o_5) = 0.1$

35. $\Pr(A) = 1/5$, $\Pr(B) = 2/15$, $\Pr(C) = 4/15$, $\Pr(D) = 6/15$

37. **(a)** 3/8 **(b)** 1/4 **(c)** 0 **(d)** 1/4

39. **(a)** 1/6 **(b)** 1/9 **(c)** 2/9

41. **(a)** 1/1024 **(b)** 5/512 **(c)** 11/1024

43. **(a)** 1/1024 **(b)** 1/1024 **(c)** 5/512 **(d)** 11/1024 **(e)** 0 **(f)** 7/128

45. **(a)** 4/15 **(b)** 11/15 **(c)** 1/1365

47. 1023/1024

49. **(a)** 4 to 3 **(b)** 3 to 2

51. **(a)** 3/8 **(b)** 15/23

16.4 Expectations

53. 82.9% **55.** 16.4 **57.** $7.50 **59.** 1.5

16.5 Measuring Risk

61. $−1.00 **63.** $\$-\frac{1}{38} \approx \$-0.0263$

65. It makes sense for Joe to purchase the extended warranty: The warranty has an expected value of $\$320 \times 0.24 + \$(-80) \times 0.76 = \$16$.

67. $525 per policy

JOGGING

69. **(a)** ${}_{35}P_3 = 39{,}270$ **(b)** ${}_{15}P_3 = 2730$ **(c)** ${}_{15}P_3 + {}_{20}P_3 = 9570$ **(d)** $39{,}270 - 9570 = 29{,}700$

71. **(a)** $\{AARRR, ARARR, ARRAR, RAARR, RARAR, RRAAR\}$
(b) $\{RRR, ARRR, RARR, RRAR, AARRR, ARARR, ARRAR, RAARR, RARAR, RRAAR\}$
(c) $\{AARRR, ARARR, ARRAR, RAARR, RARAR, RRAAR, RRAAA, RARAA, RAARA, ARRAA, ARARA, AARRA\}$

73. **(a)** 7/40 **(b)** 63/1000 **(c)** 7/120

75. If N is odd, then the probability is 0. If N is even and $K = \frac{N}{2}$, then the probability of getting K heads and K tails is ${}_NC_K/2^N$.

77. $\frac{{}_5C_1 \cdot {}_{495}C_2}{{}_{500}C_5} + \frac{{}_5C_2 \cdot {}_{495}C_1}{{}_{500}C_5} + \frac{{}_5C_3 \cdot {}_{495}C_0}{{}_{500}C_5}$

79. **(a)** $\frac{6}{6^5} \approx 0.00077$ **(b)** $\frac{5 \times 6 \times 5}{6^5} \approx 0.019$ **(c)** $\frac{2 \times 5!}{6^5} \approx 0.031$

CHAPTER 17 The Mathematics of Normality

WALKING

17.2 Normal Curves and Normal Distribution

1. **(a)** 80 lb **(b)** 80 lb **(c)** 10 lb

3. **(a)** 40 in. **(b)** 8 in. **(c)** 45.4 in. **(d)** 34.6 in.

5. **(a)** 89.6 lb **(b)** 72.8 lb

7. **(a)** 500 points **(b)** 100 points

9. 20 in. **11.** $\mu \neq M$

13. In a normal distribution, the mean μ must be at the median of the quartiles.

15. **(a)** 1 **(b)** −2 **(c)** 1.6 **(d)** −1.8

17. **(a)** 1.5 **(b)** −0.5 **(c)** −0.75 **(d)** 1.1

19. **(a)** 183.5 ft **(b)** 152.3 ft

21. **(a)** 199.1 ft **(b)** 111.74 ft

23. 17 lb. **25.** 5

27. $\mu = 52, \sigma = 16$ **29.** $\mu = 48$ in., $\sigma = 6$ in.

31. **(a)** $\mu = 92$ cm, $\sigma = 3.4$ cm **(b)** $Q_1 = 89.7$ cm, $Q_3 = 94.3$ cm

33. **(a)** 6.3 in. **(b)** 77.8 in.

35. $\mu = 69.3$ cm **37.** 59%

39. **(a)** 97.5% **(b)** 13.5%

17.3 Modeling Approximately Normal Distributions

41. **(a)** 52 points **(b)** 50% **(c)** 68% **(d)** 16%

43. **(a)** 44.6 points **(b)** 59.4 points

45. **(a)** 1900 **(b)** 1630

47. **(a)** 16th percentile **(b)** 84th percentile **(c)** 99.85th percentile

49. **(a)** 95% **(b)** 47.5% **(c)** 97.5%

51. **(a)** 13 **(b)** 80 **(c)** 250 **(d)** 420 **(e)** 488 **(f)** 499

53. **(a)** 16th percentile **(b)** 97.5th percentile **(c)** 8.8 kg

55. **(a)** 97.5th percentile **(b)** 99.85th percentile **(c)** 4.23 kg

17.4 Normality in Random Events

57. **(a)** $\mu = 1800, \sigma = 30$ **(b)** approximately 68% **(c)** approximately 34% **(d)** approximately 13.5%

59. **(a)** approximately 68% **(b)** approximately 16% **(c)** approximately 84%

61. **(a)** $\mu = 240, \sigma = 12$ **(b)** $Q_1 \approx 232, Q_3 \approx 248$ **(c)** approximately 0.95

63. **(a)** $\mu = 30, \sigma = 5$ **(b)** approximately 0.025 **(c)** approximately 0.34

JOGGING

65. 0 **67.** **(a)** 9.73 kg **(b)** 7.36 kg

69. **(a)** 60th percentile **(b)** 30th percentile

71. **(a)** 590 or 600 points **(b)** 630 or 640 **(c)** 570 or 580 points

73. **(a)** 78 points **(b)** 70.4 points **(c)** 60 points **(d)** 48.7 points

INDEX

CREDITS

CHAPTER 1
p. 2 Rich Graessle/Icon Sportswire CGV/Rich Graessle/Icon Sportswire/Newscom **p. 5** Smock John/SIPA/Newscom **p. 6** JOHN ANGELILLO/UPI/Newscom **p. 7** JINFphoto/Newscom **p. 9** Joshua Lott/The New York Times/Redux Pictures **p. 12** Pearson Education, Inc. **p. 12** Frilley/Jean Jacques/The Art Gallery Collection/Alamy **p. 14** McClatchy-Tribune/Tribune Content Agency LLC/Alamy Stock Photo **p. 22** Yan Sheng/CNImaging/Newscom

CHAPTER 2
p. 39 Presidential Election Laws, The Constitution, Article II, U S National Archieves and Records Administration **p. 42** Scott J. Ferrell/Congressional Quarterly/Alamy Stock Photo **p. 47** The Constitution of the United States: A Transcription, Article I, U S National Archieves and Records Administration **p. 50** Epa European Pressphoto Agency B.V./Alamy Stock Photo **p. 50** Ben Mattison/Yale School of Management

CHAPTER 3
p. 68 Chester Higgins/The New York Times/Redux Pictures **p. 76** Office of Communications/Princeton University **p. 78** A.M. Fink **p. 84** Peter Tannenbaum **p. 88** Album/Oronoz/Newscom

CHAPTER 4
p. 102 Graphithèque/Fotolia **p. 103** The Constitution of the United States: A Transcription, Article I, U S National Archieves and Records Administration **p. 107** Library of Congress Prints and Photographs Division [LC-DIG-det-4a26166] **p. 109** National Archives **p. 112** Balinski, Michel Louis **p. 112** Fernanda F. Young **p. 112** Library of Congress Prints and Photographs Division[LC-USZ62-117119] **p. 113** Library of Congress Prints and Photographs Division[LC-DIG-pga-06703] **p. 114** The University of Texas at Austin/The Dolph Briscoe Center for American History **p. 123** The Constitution of the United States: A Transcription, Article I, U S National Archieves and Records Administration

CHAPTER 5
p. 136 Victoria Arocho/AP Images **p. 140** Pearson Education, Inc. **p. 143** Air Rarotonga **p. 144** Auremar/Shutterstock **p. 144** Alexander Trinitatov/Shutterstock **p. 144** Nikkytok/Shutterstock **p. 144** Auremar/Shutterstock **p. 144** StockLite/Shutterstock **p. 144** Blend Images/Shutterstock **p. 144** Wavebreakmedia/Shutterstock **p. 144** Creativa Images/Shutterstock **p. 144** EDHAR/Shutterstock **p. 144** Ijansempoi/Shutterstock **p. 144** Goodluz/Shutterstock **p. 144** Goodluz/Shutterstock **p. 144** Monkey Business Images/Shutterstock **p. 144** Monkey Business Images/Shutterstock **p. 144** Dinostock/Fotolia **p. 144** Mavoimages/Fotolia **p. 144** Christine Langer-Pueschel/Shutterstock **p. 144** Imtmphoto/Shutterstock **p. 144** Auremar/123RF

CHAPTER 6
p. 172 JPL-Caltech/MSSS/NASA **p. 174** JPL/DLR/NASA **p. 174** Galileo Project/DLR/JPL/NASA **p. 174** Galileo Project/JPL/NASA **p. 174** University of Arizona/JPL/NASA **p. 174** NASA **p. 174** JPL-Caltech/NASA **p. 175** JPL Caltech/Univ. of Arizona/NASA **p. 176** ImageBROKER/Alamy Stock Photo **p. 177** Neftali/Alamy Stock Photo **p. 183** JPL/DLR/NASA **p. 184** Galileo Project/JPL/NASA

CHAPTER 7
p. 204 Kimihiro Hoshino/AFP/Getty Images/Newscom **p. 207** Kodda/Shutterstock **p. 218** Brad Loper/KRT/Newscom

CHAPTER 8
p. 226 Franck Boston/Shutterstock **p. 227** Arena Creative/Shutterstock **p. 228** Iurii/Shutterstock **p. 229** BDR/Alamy Stock Photo **p. 230** Rich Pedroncelli/File/AP Images **p. 231** Glenn Research Center/NASA

CHAPTER 9
p. 258 US Census Bureau **p. 261** Pearson Education, Inc. **p. 264** Pressmaster/Shutterstock **p. 268** Csp/Shutterstock **p. 269** Mike Nelson/EPA/Newscom **p. 274** Patricia Hofmeester/Shutterstock

CHAPTER 10
p. 290 Tiero/Fotolia **p. 291** *The Ascent of Money: A Financial History of the World* by Niall Ferguson. Published by Penguin Press, © 2008 **p. 292** Doug Raphael/Shutterstock **p. 293** Patrick T. Power/Shutterstock **p. 294** **K.** Geijer/Fotolia **p. 313** Andy Dean/Fotolia **p. 317** Mark Plumley/Shutterstock

CHAPTER 11

p. 326 Peter Parks/AFP/Getty Images **p. 327** Janaka Dharmasena/Shutterstock **p. 330** Rui Vale de Sousa/Fotolia **p. 330** Pathdoc/Fotolia **p. 332** Photobank/Fotolia **p. 333** Jitty26/Fotolia **p. 336** Appler/Fotolia **p. 336** Modify260/Fotolia **p. 336** Rossco/Fotolia **p. 339** Irmaiirma/Fotolia **p. 339** Imagewriter/Fotolia **p. 339** Aldorado/Fotolia **p. 339** Fantasy/Fotolia **p. 339** Alex Mit/Shutterstock **p. 339** SnowCrystals.com **p. 339** SnowCrystals.com **p. 340** Wendy Kaveney/Fotolia **p. 340** Rafael Ramirez Lee/Shutterstock **p. 340** Eva Madrazo/Shutterstock **p. 341** Axusha/Shutterstock **p. 341** Darla Hallmark/Shutterstock **p. 341** Peter Tannenbaum **p. 344** Janaka Dharmasena/Shutterstock

CHAPTER 12

p. 356 Nnerto/Fotolia **p. 356** Sergey Skleznev/Fotolia **p. 356** Sergey Skleznev/Fotolia **p. 361** Courtesy of John Walker/Fourmilab **p. 361** Courtesy of John Walker/Fourmilab **p. 361** Courtesy of John Walker/Fourmilab **p. 361** Courtesy of John Walker/Fourmilab **p. 362** Joingate/Shutterstock **p. 362** Sergii Figurnyi/Shutterstock **p. 362** Dario Sabljak/Shutterstock **p. 362** Zagart/Fotolia **p. 362** Valdezrl/Fotolia **p. 362** Romansl/Fotolia **p. 362** Earth as Art/USGS **p. 363** Michael Nicholson/Corbis/Getty Images **p. 367** Oliver Klimek/Shutterstock **p. 367** Destillat/Fotolia **p. 368** Four lines from *On Poetry: A Rhapsody* by Jonathan Swift, © 1733 **p. 368** Hank Morgan/Getty Images **p. 368** Rollo Silver/Peter Tannenbaum **p. 368** Georg Lehnerer/Fotolia **p. 369** Rollo Silver/Peter Tannenbaum **p. 369** Rollo Silver/Peter Tannenbaum **p. 369** Rollo Silver/Peter Tannenbaum **p. 369** Rollo Silver/Peter Tannenbaum **p. 369** Rollo Silver/Peter Tannenbaum **p. 369** Rollo Silver/Peter Tannenbaum **p. 374** Tatiana Sayig/Shutterstock **p. 374** Taras Vyshnya/Shutterstock **p. 374** Dimitri Surkov/Fotolia **p. 375** Dmitris K/Shutterstock **p. 375** Vichie81/Shutterstock **p. 375** Bill Mack/Shutterstock **p. 375** Oliver Klimek/Shutterstock **p. 375** Stephen Coburn/Shutterstock **p. 375** Polina Katritch/Shutterstock

CHAPTER 13

p. 386 McIek/Shutterstock **p. 390** Peter Tannenbaum **p. 390** Peter Tannenbaum **p. 390** E. Spek/Shutterstock **p. 390** Alex Virid/Shutterstock **p. 390** Alex Virid/Shutterstock **p. 390** Eiji Ueda/Shutterstock **p. 390** Eiji Ueda/Shutterstock **p. 392** Katusha161107/Fotolia **p. 392** Amurca/Fotolia **p. 392** Laurent Renault/Shutterstock **p. 399** Aaltair/Shutterstock **p. 399** Arto Hakola/Shutterstock **p. 399** Jim Barber/Shutterstock **p. 401** Aleksandar Mijatovic/Fotolia **p. 401** Nikhg/Fotolia **p. 401** Nikhg/Fotolia **p. 401** Nikhg/Fotolia **p. 401** Le Do/Shutterstock **p. 401** Leo Blanchette/Shutterstock

CHAPTER 14

p. 410 Valzan/Shutterstock **p. 411** From *The Adventure of the Copper Beeches* by Arthur Conan Doyle **p. 412** Steve Lovegrove/Shutterstock **p. 413** Eric Gevaert/Shutterstock **p. 415** Jugulator/Shutterstock **p. 416** Max5128/Fotolia **p. 417** The Constitution of the United States: A Transcription, Article I, U S National Archieves and Records Administration **p. 418** Kenishirotie/Fotolia **p. 421** From *Case Study I: The 1936 Literary Digest Poll* in Literary Digest. Published by Literary Digest, © 1936 **p. 423** Bettmann/UPI/Getty Images **p. 424** Mishchenko Mikhail/Shutterstock

CHAPTER 15

p. 442 Dragana Gerasimoski/Shutterstock **p. 443** Sergey Nivens/Fotolia **p. 447** Fedor A.Sidorov/Shutterstock **p. 448** Roman Sigaev/Shutterstock **p. 448** Angela Waye/Shutterstock **p. 448** Mirexon/Fotolia

CHAPTER 16

p. 472 Kena Betancur/Stringer/AFP/Getty Images **p. 473** James Steidl/Shutterstock **p. 474** Ronstik/Shutterstock **p. 474** Ronstik/Shutterstock **p. 475** Sergey Nivens/Shutterstock **p. 475** ZUMA Press, Inc./Alamy Stock Photo **p. 476** Melissa Schalke/Fotolia **p. 480** HLPhoto/Shutterstock **p. 481** Oliver Hoffmann/Shutterstock **p. 481** Cybrain/Shutterstock **p. 482** Peter Tannenbaum **p. 482** Peter Tannenbaum **p. 482** Peter Tannenbaum **p. 483** JCJG Photography/Shutterstock **p. 484** Doug Pensinger/Getty Images **p. 484** Peter Tannenbaum **p. 488** Peter Tannenbaum **p. 489** Doug Pensinger/Getty Images

CHAPTER 17

pp. 508, 514 Peter Tannenbaum

ENDSHEETS

p. FES3 Pearson Education, Inc.; Kimihiro Hoshino/AFP/Getty Images/Newscom; Nnerto/Fotolia; Sergey Skleznev/Fotolia; Sergey Skleznev/Fotolia

INDEX OF APPLICATIONS

Sports

Statistics and Demographics

Technology

Transportation